Catalytic
Asymmetric
Synthesis

Catalytic Asymmetric Synthesis

EDITED BY

Iwao Ojima

Iwao Ojima
Department of Chemistry
State University of New York at Stony Brook
Stony Brook, NY 11794-3400

This book is printed on acid-free paper. ∞

Library of Congress Cataloging-in-Publication Data

Catalytic asymmetric synthesis / edited by Iwao Ojima.
    p. cm.
  Includes bibliographical references and index.
  ISBN 1-56081-532-9 (acid-free)
  1. Asymmetric synthesis. 2. Catalysis. I. Ojima, Iwao. 1945–
QD262.c356 1993
541.3'9—dc20

                        93-19389
                          CIP

Printed in the United States of America

ISBN: 1-56081-532-9 VCH Publishers
ISBN: 3-527-89532-9 VCH Verlagsgesellschaft

Printing history:
10 9 8 7 6 5 4 3 2 1

Published jointly by

VCH Publishers, Inc.	VCH Verlagsgesellschaft mbH	VCH Publishers (UK) Ltd.
220 East 23rd Street	P.O. Box 10 11 61	8 Wellington Court
New York, N.Y. 10010-4606	69451 Weinheim, Germany	Cambridge CB1 1HZ
		United Kingdom

Preface

Biological systems, in most cases, recognize the members of a pair of enantiomers as different substances, and the two enantiomers will elicit different responses. Thus, one enantiomer may act as a very effective therapeutic drug whereas the other enantiomer is highly toxic. The sad example of thalidomide is well known. It is the responsibility of synthetic chemists to prevent a repetition of the thalidomide tragedy by providing highly efficient and reliable methods for the synthesis of desired compounds in an enantiomerically pure state—that is, with 100% enantiomeric excess (% ee). It has been shown for many pharmaceuticals that only one enantiomer contains all the desired activity, and the other is either totally inactive or toxic. Recent rulings of the Food and Drug Administration (FDA) in the United States clearly reflect the current situation in "chiral drugs": pharmaceutical industries will have to provide rigorous justification to obtain the FDA's approval of racemates. There are several methods to obtain enantiomerically pure materials, which include classical optical resolution via diastereomers, chromatographic separation of enantiomers, enzymic resolution, chemical kinetic resolution, and asymmetric synthesis.

The importance and practicality of asymmetric synthesis as a tool to obtain enantiomerically pure or enriched compounds has been fully acknowledged by chemists in synthetic organic chemistry, medicinal chemistry, agricultural chemistry, natural products chemistry, the pharmaceutical industries, and the agricultural industries. This prominence is due to the explosive development of newer and more efficient methods during the last decade.

This book describes recent advances in catalytic asymmetric synthesis with brief summaries of earlier achievements as well as general discussions of the reactions. A book reviewing this topic, *Asymmetric Synthesis,* Vol. 5—*Chiral Catalysis,* edited by J. D. Morrison and published by Academic Press in 1985, compiles important contributions through 1982. Another book, *Asymmetric Catalysis*, edited by B. Bosnich and published by Martinus Nijhoff in 1986, also concisely covers contributions up to early 1984. In 1971 Prentice Hall, Inc. published an excellent book *Asymmetric Organic Reactions*, in which J. D. Morrison and H. S. Mosher reviewed all earlier important work on the subject and compiled nearly 850 relevant publications through 1968, including some papers published in 1969. In the early 1980s, a survey of publications dealing with "asymmetric synthesis" (in a broad sense) indicated that the total number of papers in this area of research published in the 10 years after the book by Morrison and Mosher (i.e., 1971–1980) was almost the same as that of all the papers published before 1971. This doubling of output clearly indicates the attention paid to this important topic in the 1970s. In the 1980s, research on "asymmetric synthesis: became even more important and popular, since enantiomerically pure compounds were required for the total synthesis of natural products, pharmaceuticals, and agricultural agents. It would not be an exaggeration to say that the number of publications on asymmetric synthesis has increased exponentially every year.

Among the types of asymmetric reaction, the most desirable and the most challenging is *catalytic* asymmetric synthesis, since one chiral catalyst molecule can create millions of chiral product molecules, just as enzymes do in biological systems. Among the significant achievements in basic research, asymmetric hydrogenation of dehydroamino acids, as described in the ground-breaking work by W. S. Knowles et al., the Sharpless epoxidation described by K. B. Sharpless et al., and the second-generation asymmetric hydrogenation processes developed by R. Noyori et al. deserve particular attention because of the tremendous impact these processes have had in synthetic organic chemistry. Catalytic asymmetric synthesis often has significant economic advantages over stoichiometric asymmetric synthesis for industrial scale production of enantiomerically pure compounds. In fact, a number of catalytic asymmetric reactions, including the "Takasago process" (asymmetric isomerization), the "Sumitomo process" (asymmetric cyclopropanation), and the "Arco process" (asymmetric Sharpless epoxidation) were commercialized in the 1980s. These supplement the epoch-making "Monsanto process" (asymmetric hydrogenation) established in the early 1970s. As this book will uncover, other catalytic asymmetric reactions have high potential as commercial processes. Extensive research on new and effective catalytic asymmetric reactions will surely continue beyond the year 2000, and catalytic asymmetric processes promoted by man-made chiral catalysts will become mainstream chemical technology in the twenty-first century.

This book covers the following catalytic asymmetric reactions: asymmetric hydrogenation (Chapter 1), isomerization (Chapter 2), cyclopropanation (Chapter 3), oxidations (epoxidation of allylic alcohols as well as unfunctionalized olefins, oxidation of sulfides, and dihydroxylation of olefins: Chapter 4), carbonylations (Chapter 5), hydrosilylation (Chapter 6), carbon–carbon bond forming reactions (allylic alkylation, Grignard cross-coupling, and aldol reactions: Chapter 7), phase transfer reactions (Chapter 8), and Lewis acid catalyzed reactions (Chapter 9). The authors of the chapters are all world leaders in this field, who outline and discuss the essence of each catalytic asymmetric reaction. In addition, a convenient list of the chiral ligands appearing in this book, with citation of relevant references, is provided as an Appendix.

This book will serve as an excellent reference book for graduate students as well as chemists at all levels in both academic and industrial laboratories.

Iwao Ojima
March 1993

Contents

1. Asymmetric Hydrogenation   1
Hidemasa Takaya, Tetsuo Ohta, and Ryoji Noyori

1.1 Chiral Transition Metal Catalysts   1
1.2 Asymmetric Hydrogenation of Olefins   6
1.3 Asymmetric Hydrogenation of Ketones   20
1.4 Asymmetric Hydrogenation of Imines   31
1.5 Conclusion   32
References   33

2. Asymmetric Isomerization of Allylamines   41
Susumu Akutagawa and Kazuhide Tani

2.1 Introduction   41
2.2 Asymmetric Isomerization of Allylamine   42
2.3 Isomerization Process Development   46
2.4 Commercial Manufacture   51
2.5 Scope and Limitations   54
2.6 Conclusion   59
References   60

ix

3. Asymmetric Cyclopropanation   63
Michael P. Doyle

3.1 Introduction   63
3.2 Asymmetric Intermolecular Cyclopropanation   67
3.3 Asymmetric Intramolecular Cyclopropanation   89
3.4 Intermolecular Cyclopropanation of Alkynes   94
3.5 Conclusion   96
References   97

4. Asymmetric Oxidation   101
4.1 Catalytic Asymmetric Epoxidation of Allylic Alcohols   103
Roy A. Johnson and K. Barry Sharpless
4.1.1 Introduction   103
4.1.2 Fundamental Elements of Titanium Tartrate Catalyzed Asymmetric Epoxidation   104
4.1.3 Reaction Variables for Titanium Tartrate Catalyzed Asymmetric Epoxidation   108
4.1.4 Sources of Allylic Alcohols   112
4.1.5 Asymmetric Epoxidation by Substrate Structure   114
4.1.6 Mechanism of Titanium Tartrate Catalyzed Asymmetric Epoxidation   144
4.1.7 Other Asymmetric Epoxidations and Oxidations Catalyzed by Titanium Complexes   148
4.1.8 Conclusion   152
References   152

4.2 Asymmetric Catalytic Epoxidation of Unfunctionalized Olefins   159
Eric N. Jacobson
4.2.1 Introduction   159
4.2.2 Transition Metal Based Catalysts   160
4.2.3 Asymmetric Epoxidation with Organic Oxidants   193
4.2.4 Conclusion   199
References   199

4.3 Asymmetric Oxidation of Sulfides   203
Henri B. Kagan
4.3.1 Introduction   203
4.3.2 Oxidation in the Presence of Chiral Titanium Alcoholates   204

4.3.3 Chiral Titanium–Schiff Base Catalysts   212
4.3.4 Chiral Vanadium (IV)–Schiff Base Catalysts   214
4.3.5 Iron– or Manganese–Porphyrin Catalysts   214
4.3.6 Heterogeneous Catalysts   216
4.3.7 Chiral Flavins as the Catalysts   217
4.3.8 Template Effect   218
4.3.9 Enzymatic Reactions   220
4.3.10 Conclusion   223
References   223

4.4 Catalytic Asymmetric Dihydroxylation   227
Roy A. Johnson and K. Barry Sharpless
4.4.1 Introduction   227
4.4.2 General Features of Osmium-Catalyzed Asymmetric
      Dihydroxylation   233
4.4.3 Catalytic Asymmetric Dihydroxylations by Olefin
      Substitution Pattern   245
4.4.4 Diol Activation   261
4.4.5 Conclusion   266
References   268

5. Asymmetric Carbonylation   273
Giambattista Consiglio

5.1 Introduction   273
5.2 Enantioface Discriminating Carbonylation Reactions of Olefinic
    Substrates   275
5.3 Enantiomer Discriminating Carbonylation Reactions   291
5.4 Enantiotopic Group Discriminating Synthesis of Lactones   294
5.5 Conclusion and Outlook   295
References   296

6. Asymmetric Hydrosilylation   303
Henri Brunner, Hisao Nishiyama, and Kenji Itoh

6.1 Introduction   303
6.2 Asymmetric Hydrosilylation of Ketones   304
6.3 Asymmetric Hydrosilylation of α,β-Unsaturated Ketones   312
6.4 Asymmetric Hydrosilylation of Keto Esters   312
6.5 Asymmetric Hydrosilylation of Imines   314
6.6 Asymmetric Hydrosilylation of Olefins   315
6.7 Asymmetric Synthesis of Chiral Silicon Compounds   319
6.8 Conclusion   320
References   320

7. Asymmetric Carbon–Carbon Bond Forming Reactions   323

7.1 Asymmetric Allylic Substitution and Grignard Cross-Coupling   325

Tamio Hayashi

7.1.1 Asymmetric Allylic Substitution Reactions Catalyzed by Palladium and Nickel Complexes   325

7.1.2 Asymmetric Cross-Coupling Catalyzed by Nickel and Palladium Complexes   350

7.1.3 Conclusion   361

References   363

7.2 Asymmetric Aldol Reactions   367

Masaya Sawamura and Yoshihiko Ito

7.2.1 Introduction   367

7.2.2 Gold-Catalyzed Asymmetric Aldol Reaction of α-Isocyanocarboxylates   367

7.2.3 Asymmetric Nitroaldol Reaction   377

7.2.4 Lewis Acid Catalyzed Asymmetric Aldol Reaction   379

7.2.5 Conclusion   386

References   387

8. Asymmetric Phase Transfer Reactions   389

Martin J. O'Donnell

8.1 Introduction   389

8.2 The Phase Transfer System   389

8.3 Phase Transfer Reactions   395

8.4 Conclusion and Future Prospects   404

References   405

9. Asymmetric Reactions with Chiral Lewis Acid Catalysts   413

Keiji Maruoka and Hisashi Yamamoto

9.1 Introduction   413

9.2 Chiral Aluminum Reagents   415

9.3 Chiral Titanium Reagents   421

9.4 Chiral Boron Reagents   427

9.5 Chiral Lewis Acids of Other Metals   438

9.6 Conclusion   438

References   438

Epilogue   441
Iwao Ojima

Appendix: List of Chiral Ligands   445

Index   471

1

Asymmetric Hydrogenation

Hidemasa Takaya
Tetsuo Ohta

Division of Material Chemistry
Faculty of Engineering, Kyoto University
Kyoto, Japan

Ryoji Noyori

Department of Chemistry
Nagoya University
Nagoya, Japan

The first enantioselective hydrogenation of unsaturated compounds appeared in the late 1930s using metallic catalysts deposited on chiral supports [1]. This method attracted much attention from a synthetic point of view when its enantioselectivity exceeded 60% ee in the middle 1950s [2]. Biopolymers such as polypeptides, polysaccharides, and cellulose have been used as chiral supports. Much later, in 1968, Knowles and Horner independently reported homogeneous asymmetric hydrogenation with Rh complexes bearing chiral tertiary phosphines [3,4]. Recent developments in homogeneous asymmetric hydrogenation by means of manufactured transition metal catalysts are remarkable. Their efficiencies are often comparable or even superior to those of biocatalysts, and now this methodology is becoming one of the most attractive and useful synthetic approaches to optically active organic compounds. Many excellent reviews on this field have been published [5]. This chapter reviews asymmetric hydrogenation catalyzed by homogeneous transition metal catalysts by putting the emphasis on recent advances. Asymmetric hydrogenation with chiral heterogeneous catalysts and biocatalysts is not included.

1.1. Chiral Transition Metal Catalysts

1.1.1. Chiral Ligands

In the early 1970s, Kagan, Knowles, and others developed Rh(I) complexes of chiral chelating diphosphines, **1** and **2**, respectively, and had great success in the

Chart 1

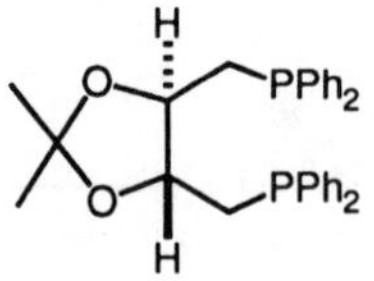

(R,R)-DIOP (1)[6]

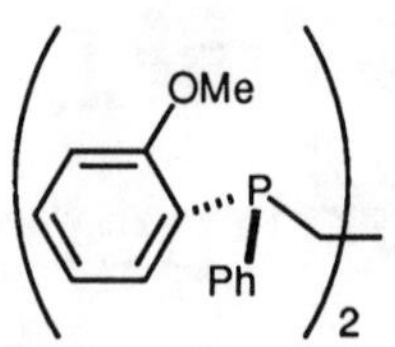

(R,R)-DIPAMP (2)[7]

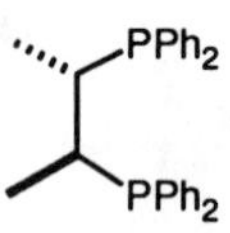

(S,S)-CHIRAPHOS (3)[10]

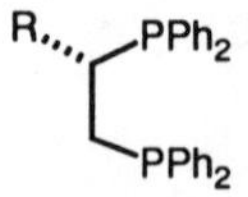

R = Me: (S)-PROPHOS (4)[11]
R = Ph: (S)-PHENPHOS (5)[12]
R = c-C_6H_{11}: (S)-CYCPHOS (6)[13]

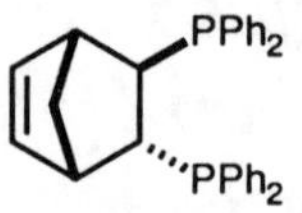

(R,R)-NORPHOS (7)[14]

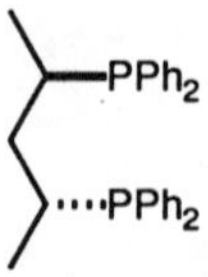

(R,R)-BDPP (8)[15]

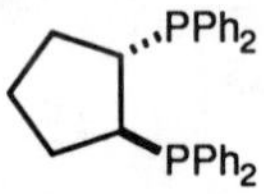

(S,S)-DPCP (9)[16]

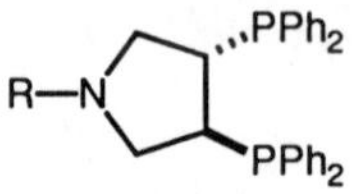

(S,S)-PYRPHOS (10)[17]

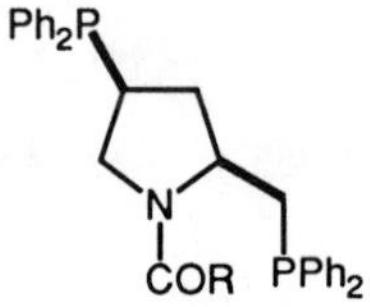

R = t-BuO: (S,S)-BPPM (11)[18]
R = R'NH: (S,S)-R'-CAPP (12)[19]
R' = C_6H_3-3,4-Cl_2;
C_6H_4-4-Br

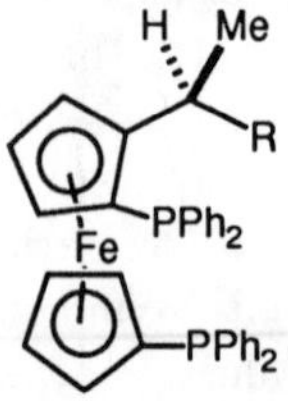

R = NMe$_2$: (R,S)-BPPFA (13a)[9]
R = OH: (R,S)-BPPFOH (13b)

R = N(Me)CH$_2$CH$_2$N⟨piperidine⟩ (13c)

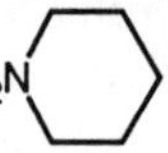

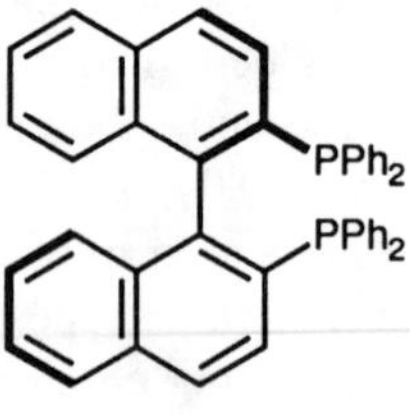

(R)-BINAP (14)[20]

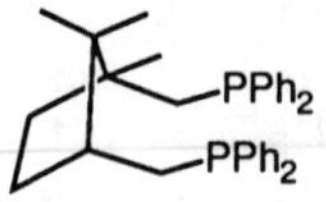

CAMPHOS (15)[21]

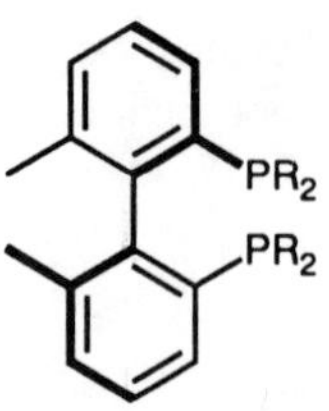

R = Ph: (R)-BIPHEMP (16)[22]
R = c-C_6H_{11}: (R)-BICHEP (17)[23]

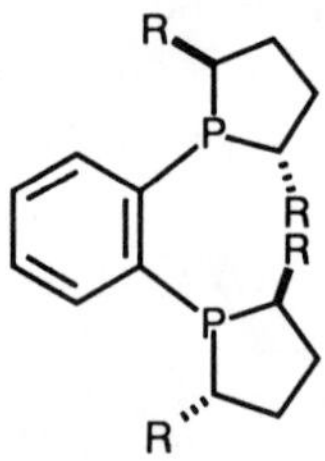

R = Me: Me-DuPHOS (18a)[24]
R = Et: Et-DuPHOS (18b)

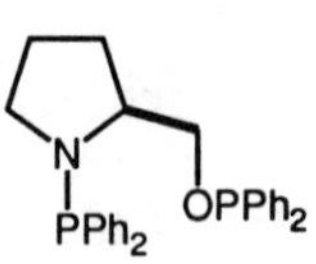

ProNOP (19)[25]

asymmetric hydrogenation of α-acylaminoacrylic acids [6,7]. Since then, a great number of optically pure diphosphines have been synthesized and used as chiral ligands [8]. Some typical chiral diphosphines (**1–19**) are listed in Chart 1. Ferrocenyl phosphines **13** are also important chiral ligands [9]. Chiral bidentate ligands such as diphosphinites [5b,26], bis(aminophosphine) [5b,27], and aminophosphine–phosphinites [5b,28] have also been developed [5b]. The Rh, Ru, and Ir complexes of these ligands have been used as catalysts for asymmetric hydrogenation of C=C, C=O, and C=N bonds. Much information on the catalytic activities and stereoselectivities of the chiral catalysts has been accumulated, but the choice of catalyst and reaction conditions still must be made empirically for an individual substrate.

Although many chiral dinitrogen ligands have also been developed, only a few of them are used for asymmetric hydrogenation. The Co complex of **20** was used for the asymmetric hydrogenation of α,β-unsaturated esters and amides in combination with $NaBH_4$ as a reducing agent [29]. The reduction of simple ketones by asymmetric transfer hydrogenation from 2-propanol was catalyzed by Ir complexes of **21** [30] and **22** [31].

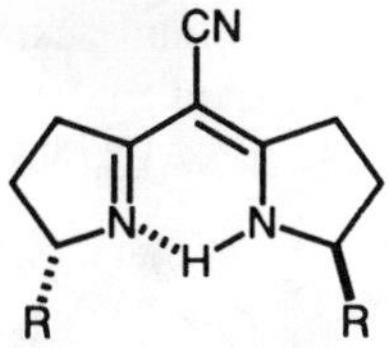

R = $CH_2OSiMe_2(t\text{-Bu})$ (**20**)[29] 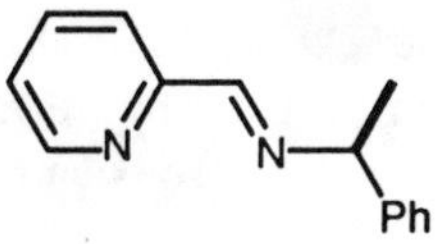   (S)-PPEI (**21**)[30] 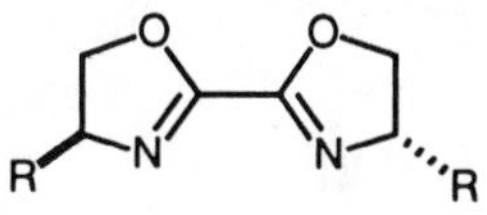   R = i-Pr (**22**)[31]

Cyclopentadienyl ligands are promising as chiral auxiliaries for the asymmetric hydrogenation and polymerization of simple olefins. Several chiral cyclopentadienes have been prepared. Cyclopentadiene **23** with C_2 symmetry was used for the Ti-catalyzed asymmetric hydrogenation of prochiral olefinic hydrocarbons [32].

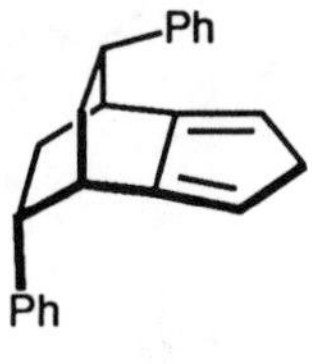

(**23**)[32]

1.1.2. Chiral Catalysts

Rhodium complexes of chiral diphosphines are usually prepared *in situ* by mixing [RhCl(diene)]$_2$ or [Rh(diene)$_2$]Y (Y = BF_4 and ClO_4) and a diphosphine, and then used directly as a catalyst. However, one must be careful in the preparation of catalysts, since rhodium species without phosphine ligands often work as catalysts for hydrogenation. In the case of the BINAP ligand, even the isolated complex [Rh(BINAP)(NBD)]ClO_4 (NBD = norbornadiene) formed two catalytically active species upon treatment with H_2 in methanol. One of them, [Rh(BINAP)(MeOH)$_2$]ClO_4, showed high enantioselectivities in the hydrogenation of N-acylaminoacrylic acids, while the other dimeric species was found to be a poor catalyst [33].

In the late 1970's, James reported his pioneering work on the use of chiral Ru(II) complexes as catalysts for asymmetric hydrogenation [34]. The Ru complexes, $Ru_2Cl_4(DIOP)_3$, RuHCl(DIOP)$_2$, and RuCl$_2$(DIOP)$_2$, were synthesized by a ligand substitution reaction of RuCl$_2$(PPh$_3$)$_3$ and excess DIOP, and used as catalysts for asymmetric hydrogenation.

The first BINAP–Ru(II) complex, **24**, was synthesized by the Tokyo University group according to equation 1 [35]. Mononuclear dicarboxylate complexes **25** are derived from **24** in 71–87% yields by treatment of **24** (or its derivatives) with a sodium carboxylate in t-butanol at 80°C (eq. 2) [36]. The anion exchange reaction can be performed in two phases with (PhCH$_2$NEt$_3$)Br as a phase transfer catalyst. The acetate complex **25a** was also prepared by the reaction of [RuCl$_2$(benzene)]$_2$ or RuCl$_2$(SbPh$_3$)$_3$ with BINAP in DMF followed by treatment with sodium acetate [37] or by the reaction of [Ru(OAc)$_2$(COD)] (COD = 1,5-cyclooctadiene) with BINAP [38]. A catalyst system obtained by treatment of Ru$_2$Cl$_4$(BINAP)$_2$(NEt$_3$) (**24**) with washed Dowex-50 resin has also been reported [39].

$$[RuCl_2(COD)]_n + (S)\text{-BINAP} \xrightarrow[\text{toluene}]{Et_3N} Ru_2Cl_4[(S)\text{-BINAP}]_2\cdot(NEt_3) \qquad (1)$$
$$(S)\text{-}\mathbf{24}$$

$$(S)\text{-}\mathbf{24} \xrightarrow[\substack{\text{or} \\ RCO_2Na, H_2O\text{—}CH_2Cl_2, \\ (PhCH_2NEt_3)Br}]{RCO_2Na,\ t\text{-BuOH}} \qquad (2)$$

$$\Lambda\text{-}(S)\text{-}\mathbf{25}$$

a: R = Me; Ar = Ph   **d**: R = Me; Ar = C$_6$H$_4$-4-Cl
b: R = Me; Ar = C$_6$H$_4$-4-Me   **e**: R = Me; Ar = C$_6$H$_4$-4-F
c: R = Me; Ar = C$_6$H$_4$-4-OMe   **f**: R = t-Bu; Ar = Ph

Addition of 2 equiv. of HX (HCl, HBr) or Me$_3$SiI to the acetate complex **25a** affords a mixture of halogen-containing species, defined by the empirical formula [RuCl$_2$(BINAP)], which are excellent catalysts for the asymmetric hydrogenation of functionalized ketones [40]. Several new halogen containing cationic BINAP–Ru(II) complexes have also been synthesized [41]. Treatment of **26** or **28** with one equiv. of (S)-BINAP gives (S)-**27** and (S)-**29**, respectively (eqs. 3 and 4). The iodide complex **27c** is prone to liberate benzene, while the corresponding p-cymene complex (S)-**29c** is stable enough to be isolated in pure form. The chloride ion of (S)-**27a** can easily be replaced by BF_4^- and BPh_4^-, by treatment with $AgBF_4$ in dichloromethane and with $NaBPh_4$ in methanol, to give (S)-**27d** and (S)-**27e**, respectively. A complex prepared *in situ* by simple mixing of **28c** and ligand **14** is also usable for the asymmetric hydrogenation of ketones [41]. BINAP–$RuCl_2$ complexes that are effective as catalysts for the hydrogenation of functionalized ketones are also formed when $[RuCl_2(benzene)]_2$ or $RuCl_2(SbPh_3)_3$ is treated with BINAP in DMF [42].

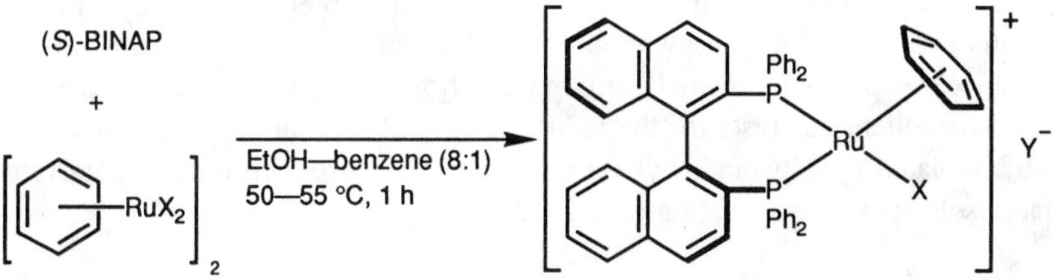

(S)-BINAP

+

26

a: X = Cl
b: X = Br
c: X = I

EtOH—benzene (8:1)
50—55 °C, 1 h

(S)-**27**

a: X = Y = Cl   90% yield
b: X = Y = Br   94%
c: X = Y = I   52%
d: X = Cl, Y = BF_4
e: X = Cl, Y = BPh_4

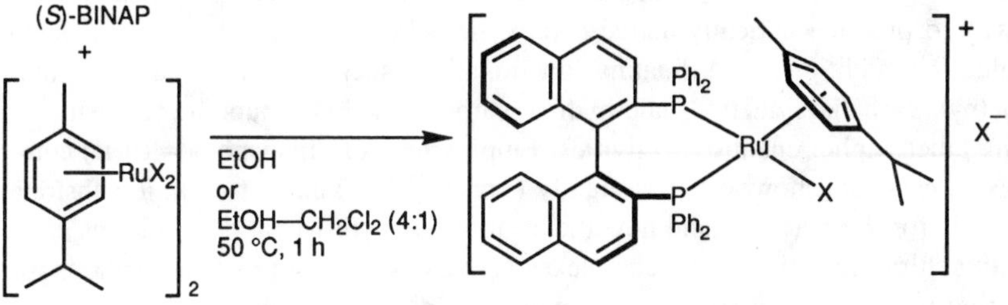

(3)

(S)-BINAP

+

28

a: X = Cl
b: X = Br
c: X = I

EtOH
or
EtOH—CH_2Cl_2 (4:1)
50 °C, 1 h

(S)-**29**

a: X = Cl   94% yield
b: X = Br   97%
c: X = I   94%

(4)

Recently, mononuclear complexes [**30**: R = Me; P—P = DIOP (**1**), CHIRAPHOS (**3**), NORPHOS (**7**), BINAP (**14**); **31**: P—P = DIOP (**1**), BPPM (**11**), BINAP (**14**), etc.; **32**: P—P, BIPHEMP (**16**), BINAP (**14**)] were prepared from chiral diphosphines and Ru(2-methallyl)$_2$(COD) [43], Ru(allyl)(acac-F_6)(COD) (acac-F_6 = hexafluoropentanedionate) [44], or [Ru(OCOCF$_3$)$_2$(COD)]$_2$ [38]. BICHEP–Ru complexes such as [RuX(BICHEP)(p-cymene)]X (X = Cl, I) and Ru(OCOMe)$_2$(BICHEP) have also been prepared [45].

30   **31**   **32**

P—P = diphosphine

Complexes of Ir and Co with chiral ligands are usually prepared in situ and used for asymmetric hydrogenations without isolation. For example, the semicorrin–cobalt complexes prepared in situ by mixing $CoCl_2$ with slight excess of semicorrin ligands are excellent catalysts for the reduction of α,β-unsaturated esters and amides [46]. Isolated cyclopentadienyl–Ti and –Sm complexes are used as catalysts for enantioselective hydrogenation of simple olefins [32, 105].

1.2. Asymmetric Hydrogenation of Olefins

Acrylic acids were chosen as substrates in the early studies on the asymmetric hydrogenation of olefins. Reduction of (E)-3-phenyl-2-butenoic acid, with neo-menthyldiphenylphosphine–rhodium complex as catalyst, attained 61% ee. This example showed that the asymmetric hydrogenation of acrylic acids was a promising way to produce optically active carboxylic acids [47]. However, it was soon revealed that additional coordinating functionalities such as amido, carboxyl, amidomethyl, carbalkoxymethyl, and hydroxycarbonylmethyl group, are requisite for getting much higher enantioselectivities. Employment of diphosphine–Ru(II) complexes as catalysts, however, has greatly expanded the kinds of olefinic substrate applicable for the reaction, and now the asymmetric hydrogenation of olefins catalyzed by chiral transition metal complexes is one of the most practical methods for preparing optically active organic compounds. Chiral cyclopentadienyl complexes of titanium and lanthanide elements have also opened a new way for asymmetric hydrogenation of simple olefins.

1.2.1. *N*-Acylaminoacrylic Acids and Related Compounds

α-Acylaminoacrylic acids (**33**) were the first olefinic substrates successfully used in the homogeneous asymmetric hydrogenation (eq. 5). Kagan reported enantioselective hydrogenation of (Z)-α-acetamidocinnamic acid with DIOP–Rh$^+$ complexes, giving *N*-acetylphenylalanine as the product with 72% ee [6]. The Monsanto group headed by Knowles developed a chiral ligand, (R)-o-anisylcyclohexylmethylphosphine [(R)-CAMP] [3] and DIPAMP (**2**) [7] and achieved very high enantioselectivities (>90% ee) in the Rh(I)-catalyzed asymmetric hydrogenation of **33**. A num-

Table 1 Enantiomeric Excess (%) (Absolute Configuration) of Products in Asymmetric Hydrogenation of α-Acylaminoacrylic Acids

	Ligand of Rh Complex					
Substrate	(R,R)-DIOP	(R,R)-DIPAMP	(S,S)-NORPHOS	(S,S)-BPPM	(S)-BINAP[a]	(S,R)-BPPFA
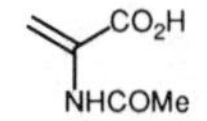 ![structure 1]	73 (R)	90 (S)	95 (R)	99 (R)	[98] (R)	76 (S)
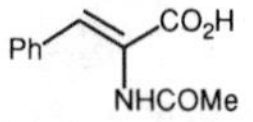 ![structure 2]	85 (R)	96 (S)	95 (R)	91 (R)	[100] (R)	93 (S)
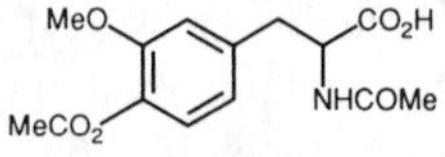 ![structure 3]	84 (R)	94 (S)	94 (R)	86 (R)	[79] (R)	86 (S)

[a] Figures in brackets indicate the results for N-benzoyl derivatives.

ber of chiral phosphine ligand–metal complexes are now known to be efficient catalysts for this type of asymmetric hydrogenation (Table 1) (see Refs. 5b, 8, 23, 24 and references cited therein). In contrast to the high ee that is obtained for (Z)-α-acylaminoacrylic acids and esters, the asymmetric hydrogenation of the corresponding (E)-isomers usually proceeds very slowly with poor enantioselectivity, although some alkyl-substituted (E)-α-acylaminoacrylic acids are hydrogenated to give the products with high enantiomeric purities [48]. Recently, very high catalytic activities for (Z)-isomers were achieved by the PYROPHOS–Rh [17], PPCP–Rh [49], BICHEP–Rh complexes [23, 50], etc. [51]. N-Acylaminoacrylic acids, such as (Z)-2-acetamido-4-(methylhydroxyphosphinyl)but-2-enoic acid [52] and methyl (Z)-2-acetamido-3-(3-pyridyl)acrylate [53], were also reduced to give the products with 91 and 99% ee, respectively.

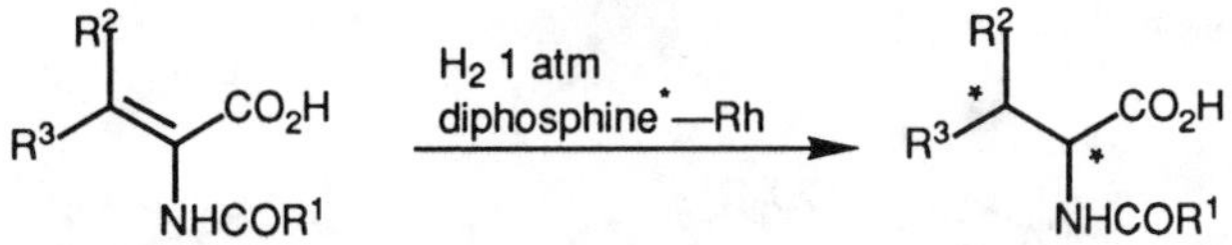

$$(5)$$

With a water-soluble catalyst, [Rh(COD){(3R,4R)-3,4-bis(diphenylphosphino)-1,1-dimethylpyrrolidinium-P,P'}] · 2BF$_4$, the sodium salt of α-acetamidocinnamic acid was hydrogenated in aqueous media to give the product with 90% ee [54–56].

The mechanism of the reaction has been elucidated by detailed studies on the catalytic cycle [57–59]. Halpern and coworkers determined the structure of a key intermediate, the catalyst–substrate adduct [Rh((S,S)-CHIRAPHOS)(EAC)]$^+$, in the asymmetric hydrogenation of ethyl 2-acetamidocinnamate (EAC) by X-ray analysis and demonstrated that the minor diastereomeric catalyst–substrate adduct is more reactive (at least 10^3 times) toward H$_2$ than the major one and thus gives the predominant product.

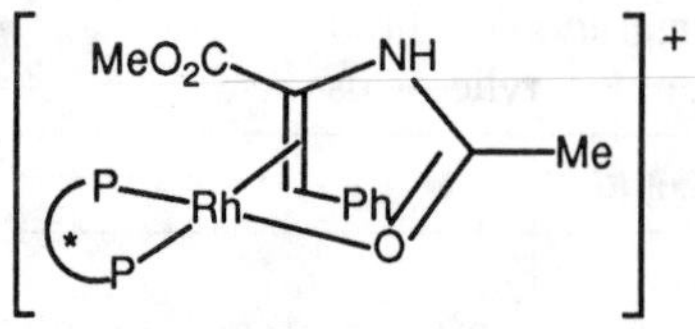

Minor diastereomeric adduct 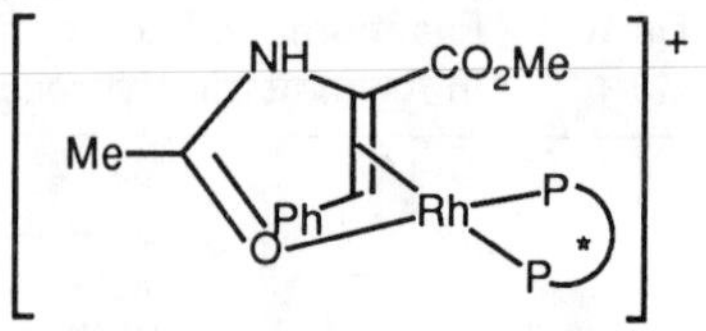   Major diastereomeric adduct

$[Rh((S,S)\text{-CHIRAPHOS})(EAC)]^+$

EAC = ethyl (Z)-α-acetamidocinnamate

The origin of enantioselectivity in the asymmetric hydrogenation of *N*-acyl-aminoacrylic acids has been discussed based on the structures of cationic rhodium complexes with chiral diphosphines determined by *X*-ray crystallography [60–63]. Dissymmetric arrangement of phenyl rings on phosphorus atoms seems to play an important role.

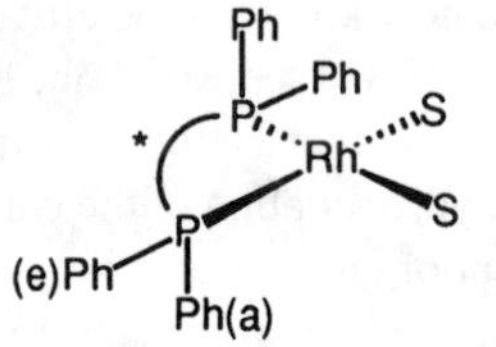

P chirality 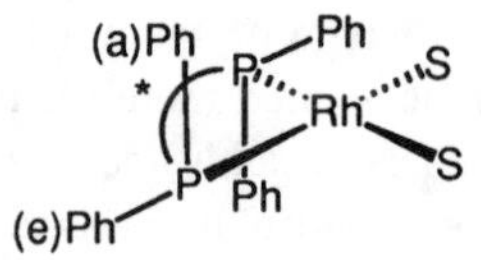   **M** chirality

BINAP–Ru(II) complexes are also efficient catalysts for the asymmetric hydrogenation of *N*-acylaminoacrylic acids and their derivatives [35]. Compound **35** was hydrogenated to **36** with 79–92% ee by using Ru complexes bearing (R)-BINAP [35,64]. It should be noted that with $[Rh((R)\text{-BINAP})(MeOH)_2]ClO_4$ as the catalyst [33,65], the opposite enantioface selection was observed, giving the other enantiomer **37** with 92–100% ee.

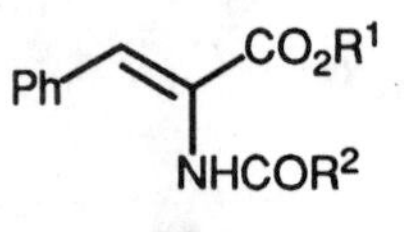

35 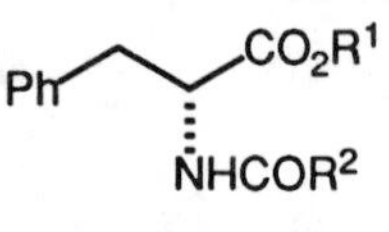   **36** 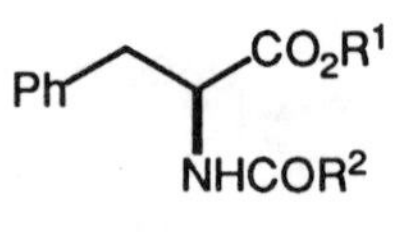   **37**

(E)-β-*N*-Acylaminoacrylic acid derivatives were hydrogenated with BINAP–Ru(II) complexes to give the products with up to 96% ee [66]. The (Z)-isomers are more reactive, but enantioselectivities are poor. Here again the enantioface selection exhibited by BINAP–Ru(II) complexes was opposite to that observed in the catalysis of BINAP–Rh(I) complexes.

The hydroxycarbonyl group in α-acylaminoacrylic acids can be replaced by other electron-withdrawing groups such as alkoxycarbonyl, carbamoyl, keto, and cyano groups [5b]. These substrates **38** are reduced to **39** with high enantioselectivities (eq. 6).

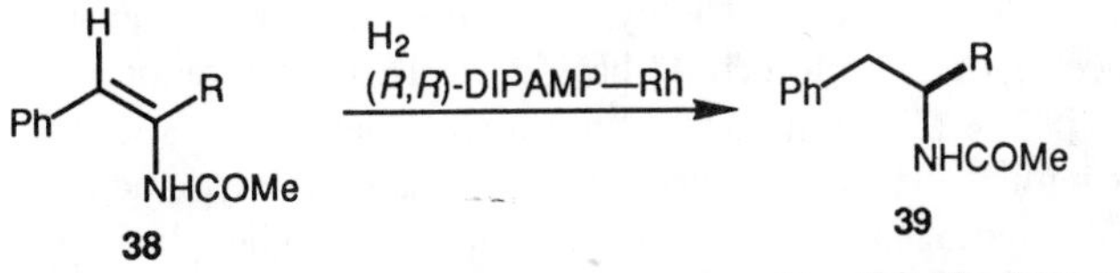

$$(6)$$

R = CO$_2$Me; 95% ee
CONH$_2$; 94% ee
COPh; 85% ee
CN; 89% ee (N-benzoyl)

Dipeptides and oligopeptides have been obtained with high enantiomeric purities by asymmetric hydrogenation of the corresponding dehydropeptides catalyzed by Rh complexes [67]. This methodology provides an effective and convenient method for the preparation of peptides. Diastereoselective hydrogenation has been carried out with diphosphine–Rh and –Ru complexes [68,69]. Carbamate-directed stereoselective hydrogenation and kinetic resolution of N-carbalkoxy-2-(aminoalkyl)acrylates were also reported [68].

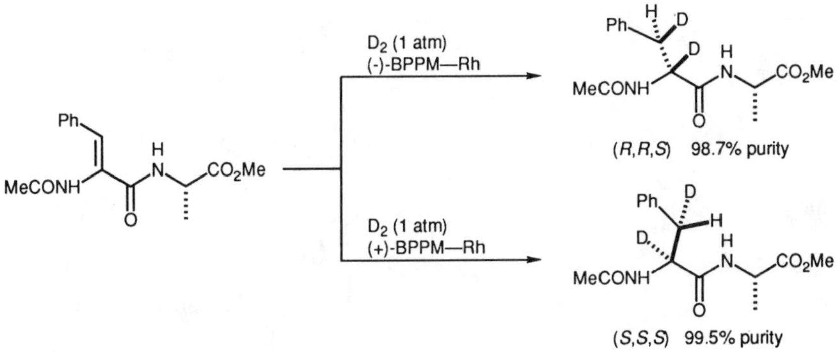

1.2.2. Enamides

Acyclic enamides **40** without other functionality were hydrogenated to **41** by using DIOP–Rh(I) complexes with up to 92% enantioselectivity (eq. 7) [70].

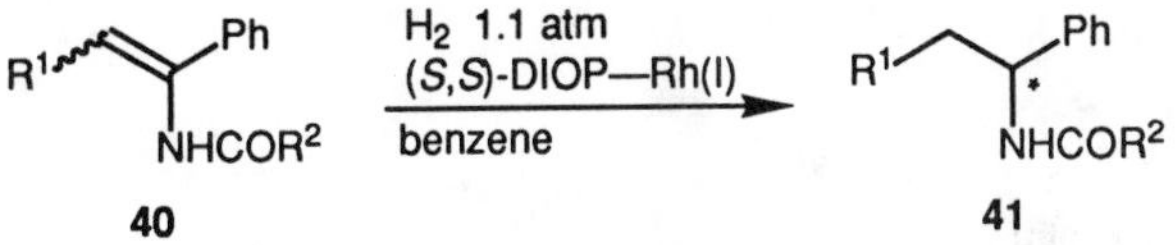

$$(7)$$

R^1	R^2	%ee
(Z)-Me	Me	92%
(Z)-Me	i-Pr	90%
Et	Me	90%
(Z)-c-C$_6$H$_{11}$	Me	82%

1-Substituted tetrahydroisoquinolines of type **43** are an important class of physiologically active compounds, which also serve as key intermediates for the prepara-

tion of a variety of isoquinoline alkaloids. The hydrogenation of (Z)-enamide substrates **42** in the presence of (R)-**25** afforded **43** having the 1R configuration with
95–100% ee (eq. 8) [71]. The (E)-isomers of **42** are inert under the same conditions. The Ru complex with BIPHEMP showed almost the same catalytic activity
and stereoselectivity in the hydrogenation of N-formyl-1-(4-methoxyphenylmethyli-
dene)-1,2,3,4,5,6,7,8-tetrahydroisoquinoline [38].

This enantioselective reaction, followed by removal and/or modification of the
N-acyl group, leads easily to tetrahydropapaveline (**44**), laudanosine (**45**), norreticurine (**48**), trethoquinol (**49**), and salsolidine (**50**) with high enantiomeric purities.

This method has been further applied to substrate **51**, establishing a general route
to benzomorphans and morphinans based on asymmetric catalysis (eq. 9) [72].

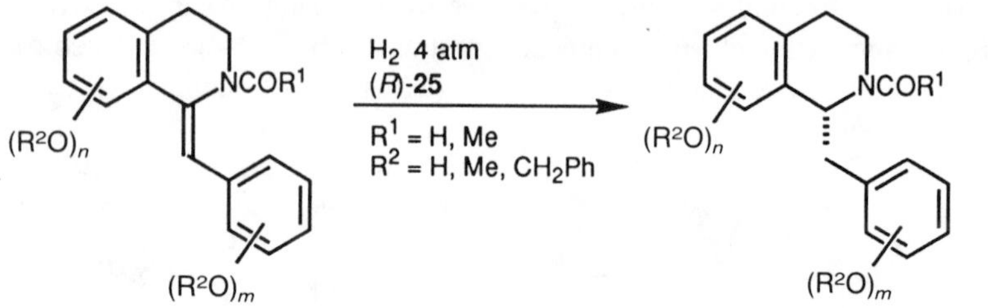

$$\text{(8)}$$

$$R^1 = H, Me$$
$$R^2 = H, Me, CH_2Ph$$

42

43
95—100% ee

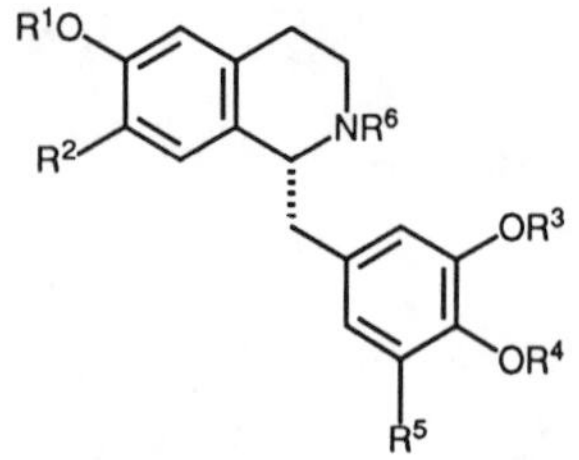

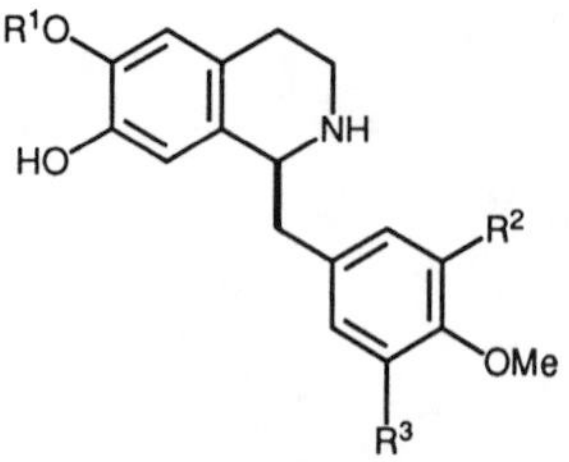

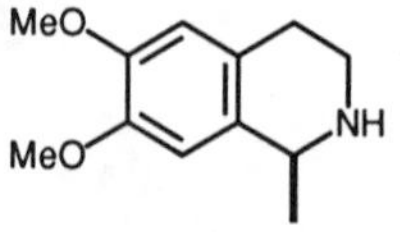

	R^1	R^2	R^3	R^4	R^5	R^6	% ee
44	Me	OMe	Me	Me	H	H	100
45	Me	OMe	Me	Me	H	Me	100
46	Me	H	CH$_2$Ph	Me	OCH$_2$Ph	H	97
47	Me	H	H	Me	H	H	95

	R^1	R^2	R^3	% ee
48	Me	OH	H	96
49	H	OMe	OMe	100

50   97% ee

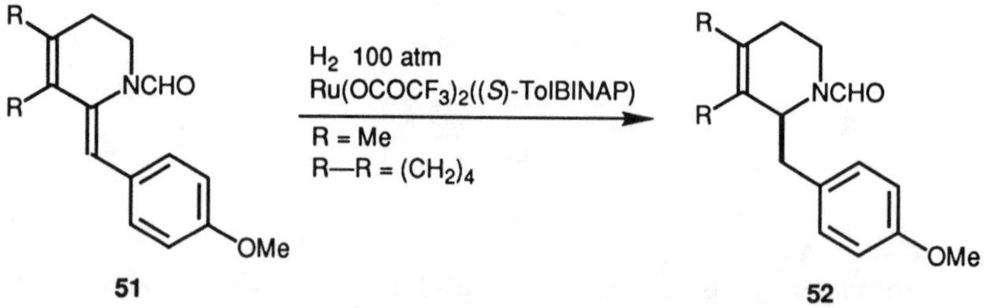

H$_2$ 100 atm
Ru(OCOCF$_3$)$_2$((S)-TolBINAP)

R = Me
R—R = (CH$_2$)$_4$

$$\text{(9)}$$

51

52
97% ee

1.2.3. Enol Derivatives

Enol esters are interesting substrates for asymmetric hydrogenation because the products can easily be converted to optically active alcohols. Several enol esters have been used in asymmetric hydrogenation catalyzed by Rh complexes with chiral diphosphine ligands such as DIPAMP [5b,73], PROPHOS [74], and DIOP [75]. Substrates, such as ethyl 2-oxo-3-phenylpropanoate enol acetate (DIPAMP–Rh, 95% ee [5b]), 2,4-pentanedione enol acetate (DIPAMP–Rh, 90% ee [73]), and ethyl acetoacetate enol acetate (DIPAMP–Rh, 89% ee [73]; PROPHOS–Rh, 81% ee [74]) which have an additional functional group, are hydrogenated with high enantioselectivities. Simple enols are difficult to hydrogenate to give products with high enantiomeric excess, except for the case of acetophenone enol diphenylphosphinate (DIOP–Rh, 80% ee [75]). Recently, chiral diphopholanes were reported to be excellent ligands for the asymmetric hydrogenation of enol acetates **53** to acetate **54** (eq. 10). For example, trifluoroacetone enol acetate and 1-naphthyl methyl ketone enol acetate are hydrogenated by using a complex of rhodium and the diphospholanes (**18** and **55**) to give the corresponding esters with 95 and 94% ee, respectively [24b].

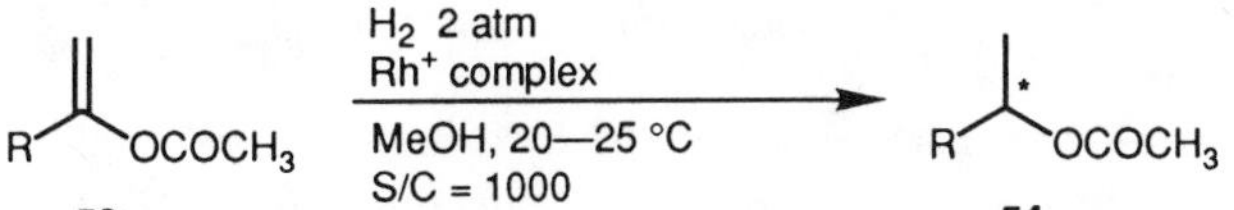

$$\text{(10)}$$

R	Ligand[b]	% ee	Config.
Ph	(S,S)-18a	89	S
C₆H₄-4-NO₂	(R,R)-55	90	R
C₆H₄-3-Cl	(S,S)-18b	91	a
1-Naphthyl	(R,R)-55	94	R
CO₂Et	(R,R)-18b	>99	R
CF₃	(R,R)-55	>95	R

a Configuration was not determined.

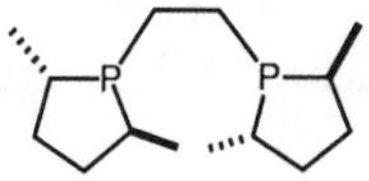

55

Enol lactone **56** was hydrogenated smoothly in the presence of BINAP–Ru(II) complexes to give lactone **57** with 94% ee (eq. 11) [76].

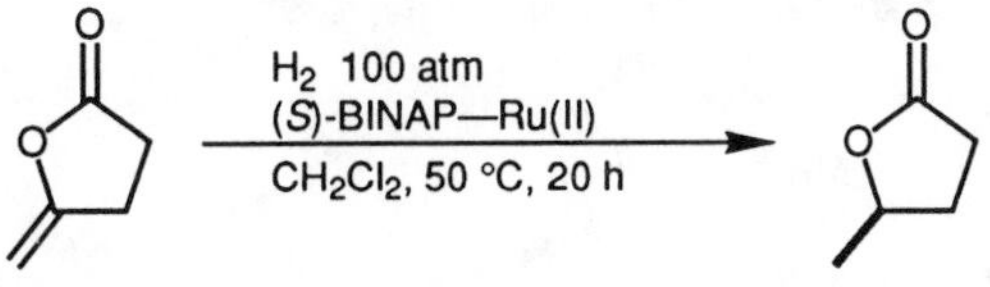

$$\text{(11)}$$

56

57
94% ee

1.2.4. α,β-Unsaturated Esters, Amides, and Ketones

There are only a few examples of asymmetric hydrogenation with high enantioselectivity for olefins without an amide or hydroxycarbonyl group. Methyl itaconate was hydrogenated by using diphosphine–Rh catalysts with high enantioselectivity (BICHEP–Rh, > 99% ee [23,50]; DIPAMP–Rh, 88% ee [77]; BPPM derivatives–Rh, 93% ee [78]). A triphospholane–Rh complex was also usable for the asymmetric hydrogenation of methyl itaconate (94% ee) [79].

Recently, α,β-unsaturated carboxylates **58** were reduced to **59** with sodium borohydride in the presence of semicorrin **20**–Co catalyst with up to 96% ee (eq. 12). The (E)- and (Z)-isomers afforded products with opposite configurations. The isolated double bonds in the substrates remained untouched [80].

$$R \diagup \diagdown CO_2Et \ + \ NaBH_4 \ \xrightarrow[\text{EtOH/DMF, 25 °C}]{\substack{\text{1 mol% of CoCl}_2 \\ \text{1.2 mol% of ligand 20}}} \ R \diagup CO_2Et \qquad (12)$$

58   **59**

R = (CH₂)₂Ph;   94% ee
R = CH₂CH=CMe₂;   94% ee
R = i-Pr;   96% ee
R = Ph;   81% ee

α,β-Unsaturated amides **60** were also reduced to **61** by using the same catalyst system described above (see eq. 13). For example, N,N-dimethyl-3-methyl-(E)-5-phenyl-2-butenamide was reduced at room temperature to the corresponding saturated amide with 98.7% ee [81]. When the amount of catalyst was decreased from 2 mol % to 0.1 % at constant substrate and borohydride concentrations, the reaction rate remained essentially the same, whereas the enantiomeric excess increased from 96.5 to 98.7% ee. However, a very low catalyst concentration (e.g., substrate/catalyst ratio of 10,000:1) caused a substantial decrease in the reaction rate. The enantioselectivity also decreased to some extent, but stayed above 95% ee. In keeping with the results obtained for the corresponding esters [80], the (E)- and (Z)-isomers were converted to the products with opposite absolute configuration.

$$R^2 \diagup\diagdown^{R^1} CONHR^3 \ + \ NaBH_4 \ \xrightarrow[\text{EtOH/diglyme}]{\substack{\text{1 mol% of CoCl}_2 \\ \text{1.2 mol% of ligand 20}}} \ R^2 \diagup^{R^1} CONHR^3 \qquad (13)$$

60   **61**

R¹	R²	R³	% ee	Config.
Me	(CH₂)₂Ph	H	96.6	R
Me	(CH₂)₂Ph	Me	98.7	R
Me	(CH₂)₂Ph	i-Pr	95.4	R
(CH₂)₂Ph	Me	H	95.0	S
(CH₂)₂Ph	Me	Me	97.2	S
Me	c-C₆H₁₁	Me	98.9	a
Me	Ph	Me	92.4	S
c-C₆H₁₁	Me	Me	97.2	a

a Configuration was not determined.

With the catalysts derived from (S,S)-1,2-bis(diphenylphosphinomethyl)cyclo-butane and RhH(CO)(PPh$_3$)$_3$ or rhodium carbonyls, the α,β-unsaturated aldehydes neral and geranial were hydrogenated to (R)- or (S)-citronellal with 79% and 60% ee, respectively [82].

Cyclic α,β-unsaturated ketones such as isophorone and 2-methyl-2-cyclohexe-none have been hydrogenated by the use of ruthenium hydrides coordinated with chiral diphosphines to give the saturated ketones with up to 62% ee, though conver-sions are not satisfactory [83].

Asymmetric hydrogenation of 2-alkylidene-γ-butyrolactones (62a,b) and 2-pen-tylidenecyclopentanone (62c) catalyzed by BINAP–Ru(II) complexes affords the corresponding γ-butyrolactones (63a,b) and cyclopentanone (63c) with 95–98% ee, respectively (eq. 14) [76]. Asymmetric hydrogenation of (E)- and (Z)-2-propyl-idene-γ-butyrolactones catalyzed by the same catalyst gave the products of the same absolute configuration with almost equal enantioselectivities; that is, olefin geome-try does not affect the stereochemistry and enantioselectivity.

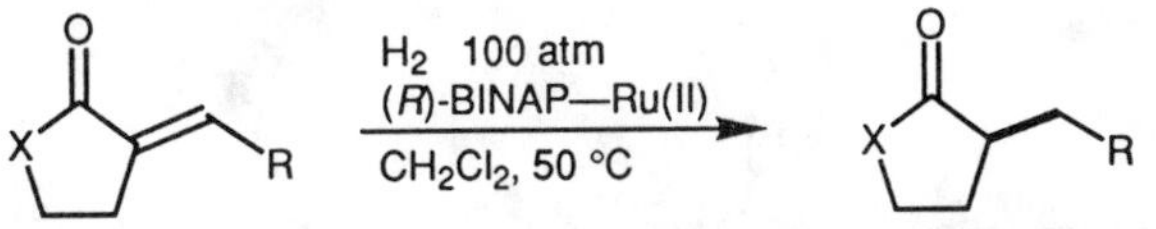

$$ (14) $$

a: X = O, R = H; 95% ee
b: X = O, R = Et; 95% ee
c: X = CH$_2$, R = n-Bu; 98% ee

1.2.5. Unsaturated Carboxylic Acids

Itaconic acid (64) has been reduced to (S)-2-methylsuccinic acid (65) with 95% ee in the presence of (S,S)-R-CAPP–Rh$^+$ complex (eq. 15) [84]. When BICHEP–Rh$^+$ complex was used as the catalyst, the reaction completed in 5 minutes and 65 was obtained with 93% ee under mild conditions (S/C = 10,000, H$_2$ 5 atm, room temperature) [50]. Asymmetric reduction of itaconic acid by transfer hydrogena-tion, with the catalyst derived from [RhCl(COD)]$_2$ and BPPM and triethylam-monium formate as the hydrogen source, also proceeds smoothly to give 65 with 84% ee [85]. Several new ligands such as MOD-DIOP, having electron-donating substituents on the phenyl moieties, were developed, and their complexes showed better enantioselectivities than those obtained by using DIOP–Rh complex in the asymmetric hydrogenation of itaconic acid, methyl itaconate, dimethyl itaconate, and their derivatives [86,87].

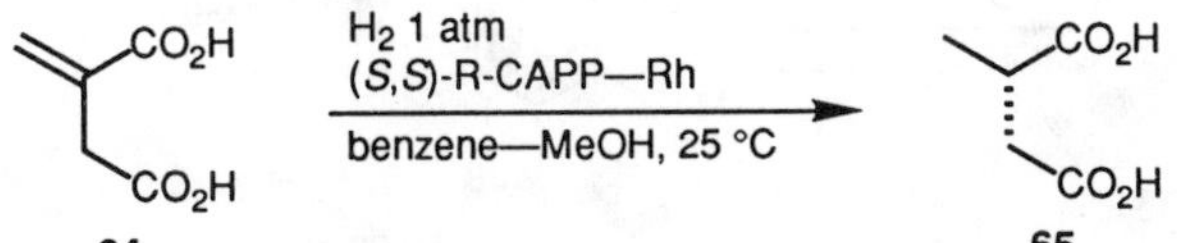

$$ (15) $$

95% ee

BINAP–Ru complexes are also efficient catalysts for the asymmetric hydrogenation of itaconic acid (88% ee) and derivatives (75–94% ee) [35b,88]. Asymmetric hydrogenation of a conjugated diacid, 2,3-dimethylidenesuccinic acid, catalyzed by BINAP–Ru complex, afforded the corresponding saturated diacid with 98% diastereoselectivity (de) and 96% ee [89].

The complexes of type **25** are excellent catalysts for the asymmetric hydrogenation of acrylic acids (**66**) having only carboxylic acid functionality (eq. 16) [90]. The enantioselectivity is highly dependent on the structure of the substrate and the reaction conditions, including initial pressure of hydrogen. As shown below, various oxygen-functionalized unsaturated carboxylic acids can be used as substrates. Asymmetric hydrogenation of **67** catalyzed by (*S*)-**25a** afforded (*S*)-naproxen (**68**), a useful anti-inflammatory agent, with 97% ee in 92% yield (eq. 17). Certain β,γ-unsaturated carboxylic acids were also hydrogenated with 81–88% ee. Other BINAP–Ru(II) complexes such as Ru(2-methallyl)$_2$(BINAP) (**30**) [43] and Ru(allyl)(acac-F$_6$)(BINAP) (**31**) [44] were also used for the asymmetric hydrogenation of tiglic acid.

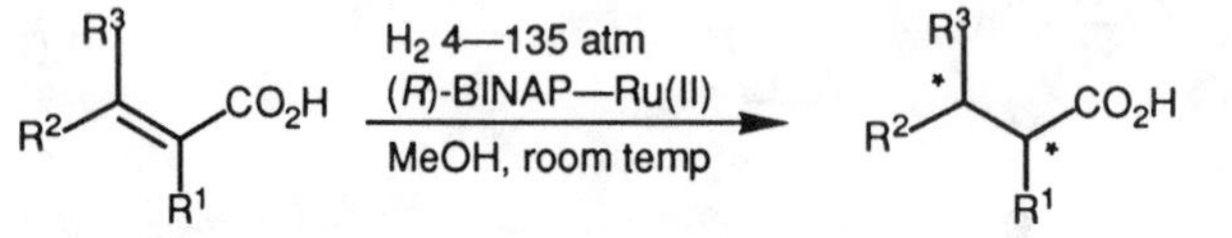

(16)

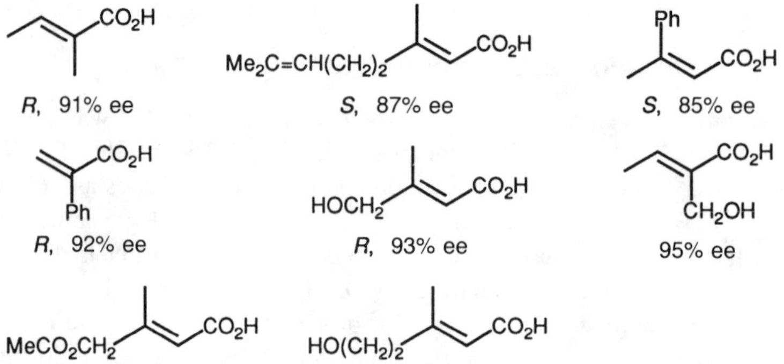

R, 91% ee

S, 87% ee

S, 85% ee

R, 92% ee

R, 93% ee

95% ee

R, 95% ee

S, 93% ee

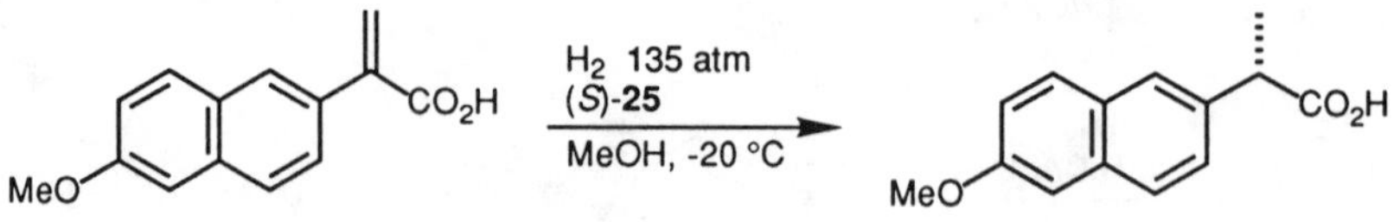

(17)

67

68   97% ee
(*S*)-Naproxen

Recently, very high enantioselectivity has been accomplished in the asymmetric hydrogenation of trisubstituted acrylic acids **69** catalyzed by a cationic **13c**–Rh complex (eq. 18) [91].

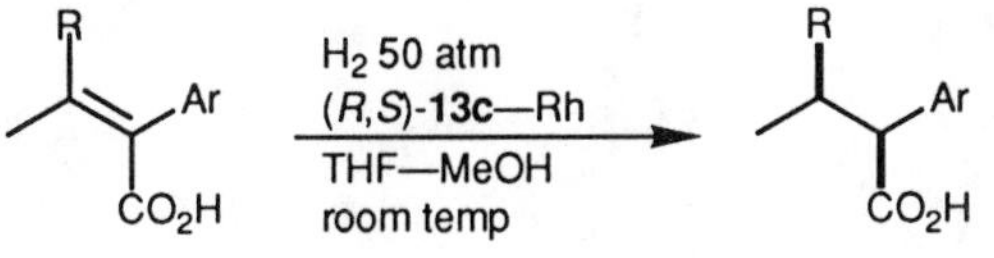

(18)

R	Ar	% ee
Me	Ph	98.4
Me	C$_6$H$_4$-4-Cl	97.4
Me	C$_6$H$_4$-4-OMe	96.7
Me	2-naphthyl	97.3
Et	Ph	97.3
Ph	Ph	92.1

Some experimental evidence for mechanisms has been obtained with complex **25**. Deuterium incorporation experiments for the asymmetric hydrogenation of acrylic acid derivatives have shown that α-hydrogens in the products usually originate from gaseous hydrogen, while most of β-hydrogens (50–100%) come from protic solvents or substrates. The overall stereochemistry of the hydrogen addition is perfectly cis in the deuteration of tiglic acid and cinnamic acid (eq. 19) [92]. These experiments suggest that the reaction of acrylic acids proceeds through a mechanism involving a monohydride complex of type **70**. Ashby and Halpern have clarified the kinetic aspects of this reaction and have suggested that the reaction intermediate is anionic monohydride ruthenium species **71** [93].

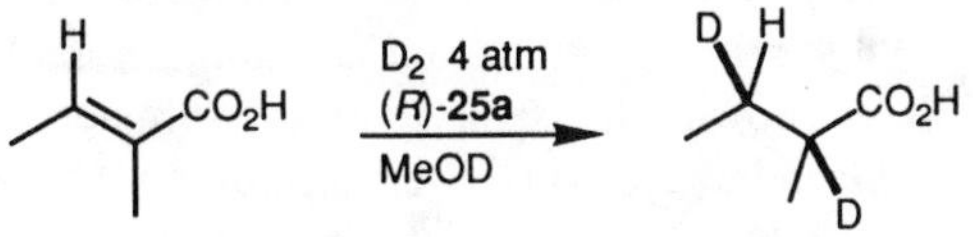

(19)

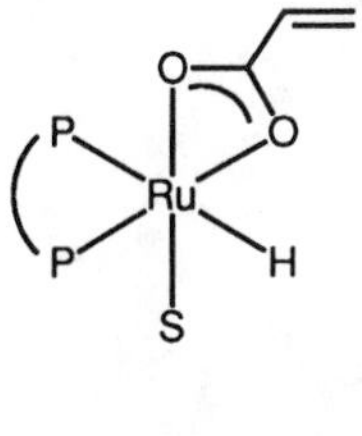

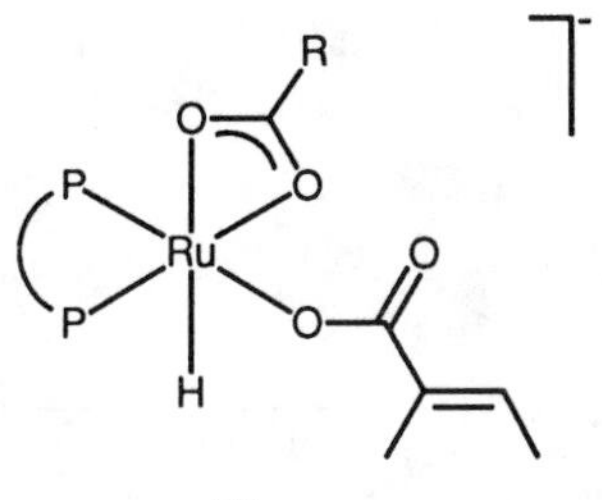

1.2.6. Allylic and Homoallylic Alcohols

Cationic phosphine–Rh and –Ir complexes have been used for diastereoselective hydrogenation of chiral allylic and homoallylic alcohols [94]. Here, the preexisting

Table 2 Asymmetric Hydrogenation of Geraniol (**72**) and Nerol (**73**)

Substrate	Catalyst	S/C	% op[a]	% ee[b]	Config.
72	Ru(OCOMe)₂((S)-BINAP) [(S)-**25a**]	530	98	96	R
72	Ru(OCOMCMe₃)₂((S)-BINAP) [(S)-**25f**]	500		98	R
72	Ru(OCOMe)₂((S)-TolBINAP) [(S)-**25b**]	10,000	99	96	R
72	Ru(OCOCF₃)₂((S)-BINAP)	50,000	96		R
72	Ru(OCOCF₃)₂((S)-TolBINAP)	50,000	97		R
73	[(R)-**25a**]	540		98	R
73	[(S)-**25b**]	570		98	S

[a] Determined by optical rotation.
[b] Determined by high performance liquid chromatography.

chirality at the sp^3-hybridized carbons induces new asymmetry on the neighboring olefinic diastereofaces through coordination of the hydroxyl group to the transition metals. High enantioselectivity, however, has not been obtained for the asymmetric hydrogenation of prochiral allylic alcohols catalyzed by Rh complexes. For example, the reaction of geraniol (**72**) catalyzed by the (S)-CyBINAP–Rh complex gave (S)-citronellol with 66% ee [95].

On the other hand, the asymmetric hydrogenation of prochiral allylic and homo-allylic alcohols catalyzed by Ru(II) complexes **25** afforded the products with exceptionally high enantioselectivities [96]. Geraniol (**72**) and nerol (**73**) were hydrogenated in methanol at room temperature under an initial hydrogen pressure of 90–100 atm to give citronellol (**74**) with 96–99% ee in nearly quantitative yield (Scheme 1 and Table 2). This catalyst is extremely efficient, so that the substrate/catalyst mole ratio (S/C ratio) of 50,000 is practical. The allylic and nonallylic double bonds in the substrates can clearly be differentiated to give the product contaminated with less than 0.5% dihydrocitronellol. Asymmetric hydrogenation of homogeraniol gave 4,8-dimethyl-7-nonenol with 92% ee, showing the same sense of asymmetric induction. The Ru complexes of BIPHEMP and its derivative were also used for this hydrogenation [38]. With Ru(OCOCF₃)₂(BIPHEMP), **72** and tetrahydrofarnesol were hydrogenated with 98.2 and 98.7% ee, respectively.

Scheme 1

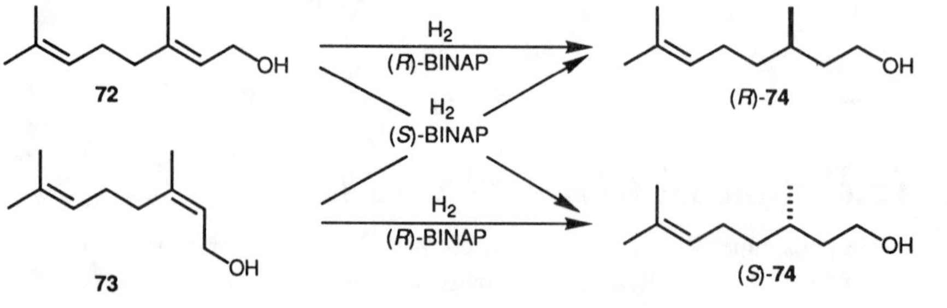

As shown in Scheme 2, this method has been successfully applied to the synthesis of (3*R*,7*R*)-3,7,11-trimethyldodecanol (**78**), a key intermediate for the synthesis of α-tocopherol (vitamin E) (**75**). The synthesis of **78** starts from citronellal (**76**), whose asymmetric synthesis has already been established for the industrial production of (−)-menthol. Conversion of **76** to **77** followed by asymmetric hydrogenation catalyzed by the BINAP–Ru complex affords the desired alcohol **78** with 99% diastereoselectivity [96].

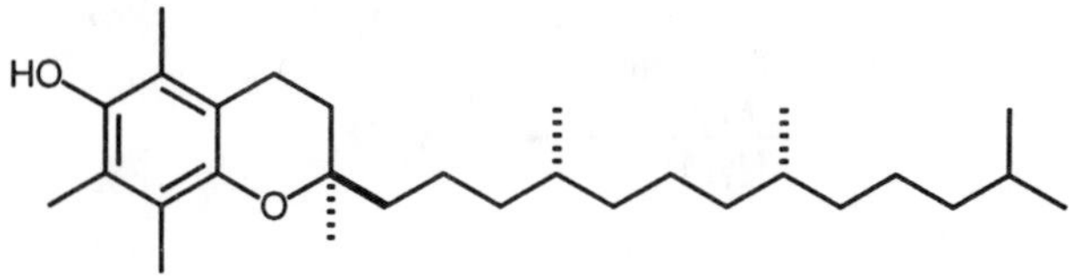
HO
O

75

Scheme 2

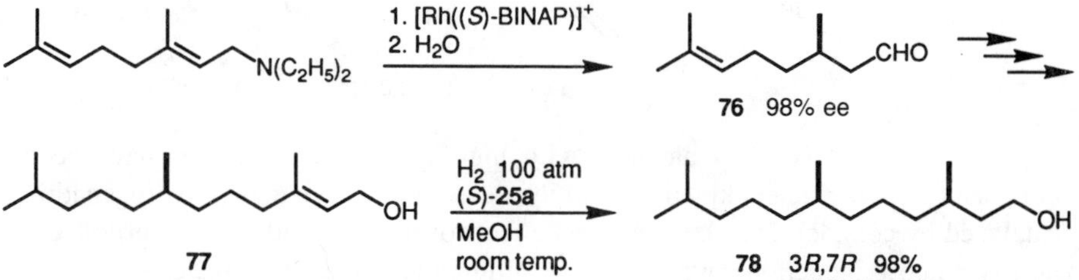
N(C2H5)2
1. [Rh((S)-BINAP)]+
2. H2O
CHO
OH
H2 100 atm
(S)-25a
MeOH
room temp.
OH

76 98% ee

77

78 3*R*,7*R* 98%
3*S*,7*R* 1%
3*R*,7*S* 1%

This asymmetric hydrogenation of allylic alcohols with the use of BINAP–Ru(II) was successfully applied to the synthesis of dolichols (**80**) via plant polyrenols **79** (eq. 20) [97].

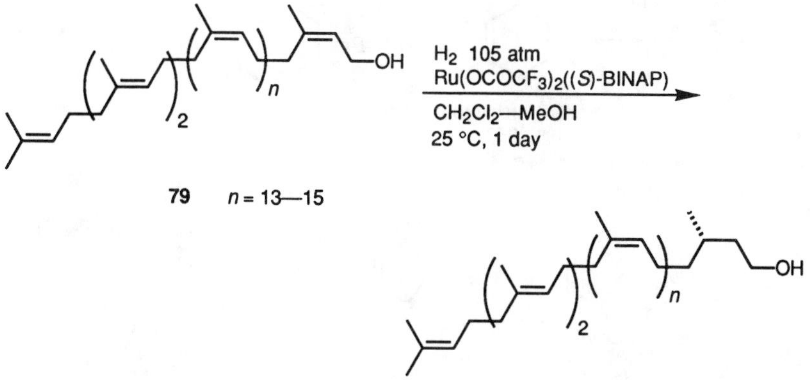
OH
n
2
H2 105 atm
Ru(OCOCF3)2((S)-BINAP)
CH2Cl2—MeOH
25 °C, 1 day
OH
n
2

79 *n* = 13—15

(20)

80 >95% optical purity

Compound β-**82**, an important intermediate in the synthesis of 1β-methylcarbapenems, was prepared from **81** using asymmetric hydrogenation catalyzed by BINAP–Ru(II) complexes [98]. With the use of $Ru(OCOMe)_2$((*R*)-TolBINAP) [(*R*)-**25b**], the ratio of β- to α-**82** went as high as 99.9:0.1. Only moderate selectivity (β/α = 22:78) was obtained when the enantiomeric (*S*)-**25b** was used. These results clearly indicate that effective double-asymmetric induction for the matching pair is operative in the former case.

81

TMBDS = $SiMe_2$(*t*-Bu)

β-**82** (99.9:0.1) α-**82**

(21)

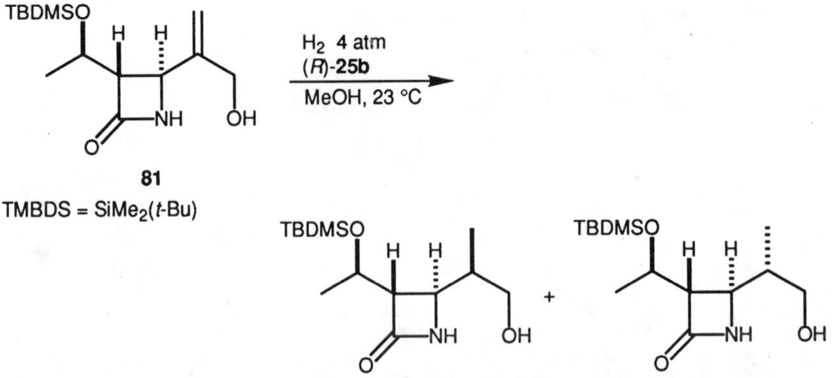

Chiral allylic secondary alcohols with high enantiomeric purities have been obtained through efficient kinetic resolution occurring in asymmetric hydrogenation catalyzed by complex **25**. The combined effects of intramolecular and intermolecular asymmetric induction gave up to 76:1 differentiation between the enantiomeric unsaturated alcohols **83** (eq. 22) [99]. This method was applied to a practical kinetic resolution of 4-hydroxy-2-cyclopentenone (**84**). Asymmetric hydrogenation of racemic substrate **84** with 0.1 mol % of (*S*)-**25** at 4 atm proceeded with k_f/k_s 11 and, at 60% conversion, afforded slow-reacting (*R*)-**84** with 98% ee.

83   **83** (recovered) 49%, 95% ee k_f/k_s = 76   trans:cis = 66:1

(22)

(±)-**84**   (*R*)-**84**

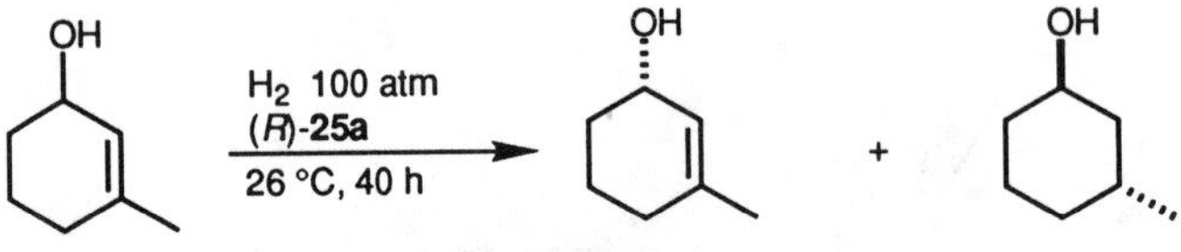

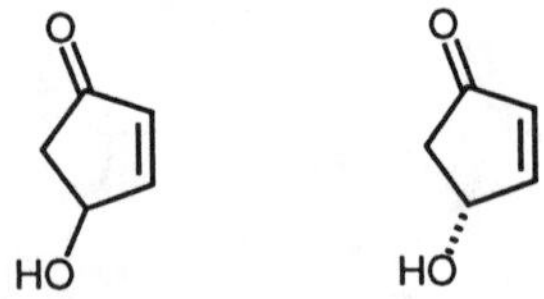

1.2.7. Simple Olefins

Asymmetric hydrogenation of simple olefins with high enantiomeric excesses has not been attained using chiral Rh and Ru catalysts [100]. Fused cyclopentadienyl ligands such as **23** with C_2 symmetry have been successfully used in the enantioselective titanocene-catalyzed hydrogenation of simple alkenes [32,101–103]. In this reaction, 2-phenyl-1-butene (**85**) was hydrogenated to give **86** with 95% ee at −75°C [32]. A chiral Ti complex coordinated with ethylenebis(tetrahydro-1-indenyl) catalyzed the asymmetric deuteration of styrene at room temperature to give ethylbenzene-1,2-d_2 with 65% ee [104]. Samarium complexes have also been used as catalysts for asymmetric hydrogenation of **85** [105]. The chiral complex **88** brought about high enantioselectivity and high catalytic activity (e.g., its turnover number reached 20,000/h at room temperature).

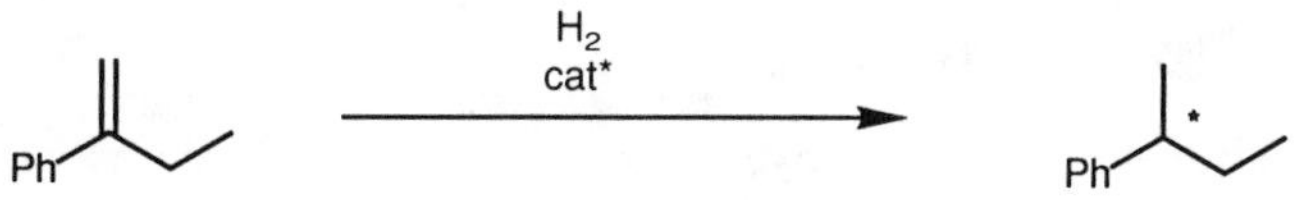

(23)

Catalyst (cat*)	Conditions	Product
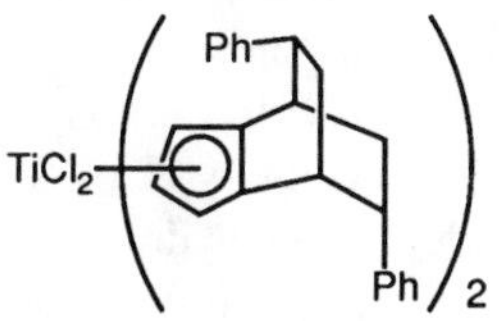 **87** + n-BuLi	-75 °C	95% ee (S)
70/30 (S)/(R)-**88b**[a,b]	-78 °C	96% ee (S)
70/30 (S)/(R)-**88b**[a,b]	25 °C	64% ee (S)
(R)-**88a**[a,b]	25 °C	71% ee (R)

[a]

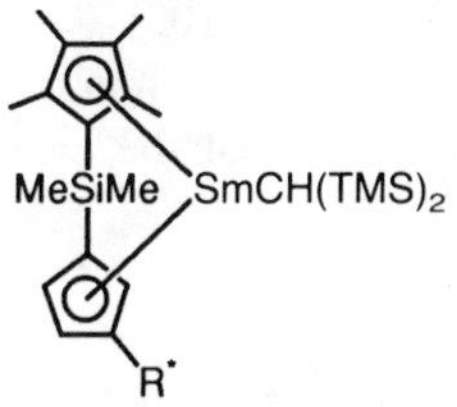

88a R* = (+)-neomenthyl
88b R* = (-)-menthyl

[b] (S) and (R) represent the chirality of cp ligand.

1.3. Asymmetric Hydrogenation of Ketones

Optically active secondary alcohols with a neighboring functional group are extremely useful starting materials for the syntheses of various biologically active compounds. Consequently, the asymmetric hydrogenation of functionalized ketones catalyzed by chiral transition metal complexes has attracted much interest. Diphosphine complexes of Rh and Ru have been used as catalysts for this reaction. Diphosphine–Ru complexes are also excellent hydrogenation catalysts for dynamic kinetic resolution of chiral α-substituted β-keto esters. High enantioselectivity in the hydrogenation of simple ketones has been difficult to attain with conventional chiral diphosphine complexes. Recently, however, some complexes of Rh and Ir with chiral nitrogen ligands were shown to be excellent catalysts for enantioselective transfer hydrogenation of alkyl aryl ketones.

1.3.1. Functionalized Ketones

Complexes **25** can catalyze the asymmetric hydrogenation of α-amino ketones **89** to give amino alcohols **90** with high enantioselectivity (eq. 24) [106]. Although other functionalized ketones such as β-keto ester **91** cannot be reduced by using **25**, catalytic systems empirically described as [RuX$_2$(BINAP)], which are derived from **25a** and 2 equiv. of HX (X = Cl, Br) or Me$_3$SiI (eq. 24) [106, 107] as well as the complexes **24** [106–108], **27** and **29** [41], are extremely efficient catalysts for the asymmetric hydrogenation of ketones bearing various functionalities including dialkylamino, hydroxyl, alkoxyl, siloxyl, keto, alkoxycarbonyl, alkylthiocarbonyl, dialkylaminocarbonyl, and carboxyl, to give the corresponding secondary alcohols with exceptionally high enantioselectivities (~100% ee) (Table 3). The hydrogenation proceeds smoothly in methanol at room temperature with an initial hydrogen pressure of 3–100 atm. The catalytic systems prepared in situ from complexes **26**, **28**, or Ru$_2$Cl$_4$(COD)$_2$(MeCN) and BINAP ligands can be used without loss of enantioselectivities. Asymmetric hydrogenation of prochiral 4-oxoalkanoic esters and o-acylbenzoates catalyzed by BINAP–Ru(II) gives the corresponding alcohols with enantiomeric excesses exceeding 98%, which are easily converted to γ-lactones [109]. Asymmetric hydrogenation of prochiral, symmetric α- and β-diketones afforded a mixture of dl- and meso-diols (eq. 26). The enantioselectivities for the former products (R,R)-dl-**94** were very high (99–100% ee) [106,108]. The stereochemistry of the hydrogenation suggests that a simultaneous coordination of the carbonyl oxygen and heteroatom Y to Ru atom is important at the enantioface differentiation step (eq. 27).

$$\text{(24)}$$

R	Catalyst	% ee
Me	(S)-**25a**	96
i-Pr	(S)-**25a**	95
Ph	(S)-**25b**	93

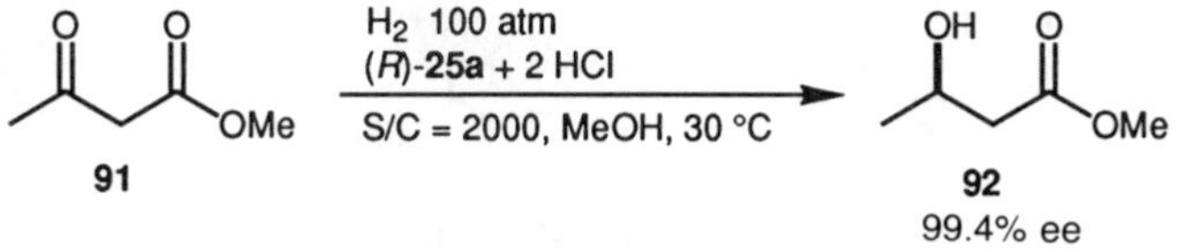

$$\text{91} \xrightarrow[\text{S/C = 2000, MeOH, 30 °C}]{\substack{H_2 \ 100 \ atm \\ (R)\text{-}25a + 2 \ HCl}} \text{92} \qquad (25)$$

91 → 92, 99.4% ee

$$\text{93} \xrightarrow[\text{20—50 °C}]{\substack{H_2 \ 50\text{—}72 \ atm \\ (R)\text{-BINAP—Ru}}} (R,R)\text{-}dl\text{-94} + meso\text{-94} \qquad (26)$$

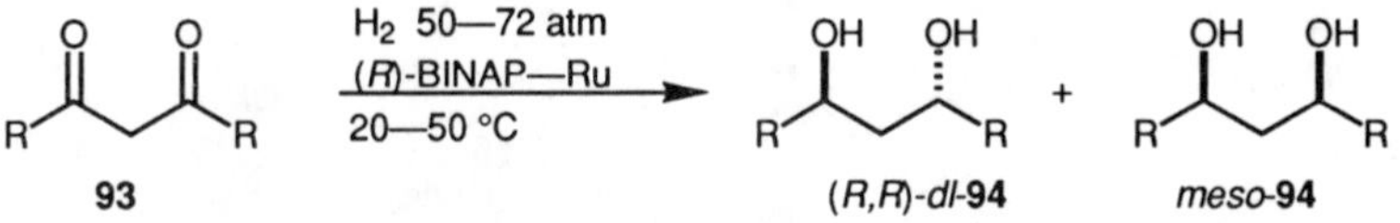

R	dl:meso	% ee
Me	99:1	100
Et	98:2	96

$$\xrightarrow[]{\substack{H_2 \\ (S)\text{-BINAP—Ru(II)}}} \qquad (27)$$

X = Y, C–Y, C–C–Y
Y = O, N, Br, etc.
C = sp^2 or nonstereogenic sp^3 carbon

The chemical and catalytic behaviors of complexes 27 and 29 have been extensively studied [110]. The arene ligands of 27 and 29 are easily liberated under the catalytic conditions to afford coordinatively unsaturated species that exhibit sufficient catalytic activity and enantioselectivity in the hydrogenation of a number of unsaturated substrates [41]. When methanol was used as a solvent, the reduction of 91 was fast, though a small amount of dimethyl acetal was formed as a by-product. The formation of such by-product can be avoided if dichloromethane or aqueous methanol is used.

Ruthenium complexes bearing BIPHEMP [38] and its derivatives [111] were also good catalysts for the asymmetric hydrogenation of β-keto esters.

In some cases, hydrogenation at high temperatures shortens the reaction time without significant loss of the enantiomeric excess of the products [42]. Compound 95, a ketone with both an ester and a halogen functionality, afforded upon hydrogenation under the standard conditions (ethanol as solvent, room temperature, 100 atm of H_2, 25–40 h) the desired alcohol with less than 70% ee. A surprisingly high enantioselectivity, however, was obtained at higher temperatures and a short reaction time (eq. 28) [112]. Compound 96 can easily be converted to carnitine. Chemoselective hydrogenation of the carbonyl moiety in β-keto ester 97 with an isolated trisubstituted or disubstituted C=C bond was also easily achieved (eq. 29) [39]. A key intermediate for FK-506, a macrolide antibiotic agent, was prepared by this method [113].

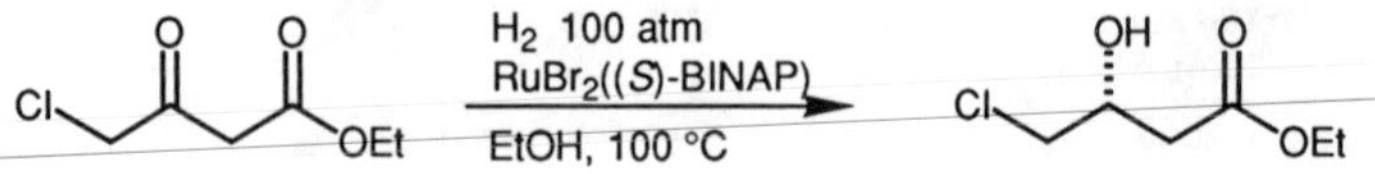

$$(28)$$

95 → **96**
97% yield, 97% ee

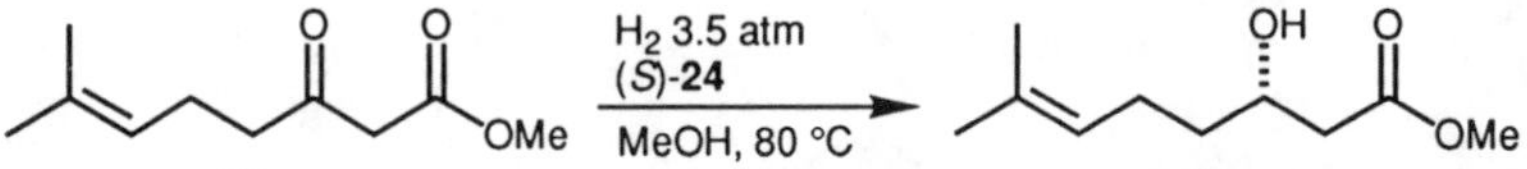

$$(29)$$

97 → **98**
96% yield, 98% ee

Diastereoselective hydrogenation of *N*-protected γ-amino-β-keto esters catalyzed by BINAP–Ru(II) complexes provides an efficient route to the statine series with high enantiomeric purities which are important components of protease inhibitors (eq. 30). When [RuBr$_2$((*R*)-BINAP)] was used as catalyst for the hydrogenation of **99, 100** was obtained in very high yields with high threo selectivities and excellent enantioselectivities (eq. 30), while with [RuBr$_2$((*S*)-BINAP)] *erythro*-**100** (> 98% ee) was obtained as the predominant product (threo/erythro ratio = 9 : 91) [114].

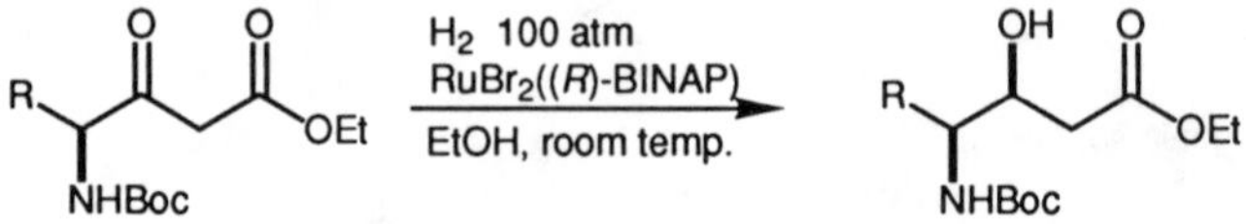

$$(30)$$

99 → **100**
>92% yield, 97—100% ee
threo:erythro = >99:1

For the asymmetric hydrogenation of α-keto carboxylic acid derivatives **101**, BICHEP–Ru complexes are efficient catalysts, which give **102** with up to 99% ee (eq. 31) [45]. High enantioselectivity was also obtained in the hydrogenation of 1,3-dialkoxyacetone **103** catalyzed by (*S*)-**24** (eq. 32) [115].

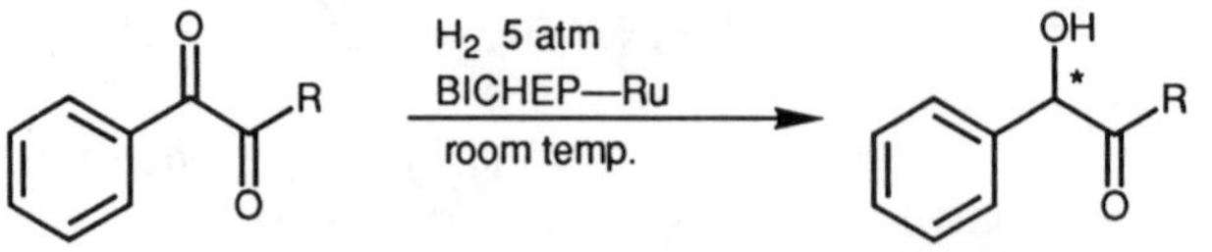

$$(31)$$

101 → **102**

R = OMe   >99% ee
R = NH-*t*-Bu   93% ee
R = NHCH$_2$Ph   88% ee

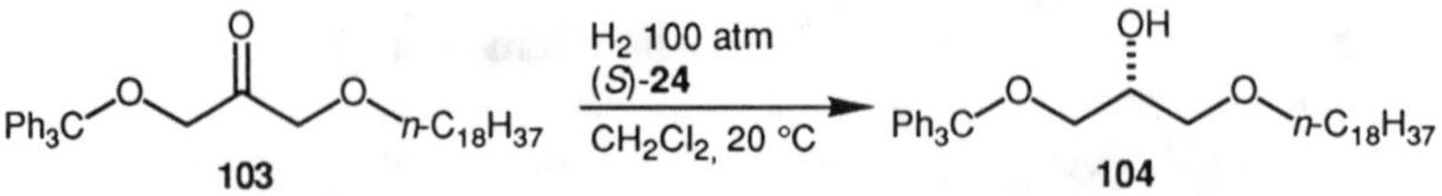

$$(32)$$

103 → **104**
>70% yield, >96% ee

Table 3 Asymmetric Hydrogenation of Functionalized Ketones (RCOR^1)
Catalyzed by Ru Complexes

R	R^1	Catalyst	Product % ee	Config.	Ref.
Me	CH$_2$NMe$_2$	(S)-**25a**	96	S	106
Me	CH$_2$NMe$_2$	(S)-**29c**	99	S	41
i-Pr	CH$_2$NMe$_2$	(S)-**25a**	95	S	106
Ph	CH$_2$NMe$_2$	(S)-**106**	95	S	106
Me	CH$_2$OH	(R)-**105**	92	R	106
Me	CH$_2$OH	(R)-**107**	91	R	42
Me	(CH$_2$)$_2$OH	(R)-**105**	98	R	106
Me	(CH$_2$)$_2$OH	(R)-**106**	98	R	42
Me	CH$_2$CO$_2$Me	(R)-**105**	>99	R	107
Me	CH$_2$CO$_2$Me	(R)-**106**	>99	R	107
Me	CH$_2$CO$_2$Me	(S)-**27a**	98	S	41
Me	CH$_2$CO$_2$Me	(S)-**27c**	97	S	41
Me	CH$_2$CO$_2$Me	(S)-**29c**	99	S	41
Me	CH$_2$CO$_2$Et	(R)-**106**	>99	R	109
Me	CH$_2$CO$_2$Et	(R)-**107**	99	R	42
Me	CH$_2$CO$_2$Et	(S)-**112**	100	S	111
Me	CH$_2$CO$_2$Et	(S)-**113**	95	S	111
Me	CH$_2$CO$_2$Et	(R)-**114**	100	R	111
Me	CH$_2$CO$_2$-i-Pr	(R)-**106**	98	R	107
Me	CH$_2$CO$_2$-t-Bu	(R)-**105**	98	R	107
Et	CH$_2$CO$_2$Me	(R)-**106**	100	R	107
n-Pr	CH$_2$CO$_2$Me	(S)-**24** + Dowex-50	98	S	39
n-Bu	CH$_2$CO$_2$Me	(S)-**24** + Dowex-50	97	S	39
n-Bu	CH$_2$CO$_2$Me	(S)-**105**	98	S	107
i-Pr	CH$_2$CO$_2$Me	(R)-**105**	>99	S	107
n-C$_{11}$H$_{23}$	CH$_2$CO$_2$Me	(R)-**115**	96	R	38
n-C$_{11}$H$_{23}$	CH$_2$CO$_2$Me	(R)-**116**	95	R	38
n-C$_{11}$H$_{23}$	CH$_2$CO$_2$Me	(R)-**117**	97	R	38
n-C$_{11}$H$_{23}$	CH$_2$CO$_2$Me	(R)-**118**	97	R	38
n-C$_{11}$H$_{23}$	CH$_2$CO$_2$Me	(R)-**119**	97	R	38
(CH$_2$)$_2$CH=CMe$_2$	CH$_2$CO$_2$Me	(S)-**24**	98	S	39
(CH$_2$)$_2$CH=CHMe	CH$_2$CO$_2$Me	(S)-**24**	98	S	39
Ph	CH$_2$CO$_2$Et	(R)-**106**	85	S	107
CH$_2$Cl	CH$_2$CO$_2$Et	(S)-**25**	97	R	112
CH$_2$Cl	CH$_2$CO$_2$Et	(S)-**107**	93	R	42
CH$_2$OSi(i-Pr)$_3$	CH$_2$CO$_2$Et	(S)-**106**	95	R	106
(CH$_2$)$_2$OCH$_2$Ph	CH$_2$CO$_2$Et	(S)-**106**	99	S	106
(CH$_2$)$_2$OPMB	CH$_2$CO$_2$Me	(S)-**24**	~100	S	113
Me	(CH$_2$)$_2$CO$_2$Et	(S)-**105**	99.5	S	109
Et	(CH$_2$)$_2$CO$_2$Et	(R)-**105**	98	R	109
n-Octyl	(CH$_2$)$_2$CO$_2$Et	(R)-**105**	98	R	109
Me	C$_6$H$_4$-2-CO$_2$Et	(S)-**105**	97	S	109
Me	CH$_2$CONMe$_2$	(S)-**106**	96	S	106
Me	CH$_2$COSEt	(R)-**105**	93	R	106
Me	C$_6$H$_4$-2-Br	(R)-**24**	92	R	106
Me	C$_6$H$_4$-2-Br	(R)-**106**	92	R	106

(*Continued*)

Table 3 (*Continued*)

R	R¹	Catalyst	Product % ee	Config.	Ref.
Me	C_6H_4-2-Br	(S)-**107**	96	S	42
Me	COMe	(S)-**106**	100ᵃ	S,S	106
Me	CH_2COMe	(R)-**105**	100ᵇ	R,R	106
Me	CH_2COMe	(R)-**24**	>99ᶜ	R,R	108
Me	CHMeCOMe	(S)-**105**	99ᵇ	S,S	106
Me	CH_2COEt	(R)-**24**	94ᵈ	R,R	108
Me	CH_2CO-*i*-Pr	(R)-**24**	98ᵉ	R,R	108
Me	$CH_2COCHMeEt$	(R)-**24**	98ᶠ	R,R	108
Et	CH_2COEt	(R)-**24**	96ᵍ	R,R	108
Me	CH_2COPh	(R)-**24**	98ʰ	R	108
Me	CH_2COPh	(R)-**24**	99ⁱ	R,R	108
Me	CO_2Me	(R)-**105**	83	R	106
Me	CO_2Me	(S)-**27a**	88	S	116
i-Pr	CO_2Me	(S)-**27a**	90	S	116
c-C_6H_{11}	CO_2Me	(S)-**27a**	90	S	116
Ph	CO_2Me	(S)-**27a**	89	S	116
Ph	CO_2Me	(S)-**108**	91	S	45
Ph	CO_2Me	(S)-**109**	>99	S	45
C_6H_4-4-Cl	CO_2Me	(S)-**27a**	93	S	116
C_6H_4-4-Me	CO_2Me	(S)-**27a**	93	S	116
C_6H_4-4-OMe	CO_2Me	(S)-**27a**	86	S	116
C_6H_4-4-NO_2	CO_2Me	(S)-**27a**	88	S	116
Ph	$CONHCH_2Ph$	(R)-**110**	88	S	45
Ph	CONH-*t*-Bu	(R)-**110**	93	R	45
Ph	CONH-*t*-Bu	(R)-**111**	88	R	45
CH_2O-*n*-$C_{18}H_{37}$	CH_2OCPh_3	(S)-**24**	>96	R	115
CH_2OCH_2Ph	CH_2OCPh_3	(S)-**24**	88	R	115

105: $RuCl_2$(BINAP)
106: $RuBr_2$(BINAP)
107: [$RuCl_2$(benzene)]₂ +
    BINAP in DMF
108: [RuCl(TolBINAP)
    (benzene)]Cl
109: [RuI(BICHEP)(*p*-cymene)]I

110: [RuCl(BICHEP)
    (*p*-cymene)]Cl
111: Ru(OCOMe)₂(BICHEP)
112: [RuI_2(benzene)]₂ +
    FUPMOP
113: [RuI_2(benzene)]₂ + BIFUP
114: [RuI_2(benzene)]₂ +
    BIMOP

115: Ru_2Cl_4(COD)₂(MeCN) + BINAP
116: Ru_2Cl_4(COD)₂(MeCN) +
    TolBINAP
117: Ru_2Cl_4(COD)₂(MeCN) +
    MeO-BIPHEMP
118: Ru_2Cl_4(COD)₂(MeCN) +
    BIPHEMP
119: Ru(OAc)₂(BIPHEMP) + 2HCl

ᵃ *dl*/*meso* = 26 / 76.
ᵇ *dl*/*meso* = 99 / 1.
ᶜ *dl*/*meso* = 99 / 1.

ᵈ *dl*/*meso* = 94 / 6.
ᵉ *dl*/*meso* = 97 / 3.
ᶠ *dl*/*meso* = 91 / 9.

ᵍ *dl*/*meso* = 98 / 2.
ʰ diol/hydroxyketone = 2 / 98.
ⁱ At 100 °C, *dl*/*meso*/hydroxyketone
    = 89 / 9 / 2.

Amino ketone hydrochlorides were hydrogenated with high enantioselectivities by using **13b**–Rh (eq. 33) [117]. Methylaminomethyl 3,4-dihydroxyphenyl ketone hydrochloride (**120a**), upon hydrogenation catalyzed by an (R,S)-**13b**–Rh complex, gave epinephrine hydrochloride (**121a**) with up to 95% ee. DIOP is also a good ligand for the reduction of α-amino ketones [118].

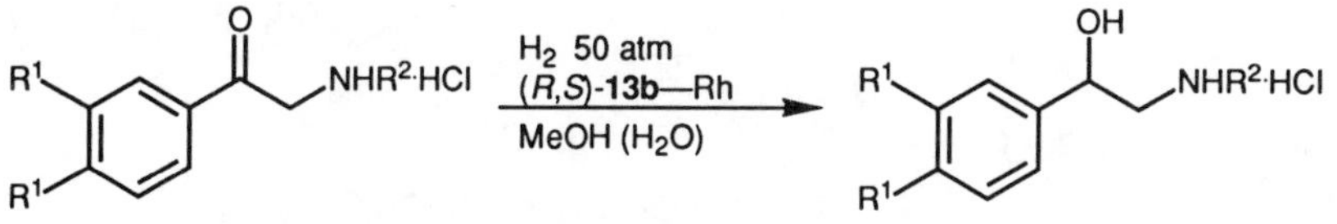

$$
\textbf{120} \xrightarrow[\text{MeOH (H}_2\text{O)}]{\substack{\text{H}_2 \text{ 50 atm} \\ (R,S)\text{-13b—Rh}}} \textbf{121}
\tag{33}
$$

a: R^1 = OH; R^2 = Me   95% ee
b: R^1 = OMe; R^2 = H   92% ee

For the hydrogenation of α-dimethylaminoacetophenone hydrochloride, a very high substrate/catalyst ratio (100,000) has been attained with the use of new ligands prepared on the basis of the "respective control concept" to give the corresponding amino alcohol hydrochloride with up to 96% ee [119–123]. This method was applied to the synthesis of (S)-1-amino-3-aryloxy-2-propanol derivatives such as (S)-propranolol hydrochloride (**122**), in which 3-aryloxy-2-oxo-1-propylamine derivatives were hydrogenated using MCCPM–Rh as the catalyst (eq. 34) [119]. Carnitine (**123**) (eq. 35) [120], mephenoxalone (**124**) [121a], fluoxetine hydrochloride (**125**) [121b], levamisole (**126**) [122], and phenylephrine hydrochloride (**127**) [123] were also synthesized.

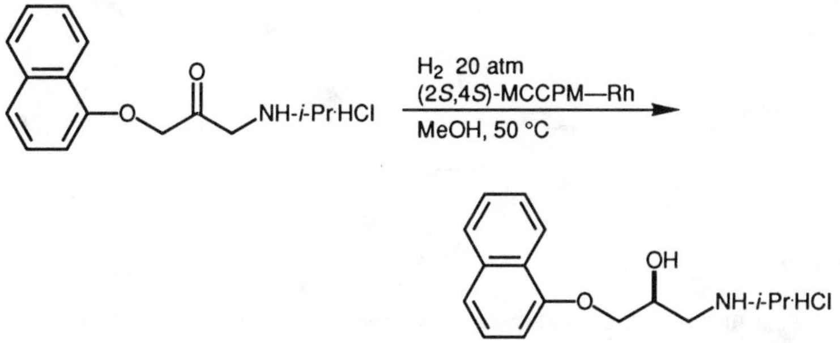

$$\tag{34}$$

122 (S)-Propranolol hydrochloride
90.8% ee

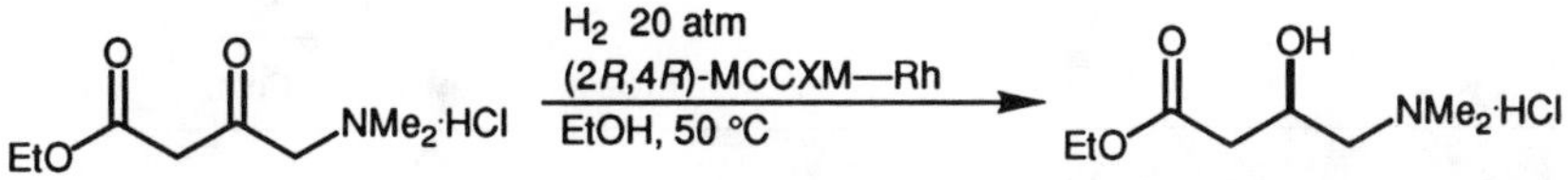

$$\tag{35}$$

(R)-Norcarnitine hydrochloride
85.0% ee

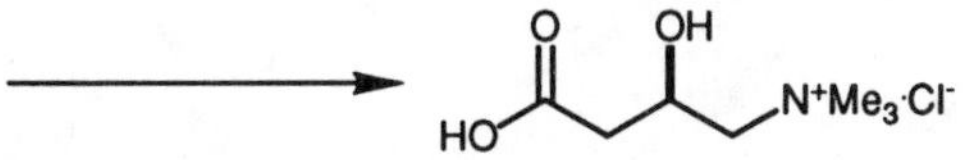

123
(R)-Carnitine hydrochloride

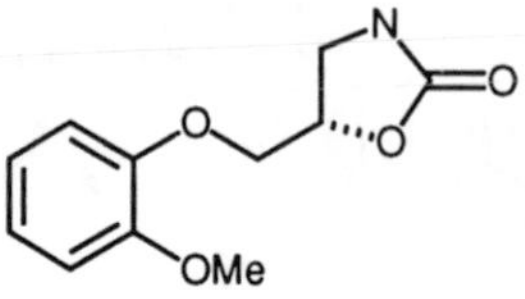

124

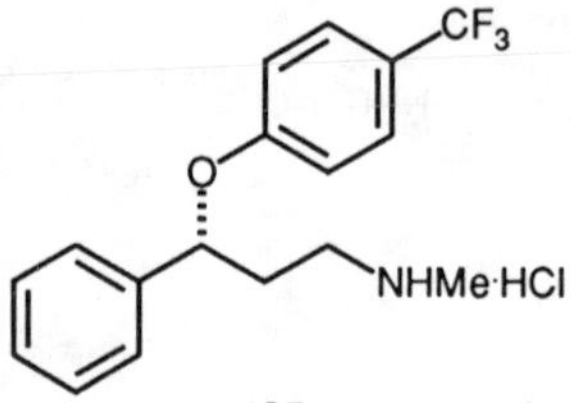

125

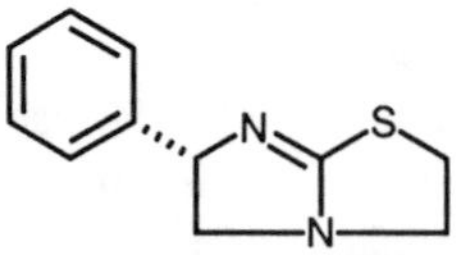

126

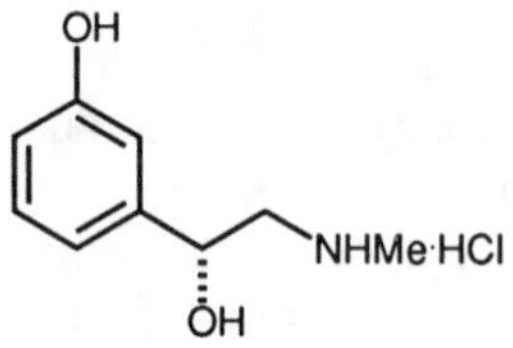

127

α-Keto esters are reduced by Rh catalysts bearing BCPM or MCCPM (ketopan-tolactone, 92% ee, [124,125]; methyl pyruvate, 87% ee [126]). Pantolactone (**129**) was also obtained by the hydrogenation of ketopantolactone (**128**) with the use of BPPM–Rh [124], or alkyl-isoAlaNOP–Rh [127] as catalysts (eq. 36).

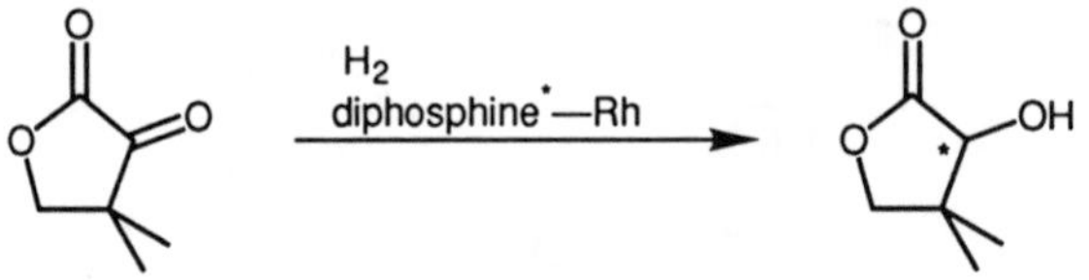

$$(36)$$

128   **129**

Table 4 shows a number of Rh-catalyzed asymmetric hydrogenation of ketones.

Table 4 Asymmetric Hydrogenation of Functionalized Ketones (RCOR¹)
Catalyzed by Rh Complexes

			Product		
R	R¹	Ligand	% ee	Config.	Ref.
Me	CH₂NMe₂ · HCl	(2S,4S)-**130**	86	S	119
CH₂Ph	CH₂NEt₂ · HCl	(2S,4S)-**131**	91	S	119
Ph	CH₂NH₂ · HCl	(2S,4S)-**131**	81	S	122
Ph	CH₂NHMe · HCl	(2S,4S)-**131**	81	S	122
Ph	CH₂NHCH₂Ph · HCl	(2S,4S)-**131**	87	S	122
Ph	CH₂NHCH₂Ph · HCl	(2S,4S)-**130**	93	S	122
Ph	CH₂NEt₂	(R,R)-**1**	93	(+)	118
Ph	CH₂NEt₂ · HCl	(2S,4S)-**130**	97	S	122
Ph	CH₂NEt₂ · HCl	(2S,4S)-**131**	93	S	122

(Continued)

Table 4   (*Continued*)

R	R^1	Ligand	% ee	Config.	Ref.
			Product		
Ph	CH$_2$NMeCH$_2$Ph · HCl	(2S,4S)-**131**	85	S	122
Ph	CH$_2$NMeCH$_2$Ph · HCl	(2S,4S)-**130**	91	S	122
C$_6$H$_3$-3,4-(OH)$_2$	CH$_2$NHMe · HCl	(R,S)-**13b**	95	R	117
C$_6$H$_3$-3,4-(OMe)$_2$	CH$_2$NH$_2$ · HCl	(R,S)-**13b**	92	R	117
C$_6$H$_4$-3-OCH$_2$Ph	CH$_2$NMeCH$_2$Ph · HCl	(2S,4S)-**130**	85	S	123
C$_6$H$_4$-3-OCH$_2$Ph	CH$_2$NMeCH$_2$Ph · HCl	(2S,4S)-**133**	85	S	123
2-Naphthyl	CH$_2$NEt$_2$	(R,R)-**1**	95	(+)	118
CH$_2$OPh	CH$_2$NHCHMe$_2$ · HCl	(2S,4S)-**130**	87	S	119
CH$_2$OPh	CH$_2$NHCH$_2$Ph · HCl	(2S,4S)-**130**	97	S	119
CH$_2$O-(C$_6$H$_3$-3,5-Me$_2$)	CH$_2$NHCH$_2$Ph · HCl	(2S,4S)-**130**	95	S	119
CH$_2$OC$_6$H$_4$(CH$_2$)$_2$OMe	CH$_2$NH-i-Pr · HCl	(2S,4S)-**130**	93	S	119
CH$_2$O(1-Naphthyl)	CH$_2$NH-i-Pr · HCl	(2S,4S)-**130**	91	S	119
CH$_2$CO$_2$Et	CH$_2$NMe$_2$ · HCl	(2S,4S)-**135**	85	S	120
CH$_2$CO$_2$Et	CH$_2$NMe$_2$ · HCl	(2S,4S)-**130**	83	S	120
Ph	(CH$_2$)$_2$NMeCH$_2$Ph · HCl	(2S,4S)-**130**	91	R	121
Ph	(CH$_2$)$_2$NMeCH$_2$Ph · HCl	(2S,4S)-**135**	82	R	121
Ph	(CH$_2$)$_2$HMeCH$_2$Ph · HCl	(2S,4S)-**131**	83	R	121
—CMe$_2$CH$_2$OCO—		(S,S)-**11**	87	R	124
—CMe$_2$CH$_2$OCO—		.(S,S)-**136**	81	R	127
—CMe$_2$CH$_2$OCO—		(S)-**137**	80	S	127
—CMe$_2$CH$_2$OCO—		(S)-**138**	89	S	127
—CMe$_2$CH$_2$OCO—		(S)-**139**	89	S	127
—CMe$_2$CH$_2$OCO—		(2S,4S)-**131**	92	R	125
Me	CO$_2$Me	(2S,4S)-**130**	87	R	126
Me	CO$_2$Me	(2S,4S)-**132**	85	R	126
Me	CO$_2$Me	(2S,4S)-**133**	84	R	126
Me	CO$_2$Me	(2S,4S)-**134**	86	R	126
Me	CH$_2$CO$_2$NHEt$_3$ · HCl	(R,S)-**13b**	83	R	117

	R^1	R^2	
130	NHMe	Ph	MCCPM
131	O-t-Bu	Ph	BCPM
132	NHPh	Ph	PCCPM
133	OMe	Ph	MCPM
134	OPh	Ph	PCPM
135	NHMe	C$_6$H$_3$-3,5-Me$_2$	MCCXM

136   R^1 = Cp, R^2 = Cy   (S)-Cp,Cy-ProNOP

137   R^1 = R^2 = Cy   (S)-Cy-isoAlaNOP
138   R^1 = R^2 = Cp   (S)-Cp-isoAlaNOP
139   R^1 = Cp, R^2 = Cy   (S)-Cp,Cy-isoAlaNOP

1.3.2. α-Substituted β-Keto Esters: Dynamic Kinetic Resolution

β-Keto esters (**140**), with a substituent at the α-position, are chiral compounds, but their racemization occurs very rapidly. If hydrogenation of such chirally labile substrates occurs stereoselectively via "dynamic kinetic resolution," only one diastereomer will be formed with high enantioselectivity. The BINAP—Ru catalysts can realize this type of hydrogenation (eq. 37) [128]. Very high *syn* selectivities have been observed for alicyclic keto esters when the reactions were carried out in dichloromethane.

R¹	R²	Solvent	syn:anti	Syn Isomer (% ee)
Me	Me	MeOH	51:49	96
Me	NHCOMe	CH₂Cl₂	99:1	98
(methylenedioxyphenyl)	NHCOMe	CH₂Cl₂	99:1	94
(methylenedioxyphenyl)	NHCO₂CH₂Ph	CH₂Cl₂	99:1	92

A similar asymmetric hydrogenation of (±)-**142** with (*R*)-BINAP–Ru(II) gives predominantly the syn alcohol **143** (eq. 38) [128]. Compound *syn*-(2*S*,3*R*)-**143** is an important intermediate for the synthesis of carbapenem antibiotics, because it can be converted to the acetate **145** via **144** in high yield (eq. 39) [129]. Catalytic conditions that afford *syn*-(2*S*,3*R*)-**143** selectively have been extensively studied by using various BINAP–Ru(II) complexes [130]. The diastereoselectivities are highly dependent on the kinds of solvent and halide anions in the BINAP–Ru(II) complexes as well as substituents on the four phenyl rings of BINAP ligands. Among the BINAP–Ru(II) halide complexes, **29c** bearing iodide anions gave the highest diastereoselectivity. Representative results are shown in Table 5. In general, much lower diastereoselectivity was obtained in methanol than in dichloromethane, though hydrogenation proceeds much faster in methanol than in dichloromethane. It has also been revealed that the introduction of methyl substituents at the 3- and 5-positions of the phenyl rings of BINAP results in higher diastereoselectivity, while a substituent such as Me, *t*-Bu, MeO, F, Cl, or CF₃ at the 4-position does not exert a remarkable effect on diastereoselectivity. Replacement of the phenyl rings of BINAP by cyclohexyl or cyclopentyl groups resulted in total loss of catalytic activity. The highest diastereoselectivity (98% de) has been obtained with the (*R*)-3,5-(*t*-Bu)₂BINAP–[RuI₂(*p*-cymene)]₂ system as the catalyst. The effect of iodide ion on stereoselectivity has also been observed in the hydrogenation of racemic **146** and **148** (eqs. 40 and 41) [131,132].

Table 5 Steroselectivities of the Hydrogenation of (±)-**142**
Catalyzed by BINAP–Ru(II) Complexes

		Product		
Catalyst[a]	Solvent	% de	% ee	Config.
[RuCl((R)-BINAP)(benzene)]Cl	CH$_2$Cl$_2$	74	90	(2S,3R)
[RuCl((R)-BINAP)(benzene)]Br	CH$_2$Cl$_2$	79	98	(2S,3R)
[RuI((R)-BINAP)(p-cymene)]I	CH$_2$Cl$_2$–MeOH	84	99	(2S,3R)
[RuI((S)-m-TolBINAP)(p-cymene)]I	MeOH	67	91	(2R,3S)
[RuI((S)-m-XylylBINAP)(p-cymene)]I	MeOH	73	91	(2R,3S)
[RuI((S)-m-XylylBINAP)(p-cymene)]I	CH$_2$Cl$_2$–MeOH	91	98	(2R,3S)
[RuI((S)-m-XylylBINAP)(p-cymene)]I	CH$_2$Cl$_2$	95	99	(2R,3S)
[RuI$_2$(p-Cymene)]$_2$ + (R)-3,5-(t-Bu)$_2$BINAP	CH$_2$Cl$_2$–MeOH	98	99	(2S,3R)

[a]   m-Tol: [Me]   m-Xylyl: [Me, Me]   3,5-(t-Bu)$_2$:   [t-Bu, t-Bu]

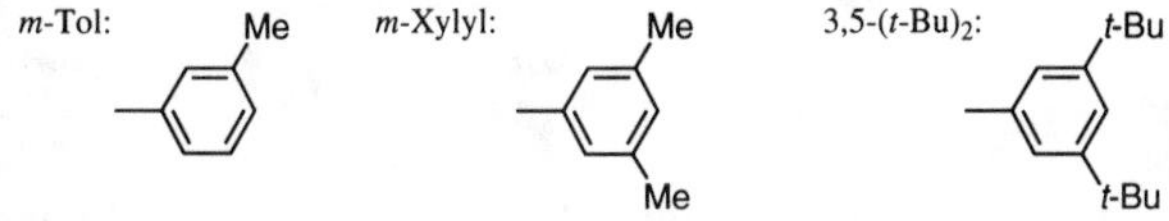

[b]   CH$_2$Cl$_2$ + 0.5% H$_2$O.
[c]   CH$_2$Cl$_2$/MeOH = 7 / 1.

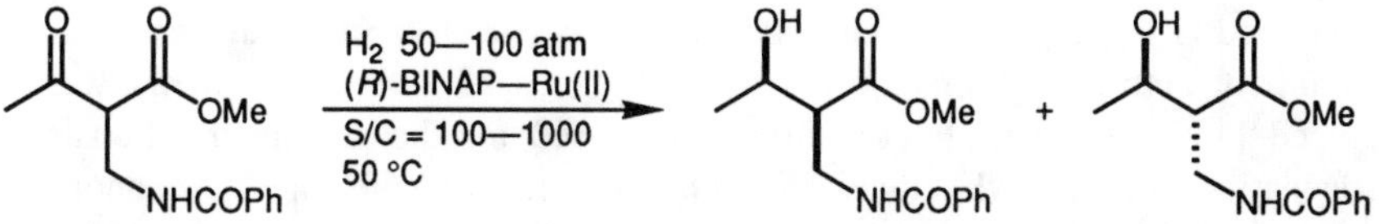

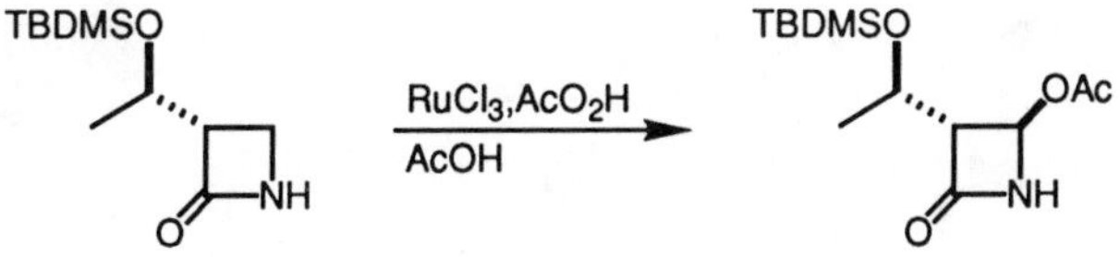

$$(38)$$

$$(39)$$

144   145
TBDMS = SiMe$_2$(t-Bu)

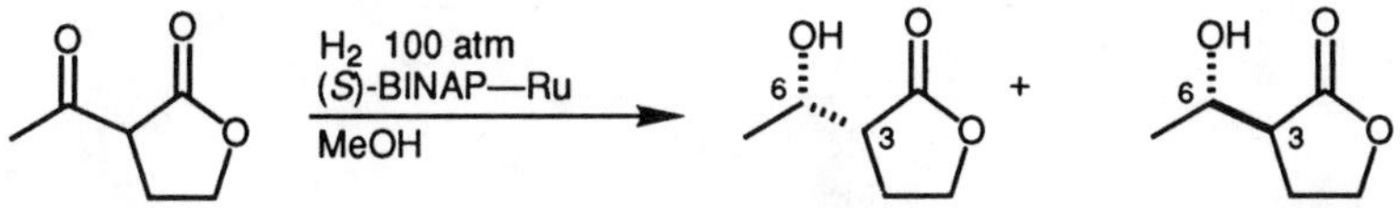

$$(40)$$

	Config.	% ee	syn:anti
(R)-**27a**	syn (3S,6R)	93	97.5:92.5
(S)-**29c**	syn (3R,6S)	97	99:1

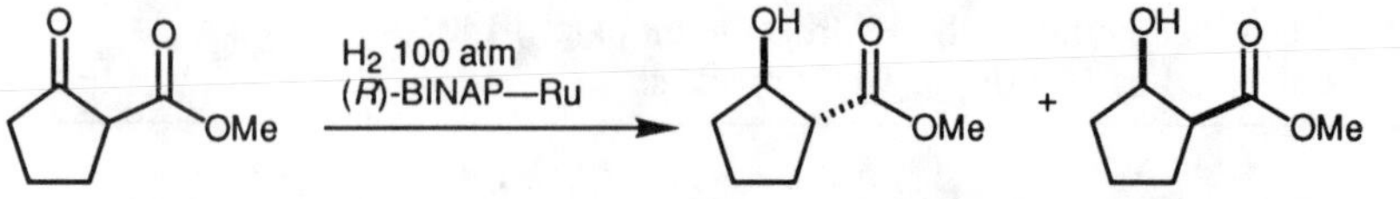

$$(41)$$

Cat.	S/C	Solvent	Product Config.		% ee	syn:anti
(R)-27a	1170	MeOH	anti	(1R,2R)	88	18.5:81.5
		CH$_2$Cl$_2$	anti	(1R,2R)	92	1:99
(S)-29c	2500	MeOH	anti	(1S,2S)	99	4:96

Recently, a highly stereoselective hydrogenation of α-acylamino-β-keto ester was attained with a CHIRAPHOS–Ru(II) complex and the reaction was applied to the synthesis of *l*-threonine and allothreonine [133].

1.3.3. Simple Ketones

Highly enantioselective hydrogenation of ketones without any other functionality is still difficult to realize. Recently, transfer hydrogenation of simple ketones with 2-propanol catalyzed by Ir complexes with dinitrogen ligands was reported. Complex [IrI((S)-PPEI)(COD)] is a very good catalyst precursor for the asymmetric transfer hydrogenation of *t*-butyl phenyl ketone and enantioselectivity up to 84% ee was obtained [134]. In this reaction, additives such as KOH, NaI, and even air are important for getting high enantioselectivity. Ligands **22** with C_2 symmetry were developed [135] and their Ir complexes were used for the asymmetric transfer hydrogenation of alkyl aryl ketones **150** in the presence of KOH and 2-propanol to give the products **151** with up to 91% ee (at 70% conversion; 88% ee at 93% conversion) (eq. 42) [31].

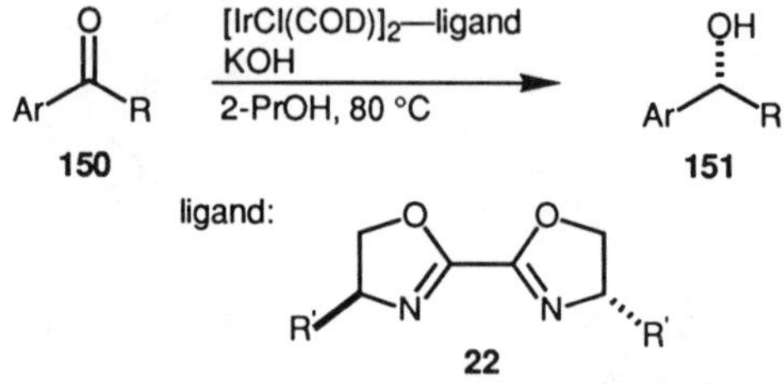

$$(42)$$

Ar	R	R'	% ee
Ph	Me	CH$_2$Ph	47
Ph	Me	*i*-Pr	58
C$_6$H$_4$-4-OMe	Me	*i*-Pr	57
2-Naphthyl	Me	*i*-Pr	63
Ph	*i*-Pr	*i*-Pr	91

The catalysts prepared in situ from [RhCl(diene)]$_2$ (diene = 1,5-hexadiene or 1,5-cyclooctadiene) and 3-alkylphenanthrolines display very high catalytic activity in the transfer hydrogenation of acetophenone. Turnover rates up to 10,000 cycles/hour have been recorded in 2-propanol solution at 83°C in the presence of KOH as a promoter, and asymmetric inductions up to 65% ee are obtained with (−)-(S)-3-(1,2,2-trimethylpropyl)-1,10-phenanthroline as ligand [136].

1.4. Asymmetric Hydrogenation of Imines

A relatively small number of examples have appeared on the enantioselective hydrogenation of Schiff bases, though the resulting optically active amines are synthetically important. For the asymmetric hydrogenation of acetophenonebenzylimine (**152**), the catalytic system prepared in situ from [RhCl(diene)]$_2$ and BDPP was used in methanol at 0°C to give the product (**153**) with 83% ee [137]. This imine was also hydrogenated by using the system derived from [RhCl(NBD)]$_2$ and CYCPHOS as catalyst to give **153** with 67% ee. When one equivalent of KI was added to this catalyst system, the enantioselectivity increased to 79% ee [138]. In the asymmetric hydrogenation of p-methoxyacetophenonebenzylimine at −25°C with KI as cocatalyst, 91% ee was achieved.

In some cases, the use of water-soluble catalysts afforded a high enantiomeric excess [139], and a number of chiral sulfonated diphosphines have been tested as ligands (eq. 43). High enantiomeric excesses (acetophenonebenzylimine, > 95% ee) were achieved by using [RhCl(NBD)]$_2$ with a mixture of mono-, di-, and trisulfonated BDPP at 20°C in an H$_2$O–ethyl acetate mixture under H$_2$ atmosphere.

$$\text{(43)}$$

Ar	Conditions	% ee
Ph	[RhCl(NBD)]$_2$ + BDPP + NEt$_3$ MeOH, 0 °C, 6 h	83%
C$_6$H$_4$-4-OMe	[RhCl(NBD)]$_2$ + CYCPHOS + KI benzene/MeOH, -25 °C, 144 h	91%
Ph	[RhCl(NBD)]$_2$ + sulfonated BDPP H$_2$O/EtOAc, 20 °C	>95%

Diphosphine–Ir complexes are also good catalysts for the asymmetric hydrogenation of imines. The hydrogenation of imine **154** in the presence of [IrCl(COD)]$_2$/BDPP catalyst and KI gave **155** with 84% ee [140]. With the use of [IrI$_2$H(BDPP)], cyclic imine **156** was hydrogenated to **157** with 80% ee [141]. Interestingly, the latter complex showed low enantioselectivity in the hydrogenation of **154** (34% ee).

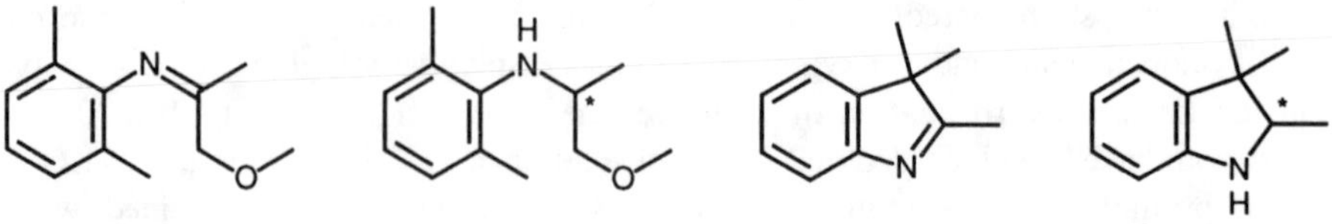

154	**155**	**156**	**157**

Imine **158** derived from saccharine was hydrogenated in the presence of (*R*)-**24** to give (*R*)-sultam (**159**) in 84% yield with >99% ee.

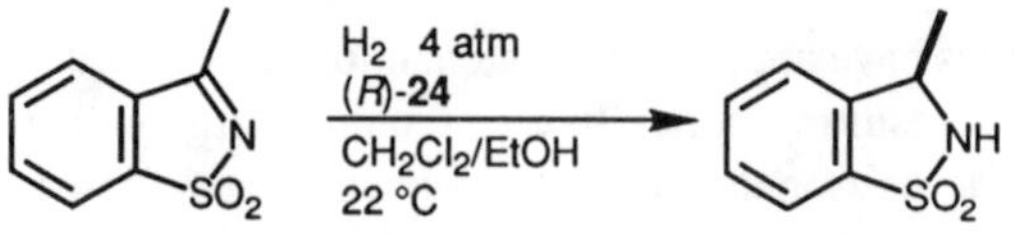

(44)

Quite recently, a highly enantioselective hydrogenation of C=N bonds (up to 98% ee) was reported with new Ti and Rh catalysts [143,144].

1.5. Conclusion

Homogeneous asymmetric hydrogenation, which originated in late 1960s, is now regarded as a powerful tool to prepare optically active organic molecules. At early stages of the investigation, however, substrates to be hydrogenated by chiral phosphine–Rh(I) complexes were limited to certain functionalized olefins such as α-acylaminoacrylic acids. A new generation of asymmetric hydrogenation started in the mid-1980s, when chiral diphosphine–Ru(II) catalysts were devised. The Ru(II) catalyst systems have expanded substrates to a variety of functionalized olefins and ketones for which high enantioselectivities and sufficiently high conversions had not been attained by conventional Rh(I) complexes. Recently, chiral complexes of Group 4 transition metals and lanthanide metals have successfully been used for the asymmetric hydrogenation of simple olefins. Now one can anticipate developments of efficient new catalytic systems for the asymmetric hydrogenation of simple ketones, imines, and related substrates with the use of complexes of Ir(I) and other transition metals.

Although most of the mechanisms of asymmetric hydrogenation and the factors controlling enantioselectivities remain to be elucidated, their synthetic applicability seems to be rapidly expanding. Purely chemical processes are clean, operationally simple, economical, and adaptable for a large-scale production. Sometimes, they can provide products with enantiomeric purities even higher than those of natural origins. Moreover, both *R* and *S* enantiomers are obtainable with equal ease by variation of substrate geometry or catalyst handedness. Thus, chemical asymmetric hydrogenation processes are often superior to any other methods. All these facts lead us to anticipate that asymmetric hydrogenation catalyzed by transition metal complexes will be widely used in the near future on the industrial scale to produce not only physiologically active compounds and their synthetic intermediates, but also new materials such as color liquid crystals, nonlinear optics, and biodegradable plastics.

References

1. Recent review: Blaser, H.-U. *Tetrahedron: Asymmetry,* **1991,** *2,* 843.

2. Izumi, Y. *Adv. Catal.* **1983,** *32,* 215.

3. Knowles, W. S.; Sabacky, M. J. *J. Chem. Soc., Chem. Commun.* **1968,** 1445.

4. Horner, L.; Siegel, H.; Büthe, H. *Angew. Chem., Int. Ed. Engl.* **1968,** *7,* 942.

5. Reviews: (a) Kagan, H. B. in *Asymmetric Synthesis,* Morrison, J. D. (Ed.), Academic Press: Orlando, FL, **1985;** Vol. 5, p. 1. (b) Koenig, K. E. In *Asymmetric Synthesis,* Morrison, J. D. (Ed.); Academic Press: Orlando, FL, **1985;** Vol. 5, p. 71. (c) Scott, J. W.; Valentine, D. *Science,* **1974,** *184,* 4140. (d) Morrison, J. D.; Melser, W. F.; Hathaway, S. In *Catalysis in Organic Chemistry,* Rylander, P. N.; Greenfield, H. (Eds.); Academic Press: Orlando, FL, **1976;** p. 229. (e) Caplar, V.; Comisso, G.; Sunjic, V. *Synthesis,* **1981,** 85. (f) Valentine, D.; Scott, J. W. *Synthesis,* **1978,** 329. (g) Kozikowski, A. P.; Watter, H. F. *Synthesis,* **1976,** 561. (h) Maugh, T. H. *Science,* **1983,** *221,* 351. (i) Mosher, H. S.; Morrison, J. D. *Science,* **1983,** *221,* 1013. (j) Knowles, W. S.; Christopfel, W. C.; Koenig, K. E.; Hobbs, C. F. In *Catalytic Aspects of Metal Phosphine Complexes,* Alyea, E. C.; Meek, D. W. (Eds.); American Chemical Society: Washington, DC, **1982;** p. 325. (k) Merrill, R. E. *ChemTech,* **1981,** 118. (l) Knowles, W. S. *Acc. Chem. Res.* **1983,** *16,* 106. (m) Drauz, K.; Kleeman, A.; Martens, J. *Angew. Chem., Int. Ed. Engl.* **1982,** *21,* 584. (n) Kagan, H. B. *Bull. Chim. Soc. Fr.* **1988,** 846. (o) Noyori, R.; Kitamura, M. In *Modern Synthetic Methods,* Schefford, R. (Ed.); Springer-Verlag: Berlin, **1989;** Vol. 5, p. 115. (o) Brunner, H. *Synthesis,* **1988,** 645. (p) Noyori, R. *Science,* **1990,** *248,* 1194. (q) Ojima, I.; Clos, N.; Bastos, C. *Tetrahedron,* **1989,** *45,* 6901. (r) Blyston, S. L. *Chem. Rev.* **1989,** *89,* 1663. (s) Kagan, H. B.; Sasaki, M. In *The Chemistry of Organophosphorous Compounds,* Hartley, F. R. (Ed.); Wiley: New York, **1990;** Vol. 1, Chapter 3, p. 53.

6. Kagan, H. B.; Dang, T.-P. *J. Am. Chem. Soc.* **1972,** *94,* 6429.

7. (a) Knowles, W. S.; Sabacky, M. J.; Vineyard, B. D. *J. Chem. Soc., Chem. Commun.* **1972,** 10. (b) Vineyard, B. D.; Knowles, W. S.; Sabacky, M. J.; Bachman, G. L.; Weinkauff, D. J. *J. Am. Chem. Soc.* **1977,** *99,* 5946.

8. Brunner, H. *Top. Stereochem.* **1988,** *18,* 129.

9. (a) Hayashi, T.; Kumada, M. In *Fundamental Research in Homogeneous Catalysis,* Ishii, Y.; Tsutsui, M. (Eds.); Plenum: New York, **1978;** Vol. 2, p. 159. (b) Hayashi, T.; Mise, T.; Fukushima, M.; Kagotani, M.; Nagashima, N.; Hamada, Y.; Motsumoto, A.; Kawakami, S.; Konishi, M.; Yamamoto, K.; Kumada, M. *Bull. Chem. Soc. Jpn.* **1980,** *53,* 1138. (c) Hayashi, T.; Kumada, M. *Acc. Chem. Res.* **1982,** *15,* 395. (d) Hayashi, T.; Konishi, M.; Fukushima, M.; Mise, T.; Kagotani, M.; Tajika, M.; Kumada, M. *J. Am. Chem. Soc.* **1982,** *104,* 180.

10. Fryzuk, M. D.; Bosnich, B. *J. Am. Chem. Soc.* **1977,** *99,* 6262.

11. Fryzuk, M. D.; Bosnich, B. *J. Am. Chem. Soc.* **1978,** *100,* 5491.

12. King, R. B.; Bakos, J.; Hoff, C. D.; Markó, L. *J. Org. Chem.* **1979,** *44,* 1729.

13. (a) Riley, D. P.; Shumate, R. E. *J. Org. Chem.* **1980,** *45,* 5187. (b) Oliver, J. D.; Riley, D. P. *Organometallics,* **1983,** *2,* 1032.

14. Brunner, H.; Pieronczyk, W.; Schönhammer, B.; Streng, K.; Bernal, I.; Korp, J. *Chem. Ber.* **1981,** *114,* 1137.

15. McNeil, P. A.; Roberts, N. K.; Bosnich, B. *J. Am. Chem. Soc.* **1981,** *103,* 2280.

16. Allen, D. L.; Gibson, V. C.; Green, M. L. H.; Skinner, J. F.; Bashkin, J.; Grebenik, P. D. *J. Chem. Soc., Chem. Commun.* **1983**, 895.

17. Nagel, U.; Kinzel, E.; Andrade, J.; Prescher, G. *Chem. Ber.* **1986**, *119*, 3326.

18. Achiwa, K. *J. Am. Chem. Soc.* **1976**, *98*, 8265.

19. Ojima, I.; Yoda, N. *Tetrahedron Lett.* **1980**, *21*, 1051.

20. (a) Miyashita, A.; Yasuda, A.; Takaya, H.; Toriumi, K.; Ito, T.; Souchi, T.; Noyori, R. *J. Am. Chem. Soc.* **1980**, *102*, 7932. (b) Miyashita, A.; Takaya, H.; Souchi, T.; Noyori, R. *Tetrahedron,* **1984**, *40*, 1245. (c) Takaya, H.; Mashima, K.; Koyano, K.; Yagi, M.; Kumobayashi, H.; Taketomi, T.; Akutagawa, S.; Noyori, R. *J. Org. Chem.* **1986**, *51*, 629. (d) Takaya, H.; Akutagawa, S.; Noyori, R. *Org. Synth.* **1988**, *67*, 20.

21. Morrison, J. D.; Masler, W. F.; Neuberg, M. K. *Adv. Catal.* **1976**, *25*, 81.

22. Schmid, R.; Cereghetti, M.; Heiser, B.; Schönholzer, P.; Hansen, H.-J. *Helv. Chim. Acta,* **1988**, *71*, 897.

23. Miyashita, A.; Karino, H.; Shimamura, J.-I.; Chiba, T.; Nagano, K.; Nohira, H.; Takaya, H. *Chem. Lett.* **1989**, 1849.

24. (a) Burk, M. J.; Feaster, J. E.; Harlow, R. L. *Organometallics,* **1990**, *9*, 2653. (b) Burk, M. J. *J. Am. Chem. Soc.* **1991**, *113*, 8518.

25. Karim, A.; Mortreux, A.; Petit, F. *J. Organomet. Chem.* **1986**, *312*, 375.

26. (a) Selke, R.; Pracejus, H. *J. Mol. Catal.,* **1986**, *37*, 213. (b) Habus, I.; Raza, Z.; Sunjic, V. *J. Mol. Catal.* **1987**, *42*, 173.

27. Fiorini, M.; Giongo, G. M. (a) *J. Mol. Catal.* **1979**, *5*, 303; (b) *J. Mol. Catal.* **1980**, *7*, 411.

28. Cesarotti, E.; Chiesa, A.; Ciani, G.; Sironi, A. *J. Organomet. Chem.* **1983**, *251*, 79.

29. (a) Fritshi, H.; Leutenegger, U.; Pfaltz, A. *Angew. Chem., Int. Ed. Engl.* **1986**, *25*, 1005. (b) Fritschi, H.; Leutenegger, U.; Siegmann, K.; Pfaltz, A.; Keller, W.; Kratky, C. *Helv. Chim. Acta,* **1988**, *71*, 1541.

30. Nerdel, F.; Becker, K.; Kresze, G. *Chem. Ber.* **1956**, *12*, 2862.

31. Müller, D.; Umbricht, G.; Weber, B.; Pfaltz, A. *Helv. Chim. Acta,* **1991**, *74*, 232.

32. Halterman, R. L.; Vollhardt, K. P. C.; Welker, M. E.; Bläser, D.; Boese, R. *J. Am. Chem. Soc.* **1987**, *109*, 8105.

33. Miyashita, A.; Yasuda, A.; Takaya, H.; Toriumi, K.; Ito, T.; Noyori, R. *J. Am. Chem. Soc.* **1980**, *102*, 7932.

34. (a) James, B. R.; McMillan, R. S.; Morris, R. H.; Wang, D. K. W. *Adv. Chem. Ser.* **1978**, *167*, 122. (b) Ball, R. G.; James, B. R.; Trotter, J.; Wang, D. K. W. *J. Chem. Soc., Chem. Commun.* **1979**, 460.

35. (a) Ikariya, T.; Ishii, Y.; Kawano, H.; Arai, T.; Saburi, M.; Yoshikawa, S.; Akutagawa, S. *J. Chem. Soc., Chem. Commun.* **1985**, 922. (b) Kawano, H.; Ikariya, T.; Ishii, Y.; Saburi, M.; Yoshikawa, S.; Uchida, Y.; Kumobayashi, H. *J. Chem. Soc., Perkin Trans.* 1, **1989**, 1571.

36. Ohta, T.; Takaya, H.; Noyori, R. *Inorg. Chem.* **1988**, *27*, 566.

37. Takaya, H.; Ohta, T.; Inoue, S.; Tokunaga, M.; Kitamura, M.,; Noyori, R. *Org. Synth.* in press.

38. Heiser, B.; Broger, E. A.; Crameri, Y. *Tetrahedron: Asymmetry,* **1991**, *2*, 51.

39. Taber, D. F.; Silverberg, L. J. *Tetrahedron Lett.* **1991**, *32*, 4227.

40. Kitamura, M.; Ohkuma, T.; Inoue, S.; Sayo, N.; Kumobayashi, H.; Akutagawa, S.; Ohta, T.; Takaya, H.; Noyori, R. *J. Am. Chem. Soc.* **1988**, *110*, 629.

41. Mashima, K.; Kusano, K.; Ohta, T.; Noyori, R.; Takaya, H. *J. Chem. Soc., Chem. Commun.* **1989**, 1208.

42. Kitamura, M.; Tokunaga, M.; Ohkuma, T.; Noyori, R. *Tetrahedron Lett.* **1991**, *32*, 4163.

43. Jenet, J. P.; Mallart, S.; Pinel, C.; Juge, S.; Laffitte, J. A. *Tetrahedron: Asymmetry*, **1991**, *2*, 43.

44. (a) Alcock, N. W.; Brown, J. M.; Rose, M.; Wienand, A. *Tetrahedron: Asymmetry*, **1991**, *2*, 47. (b) Brown, J. M.; Brunner, H.; Leitner, W.; Rose, M. *Tetrahedron: Asymmetry*, **1991**, *2*, 331.

45. Miyashita, A.; Chiba, T.; Nohira, H.; Takaya, H. *Abstract of 38th Symposium on Organometallic Chemistry*, Kyoto, 1991, B216.

46. Lentenegger, U.; Madin, A.; Pfaltz, A. *Angew. Chem., Int. Ed. Engl.* **1989**, *28*, 60.

47. Morrison, J. D.; Burnett, R. E.; Aguiar, A. M.; Morrow, C. J.; Phillips, C. *J. Am. Chem. Soc.* **1971**, *93*, 1301.

48. Scott, J. W.; Keith, D. D.; Nix, Jr., G.; Parrish, D. R.; Remington, S.; Roth, G. P.; Townsend, J. M.; Valentine, Jr., D.; Young, R. *J. Org. Chem.* **1981**, *46*, 5086.

49. Inoguchi, K.; Achiwa, K. *Synlett*, **1991**, 49.

50. Chiba, T.; Miyashita, A.; Nohira, H.; Takaya, H. *Tetrahedron Lett.* **1991**, *32*, 4745.

51. Takahashi, H.; Achiwa, K. *Chem. Lett.* **1989**, 305.

52. Zeiss, H.-J. *J. Org. Chem.* **1991**, *56*, 1783.

53. Bozell, J. J.; Vogt, C. E.; Gozum, J. *J. Org. Chem.* **1991**, *56*, 2584.

54. Nagel, U.; Kinzel, E. *Chem. Ber.* **1986**, *119*, 1731.

55. (a) Sinou, D. *Bull. Soc. Chim. Fr.* **1987**, 480. (b) Alario, F.; Amrani, Y.; Collenille, Y.; Dang, T. P.; Jenck, J.; Morel, D.; Sinou, D. *J. Chem. Soc., Chem. Commun.* **1986**, 202. (c) Amrani, Y.; Lecomte, L.; Sinou, D.; Bakos, J.; Tóth, I.; Heil, B. *Organometallics*, **1989**, *8*, 542.

56. (a) Tóth, I.; Hanson, B. E. *Tetrahedron: Asymmetry*, **1990**, *1*, 895. (b) Tóth, I.; Hanson, B. E.; Davis, M. E. *Tetrahedron: Asymmetry*, **1990**, *1*, 913.

57. (a) Halpern, J. *Pure Appl. Chem.* **1983**, *55*, 99. (b) Landis, C. R.; Halpern, J. *J. Am. Chem. Soc.* **1987**, *109*, 1746. (c) Halpern, J. In *Asymmetric Catalysis*, Morrison, J. D. (Ed.); Academic Press: Orlando, FL, **1985**; Vol. 5, p. 41.

58. (a) Brown, J. M.; Chaloner, P. A. *J. Am. Chem. Soc.* **1980**, *102*, 3040. (b) Brown, J. M.; Chaloner, P. A.; Morris, G. A. *J. Chem. Soc., Chem. Commun.* **1983**, 644. (c) Brown, J. M.; Canning, L. R.; Downs, A. J.; Forster, A. M. *J. Organomet. Chem.* **1983**, *255*, 103. (d) Brown, J. M.; Evans, P. L. *Tetrahedron Lett.* **1988**, *44*, 4905.

59. Ojima, I.; Kogure, T.; Yoda, N. *J. Org. Chem.* **1980**, *45*, 4728.

60. Knowles, W. S.; Vineyard, B. D.; Sabacky, M. J.; Stults, B. R. In *Fundamental Research in Homogeneous Catalysis*, Tsutsui, M. (Ed.); Plenum: New York, **1979**; Vol. 3, p. 537.

61. Kashiwabara, K.; Hanaki, K.; Fujita, J. *Bull. Chem. Soc. Jpn.* **1980**, *53*, 2275.

62. Sakuraba, S.; Morimoto, T.; Achiwa, K. *Tetrahedron: Asymmetry*, **1991**, *2*, 597.

63. Nagel, U.; Rieger, B. *Organometallics*, **1989**, *8*, 1534.

64. Noyori, R.; Ikeda, T.; Ohkuma, T.; Widhalm, M.; Kitamura, M.; Takaya, H.; Akutagawa, S.; Sayo, N.; Saito, T.; Taketomi, T.; Kumobayashi, H. *J. Am. Chem. Soc.* **1989**, *111*, 9134.

65. Miyashita, A.; Takaya, H.; Souchi, T.; Noyori, R. *Tetrahedron,* **1984,** *40,* 1245.

66. Lubell, W. D.; Kitamura, M.; Noyori, R. *Tetrahedron: Asymmetry,* **1991,** *2,* 543.

67. (a) Onuma, K.; Ito, T.; Nakamura, A. *Chem. Lett.* **1980,** 481. (b) Meyer, D.; Poulin, J.-C.; Kagan, H. B.; Levine-Pinto, H.; Morgat, J.-L.; Fromageot, P. *J. Org. Chem.* **1980,** *45,* 4680. (c) Kleeman, A.; Martens, J.; Samson, M.; Bergstein, W. *Synthesis,* **1981,** 740. (d) Ojima, I.; Yatabe, M. *Chem. Lett.* **1982,** 1335. (e) Ojima, I.; Yoda, N. *Tetrahedron Lett.* **1982,** *23,* 3913. (f) Ojima, I.; Yoda, N.; Yatabe, M. *Tetrahedron Lett.* **1982,** *23,* 3917. (g) Ojima, I.; Kogure, T.; Yoda, N.; Suzuki, T.; Yatabe, M.; Tanaka, T. *J. Org. Chem.* **1982,** *47,* 1329. (h) Levine-Pinto, H.; Margat, J. L.; Fromagoet, P.; Meyer, D.; Poulin, J. P.; Kagan, H. B. *Tetrahedron,* **1982,** 481. (i) Poulin, J.-C.; Kagan, H. B. *J. Chem. Soc., Chem. Commun.* **1982,** 1261. (j) Ojima, I.; Yoda, N.; Yatabe, M.; Kogure, T. *Tetrahedron,* **1984,** *40,* 1255. (k) Yamagishi, T.; Ikeda, S.; Yatagai, M.; Yamaguchi, M.; Hida, M. *J. Chem. Soc., Perkin Trans. 1,* **1988,** 1787.

68. Takagi, M.; Yamamoto, K. *Tetrahedron,* **1991,** *47,* 8869.

69. (a) Brown, J. M.; James, A. P.; Prior, L. M. *Tetrahedron Lett.* **1987,** *28,* 2179. (b) Brown, J. M.; Evans, P. L.; James, A. P. *Org. Synth.* **1989,** *68,* 64.

70. (a) Kagan, H. B.; Langloirs, N.; Dang, T. P. *J. Organomet. Chem.* **1975,** *96,* 353. (b) Sinou, D.; Kagan, H. B. *J. Organomet. Chem.* **1976,** *114,* 325.

71. Noyori, R.; Ohta, M.; Hsiao, Yi; Kitamura, M.; Ohta, T.; Takaya, H. *J. Am. Chem. Soc.* **1986,** *108,* 7117.

72. Kitamura, M.; Hsiao, Y.; Noyori, R.; Takaya, H. *Tetrahedron Lett.* **1987,** *28,* 4829.

73. Koenig, K. E.; Bachman, G. L.; Vineyard, B. D. *J. Org. Chem.* **1980,** *45,* 2362.

74. Merrill, R. E. *ChemTech,* **1981,** 118.

75. Hayashi, T.; Kanehira, K.; Kumada, M. *Tetrahedron Lett.* **1981,** *22,* 4417.

76. Ohta, T.; Miyake, T.; Seido, N.; Kumobayashi, H.; Akutagawa, S.; Takaya, H. *Tetrahedron Lett.* **1992,** *33,* 635.

77. Cristopfel, W. C.; Vineyard, B. D. *J. Am. Chem. Soc.* **1979,** *101,* 4406.

78. Inoguchi, K.; Morimoto, T.; Achiwa, K. *J. Organomet. Chem.* **1989,** *370,* C9.

79. Burk, M. J.; Feaster, J. E.; Harlow, R. L. *Tetrahedron: Asymmetry,* **1991,** *2,* 569.

80. Leutenegger, U.; Madin, A.; Pfaltz, A. *Angew. Chem., Int. Ed. Engl.* **1989,** *28,* 60.

81. Matt, P.; Pfaltz, A. *Tetrahedron: Asymmetry,* **1991,** *2,* 691.

82. Dang, T.-P.; Aviron-Violet, P.; Colleuille, J. Y.; Varagnat, Y. C. *J. Mol. Catal.* **1982,** *16,* 51.

83. (a) Massonneau, V.; Maux, P. L.; Simonneaux, G. *Tetrahedron Lett.* **1986,** *27,* 5497. (b) Massoneau, V.; Maux, P. L.; Simonneaux, G. *J. Organomet. Chem.* **1987,** *327,* 269.

84. (a) Achiwa, K. *Tetrahedron Lett.* **1978,** *17,* 1475. (b) Jendralla, H. *Tetrahedron Lett.* **1991,** *32,* 3671.

85. Brunner, H.; Leitner, W. *Angew. Chem., Int. Ed. Engl.* **1988,** *27,* 1180.

86. Morimoto, T.; Chiba, M.; Achiwa, K. (a) *Tetrahedron Lett.,* **1989,** *30,* 735; (b) *Tetrahedron Lett.* **1990,** *31,* 261.

87. (a) Takahashi, H.; Achiwa, K. *Chem. Lett.* **1987,** 1921. (b) Takahashi, H.; Yamamoto, N.; Takeda, H.; Achiwa, K. *Chem. Lett.* **1989,** 559.

88. (a) Kawano, H.; Ishii, Y.; Ikariya, T.; Saburi, M.; Yoshikawa, S.; Uchida, Y.; Ku *Tetrahedron Lett.* **1987,** *28,* 1905. (b) Shao, L.; Miyata, S.; Muramatsu, H.; Kaw Y.; Saburi, M.; Uchida, Y. *J. Chem. Soc., Perkin Trans. 1,* **1990,** 1441.

89. Muramatsu, H.; Kawano, H.; Ishii, Y.; Saburi, M.; Uchida, Y. *J. Chem. Soc., Chem. Commun.* **1989,** 769.

90. Ohta, T.; Takaya, H.; Kitamura, M.; Nagai, K.; Noyori, R. *J. Org. Chem.* **1987,** *52,* 3174.

91. (a) Hayashi, T.; Kawamura, N.; Ito, Y. *J. Am. Chem. Soc.,* **1987,** *109,* 7876. (b) Hayashi, T.; Kawamura, N.; Ito, Y. *Tetrahedron Lett.* **1988,** *29,* 5969.

92. Ohta, T.; Takaya, H.; Noyori, R. *Tetrahedron Lett.* **1990,** *31,* 7189.

93. (a) Ashby, M. T.; Halpern, J. *J. Am. Chem. Soc.* **1991,** *113,* 589. (b) Ashby, M. T.; Khan, M. A.; Halpern, J. *Organometallics,* **1991,** *10,* 2011.

94. Brown, J. M. *Angew. Chem., Int. Ed. Engl.* **1987,** *26,* 190, and references cited therein.

95. Inoue, S.-I.; Osada, M.; Koyano, K.; Takaya, H.; Noyori, R. *Chem. Lett.* **1985,** 1007.

96. Takaya, H.; Ohta, T.; Sayo, N.; Kumobayashi, H.; Akutagawa, S.; Inoue, S.; Kasahara, I.; Noyori, R. *J. Am. Chem. Soc.* **1987,** *109,* 1596, 4129.

97. (a) Imperiali, B.; Zimmerman, J. W. *Tetrahedron Lett.* **1988,** *29,* 5343. (b) Kitamura, M.; Noyori, R. Unpublished results.

98. Kitamura, M.; Nagai, K.; Hsiao, Y.; Noyori, R. *Tetrahedron Lett.* **1990,** *31,* 549.

99. Kitamura, M.; Kasahara, I.; Manabe, K.; Noyori, R.; Takaya, H. *J. Org. Chem.* **1988,** *53,* 708.

100. (a) Tanaka, M.; Ogata, I. *J. Chem. Soc., Chem. Commun.* **1975,** 735. (b) Kawabata, Y.; Tanaka, M.; Ogata, I. *Chem. Lett.* **1976,** 1213. (c) Schofield, P. A.; Adams, H.; Bailey, N. A.; Cesarotti, E.; White, C. *J. Organomet. Chem.* **1991,** *412,* 273.

101. Halterman, R. L.; Vollhardt, K. P. C. *Organometallics,* **1988,** *7,* 883.

102. Paquette, L. A.; McKinney, J. A.; McLaughlin, M. L.; Rheingold, A. L. *Tetrahedron Lett.* **1986,** *27,* 5599.

103. Gallucci, J. C.; Gautheron, B.; Gugelchuk, M.; Meunier, P.; Paquette, L. A. *Organometallics,* **1987,** *6,* 15.

104. Waymouth, R.; Pino, P. *J. Am. Chem. Soc.* **1990,** *112,* 4911.

105. Conticello, V. P.; Brard, L.; Giardello, M. A.; Tsuji, Y.; Sabat, M.; Stern, C. L.; Marks, T. J. *J. Am. Chem. Soc.* **1992,** *114,* 2761.

106. Kitamura, M.; Okuma, T.; Inoue, S.; Sayo, N.; Kumobayashi, H.; Akutagawa, S.; Ohta, T.; Takaya, H.; Noyori, R. *J. Am. Chem. Soc.* **1988,** *110,* 629.

107. Noyori, R.; Okuma, T.; Kitamura, M.; Takaya, H.; Sayo, N.; Kumobayashi, H.; Akutagawa, S. *J. Am. Chem. Soc.* **1987,** *109,* 5856.

108. Kawano, H.; Ishii, Y.; Saburi, M.; Uchida, Y. *J. Chem. Soc., Chem. Commun.* **1988,** 87.

109. Ohkuma, T.; Kitamura, M.; Noyori, R. *Tetrahedron Lett.* **1990,** *31,* 5509.

110. Mashima, K.; Hino, T.; Takaya, H. (a) *Tetrahedron Lett.* **1991,** *32,* 3101. (b) *J. Chem. Soc., Dalton Trans.* **1992,** 2099.

111. Murata, M.; Morimoto, T.; Achiwa, K. *Synlett,* **1991,** 827.

112. Kitamura, M.; Ohkuma, T.; Takaya, H.; Noyori, R. *Tetrahedron Lett.* **1988,** *29,* 1555.

113. Jones, A. B.; Yamaguchi, M.; Patten, A.; Danishefsky, S. J.; Ragan, J. A.; Smith, D. B.; Schreiber, S. L. *J. Org. Chem.* **1989,** *54,* 19.

114. Nishi, T.; Kitamura, M.; Ohkuma, T.; Noyori, R. *Tetrahedron Lett.* **1988,** *29,* 6327.

115. Cesarotti, E.; Mauri, A.; Pallavicini, M.; Villa, L. *Tetrahedron Lett.* **1991,** *32,* 4381.

116. Mashima, K.; Sato, N.; Kusano, K.; Nozaki, K.; Takaya, H. Unpublished results.

117. (a) Hayashi, T.; Katsumura, A.; Konishi, M.; Kumada, M. *Tetrahedron Lett.* **1979,** 425. (b) Hayashi, T.; Mise, T.; Kumada, M. *Tetrahedron Lett.* **1976,** 4351.

118. Toros, S.; Kollar, L.; Heil, B.; Markó, L. *J. Organomet. Chem.* **1982,** *232,* C17.

119. Takahashi, H.; Sakuraba, S.; Takeda, H.; Achiwa, K. *J. Am. Chem. Soc.* **1990,** *112,* 5876.

120. Takeda, H.; Hosokawa, S.; Aburatani, M.; Achiwa, K. *Synlett,* **1991,** 193.

121. (a) Takahashi, H.; Sakuraba, S.; Achiwa, K. *Abstracts of 36th Organometallic Chemistry,* Tokyo, **1989,** B107. (b) Sakuraba, S.; Achiwa, K. *Synlett,* **1991,** 689.

122. Takeda, H.; Tachinami, T.; Aburatani, M.; Takahashi, H.; Motimoto, T.; Achiwa, K. *Tetrahedron Lett.* **1989,** *30,* 363.

123. Takeda, H.; Tachinami, T.; Aburatani, M.; Takahashi, H.; Morimoto, T.; Achiwa, K. *Tetrahedron Lett.* **1989,** *30,* 367.

124. (a) Achiwa, K.; Kogure, T.; Ojima, I. *Tetrahedron Lett.* **1977,** 4431. (b) Ojima, I., Kogure, T.; Terasaki, T.; Achiwa, K. *J. Org. Chem.* **1978,** *43,* 3444.

125. (a) Takahashi, H.; Hattori, M.; Chiba, M.; Morimoto, T.; Achiwa, K. *Tetrahedron Lett.* **1986,** *27,* 4477. (b) Morimoto, T.; Takahashi, H.; Fujii, K.; Chiba, M.; Achiwa, K. *Chem. Lett.* **1986,** 2061.

126. Takahashi, H.; Morimoto, T.; Achiwa, K. *Chem. Lett.* **1987,** 855.

127. (a) Hatat, C.; Karim, A.; Kokel, N.; Mortreux, A.; Petit, F. *Tetrahedron Lett.* **1988,** *29,* 3675. (b) Hatat, C.; Kokel, N.; Mortreux, A.; Petit, F. *Tetrahedron Lett.* **1990,** *31,* 4139.

128. Noyori, R.; Ikeda, T.; Ohkuma, T.; Widhelm, M.; Kitamura, M.; Takaya, H.; Akutagawa, S.; Sayo, N.; Saito, T.; Taketomi, T.; Kumobayashi, H. *J. Am. Chem. Soc.* **1989,** *111,* 9134.

129. Murahashi, S.-I.; Naota, T.; Kuwabara, T.; Saito, T.; Kumobayashi, H.; Akutagawa, S. *J. Am. Chem. Soc.* **1990,** *112,* 7820.

130. Mahsima, K.; Matsumura, Y.-I.; Kusano, K.; Kumobayashi, H.; Sayo, N.; Hori, Y.; Ishizaki, T.; Akutagawa, S.; Takaya, H. *J. Chem. Soc., Chem. Commun.* **1991,** 609.

131. Kitamura, M.; Ohkuma, T.; Tokunaga, M.; Noyori, R. *Tetrahedron: Asymmetry,* **1990,** *1,* 1.

132. Mashima, K.; Kusano, K.; Takaya, H. Unpublished results.

133. Genet, J. P.; Pinel, C.; Mallart, S.; Juge, S.; Thorimbert, S.; Laffitte, J. A. *Tetrahedron: Asymmetry,* **1991,** *2,* 555.

134. Zassinovich, G.; Bettella, R.; Mestroni, G.; Bresciani-Pahor, N.; Geremia, S.; Randaccio, L. *J. Organomet. Chem.* **1989,** *370,* 187.

135. Bolm, C. *Angew. Chem., Int. Ed. Engl.* **1991,** *30,* 542. See also references cited therein.

136. Gladiali, S.; Pinna, L.; Deloga, G.; Martin, S. D.; Zassinovich, G.; Mestroni, G. *Tetrahedron: Asymmetry,* **1990,** *1,* 635.

137. Bakos, J.; Tóth, I.; Heil, B.; Szalontai, G.; Párkányi, L.; Fülöp, V. *J. Organomet. Chem.* **1989,** *370,* 263.

138. (a) Kang, G.-J.; Cullen, W. R.; Fryzuk, M. D.; James, B. R.; Kutney, J. P. *J. Chem. Soc., Chem. Commun.* **1988,** 1466. (b) Becalski, A. G.; Cullen, W. R.; Fryzuk, M. D.; James, B. R.; Kang, G.-J.; Rettig, S. J. *Inorg. Chem.* **1991,** *30,* 5002.

139. Bakos, J.; Heil, B.; Lecomte, L.; Sinou, D. *Abstract of 5th OMCOS,* Florence, Italy, **1989,** PS1–36.

140. Spindler, F.; Pugin, B.; Blaser, H.-U. *Angew. Chem., Int. Ed. Engl.* **1990,** *29,* 558.

141. Chan, Y. N. C.; Osborn, J. A. *J. Am. Chem. Soc.* **1990,** *112,* 9400.

142. Oppolzer, W.; Wills, M.; Starkemann, C.; Bernardinelli, G. *Tetrahedron Lett.* **1990,** *31,* 4117.

143. Willoughby, C. A.; Buchwald, S. L. *J. Am. Chem. Soc.* **1992,** *114,* 7562.

144. Burk, M. J.; Feaster. J. E. *J. Am. Chem. Soc.* **1992,** *114,* 6266.

2

Asymmetric Isomerization of Allylamines

Susumu Akutagawa

Takasago Research Institute, Inc.

Kazuhide Tani

Department of Chemistry, Faculty of Engineering Science
Osaka University

2.1. Introduction

Olefinic double-bond isomerization is probably one of the most commonly observed and well-studied reactions that uses transition metals as catalysts [1]. However, prior to our first achievement of asymmetric isomerization of allylamine by optically active Co(I) complex catalysts [2], there were only a few examples of catalytic asymmetric isomerization, and these were characterized by very low asymmetric induction ($< 4\%$ ee) [3]. In 1978 we reported that an enantioselective hydrogen migration of a prochiral allylamine such as *N,N*-diethylgeranylamine (**1**) or *N,N*-diethylnerylamine (**2**) gave optically active citronellal (*E*)-enamine **3** with about 32% ee utilizing Co(I)–DIOP [DIOP = 2,3-*O*-isopropylidene-2,3-dihydroxy-1,4-bis(diphenylphosphino)butane] complexes as the catalyst (eq. 1).

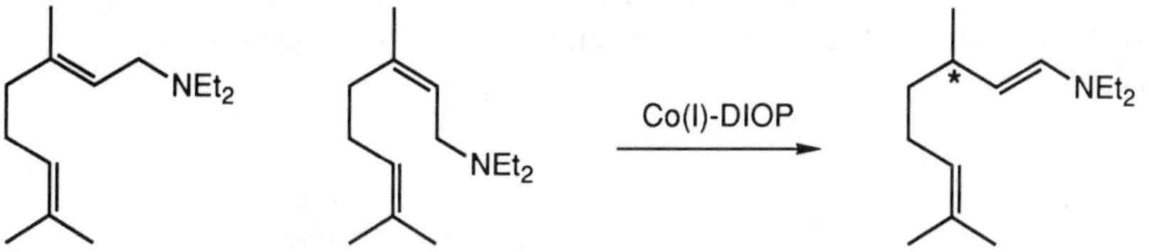

$$ \text{(1)} $$

| **1** | **2** | **3** |

Optically active citronellal, which can be quantitatively obtained from the optically active citronellal enamine, is a useful intermediate for the synthesis of many optically active natural products. Thus, the discovery of the asymmetric isomerization, although both chemo- and enantioselectivities were insufficient, opened the door to the synthesis of optically active terpenoids. A great improvement in the chemo- and enantioselectivities for the reaction in equation 1 (up to 99% yield for **3** and 98% ee) has been achieved by employing cationic Rh(I)–BINAP (BINAP = 2,2′-bis(diphenylphosphino)-1,1′-binaphthyl) complexes, [Rh(BINAP)(L$_2$)]$^+$(L = diene or solvent), as the catalyst precursor. Further studies of the Rh(I)–BINAP species led to the discovery of the thermally stable [Rh(BINAP)$_2$]$^+$ complex, which is suitable for industrial uses, achieving a very high "turnover number" (TON: enamine moles produced by one mole of catalyst during 18 h) ($\sim$3 × 10^5) without impairing chemo- and enantioselectivities. Since 1983 Takasago International Corporation has been successfully manufacturing (−)-menthol on a commercial basis ($\sim$1500 t/year) using this asymmetric isomerization process, the "Takasago process."

This chapter describes the discovery of the new asymmetric isomerization process, improvements in the catalyst, some practical aspects, and the scope and limitation of the process, with emphasis on developments during commercial manufacturing, as review articles on this subject [4] have been published.

2.2. Asymmetric Isomerization of Allylamines

Although a variety of transition metal complexes have been reported to be active catalysts for the migration of inner double bonds to terminal ones in functionalized allylic systems (eq. 2) [5], prochiral allylic compounds with a multisubstituted olefin (R^1, R^2 ≠ H in eq. 2) are not always susceptible to catalysis or show only a low reactivity [1 d]. There are merits in choosing allylamines **1** and **2** as the substrates for enantioselective isomerization: (1) optically pure citronellal, which is an important starting material for optically active terpenoids such as (−)-menthol cannot be obtained directly from natural sources [6], and (2) both (E)-allylamine **1** and (Z)-allylamine **2** can be prepared in reasonable yields from myrcene or isoprene, respectively. The (E)-allylamine **1** is obtained from the reaction of myrcene and diethylamine in the presence of lithium diethylamide under argon in an almost quantitative yield (eq. 3) [7]. The (Z)-allylamine **2** can also be prepared with high selectivity ($\sim$90%) by lithium-catalyzed telomerization of isoprene using diethylamine as a telomer (eq. 4) [8]. Thus, either natural or petroleum resources can be selected.

R^2∕∿∖∖X ⟶ R^2∕∿∖∖X   (2)

X=OH, OR, NRR', etc.

Table 1 Metal Catalyzed Isomerization of Allylamine*

Catalyst	Substrate	Product (Selectivity%)
Cp_2TiCl_2 / ^{i}PrMgBr	1	**4** (100)
$CoH(N_2)(PPh_3)_3$	1	**3** (85), **4** (15)
$Co(acac)_2$ / PPh_3 / ^{i}Bu$_2$AlH	2	**3** (81), **4** (19)
$Co(acac)_2$ / PPh_3 / ^{i}Bu$_2$AlH	5	**6** (~100)
$[Rh(PPh_3)_2(COD)]^+$	1	**3** (~100)
$[Rh(DIPHOS)(Solvent)_n]^+$	1	**3** (>96)
$[Rh(BINAP)(COD)]^+$	1	**3** (>96)

* Reaction was carried out in THF at 60°C under N_2, [Substrate]/[Catalyst] 100, for 15hrs.

Migration of the multisubstituted inner double bonds of **1** or **2** has been found to be effected by various metal catalysts (Table 1). As Table 1 shows, the product selectivity varies dramatically with the catalyst employed. With the exception of cationic Rh(I)–tertiary phosphine complexes, the selectivity for the desired citronellal enamine **3** is not sufficient. In the case of titanium and cobalt complexes, the 6-double bond also migrates to give a considerable amount of the undesired conjugated dienamine **4**. A secondary allylamine, N-cyclohexylgeranylamine (**5**), is an effective substrate that can be isomerized cleanly to give the corresponding imine **6** even with Co(I) catalysts. Cationic rhodium(I) complexes with phosphine or

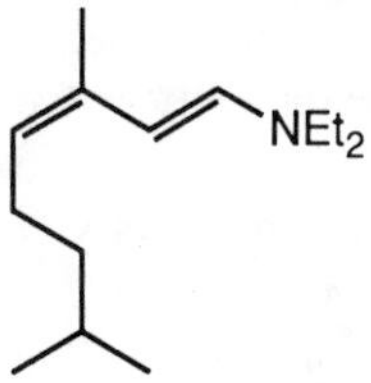

| **4** | **5** | **6** | **7** |

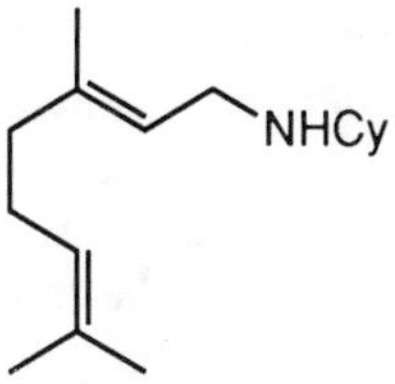

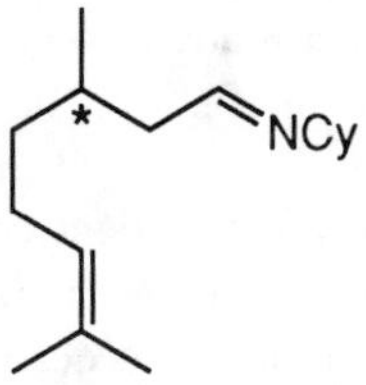

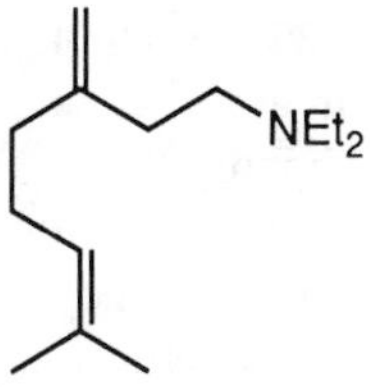

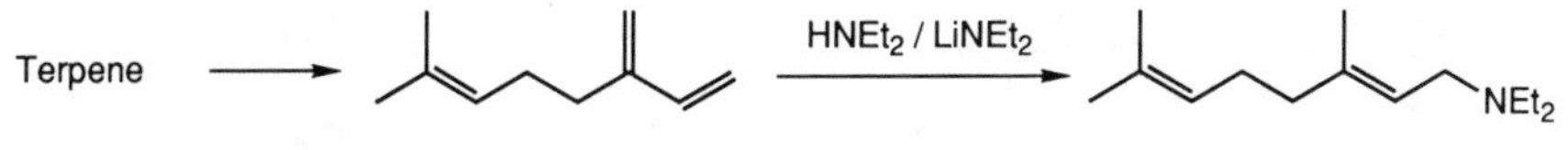

Terpene $\longrightarrow$ [structure] $\xrightarrow{\text{HNEt}_2 \text{ / LiNEt}_2}$ [structure] NEt$_2$ **(3)**

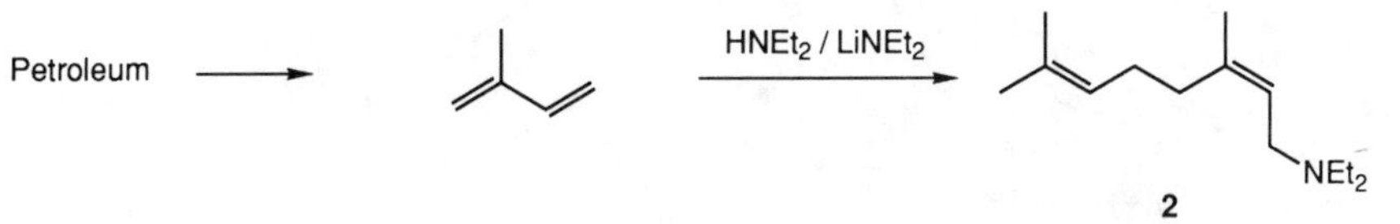

1

Petroleum $\longrightarrow$ [structure] $\xrightarrow{\text{HNEt}_2 \text{ / LiNEt}_2}$ [structure] NEt$_2$ **(4)**

2

bisphosphine are very effective catalysts for the isomerization of **1** and **2** to give the same (E)-enamine **3**; the undesired dienamine **4** is virtually absent.

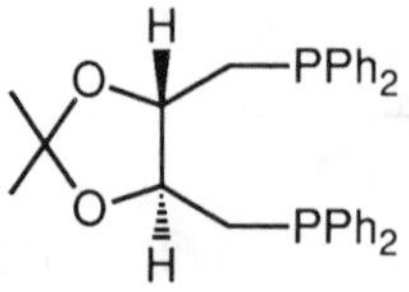

(+)-DIOP 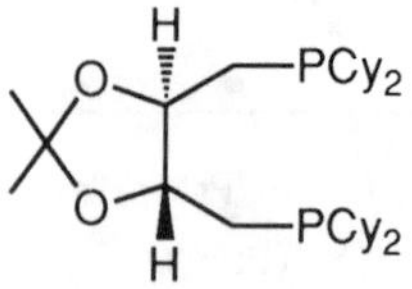   (-)-CyDIOP

The first asymmetric isomerization of an allylamine was achieved with the optically active cobalt catalysts as mentioned above. The isomerization of **1** with Co(acac)$_2$/(+)-DIOP/iBu$_2$AlH catalyst gave (3R)-**3** with 32% ee (39% chemical yield) accompanied by a considerable amount of undesired dienamine **4**. Although with the same catalyst system a higher enantioselectivity (57% ee) and chemical yield (60%) were obtained by using the secondary allylamine **5** as the substrate, these were insufficient for a commercial process. As mentioned earlier we have found that the cationic Rh(I)–phosphine complexes, which are effective as hydrogenation catalysts, showed very high selectivities and catalytic activities in the isomerization of **1** or **2** to the (E)-enamine **3**. Since many chiral diphosphines had already been developed, several typical optically active diphosphines were tested as the chiral ligand for asymmetric isomerization. Results are listed in Table 2. Among the DIOP-based ligands, only CyDIOP gave an exceptionally high enantioselectivity of 77% ee, but the catalytic activity was not sufficient with such a diphosphine where all the phosphorus substituents were aliphatic groups. The DIOP complex showed a fairly high catalytic activity but only a low asymmetric induction of about 22% ee; DIOP was also known to give low enantioselectivities in the asymmetric hydrogenation of some olefins [9]. In contrast to these conventional optically active diphosphines, BINAP, in which all the substituents on both phosphorus atoms are aromatic groups, showed excellent enantioselectivity as well as chemoselectivity and catalytic activity for the isomerization. One of the favorable features of this asymmetric isomerization is the desirable stereochemical correlation between the substrate geometries, product (E)-enamine configurations, and the BINAP chirality, as shown in Scheme 1. This correlation is established with an almost perfect enantioselectivity and a practically quantitative chemical yield in all routes.

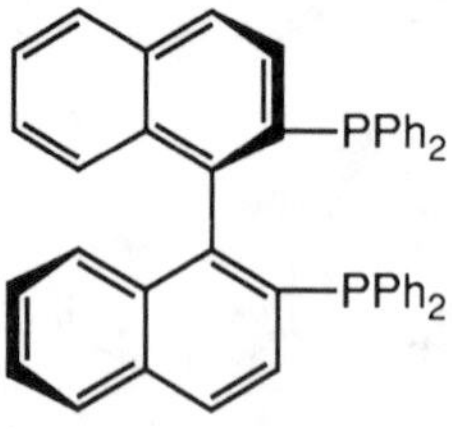

(R)(+)-BINAP 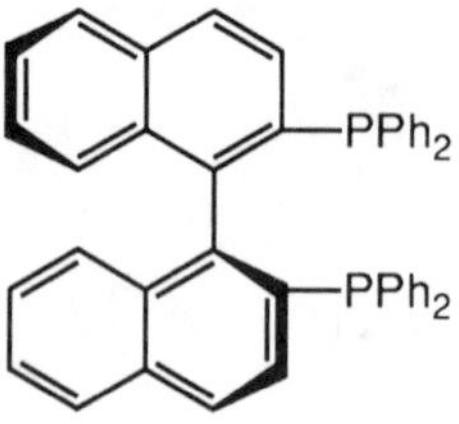   (S)(-)-BINAP

Table 2 Asymmetric Isomerization of Allylamine*

Catalyst	Substrate	Conversion(%)	Product (Selectivity%)	%ee (Configuration)
Co(acac)$_2$ / (+)-DIOP / ^{i}Bu$_2$AlH	1	45	3 (87), 4 (13)	35(**R**)
Co(acac)$_2$ / (+)-DIOP / ^{i}Bu$_2$AlH	5	62	6 (97)	57(**R**)
[Rh{(+)-DIOP}(COD)]$^+$	1	71	3 (100)	22(**R**)
[Rh{(−)-CyDIOP}(Solvent)$_2$]$^+$	1	18	3 (80)	77(**S**)
[Rh{(**R**)-BINAP}(COD)]$^+$	1	100	3 (100)	97(**S**)
[Rh{(**R**)-BINAP}(COD)]$^+$	2	97	3 (100)	96(**R**)
[Rh{(**S**)-BINAP}(COD)]$^+$	1	100	3 (100)	97(**R**)
[Rh{(**S**)-BINAP}(COD)]$^+$	5	100	6 (100)	98(**R**)

* Reaction was carried out in THF at 60°C with the [substrate]/[catalyst] ratio of 100 for 15h.

45

Scheme 1

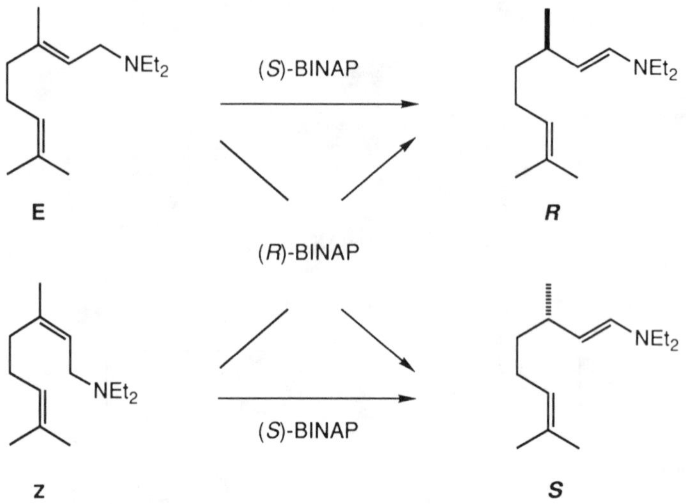

Since BINAP enantiomers as well as the both (*E*)- and (*Z*)-allylamine substrates are easily obtained, this stereochemical relation provides the following economical advantages: (1) *the option of taking the starting material either from a natural resource (renewable terpene) or from petroleum,* and (2) *easy access to both enantiomers of citronellal from a single intermediate.*

2.3. Isomerization Process Development

In spite of the remarkable high selectivities, the drawback of the original asymmetric isomerization process was the high price of the Rh–BINAP catalyst (cf. $RhCl_3$, \$42.50/500 mg, (+)- or (−)-BINAP, \$25.40/100 mg: *Aldrich,* 1991). For the industrial production of (+)-citronellal in 1000 t/year scale, we had to raise the TON to an optimum level > 50,000) while maintaining high regio- and stereo selectivities. The following process development has enabled us to fulfill these requirements.

2.3.1. Great Improvement in the Procedure for Synthesizing Optically Active BINAP

The chiral ligand BINAP was originally prepared from 2,2′-binaphthol and resolved by complexation with an optically active palladium complex [10]. A new method starting from 2-naphthol was developed (Scheme 2) [11]. In this method optical resolution was achieved at the stage of BINAP dioxide (BINAPO) using inexpensive optically active acids such as camphorsulfonic acid.

Scheme 2

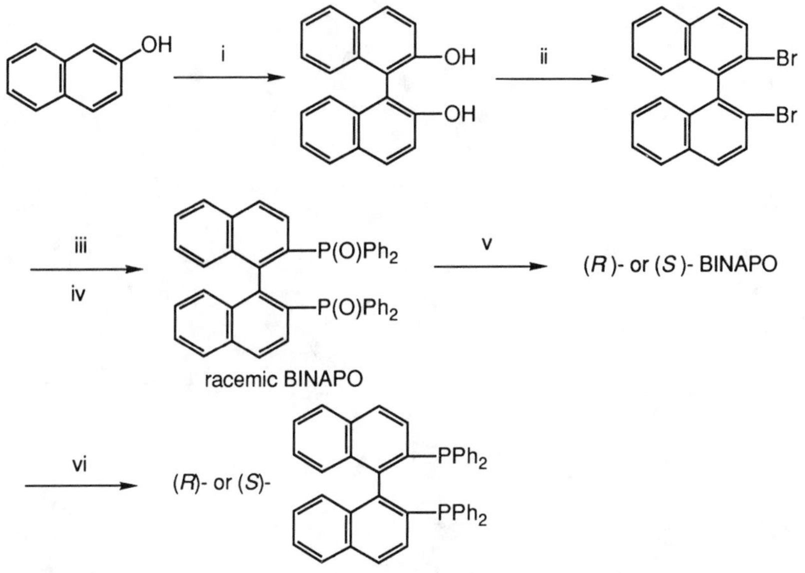

i : FeCl$_3$ ii : Ph$_3$PBr$_2$ iii : Mg / THF
iv : Ph$_2$P(O)Cl   v : Optical resolution   vi : HSiCl$_3$ / NEt$_3$
                          with camphorsulfonic acid

2.3.2. Removal of Catalyst Inhibitors

The Rh–BINAP catalysts are very sensitive to impurities such as oxygen, moisture, and carbon dioxide. If an excess of water ($[H_2O]/[Rh] = 15$) is present in the reaction mixture, the isomerization is stopped after a few turnovers with precipitation of air-stable red-brown crystals, which were found to be [{Rh(BINAP)}$_3$(μ_3-OH)$_2$]ClO$_4$. X-Ray analysis of this complex (H$_2$O was replaced by D$_2$O) revealed a unique structure of a triangular Rh$_3$ core capped with two triply bridging OH groups (Fig. 1). This Rh(I)–trinuclear complex was totally inactive as an isomerization catalyst, which suggests a catalyst deactivation mechanism [12]. The effect of several additives on the isomerization of **2** with [Rh(BINAP)(COD)]$^+$ has been examined, and the results are summarized in Table 3.

Although donor substances retard the reaction rate, significant deactivation of the catalyst by the conjugated dienamine **4** was observed. In connection with this, we have found that an isomeric dienamine of the substrate 2-[2-(N,N-diethylamino)ethyl]-6-methyl-1,5-heptadiene (**7**), which is always present in small amount (0.5–0.7%) in the crude commercial products **1** and **2**, also acts as a deactivator (see Table 4). Thus, a careful and exhaustive pretreatment of the substrate and the reaction system are necessary to attain a high TON.

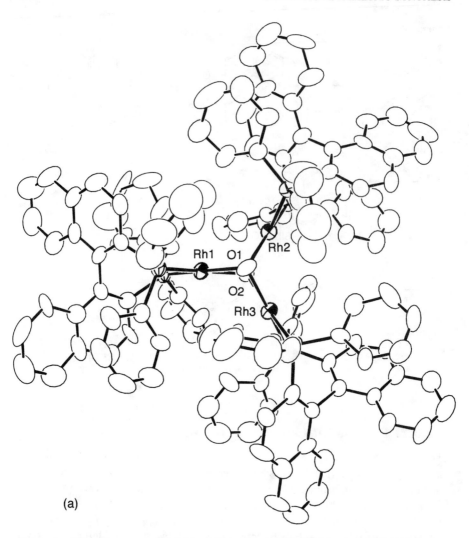

(a)

Figure 1 Molecular structure of the cationic part in [{Rh((R)(+)–BINAP)}₃(OD)₂]ClO₄: ORTEP drawing (a) and space-filling representations of the side view (b) and top view (c). Rh–Rh = 3.086–3.102 Å, Rh–P = 2.194–2.222 Å, Rh–O = 2.122–2.163 Å; Rh–Rh–Rh = 59.75–60.25°.

2.3.3. Development of New Catalyst Systems

As the catalyst precursor for the present isomerization, [Rh(BINAP)(COD)]⁺ was conveniently employed in the original process. However, further improvement in TON was achieved by the discovery of a new catalyst, [Rh(BINAP)₂]⁺ (**8**) [13]. It was a surprise for us to find that the bis(BINAP) complex showed considerable catalytic activity at a high temperature (> 80°C), because [Rh(DIPHOS)₂]ClO₄ was

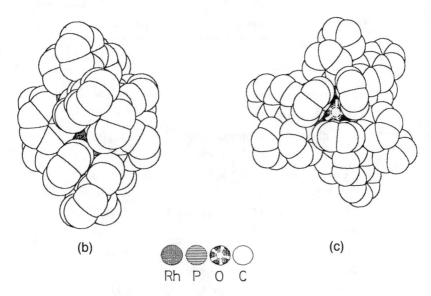

(b)   (c)

Rh P O C

Figure 1 (*continued*)

completely inactive for the isomerization of **1** and **2**, even at 120°C. The enantioselectivity was maintained as high (> 96% ee) as that obtained with [Rh(BINAP)(COD)]⁺ at 40°C. By employing the bis(BINAP) complex, it became feasible to reuse the catalyst and to raise the TON above 400,000. Progress in the improvement of TON is summarized in Table 4. An X-ray structural analysis of complex **8** [13] indicated the existence of significant distortion toward a tetrahedron and a considerable elongation of Rh–P distances (2.368(6)–2.388(6)Å) due to steric con-

Table 3   Effect of Additive on the Isomerization of **2** with Rh-BINAP Catalyst

Substrate	Additive to $[Rh\{(+)\text{-}BINAP\}(COD)]^+$	Additive / Catalyst	k_{obs} a) mol / l / min
2 (structure) b)	none	---	37.
	NEt₃	4	20.
	COD	2	8.5
	dienamine(**4**)	2	1.8

a) THF, $[Rh\{(+)\text{-}BINAP\}(COD)]^+$, [S] = 0.24 mol /$l$, [S] / [C] = 100, 60°C

b) Commercial production of **2** always accompanies 0.5~0.7% of **7**

Table 4   Advancement of TON for the Isomerization of **1**

Improvement	TON
[Rh(BINAP)(COD)]$^+$, 1 without special pretreatment	100
[Rh(BINAP)(COD)]$^+$, 1 treated with Vitride®	1,000
[Rh(BINAP)(COD)]$^+$, 1 after removal of amine isomer **7**	8,000
Reuse of **8** (10% loss)	80,000
Reuse of **8** (2% loss*)	400,000

* Owing to total quality control

gestion. The observed Rh–P distance is perhaps the longest among the known bisphosphine–Rh(I) complexes. These features may explain the unusual catalytic activity of **8**.

The stereochemical correlation observed with the [Rh(BINAP)$_2$]$^+$ catalyst was the same as that obtained with [Rh(BINAP)(COD)]$^+$ or [Rh(BINAP)(solvent)$_n$]$^+$ (Scheme 1), implying that an identical catalyst species determines the enantioselection. This requires dissociation of one of the BINAP ligands of [Rh(BINAP)$_2$]$^+$ during the catalysis. Supporting evidence for the dissociation of BINAP from [Rh(BINAP)$_2$]$^+$ has been obtained from a ^{31}P NMR study of the reaction mixture of **8** with a pseudosubstrate, triethylamine, at 90°C [14]. The greatest advantage of using [Rh(BINAP)$_2$]$^+$ as the catalyst is the resulting thermal and chemical stability, which enabled the reuse of the recovered catalyst. A further improvement of the catalyst was realized by employing the *p*-TolBINAP ligand [11b]. The new catalyst,

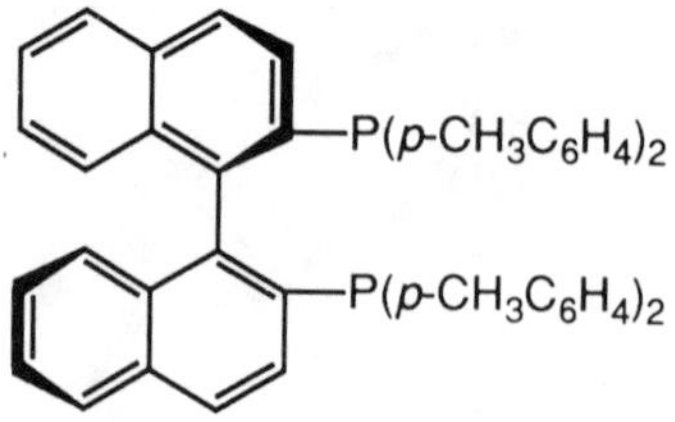

(*R*)(+)-*p*-Tol BINAP

[Rh(p-TolBINAP)$_2$]$^+$, also showed excellent stability, allowing multiple reuses of the recovered catalyst. Both chemo- and enantioselectivities for the isomerization of **1** and **2** also remained high enough (> 97% ee) with the new catalyst. Another merit found in the new catalyst was a better solubility in organic solvent, a valuable feature in industrial processes.

2.4. Commercial Manufacture

As described above, hydrolysis of the optically active enamine **3** proceeds without racemization and produces an optically active aldehyde, citronellal, with a very high optical purity (> 98% ee). The optical purity of citronellal available from natural sources is known to be no more than 80% ee [5]. the present asymmetric isomerization of the allylamine **1** is utilized as the key step for the industrial production of (−)-menthol (Scheme 3).

Scheme 3

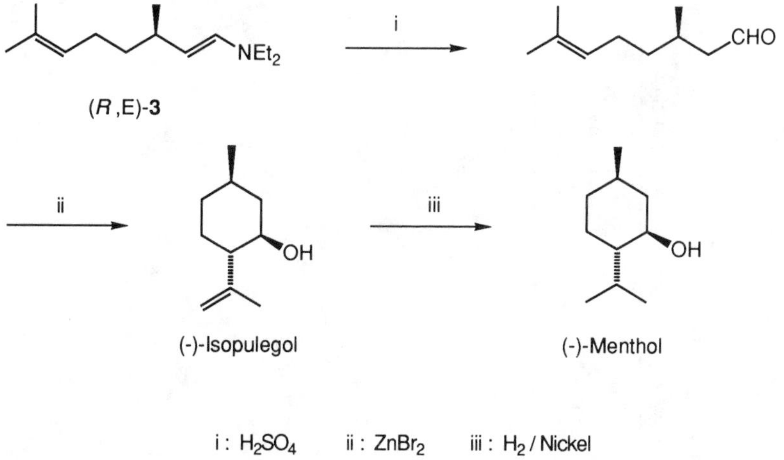

(*R*,E)-**3**

(-)-Isopulegol

(-)-Menthol

i : H$_2$SO$_4$ ii : ZnBr$_2$ iii : H$_2$ / Nickel

The ZnBr$_2$-catalyzed cyclization of (+)-citronellal, an intramolecular ene reaction, proceeds stereospecifically to give almost quantitatively (−)-isopulegol, where all the substituents are in the equatorial position. This forms a contrast to ordinary Lewis acid catalyzed cyclization of citronellal, which gives only 65% of (−)-isopulegol. The enantiomeric purity of (−)-isopulegol was raised to 100% ee by recrystallization at −50°C, and each step proceeds with high chemical yield (> 92%), therefore the synthesis of pure (−)-menthol by the present procedure became a commercial process. With an estimated worldwide consumption of 4500 tons per year, (−)-menthol is widely used in many consumer products including cigarettes, chewing gum, toothpaste, and pharmaceutical products. At present, natural (−)-menthol is obtained mostly from *Mentha arvenis* cultivated in China, while synthetic materials are produced in Germany, the United States, and Japan. Takasago's

Table 5 Optically Active Terpenoids Produced Based on the Asymmetric Isomerization of Allylamines by Takasago

Name	Formula	Chem. Purity (%)	%ee	Production (tons/year)	Use
(+)-Citronellal	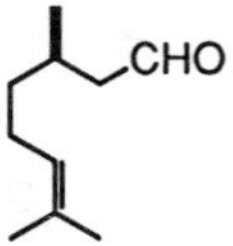 (structure, CHO)	98.0	97±1	1,500	Intermediate
(-)-Isopulegol	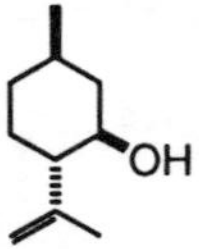 (structure, OH)	100	100	1,100	Intermediate
(-)-Menthol	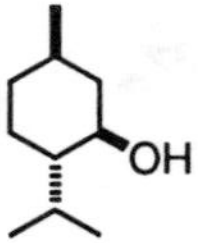 (structure, OH)	100	100	1,000	Pharmaceuticals Tobacco Household Products
(-)-Citronellol	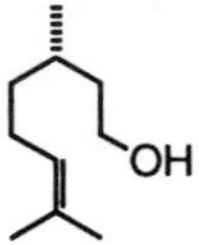 (structure, OH)	99	98	20	Fragrances

Table 5 (continued)

Name	Formula	Chem. Purity (%)	%ee	Production (tons/year)	Use
(+)-Citronellol	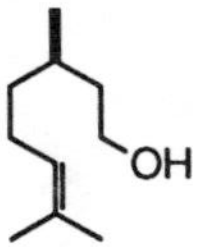	99	98	20	Fragrances
(-)-7-Hydroxy-citronellal	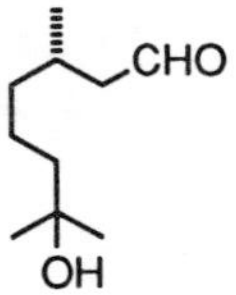	99	98	40	Fragrances
S-7-Methoxy-citronellal	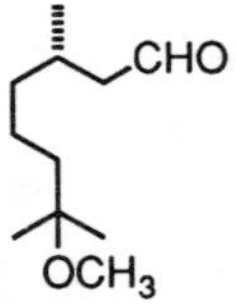	99	98	10	Insect Growth Regulator
S-3,7-Dimethyl-1-octanal	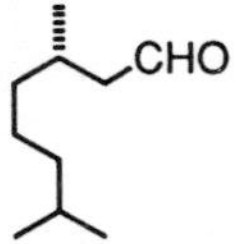	99	98	7	Insect Growth Regulator

menthol synthesis is now yielding the isomer with (1*R*, 3*R*, 4*S*)-configuration, the only useful isomer, among the eight possible isomers, in its enantiomerically pure form without any optical resolution process. Since 1984 Takasago has also been producing a number of optically active terpenoids beside (−)-menthol based on the asymmetric isomerization of allylamines (Table 5).

2.5. Scope and Limitations

The present enantioselective isomerization process requires prochiral allylamines free of the geometrical isomer. If such an allylamine is at hand, high asymmetric induction can be realized for a wide range of tertiary or secondary amines with alkyl substituents on the nitrogen atom. Thus, with [Rh{(+)−BINAP}(COD)]$^+$ the secondary allylamine **5** gives quantitatively the corresponding (*S*)-imine **6** with 96% ee. Allylamines with a styrene-type conjugation (**9**) (though slowly reacting substrates) and those having a hydroxy group (**10**) and a methoxy group (**11**) also act as effective substrates to give the corresponding enamines with high enantiomeric purities (> 95% ee) (see Table 5). For all these substrates the stereochemical relationship shown in Scheme 1 remain unaffected. Asymmetric isomerization has successfully been applied to the synthesis of the optically active α-tocopherol side chain [15]. Thus the C-15 allylamine (*E*)(7*R*)-1-dimethylamino-3,7,11-trimethyl-2-dodecene (**12**) was isomerized to the corresponding C-15 (*R*,*R*)-enamine (**13**) with a very high enantiomeric purity (97% ee) in excellent chemical yield (98%) (eq. 5).

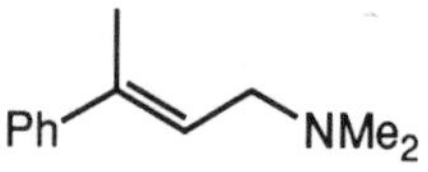

| 9 | 10 | 11 |

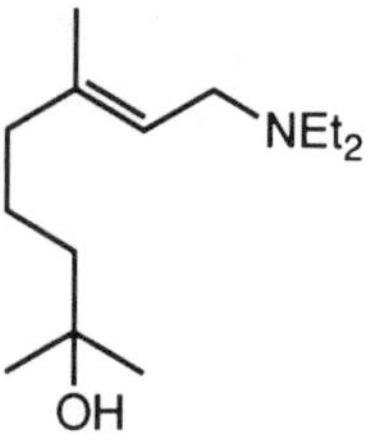

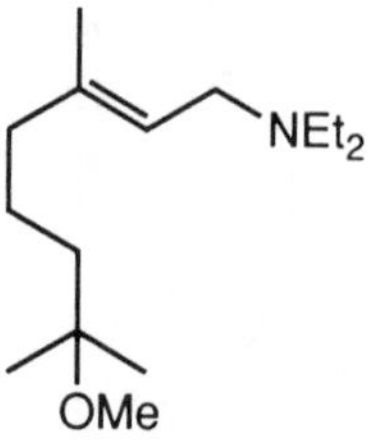

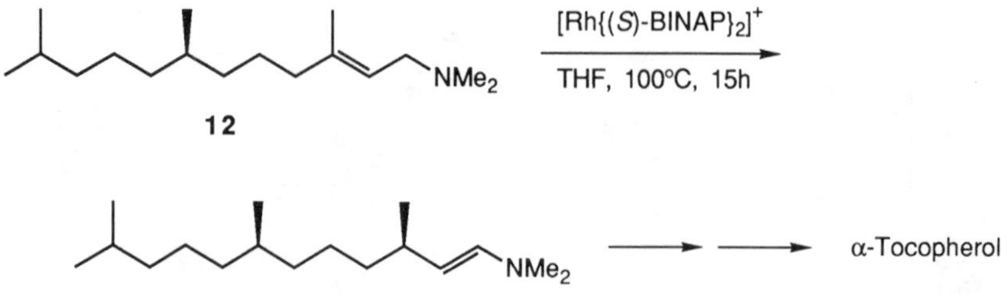

[Rh{(*S*)-BINAP}$_2$]$^+$

THF, 100°C, 15h

12

α-Tocopherol   (5)

13

The basicity of the amine nitrogen appears to be an important factor for an effective asymmetric induction. Phenyl substituents on the nitrogen atom greatly retard the reaction rate. Thus, N-phenyl- and N,N-diphenylgeranylamine are inert at 40°C and 24 hours reaction time. There are a few characteristic features worth noting. If an allylamine is secondary, the product is the corresponding imine, a more stable valence tautomer of the enamine, which cannot be detected in the reaction mixture. The exclusive formation of an (E)-enamine regardless of the double-bond geometry of the starting substrate is another noticeable feature of isomerization. Homoallylamines such as **7** or N,N-dimethyl-3-butenylamine (**14**) are inactive substrates. Besides acyclic allylamines, cyclic allylamine **15**, which should yield the (Z)-enamine **16** on isomerization, is also an active substrate. We have examined the isomerization of several tetrahydropyridine derivatives (**17a–e**) with the Rh–

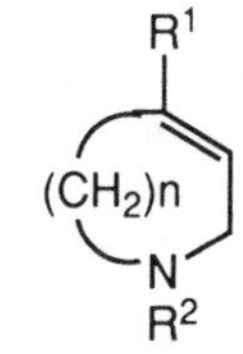

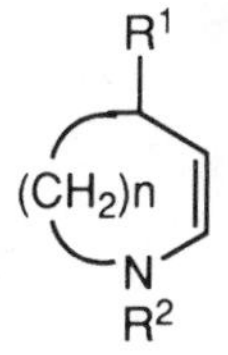

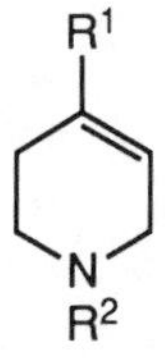

a	R^1=H,	R^2=Me
b	R^1=H,	R^2=CH$_2$Ph
c	R^1=Et,	R^2=Me
d	R^1=Me,	R^2=CH$_2$Ph
e	R^1=Me,	R^2=H

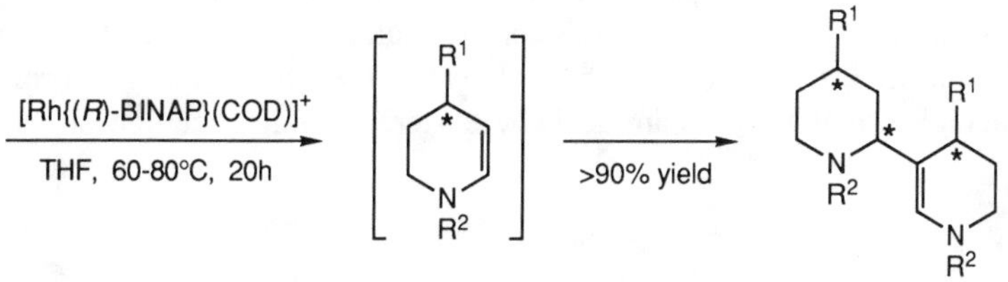

$$(6)$$

c R^1=Et, R^2=Me
$[\alpha]_D^{22}$+66.4° (CHCl$_3$)

d R^1=Me, R^2=CH$_2$Ph
$[\alpha]_D^{22}$+35.9° (CHCl$_3$)

BINAP catalyst [15]. In these reactions, however, the corresponding (Z)-enamines of type **16** are not isolated as monomeric forms, only the dimer **18** (from the secondary allylamine) or trimer **20** (from the tertiary allylamine) has been isolated in optically active forms when prochiral substrates are employed (eqs. 6 and 8). The configuration and the enantiomeric purity of these enamines (**18** and **20**) have not been determined. The achiral system **17a,b** yielded only racemic dimers **19a,b** (eq. 7). Thus, it is likely that the oligomerization of the enamine intermediate is a noncatalyzed thermal process.

$$
\textbf{17a,b} \xrightarrow[\text{THF, 60°C, 20h}]{[\text{Rh}\{(R)\text{-BINAP}\}(\text{COD})]^+} \quad \left[\underset{R^2}{\underset{|}{\overset{\frown}{N}}} \right] \xrightarrow{>90\% \text{ yield}} \quad \underset{19}{\text{structure}} \tag{7}
$$

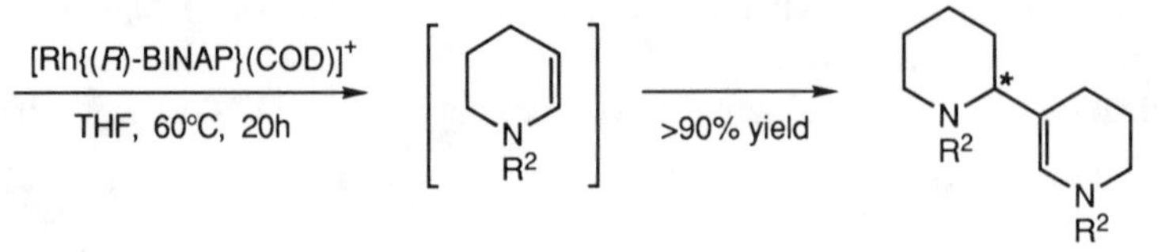

a R^2=Me, $[\alpha]_D$ 0°

b R^2=CH$_2$Ph, $[\alpha]_D$ 0°

$$
\textbf{17e} \xrightarrow[\text{THF, 100°C, 60h}]{[\text{Rh}\{(R)\text{-BINAP}\}(\text{COD})]^+} \quad \left[\underset{N}{\overset{\text{Me}}{\text{structure}}} \right] \xrightarrow{26\%} \quad \textbf{20} \tag{8}
$$

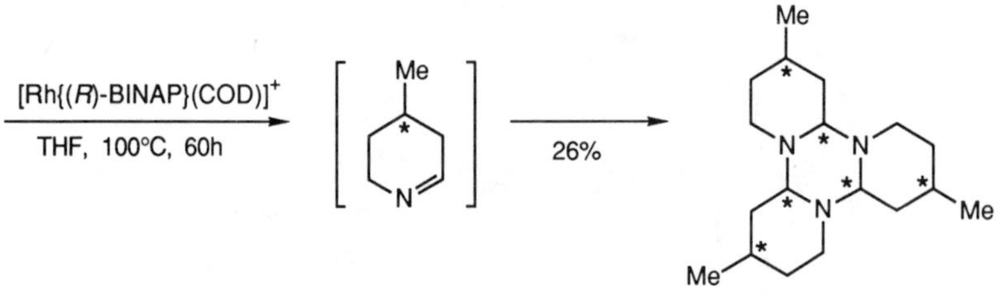

20 $[\alpha]_D^{22}$ +14.7° (CHCl$_3$)

Allylamides are slow-reacting substrates [16]. The isomerization of N-acetylger-anylamide (**21**) and N-acetylnerylamide (**22**) with [Rh{(+)–BINAP}(COD)]$^+$ needs a temperature of 150°C to give the corresponding allylamide **23** with high enantiomeric purity (> 95% ee), and the yield is low (< 30%) due to the formation of a considerable amount of dienamide **24**. The stereochemical relationship is the same

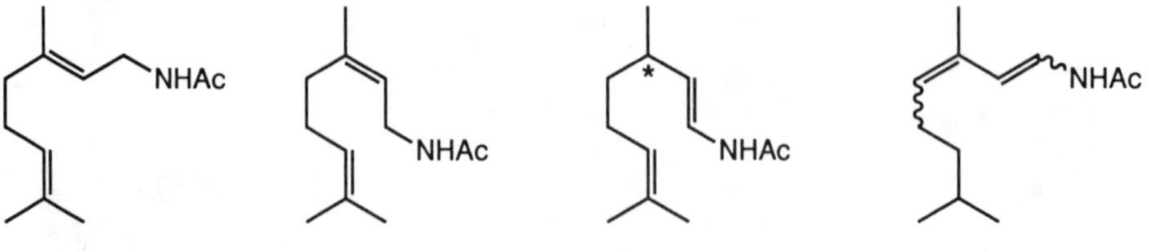

21   **22**   **23**   **24**

as that found with allylamine (see Scheme 1). A cyclic allylamide (**25**) is also isomerized selectively at 150°C to the corresponding enamide (**26**) with high enantiomeric purity (98% ee)(eq. 9).

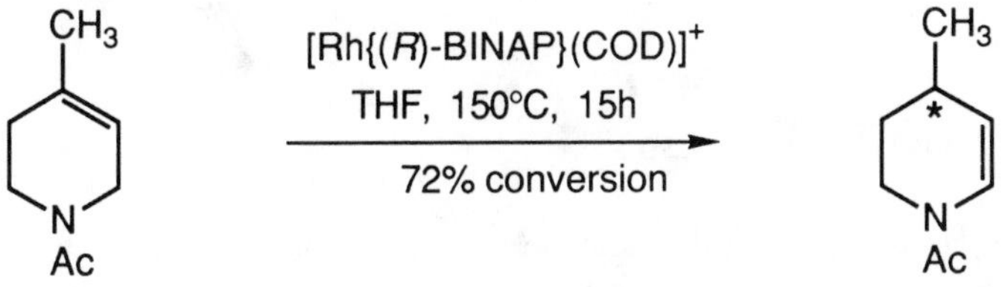

$$[Rh\{(R)\text{-BINAP}\}(COD)]^+$$
THF, 150°C, 15h
72% conversion

(9)

98%ee

25 **26**

The enantioselective isomerization of the prochiral alcohols **27** and **28** has also been achieved with 1 mol % of Rh–BINAP catalyst (eqs. 10 and 11) [17]. However, neither chemo- nor enantioselectivity was sufficiently high, though the enantiomeric excesses of the products were much higher than the values hitherto reported [3a]. The formation of the S-configuration at C(3) of the aldehyde produced from the E-double bond of the starting alcohol is again the same as the situation observed for the isomerization of allylamine (Scheme 1).

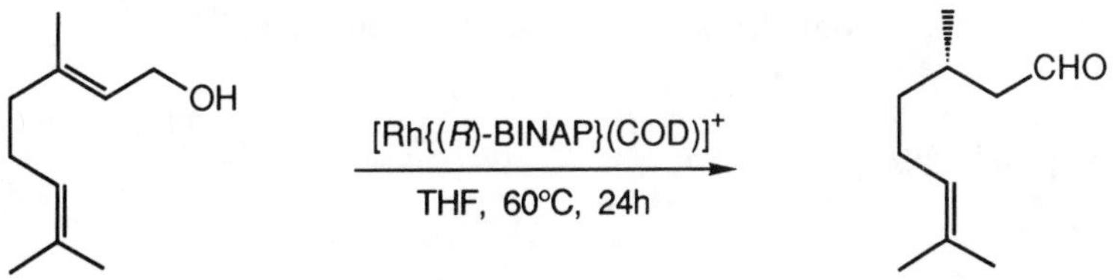

$$[Rh\{(R)\text{-BINAP}\}(COD)]^+$$
THF, 60°C, 24h

(10)

27 70% yield
37%ee

$$[Rh\{(R)\text{-BINAP}\}(COD)]^+$$
THF, 60°C, 24h

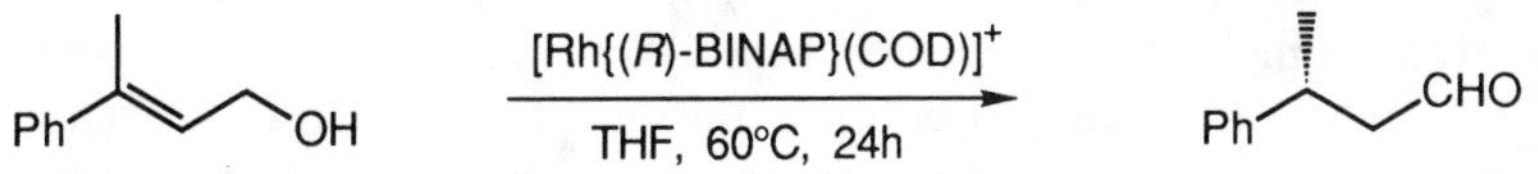

(11)

28 47% yield
53%ee

The isomerization of cyclic allyl alcohols to produce ketones proceeds more cleanly [17]. Effective kinetic resolution of racemic cyclic allylic alcohols has been reported [18]. The isomerization of racemic 4-hydroxy-2-cyclopentanone (**29**) in the presence of 0.5 mol % of $[Rh\{(R)\text{–BINAP}\}(MeOH)_2]^+$ in THF proceeded with 5:1 enantiomeric discrimination at 0°C to give 1,3-cyclopentadione (**31**) via enol ketone **30**, leaving the R-starting allylic alcohol (91% ee and 27% recovery yield) at 72% conversion after 14 days (eq. 12). (R)-4-Hydroxy-2-cyclopentenone is a key building block for prostaglandin synthesis [19].

Recently it has been reported that the catalytic isomerization of allylic alcohols effected by [Rh(diphosphine)(solvent)$_2$]$^+$ at 25°C yields synthetically useful quantities of the corresponding simple enols and that the transformation of allylic alcohols to enols and thereby to ketonic products proceeds catalytically via hydrido-π-allylic and hydrido-π-oxyallylic intermediates, respectively [20]. Consistently, enantioselection has been observed in the process of conversion of a prochiral enol to a chiral aldehyde. Thus, the prochiral substrate 32 is transformed to the optically active aldehyde 34 with 18% ee by using [Rh(BINAP)]$^+$ catalyst (eq. 13). Accordingly, this isomerization proceeds via a different mechanism from that of the isomerization of allylamine. For the reaction mechanism of the asymmetric isomerization of allylamine, we propose a unique, nitrogen-triggered mechanism, where the nitrogen coordination of the allylamine to the Rh$^+$ center plays an important role in the reaction process as well as in the chiral recognition [4b, 14]. A proposed mechanism for the isomerization of allylamine is illustrated in Scheme 4.

Regarding the catalyst precursor, among cationic Rh(I)–bisphosphane complexes the solvent complex [Rh(BINAP)(solvent)n]$^+$ is the most active. For example, [Rh(BINAP)(MeOH)$_2$] ClO$_4$ is effective for the isomerization of 1 even at temperatures below -20°C, but it is very sensitive to impurities. The diene complex [Rh(BINAP)(diene)]$^+$ is most conveniently used in laboratory scale experiments and is sufficiently active around ambient temperature. The bis(BINAP) complex [Rh(BINAP)$_2$]$^+$ is a less active catalyst and needs temperatures above 90°C for practical use, but this complex is the most robust and the most adequate for industrial purposes, as mentioned above. All three catalyst types show equally high chemo- and stereoselectivity with the same stereochemical relationship for the isomerization. In contrast to the very active cationic Rh(I) complexes, the neutral Rh(I) complex, "Rh(BINAP)Cl" prepared in situ from [Rh(diene)Cl]$_2$ with two molecules of the BINAP ligand, was not effective for the isomerization of allylamine. This may imply that the Lewis acidity of the rhodium metal is an important factor for catalytic activity. Peraryldiphosphine ligands, where all substituents on the phosphorus atoms are aromatic groups, form the most active catalysts.

Scheme 4 A nitrogen-triggered mechanism for the isomerization of allylamine with cationic Rh(I)–BINAP complexes.

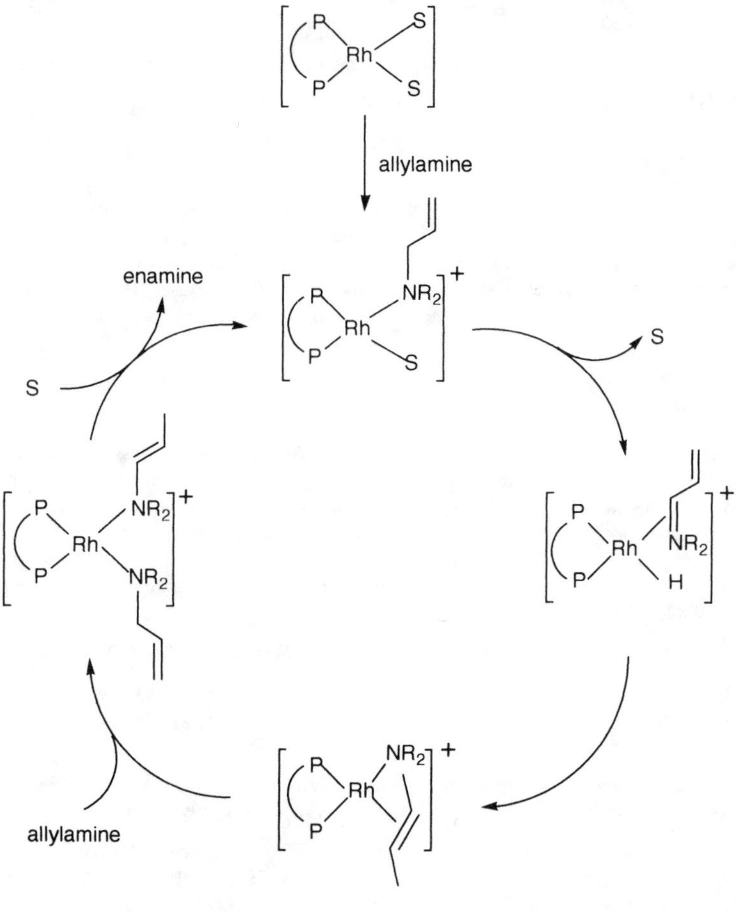

S = solvent, P⌒P = BINAP

Besides BINAP or p-TolBINAP, optically active peraryldiphosphines with axial chirality based on the biphenyl groups (6,6′-dimethylbiphenyl-2,2′-diyl)bis(diphenylphosphine) and its analog are also effective ligands for the asymmetric isomerization as expected [21].

2.6. Conclusion

During the past decade, metal-catalyzed asymmetric reactions have become one of the indispensable synthetic methodologies both in academic and industrial fields. The asymmetric isomerization of allylamine to an optically active enamine is a typical example of the successful application of basic research to an industrial process. We believe that Takasago's successful development of large-scale asymmetric catalysis will have a great impact on both synthetic chemistry and the fine chemical industries. The rhodium–BINAP catalysts, though very expensive, have become one of the cheapest catalysts in the chemical industry through extensive process development.

Acknowledgment

The authors express their deepest gratitude to the following people for their collaboration: Professors J. Tanaka (Shizuoka University) and S. Watanabe (Chiba University)—terpenoid amine synthesis; Professors R. Noyori (Nagoya University) and H. Takaya (Kyoto University)—BINAP synthesis; Professor S. Otsuka and Dr. T. Yamagata (Osaka University) and Dr. H. Kumobayashi (Takasago)—asymmetric isomerizations; and T. Sakaguchi, M. Yagi, H. Nagashima, and N. Murakami (Takasago)—process development.

References

1. (a) Jolly, P. W.; Wilke, G. *The Organic Chemistry of Nickel;* Academic Press: Orlando, FL, **1975;** Vol. 2, Chapter 1. (b) Houghton, R. P. *Metal Complexes in Organic Chemistry;* Cambridge University Press: London, New York, **1979;** p. 258. (c) Parshall, G. W. *Homogeneous Catalysis;* Wiley: New York, **1980;** Chapter 3.3. (d) Davies, S. G. *Organotransition Metal Chemistry: Applications to Organic Synthesis;* Pergamon: Oxford, **1982;** Chapter 7. (e) Colquhoun, H. M.; Holton, J.; Thompson, D. J.; Twigg, M. V. In *New Pathways for Organic Synthesis;* Plenum: New York, **1984;** Chapter 5.

2. Kumobayashi, H.; Akutagawa, S.; Otsuka, S. *J. Am. Chem. Soc.* **1978,** *100,* 3949.

3. (a) Botteghi, C.; Giacomelli, G. *Gazz. Chim. Ital.* **1976,** *106,* 1131. (b) Carlini, C.; Politi, D.; Ciardelli, F. *J. Chem. Soc., Chem., Commun.* **1970,** 1260. (c) Giacomelli, G.; Bertero, L.; Lardicci, L.; Menicagli, R. *J. Org. Chem.* **1981,** *46,* 3707.

4. (a) Otsuka, S.; Tani, K. In *Asymmetric Synthesis;* Morrison, J. D. (Ed.); Academic Press: Orlando, FL **1985;** Vol. 5, Chapter 6. (b) Otsuka, S.; Tani, K. *Synthesis,* **1991,** 665.

5. For examples, see (a) Baudry, D.; Ephritikhine, M.; Felkin, H. *Nouveau J. Chim.,* **1978,** *2,* 355. (b) Golborn, P.; Scheinmann, F. *J. Chem. Soc. Perkin Trans. 1,* **1973,** 2870. (c) Baudry, D.; Ephritikhine, M.; Felkin, H. *J. Chem. Soc. Chem. Commun.,* **1978,** 694. (d) Stille, J. K.; Becker, Y. *J. Org. Chem.* **1980,** *45,* 2139. (e) Suzuki, H.; Koyama, Y.; Moro-Oka, Y.; Ikawa, T. *Tetrahedron Lett.* **1979,** 1415.

6. Sully, B. D.; Williams, P. L. *Perfum. Essent. Oil Rec.* **1968,** *59,* 365.

7. (a) Takabe, K.; Katagiri, T.; Tanaka, J. *Bull. Chem. Soc. Jpn.* **1973,** *46,* 222. (b) Fujita, T.; Suga, K.; Watanabe, S. *Chem. Ind. (London)* **1973,** 231. (c) Takabe, K.; Katagiri, T.; Tanaka, J.; Fugita, T.; Watanabe, S.; Suga, K. *Org. Synth.* **1989,** *67,* 44.

8. (a) Takabe, K.; Katagiri, T.; Tanaka, J. *Chem. Lett.* **1977,** 1025. (b) Takabe, K.; Katagiri, T.; Tanaka, J. *Tetrahedron Lett.* **1972,** 4009. (c) Takabe, K.; Yamada, T.; Katagiri, T.; Tanaka, J. *Org. Synth.* **1989,** *67,* 48.

9. Koenig, K. E. In *Asymmetric Synthesis;* Morrison, J. D. (Ed.); Academic Press: Orlando, FL **1985;** Vol. 5, Chapter 3.

10. (a) Miyashita, A.; Takaya, H.; Souchi, T.; Noyori, R. *Tetrahedron,* **1984,** *40,* 1245. (b) Miyashita, A.; Yasuda, A.; Takaya, H.; Toriumi, K.; Ito, T.; Souchi, T.; Noyori, R. *J. Am. Chem. Soc.* **1980,** *102,* 7932.

11. (a) Takaya, H.; Akutagawa, S.; Noyori, R. *Org. Synth.* **1989,** *67,* 20. (b) Takaya, H.; Mashima, K.; Koyano, K.; Yagi, M.; Kumobayashi, H.; Taketomi, T.; Akutagawa, S.; Noyori, R. *J. Org. Chem.* **1986,** *51,* 629.

12. Yamagata, T.; Tani, K.; Tatsuno, Y.; Saito, T. *J. Chem. Soc., Chem. Commun.* **1988,** 466.

13. Tani, K.; Yamagata, T.; Tatsuno, Y.; Yamagata, Y.; Tomita, K.; Akutagawa, S.; Kumobayashi, H.; Otsuka, S. *Angew. Chem.* **1989,** *85,* 232; *Angew. Chem., Int. Ed. Engl.* **1985,** *24,* 217.

14. Inoue, S.-I.; Takaya, H.; Tani, K.; Otsuka, S.; Saito, T.; Noyori, R. *J. Am. Chem. Soc.* **1990,** *112,* 4897.

15. Takabe, K.; Uchiyama, Y.; Okisaka, K.; Yamada, T.; Katagiri, T.; Okazaki, T.; Oketa, Y.; Kumobayashi, H.; Akutagawa, S. *Tetrahedron Lett.* **1985,** *26,* 5153.

16. Tani, K. Unpublished results.

17. Tani, K. *Pure Appl. Chem.* **1985,** *57,* 1845.

18. Kitamura, M.; Manabe, K.; Noyori, R.; Takaya, H. *Tetrahedron Lett.* **1987,** *28,* 4719.

19. (a) Noyori, R.; Suzuki, M. *Angew. Chem., Int. Ed. Engl.* **1984,** *23,* 847. (b) Suzuki, M.; Yanagisawa, A.; Noyori, R. *J. Am. Chem. Soc.* **1985,** *107,* 3348.

20. Bergens, S. T.; Bosnich, B. *J. Am. Chem. Soc.* **1991,** *113,* 958.

21. Schmid, R.; Cereghetti, M.; Heiser, B.; Schönholzer, P.; Hansen, H.-J. *Helv. Chim. Acta,* **1988,** *71,* 897.

Asymmetric Cyclopropanation

Michael P. Doyle

Department of Chemistry
Trinity University
San Antonio, Texas

3.1. Introduction

The first chiral transition metal catalyst designed for an enantioselective transformation was applied to the reaction between a diazo ester and an alkene to form cyclopropanes [1]. In that application Nozaki and coworkers employed a Schiff base–copper(II) complex (**1**), whose chiral ligand was derived from α-phenethylamine, to catalyze the cyclopropanation of styrene with ethyl diazoacetate (eq. 1) [2]. Although enantiomeric excesses were low, the principle of chiral catalyst development for generating and distinguishing between diastereomeric transition states involving transition metals was established. Following from that first report in 1966, elaboration on ligand designs for chiral Schiff base–copper(II) complexes (**2**) by Aratani and coworkers [3–6] have led to high enantioselectivities in selected intermolecular cyclopropanation reactions (eq. 2–4), but unlike chiral catalysts that are effective for hydrogenation reactions [7], their development was slow, hindered in part by the complexity of the transformation and by limited potential applications. However, recent advances in ligand design for transition metals that are effective for

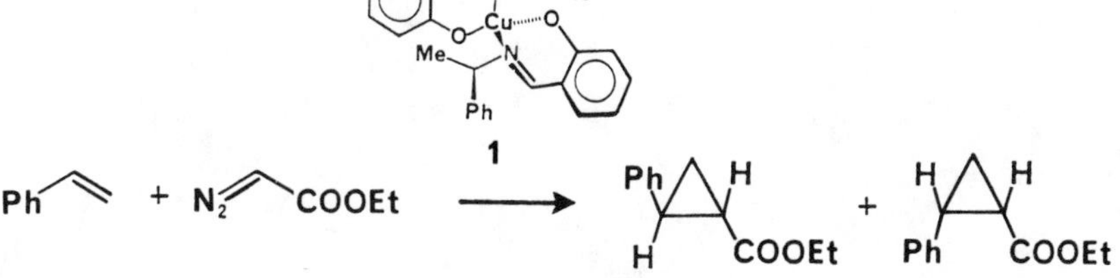

$$Ph \diagdown \diagup + N_2 \diagdown \diagup COOEt \longrightarrow \underset{H}{\overset{Ph}{\diagup}} \triangle \underset{COOEt}{\overset{H}{\diagdown}} + \underset{Ph}{\overset{H}{\diagup}} \triangle \underset{COOEt}{\overset{H}{\diagdown}} \qquad (1)$$

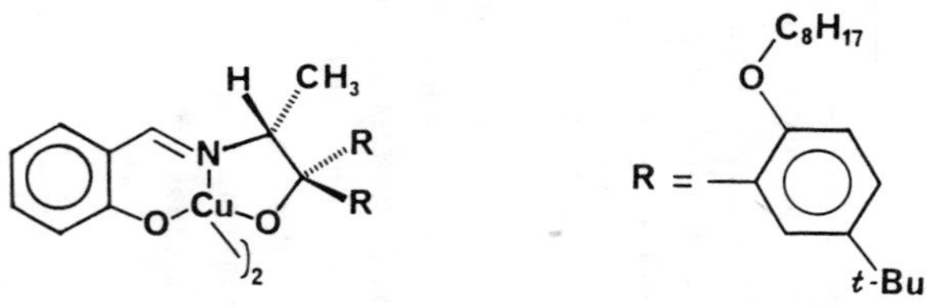

$R = $ (4-t-Bu-2-$C_8H_{17}O$-phenyl)

(R)-2

Me$_2$C=CH–C(Me)=CH$_2$ + N$_2$=CH–COOR' $\xrightarrow{\text{(R)-2}}$

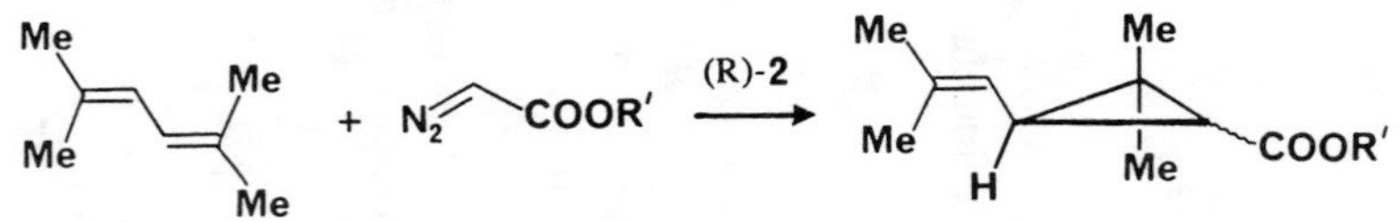

R' = Et: 68% ee (trans), 62% ee (cis)
R' = t-Bu: 75% ee (trans), 46% ee (cis)
R' = l-Menthyl: 94% ee (trans), 46% ee (cis)

(2)

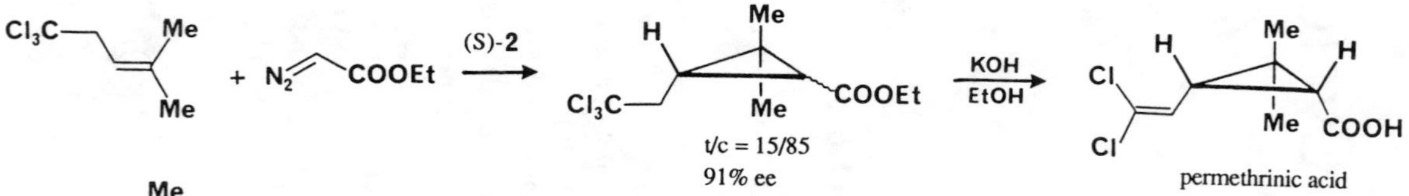

Cl$_3$C–CH$_2$–C(Me)=CH–... + N$_2$=CH–COOEt $\xrightarrow{\text{(S)-2}}$

$t/c = 15/85$
91% ee

$\xrightarrow[\text{EtOH}]{\text{KOH}}$ permethrinic acid

(3)

Me$_2$C=CH$_2$ + N$_2$=CH–COOEt $\xrightarrow{\text{(R)-2}}$

92% ee

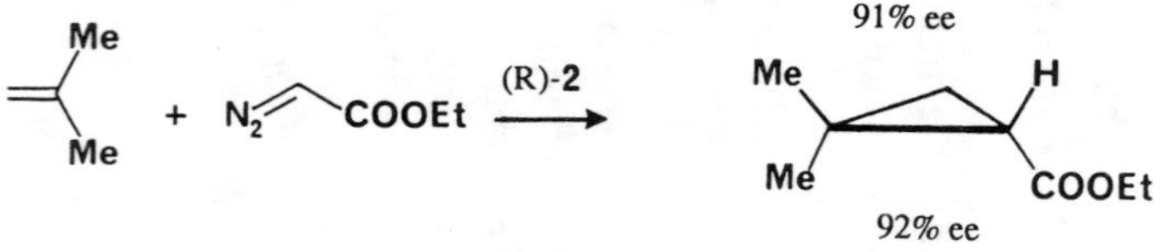

(4)

64

cyclopropanation reactions [8,9] have intensified efforts directed toward high enantioselectivity and have led to a resurgence of interest in chiral cyclopropane compounds and their uses [10]. Today, several catalyst designs are available that provide exceptional enantiocontrol in both inter- and intramolecular cyclopropanation reactions [11].

Cyclopropane formation occurs from reactions between diazo compounds and alkenes, catalyzed by a wide variety of transition metal compounds [12], that involve the addition of a carbene entity to a carbon–carbon double bond. This transformation is stereospecific and generally occurs with "electron-rich" alkenes, including substituted olefins, dienes, and vinyl ethers, but not α,β-unsaturated carbonyl compounds or nitriles [13–15]. Relative reactivities portray a highly electrophilic intermediate and an "early" transition state for cyclopropanation reactions [16,17], accounting in part for the relative difficulty in controlling selectivity. For intermolecular reactions, the formation of geometrical isomers, of regioisomers from reactions with dienes, and of enantiomers must all be taken into account.

Catalysts that promote the extrusion of dinitrogen from diazo compounds are effective for cyclopropanation reactions. Like electrophilic addition to diazo compounds [18] from which diazonium ions and, subsequently, carbocations are generated (eq. 5), transition metal compounds that can act as electrophiles are potentially effective catalysts.

$$RCH{=}N_2 + E^+ \rightarrow E{-}\overset{\overset{\displaystyle N_2^+}{\displaystyle |}}{C}HR \rightarrow E{-}\overset{+}{C}HR + N_2 \qquad (5)$$

These compounds possess an open coordination site, which allows the formation of a diazo carbon–metal bond with a diazo compound and, after loss of dinitrogen, affords a metal carbene (Scheme 1). Lewis bases (B:) that can occupy the open coordination site inhibit catalytic activity. The electrophilic nature of the metal carbene is seen in its subsequent reactions with nucleophiles (S:), which occur with the transfer of the carbene entity from the metal to a nucleophile without ever having generated or transformed an actual "free" carbene. Nucleophiles for these carbene transfer reactions are extensive and include, in addition to olefins or acetylenes, Z–H bonds (Z = C, N, O, Si, S) for insertion reactions [13–15] and heteroatoms (O, S, N, I, Br, Cl) for ylide generation [19].

Scheme 1

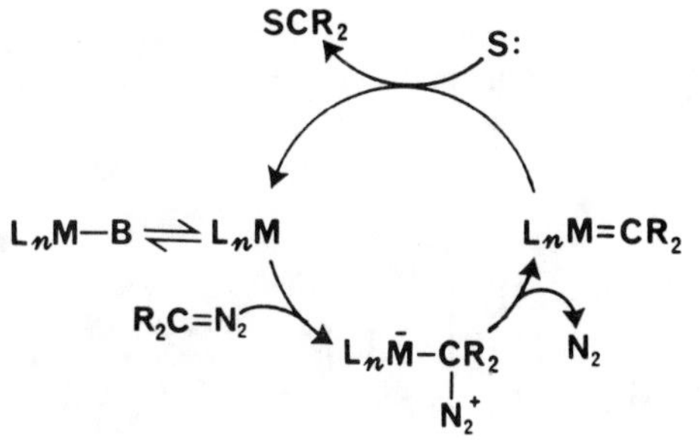

Among transition metal compounds that are effective for cyclopropanation reactions, those of copper and rhodium have received the greatest attention [13–15]. Copper catalysis for reactions of diazo compounds with olefins has been known for more than 80 years [20], but rhodium catalysis, in the form of dirhodium(II) tetraacetate, was not reported until the 1970s [21]. Although metal carbene intermediates with catalytically active copper or rhodium compounds have not yet been observed, correlations in relative reactivities and diastereomeric selectivities for cyclopropanation between pentacarbonyltungsten carbenes [$(CO)_5W=CHR$; R = Ph, COOEt] and those issued from dirhodium(II) tetrakis(acetate), $Rh_2(OAc)_4$ [17,22], strongly suggest their involvement in catalytic reactions. Apparent trapping of a rhodium carbene or its diazonium ion precursor by iodide [23] from the reaction of Rh(TTP)I with ethyl diazoacetate (eq. 6, TTP = teta-*p*-tolylporphyrin) also supports the catalytic cycle that is outlined in Scheme 1.

$$(TTP)RhI \xrightarrow{\text{N}_2\text{CHCOOEt}} \left[(TTP)Rh\!-\!\overset{\overset{+}{\text{N}_2}}{\underset{|}{\text{CHCOOEt}}} \right] \xrightarrow{-\text{N}_2} [(TTP)Rh\!=\!CHCOOEt] \quad (6)$$

$$LnM\!=\!CR_2 \leftrightarrow L_n\overset{-}{M}\!-\!\overset{+}{C}R_2$$

$$\textbf{3a} \qquad\qquad \textbf{3b}$$

Two contributing resonance structures (**3a** and **3b**), in the formalism of ylide structures, can be used to describe metal carbene intermediates. The highly electrophilic character of carbene intermediates derived from copper and rhodium catalysts suggests that the contribution from the metal-stabilized carbocation form **3b** is important in the overall evaluation of the reactivities and selectivities of these metal carbene intermediates. Emphasis on the metal carbene structure **3a** has led to the proposal that cyclopropane formation from reactions with alkenes occurs through the intervention of a metallocyclobutane (eq. 7) [24]. The metal-stabilized carbocation structure **3b** is consistent with the suggestion that L_nM dissociates from the carbene as bond formation occurs between the carbene and the reacting alkene (eq. 8) [13,17].

$$\begin{array}{c} ACH\!=\!CH_2 \\ + \\ LnM\!=\!CR_2 \end{array} \rightarrow \left[\begin{array}{c} ACH\!-\!CH_2 \\ |\qquad\quad| \\ L_nM\!\!-\!\!-\!CR_2 \end{array} \right] \rightarrow L_nM \;+\; \begin{array}{c} C \\ \diagup\;\diagdown \\ ACH\!-\!CR_2 \end{array} \quad (7)$$

$$H_2C=CHA \atop {+ \atop L_nM\overset{+}{-}CR_2}} \quad \left[L_nM----\overset{\overset{HC--CHA}{|}}{\underset{R_2}{C}}\right]^{\ddagger} \quad L_nM + {H_2 \atop {C \atop {/ \backslash \atop ACH-CR_2}}} \qquad (8)$$

Diazo carbonyl compounds, especially diazo ketones and diazo esters [25], are the most suitable substrates for cyclopropanation reactions catalyzed by copper or rhodium compounds. Diazoalkanes are less useful owing to more pronounced "carbene dimer" formation that competes with cyclopropanation [17]. This competing reaction occurs by electrophilic addition of the metal-stabilized carbocation to the diazo compound followed by dinitrogen loss and formation of the alkene product that occurs with regeneration of the catalytically active metal complex (eq. 9) [26].

$$L_nM=CHR + RCH=N_2 \rightarrow \left[\overset{L_n\bar{M}-CHR}{\underset{RCH-N_2^+}{|}} \right] \rightarrow L_nM + RCH=CHR + N_2 \qquad (9)$$

That the transition metal is involved in the product-determining step for carbene dimer formation is suggested by the predominance of the cis geometry for the olefin product [26,27]. Among the transition metal catalysts that have been employed for cyclopropanation, only those of palladium(II) are productive with diazomethane [28], and this may be the result of a mechanism whereby the palladium-coordinated alkene undergoes electrophilic addition to diazomethane (*vide infra*: Scheme 5) rather than by a metal carbene transformation.

The development of chiral transition metal catalysts for the asymmetric synthesis of cyclopropanes has taken place during the evolution of mechanistic understanding for metal carbene generation and for cyclopropane formation. Concepts placed in the design of chiral catalysts have been influenced by these mechanistic developments, and further refinements can be expected.

3.2. Asymmetric Intermolecular Cyclopropanation

3.2.1. Chiral Salicylaldimine Copper Catalysts

Although the degree of enantioselection for Nozaki's copper(II)–Schiff base catalyst (**1**) was low [2,13], extensive systematic screening of salicylaldimine ligands by Aratani and coworkers nearly 10 years later brought about design changes in the chiral salicylaldimine copper catalysts that gave dramatic improvements in optical yields for selected intermolecular cyclopropanation reactions [3,4]. The ligands offering the greatest advantage were those prepared from salicyclaldehyde and amino alcohols derived from alanine [3]. Ligands derived from phenylalanine, valine, and leucine were also evaluated [6], but lower enantioselectivities were obtained from cyclopropanation reactions with 2,5-dimethyl-2,4-hexadiene (eq. 2).

When the carbinol substituents (R) were the bulky 5-*tert*-butyl-2-(*n*-octyl-oxy)phenyl group, optimum enantioselectivities were achieved with the catalytic use of the corresponding copper(II) complex (**4**), in both enantiomeric forms, for intermolecular cyclopropanation reactions. Specific applications of the Aratani catalysts have included the synthesis of chrysanthemic acid esters (eq. 2) and a precursor to permethrinic acid (eq. 3), both potent units of pyrethroid insecticides, and for the commercial production of ethyl (*S*)-2,2-dimethylcyclopropanecarboxylate (eq. 4), the "Sumitomo process," which is employed for the construction of cilastatin, an in vivo stabilizer of the antibiotic imipenem. Several other uses of these catalysts and their derivatives for cyclopropanation reactions have been reported recently albeit, in most cases, with only moderate enantioselectivities [29–32].

(R)-4

Enantioselection with the Aratani catalysts is influenced by the steric bulk of the R group and by the size of the alkyl substituent derived from the amino alcohol (A = CH_3 > CH_2Ph > *i*-Pr > *i*-Bu). In addition, the diazo ester has a significant influence on both enantioselectivity and on the trans/cis ratio of the resulting cyclopropane product [4]. As reported in Table 1 for the production of chrysanthemic acid esters, increasing the size of the diazo ester substituent increases the percentage

Table 1 Cyclopropanation of 2,5-Dimethyl-2,4-hexadiene with Diazo Esters Catalyzed by the Aratani Catalysts[a]

					% ee[c]	
R'	A	$N_2CHCOOR$*:R*[b]	Yield (%)	trans:cis	trans	cis
n-Octyl	CH_3	CH_2CH_3	54	51:49	68	62
n-Octyl	CH_3	Cyclohexyl	71	58:42	70	58
n-Octyl	CH_3	$C(CH_3)_3$	74	75:25	75	46
n-Octyl	CH_3	1-Adamantyl	82	84:16	85	46
n-Octyl	CH_3	$C(i\text{-}Pr)_2CH_3$	64	92:8	88	*d*
n-Octyl	CH_3	*d*-Menthyl	64	72:28	90	59
n-Octyl	CH_3	*l*-Menthyl	*d*	93:7	94	46
n-Butyl	CH_3	*l*-Menthyl	67	89:11	87	25
n-Heptyl	$PhCH_2$	*l*-Menthyl	42	91:9	86	22

[a] Results for (*R*)-**2** are reported.
[b] *l*-Menthyl = (1*R*,2*S*,5*R*)-2-isopropyl-5-methylcyclohexyl.
[c] % de for menthyl esters.
[d] Not reported.

Source: Reference 4.

of cyclopropane product having the trans geometry and also increases the enantiomeric excess of this product. Enantioselectivity for the cis isomer decreases with increasing size of the diazoacetate, at least for the examples reported in Table 1. A higher enantioselectivity for the trans isomer and a lower enantioselectivity for the cis isomer are obtained with *l*-menthyl diazoacetate and (*R*)-**4**, which also produces a higher trans/cis product ratio.

With a series of substituted ethylenes, diastereoselectivities for cyclopropanation with chiral menthyl diazoacetates generally increase with increasing substitution about the carbon–carbon double bond (Table 2). However, structural effects from the olefin on stereoselection are relatively minor, and only with trisubstituted olefins do diastereomeric excesses (de) exceed 90%. 5-Halo-2-pentenes (Table 2: entries 11 and 12) are exceptional in this series with both an inverted preference for formation of the *cis*-cyclopropane derivative and the high diastereoselectivity for this stereoisomer. Only one other example has been reported to have this unexpectedly high stereochemical preference (eq. 10) [30]. This cis selectivity may be the result of a dipolar interaction between the polar substituent of the reacting alkene and the metal carbene, but no other examples show the relative independence of enantioselectivity on the diazoacetate (e.g., entry 11 of Table 2, *l*-menthyl vs. ethyl in eq. 3) as do these halogenated alkenes.

(10)

	% ee		
	trans:cis	cis	trans
R = *l*-menthyl	44:56	>95	>95
Et	37:63	80	29

With modified Aratani catalysts (R = Ph and A = CH$_2$Ph) Reissig observed moderate enantioselectivities (30–40% ee's for the *trans*-cyclopropane isomer) for reactions between trimethylsilyl vinyl ethers and methyl diazoacetate [29], but vinyl ethers are the most reactive olefins toward cyclopropanation and also the least selective [33,34]. Other chiral Schiff bases have been examined for enantioselection using the in situ method for catalyst preparation that was pioneered by Brunner, but enantioselectivities were generally low [35].

According to Aratani, the results obtained are consistent with a reduced mononuclear copper(I) complex, existing in a tetrahedral configuration, that forms a metal carbene intermediate [6]. Alkene addition occurs by displacement of an oxygen ligand, which allows the production of a metallacyclobutane intermediate that collapses to cyclopropane product with regeneration of the catalytically active copper compound (Scheme 2). Approach of the alkene to the carbene occurs from

Table 2   Diastereoselective Cyclopropanation of Alkenes with *l*-Menthyl Diazoacetate (*l*-MDA) Catalyzed by the Aratani Catalyst 4 (A = CH_3)[a]

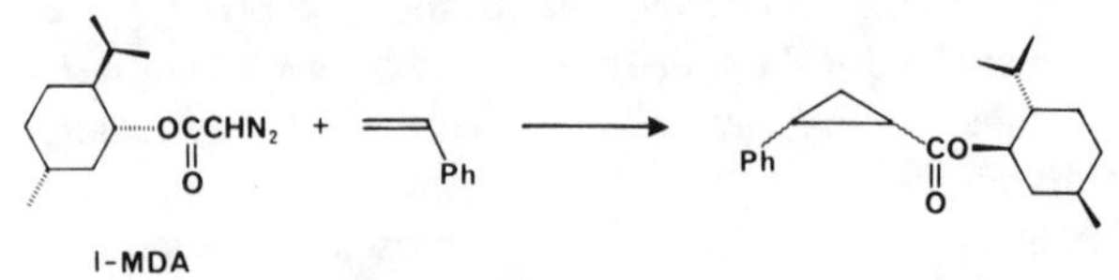

				% de[b]	
Entry	Alkene	Catalyst Configuration	trans:cis	trans	cis
1	Styrene	R	86:14	69 (1S,2S)	54 (1S,2R)
2		S	82:18	81 (1R,2R)	78 (1R,2S)
3	1-Octene	R	83:17	76 (1S,2S)	46 (1S,2S)
4		S	78:22	84 (1R,2R)	64 (1R,2R)
5	*trans*-4-Octene	R		82 (2S,3S)	
6		S		84 (2R,3R)	
7	α-Methylstyrene	R	60:40	68	86
8		S	64:36	58	68
9	1,1-Diphenylethylene	R		75 (1S)	
10		S		64 (1R)	
11	$Cl_3CCH_2CH{=}CMe_2$[c]	S	15:85	19 (1R,2R)	93 (1R,2S)
12	$BrCH_2CH_2CH{=}CMe_2$	S	17:83	23 (1R,2R)	95 (1R,2S)
13	$Me_2C{=}CH{-}CH{=}CMe_2$	R	93:7	94 (1S,2S)	46 (1S,2R)

[a]   *l*-Menthyl = (1R,2S,5R)-2-isopropyl-5-methylcyclohexyl.
[b]   Configuration in parentheses.
[c]   With ethyl diazoacetate the same trans:cis product ratio was obtained (eq. 3): cis, 91% ee; trans, 11% ee.

Source: Reference 5.

Scheme 2

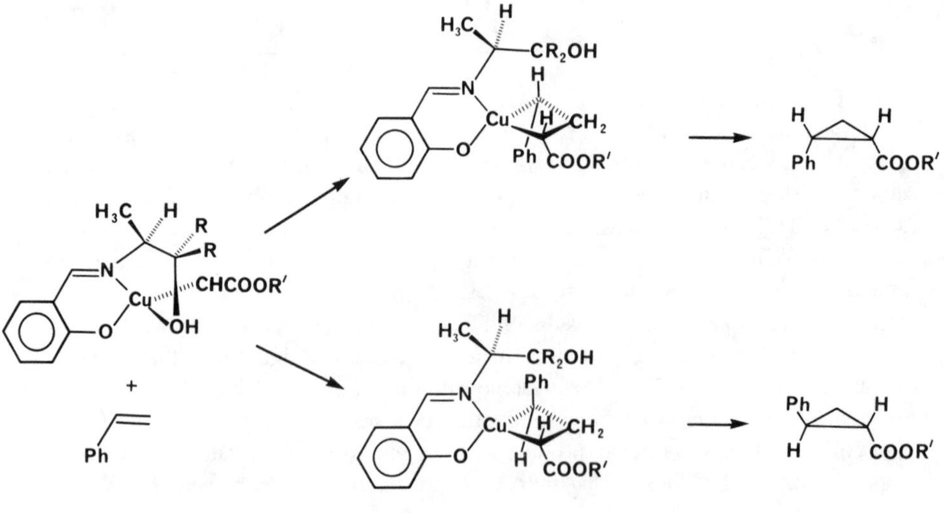

the less hindered side. However, although this explanation predicts the observed predominant cyclopropane configuration, it does not adequately account for the preferential trans stereochemistry that is observed in most cases (Table 2). An alternative explanation, based on metal carbenes as metal-stabilized carbocations, has also been proposed to explain these results [11], and its predictions better represent observed selectivities (Scheme 3). In this description the *p*-orbital of the carbene carbon is projected in front of the copper nucleus, and its orientation is selected to minimize steric interactions between the ligand's bulky 5-*tert*-butyl-2-octyloxyphenyl group and the carbene's ester substituent (R). In the two limiting

Scheme 3

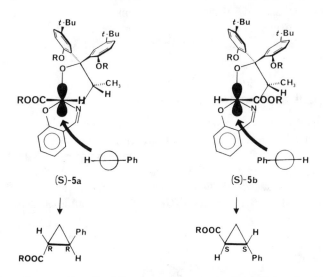

(S)-5a   (S)-5b

conformations that are described for the catalyst in the *S*-configuration (**5a** and **5b**), the preferred conformation is the one in which the ester substituent of the carbene entity is oriented to the side away from the sterically more crowded ligand region. The hydrogen that protrudes from the ligand toward the carbene substituents limits rotation and governs, in large part, the preference for conformation **5a** over **5b**. Addition of an alkene occurs preferentially from the side opposite to the ligand's bulky 5-*tert*-butyl-2-octyloxyphenyl group. Cyclopropanation takes place by approach of the olefin's π-bond to the carbene's *p*-orbital so that the alkene carbon bearing the substituent (Ph of the Newman projection in Scheme 3) is projected away from the metal in the manner portrayed in equation 8. This model, which in Scheme 3 focuses on the formation of *trans*-2-phenylcyclopropanecarboxylates, explains the enantiomeric preferences of the Aratani catalysts (Table 2) as well as the influence of the size of the ester substituent on stereoselectivity (Table 1) for most examples reported thus far. The preferred configurations for the trans and cis isomers of 2-phenylcyclopropanecarboxylate esters both arise from approach of the alkene to the same conformation of the metal carbene.

The methyl group at the chiral center of the ligand controls, at least in part, the direction of approach of the diazo ester to the electrophilic copper. By inhibiting approach of the diazo ester to the "backside" of the copper, the substituent on the chiral center effectively limits metal carbene formation that would have resulted in the mirror image configurations of those cyclopropane products that are preferentially formed. That the *trans*-cyclopropane isomer is formed with a higher enantiomeric purity than the cis isomer for reactions with monosubstituted alkenes may be a result of the closer approach by the olefin to the carbene carbon if the distinguishing olefin substituent and the ester substituent are anti to each other. The influence of olefin substitution (tri- > di- > mono-) on enantioselectivity is, as inferred by Aratani [6], probably a function of the degree to which bond formation has occurred in the transition state for electrophilic addition. The absence of restricted access to one side of the carbene carbon allows relatively unimpeded approach even by trisubstituted olefins to the carbene center, and this structural feature is absent in other catalyst designs which, because of impeded approach, are more limited in their applicability.

3.2.2. Chiral Semicorrin/Bisoxazoline Copper Catalysts

The most significant recent advances in the development of chiral copper catalysts for enantioselective cyclopropanation reactions were initiated by Pfaltz with his report in 1986 [8] of semicorrin copper(II) complexes and extended by Pfaltz [36], Masamune [37], and Evans [38] with bisoxazoline copper compounds. Of the catalysts with chiral semicorrin ligands that were initially evaluated [39], the copper complex (**6**) in which R is a bulky $C(CH_3)_2OH$ group afforded the highest selec-

6

tivities [40] (L = a second semicorrin ligand for the Cu(II) complex or a carbene ligand in a Cu(I) complex). For the cyclopropanation of two monosubstituted alkenes, styrene and 1-heptene, enantioselectivities were significantly higher than those obtained with the use of the Aratani catalysts (Table 3). As with the Aratani catalysts, enantioselectivities for cyclopropane formation with **6** are responsive to the steric bulk of the diazo ester, are higher for the trans isomer than for the cis form, and are influenced by the absolute configuration of a chiral diazo ester (*d*- and *l*-menthyl diazoacetate), although not to the same degree as are reported for **4** in Tables 1 and 2. 1,3-Butadiene and 4-methyl-1,3-pentadiene, whose higher reactivities result in higher product yields than terminal alkenes, form cyclopropane products with 97% ee in reactions with *d*-menthyl diazoacetate (eq. 11). Regiocontrol is complete, but diastereocontrol is only moderate.

Table 3 Enantioselective Cyclopropanation of Alkenes with Diazo Compounds
Catalyzed by the Pfaltz Semicorrin Catalyst **6**

Alkene	$N_2CHCOOR^a$:R	trans:cis	% ee[b] trans	cis
Styrene	*l*-Menthyl	85:15	91 (1*S*,2*S*)	90 (1*S*,2*R*)
Styrene	*d*-Menthyl	82:18	97 (1*S*,2*S*)	95 (1*S*,2*R*)
Styrene	*tert*-Butyl	81:19	93 (1*S*,2*S*)	93 (1*S*,2*R*)
Styrene	Ethyl	73:27	92 (1*S*,2*S*)	79 (1*S*,2*R*)
1-Heptene	*d*-Menthyl	82:18	92 (1*S*,2*S*)	92 (1*S*,2*R*)
1,3-Butadiene	*d*-Menthyl	63:37	97 (1*S*,2*S*)	97 (1*S*,2*R*)
4-Methyl-1,3-butadiene	*d*-Menthyl	63:37	97 (1*S*,2*S*)	97 (1*S*,2*R*)

[a] *d*-Menthyl = (1*S*,2*R*,5*S*)-2-isopropyl-5-methylcyclohexyl.
[b] Absolute configuration of cyclopropane product in parentheses.

$$R = d\text{-menthyl}$$

(11)

67%   33%
97% ee   97% ee

Whereas the (semicorrinato)copper catalysts are exceptionally effective with monosubstituted alkenes, their enantiocontrol with 1,2-di- and trisubstituted alkenes is less effective than the control achieved with the use of the Aratani catalysts. Chrysanthemic acid esters, for example, are prepared in much lower yields and with significantly reduced enantiocontrol with the Pfaltz catalysts than with the Aratani catalysts [41]. In addition, alkenes less reactive than conjugated olefins like styrene or 1,3-butadiene are sluggish in their reactions with the intermediate metal carbene; for example, cyclopropanation of 1-heptene gave only a 30% product yield whereas with styrene or conjugated dienes, cyclopropanation yields were more than double this value.

Pfaltz has described the active catalyst as a copper(I) derivative with only one semicorrin ligand, and he has provided a mechanistic description that portrays rational transition state geometries [40]. Using the shorthand notation employed earlier to describe enantioselection with the Aratani catalysts, the effectiveness of the Pfaltz catalyst can be seen through the depictions in Scheme 4. The preferred conformation for the metal carbene is that in which carbene substituents are oriented to minimize steric interaction with the semicorrin ligand's substituents (R) which, in turn, maximizes interaction with the approaching alkene. Four limiting conformations can be envisioned in this front view of the reacting system (**7a–d**), of which **7b** and **7d**, the two that have the olefin substituent Ph on the same side as the ligand's R-group, are less stable. Because of the C_2 symmetry of the semicorrin ring, the same enantiomers are formed from these models whether the COZ substituent is

Scheme 4

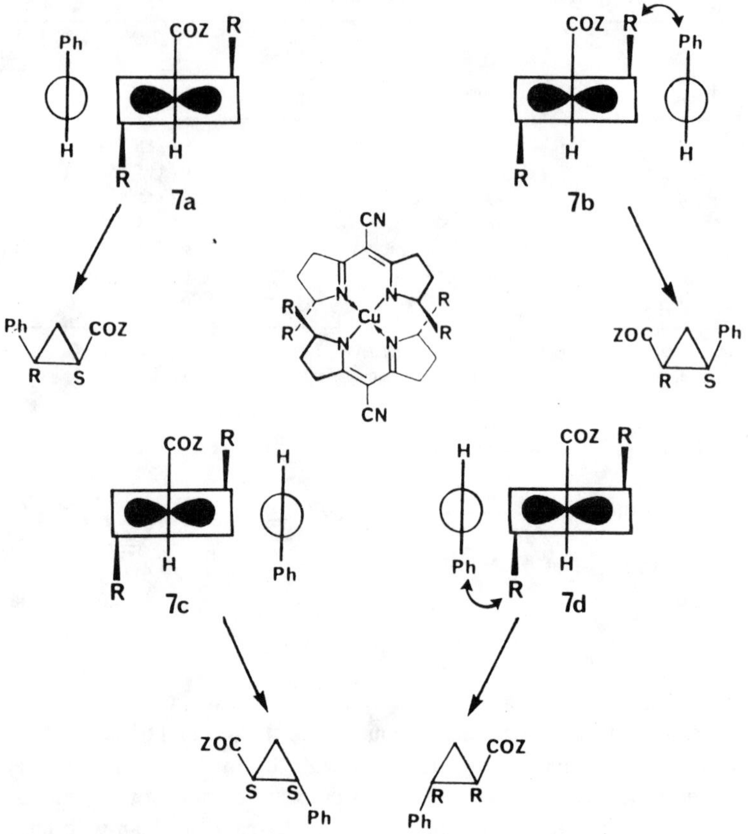

positioned up (Scheme 4) or down. These models suggest that the observed high enantioselectivities arise from interactions of R with olefin substituents, and they imply that monosubstituted alkenes should be preferred substrates. Furthermore, increasing the size of the ester group (COZ) probably changes the orientation of the carbene's carbonyl substituent relative to the semicorrin ring and magnifies the influence of the semicorrin R-group on enantioselection.

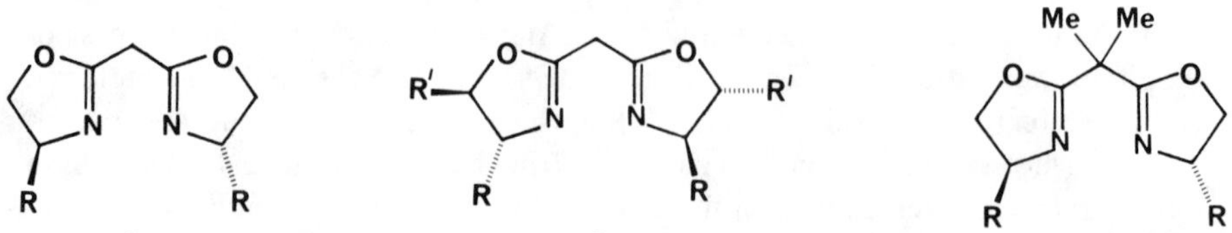

Copper complexes with bisoxazoline ligands **8** and **9** that were first reported by Masamune and coworkers [37] have incited considerable interest because of the

Table 4 Enantioselective Cyclopropanation of Styrene with Diazo Esters Using Bisoxazoline CopperCatalysts

Ligand	R	R'	trans:cis	trans[a]	cis[b]	Ref.
	$N_2CHCOOR'$			% ee		
8	Ph	Et	70:30	60	52	37
8	$PhCH_2$	Et	71:29	36	15	37
8	i-Pr	Et	71:29	46	31	37
8	i-Pr	Et	64:36	64	48	38
8	t-Bu	Et	75:25	90	77	37
8	t-Bu	Et	77:23	98	93	38
8	t-Bu	d-Menthyl	84:16	98	80	37
8	t-Bu	l-Menthyl	84:16	98	96	37
8	t-Bu	l-Menthyl	87:13	96	97	36
8	$Me_2C(OH)$[c]	d-Menthyl	83:17	90[d]	90[e]	36
9	Me	Et	71:29	28[d]	30[e]	37
10	t-Bu	Et	73:27	99	97	38
10	t-Bu	t-Bu	81:19	96	93	38
10	t-Bu	BHT[f]	94:6	99		38

[a] Product from styrene has the $(1R,2R)$-configuration.
[b] Product from styrene has the $(1R,2S)$-configuration.
[c] Catalyst has opposite configuration to **8** (R = t-Bu).
[d] $(1S,2S)$-configuration.
[e] $(1S,2R)$-configuration.
[f] 2,6-Di-$tert$-butyl-4-methylphenyl.

exceptional enantiocontrol that can be achieved with their use as catalysts for cyclopropanation reactions. Concurrent investigations by Evans [38], who added **10**, and Pfaltz [36], who investigated a similar series, have established the bisoxazoline ligands to be suitable alternatives to semicorrin ligands for copper in creating a highly enantioselective environment for intermolecular cyclopropanation. For the first time, diazoacetates with ester substituents as small as ethyl could be used to achieve greater than 90% ee in reactions with styrene (Table 4).

As expected, increasing the size of the R group increases enantioselection, and the buttressing effect on the bisoxazoline ring caused by the geminal dimethyl group in **10** provides further enhancement of enantiocontrol. However, as can be seen from the results in Table 4, the ligand's R substituent has only a minor influence on the trans/cis ratio of cyclopropane products. To increase product diastereoselectivity, Evans [38] increased the size of the ester substituent from ethyl to $tert$-butyl and then to the bulky BHT ester, which had been reported by Doyle [42] to provide exceptionally high diastereocontrol in catalytic cyclopropanation reactions. However, the disadvantage of these BHT esters is their low reactivity to chemical reduction and unreactivity toward hydrolysis.

Applications of these catalysts to alkenes other than styrene have demonstrated the potential generality of their uses for asymmetric intermolecular cyclopropanation (Table 5). The trans/cis (or E/Z) ratios of cyclopropane products are higher than those obtained with styrene, and diastereoselectivities or enantioselectivities for the

Table 5   Enantioselective Cyclopropanation of Alkenes with Diazo Esters
Using Bisoxazoline Copper Catalysts

Alkene	Ligand[a]	$N_2CHCOOR':R'$	Yield (%)	trans:cis	% ee[b] trans	cis	Ref.
1-Octene	8	*l*-Menthyl	76	94:6	99	30	37
		d-Menthyl	72	90:10	75	45	37
α-Methylstyrene	8	*l*-Menthyl	78	89:11	92	79	37
		d-Menthyl	72	85:15	83	77	37
trans-4-Octene	8	*l*-Menthyl	52	—	88		37
2,3,3-Trimethyl-1-1-butene	8	*l*-Menthyl	60	95:5	80	91	37
		d-Menthyl	55	98:2	77	n.d.[c]	37
Isobutylene	10	Et	91	—	>99		38
1,1-Diphenylethene	10	Et	75	—	99		38

[a]   R = *t*-Bu.
[b]   % de values for reactions with menthyl diazoacetate.
[c]   Not determined.

trans (*E*) isomer are generally greater than 80%. Although these reactions are
generally performed with 1.0 mol % of catalyst, Evans has optimized the cyclopropanation of isobutylene to a 0.25 mole scale [38], using only 0.1 mol % of
catalyst, and obtained a 91% yield of the (*S*)-cyclopropane enantiomer with greater
than 99% ee—significantly greater than for the same reaction performed with the
Aratani catalyst (eq. 4) [6].

11a	R = CH$_2$OSiMe$_2$(*t*-Bu)
11b	R = CMe$_2$OSiMe$_3$
11c	R = CMe$_2$OSiMe$_2$(*t*-Bu)

More recently Pfaltz and coworkers have prepared a series of 5-aza-semicorrin
ligands (**11**) and examined the catalysts formed from these ligands with copper(I)
triflate in comparison with **6** and **8** for cyclopropanation of styrene [43]. Enantioselectivities for Cu(I)OTf/**11** fall between those obtained from use of the Evans
catalyst Cu(I)OTf/**10** and those from Pfaltz's **6** and Masamune's **8** (Table 6). Cyclopropane product yields are generally in the range of 75–90%. Catalyst preparation
includes the addition of copper(I) triflate to a slight excess of the chiral ligand in
1,2-dichloroethane. The resulting suspension is stirred and then filtered through
glass wool. The green filtrate is used directly for cyclopropanation.

Copper catalysts for metal carbene transformations, as originally established by
Solomon and Kochi [44] for copper triflate, are active as copper(I) complexes, not
copper(II). Although there has been some disagreement with this proposition for
some simple copper compounds [45], bisoxazoline, semicorrin, and even the Aratani catalysts are active only when copper is in its +1 oxidation state [6,8,37,38].

Table 6 Enantioselective Cyclopropanation of Styrene
with 5-Aza-semicorrin Copper Catalysts

Catalyst	Diazoacetate R	Yield (%)	trans:cis	% ee[a] trans	cis
11a/Cu(I)OTf	Et	40	75:25	66 (1S,2S)	43 (1S,2R)
11b/Cu(I)OTf	Et	80	75:25	94 (1S,2S)	68 (1S,2R)
	t-Bu	87	86:14	96 (1S,2S)	90 (1S,2R)
	d-Menthyl	89	84:16	98 (1S,2S)	99 (1S,2R)
11c/Cu(I)OTf	Et	45	77:23	95 (1S,2S)	90 (1S,2R)
	t-Bu	75	81:19	94 (1S,2S)	95 (1S,2R)
	d-Menthyl	75	84:16	98 (1S,2S)	99 (1S,2R)

[a] Absolute configuration of cyclopropane product in parentheses.

Source: Reference 43.

The chiral copper(I) catalysts have been produced from the corresponding chiral copper(II) complex by reduction with phenylhydrazine [6,8,37] prior to use (eq. 12) or generated in situ by ligand replacement from copper(I) triflate [38] or copper(I) *tert*-butoxide [8] (eq. 13, X = OTf or O-t-Bu). The method of catalyst preparation

$$CuL_2^* \xrightarrow{\text{PhNHNH}_2} CuL^* + L^*H \qquad (12)$$

$$CuX + L^*H \longrightarrow CuL^* + HX \qquad (13)$$

does influence enantioselectivity. With **8** (R = t-Bu), for example, the catalyst generated according to eq. 12 [37] provided 90 and 77% ee for the *trans*- and *cis*-2-phenylcyclopropanecarboxylate esters, respectively (Table 4), whereas the presumed same catalyst generated in situ (eq. 13, X = OTf) gave 98 and 93% ee for the same *trans*- and *cis*-products, respectively [38]. Similar results were observed by Pfaltz [8], but the catalyst generated in situ gave lower enantioselectivity than that produced by phenylhydrazine reduction. Obviously this facet of catalyst generation requires more study.

Applications of these catalysts beyond styrene suggest their potential for highly enantioselective intermolecular cyclopropanation (Table 5). Whether bisoxazoline ligands **8** are better than the corresponding semicorrin ligands or not is yet to be determined. However, bisoxazoline ligands **10** are clearly capable of exceptionally high enantioselection. Furthermore, Cu(OTf)/**10** (R = t-Bu) has been reported by Evans [38] to be capable of enantioselective aziridination (eq. 14), further expanding the potential of this catalytic methodology.

$$PhCH{=}CH_2 \ + \ PhI{=}NTs \ \xrightarrow{\text{Cu(OTf)/10}} \ \underset{Ph}{\overset{\overset{\displaystyle Ts}{\overset{|}{N}}}{\triangle}} \ + \ PhI \qquad (14)$$

3.2.3. Chiral Bis(1,2-dioximato)cobalt(II) Catalysts

In 1974, before Aratani had introduced his chiral salicylaldimine copper catalysts, Nakamura and Otsuka reported chiral bis(α-camphorquinonedioximato)cobalt(II) **12** and related complexes to be effective enantioselective cyclopropanation catalysts [46], and they fully described the preparation, characteristics, and uses of these catalysts in 1978 [47,48]. Enantioselectivities as high as 88% ee were achieved for the cyclopropanation of styrene with neopentyl diazoacetate, and chemical yields greater than 90% were obtained in several cases. Dioximatocobalt(II) catalysts are unusual in their ability to catalyze cyclopropanation reactions that occur with conjugated olefins (e.g., styrene, 1,3-butadiene, and 1-phenyl-1,3-butadiene) and certain α,β-unsaturated esters (e.g., methyl α-phenylacrylate), but not with simple olefins and vinyl ethers. In this regard they do not behave like metal carbenes formed with copper or rhodium catalysts that are characteristically electrophilic in their reactions toward alkenes (vinyl ethers > dienes > simple olefins >> α,β-unsaturated esters) [13,14], and this divergence has not been adequately explained. However, despite their ability to attain high enantioselectivities in cyclopropanation reactions with ethyl diazoacetate and other diazo esters, no additional details concerning these cobalt(II) catalysts have been published since the initial reports by Nakamura and Otsuka.

12

Although enantioselectivities are relatively high (75% ee for ethyl (1S,2S)-*trans*-2-phenylcyclopropanecarboxylate and 67% ee for the (1S,2R)-*cis*-isomer from the reaction of styrene with ethyl diazoacetate), diastereoselectivities are low. However, at least for reactions with styrene, the trans/cis product ratio changes markedly with small changes in the ester group of diazoacetates [48], and enantioselectivities are similarly responsive (Table 7). Enantioselectivities increase with decreasing temperature, but they level off below 0°C. Several other dioximatocobalt(II) catalysts were also used, but none of them gave enantioselectivities as high as those obtained with **12**.

In reactions of ethyl diazoacetate with *cis*-1,2-dideuteriostyrene, epimerization of deuterium in the cyclopropane product is observed (eq. 15). Free rotation about the single bond between C-1 and C-2 of the original 1,2-dideuteriostyrene in a reaction intermediate is inferred. (No such isomerization occurred using the same alkene with the Aratani catalyst [32].) Furthermore, the presence of a base, formulated as an axial ligand of cobalt, influences enantioselection and diastereoselection, and a significant decrease in enantioselectivity is normally observed in reactions performed in the presence of most pyridine bases [48].

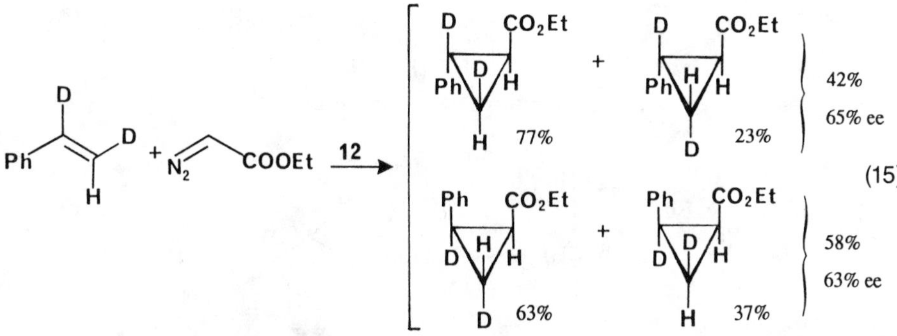

$$(15)$$

Slow evolution of nitrogen occurs when ethyl diazoacetate is treated with the chiral cobalt(II) catalyst in the absence of styrene [48]. However, marked acceleration occurs upon addition of styrene. The reaction rate depends on the concentrations of both catalyst and diazoacetate in neat styrene, and as the styrene concentration is decreased its rate dependence approaches first order. The extensive data accumulated by Nakamura and Otsuka, although interpreted by the intervention of metal carbene and metallacyclobutane intermediates (eq. 7), can also be rationalized by the mechanism in Scheme 5 (E = ester).

Coordination of the chiral cobalt(II) catalyst with the alkene may activate the alkene for electrophilic addition to the diazo compound. Subsequent ring closure can be envisioned to occur via a diazonium ion intermediate (13), without involving a metal carbene intermediate at any stage. Such a mechanism accounts for the rate acceleration caused by styrene and the isomerization observed in cyclopropane formation with cis-1,2-dideuteriostyrene; it also may explain the unusual reactivities of these catalysts toward even normally unreactive α,β-unsaturated esters. Association with the diazo ester provides a separate pathway for decomposition that, if binding occurs to the terminal nitrogen of the diazo compound (14), would not

Table 7 Enantioselective Cyclopropanation of Alkenes with Diazo Esters Using Bis(α-camphorquinone-dioximato)cobalt(II) (12)[a]

Alkene	$N_2CHCOOR'$ R'=	trans:cis	Yield (%)	% ee trans	cis
PhCH=CH$_2$	Me	41:59	94	61	—
	Et	46:54	92	75	67
	i-Bu	48:52	94	80	—
	i-Pr	53:47	91	84	—
	Cyclohexyl	59:41	72	78	—
	Neopentyl	70:30	87	88	81
Ph$_2$C=CH$_2$	Et	—	95	70	
H$_2$C=C(Ph)COOMe	Et	—	92	71	37

[a] Reactions performed in neat olefin below 5°C using 2–3 mol % of catalyst.

Source: References 46 and 47.

Scheme 5

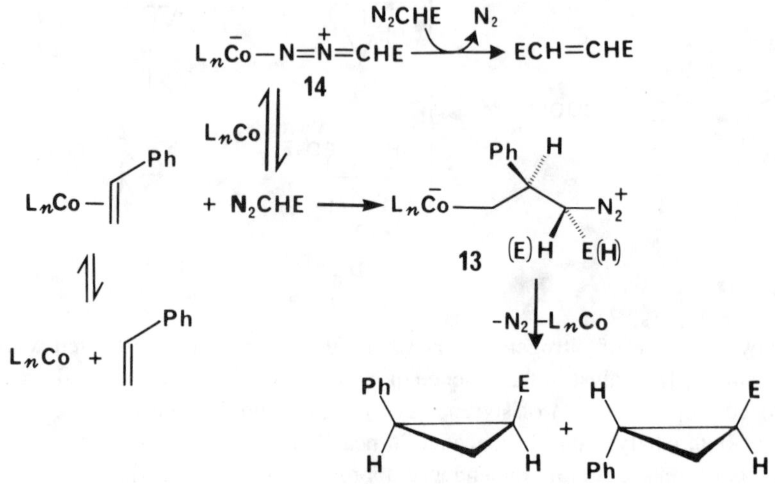

necessitate metal carbene intermediates on the way to maleate and fumarate products.

Reactions of transition metal coordinated olefins with diazo compounds as a route to cyclopropane products have not yet been rigorously established. Catalysts that should be effective in this pathway are those that are more susceptible to olefin coordination than to association with a diazo compound and also those whose coordinated alkene is sufficiently electrophilic to react with diazo compounds, especially diazomethane. Palladium(II), platinum(II), and cobalt(II) compounds [12] appear to be capable of promoting olefin-coordination-induced cyclopropanation reactions, but further investigations will be required to unravel this mechanistic possibility.

3.2.4. Chiral Dirhodium(II) Catalysts

Rhodium(II) carboxylates, especially $Rh_2(OAc)_4$, have emerged as the most generally effective catalysts for carbenoid transformations [13–15] and, because of this, there is growing interest in the design and development of dirhodium(II) complexes that possess chiral ligands. The first of these applied to cyclopropanation reactions was the chiral rhodium(II) carboxylate derivates developed by Brunner [49], who prepared 13 chiral dirhodium(II) tetrakis(carboxylate) derivatives (**15**) from enantiomerically pure carboxylic acids $R^1R^2R^3CCOOH$ with substituents that were varied from H, Me, and Ph to OH, NHAc, and CF_3. However, reactions performed between ethyl diazoacetate and styrene yielded cyclopropane products with less than 12% ee. The situation was analogous to that encountered by Nozaki [2] in the first applications of chiral Schiff base–copper(II) catalysts.

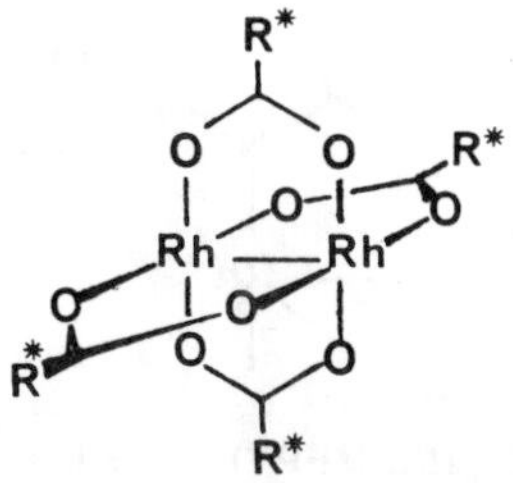

15

Rhodium(II) carboxylates are structurally well defined, having D_{2h} symmetry [50], with axial coordination sites at which carbene formation occurs in reactions with diazo compounds. With chiral dirhodium(II) carboxylates the chiral center is relatively far from the carbene center in the metal carbene intermediate. As demonstrated by Callot and Metz [51] for 2,6-disubstituted benzoate derivatives of dirhodium(II), effective control of selectivity is possible only when the carboxylate substituents can restrict the orientation of the carbene about the metal center. That enantioselection is observed with the chiral rhodium(II) carboxylates is encouraging, and further efforts are warranted.

In contrast to rhodium(II) carboxylates, rhodium(II) carboxamides provide placement of a chiral center adjacent to nitrogen in close proximity to the carbenoid center. However, unlike the rhodium(II) carboxylates, for which only one isomer derived from the ligand is possible, rhodium(II) carboxamides can conceivably be constructed from chiral carboxamides in any one or all of four possible configurations. When *N*-phenylacetamide is used as the ligand, the isomer (**16**) possessing the four bridging amide ligands in which two oxygen and two nitrogen donor atoms are bonded to each rhodium and the nitrogen donor atoms are in a cis arrangement is preferred among other possible arrangements [52]. For acetamide or trifluoroacetamide ligand **16** is the sole complex isolated [53]. Rhodium(II) carboxamides are less reactive toward diazo compounds but provide higher selectivities in several carbenoid reactions than do rhodium(II) carboxylates [13,42,54].

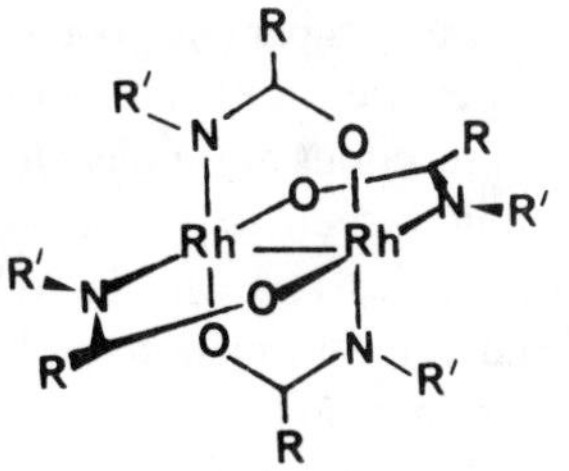

16

The preparation and use of chiral dirhodium(II) methyl 2-pyrrolidone-5-carboxylate (**17**) and 4-alkyloxazolidinones (**18–20**) for intermolecular cyclopropanation reactions have recently been reported [9]. These catalysts are prepared in high yield

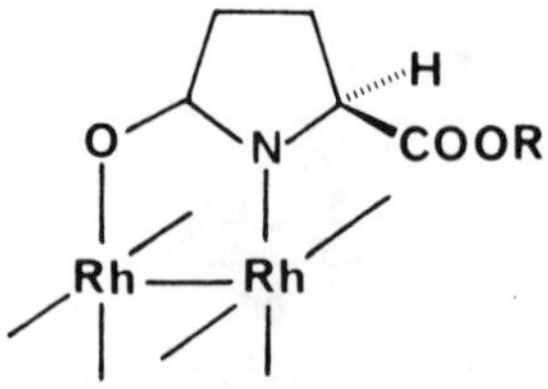

Rh₂(5S-MEPY)₄, R = Me

17

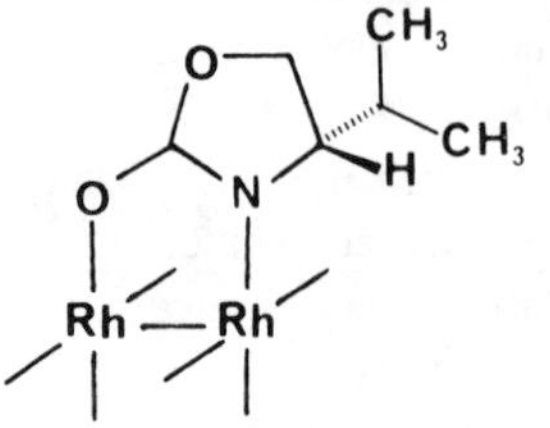

Rh₂(4S-IPOX)₄

18

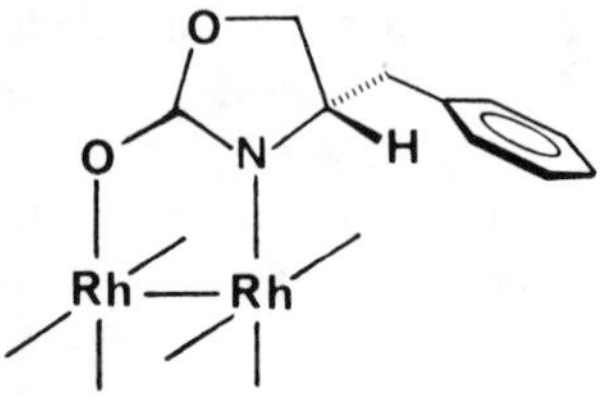

Rh₂(4S-BNOX)₄

19

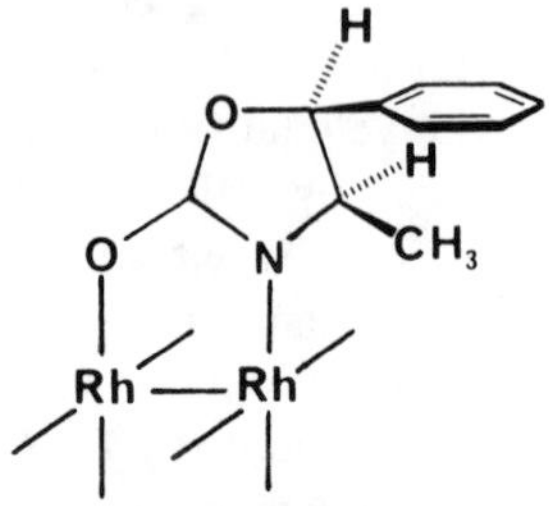

Rh₂(4R-MPOX)₄

20

and in only the cis geometry by ligand exchange in refluxing chlorobenzene with rhodium(II) acetate. Their design places the chiral center adjacent to the rhodium-bound nitrogen and, like the salicylaldimine and semicorrin/bisoxazoline copper catalysts, interactions with the protruding alkyl or carboxylate substituents control enantioselectivity. These dirhodium(II) catalysts block approach of the alkene in a manner that is significantly different from previous catalyst designs so that diastereoselectivities for the *cis*-cyclopropane derivative from reactions of menthyl diazoacetates with styrene are greater than those for the *trans*-cyclopropane derivative (Table 8). Product yields are comparable to those obtained with the chiral catalysts, but diastereoselectivities are lower. Furthermore, with *d*- and *l*-menthyl diazoacetates, cyclopropanation reactions are subject to exceptionally large "double diastereoselectivity" not previously seen to the same degree with chiral copper catalysts (compare Table 8 with Tables 1, 3, and 4). Major competing reactions are those due to carbene dimer and azine formation, and with menthyl diazoacetates intramolecular carbon–hydrogen insertion also occurs [54a].

The significant increase in diastereoselection for cyclopropanation of styrene with menthyl diazoacetates with the use of Rh₂(5S-MEPY)₄ and Rh₂(5R-MEPY)₄ catalysts has been attributed [11] to the ability of their carboxylate substituents to orient and stabilize the rhodium-bound carbene and thereby direct incoming nucleophiles to attack the carbene from the side opposite to the stabilizing substituents (Scheme 6). Here electronic interaction of the *p*-orbital of the metal carbene with the

Table 8 Enantioselective Cyclopropanation of Styrene with *d*- and *l*-Menthyl Diazoacetate Using Dirhodium(II) Oxazolidinones and Methyl 2-Pyrrolidone-5-carboxylate

Catalyst	Menthyl Diazoacetate	trans:cis	% de[b] trans	% de[b] cis
$Rh_2(OAc)_4$	*l*	68:32	5 (1*R*,2*R*)	13 (1*R*,2*S*)
$Rh_2(OAc)_4$	*d*	55:45	9 (1*S*,2*S*)	13 (1*S*,2*R*)
$Rh_2(4S\text{-IPOX})_4$ **(18)**	*l*	70:30	34 (1*R*,2*R*)	56 (1*R*,2*S*)
$Rh_2(4S\text{-IPOX})_4$ **(18)**	*d*	75:25	2 (1*R*,2*R*)	4 (1*R*,2*S*)
$Rh_2(4S\text{-BNOX})_4$ **(19)**	*l*	61:39	34 (1*R*,2*R*)	63 (1*R*,2*S*)
$Rh_2(4S\text{-BNOX})_4$ **(19)**	*d*	63:37	4 (1*R*,2*R*)	24 (1*R*,2*S*)
$Rh_2(4R\text{-BNOX})_4$ **(*R*-19)**	*l*	62:38	4 (1*S*,2*S*)	25 (1*S*,2*R*)
$Rh_2(4R\text{-BNOX})_4$ **(*R*-19)**	*d*	67:33	34 (1*S*,2*S*)	62 (1*S*,2*R*)
$Rh_2(4R\text{-MPOX})_4$ **(20)**	*l*	71:29	4 (1*R*,2*R*)	4 (1*R*,2*S*)
$Rh_2(4R\text{-MPOX})_4$ **(20)**	*d*	77:23	23 (1*S*,2*S*)	20 (1*S*,2*R*)
$Rh_2(5S\text{-MEPY})_4$ **(17)**	*l*	67:33	56 (1*S*,2*S*)	79 (1*S*,2*R*)
$Rh_2(5S\text{-MEPY})_4$ **(17)**	*d*	67:33	48 (1*S*,2*S*)	86 (1*S*,2*R*)
$Rh_2(5R\text{-MEPY})_4$ **(*R*-17)**	*l*	67:33	48 (1*R*,2*R*)	85 (1*R*,2*S*)
$Rh_2(5R\text{-MEPY})_4$ **(*R*-17)**	*d*	67:33	55 (1*R*,2*R*)	78 (1*R*,2*S*)

[a] *d*-Menthyl = (1*S*,2*R*,5*S*)-2-isopropyl-5-methylcyclohexyl.

[b] Absolute configuration of cyclopropane product in parentheses.

Source: Reference 53.

Scheme 6

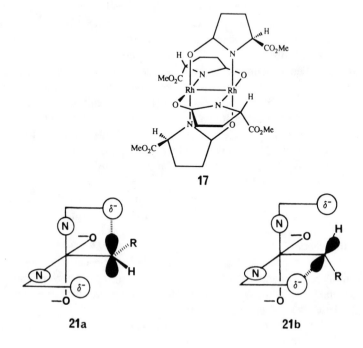

17

21a   **21b**

Table 9 Diastereoselective Cyclopropanation of Alkenes with d-Menthyl Diazoacetate Using $Rh_2(5S\text{-MEPY})_4$

		% de	
Alkene	trans:cis	trans	cis
Styrene	67:33	48	86
α-Methylstyrene	60:40	60	n.d.[a]
3,3-Dimethyl-1-butene	71:29	65	91
Ethyl vinyl ether	63:37	47	60

[a] Not determined.

polar carboxylate substituent of the ligand (δ^-) provides the required stabilization and orientation. Because two nitrogens of the carboxamide ligand place two polar substituents in a cis geometry on each face of the dirhodium(II) compound, two limiting carbene orientations are possible (**21a** and **21b**). The carbene substituent R is assumed to take the optimum electronic or less sterically congested position.

The success of this approach to the design of effective chiral rhodium(II) carboxamide catalysts can be seen from the enhanced enantioselectivities achieved with the use of $Rh_2(5S\text{-MEPY})_4$ in intermolecular cyclopropanation reactions of d-menthyl diazoacetate with monosubstituted olefins (Table 9). The basis for this selectivity enhancement over that achieved with chiral dirhodium(II) oxazolidinones is explained through the mechanistic interpretation advanced in Scheme 7. These projections portray the carbene bonded to the dirhodium framework (back) with the chiral ligand's carboxylate substituents in the spacial positions that models suggest would be encountered by carbene substituents. The front view of the reacting metal carbene is depicted in two limiting conformations with the larger carbene substituents (R) located preferentially at the less sterically congested side. The attachment ⊘ to ⑧ describes the direction of the carboxylate moiety away from the pyrrolidone ligand: left and down for $Rh_2(5S\text{-MEPY})_4$, right and up for $Rh_2(5R\text{-MEPY})_4$. The ligand carbonyl group that stabilizes the electrophilic carbene center enforces nucleophilic attack of the alkene from the backside. Styrene, as the reacting alkene, is aligned with the carbene to form either the trans- or cis-disubstituted cyclopropane product. Of the two possible conformations directed to the formation of either cis- or trans-cyclopropane isomer, the one in which the R group (R = COZ) lies on the same side as the ligand's carboxylate group is less stable because of steric and/or electronic repulsions. The preferential formation of the (1S,2S)- and (1S,2R)-enantiomers with $Rh_2(5S\text{-MEPY})_4$ and the (1R,2R)- and (1R,2S)-enantiomers with $Rh_2(5R\text{-MEPY})_4$ is predicted by this model.

According to this model, increasing the steric bulk of the carbene substituent R is predicted to increase enantioselectivity for formation of the cis-cyclopropane isomer without similarly affecting the selectivity for the trans-cyclopropane isomer, and indeed this is what is observed (Table 10).

Scheme 7

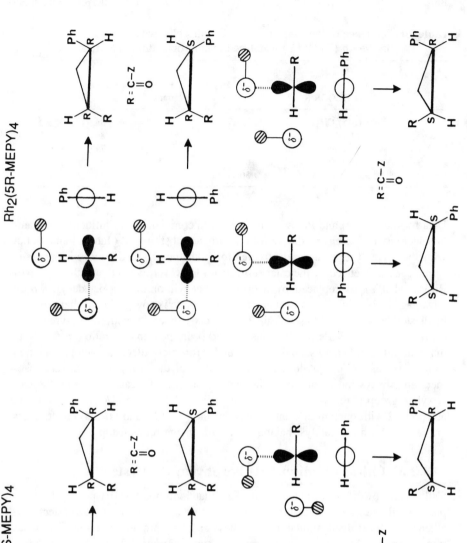

Rh₂(5S-MEPY)₄

Rh₂(5R-MEPY)₄

85

Table 10 Influence of the Diazo Ester on Enantioselectivity
in the $Rh_2(5S\text{-MEPY})_4$-catalyzed Cyclopropanation of Styrene

		% ee	
N₂CHCOOR:R	trans:cis	trans	cis
$C(i\text{-Pr})_2CH_3$	73:27	67	83
d-Menthyl	67:33	48	86
l-Menthyl	67:33	56	79
t-Bu	60:40	57	73
Et	56:44	58	33

Data for the Aratani and Pfaltz catalysts show a complementary influence on enantioselectivity by increasing the size of the carbene substituent R (Tables 1 and 3), but
with these catalysts enantioselectivities for both the *cis*- and *trans*-cyclopropane
isomers are influenced by the ester substituent. However, when the methyl group of
$Rh_2(5S\text{-MEPY})_4$ is replaced by neopentyl, benzyl, or even *n*-octadecyl, whose
advantage is its solubility in pentane or hexane, there is no significant change either
in diastereoselectivity (78–86% de for the cis isomer formed from styrene and
d-MDA and 44–53% de for the trans isomer) or in the trans/cis ratio ($\pm 2\%$). Note
that with all chiral catalysts employed thus far for intermolecular cyclopropanation
using diazo esters, the predominant enantiomers of the cis(syn) or trans(anti) isomers are always those for which the configuration at the carbon bearing the carboxylate group is the same (*R* or *S*). Thus, the (1*R*,1*S*)- and (1*R*,2*R*)-enantiomers
are favored with one chiral catalyst whereas the (1*S*,2*R*)- and (1*S*,2*S*)-enantiomers
are favored with the catalyst having the mirror image relationship.

3.2.5. Chiral Rhodium(III) Porphyrin Catalysts

Callot and coworkers established in 1982 that iodorhodium(III) porphyrin complexes could be used as cyclopropanation catalysts for reactions of diazo esters with
alkenes; with *cis*-disubstituted alkenes these catalysts provide preferential production of cis(syn)-disubstituted cyclopropanes (syn/anti up to 3.3 with 1,4-cyclohexa-
diene) [55]. More recently, chiral porphyrins have been designed and prepared by
Kodadek and coworkers [56], and their iodorhodium(III) complexes have been
examined for asymmetric induction in catalytic cyclopropanation reactions [57,58].
The intent here has been to affix chiral attachments onto the four porphyrin positions
that are occupied in tetraphenylporphyrin by a phenyl group. Iodorhodium(III)
catalysts with chiral binaphthyl (**22**, called "chiral wall" porphyrin [57]) and the
structurally analogous chiral pyrenylnaphthyl (**23**, called "chiral fortress" porphyrin
[58]) have been prepared, but their ability to control enantioselectivity is low to
moderate (Table 11) with ethyl diazoacetate. However, they have an exceptionally
strong propensity for the production of the cis(syn) stereoisomer to an even greater
extent than that reported by Callot and coworkers for the corresponding tetra-

Table 11 Enantioselective Cyclopropanation of Alkenes with Ethyl Diazoacetate Using Iodorhodium(III)–Porphyrin Catalysts

| Catalyst | Alkene | cis(syn):trans(anti) | % ee | | Temp. (C°) |
			cis(syn)	trans(anti)	
22	Styrene	70:30	10[a]	n.d.[b]	0
23		71:29	15	n.d.[b]	25
22	(Z)-1-Phenylpropene	89:11	20	50	0
23		93:7	0	20	0
23		84:16	25	20	25
22	3-Phenyl-2-propene	81:19	45	60	0
23		80:20	20	40	0
23		50:50	10	10	25
23	Ethyl vinyl ether	45:55	15	10	25

[a] (1S,5R)-configuration.
[b] Not determined.

Source: References 57 and 58.

mesitylporphyrin catalyst [55]. The oxidation state of the catalytically active rhodium compound is unknown and could be rhodium(II).

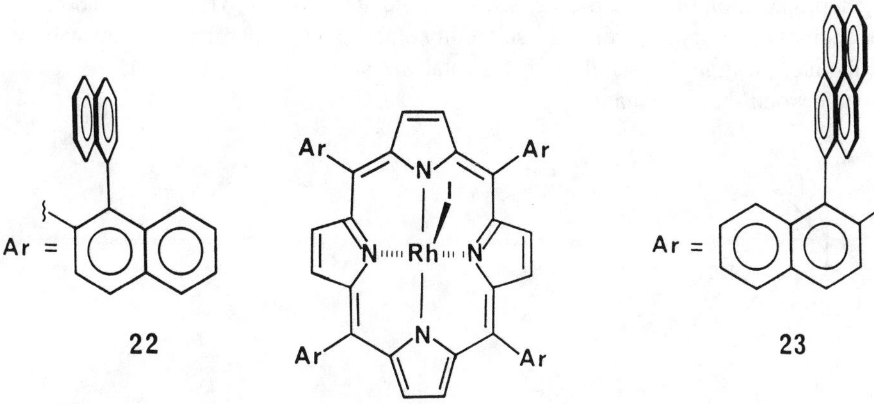

3.2.6. Other Chiral Catalysts

Although chiral copper salicylaldimine and semicorrin or bisoxazoline catalysts have provided high enantiocontrol in cyclopropanation reactions, a large number of chiral ligands for copper have proved to be less effective. Copper(II) tartrate has been used for the enantioselective cyclopropanation of 3-methoxystyrene by 4-bromo-1-diazo-2-butanone (46% ee for the trans isomer) [59], but no further reports on the selectivity of this catalyst have been published. The binaphthyl-*o,o'*-diamine Schiff base **24** has been used as a ligand for copper, but is relatively ineffective for asymmetric cyclopropanation of 1,1-diphenylethylene by ethyl diazoacetate (24% ee) [35].

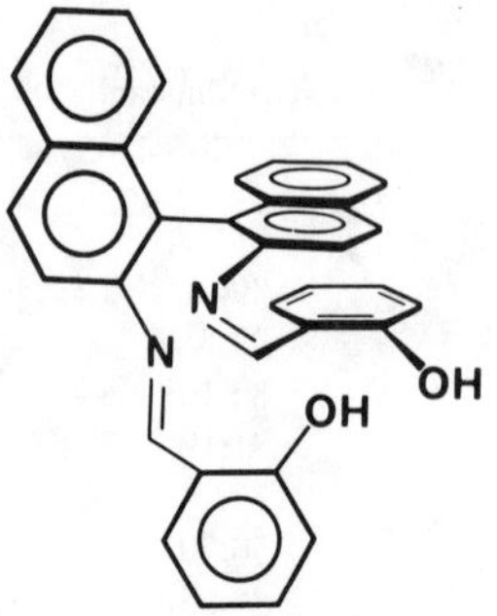

24

As the most enantioselective process published to date, Matlin and coworkers have communicated [60] enantioselectivity of 100% ee for the cyclopropanation of styrene with 2-diazodimedone using copper(II) chelates (**25**) possessing trifluoroacetyl-(+)-camphor ligands (eq. 16). However, Dauben has reported that a camphor catalyst of this type was ineffective for asymmetric induction in the intramolecular cyclopropanation of 1-diazo-6-hepten-2-one (eq. 18, 2% ee) [61], and the absence of additional reports regarding the suitability of **25** for asymmetric cyclopropanation calls into question the viability of this catalyst, at least for cyclopropanation reactions beyond that of equation 16.

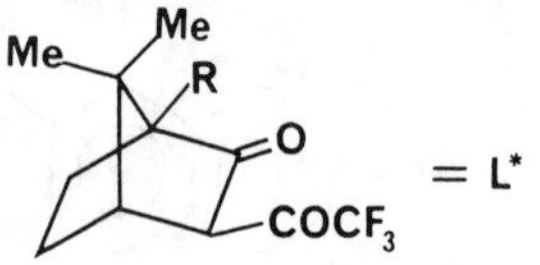

25

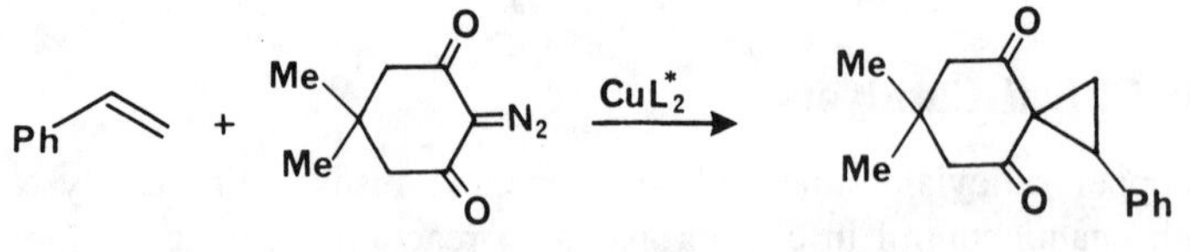

$$\tag{16}$$

R = Me	92% ee
R = H$_2$C=CH	100% ee

With Schiff bases derived from pyridine-2-carboxaldehyde and amino sugars as chiral ligands for copper, low enantiocontrol is observed for intermolecular cyclopropanation reactions of ethyl diazoacetate with olefins leading to 2,2-dimethyl-3-vinyl-2-cyclopropanecarboxylates [31,62]. In these cases the configuration of the cyclopropane product is dependent on the configuration at C-2 of the 2-amino sugar (e.g., 2-amino-D-glucopyranoside versus 2-amino-D-allopyranoside for predominant R or S product configuration). However, enantioselectivities in these cases are so low that predictions regarding configurational preferences are unreliable.

Other transition metal compounds capable of serving as enantioselective cyclopropanation catalysts, especially those of palladium or platinum [12], have not yet been effectively developed.

3.3. Asymmetric Intramolecular Cyclopropanation

Intramolecular cyclopropanation reactions of alkenyl diazo carbonyl compounds are among the most useful catalytic metal carbene transformations, and the diversity of their applications for organic syntheses is substantial [13–15,63]. Catalytic asymmetric reactions, however, have only recently been reported. An early application of the Aratani catalyst **4** (A = PhCH$_2$) to the asymmetric synthesis of dihydrochrysanthemolactone (eq. 17, in refluxing cyclohexane) showed low enantiocontrol (23% ee) and relatively low yield (47%) [64]. Dauben and coworkers have more

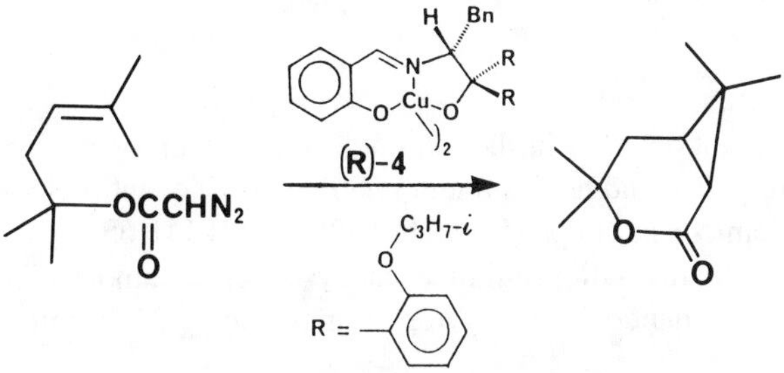

$$(17)$$

recently examined enantioselectivity in the intramolecular cyclopropanation of 1-diazo-5-hexen-2-one (eq. 18, R^1 = R^2 = H) and its homolog (eq. 19) using a similar Aratani catalyst [61]. Product yields and enantioselectivities of these reactions depend on the reagents and conditions employed. The best results (77% ee for eq. 18, 34% ee for eq. 19) were obtained upon treatment of **4** (A = PhCH$_2$) with 0.25 equiv. of DIBAH in benzene. Other diazo carbonyl compounds, including a β-keto-α-diazo ester and α-diazo esters, underwent intramolecular cyclopropanation in moderate to good yields, but with low enantiocontrol. For example, diazo esters **28** (n = 1, 2) were converted to **29** (n = 1, 2) in 0 and 25% ee, respectively, upon treatment with the chiral Schiff base catalyst. In these reactions, however, 25 mol % of DIBAH-treated catalyst was required to avoid catalyst poisoning by the [2 + 3] pyrazoline cycloadduct that was formed as a by-product.

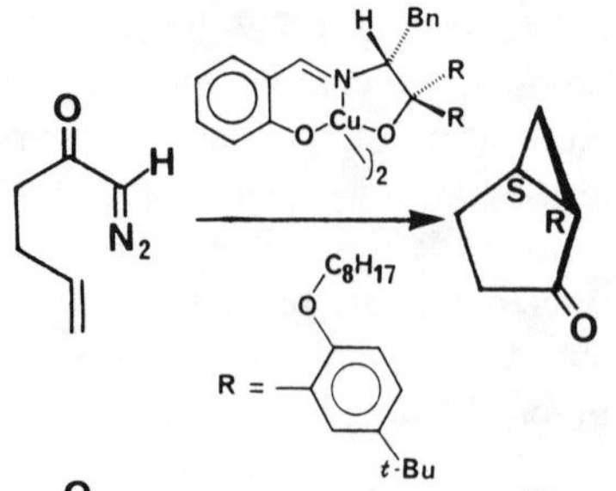

(18)

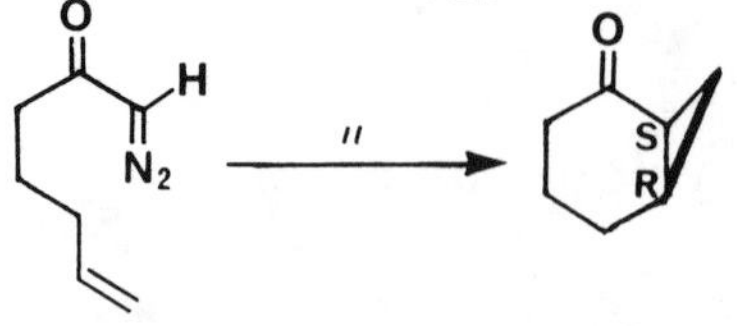

(19)

26   **27**

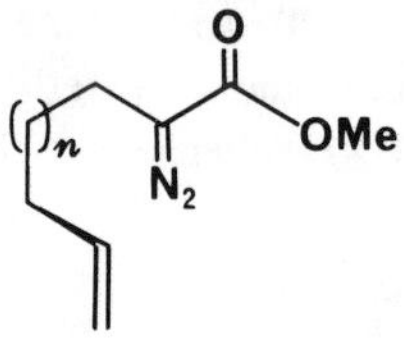

28

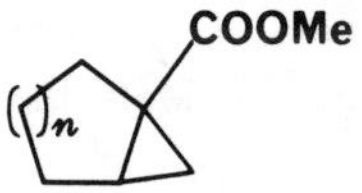

29

Pfaltz has also examined enantiocontrol in the intramolecular cyclopropanation of diazo ketones **26** and **30** (eq. 20), and he has found relatively high enantioselectivities with the use of his semicorrin copper(I) catalyst (Table 12) [41,65]. This catalyst is obviously superior to the salicylaldimine catalysts for intramolecular cyclopropanation reactions, and enantioselectivity for the reaction of **26** is better than that for **30**.

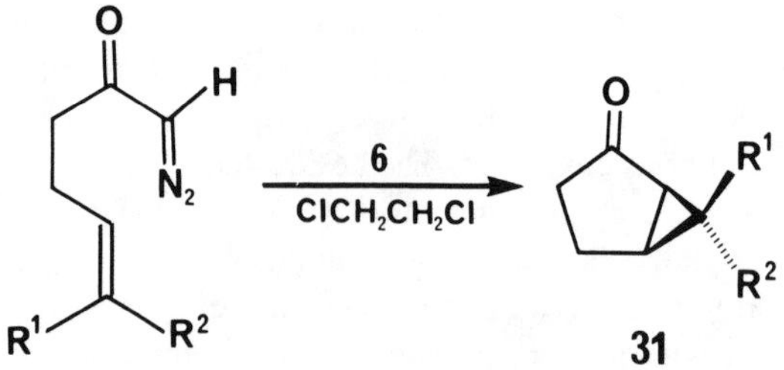

(20)

Doyle, Martin, Müller, and coworkers have reported excellent enantioselectivity for intramolecular cyclopropanation of a series of allyl diazoacetates (eq. 21) using dirhodium(II) tetrakis(methyl 2-pyrrolidone-5-carboxylates), $Rh_2(MEPY)_4$ in either

Table 12 Enantioselective Intramolecular Cyclopropanation of 1-Diazo-*n*-alken-2-ones with Aratani and Pfaltz Catalysts

Diazo Ketone	Catalyst	Catalyst (mol %)	Temp. (C°)	Product	% ee
30 ($R^1 = R^2 = H$)	**4** (A = PhCH$_2$)[a]	3	25	**31** ($R^1 = R^2 = H$)	77
	6 (R = Me$_2$C(OH))	3	25		76
30 ($R^1 = R^2 = CH_3$)	**6** (R = Me$_2$C(OH))	3	25	**31** ($R^1 = R^2 = CH_3$)	68
30 ($R^1 = H$, $R^2 = H_2C{=}CH$)	**7** (R = Me$_2$C(OH))	1	25	**31** ($R^1 = H$; $R^2 = H_2C{=}CH$)	24
26	**4** (A = PhCH$_2$)[a]	3	−10	**27**	34
	4 (A = PhCH$_2$)[a]	3	25		22
	6 (R = Me$_2$C(OH))	3	25		92

[a] Pretreated with 0.75 mol % DIBAH.

Source: References 61 and 65.

Table 13 Enantioselective Intramolecular Cyclopropanation of Allyl Diazoacetates Catalyzed by $Rh_2(5S\text{-MEPY})_4$

Cpd	R^1	R^2	Yield (%)	% ee	Cpd	R^1	R^2	Yield %	% ee
33a	H	H	74	88	33f	CH_3	CH_3	82	98
33b	H	Ph	45	≥94	33g	Ph	H	59	65
33c	H	CH_2CH_3	88	≥94	33h	$CH_2CH_2CH_3$	H	74	75
33d	H	CH_2Ph	80	≥94	33i	H	$C_6H_{11}CH_2$	45	68
33e	H	$Sn(n\text{-Bu})_3$	78	≥94	33j	H	$(CH_3)_2CHCH_2$	29	72

their *R*- or *S*- configurations (Table 13) [66]. cis-Disubstituted olefins generally lead to higher enantioselectivities than do the corresponding trans-disubstituted olefins, even with substituents as bulky as tri-*n*-butyltin, and virtually complete enantio-control has been realized for five different substrates. Produce yields are high except for **33i** and **33j**, where steric factors appear to limit olefin approach to the metal carbene center.

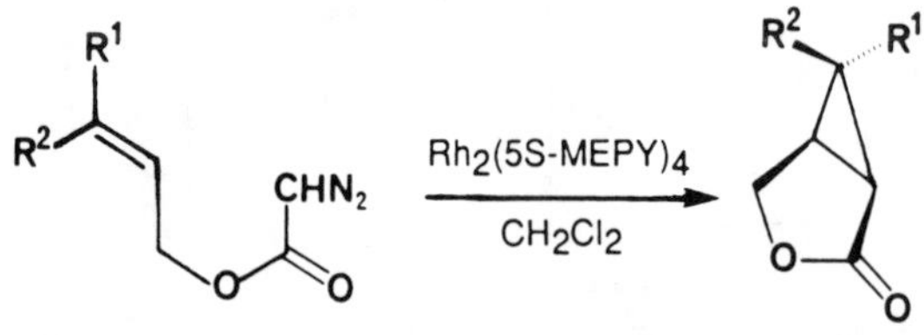

$$(21)$$

Chiral rhodium(II) oxazolidinones $Rh_2(BNOX)_4$ and $Rh_2(IPOX)_4$ were not as effective as $Rh_2(MEPY)_4$ for enantioselective intramolecular cyclopropanation, even through the steric bulk of their chiral ligand attachments (COOMe vs. *i*-Pr or CH_2Ph) are similar. Significantly lower yields and lower enantiomeric excesses resulted from dinitrogen extrusion from **34** catalyzed by either $Rh_2(4S\text{-IPOX})_4$, $Rh_2(4S\text{-BNOX})_4$, or $Rh_2(4R\text{-BNOX})_4$ (eq. 22). In addition, the formation of bu-

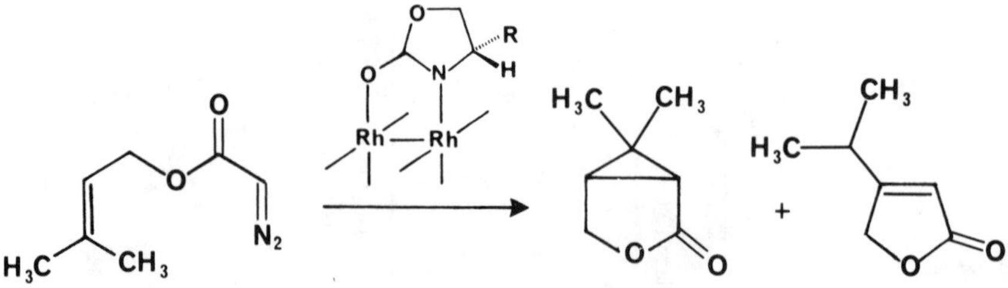

$$(22)$$

	35	36
$Rh_2(4S\text{-IPOX})_4$	44% (43% ee)	5%
$Rh_2(4S\text{-BNOX})_4$	52% (55% ee)	11%
$Rh_2(4R\text{-BNOX})_4$	50% (56% ee)	12%

Scheme 8

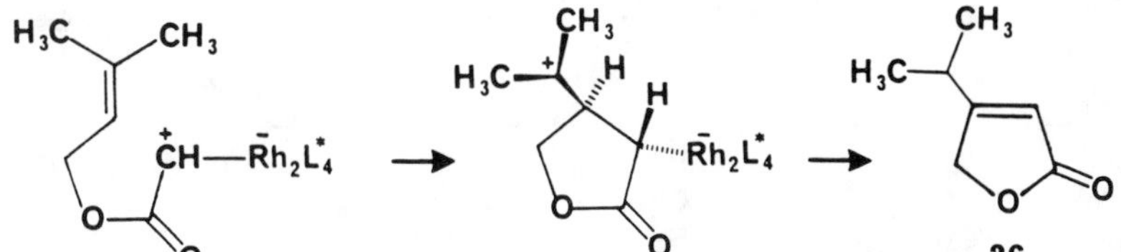

Rh₂(4S-BNOX)₄	12%
Rh₂(4R-BNOX)₄	11%
Rh₂(4S-IPOX)₄	5%
Rh₂(5S-MEPY)₄	< 1%

tenolide **36** (Scheme 8) was substantial in reactions performed with Rh₂(BNOX)₄ and Rh₂(IPOX)₄ but was only a minor constituent (≤ 1%) from reactions catalyzed by Rh₂(5S-MEPY)₄: The product **36** arises from formal carbenium ion addition to the double bond followed by 1,2-hydrogen migration and dissociation of Rh₂L*₄. This difference can be attributed to the ability of the carboxylate substituents to stabilize the carbocation form of the intermediate metal carbene, thus limiting the Rh₂(5S-MEPY)₄-catalyzed reaction to concerted carbene addition to the carbon–carbon double bond.

The directional orientation of chiral ligand substituents on Rh₂(5S-MEPY)₄ establishes relatively unimpeded pathways for intramolecular cyclization. Of the two possible configurations molecular modeling indicates that **37** is preferred. According to this model Z-olefins should afford higher enantioselectivities than E-olefins, whose substituent is buttressed against the ligand's carboxylate group. The absolute configuration of the bicyclic cyclopropane product is consistent with this interpretation.

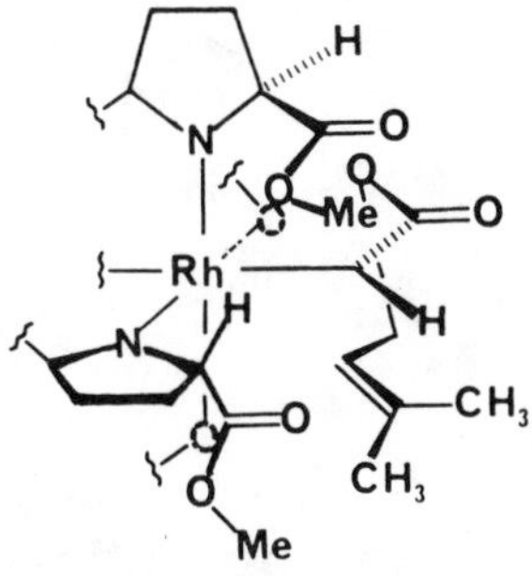

37

Intramolecular cyclopropanation of the next higher homologs of the allyl diazoacetates (eq. 23) catalyzed by Rh₂(MEPY)₄ also give high enantiomeric purities for the addition product, and isolated yields are also high (Table 14) [67]. Once again, cis-disubstituted olefins lead to higher enantioselectivities than do trans-disubstituted olefins, but here the differences are not as great as they were with allyl

Table 14 Enantioselective Intramolecular Cyclopropanation of Homoallyl
Diazoacetates Catalyzed by $Rh_2(5S\text{-MEPY})_4$

Cpd	R^1	R^2	R^3	Yield (%)	% ee
39a	H	H	H	80	71
39b	CH_3	CH_3	H	74	77
39c	H	H	CH_3	76	79
39d	CH_3CH_2	H	H	80	90
39e	H	CH_3CH_2	H	65	82
39f	Ph	H	H	73	88
39g	H	Ph	H	55	73
39h	Cyclohexyl	H	H	77	80
39i	$PhCH_2$	H	H	68	80
39j	TMS	H	H	65	86

diazoacetates. Although enantiocontrol is generally more pronounced for allylic diazoacetates than for their homoallylic counterparts, there are exceptions (e.g., **33h** vs. **39e**) that imply a greater flexibility for cyclization to **39** than to **33**. The composite data suggest that chiral dirhodium(II) carboxamide catalysts are superior to chiral copper catalysts for intramolecular cyclopropanation reactions, but a conclusion must await more detailed comparisons with the same substrates. The design of the $Rh_2(MEPY)_4$ catalysts, with the cis arrangement of its chiral attachments, favors intramolecular transformations by keeping adjacent quadrants on the catalyst face open for cyclization.

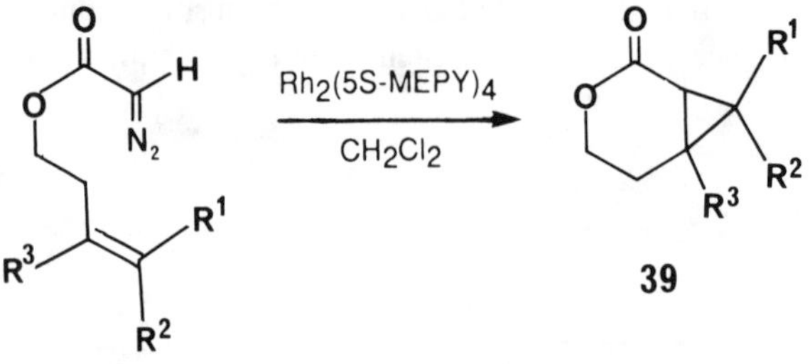

(23)

3.4. Intermolecular Cyclopropenation of Alkynes

Functionalized cyclopropenes are viable synthetic intermediates, but their applications [68,69] to a wide variety of carbocyclic and heterocyclic systems have been largely ignored because of the relative inaccessibility of these strained compounds. However, recent advances in the synthesis of cyclopropenes, particularly through rhodium(II) carboxylate catalyzed decomposition of diazo esters in the presence of

Table 15 Enantioselective Cyclopropanation of Alkenes Catalyzed by $Rh_2(5R\text{-MEPY})$

Diazoacetate:R	1-Alkyne:R'	Cyclopropene	Yield (%)	% ee
Et	MeOCH$_2$	**40a**	73	69
t-Bu		**40b**	56	78
d-Menthyl		**40c**	43	98
l-Menthyl		**40d**	45	43
Et	n-Bu	**41a**	70	54
t-Bu		**41b**	69	53
d-Menthyl		**41c**	46	86
l-Menthyl		**41d**	46	20
Et	t-Bu	**42a**	85	57
t-Bu		**42b**	57	70
d-Menthyl		**42c**	51	77
l-Menthyl		**42d**	50	56

Source: Reference 81.

alkynes [70–73], has made available an array of stable 3-cyclopropenecarboxylate esters. Previously, copper catalysts provided low to moderate yields of cyclopropenes in reactions of diazo esters with disubstituted acetylenes [74], but the higher temperatures required for these carbenoid reactions often led to thermal or catalytic ring opening and products derived from vinylcarbene intermediates [75–78]. Potential uses of the cyclopropene ring as a template in enantiocontrolled syntheses have been recognized, but until recently synthetic chiral cyclopropene derivatives have been accessible only through resolution [79] or from natural products [80].

With ethyl diazoacetate and representative alkynes, use of catalytic amounts of $Rh_2(5R\text{-MEPY})_4$ results in the formation of ethyl cyclopropene-3-carboxylates (eq. 24) with 54–69% ee in good yields (70–85%) [81]. Virtually identical results, except for absolute configuration, are obtained with the use of $Rh_2(5S\text{-MEPY})_4$.

$$H\text{-}C\equiv C\text{-}R \ + \ N_2CHCOOR' \xrightarrow[CH_2Cl_2]{Rh_2(5R\text{-MEPY})_4} \qquad\qquad (24)$$

$$
\begin{array}{ll}
\textbf{40} & R = MeOCH_2 \\
\textbf{41} & R = n\text{-Bu} \\
\textbf{42} & R = t\text{-Bu}
\end{array}
\qquad
\begin{array}{ll}
\textbf{a} & R' = Et \\
\textbf{b} & R' = t\text{-Bu} \\
\textbf{c} & R' = d\text{-menthyl} \\
\textbf{d} & R' = l\text{-menthyl}
\end{array}
$$

With d-menthyl diazoacetate and the same series of alkynes, selectivities as high as 98% de have been achieved (Table 15). Enantioselectivities increase with the steric size of the diazo ester, and the polarity of the alkyne substituent also appears to influence enantiocontrol. The fact that enantiomeric purities of cyclopropenes from reactions with propargyl methyl ether are higher than those from reactions with 1-hexyne and 3,3-dimethyl-2-butyne suggests that polar interactions of the methoxy substituent of this alkyne with ligands of the catalyst may be operative. Indeed, cyclopropenation of propargyl acetate by l-menthyl diazoacetate in the presence of Rh$_2$(5S-MEPY)$_4$ also resulted in 98% de for the cyclopropenecarboxylate product.

The use of Rh$_2$(5R-MEPY)$_4$ and Rh$_2$(5S-MEPY)$_4$ for reactions with menthyl diazoacetates (MDA) produces an enormous "double diastereoselection" not previously observed to the same degree in cyclopropanation reactions. With methyl propargyl ether, for example, Rh$_2$(5R-MEPY)$_4$ catalyzed reactions of d-MDA yield **40c** with 98% de but l-MDA produces **40d** with only 40% de; with Rh$_2$(5S-MEPY)$_4$, l-MDA gives the higher selectivity (98% de) than does d-MDA (43% de). Similar results are obtained from reactions of MDA with 1-hexyne and 3,3-dimethyl-1-propyne. The diazocarboxylate substituent obviously plays a critical role in establishing the more effective carbene orientation for addition to the alkyne.

Rhodium(II) carboxamide catalysts derived from 4(R)-benzyloxazolidinone (4R-BNOXH) and 4(S)-isopropyloxazolidinone (4S-IPOXH) provided only a fraction of the enantioselection obtained with Rh$_2$(MEPY)$_4$ catalysts. Whereas cyclopropenation of 1-hexyne with ethyl diazoacetate in the presence of Rh$_2$(5R-MEPY)$_4$ resulted in **41a** with 54% ee, Rh$_2$(4R-BNOX)$_4$ gave the same compound with 5% ee, and Rh$_2$(4S-IPOX)$_4$ provided only 6% ee. Preliminary results with the semicorrin copper catalyst **6** (R = CH$_2$OSi(t-Bu)Me$_2$) on the reaction between ethyl diazoacetate with 1-hexyne showed only 10% ee and low product yield.

3.5. Conclusion

Significant progress has been made recently in the design and development of chiral transition metal catalysts for metal carbene transformations. The major focus of these efforts has been on intermolecular cyclopropanation reactions, but recent results have demonstrated highly enantioselective intramolecular counterparts. Both copper and rhodium compounds have been effectively employed with chiral salicylaldimine (**4**), semicorrin (**6** and **11**), and bisoxazoline (**8–10**) ligands for copper and chiral 2-pyrrolidone-5-carboxylate (**17**) ligands for dirhodium(II) offering the highest enantioselection, often exceeding 95% ee in selected examples. Extensions to other metal carbene transformations, particularly carbon–hydrogen insertion, have only recently been undertaken with dirhodium(II) catalysts [82–84], but enantioselectivities greater than 90% ee have already been achieved in selected cases [84]. Accordingly, new applications of asymmetric carbene reactions will be explored extensively in the near future.

Acknowledgment

Support from the National Science Foundation, National Institutes of Health (GM-46503), and the Robert A. Welch Foundation for investigations of the design and development of chiral dirhodium(II) carboxamide catalysts for asymmetric cyclopropanation and related transformations is gratefully acknowledged.

References

1. Noyori, R. *Science,* **1990,** *248,* 1194.

2. Nozaki, H.; Moriuti, S.; Takaya, H.; Noyori, R. *Tetrahedron Lett.* **1966,** 5239.

3. Aratani, T.; Yoneyoshi, Y.; Nagase, T. *Tetrahedron Lett.* **1975,** 1707.

4. Aratani, T.; Yoneyoshi, Y.; Nagase, T. *Tetrahedron Lett.* **1977,** 2599.

5. Aratani, T.; Yoneyoshi, Y.; Nagase, T. *Tetrahedron Lett.* **1982,** *23,* 685.

6. Aratani, T. *Pure Appl. Chem.* **1985,** *57,* 1839.

7. Takaya, H.; Ohta, T.; Noyori, R. In *Catalytic Asymmetric Synthesis;* Ojima, I. (Ed.); VCH Publishers: New York, 1992; Chapter 1.

8. Fritschi, H.; Leuteneggar, U.; Pfaltz, A. *Angew. Chem., Int. Ed. Engl.* **1986,** *25,* 1005.

9. Doyle, M. P.; Brandes, B. D.; Kazala, A. P.; Pieters, R. J.; Jarstfer, M. B.; Watkins, L. M.; Eagle, C. T. *Tetrahedron Lett.* **1990,** *31,* 6613.

10. Salaün, J. *Chem. Rev.* **1989,** *89,* 1247.

11. Doyle, M. P. *Recl. Trav. Chim. Pays-Bas,* **1991,** *110,* 305.

12. (a) Tamblyn, W. H.; Hoffman, S. R.; Doyle, M. P. *J. Organometl. Chem.* **1981,** *216,* C64. (b) Hanks, T. W.; Jennings, P. W. *J. Am. Chem. Soc.* **1987,** *109,* 5023. (c) Tomilov, Y. V.; Kostitsyn, A. B.; Shulishov, E. V.; Nefedov, O. M. *Synthesis,* **1990,** 246.

13. (a) Doyle, M. P. *Chem. Rev.* **1986,** *19,* 348. (b) Doyle, M. P. *Acc. Chem. Res.* **1986,** *19,* 348.

14. Maas, G. *Top. Curr. Chem.* **1987,** *137,* 75.

15. Adams, J.; Spero, D. M. *Tetrahedron,* **1991,** *47,* 1765.

16. Anciaux, A. J.; Hubert, A. J.; Noels, A. F.; Petiniot, N.; Teyssié, Ph. *J. Org. Chem.* **1980,** *45,* 695.

17. Doyle, M. P.; Griffin, J. H.; Bagheri, V.; Dorow, R. L. *Organometallics,* **1984,** *3,* 53.

18. More O'Ferrall, R. A. *Adv. Phys. Org. Chem.* **1967,** *5,* 331.

19. Padwa, A.; Hornbuckle, S. F. *Chem. Rev.* **1991,** *91,* 263.

20. Silberrad, O.; Roy, C. S. *J. Chem. Soc.* **1906,** *89,* 179.

21. (a) Paulissen, R.; Reimlinger, H.; Hayez, E.; Hubert, A. J.; Teyssie, P. *Tetrahedron Lett.* **1973,** 2233. (b) Hubert, A. J.; Noels, A. F.; Anciaux, A. J.; Teyssie, P. *Synthesis,* **1976,** 600.

22. Doyle, M. P.; Griffin, J. H.; Conceicao, J. da *J. Chem. Soc., Chem. Commun.* **1985,** 328.

23. Maxwell, J.; Kodadek, T. *Organometallics,* **1991,** *10,* 4.

24. Brookhart, M.; Studabaker, W. B. *Chem. Rev.* **1987,** *87,* 411.

25. Regitz, M.; Maas, G. *Aliphatic Diazo Compounds—Properties and Synthesis;* Academic Press: Orlando, FL, 1987.

26. Shankar, B. K. R.; Shechter, H. *Tetrahedron Lett.* **1982,** *23,* 2277.

27. (a) Doyle, M. P.; High, K. G.; Oon, S.-M.; Osborn, A. K. *Tetrahedron Lett.* **1989,** *30,* 3049. (b) Müller, P.; Pautex, N.; Doyle, M. P.; Bagheri, V. *Helv. Chim. Acta,* **1990,** *73,* 1233.

28. (a) Suda, M. *Synthesis,* **1981,** 714. (b) Vallgarda, J.; Hacksell, U. *Tetrahedron Lett.* **1991,** *32,* 5625.

29. Kunz, T.; Reissig, H.-U. *Tetrahedron Lett.* **1989,** *30,* 2079.

30. Becalski, A.; Cullen, W. R.; Fryzuk, M. D.; Herb, G.; James, B. R.; Kutney, J. P.; Piotrowska, K.;
     Tapiolas, D. *Can. J. Chem.* **1988,** *66,* 3108.

31. Laidler, D. A.; Milner, D. J. *J. Organomet. Chem.* **1984,** *270,* 121.

32. Baldwin, J. E.; Barden, T. C. *J. Am. Chem. Soc.* **1984,** *106,* 6364.

33. Doyle, M. P.; Dorow, R. L.; Buhro, W. E.; Griffin, J. H.; Tamblyn, W. H.; Trudell, M. L.
     Organometallics, **1984,** *3,* 44.

34. Doyle, M. P.; van Leusen, D. *J. Org. Chem.* **1982,** *47,* 5326.

35. Brunner, H.; Miehling, W. *Monatsh. Chem.* **1984,** *115,* 1237.

36. Müller, D.; Umbricht, G.; Weber, B.; Pfaltz, A. *Helv. Chim. Acta,* **1991,** *74,* 232.

37. Lowenthal, R. E.; Abiko, A.; Masamune, S. *Tetrahedron Lett.* **1990,** *31,* 6005.

38. Evans, D. A.; Woerpel, K. A.; Hinman, M. M. *J. Am. Chem. Soc.* **1991,** *113,* 726.

39. Fritschi, H.; Leutenegger, U.; Siegmann, K.; Pfaltz, A.; Keller, W.; Kratky, Ch. *Helv. Chim. Acta,*
     1988, *71,* 1541.

40. Fritschi, H. Leutenegger, U.; Pfaltz, A. *Helv. Chim. Acta,* **1988,** *71,* 1553.

41. Pfaltz, A. *Mod. Synth. Methods,* **1989,** *5,* 199.

42. Doyle, M. P.; Bagheri, V.; Wandless, T. J.; Harn, N. K.; Brinker, D. A.; Eagle, C. T.; Loh, K.-L.
     J. Am. Chem. Soc. **1990,** *112,* 1906.

43. Leutenegger, U.; Umbricht, G.; Fahrni, C.; von Matt, P.; Pfaltz, A. *Tetrahedron,* **1992,** *48,* 2143.

44. Solomon, R. G.; Kochi, J. K. *J. Am. Chem. Soc.* **1973,** *95,* 3300.

45. Wulfman, D. S.; Linstrumelle, G.; Cooper, C. F. In *The Chemistry of Diazonium and Diazo
     Groups;* Patai, S., Ed.; Wiley: New York, 1978; Part 2, Chapter 18.

46. Tatsuno, Y.; Konishi, A.; Nakamura, A.; Otsuka, S. *J. Chem. Soc., Chem. Commun.* **1974,** 588.

47. Nakamura, A.; Konishi, A.; Tatsuno, Y.; Otsuka, S. *J. Am. Chem. Soc.* **1978,** *100,* 3443, 6544.

48. Nakamura, A.; Konishi, A.; Tsujitani, R.; Kudo, M.; Otsuka, S. *J. Am. Chem. Soc.* **1978,** *100,*
     3449.

49. Brunner, H.; Kluschanzoff, H.; Wutz, K. *Bull. Chem. Soc. Belg.* **1989,** *98,* 63.

50. (a) Cotton, F. A.; Walton, R. A. *Multiple Bonds Between Metal Atoms;* Wiley: New York, 1982;
     Chapter 7. (b) Felthouse, T. R. *Prog. Inorg. Chem.* **1982,** *29,* 73. (c) Boyar, E. B.; Robinson, S. D.
     Coord. Chem. Rev. **1983,** *50,* 109.

51. Callot, H. J.; Metz, F. *Tetrahedron,* **1985,** *41,* 4495.

52. Bear, J. L.; Zhu, T. P.; Malinski, T.; Dennis, A. M.; Kadish, K. M. *Inorg. Chem.* **1984,** *23,* 674.

53. (a) Ahsan, M. Q.; Bernal, I.; Bear, J. L. *Inorg. Chem.* **1986,** *25,* 260. (b) Ahsan, M. Q.; Bernal, I.;
     Bear, J. L. *Inorg. Chim. Acta,* **1986,** *115,* 135.

54. (a) Doyle, M. P.; Bagheri, V.; Pearson, M. M.; Edwards, J. D. *Tetrahedron Lett.* **1989,** *30,* 7001.
     (b) Doyle, M. P.; Taunton, J.; Pho, H. Q. *Tetrahedron Lett.* **1989,** *30,* 5397. (c) Doyle, M. P.;
     Pieters, R. J.; Taunton, J.; Pho, H. Q.; Padwa, A.; Hertzog, D. L.; Precedo, L. *J. Org. Chem.*
     1991, *56,* 820.

55. Callot, H. J.; Metz, F.; Piechoki, C. *Tetrahedron,* **1982,** *38,* 2365.

56. O'Malley, S.; Kodadek, T. *J. Am. Chem. Soc.* **1989,** *111,* 9116.

57. O'Malley, S.; Kodadek, T. *Tetrahedron Lett.* **1991,** *32,* 2445.

58. O'Malley, S.; Kodadek, T. *Organometallics,* **1992,** *11,* 2299.

59. (a) Daniewski, A. R.; Kowalczyk-Przewloka, T. *Tetrahedron Lett.* **1982,** *23,* 2411. (b) Daniewski, A. R.; Kowakzyk-Przewloka, T. *J. Org. Chem.* **1985,** *50,* 2976.

60. Matlin, S. A.; Lough, W. J.; Chan, L.; Abram, D. M. H.; Zhou, Z. *J. Chem. Soc., Chem. Commun.* **1984,** 1038.

61. Dauben, W. G.; Hendricks, R. T.; Luzzio, M. J.; Ng, H. P. *Tetrahedron Lett.* **1990,** *31,* 6969.

62. (a) Holland, D.; Laidler, D. A.; Milner, D. J. *J. Mol. Catal.* **1981,** *11,* 119. (b) Holland, D.; Laidler, D. A.; Milner, D. J. *Inorg. Chim. Acta,* **1981,** *54,* L21.

63. Burke, S. T.; Grieco, P. A. *Org. Reactions (N.Y.)* **1979,** *26,* 361.

64. Hirai, H.; Matsui, M. *Agric. Biol. Chem.* **1974,** *40,* 169.

65. Pfaltz, A. Private communication on results by C. Pigué and B. Fähndrich.

66. Doyle, M. P.; Pieters, R. J.; Martin, S. F.; Austin, R. E.; Oalmann, C. J.; Müller, P. *J. Am. Chem. Soc.* **1991,** *113,* 1423.

67. Martin, S. F., Oalmann, C. J.; Liras, S. *Tetrahedron Lett.* **1992,** *33,* 6727.

68. Baird, M. S. *Top. Curr. Chem.* **1988,** *144,* 137.

69. Binger, P.; Buch, H. M. *Top. Curr. Chem.* **1986,** *135,* 77.

70. Protopopova, M. N.; Shapiro, E. A. *Russ. Chem. Rev.* **1989,** *58,* 667.

71. Petiniot, N.; Anciaux, A. J.; Noels, A. F.; Hubert, A. J.; Teyssie, P. *Tetrahedron Lett.* **1978,** 1239.

72. (a) Dowd, P.; Schappert, R.; Garnver, P.; Go, C. L. *J. Org. Chem.* **1985,** *50,* 44. (b) Cho, S. H.; Liebeskind, L. S. *J. Org. Chem.* **1987,** *52,* 2631. (c) O'Bannon, P. E.; Dailey, W. P. *J. Org. Chem.* **1991,** *56,* 2258.

73. Shapiro, E. A.; Romanova, T. N.; Dolgii, I. E.; Nefedov, O. M. *Izv. Alkad. Nauk SSR, Ser. Khim.* **1984,** 2535.

74. Maier, G.; Hoppe, M.; Reisenauer, H. P.; Kruger, C. *Angew. Chem., Int. Ed. Engl.* **1982,** *21,* 437.

75. Wenkert, E.; Alonso, M. E.; Buckwalter, B. L.; Sanchez, E. L. *J. Am. Chem. Soc.* **1983,** *105,* 2021.

76. Müller, P.; Pautex, N.; Doyle, M. P.; Bagheri, V. *Helv. Chim. Acta,* **1990,** *73,* 1233.

77. Padwa, A.; Chiacchio, U.; Garreau, Y.; Kassir, J. M.; Krumpe, K. E.; Schoffstall, A. M. *J. Org. Chem.* **1990,** *55,* 414.

78. Hoye, T. R.; Dinsmore, C. J.; Johnson, D. S.; Korkowski, P. F.; *J. Org. Chem.* **1990,** *55,* 4519.

79. (a) Pincock, J. A.; Moutsokapas, A. A. *Can. J. Chem.* **1977,** *55,* 979. (b) Aldulayymi, J. R.; Baird, M. S. *Tetrahedron,* **1990,** *46,* 5703.

80. Doss, G. A.; Djerassi, C. *J. Am. Chem. Soc.* **1988,** *110,* 8124.

81. Protopopova, M. N.; Doyle, M. P.; Müller, P.; Ene, D. *J. Am. Chem. Soc.* **1992,** *114,* 2755.

82. Hashimoto, S.; Watanabe, N.; Ikegami, S. *Tetrahedron Lett.* **1990,** *31,* 5173.

83. Kennedy, M.; McKervey, M. A.; Maguire, A. R.; Roos, G. H. P. *J. Chem. Soc., Chem. Commun.* **1990,** 361.

84. Doyle, M. P.; van Oeveren, A.; Westrum, L. J.; Protopopova, M. N.; Clayton, Jr., T. W. *J. Am. Chem. Soc.* **1991,** *113,* 8982.

Asymmetric Oxidation

4.1 Catalytic Asymmetric Epoxidation
of Allylic Alcohols
Roy A. Johnson and K. Barry Sharpless

4.2 Asymmetric Catalytic Epoxidation
of Unfunctionalized Olefins
Eric N. Jacobsen

4.3 Asymmetric Oxidation of Sulfides
Henri B. Kagan

4.4 Asymmetric Dihydroxylation
Roy A. Johnson and K. Barry Sharpless

4.1

Catalytic Asymmetric Epoxidation of Allylic Alcohols

Roy A. Johnson

The Upjohn Company
Kalamazoo, Michigan

K. Barry Sharpless

The Scripps Research Institute
La Jolla, California

4.1.1. Introduction

Efforts to achieve asymmetric induction in the epoxidation of olefins commenced in 1965 with a report by Henbest that a low level of enantioselectivity (8%) was achieved in epoxidations using percamphoric acid [1]. A useful level of asymmetric induction remained an elusive goal for 15 years until Katsuki and Sharpless reported that the combination of a titanium (IV) alkoxide, an optically active tartrate ester, and *t*-butyl hydroperoxide was capable of epoxidizing a wide variety of allylic alcohols in good yield and with an enantiomeric excess usually greater than 90% [2]. The titanium–tartrate complex was shown to perform as a true catalyst; the most reactive allylic alcohols were completely oxidized when only 5–10 mol % of the complex was used for the reaction and the level of enantioselectivity achieved under these catalytic conditions was within 1 or 2% of that obtained when a stoichiometric amount of the complex was used. However, with many slower reacting allylic alcohols, a complete reaction was difficult to achieve under catalytic conditions and, as a consequence, a stoichiometric quantity of the complex was used for most asymmetric epoxidations before 1986. In 1986, the addition of activated molecular sieves to the asymmetric epoxidation process was found to have an extremely beneficial effect on the reaction, such that nearly all epoxidations are completed efficiently with only 5–10 mol % of the catalyst [3,4]. In addition to an economy of reagents, the advantages of the catalytic reaction include mild conditions, easier isolations, increased yields especially of very reactive epoxy alcohol products, and application of in situ derivatization of the product. These aspects of the reaction are all discussed further in this chapter.

Success in the use of titanium tartrate catalyzed asymmetric epoxidation depends on the presence of the hydroxyl group of the allylic alcohol. The hydroxyl group enhances the rate of the reaction, thereby providing selective epoxidation of the allylic olefin in the presence of other olefins; it also is essential for the achievement of asymmetric induction. The role played by the hydroxyl group in this reaction is described in a later section of this chapter. The need for a hydroxyl group necessarily limits the scope of this asymmetric epoxidation to a fraction of all olefins. Fortunately, allylic alcohols are easily introduced into synthetic intermediates and are very versatile in organic synthesis. The titanium tartrate catalyzed asymmetric epoxidation of allylic alcohols has enjoyed extensive application as documented in the literature and in this review. The development of methods aimed at catalytic asymmetric epoxidation of unfunctionalized olefins is described in Chapter 4.2, while the catalytic asymmetric dihydroxylation of olefins, which provides an alternate method for olefin functionalization, is described in Chapter 4.4.

The literature was reviewed through early 1990 for the purposes of preparing this chapter. The following sections of this review are reprinted with only very minor changes from an earlier version [5]. Other reviews have covered various aspects of asymmetric epoxidation, including an extensive discussion of the mechanism of the reaction [6], synthetic applications through 1984 [7], a thorough compilation of uses through 1987 [8], and an excellent review of synthetic applications utilizing nonracemic glycidol and related 2,3-epoxy alcohols [9]. Use of enantomerically pure epoxy alcohols in the synthesis of sugars and other polyhydroxylated compounds [10] and for the preparation of various synthetic intermediates has been reviewed [11]. A personal account of the discovery of titanium-catalyzed asymmetric epoxidation has been recorded [12]. A comprehensive review of titanium-catalyzed asymmetric epoxidation is planned [13].

4.1.2. Fundamental Elements of Titanium Tartrate Catalyzed Asymmetric Epoxidation

The essence of titanium-catalyzed asymmetric epoxidation is illustrated in Figure 1. As Figure 1 shows, the four essential components of the reaction are the allylic alcohol substrate, a titanium(IV) alkoxide, a chiral tartrate ester, and an alkyl hydroperoxide. The asymmetric complex formed from these reagents delivers the peroxy oxygen to one face or the other of the allylic alcohol depending on the absolute configuration of the tartrate used. If D-(−)-tartrate is used, oxygen delivery will be from the top face of the allylic alcohol when drawn in the orientation shown in Figure 1, and if L-(+)-tartrate is used, oxygen delivery will be from the bottom face. The enantioselectivity of this reaction approaches 100% as measured by the enantiomeric purities of the epoxy alcohol products. An enantiomeric excess (ee) of 94%, a degree of enantiomeric purity attained in many of the epoxide products, reflects an enantioselectivity of 97:3 for epoxidation of one face of the allylic alcohol over the other.

The enantioselectivity principles portrayed in Figure 1 have been followed with-

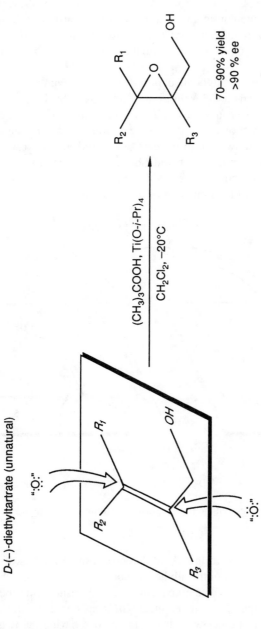

Figure 1 Enantiofacial selectivity in the epoxidation of prochiral allylic alcohols with titanium/tartrate/TBHP.

out exception in all epoxidations of prochiral allylic alcohols reported to date, and one may use these principles to assign absolute configurations to the epoxy alcohols prepared by the method. On the other hand, epoxidation of allylic alcohols with chiral substituents at C-1, C-2, and/or C-3 does not always follow these principles, and assignment of absolute configuration to the products must be made with care. Even in the latter cases, reliable assignments usually can be made if the outcome (diastereomeric ratio) of epoxidation with both the (+) and (−)-tartrate ester ligands is compared.

A structural variant of the allylic alcohol not shown in Figure 1 is encountered when a substituent is placed on C-1 as illustrated in Figure 2. Such an allylic alcohol is a racemate (unless it has been resolved) in which one enantiomer will have the R group oriented in the direction of oxygen delivery while the other enantiomer will have the R group oriented away from the direction of oxygen delivery. The enantioselective principles of asymmetric epoxidation remain in force for epoxidation of this type of substrate, but now oxygen is delivered at different rates to the two enantiomers depending on the orientation of the R group.

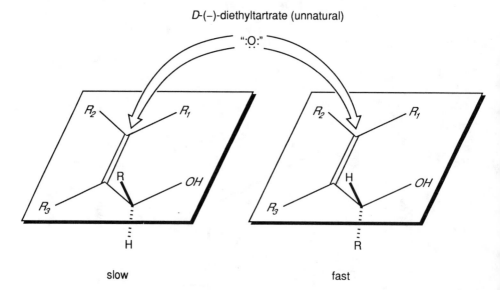

Figure 2 Diastereofacial selectivity in the epoxidation of 1-substituted allylic alcohols with titanium/tartrate/TBHP.

Experimental results have shown that the difference in these rates is of sufficient magnitude that one enantiomer of the allylic alcohol will remain largely unoxidized while the other undergoes complete epoxidation, the net result being the achievement of a kinetic resolution of the enantiomers [14]. Experience has further shown that the slow-reacting enantiomer will always be the one having the R group oriented in the direction of "oxygen" delivery. For the example illustrated in Figure 2, the titanium/D-(−)-diethyl tartrate complex will deliver oxygen to the top face in

Table 1 Compatibility of Functional Groups with the
Asymmetric Epoxidation Reaction

Compatible Functional Groups		Incompatible Groups
Acetals, ketals	Nitriles	Amines (most)
Acetylenes	Nitro	Carboxylic Acids
Alcohols (remote)	Olefins	Mercaptans
Aldehydes	Pyridines	Phenols (most)
Amides	Silyl ethers	Phosphines
Azides	Sulfones	
Carboxylic Esters	Sulfoxides	
Epoxides	Tetrazoles	
Ethers	Ureas	
Hydrazides	Urethans	
Ketones		

preference to the bottom face of the substrate in accordance with the rules implied in Figure 1, and this delivery will be more rapid when the R group is oriented toward the bottom face of the molecule. Opposite results will be obtained with the titanium/L-(+)-diethyl tartrate complex. Additional details for using this reaction in the kinetic resolution mode may be found in a later section.

An important aspect of asymmetric epoxidation not shown in Figures 1 and 2 is the fact that the allylic alcohol is coordinated to titanium as the alkoxide during the epoxidation process (see Section 4.3.1). Not only does this coordination play a key role in orientation of the allylic alcohol during the epoxidation process, but it also accounts for the selectivity of the process for allylic and homoallylic alcohols in preference to nearly all other olefins. This effect is most clearly seen in comparison of allylic alcohols with the analogous allylic ethers. The latter are essentially unchanged by the $Ti(OR)_4$/tartrate/TBHP system during the same time required for epoxidation of the allylic alcohol. The $Ti(OR)_4$/tartrate/TBHP reagent thereby exhibits selectivity for allylic and homoallylic alcohols while being compatible with other olefinic groups. Use of the reagent is compatible with many other functional groups as well (see Table 1).

An important improvement in the asymmetric epoxidation process is the finding, reported in 1986, that by adding molecular sieves to the reaction medium virtually all reactions can be performed with a catalytic amount (5–10 mol %) of the titanium–tartrate complex [3]. Previously, only a few structural classes of allylic alcohols were efficiently epoxidized by less than stoichiometric amounts of the complex, and most reactions were routinely performed with stoichiometric quantities of the reagent.

In situ derivatization of the crude epoxy alcohol product becomes a viable alternative to isolation when 5–10 mol % of catalyst is used for the epoxidation. This procedure is especially useful when the product is reactive or is difficult to isolate because of solubility in an aqueous extraction phase [15,16]. Low molecular weight epoxy alcohols such as glycidol are readily extracted from the reaction mixture after conversion to ester derivatives such as the *p*-nitrobenzoate or 3-nitrobenzenesulfonate [4,17]. This derivatization not only facilitates isolation of the product but also preserves the epoxide in a synthetically useful form.

4.1.3. Reaction Variables for Titanium Tartrate Catalyzed Asymmetric Epoxidation

This section presents a summary of the currently preferred conditions for performing titanium-catalyzed asymmetric epoxidations and is derived primarily from the detailed account of Gao et al. [4]. We wish to draw the reader's attention to several aspects of the terminology used here and throughout this chapter. The terms titanium–tartrate *complex* and titanium–tartrate *catalyst* are used interchangeably. The term *stoichiometric reaction* refers to the use of the titanium–tartrate complex in a stoichiometric ratio (100 mol %) relative to the substrate (allylic alcohol). The term *catalytic reaction* (or *quantity*) refers to the use of the titanium–tartrate complex in a catalytic ratio (usually 5–10 mol %) relative to the substrate.

4.1.3.1. Stoichiometry

Two aspects of stoichiometry are important in asymmetric epoxidation: one is the ratio of titanium to tartate used for the catalyst and the other is the ratio of catalyst to substrate. With regard to the catalyst, it is crucial to obtaining the highest possible enantioselectivity that at least a 10% excess of tartrate ester to titanium(IV) alkoxide be used in all asymmetric epoxidations. This is important whether the reaction is being done with a stoichiometric or a catalytic quantity of the complex. There appears to be no need to increase the excess of tartrate ester beyond 10–20% and, in fact, a larger excess has been shown to slow the epoxidation reaction unnecessarily [4].

The second stoichiometry consideration is the ratio of catalyst to substrate. As noted in the preceding section, virtually all asymmetric epoxidations can be performed with a catalytic amount of titanium–tartrate complex if molecular sieves are added to the reaction milieu. A study of catalyst/substrate ratios in the epoxidation of cinnamyl alcohol revealed a significant loss in enantioselectivity (Table 2) below the level of 5 mol % catalyst. At this catalyst level, the reaction rate also decreases, with the consequence that incomplete epoxidation of the substrate may occur. Presently, the recommended catalyst stoichiometry is from 5% Ti and 6% tartrate ester to 10% Ti and 12% tartrate ester [4].

4.1.3.2. Concentration

The concentration of substrate used in the asymmetric epoxidation must be given consideration because competing side reactions may increase with increased reagent concentration. The use of catalytic quantities of the titanium–tartrate complex has greatly reduced this problem. The epoxidation of most substrates under catalytic conditions may be performed at a substrate concentration up to 1 M. By contrast,

Table 2 Dependence of Enantioselectivity on Catalyst Stoichiometry

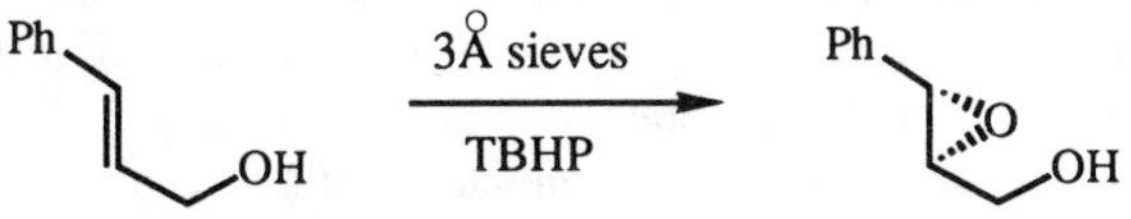

Entry	Ti(O-i-Pr)$_4$, mole %	(+)-DIPT mole %	% ee
1	5.0	6.0	92
2	4.0	5.2	87
3	2.0	2.5	69

epoxidations using stoichiometric amounts of complex are best run at substrate concentrations of 0.1 M or lower. Even with catalytic amounts of the complex, a concentration of 0.1 M may be maximal for substrates such as cinnamyl alcohol, which produce sensitive epoxy alcohol products [4].

4.1.3.3. Preparation and Aging of the Catalyst

Proper preparation of the catalyst is essential for optimal reaction rates and enantioselectivity. The preparation and storage of stock solutions of the titanium–tartrate catalyst should not be attempted because the complex is not sufficiently stable for long-term storage. Best results are obtained when the catalyst is prepared by mixing the titanium(IV) alkoxide and the tartrate in solvent at −20°C, adding either TBHP *or* the allylic alcohol, and aging the system at this temperature for 20–30 minutes. This aging period is critical to the success of the reaction and must not be eliminated. On the rare occasion that a bulky titanium(IV) alkoxide such as the *t*-butoxide is used, the aging period should be increased to one hour [18]. After the aging period, the temperature is adjusted to the desired level and the last reagent, either the allyl alcohol or the hydroperoxide, is added.

4.1.3.4. Oxidant and Epoxidation Solvent

t-Butylhydroperoxide (TBHP) is used as the oxidant for nearly all titanium-catalyzed asymmetric epoxidations. Exceptions are for allyl alcohol and methallyl alcohol, where cumylhydroperoxide is used to advantage for the epoxidation [4]. Cumylhydroperoxide can be used for other epoxidations and is reported to result in slightly faster reaction rates than are observed with TBHP [4]. Trityl hydroperoxide can also serve as an effective replacement for TBHP [6]. TBHP is generally preferred, however, since product isolation is significantly easier when this oxidant is used. The most economical source of TBHP is the commercially available 70% solution in water in which case steps must be taken to obtain anhydrous material. Detailed instructions for obtaining dry solutions of TBHP have been published elsewhere [3,4]. For smaller laboratory scale reactions, anhydrous solutions of TBHP in 2,2,4-trimethylpentane (isooctane) are available commercially. Storage of TBHP solutions over molecular sieves is not recommended, but brief drying over sieves (ca. 30 min) of the required amount of the solution just before use is a good practice.

Since the preparation and storage of stock quantities of TBHP is a convenient way to deal with this reagent, compatibility with the solvent is essential. Much care has gone into finding the optimum solvent for storage of TBHP, and recommendations have changed as additional experience has been gained. The current solvent of choice is isooctane, with the favored alternates being dichloromethane or toluene [4]. Dichloroethane should not be used [19]. Dichloromethane solutions of TBHP require storage at 0°C, and toluene solutions occasionally develop a contaminant that inhibits the catalytic reaction. Safety considerations (chance of slight pressurization) make high density polyethylene bottles preferable to glass bottles for storage of TBHP solutions. However, both dichloromethane and toluene, but not isooctane, permeate through such bottles, with the result that the concentration of the contents slowly changes with time. If the published instructions [3,4] for preparation of anhydrous TBHP in isooctane are followed, a relatively concentrated solution (5–6 M) is obtained. Aliquots of this solution are briefly dried over sieves and added directly to the epoxidation reaction without concern for removal of the isooctane. The use of dilute solutions of TBHP in isooctane should be avoided, since the additional isooctane involved in transfer will have an inhibitory effect on the rate of epoxidation and can lead to solubility problems with some substrates. Solutions of 5.5 M TBHP in isooctane are available commercially and should *always* be used instead of the 3.0 M solution.

For the asymmetric epoxidation reaction, dry alcohol-free dichloromethane (the use of dichloromethane stabilized with methanol must be avoided) is usually the solvent of choice: it is inert to the reagents, has good solvent power for the components of the reaction, and supports good epoxidation rates. A fortunate consequence of the asymmetric epoxidation process is that ligation of the allylic alcohol to the titanium center aids in solubilization of the substrate. Substrates that normally may be only modestly soluble in the above-mentioned solvents will be brought into solution as they complex with the titanium–tartrate catalyst.

4.1.3.5. Tartrate Esters

Optically active tartrate esters are the source of chirality for the asymmetric epoxidation process. With a few subtle exceptions, the esters used conventionally— dimethyl (DMT), diethyl (DET), and diisopropyl tartrate (DIPT)—are equally effective at inducing asymmetry during the crucial epoxidation event. The minor exceptions that have been noted include (a) a slight improvement in enantioselectivity (from 93 to 95% ee) when changing from DIPT to DET in the epoxidation of

E-substituted allylic alcohols such as (*E*)-2-hexen-1-ol (having only a primary alkyl chain at C-3) and (b) a higher product yield (but no change in enantioselectivity) when changing from DET to DIPT in the epoxidation of allyl alcohol [4]. Other subtle variations such as these may exist, but their discovery awaits execution of the appropriate comparative experiments. If optimal conditions are desired for a specific asymmetric epoxidation, variation of the tartrate ester is likely to be a useful exercise.

In the kinetic resolution of chiral 1-substituted allylic alcohols, there clearly is benefit to be gained in the choice of tartrate ester used for the reaction. In these reactions, the efficiency of kinetic resolution increases as the size of the tartrate alkyl ester group increases. Data for DMT, DET, and DIPT are summarized below (see Table 8 [6]), and the trend shown there continues with the use of the crystalline dicyclohexyl and dicyclododecyl tartrates [4].

The nonconventional tartrate esters **1–3** have been used to probe the mechanism of the asymmetric epoxidation process [20a]. These chain-linked bistartrates when complexed with 2 equiv. of Ti(O-*t*-Bu)$_4$ catalyze asymmetric epoxidation with good enantiofacial selectivity.

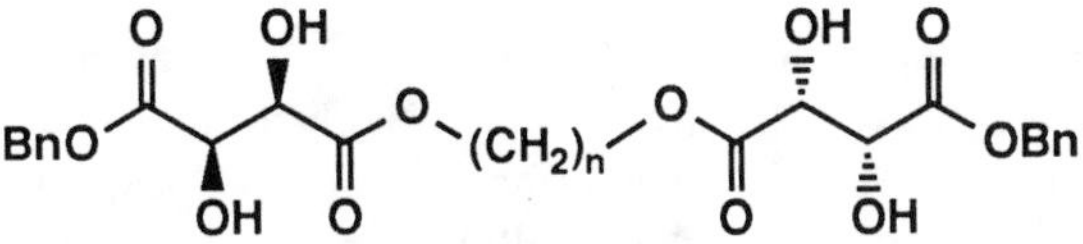

1, n = 3
2, n = 4
3, n = 5

A number of tartrate-like ligands have been studied as potential chiral auxiliaries in the asymmetric epoxidation and kinetic resolution processes [6,20b]. Although on occasion a ligand has been found that has the capability to induce high enantioselectivity into selected substrates (see Section 4.1.7.3), none has exhibited the broad scope of effectiveness seen with the tartrate esters.

Polymer-linked tartrate esters have been prepared and used for asymmetric epoxidation in efforts to simplify reaction workup procedures and to allow recycling of the chiral tartrate [21]. The tartrates were linked through an ester bond to either a hydroxymethyl or a hydroxyethyl group on the polymer backbone to form **4** and **5**, respectively.

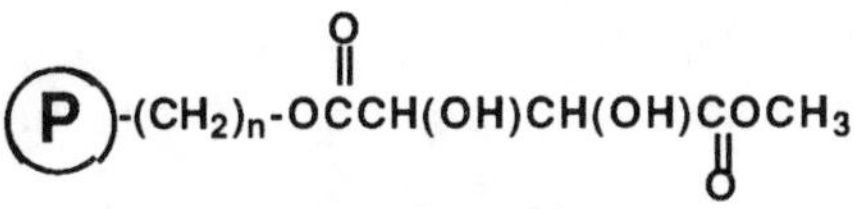

4, n = 1
5, n = 2

Epoxidation catalysts were prepared from these polymer linked tartrates by combination with 0.5 equiv. of Ti(O-*i*-Pr)$_4$, based on the weight of tartrate ester which had been added to the polymer. Epoxidation of geraniol with **4** or **5** gave epoxy alcohol with 49 and 65% ee, respectively. Recycling of the polymer-linked tartrate was possible, but the subsequent epoxidation suffered from significant loss in enantioselectivity [21].

4.1.3.6. Titanium Alkoxides

Titanium(IV) isopropoxide [*Chemical Abstracts* nomenclature: 2-propanol, titanium (4+) salt] is the titanium species of choice for preparation of the titanium–tartrate complex in the asymmetric epoxidation process. The use of titanium(IV) *t*-butoxide has been recommended for reactions in which the epoxy alcohol product is particularily sensitive to ring opening by the alkoxide [18]. The 2-substituted epoxy alcohols are one such class of compounds. Ring opening by *t*-butoxide is much slower than by *i*-propoxide. With the reduced amount of catalyst that now is sufficient for all asymmetric epoxidations, the use of Ti(O-*t*-Bu)$_4$ appears to be unnecessary in most cases, but the concept is worth noting.

4.1.3.7. Molecular Sieves

The addition of activated molecular sieves (zeolites) to the asymmetric epoxidation milieu has the beneficial effect of permitting virtually all reactions to be carried out with only 5–10 mol % of the titanium–tartrate catalyst [3,4]. Without molecular sieves, only a few of the more reactive allylic alcohols are epoxidized efficiently with less than an equivalent of the catalyst. The role of the molecular sieves is thought to be protection of the catalyst from (a) adventitious water and (b) water that may be generated in small amounts by side reactions during the epoxidation process.

There are several important guidelines to be followed in using activated molecular sieves for the asymmetric epoxidation process [4]. Stock solutions of TBHP should not be stored over molecular sieves (the sieves catalyze the slow decomposition of TBHP), but the amount of TBHP solution required for a reaction should be placed over sieves briefly (10–60 min) before use. Likewise, neither the tartrate ester nor the titanium(IV) isopropoxide should be stored over sieves. Addition of the sieves at the time of mixing the tartrate ester with the Ti(O-*i*-Pr)$_4$ followed by the normal "aging" of the catalyst is sufficient to dry these reagents, provided they initially are of good quality (see Ref. 4, p. 5771). Powdered, activated 4A molecular sieves, commercially available in preactivated form, are preferred; 3A, 4A, and 5A molecular sieves in pellet form are also effective. Only 3A sieves are effective in the case of allyl alcohol, since this substrate is small enough to be sequestered by 4A or 5A sieves. Unactivated sieves can be activated by heating at 200°C under high vacuum for at least 3 hours.

4.1.4. Sources of Allylic Alcohols

One of the amenities of present-day organic synthesis is the availability of intermediates from the many chemical supply companies. More than 100 allylic alcohols (excluding extensive listings of phorbol esters and prostaglandin structures) are offered for sale from these sources. Two concerns about such supplies should be noted. The first is the E/Z composition of acyclic allylic alcohols, which should be checked when it is not specified, and second is the optical purity of allylic alcohols offered in optically active form, which likewise should be checked.

When the allylic alcohol needed for asymmetric epoxidation is unavailable from a commercial source, reasonably general synthetic routes have been developed to allylic alcohols of several different substitution patterns. Good methods are available for the preparation of 3-substituted allylic alcohols, whereas synthesis of 2-substituted allylic alcohols is more problematic. The substrates for kinetic resolution, 1-substituted allylic alcohols, frequently can be derived by addition of alkenyl or alkynyl organometallic reagents to aldehydes followed by modification of the resulting product as required.

The Horner–Emmons addition of dialkyl carboalkoxymethylenephosphonates to aldehydes [22] has been widely used to generate α,β-unsaturated esters which, in turn, can be reduced to allylic alcohols. Under the original conditions of the Horner–Emmons reaction, the stereochemistry of the α,β-unsaturated ester is predominantly trans and therefore the *trans* allylic alcohol is obtained upon reduction. Still and Gennari have introduced an important modification of the Horner–Emmons reaction which shifts the stereochemistry of the α,β-unsaturated ester to predominantly cis [23]. Diisobutylaluminum hydride (DIBAL) has frequently been used for reduction of the alkoxycarbonyl to the primary alcohol functionality. The aldehyde needed for reaction with the Horner–Emmons reagent may be derived via Swern oxidation [24] of a primary alcohol. The net result is that one frequently sees the reaction sequence shown in equation 1 used for the net preparation of $3E$ and $3Z$ allylic alcohols.

(1)

The propargylic alcohol group may be exploited as an allylic alcohol precursor (eq. 2) and may be generated by nucleophilic addition to an electrophile [25] or by addition of a formaldehyde equivalent to a preexisting terminal acetylene group [26]. Once in place, reduction of the propargylic alcohol with lithium aluminum hydride or, preferably, with sodium bis(2-methoxyethoxy)aluminum hydride (Red-Al) [27] will produce the *trans* allylic alcohol. Alternately, catalytic reduction over Lindlar catalyst can be used to obtain the *cis* allylic alcohol [28]. The addition of other lithium acetylides to ketones produces chiral secondary alcohols, which also can be reduced by the preceding methods to the *cis* or *trans* allylic alcohols. Additional synthetic approaches to allylic alcohols may be found in the various references cited in this chapter.

$$R\text{-}X \quad + \quad HC \equiv CCH_2OR \quad \longrightarrow$$

$$R\text{-}C \equiv CCH_2OH \tag{2}$$

$$R\text{-}C \equiv CH \quad + \quad \text{"}H_2CO\text{"} \quad \longrightarrow$$

4.1.5. Asymmetric Epoxidation by Substrate Structure

The scope of allylic alcohol structures that are subject to asymmetric epoxidation was foreshadowed in the first report of this reaction. Examples of nearly all the possible substitution patterns were shown to be epoxidized in good yield and with high enantiofacial selectivity [2]. The numerous results that have appeared since the initial report have confirmed and extended the scope of the structures that have been epoxidized. This section of the chapter illustrates the structural scope without being exhaustive in coverage of the literature. Examples were chosen as much as possible from the reports in the literature that provide experimentally determined yield and % ee data. When there are limitations to the structural scope as reflected by lower enantiofacial selectivity, these cases are noted. The results presented in this section are divided according to the substitution patterns of the allylic alcohol substrates. This organization is intended to provide easy access to precedent when the synthetic chemist is contemplating asymmetric epoxidation of a new substrate.

Before commencing, the attention of the reader is drawn to the terms "enantiofacial selectivity" and "diastereoselectivity." The usage in this chapter does not conform to the strictest possible definitions of these terms. In particular, *enantiofacial selectivity* is used with reference to the selection and delivery of oxygen by the epoxidation catalyst to one face of the olefin in preference to the other. This usage extends to chiral allylic alcohols (primarily the 1-substituted allylic alcohols) when the focus of the discussion is on face selection in the epoxidation process. *Diastereoselectivity* is used in the discussion of kinetic resolution when the generation of diastereomeric compounds is emphasized.

4.1.5.1. Allyl Alcohol

Glyceraldehyde derivatives [29], asymmetrically substituted glycerol [30], and glycidol [31] are three-carbon molecules which, especially in their optically active forms, find widespread use in organic synthesis. In the past, these compounds in optically active form came almost exclusively from the degradation of natural products such as mannitol. Efficient, multistep routes from the natural products provide access to either enantiomer of these three-carbon compounds. Since the discovery of asymmetric epoxidation in 1980, the potential has existed for a convenient one-step synthesis of optically active glycidol (7) from allyl alcohol (6) [2]. However, because glycidol is one of the epoxy alcohols more sensitive to ring-opening reactions and also is a water-soluble molecule, isolation from the stoichiometric asymmetric epoxidation is difficult, and very little glycidol has been prepared in this way. Now with the use of catalytic epoxidation in the presence of molecular sieves, it is possible to isolate optically active glycidol with 88–92% ee in yields of 50–60% [4]. As a result of these improvements, both enantiomers of glycidol are available commercially.

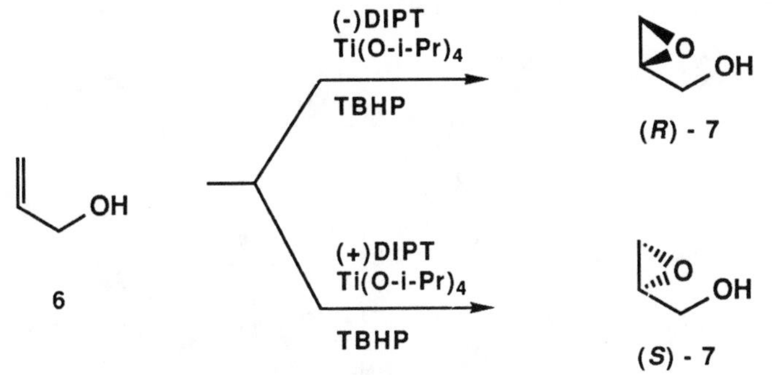

An attractive alternative to isolation of glycidol is in situ derivatization of the crude product during workup [15]. Two distinct applications of this method have been described. In the first, ring opening of glycidol (*R*)-**7** with a nucleophile such as sodium 1-naphthoxide produces an intermediate (**8**) that can be carried on to useful products—for example, for the synthesis of β-adrernergic blocking agents [15a] and antidepressants [32]. In the second, esterification of the hydroxyl group of glycidol improves the extraction of the glycidol moiety from the reaction mixture and at the same time generates a synthon in which all three carbon centers are differentiated for further reaction. Another benefit is that with certain derivatives, such as the 3-nitrobenzenesulfonate ester (**9**), recrystallization can be used to upgrade the enantiomeric purity (to > 99% ee) [4,16a].

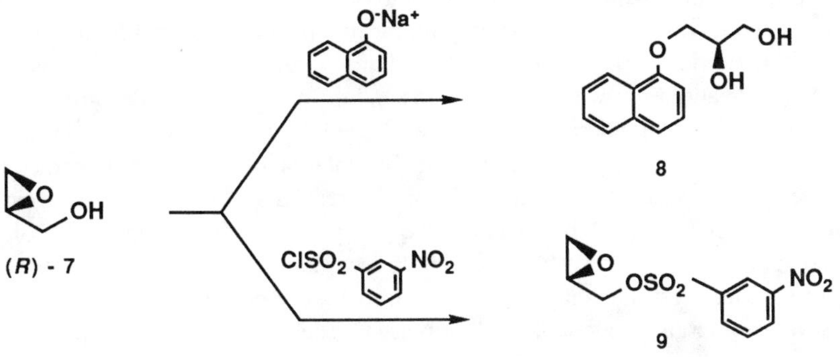

As an industrial process, production of optically active glycidol is at an early stage of development, with additional improvements and economies certain to occur. As a chemical intermediate, optically active glycidol is the most versatile epoxy alcohol prepared by asymmetric epoxidation and is poised for exploitation in organic synthesis [9,16].

4.1.5.2. 2-Substituted Allyl Alcohols

The epoxides (**11**) derived from 2-substituted allylic alcohols (**10**) are particularily susceptible to nucleophilic attack at C-3, a reaction that is promoted by titanium(IV) species [18]. When stoichiometric amounts of titanium–tartrate complex are used in these epoxidations, considerable product is lost via opening of the epoxide before it can be isolated from the reaction.

The primary nucleophilic culprit is the isopropoxide ligand of the Ti(O-i-Pr)$_4$. The use of Ti(O-t-Bu)$_4$ in place of Ti(O-i-Pr)$_4$ has been prescribed as a means to reduce this problem (the t-butoxide being a poorer nucleophile) [18]. Fortunately a better solution now exists in the form of the catalytic version of the reaction which uses only 5–10 mol % of titanium–tartrate complex and greatly reduces the amount of epoxide ring opening. Some comparisons of results from reactions run under the two sets of conditions are possible from the epoxidations summarized in Table 3 [2,4,18,33–38].

The prototype for this structural class is 2-methyl-2-propen-1-ol (methallyl alcohol), from which asymmetric epoxidation generates optically active 2- methyloxiranemethanol. Like glycidol, 2-methyloxiranemethanol has been difficult to obtain by stoichiometric asymmetric epoxidation, but with the use of the catalytic version, reasonable quantities now are produced [4] and the compound has become commercially available. In situ derivatization also can be used to recover this epoxy alcohol from the epoxidation reaction. Progress in the isolation of 2-methyloxiranemethanol is reflected in entries 1–3 of Table 3 and the results of in situ derivatization are revealed by entries 4–6. The enantiomeric purity of 2-methyloxiranemethanol produced in this way is very good (92–95% ee), and improvement to 98% ee is observed after recrystallization of the 4-nitrobenzoate derivative.

Several other allylic alcohols with primary C-2 substituents have been epoxidized with very good results (entries 7–10, 14). Epoxy alcohols have been obtained with 95–96% ee and, when the catalytic version of the reaction is used, as in entry 10, the yield is excellent. When the C-2 substituent is more highly branched, as in entries 11–13, there may be some interference with high enantiofacial selectivity by the bulky group, since the enantioselectivity in two cases (entries 11 and 12) is 86%. Another example that supports this possibility of steric interference to selective

Table 3 Epoxides from 2-Substituted Allylic Alcohols

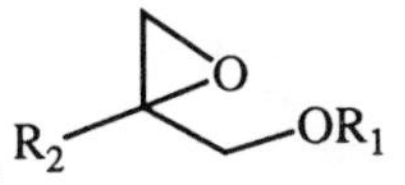

$$R_2 \diagdown \overset{O}{\diagup} \diagdown OR_1$$

Entry	Epoxide R_1	R_2	Catalyst %Ti/%Tart.	Tartrate	Yield(%)	% ee	Ref.
1	H	Me	100/100[a]	(+)-DET	-	85	33
2	H	Me	27/27	(-)-DET	32	94	34
3	H	Me	7.6/10[a]	(-)-DET	47	>95	35
4	PNB	Me	5/6	(+)-DIPT	78	92(98)[b]	4
5	Tos	Me	5/6	(+)-DIPT	69	95	4
6	Nps[c]	Me	5/6	(+)-DIPT	60	(92)[b]	4
7	H	n-Pr	4.7/5.9	(+)-DET	88	95	4
8	H	N-Nonyl	100/110	(+)-DET	53	>96	36
9	H	n-Tetradecyl	100/110[a]	(+)-DET	51	95	18
10	H	n-Tetradecyl	10/13	(+)-DET	91	96	4
11	H	i-Pr	65/120	(+)-DET	56	86	37
12	H	t-Bu	120/150[a]	(+)-DET	42	86	38
13	H	Cyclohexyl	100/100	(+)-DET	81	>95	2
14	H	-CH$_2$OBn	7.6/10[a]	(-)-DET	74	>95	35

[a]Ti(O-t-Bu)$_4$ used in this reaction. [b]Value in parenthesis is after recrystallization.
[c]Nps = 2-Naphthalenesulfonyl.

epoxidation is summarized in equation 3a [39]. In this case the optically active allylic alcohol **12**, (3R)-3,7-dimethyl-2-methylene-6-penten-1-ol, was subjected to epoxidation with both antipodes of the titanium–tartrate catalyst. With (+)-DIPT, enantiofacial selectivity was 96:4 ("matched pair" [40a]), but with (−)-DIPT selectivity fell to only 1:3 ("mismatched pair"), a further indication that a secondary C$_2$ substituent can perturb the fit of the substrate to the active catalyst species.

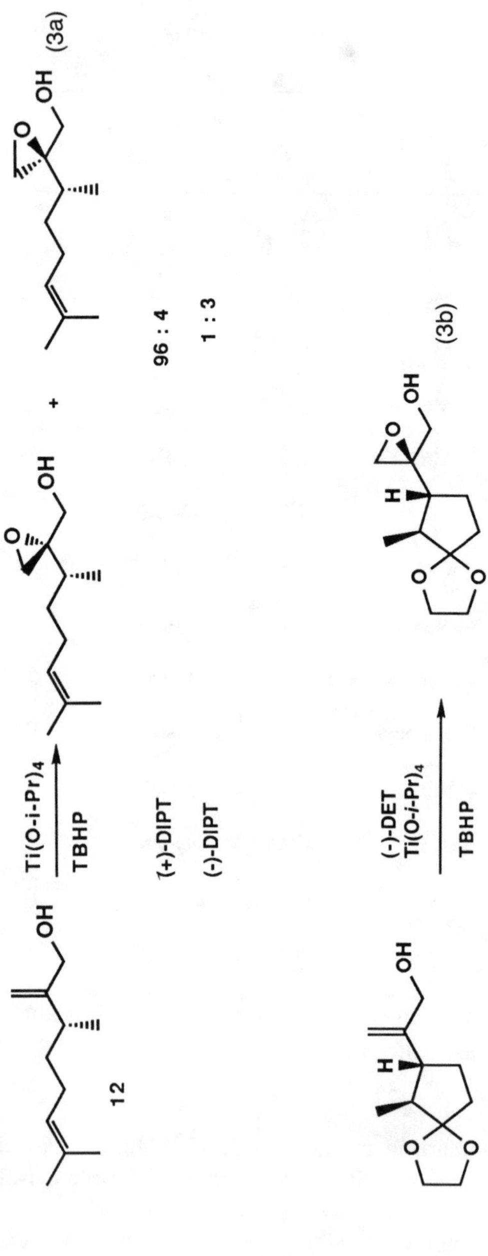
12
Ti(O-i-Pr)$_4$
TBHP
(+)-DIPT
(-)-DIPT
96 : 4
1 : 3
(3a)
(-)-DET
Ti(O-i-Pr)$_4$
TBHP
(3b)

In the epoxidation of the allylic alcohol shown in equation 3b, the epoxy alcohol is obtained in 96% yield and with a 14:1 ratio of enantiofacial selectivity [40b]. An interesting alternate route to the epoxide of entry 12 (Table 3) has been described in which 2-t-butylpropene is first converted to an allylic hydroperoxide via photooxygenation and then, in the presence of titanium–tartrate catalyst, undergoes asymmetric epoxidation (79% yield, 72% ee) [38b]. The intermediate hydroperoxide serves as the source of oxygen for the epoxidation step.

4.1.5.3. (3E)-Substituted Allyl Alcohols

Several factors contribute to the frequent use of (3E)-substituted allylic alcohols (13) for asymmetric epoxidation: (a) The allylic alcohols are easily prepared; (b) conversion to epoxy alcohol normally proceeds with good chemical yield and with better than 95% ee; (c) a large variety of functionality in the (3E) position is tolerated by the epoxidation catalyst. Representative epoxy alcohols (14) are summarized in Table 4 [2,4,18,41–53] and Figure 3 (4,54–61], with results divided arbitrarily according to whether the (3E) substituent is a hydrocarbon (Table 4) or otherwise (Fig. 3). The versatility of these and other 3-substituted epoxy alcohols for organic synthesis is illustrated with several examples in the following discussion.

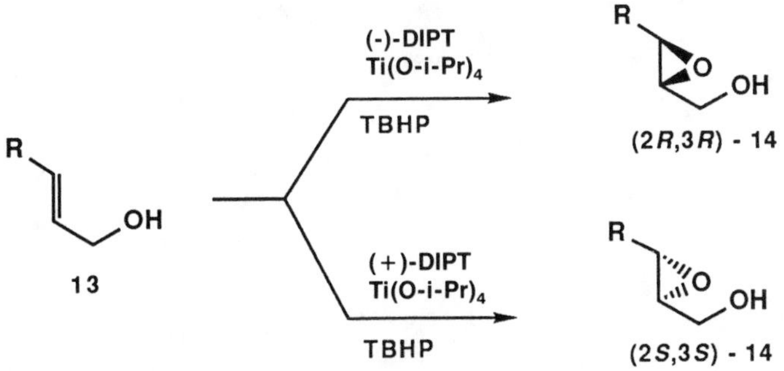

Compatibility of asymmetric epoxidation with acetals, ketals, ethers, and esters has led to extensive use of allylic alcohols containing these groups in the synthesis of polyoxygenated natural products. One such synthetic approach is illustrated by the asymmetric epoxidation of 15, an allylic alcohol derived from (S)-glyceraldehyde acetonide [59,62]. In the epoxy alcohol (16) obtained from 15, each carbon of the five carbon chain is oxygenated, and all stereochemistry has been controlled. The structural relationship of 16 to the pentoses is evident, and methods leading to these carbohydrates have been described [59,62a].

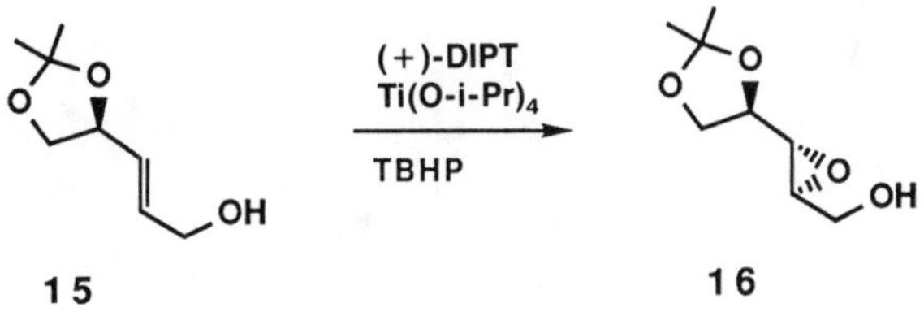

This synthetic methodology has been extended by the development of an efficient series of reactions that can transform one allylic alcohol into a second which is two carbons longer than the first. Repetition of the reaction sequence can, in principle, be continued to any desired chain length.

The key steps in this "reiterative two-carbon extension cycle" are illustrated in Figure 4, which shows a synthetic route leading from an achiral alkoxyacetaldehyde (**17**) to L-allose, one of the eight possible (L)-hexoses [63]. In practice, all eight L-hexoses were synthesized by taking advantage of branch points in the scheme and by using both antipodes of the titanium–tartrate catalyst to generate epimeric epox-

Table 4 Epoxides from (3E)-Substituted Allylic Alcohols
(Hydrocarbon Substituents)

Entry	Epoxide R	Catalyst mole %Ti/%Tart.	Tartrate	Epoxide config.	Yield(%)	% ee	Ref.
1	CH_3	100/100	(-)-DIPT	R,R	40-58	95	41-43
2	CH_3	5/6	(+)-DIPT	S,S	70	92	4
3	C_2H_5	stoich.	(-)-DIPT	R,R	80	>95	44
4	n-C_3H_7	100/104	(+)-DET	S,S	64	93	45
5	i-C_3H_7	5/6	(+)-DET	S,S	85	94	4
6	i-C_3H_7	100/104	(+)-DET	S,S	66	98	45
7	s-C_4H_9	-	(-)-DET	R,R	a	a	46
8	t-C_4H_9	120/150	(+)-DET	S,S	52	>95	38
9	CH_2=CH	5/6	(+)-DIPT	S,S	56	>91	47
10	CH_3CH=$CHCH_2$	8/10	(+)-DET	S,S	81	a	48
11	n-C_5H_{11}	a	(+)-DET	S,S	78	95	49
12	CH_2=$CH(CH_2)_3$	100/100	(+)-DET	S,S	80	>95	41
13	n-C_7H_{15}	5/7.3	(+)-DET	S,S	99	96	4
14	n-C_8H_{17}	5/6	(+)-DET	S,S	78	94	4
15	C_2H_5CH=$CHCH_2CH$=$CHCH_2$	a	(+)-DET	S,S	82	>95	46
16	C_2H_5C≡CCH_2C≡CCH_2	a	(-)-DET	R,R	76	a	50
17	n-$C_{10}H_{21}$	100/100	(+)-DET	S,S	79	>95	2
18	n-$C_{12}H_{25}$	a	(+)-DET	S,S	a	a	51
19	$C_{14}H_{29}$	a	(+)-DIPT	S,S	77	a	52
20	$C_{15}H_{31}$	120/160	(-)-DET	R,R	88	>95	53

[a]Not reported.

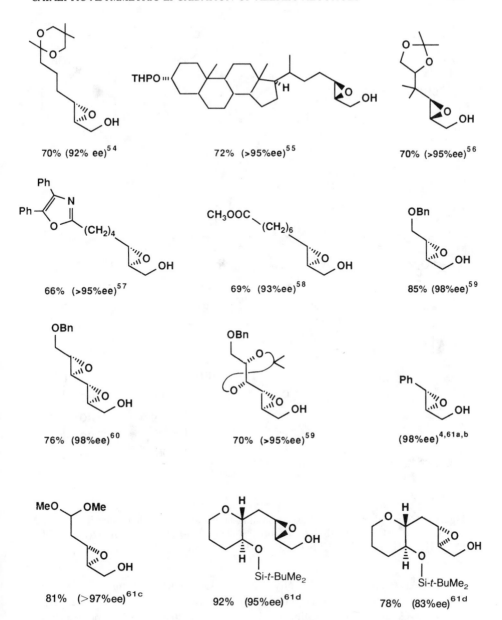

Figure 3 Epoxy alcohols from asymmetric epoxidation of (3*E*)-monosubstituted allylic alcohols.

ides. The sequence of reactions begins with the two-carbon benzyloxyacetaldeyde **17**, which can be converted to the four-carbon intermediate **18** by means of a Wittig reaction (step a). In the actual synthesis, intermediate **18** was the starting point and was obtained by an alternate method. The carboxylic acid ester of **18** is reduced

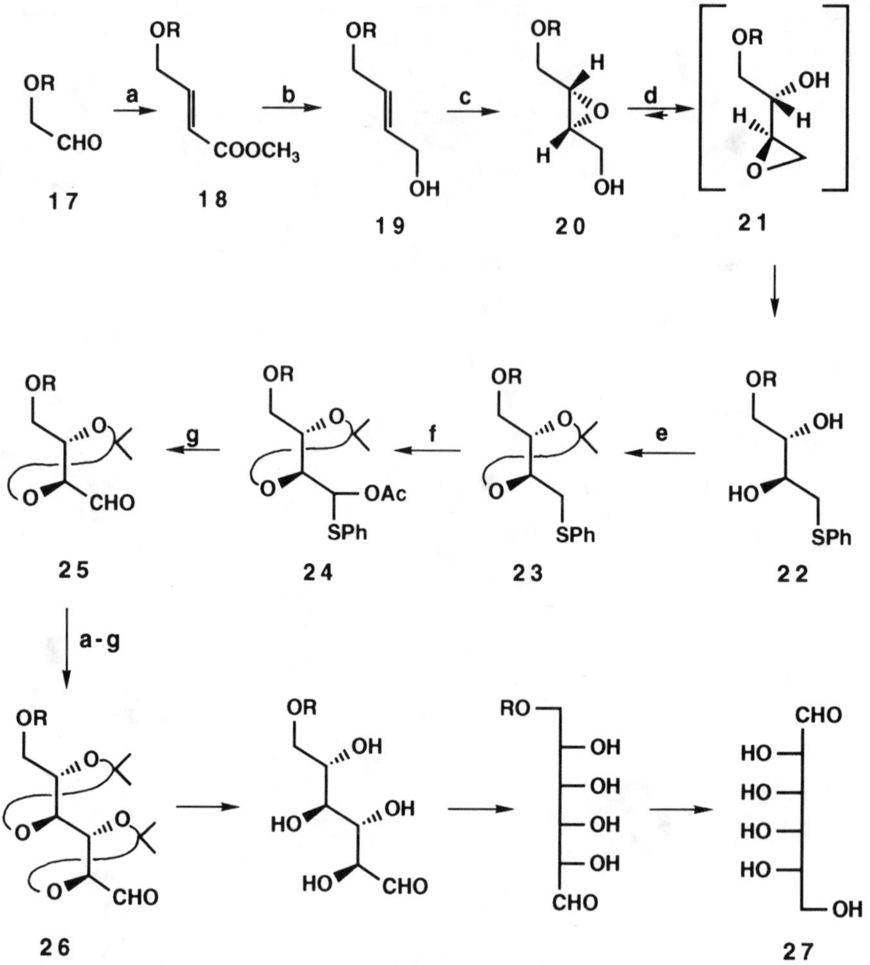

Figure 4 Synthetic route to L-allose illustrating a reiterative two-carbon extension cycle: (a) $(CH_3O)_3P(O)CHCOOCH_3$, (b) DIBAL, (c) (+)-DET/Ti(O-i-Pr)$_4$, (d) PhSH/OH$^-$, (e) CH_3 $C(=CH_2)OCH_3/H^+$, (f) m-CPBA, and (g) DIBAL (1 equiv.); H_2O.

with DIBAL (step b) to the (3E) allylic alcohol **19** which, by asymmetric epoxida-tion (step c) is converted to epoxy alcohol **20**. Base-catalyzed (Payne) rearrange-ment of **20** establishes the equilibrium shown between **20** and the 1,2-epoxy alcohol **21**. Phenylthiolate reacts regioselectively to open the 1,2-epoxide (step d), leading to the dihydroxysulfide **22**. The diol is protected by conversion (step e) to the acetonide **23**, which, upon oxidation of the sulfide to a sulfoxide followed by Pummerer rearrangement (step f), is converted to the acetoxythioacetal **24**. Reduc-tion of the latter (**24**) with one equivalent of DIBAL (step g) produces aldehyde **25**. At this point the synthetic sequence can be branched by converting a portion of the aldehyde **25** to the epimeric aldehyde (not shown) by epimerization with potassium carbonate in methanol. Both these new aldehydes can now be chain extended by repeating steps a–g, which in the case of **25** leads to the hexose derivative **26**. To obtain all eight hexoses, a further branching during the second cycle is initiated at step c with part of the material (an allylic alcohol) being subjected to asymmetric epoxidation with (−)-DET. Both these branches are carried on through step f or g, thereby producing all eight L-hexose derivatives. Deprotection of the derivatives completes the syntheses as shown for L-allose (**27**) in Figure 4.

1,2-Epoxy-3-alcohols can be derived from 2,3-epoxy-1-alcohols by the base-catalyzed Payne rearrangement as illustrated in step d of Figure 4 [59,64]. The rearrangement is completely stereospecific but, since it is reversible, it usually results in an equilibrium mixture of the two epoxy alcohols for which the relative proportions are structure dependent. Practical synthetic applications of this rearrangement therefore depend on methods that will shift the equilibrium completely in the direction desired. Nucleophiles such as thiolates and amines are sufficiently selective to react preferentially at C-1 of the 1,2-epoxy-3-alcohol and thereby shift the equilibrium completely in that direction. However, many other nucleophiles are incompatible with the reactions conditions required for the Payne rearrangement, and the approach of trapping the 1,2-epoxide cannot be used. To circumvent this problem and increase the scope of the Payne rearrangement/opening process, methods have been developed that lead to isolation of the terminal 1,2-epoxy-3-alcohols [11,65].

One method uses the 2,3-diol-1-sulfide **30**, produced by thiolate trapping of the 1,2-epoxide from the Payne rearrangement equilibrium between **28** and **29** [11,65]. The sulfide is alkylated with Me_3OBF_4 to produce a good leaving group in **31**. Then base-promoted ring closure gives the 1,2-epoxide **32** in complete preference to formation of any 1,3-oxetane. The *erythro*-epoxy alcohol precursor **31** requires sodium hydride as the base to ensure that the Payne rearrangement is not reversed back to the starting 2,3-epoxy alcohol **28**. The analogous *threo*-epoxy alcohol precursor can be closed with sodium hydroxide.

In the second method, the 2,3-epoxy-1-alcohol **28** is first converted to a mesylate (or a tosylate) and then the epoxide is opened hydrolytically with inversion at C_3 to give the diol-mesylate **33**. The slight loss of optical purity observed in this process is due to lack of complete regioselectivity for C-3 opening. Mild base is sufficient to effect ring closure of the diol-mesylate **33** to give the 1,2-epoxide **34** [11,65].

The two methods are complementary in terms of stereochemistry, such that if a 2,3-epoxy alcohol of the same absolute configuration is used to start each sequence, the *erythro*-1,2-epoxy-3-ols produced will have opposite configurations at C-2 and C-3. This is because inversion occurs at C-2 during the Payne rearrangement, while in the epoxy–mesylate opening, inversion occurs at C-3. Detailed discussions of these Payne rearrangement processes as well as of further synthetic transformations of the 1,2-epoxy alcohols have been presented elsewhere [11,65].

When two allylic alcohols are contained in a symmetrical molecule, asymmetric epoxidation proceeds with interesting consequences for stereochemical purity. The results were first described for the asymmetric epoxidation of (2Z,6E,10Z)-dodeca-2,6,10-trien-1,12-diol (**35**) [66]. The first epoxidation of **35** produces the major and minor enantiomers **36** and **37**. Since the stereogenic centers in these compounds are remote from the second allylic alcohol, each enantiomer undergoes a second epoxidation with essentially the same enantiofacial selectivity as in the first epoxidation. Three bisepoxides result, **38**, **39** (a meso compound), and **38'** (the mirror image of **38**). The overall consequence is that most of the epoxidation resulting from the undesired enantiofacial attack leads to the meso compound **39**, which is in principle separable from the major product. Very little of the mirror image compound **38'** is formed and therefore the enantiomeric purity of the major product will be very high. In the example cited, enantiomeric purity could not be determined directly but was calculated according to the expression $(A_1 + B_1)(A_2 + B_2)$, where the first term in parentheses gives the enantiofacial selectivity of the first epoxidation and the second the enantiofacial selectivity of the second epoxidation. In the example being discussed, an enantiofacial selectivity of 19:1 (90% ee) was assumed for both steps. The ratio of the three products therefore should be (19 + 1)(19 + 1) or 361:38:1 and the enantiomeric purity of **38** should be 99.45% ee [66].

Fortunately, a wide variety of functionality is compatible with the titanium–tartrate catalyst (see Table 1), but the judicious placement of functional groups relative to the allylic alcohol can lead to further desirable reactions following epoxidation. For example, in **40**, asymmetric epoxidation of the allylic alcohol is followed by intramolecular cyclization under the reaction conditions to give the tetrahydrofuran **41** [67]. Likewise, in the epoxidation of **42**, cyclization of the intermediate epoxy alcohol occurs under the reaction conditions and leads to the cyclic urethane **43** [68].

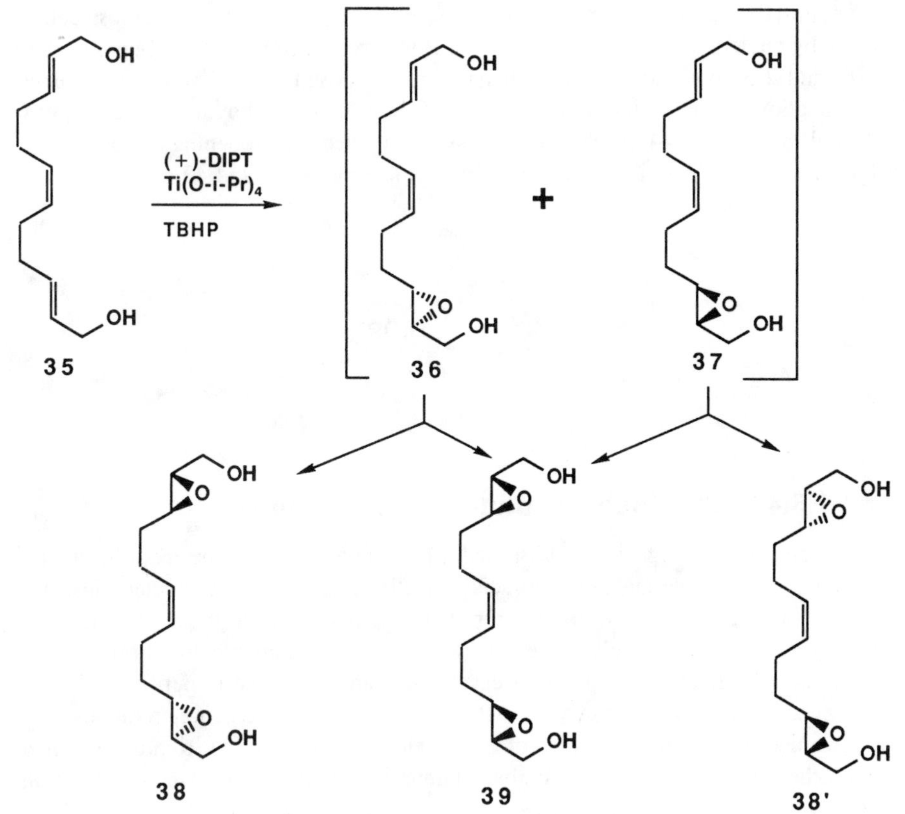
OH
OH
(+)-DIPT
Ti(O-i-Pr)$_4$
TBHP
35
OH
O
OH
36
OH
O
OH
37
OH
O
O
OH
38
OH
O
O
OH
39
OH
O
O
OH
38'

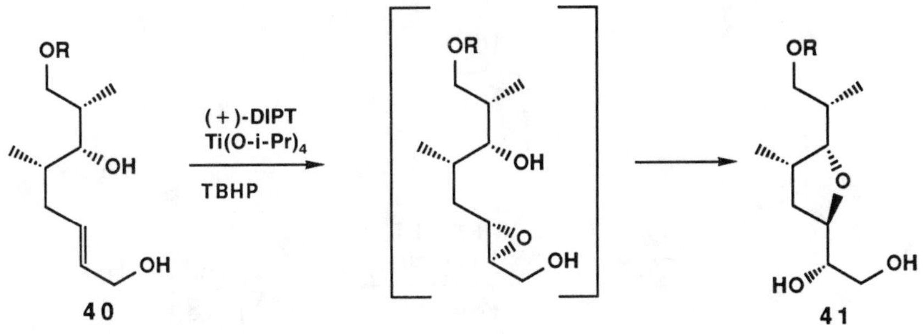
OR
OH
OH
40
(+)-DIPT
Ti(O-i-Pr)$_4$
TBHP
OR
OH
O
OH
OR
O
HO
OH
41

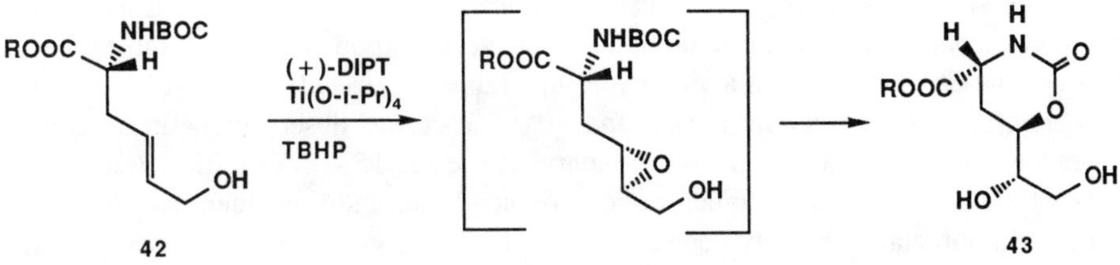
ROOC
NHBOC
H
OH
42
(+)-DIPT
Ti(O-i-Pr)$_4$
TBHP
ROOC
NHBOC
H
O
OH
H
H
N
O
ROOC
O
HO
OH
43

Titanium-(IV) isopropoxide is an effective reagent for promoting regioselective attack by nucleophiles at the 3-position of 2,3-epoxy alcohols [69], 2,3-epoxy acids [70], and 2,3-epoxy amides [70]. It has been proposed that this process involves coordination to the metal center in the bidentate manner shown for a 2,3-epoxy alcohol in structure **44**. Such titanium-assisted nucleophilic opening of epoxides is thought to play a role in the in situ reactions leading to **41** and **43**.

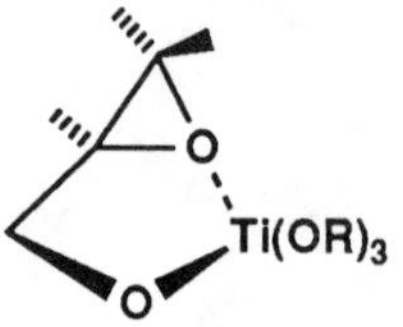

44

4.1.5.4. (3*Z*)-Monosubstituted Allyl Alcohols

Allylic alcohols having a cis-3-substituent (**45**) are the slowest to be epoxidized, and they give the most variable enantiofacial selectivity. Both these characteristics suggest that allylic alcohols of this structure have the poorest fit to the requirements of the active epoxidation catalyst. Nevertheless, asymmetric epoxidation of these substrates is still effective and in most cases gives an enantiomeric purity of at least 80% ee and often as high as 95% ee. Patience with the slower reaction rate usually is rewarded with chemical yields of epoxy alcohols comparable to those obtained with other allylic alcohols. A number of representative examples are collected in Table 5 [2,4,38,59,62a,71–78].

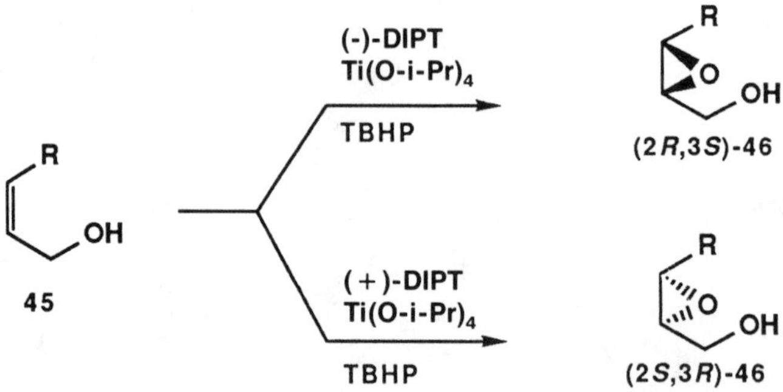

There is a rough correlation between the enantiomeric purity observed for these epoxy alcohols and the steric complexity at the α-carbon of the C-3 substituent. When the C-3 substituent is a primary group (Table 5, entries 1, 2, 4, 6–12, 19–21), enantiofacial selectivity is highest and 80–95% ee are observed for these compounds. When the substituent is secondary (entries 3, 15–18) or tertiary (entry 5), enantiofacial selectivity is much more variable. When the substituent is asymmetric, enantiofacial selectivity depends on the absolute configuration, as is evident in comparison of entries 15 with 16 and of 17 with 18 in Table 5. Epoxidation of these

Table 5 Epoxides from (3Z)-Substituted Allylic Alcohols

Entry	Epoxide R	Catalyst mole %Ti/ %Tart.	Tartrate	Epoxide config.	Yield(%)	% ee	Ref.
1	CH_3	5/6	(+)-DIPT	S,R	68	92	4
2	C_2H_5	a	(+)-DET	S,R	60	80	71
3	$CH(CH_3)_2$	a	(+)-DET	S,R	54	66	72
4	$CH_2CH(CH_3)_2$	a	(+)-DET	S,R	80	95	73
5	$C(CH_3)_3$	120/150	(+)-DET	S,R	77	25	38
6	C_9H_{19} (n)	100/100	(-)-DET	R,S	80	91	2
7	C_7H_{15} (n)	10/14	(+)-DET	S,R	74	86	4
8	C_8H_{17} (n)	5/7.4	(+)-DIPT	S,R	63	>80	4
9	$CH_2CH=CHC_5H_{11}$	110/110	(+)-DMT	S,R	70	94	74
10	$(CH_2)_3COOCH_3$	a	(+)-DET	S,R	57	95	75
11	CH_2OBn	100/100	(-)-DET	R,S	84	92	59
12	CH_2OBn	14/14	-	-	-	95	4
13	Ph	100/120	(+)-DET	S,R	61	78	76
14	$CH(CH_3)Ph$	a	(+)-DIPT	S,R	a	a	77
15	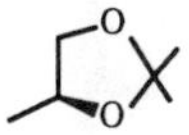	100/100	(+)-DET	S,R	55,57	93,84	59,62a
16	as above	100/100	(-)-DET	R,S	a	20	62a
17	$CH(CH_3)CH_2OBn$	100/100	(+)-DET	S,R	a	66	78a
18	as above	100/100	(-)-DET	R,S	a	0	78a
19	CH_2CH_2OBn	a	(+)-DIPT	S,R	75	92	78b
20	$C_{11}H_{23}$ (n)	120/130	(+)-DIPT	S,R	83	92	78c
21	$(CH_2CH=CH)_2CH=CH_2$	100/148	(+)-DET	S,R	59	89	78d

[a]Not reported.

chiral allylic alcohols with one antipode of catalyst yields moderate to good diastereoselectivity while with the other antipode, diastereoselectivity is virtually lacking.

4.1.5.5. (2,3E)-Disubstituted Allyl Alcohols

Extensive use in synthesis has been made of the asymmetric epoxidation of (2,3E)-disubstituted allylic alcohols. With few exceptions enantiofacial selectivity is excel-

Table 6 Epoxides from (2,3*E*)-Disubstituted Allylic Alcohols

Entry	Epoxide R$_1$	R$_2$	Tartrate	Epoxide config.	Yield(%)	% ee	Ref.
1	CH$_3$	CH$_3$	(+)-DET	2S,3S	77	94	61b
2	CH$_3$	C$_2$H$_5$	(+)-DMT	2S,3S	79	95	41,78e
3	-(CH$_2$)$_3$-		(+)-DET	2S,3S	38	>95	79
4	-(CH$_2$)$_4$-		(+)-DET	2S,3S	77	93	4
5	CH$_3$	CH$_2$OBn	(-)-DIPT	2R,3R	87	90	80
6	CH$_3$	CH(CH$_3$)CH$_2$OBn	-	-	93	>95	81
7	CH$_3$	C$_6$H$_5$	(+)-DIPT	2S,3S	79	>98	4
8	C$_6$H$_5$	C$_6$H$_5$	(+)-DET	2S,3S	70	>95[a]	2,4
9	CH$_3$	CH$_2$CH$_2$CH=CH$_2$	(+)-DET	2S,3S	71	96	82
10	CH$_3$	-CH$_2$- (aromatic OMe/MeO)	(-)-DET	2R,3R	87	>95	83
11	CH$_3$	CH$_2$CH=C(CH$_3$)$_2$	(+)-DET	2S,3S	64	>90	84a
12	CH$_3$	CH$_2$CH=C(CH$_3$)$_2$	(-)-DET	2R,3R	59	>91	84a
13	CH$_3$	(CH$_2$)$_3$OSi(CH$_3$)$_2$Bu	(-)-DET	2R,3R	89	93	84b

[a]Ee after crystallization.

lent (90–95% ee). The results for a number of epoxidations of allylic alcohols with smaller substituents are collected in Table 6 [2,4,41,61b,79–84], while a variety of other compounds with larger groups are illustrated by structures **47–60**.

The epoxy alcohol **47** is a squalene oxide analog that has been used to examine substrate specificity in enzymatic cyclizations by baker's yeast [85]. The epoxy alcohol **48** provided an optically active intermediate used in the synthesis of 3,6-epoxyauraptene and marmine [86], and epoxy alcohol **49** served as an intermediate in the synthesis of the antibiotic virantmycin [87]. In the synthesis of the three stilbene oxides **50**, **51**, and **52**, the presence of an *o*-chloro group in the 2-phenyl ring resulted in a lower enantiomeric purity (70% ee) when compared to the analogs without this chlorine substituent [88a]. The very efficient (80% yield, 96% ee) formation of **52a** by asymmetric epoxidation of the allylic alcohol precursor offers a synthetic entry to optically active 11-deoxyanthracyclinones [88b], while epoxy alcohol **52b** is one of several examples of asymmetric epoxidation used in the synthesis of brevitoxin precursors [88c]. Diastereomeric epoxy alcohols **54** and **55** are obtained in combined 90% yield (> 95% ee each) from epoxidation of the racemic alcohol **53** [89]. Diastereomeric epoxy alcohols, **57** and **58**, also are obtained with high enantiomeric purity in the epoxidation of **56** [44]. The epoxy alcohol obtained from substrate **59** undergoes further intramolecular cyclization with stereospecific formation of the cyclic ether **60** [90].

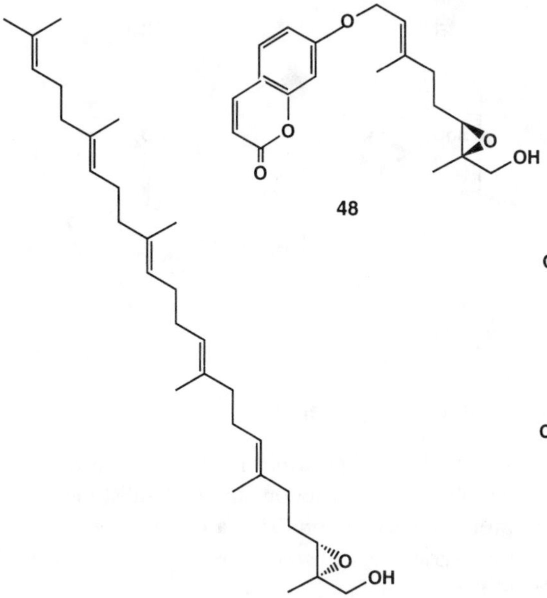

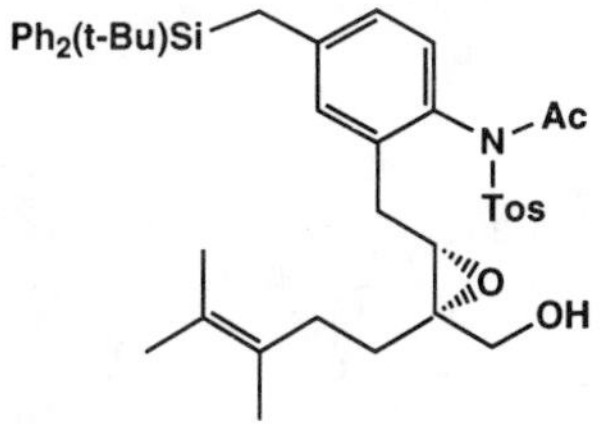

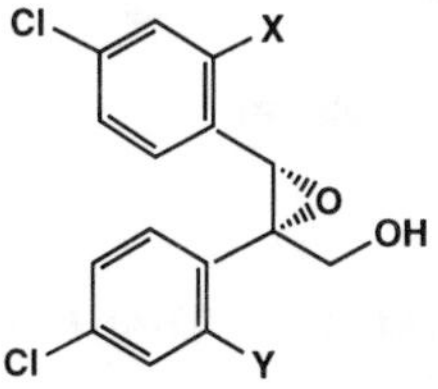

X	Y	%ee [88a]
50 H	H	>90
51 Cl	H	>90
52 Cl	Cl	70

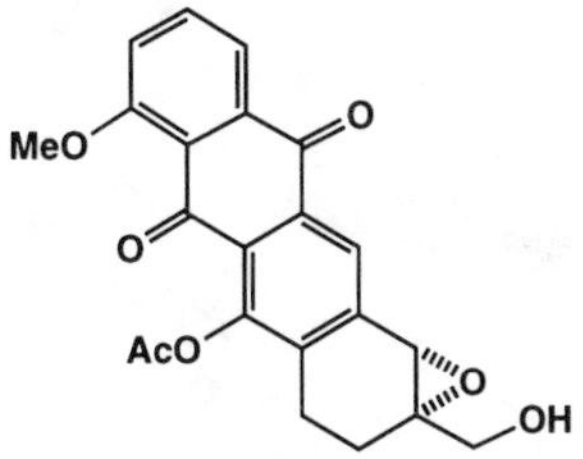

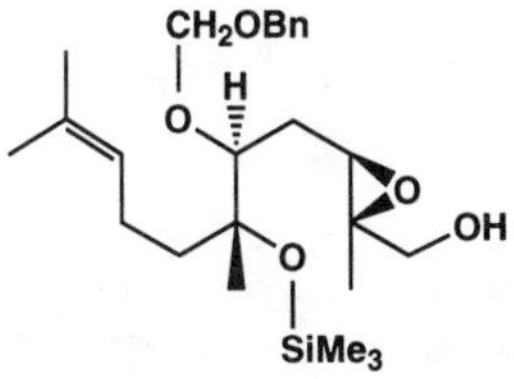

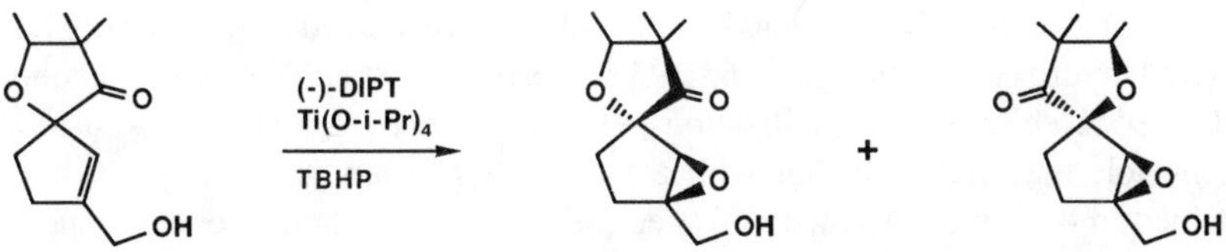

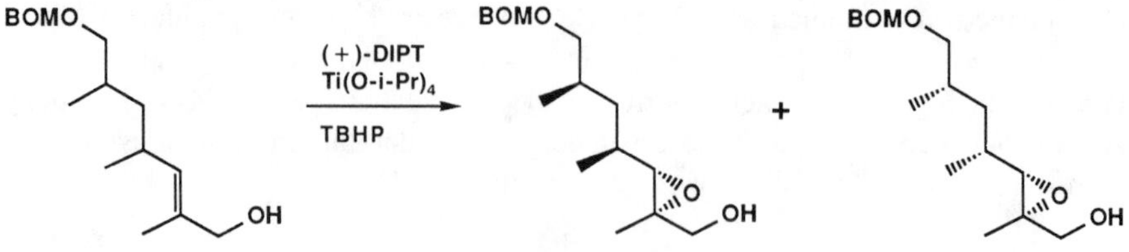

4.1.5.6. (2,3Z)-Disubstituted Allyl Alcohols

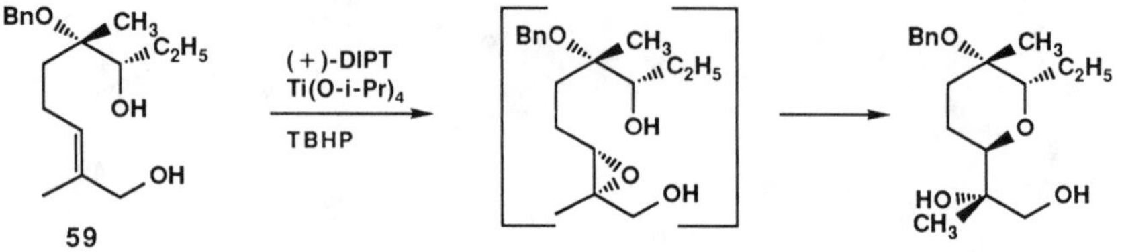

A limited number of allylic alcohols of the (2,3Z)-disubstituted type have been subjected to asymmetric epoxidation. With one exception, the C-2 substituent in these substrates has been a methyl group, the exception being a t-butyl group [38]. The (3Z)-substituents have been more varied, as illustrated by structures 61–64, which show the epoxy alcohols derived from the corresponding allylic alcohol substrates.

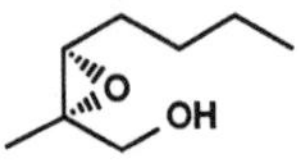

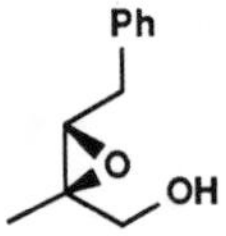

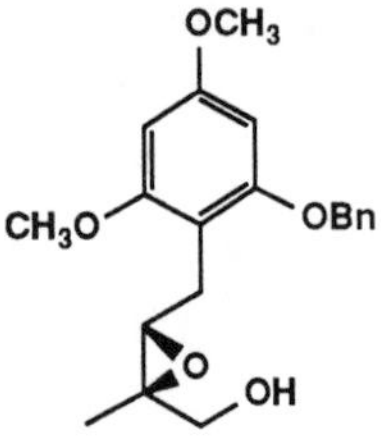

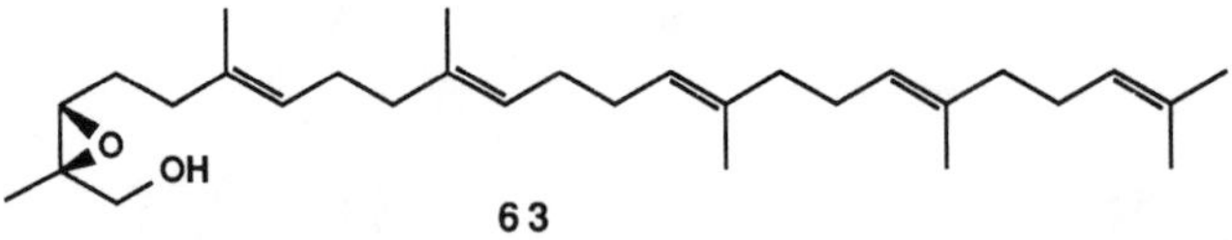

Epoxidation of (Z)-2-methyl-2-hepten-1-ol gave epoxy alcohol 61 (80% yield, 89% ee) [2], of (Z)-2-methyl-4-phenyl-2-buten-1-ol gave 62 (90%, 91% ee) [77], and of (Z)-1-hydroxysqualene gave 63 (93%, 78% ee) [85]. The epoxy alcohol 64 had > 95% ee after recrystallization [91]. In the epoxidation of (Z)-2-t-butyl-2-buten-1-ol, the allylic alcohol with a C-2 t-butyl group, the epoxy alcohol was obtained in 43% yield and with 60% ee [38]. These results lead one to expect that other 2,3Z-disubstituted allylic alcohols will be epoxidized in good yield and with enantioselectivity similar to that observed for the 3Z-monosubstituted allylic alcohols (*i.e.*, 80–95% ee).

4.1.5.7. 3,3-Disubstituted Allyl Alcohols

The 3,3-disubstituted allyl alcohols are substrates that combine a 3E-substituent with a 3Z-substituent in the same molecule. Allylic alcohols with only a 3E-substituent generally are epoxidized with excellent enantioselectivity, whereas those with only a 3Z-substituent are epoxidized with enantioselectivity in the range of 80–95% ee. In combination, many of the reported examples have a methyl substituent at the 3Z-position, and all are epoxidized with 90–95% ee (see Table 7, entries 1–4, 6) [2,4,92–97]. Only a limited number of examples with larger groups at the 3Z-position have been reported (entries 5, 7–12) and in these the enantioselectivity ranges from 84 to 94% ee.

3,3-Dimethylallyl alcohol was epoxidized with better than 90% ee (entry 1) but in low yield when a stoichiometric amount of the titanium–tartrate complex was used. However, when a catalytic amount of the complex was used and in situ derivatization employed, the *p*-nitrobenzoate (> 98% ee after recrystallization) and *p*-toluenesulfonate (93% ee) were isolated in yields of 70 and 55%, respectively.

Table 7   Epoxides from 3,3-Disubstituted Allylic Alcohols

Entry	R$_1$ (Epoxide)	R$_2$ (Epoxide)	Catalyst mol %Ti/%Tart.	Tartrate	Epoxide config.	Yield (%)	% ee	Ref.
1	CH$_3$	CH$_3$	100/100	(-)-DBT	2R	25	>90	92
2	CH$_2$CH=C(CH$_3$)$_2$	CH$_3$	200/200	(+)-DET	2S,3S	67	95	93
3	(CH$_2$)$_2$CH=C(CH$_3$)$_2$	CH$_3$	100/100	(+)-DET	2S,3S	77	95	2
4	(CH$_2$)$_2$CH=C(CH$_3$)$_2$	CH$_3$	5/7.4	(+)-DET	2S,3S	95	91	4
5	CH$_3$	(CH$_2$)$_2$CH=C(CH$_3$)$_2$	100/100	(+)-DET	2S,3R	79	94	2
6	(CH$_2$)$_2$OSi(CH$_3$)$_2$C(CH$_3$)$_3$	CH$_3$	105/157	(-)-DET	2R,3R	81	>95	94
7	CH$_3$	(CH$_2$)$_2$OSi(CH$_3$)$_2$C(CH$_3$)$_3$	a	(+)-DET	2S,3R	98	90	95
8	CH$_3$	(CH$_2$)$_2$OSi(CH$_3$)$_2$C(CH$_3$)$_3$	a	(-)-DET	2R,3S	98	86	95
9	(structure: MeO / OMe)		10/15	(+)-DET	2S,3R	97	93	96
10	(bicyclic structure)		100/110	(+)-DET	2S,3S	a	84	97a
11	(bicyclic structure)		100/110	(+)-DET	2S,3R	a	88	97a
12	(CH$_2$)$_4$OBn	CH$_2$CH(CH$_3$)CH$_2$OMEM	10/12	(-)-DIPT	2R,3S	83	91	97b

aNot reported.

Likewise, the epoxidation of geraniol with a stoichiometric amount of the complex gave epoxide (entry 3) in 77% yield (95% ee), which was improved to 95% yield (91% ee) when a catalytic amount of complex was used (entry 4).

4.1.5.8. 2,3,3-Trisubstituted Allyl Alcohols

Interesting structural diversity is present in the limited examples of trisubstituted allyl alcohols (equivalent to tetrasubstituted olefins) to which asymmetric epoxidation has been applied. The epoxides 65–70 [98–103] have been obtained from the corresponding allylic alcohols with yield and enantiomeric purity as indicated when such data have been reported. The lower enantiomeric purity observed for epoxy alcohol 69 may result from disruption of the catalyst structure by the phenolic groups or from alternate modes of binding of substrate to catalyst, again because of the phenolic groups. Phenols bind strongly to Ti(IV), which may account for the large excess (6 equiv.) of the titanium–tartrate complex that was required to achieve the yield and enantiomeric purity reported in the case of 69.

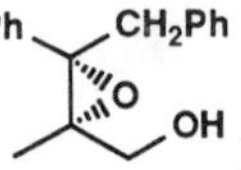

65
90% (94%ee)[98]

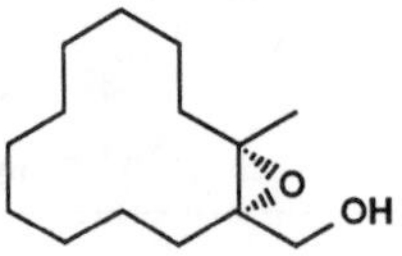

66
72% (94%ee)[99]

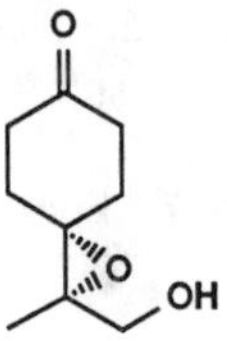

67
(>90%ee)[100]

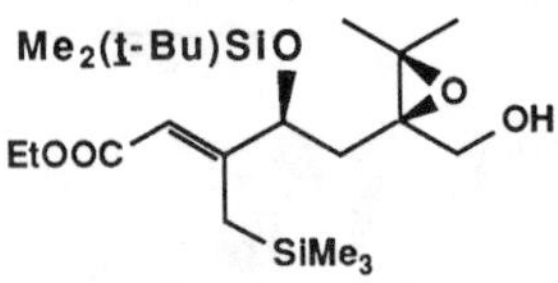

68
(>90%ee)[101]

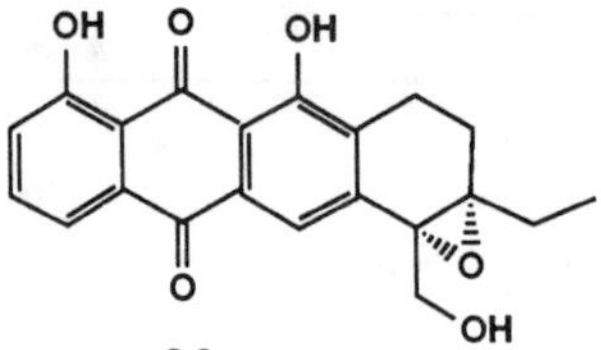

69
85% (53%ee)[102]

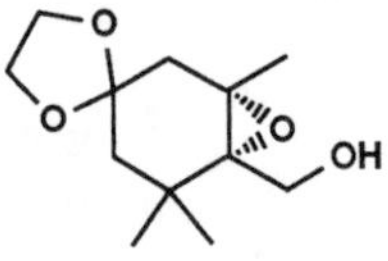

70
95% (95%ee)[103]

4.1.5.9. 1-Substituted Allyl Alcohols: Kinetic Resolution

The presence of a stereogenic center at C_1 of an allylic alcohol introduces an additional factor into the asymmetric epoxidation process in that now both enantiofacial selectivity and diastereoselectivity must be considered. It is helpful in these cases to examine epoxidation of each enantiomer of the allylic alcohol separately. Epoxidation of one enantiomer proceeds normally and produces an erythro epoxy alcohol in accord with the rules shown in Figure 1. Epoxidation of the other enantiomer proceeds at a reduced rate because contact between the C-1 substituent and the catalyst seriously impedes the necessary approach of olefin to oxidant (see

Fig. 2). The difference in epoxidation rates for the two enantiomers is usually of sufficient magnitude that either the epoxy alcohol or the recovered allylic alcohol can be produced with high enantiomeric purity. The net result is that a kinetic resolution is achieved [14]. In the case of a homochiral C-1 substituted allylic alcohol, asymmetric epoxidation will be fast and highly diastereoselective with one antipode of the titanium–tartrate catalyst but not with the other, according to the guidelines of Figure 2. Although kinetic resolution is most frequently encountered and applied to chiral C-1-substituted allylic alcohols, the rationale also is applicable to allylic alcohols with chiral substituents at other positions, examples of which have been given in several preceding subsections.

The ratio of the rates of epoxidation of the two enantiomers, k_{fast}/k_{slow}, has been defined as the relative rate (k_{rel}) and is related to both the percent conversion of allylic alcohol to epoxy alcohol and the enantiomeric purity of the remaining allylic alcohol. A mathematical relationship between these variables exists and can be represented graphically as shown in Figure 5 [14].

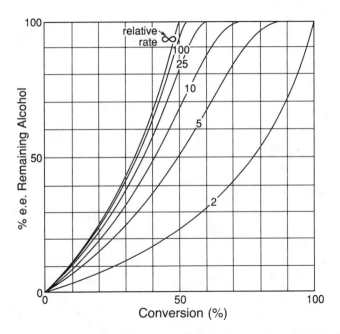

Figure 5 Dependence of enantiomeric excess on relative rate in the epoxidation of 1-substituted allylic alcohols.

If values are known for two of the three variables, the third can be predicted by use of this graph. Inspection of the graph reveals that relative rates of 25 or more are very effective for achieving kinetic resolution of 1-substituted allylic alcohols. With a relative rate of 25, the epoxidation need be carried to less than 60% conversion to achieve essentially 100% ee for the unreacted alcohol. A convenient method for limiting the extent of epoxidation to 60% is simply by controlling the amount of

Table 8 Relative Rate (k_{rel}) Data for Kinetic Resolution
of 1-Substituted Allylic Alcohols

Entry	Allylic alcohol	Reaction Time	Rel. rate at -20°C	% ee	Rel. rates at 0°C			Ref.
					DIPT	DET	DMT	
1	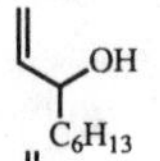 (vinyl)CH(OH)C$_6$H$_{13}$	12 da	83	>96	60			2,13
2	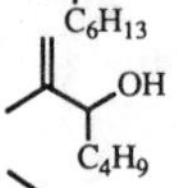 (isopropenyl)CH(OH)C$_4$H$_9$	15 hr	138	>96	96	52		2,13
3	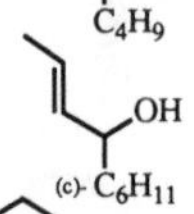 (propenyl)CH(OH)(c)-C$_6$H$_{11}$	15 hr	104	>96	74	28	15	2,13
4	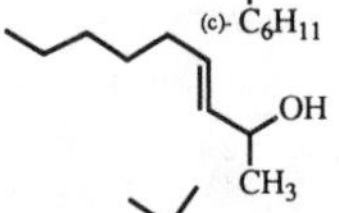 (dienyl)CH(OH)CH$_3$		160					104
5	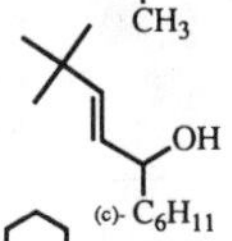 (t-Bu substituted)CH(OH)(c)-C$_6$H$_{11}$		300					104
6	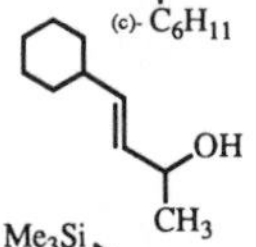 (cyclohexyl substituted)CH(OH)CH$_3$		330					104
7	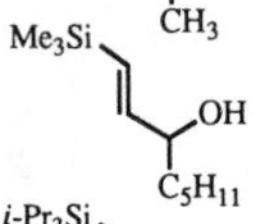 Me$_3$Si-(vinyl)CH(OH)C$_5$H$_{11}$		700					104-106
8	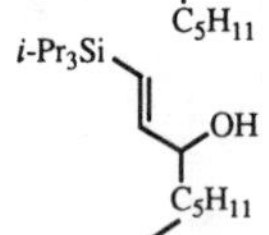 i-Pr$_3$Si-(vinyl)CH(OH)C$_5$H$_{11}$		300					104
9	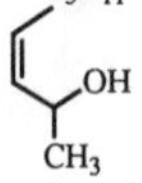 (propenyl)CH(OH)CH$_3$	6 da	20	91				2,13
10	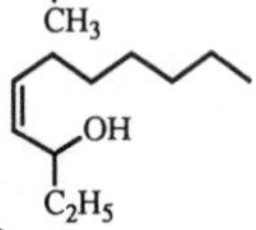 (alkenyl)CH(OH)C$_2$H$_5$	2da	16	82	13			2,13
11	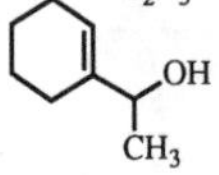 (cyclohexenyl)CH(OH)CH$_3$	15 hr	83	>96	60	38		2,13

oxidant used in the reaction. However, for some substrates (see Table 8, entries 1, 9, or 10) even k_{fast} is extremely slow and the epoxidation takes several days [2,13,104–106]. To shorten the time needed for such reactions, an alternate practice is to use an excess of oxidant and to monitor the extent of epoxidation by an appropriate analytical method. If the optically active epoxy alcohol is the desired reaction product, high enantioselectivity can be ensured by running the reaction to approximately 45% completion.

Relative rate data for the kinetic resolution/epoxidation of 1-substituted allylic alcohols of varying structure are summarized in Table 8. The k_{rel} values at −20°C for all entries in Table 8 were determined using DIPT as the chiral ligand. Additionally, for several entries (1–3, 10, 11) the dependence of k_{rel} on temperature, 0 versus $-20°C$, and on steric bulk of the tartrate ester (DIPT vs. DET vs. DMT) has been measured. Lower reaction temperature and larger tartrate ester groups both are factors that clearly increase the magnitude of k_{rel} and, therefore, improve the efficiency of the kinetic resolution process. While all the results summarized in Table 8 are from experiments in which stoichiometric quantities of titanium–tartrate complex were used, the catalytic version of the reaction also may be used for kinetic resolution [4]. When comparing results with the same tartrate ester, a slight loss in enantioselectivity is seen in the catalytic mode relative to the stoichiometric reaction. The trend toward higher enantioselectivity with bulkier tartrate esters can be used to advantage in the catalytic reaction by using dicyclohexyltartrate (DCT), which gives higher enantioselectivity than DIPT, or dicyclododecyltartrate, which gives yet higher enantioselectivity than DCT [4].

The efficiency of kinetic resolution is even greater when there is a silicon or iodo substituent in the $3E$ position of the C-1 chiral allylic alcohols. The compatibility of silyl substituents with asymmetric epoxidation conditions was first shown by the conversion of $(3E)$-3-trimethylsilylallyl alcohol into $(2R,3R)$-3-trimethylsilyloxiranemethanol in 60% yield with more than 95% ee [107a] and further exploited by the conversion of (E)-3-(triphenylsilyl)-2-[2,3-^{2}H$_2$]propenol into $(2R,3R)$-3-triphenylsilyl [2,3-^{2}H$_2$]oxiranemethanol in 96% yield with 94% ee [107b,c]. With an n-pentyl group at C-1, the k_{rel} for asymmetric epoxidation of the enantiomeric allylic alcohols is 700 (Table 8, entry 7), and both epoxy alcohol and optically active recovered allylic alcohol are obtained in 42% yield with more than 99% ee (see Table 9, entry 1). Equally good yields and enantiomeric purities are observed with other substituents in the C-1 position as is shown by entries 2–9 in Table 9 [105,108,109]. Good yields with high enantioselectivities also are reported in the kinetic resolution of $(3E)$-iodo analogs (entries 10–14) and of a $(3E)$-chloro analog (entry 15). $(3E)$-Stannyl substituents (entries 16–18) appear to be similar to carbon substituents in their effect on kinetic resolution.

The influence of both the steric and electronic properties of the silyl group on the rate of epoxidation have been examined experimentally [104]. Two different rate effects were considered. First, the overall rate of epoxidation of the silyl allylic alcohols was found to be one-fifth to one-sixth that of the similar carbon analogs. This rate difference was attributed to electronic differences between the silicon and carbon substituents. Second, the increase in k_{rel} to 700 for silyl allylic alcohols compared to carbon analogs (e.g., 104 for entry 3, Table 8) was attributed to the steric effect of the large trimethylsilyl group. As expected, when a bulky t-butyl group was placed at C-3, k_{rel} increased to 300 [104].

At the end of 1989, the number of 1-substituted allylic alcohols that had been used in kinetic resolution/asymmetric epoxidation experiments exceeded 75. In slightly more than half of these experiments, the desired product was the kinetically resolved allylic alcohol, while in the remainder the epoxy alcohol was desired. In addition to the compounds in Table 8, experimental results for other kinetically resolved alcohols are summarized in Table 10 [38,77,110–115a–d]. From these

Table 9 Kinetic Resolution of 3-Silyl-, Halo-, and Stannyl-Substituted Allylic Alcohols

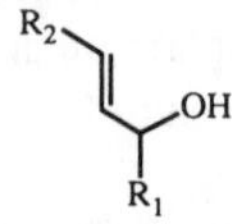

$$R_2 \diagdown \quad \diagup OH \atop R_1$$

Entry	Allylic alcohol		Allylic alcohol		Epoxy alcohol		Ref.
	R_1	R_2	Yield(%)[a]	% ee	Yield(%)[a]	% ee	
1	C_5H_{11} (n)	SiMe$_3$	42	>99	42	>99	105
2	C_3H_7 (i)	SiMe$_3$	40	>99	41	99	105b
3	C_6H_5	SiMe$_3$	44	>99	42	97	105b
4	CH_2OPh	SiMe$_3$	47	>99	46	>99	105b
5	CH_2OCH_2Ph	SiMe$_3$	43	>99	48	>99	105b
6	$CH_2CH=CHC_5H_{11}$	SiMe$_3$	44	>99	43	>99	105b
7	$CH_2CH_2OCH_2Ph$	SiMe$_3$	43	>99	45	>99	105b
8	CH_2COOBu (n)	SiMe$_3$	44	>99			105c
9	$(CH_2)_3COOCH_3$	SiMe$_3$	43	>99	45	>99	105b
10	C_5H_{11} (n)	I	49	>99	49	>99	108
11	C_2H_5	I	40	>98			108
12	$CH_2C_6H_{11}$ (c)	I	42	>99			108
13	C_5H_9 (c)	I	44	>99			108
14	C_6H_5	I	43	>98			108
15	C_5H_{11} (n)	Cl	43	>99			108
16	C_5H_{11} (n)	SnBu$_3$	40	>99	b	84	109
17	C_6H_{11} (c)	SnBu$_3$	41	>99			109
18	CH_2OPh	SnBu$_3$	40	>99			109

[a]Maximum yield is 50% [b]Not reported.

results, it appears that kinetic resolution is successful regardless of the nature of the (3E) substituent and is successful with any except the most bulky substituents at C-2.

When the allylic alcohol is the desired product of the kinetic resolution process, the accompanying epoxy alcohol also may be converted to the desired allylic alcohol by the two-step sequence shown in Scheme 1. The epoxy alcohol, after separation from the allyl alcohol, is mesylated and then subjected to reaction with sodium

Table 10 Representative Kinetic Resolutions of 1-Substituted Allylic Alcohols

$$\begin{array}{c} R_3 \diagdown \quad R_4 \\ R_2 \diagup \!\!= \!\! \diagup \!\! \overset{\displaystyle \text{OH}}{\underset{\displaystyle R_1}{|}} \end{array}$$

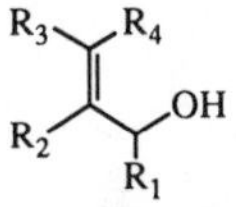

	Allylic alcohol						
Entry	R_1	R_2	R_3	R_4	Yiled(%)[a]	% ee	Ref.
1	CH_3	H	$CH_2CH=CH_2$	H	39	90	110
2	C_6H_{11} (c)	H	H	H	32	>98	111
3	C_2H_5	CH_3	H	H	b	>98	112
4	C_4H_9 (n)	H	H	H	43	>90	113
5	C_6H_3 (2,4-diCl)	C_6H_3 2,4-diCl)	H	H	42	90	114
6	C_2H_5	H	Ph	H	b	99	77
7	CH_2CH_2Ph	H	H	H	b	99	77
8	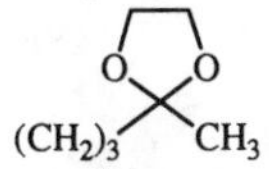 (CH$_2$)$_3$—[1,3-dioxolane]—CH$_3$	H	H	H	b	99	77
9	$CH_2\overset{O}{\overset{\|}{C}}CH=CH_2$	H	CH_3	CH_3	10	>99	77
10	$C_{12}H_{25}$ (n)	H	CH_3	H	44	97	115a
11	CH_3	$C(CH_3)_3$	H	H	b	30	38
12	$C(CH_3)_3$	H	CH_3	H	b	5	38
13	C_4H_9 (n)	H	$CH=CH_2$	H	40	90	115b
14	CH_2COOEt	CH_3	H	H	11	>95	115c
15	$C{\equiv}CC_6H_{13}$ (n)	H	H	H	35	95	115d

[a]Maximum yield is 50%.   [b]Not reported.

telluride, which effects the transformation of epoxy mesylate to the allylic alcohol with inversion at the asymmetric carbinol center [115e]. Preliminary results suggest that the rearrangement follows this pathway only when the epoxy alcohol is unsubstituted at the 3-position.

A small, structurally distinct class of 1-substituted allylic alcohols consists of those that are conformationally restricted by incorporation into a ring system. These allylic alcohols may be further subdivided into two types, depending on whether the double bond is endocyclic or exocyclic. For allylic alcohols with endocyclic double bonds, kinetic resolution gives 2-cyclohexen-1-ol (**71**) with 30% ee [14],

Scheme 1

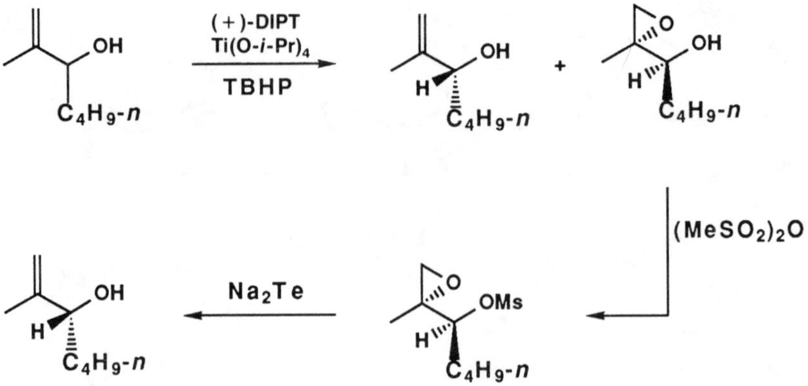

(4aS,2R)-4a-methyl-2,3,4,4a,5,6,7,8-octahydronaphthalen-2-ol (**72**) with 55% ee [116], and (R)-2-cyclohepten-1-ol (**73**) with 80% ee [14]. The epoxy alcohols (1S,2S,3R)-2,3-epoxycyclopenten-1-ol (**74**) [117], (1S,2S,4aR)-4a-decahydronaphthalen-2-ol (**75**) [116], and (1R,2S,3R)-2,3-epoxy-6-cyclononen-1-ol (**76**) [118] are obtained with 60, 61, and 90% ee, respectively.

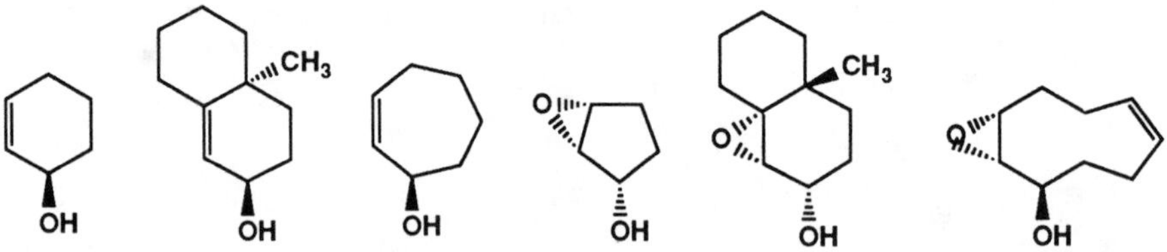

71 72 73 74 75 76

(R)-*trans*-Verbenol (**77**) is epoxidized five times as fast as (S)-*trans*-verbenol when (+)-DIPT is used in the catalyst [77]. For allylic alcohols with an exocyclic double bond, kinetic resolution gives 2-methylenecyclohexanol (**78**) with 80% ee in 46% yield when (−)-DIPT is used [119]. Epoxidation of the enantiomerically pure 4-methylene-5α-cholestan-3β-ol (**79**) is reported to be much faster with catalyst derived from (+)-DET than from (−)-DET [120].

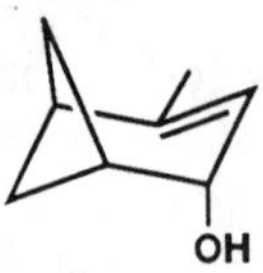

77

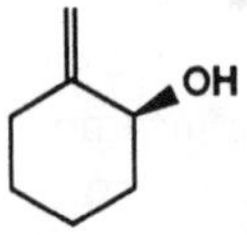

78

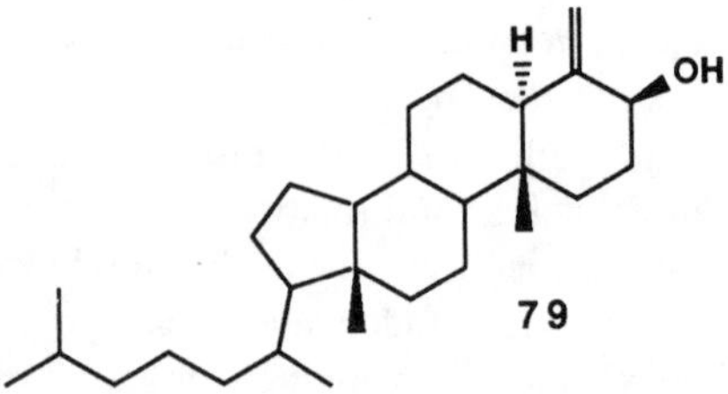

79

The variable enantioselectivities seen in these results likely stem from conformational restraints imposed by the cyclic structures, which prevent the allylic alcohols from attaining an ideal conformation for the epoxidation process (see Fig. 9, below, for the proposed ideal conformation).

One especially interesting kinetic resolution/asymmetric epoxidation substrate is (R,S)-2,4-hexadien-3-ol (**80**) [77]. The racemic diene has eight different olefinic faces at which epoxidation can occur and thereby presents an interesting challenge to the selectivity of the epoxidation catalyst. The selectivity can be tested by using slightly less than 0.5 equiv. of oxidant (because the substrate is a racemate, the maximum yield of any one product is 50%). When the reaction was run under these conditions, the only product that was formed was the (1R,2R,3R)-epoxy alcohol **81**.

Three different principles of selectivity are required to achieve this result. First, the difference in rate of epoxidation by the catalyst of a disubstituted versus a monosubstituted olefin must be such that the propenyl group is epoxidized in complete preference to the vinyl group. The effect of this selectivity is to reduce the choice of olefinic faces to the four of the two propenyl groups. Second, the inherent enantiofacial selectivity of the catalyst as represented in Figure 1 will narrow the choice of propenyl faces from four to two. Finally, the steric factor responsible for kinetic resolution of 1-substituted allylic alcohols (Fig. 2) will determine the final choice between the propenyl groups in the enantiomers of **80**. The net result is the formation of epoxy alcohol **81** and enrichment of the unreacted allylic alcohol in the (3S)-enantiomer.

trans-1,2-Dialkylcycloalkenes (**82**) have helical chirality and can be resolved if flipping of the ring from one face of the olefin to the other is restricted. When appropriately substituted, these compounds also serve as synthetic precursors to the betweenanenes. The asymmetric epoxidation approach to kinetic resolution is ideally suited for the resolution of the cycloalkenes when a hydroxymethyl group is one of the substituents on the double bond, as shown for **82**. The epoxidation of **82** with Ti(O-i-Pr)$_4$/(+)-DET and 0.6 equiv. of TBHP was complete within 10 minutes and gave resolved allylic alcohol **83** in 41% isolated yield with no detectable enantiomeric impurity and epoxy alcohol **84** in 50% yield (the maximum yields possible for **83** and **84** are 50%) [121a]. A variety of analogs of **82** including different ring sizes have been resolved by this method and have been used for the synthesis of optically active betweenanenes [121b].

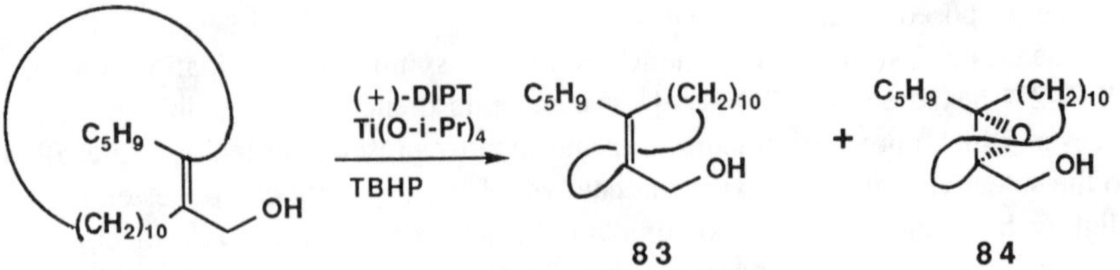

8 2   **8 3**   **8 4**

A final subclass of 1-substituted allylic alcohols is made up of carbinol derivatives having two identical olefinic substituents, the simplest example being 1,4-pentadien-3-ol or divinylcarbinol, **85**. Although these compounds per se are achiral, once they have bound to the chiral titanium complex, the two vinyl groups become stereochemically nonequivalent (diastereotopic). Asymmetric epoxidation will now occur selectively at one of the two vinyl groups, the choice being controlled by factors identical to those in effect during the kinetic resolution process. The similarity can be seen by comparison of the titanium–allylic alcohol complex portrayed in Figure 6 with the kinetic resolution process depicted in Figure 2. The pro-*S* and

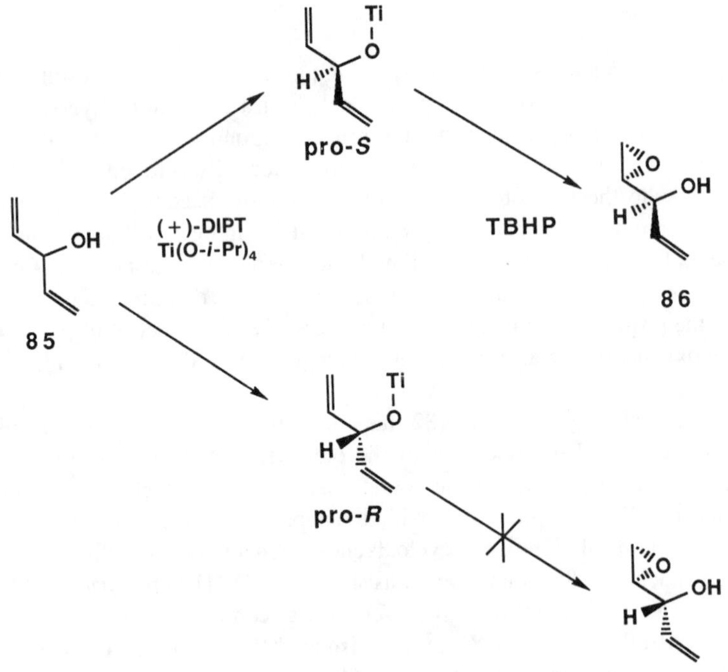

Figure 6   Asymmetric epoxidation of 1,4-pentadien-3-ol.

pro-*R* conformations shown will be sterically favored and disfavored, respectively, for the same reasons that the enantiomers of chiral C_1 allylic alcohols are distinguished during kinetic resolution. Therefore, epoxidation of **85** produces (2*R*,3*S*) epoxy alcohol **86** [122].

Further analysis of the asymmetric epoxidation of divinylcarbinol **85** including the minor products has led to recognition of the second factor that influences the optical purity of the major product **86** [26,123]. One of the minor epoxy alcohols is enantiomeric to **86** and therefore is responsible for lowering the enantiomeric purity of **86**. However, in this minor isomer the configuration of the remaining allylic alcohol group favors a rapid second epoxidation, and this isomer is quickly converted to a diepoxide. As a consequence, the enantiomeric purity of the major epoxy alcohol **86** increases as the reaction progresses. A mathematical equation relating enantiomeric purity to the various rates of epoxidation for these divinylcarbinols has been derived. This analysis can also be applied to asymmetric epoxidation of prochiral compounds such as **87** [124].

8 7

As noted earlier in this section on C-1 substituted compounds, preparation of the epoxy alcohol has been the synthetic objective nearly as often as has been the optically active allylic alcohol. The principles outlined in Figure 2 can again be used to guide the choice of tartrate ester needed in order to obtain the erythro epoxy alcohol of desired absolute configuration. By limiting the amount of oxidant (TBHP) used for the epoxidation to 0.4 equiv. (relative to substrate), optimum enantiomeric purity of the epoxy alcohol can be assured and, in most cases, will be excellent. A few representative examples of epoxides prepared in this way are summarized in Table 11 [90,105b,115c,125–130b].

In the special case in which the substrate is already enantiomerically pure (as in entry 5), it should be clear from Figure 2 that asymmetric epoxidation will be successful (with regard to diastereomeric purity) only when the choice of catalyst directs delivery of oxygen to the face of the olefin opposite that of the C-1 substituent. Such choice of catalyst is further illustrated in Scheme 2, wherein the two sequential epoxidations each proceed with better than 97% diastereoselectivity. The bisepoxide is obtained in an overall yield of 80% [130c].

4.1.5.10. 1,1-Disubstituted Allyl Alcohols

The rationale that explains the kinetic resolution of the 1-monosubstituted allylic alcohols predicts that a 1,1-disubstituted allylic alcohol will be difficult to epoxidize with the titanium–tartrate catalyst. In practice, the epoxidation of 1,1-dimethylallyl

Table 11   Epoxides from 1-Substituted Allylic Alcohols

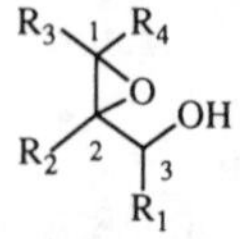

Entry	Epoxide				Tartrate	Config.[a]	Yield(%)[b]	% ee	Ref.
	R_1	R_2	R_3	R_4					
1	C_2H_5	H	H	H	(+)-DIPT	2R,3S	c	d	125
2	$C_5H_{11}(n)$	H	H	H	(-)-DIPT	2S,3R	47	91	126
3	$(CH_2)_6COOMe$	H	H	H	(+)-DIPT	2R,3S	36	>95	127
4	$CH_2C{\equiv}CSi(iPr)_3$	H	H	H	(-)-DIPT	2S,3R	40	>90	128
5[e]	C_2H_5	CH_3	H	H	(+)-DIPT	2R,3R	82	92	90
6	$CH_2CH{=}CH_2$	H	CH_3	H	(-)-DIPT	1R,2R,3R	27	>95	129a-129c
7	$CH(OBn)CH{=}CHCH_3$	H	CH_3	H	(+)-DIPT	1S,2S,3S	35	>95	130a
8	CH_3	H	$CH_2CH{=}CH_2$	H	(-)-DIPT	1R,2R,3R	40	90	129b
9	C_5H_{11}	H	$SiMe_3$	H	(+)-DIPT	1S,2S,3S	40	99	105b
10	CH_2COOEt	CH_3		H	(-)-DET	1S,2S,3R	d	>95	115c
11	CH_3	-CH_2CH=CCH_2- (Me)		H	(+)-DIPT	1S,2S,3S	37	95	130b

[a]Note that the arbitrary numbering used here may not coincide in all cases (e.g., entries 7, 10, 11) with correct Chem. Abstr. numbering. [b]Maximum yield is 50%, except for entry 5. [c]The epoxy alcohol was converted without isolation to the ethoxyethyl derivative. [d]Not reported. [e](3S)-2-Methylpent-1-en-3-ol was used as the substrate for this epoxidation.

alcohol (**88**) with a stoichiometric quantity of the titanium–tartrate complex is very slow, and no epoxy alcohol is isolated [131]. Clearly, the rate of epoxidation of this substrate is slower than the subsequent reaction(s) of the epoxide.

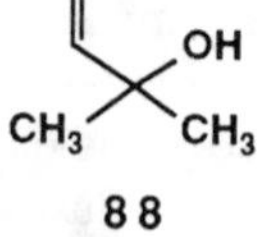

88

4.1.5.11. Homoallylic, Bishomoallylic, and Trishomoallylic Alcohols

In contrast to allylic alcohols, the asymmetric epoxidation of homoallylic alcohols shows the following three general characteristics [132]: (a) the rates of epoxidation are slower, (b) enantiofacial selectivity is reversed (i.e., oxygen is delivered to the

Scheme 2

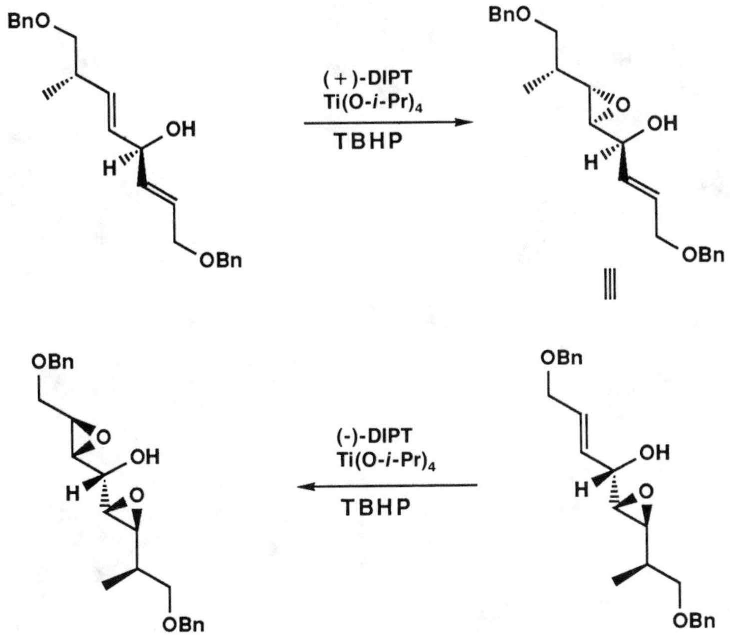

opposite face of the olefin when the same tartrate ester is used), and (c) the degree of enantiofacial selectivity is lower, with enantiomeric purities of the epoxy alcohols ranging from 20 to 55% ee. A series of seven model homoallylic alcohols, including all but one of the possible substitution patterns, has been subjected to epoxidation using the stoichiometric version of the reaction with the results providing the basis for the preceding generalizations. An analogous complex composed of zirconium(IV) isopropoxide [Zr(O-i-Pr)$_4$] and (+)-dicyclohexyltartramide has been found to catalyze asymmetric epoxidation of homoallylic alcohols with the same sense of enantiofacial selectivity as the titanium–tartrate ester complex. An improvement in enantiomeric purity was noted for epoxy alcohols derived form Z-homoallylic alcohols (to 77% ee), while other epoxy alcohols were obtained with enantiomeric purities comparable to those achieved with titanium [133].

The trishomoallylic alcohol **89** undergoes asymmetric epoxidation in a yield of 74% and with "high" diastereofacial selectivity to give **90**. Trityl hydroperoxide, which had been shown to be effective in the asymmetric epoxidation of allylic alcohols [6], was required to attain enantiofacial selectivity in the epoxidation of **89** [134a]. The titanium/tartrate/TBHP-catalyzed conversions of the bishomoallylic phenol (**90a**, $n = 1$, R = H) and the trishomoallylic analog (**90a**, $n = 2$, R = Me) into dihydrobenzofuran **90b** (22% yield, 29% ee) and dihydrobenzopyran **90c** (49 and 56% ee), respectively, is assumed to occur via the intermediate epoxides [134b]. The dihydrofuran **90b** is assigned the (2S,1'R)-configuration, whereas the configurations of the dihydropyran **90c** are unspecified.

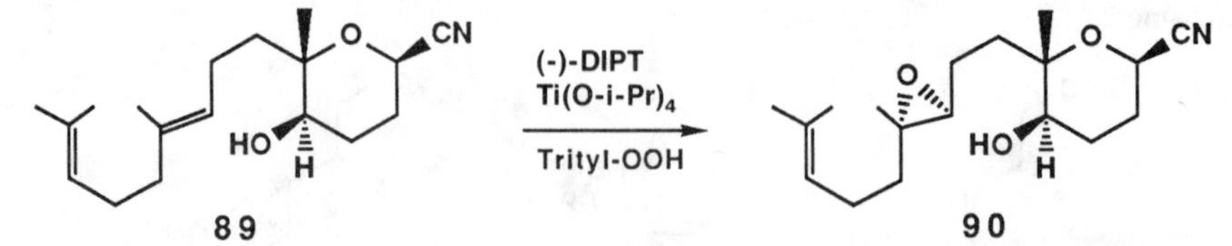

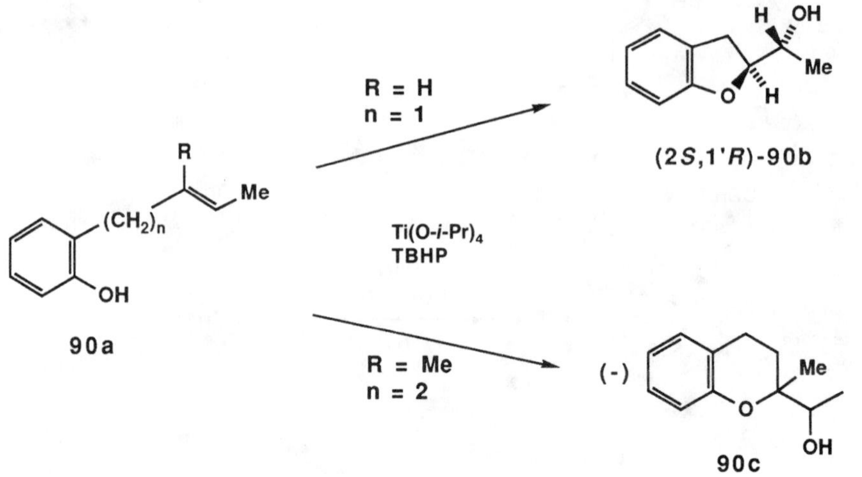

4.1.6. Mechanism of Titanium Tartrate Catalyzed Asymmetric Epoxidation

The hallmark of titanium tartrate catalyzed asymmetric epoxidation is the high degree of enantiofacial selectivity seen for a wide range of allylic alcohols. It is natural to inquire into what the mechanism of this reaction might be and what structural features of the catalyst produce these desirable results. These questions have been studied extensively, and the results have been the subject of considerable discussion [6,135,136]. For the purpose of this chapter, we review the aspects of the mechanistic–structural studies that may be helpful in devising synthetic applications of this reaction.

Of fundamental importance to an understanding of the reaction and its mechanism is the fact that in solution there is rapid exchange of titanium ligands [6,135,136]. Thus, when equimolar solutions of a titanium alkoxide and a dialkyl tartrate are mixed, the equilibrium represented by equation 4 will be quickly reached with all but the most sterically demanding alkoxides.

$$Ti(OR)_4 + \text{tartrate} \rightleftharpoons Ti(\text{tartrate})(OR)_2 + 2\,ROH \qquad (4)$$

This equilibrium is shifted far to the right because a chelating diol (i.e., the tartrate) has a much higher binding constant for titanium than do monodentate alcohols. The binding of tartrate is also enhanced by the increased acidity of its hydroxyl groups (due to the inductive effect of the esters). Spectroscopic evidence clearly reveals that two moles of free monodentate alcohol are present at equilibrium. Rapid ligand exchange continues as the hydroperoxide oxidant and the allylic alcohol substrate are added to the reaction medium. Pseudo-first-order kinetic experiments have shown a first-order rate dependence on the titanium–tartrate complex, the hydroperoxide, and the allylic alcohol and an inverse second-order dependence on the nonolefinic alcohol ligands (i.e., the isopropyl alcohol). The rate law derived from these results is expressed in equation 5. The mechanistic pathway outlined below in Figure 7 is consistent with equation 5 and clearly illustrates the ligand exchange processes essential for catalytic epoxidation.

$$\text{Rate} \;=\; k\frac{[\text{Ti(tartrate)(OR)}_2][\text{TBHP}][\text{allylic alcohol}]}{[\text{ligand alcohol}]^2} \tag{5}$$

After formation of the Ti(tartrate)(OR)$_2$ complex, the two remaining alkoxide ligands are replaced in reversible exchange reactions by the hydroperoxide (TBHP) and the allylic alcohol to give the "loaded" complex Ti(tartrate)(TBHP)(allylic alcohol). Now, in the rate-controlling step of the process, oxygen transfer from the coordinated hydroperoxide to the allylic alcohol gives the complex Ti(tartrate)(t-OBu)(epoxy alcohol). The product alkoxides are replaced by more allylic alcohol and TBHP to regenerate the "loaded" complex and complete the catalytic cycle.

An alternate mechanism invoking an ion-pair transition state assembly has been proposed to account for the enantioselectivity of the asymmetric epoxidation pro-

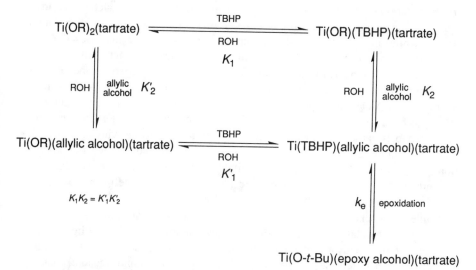

Figure 7 Ligand exchange on titanium during the asymmetric epoxidation catalytic cycle.

cess [137]. In this proposal, two additional alcohol species are required in the transition state complex. This requirement is inconsistent with the kinetic studies of this reaction that have led to the rate law expressed in equation 5 and, therefore, this proposal must be considered to be incorrect.

Much of the experimental success of asymmetric epoxidation lies in exercising proper control of equation 4 [6]. Both $Ti(OR)_4$ and $Ti(tartrate)(OR)_2$ are active epoxidation catalysts, and since the former is achiral, any contribution by that species to the epoxidation will result in loss of enantioselectivity. The addition to the reaction of more than one equivalent of tartrate, relative to titanium, will have the effect of minimizing the leftward component of the equilibrium and will suppress the amount of $Ti(OR)_4$ present in the reaction. The excess tartrate, however, forms $Ti(tartrate)_2$, which has been shown to be a catalytically inactive species and will cause a decrease in reaction rate that is proportional to the excess tartrate added. The need to minimize $Ti(OR)_4$ concentration and, at the same time, to avoid a drastic reduction in rate of epoxidation is the basis for the recommendation of a 10–20 mol % excess of tartrate over titanium for formation of the catalytic complex. After the addition of hydroperoxide and allylic alcohol to the reaction, the concentration of ROH will increase accordingly, and this will increase the leftward pressure on the equilibrium shown in equation 4. Fortunately, in most situations this shift apparently is extremely slight and is effectively suppressed by the use of excess tartrate. A shift in the equilibrium does begin to occur, however, when the reaction is run in the catalytic mode and the amount of catalyst used is less than about 5 mol % relative to allylic alcohol substrate. Loss in enantioselectivity then may be observed. This factor is the basis of the recommendation for use of 5–10 mol % of titanium–tartrate complex when using the catalytic version of asymmetric epoxidation.

Comparison of the epoxidation rates of several parasubstituted cinnamyl alcohols reveals that the olefin acts as a nucleophile toward the activated peroxide oxygen in the epoxidation reaction [136]. Relative to unsubstituted cinnamyl alcohol (relative rate = 1), an electron-withdrawing p-nitro group decreases the rate of epoxidation (0.42), while an electron-releasing group such as p-methoxy increases the rate (4.39). These results are consistent with the notion of the olefin acting as a nucleophile. Additional support for this conclusion arises from comparison of the rates for epoxidation of less substituted allylic alcohols with those for more highly substituted analogs. A clear example of this substituent effect is seen in the epoxidation of (R,S)-2,4-hexadien-3-ol (**80**), described in the preceding section, where the propenyl group is epoxidized in nearly complete preference to the vinyl group [77]. Another example is seen with the allylic–homoallylic alcohol **91**, where epoxidation occurs preferentially at the tetrasubstituted homoallylic olefin to give **92** [99]. The preferential epoxidation of the more highly substituted olefin in these compounds is consistent with a nucleophilic role for the olefin.

While the mechanistic scheme portrayed in Figure 7 provides important insight into the experimental aspects of asymmetric epoxidation, it sheds little light on the structure of the catalyst and on the features of the catalyst responsible for the concurrent high stereoselectivity and broad generality. The rapid ligand exchange, so crucial to the success of the reaction, makes characterization of the catalyst structure extremely difficult. Some reliable structural information has been obtained from spectroscopic measurements on the complex in solution [6,136b]. These data clearly support the conclusion that the major molecular species formed in solution is the dimeric composite, $Ti_2(tartrate)_2(OR)_4$. Efforts to isolate this complex, ideally as a crystalline solid, have so far been fruitless. Therefore, assignment of a structure to the dimeric complex has depended on information provided by the X-ray crystallographic structure obtained for the closely related complex, $Ti_2(dibenzyltartramide)_2(OR)_4$ [138]. The assumption of a similarity of structure for these two complexes receives some support from the fact that both catalyze the epoxidation of α-phenylcinnamyl alcohol with the same enantiofacial selectivity. From this analogy, the structure shown in Figure 8 has been proposed for the $Ti_2(tartrate)_2(OR)_4$ complex. This structure has a C_2 axis of symmetry with the two titanium atoms in identical stereochemical environments. To account for the sameness of all the tartrate ester groups in the room temperature NMR spectrum, a fluxional equilibrium between the two structurally degenerate complexes shown in Figure 8 has been proposed. Catalysis of the epoxidation process is thought to involve only one of the two titanium atoms, but the possibility that both are required has not yet been ruled out.

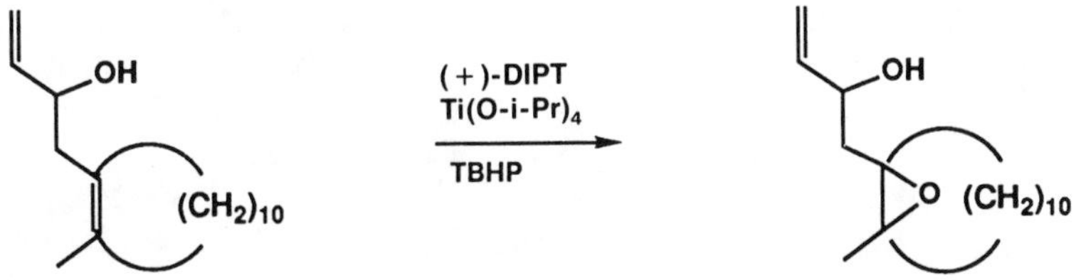

"Loading" of the catalyst with hydroperoxide and substrate can now be considered in terms of the proposed structure [6]. Orientation of these two ligands on the catalyst becomes a crucial issue. Three coordination sites, two axial and one equatorial, become available by exchange of two isopropoxides and dissociation of the coordinated ester carbonyl group. These processes can occur with minimal perturbation of the remaining catalyst structure. The three coordination sites are in a semicircular (i.e., meridional) array around one edge of the catalyst surface. In the reactive

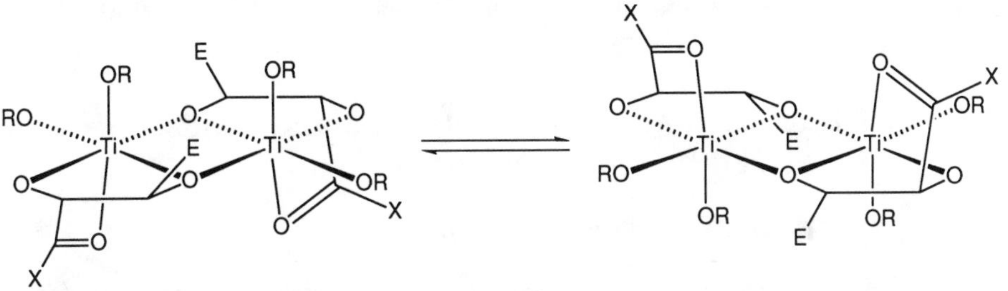

Figure 8 Fluxional equilibrium proposed for the titanium–tartrate complex in solution.

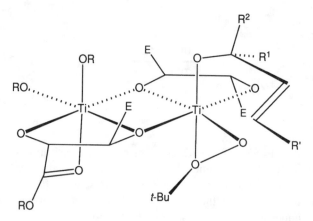

Figure 9 Proposed structure of "loaded" catalyst at the time of oxygen transfer.

mode, coordination of the hydroperoxide is assumed to be bidentate by analogy to the precedent of bidentate TBHP coordination to vanadium [6,139]. The hydroperoxide must occupy the equatorial and one of the two available axial coordination sites, with the allylic alcohol in the remaining axial site. To achieve the necessary proximity for transfer of oxygen (the distal peroxide oxygen is assumed to be transferred) to the olefin, the distal oxygen is placed in the equatorial site (Fig. 9) and the proximal oxygen is placed in the axial site. The axial site on the lower face of the complex (as drawn in Fig. 9) is chosen for the peroxide because of the larger steric demands of the *t*-butyl, or especially of the trityl group when trityl hydroperoxide is used, in comparison to the allylic alcohol.

The allylic alcohol binds to the remaining axial coordination site, where stereochemical and stereoelectronic effects dictate the conformation shown in Figure 9 [6]. The structural model of catalyst, oxidant, and substrate shown in Figure 9 illustrates a detailed version of the formalized rule presented in Figure 1. Ideally, all observed stereochemistry of epoxy alcohol and kinetic resolution products can be rationalized according to the compatibility of their binding with the stereochemistry and stereoelectronic requirements imposed by this site [6]. A transition state model for the asymmetric epoxidation complex has been calculated by a frontier orbital approach and is consistent with the formulation portrayed in Figure 9 [140].

4.1.7. Other Asymmetric Epoxidations and Oxidations Catalyzed by Titanium Complexes

4.1.7.1. Dititanium–Ditartrate Complex

The discussion to this point has focused entirely on the epoxidation of allylic (and homoallylic) alcohols catalyzed by the Ti(OR)$_2$(tartrate) complex. The role of the olefin as a nucleophile toward the activated peroxide oxygen in this reaction has been established (see discussion of mechanism). If the olefin of the allylic alcohol is replaced by another nucleophilic group then, in principle, oxidation of that group may occur (eq. 6) [141].

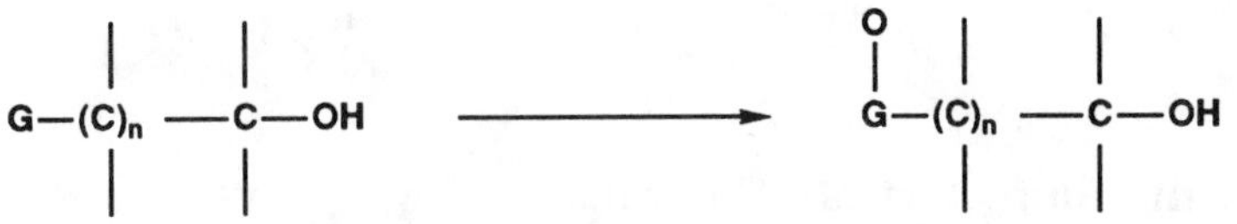

$$G-(C)_n-\overset{\displaystyle |}{\underset{\displaystyle |}{C}}-OH \longrightarrow \overset{\displaystyle O}{G}-(C)_n-\overset{\displaystyle |}{\underset{\displaystyle |}{C}}-OH \qquad (6)$$

In practice, oxidations of this type have been observed and generally have been carried out with a substrate bearing a racemic secondary alcohol so that kinetic resolution is achieved. While these oxidations are not strictly within the scope of this chapter, they are summarized briefly in equations 7–9 to acquaint the reader with other potential uses for the titanium–tartrate catalytic complex. In the kinetic resolutions shown in Equations 7 and 8, the oxidations are controlled by limiting the amount of oxidant used to 0.6 equiv. Only modest resolution was attained for the acetylenic alcohol (eq. 7, 21% ee) [77] and the allenic alcohol (eq. 8, 40% ee) [77]. Resolutions of the furanols [142] or the thiophene alcohols [143] of equation 9 generally are excellent (90–98% ee except when R_1 is a t-butyl group). Only in the kinetic resolution of the furanols has the oxidation product been identified and, in that case, is a dihydropyranone.

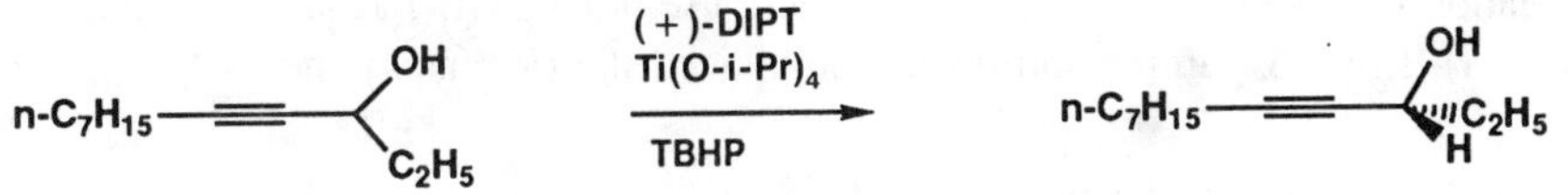

$$(7)$$

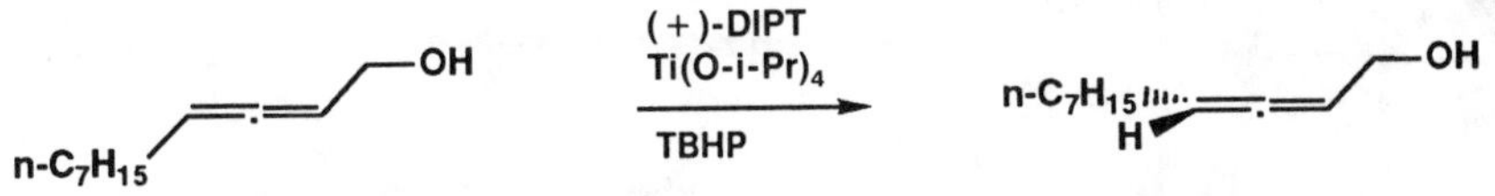

$$(8)$$

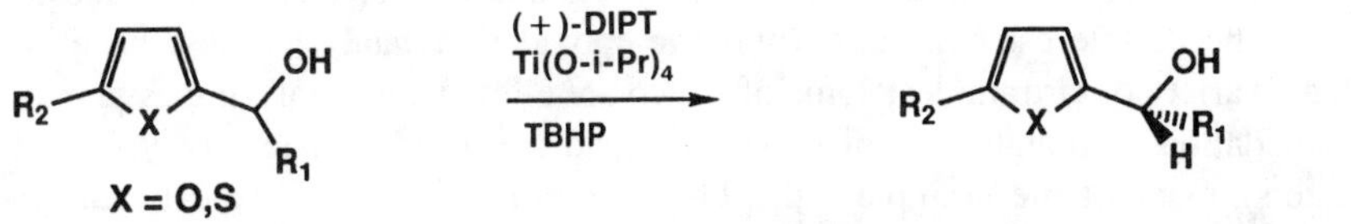

$$(9)$$

The asymmetric epoxidation of an allylic alcohol in which the carbinol has been replaced by a silanol has been described [144]. As shown in equation 10 [144], (3E)-phenylethenyldimethylsilanol is converted to an epoxysilanol in 50% yield with 85–95% ee. Note that here the longer Si—C bonds appear to overcome the restriction to epoxidation associated with a fully substituted C-1 atom in the allylic alcohol series. Fluoride cleavage of the silanol group gives (S)-styrene oxide.

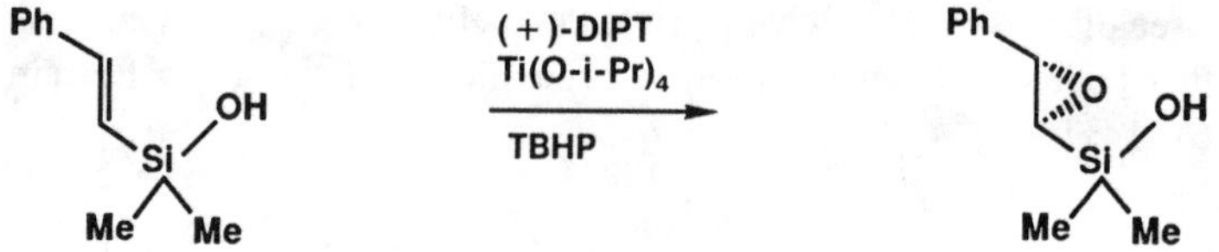

(10)

4.1.7.2. Dititanium–Tartrate Complex

The β-hydroxy amines are a class of compounds falling within the generic definition of equation 6. When the alcohol is secondary, the possibility for kinetic resolution exists if the titanium–tartrate complex is capable of catalyzing the enantioselective oxidation of the amine to an amine oxide (or other oxidation product). The use of the "standard" asymmetric epoxidation complex (i.e., $Ti_2(tartrate)_2$) to achieve such a enantioselective oxidation was unsuccessful. However, modification of the complex so that the stoichiometry lies between Ti_2 (tartrate)$_1$ and $Ti_2(tartrate)_{1.5}$ leads to a very successful kinetic resolutions of β-hydroxyamines. A representative example is shown in equation 11 [141b,c]. The oxidation and kinetic resolution of more than 20 secondary β-hydroxyamines [141,145a] provides an indication of the scope of the reaction and of some structural limitations to good kinetic resolution. These results also show a consistent correlation of absolute configuration of the resolved hydroxyamine with the configuration of tartrate used in the catalyst. This correlation is as shown in equation 11, where use of (+)-DIPT results in oxidation of the (S)-β-hydroxyamine and leaves unoxidized the (R)-enantiomer.

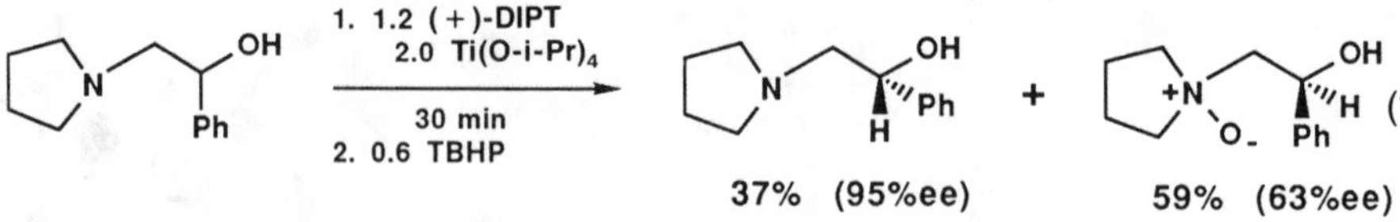

37% (95%ee)   59% (63%ee)

4.1.7.3. Titanium–Tartramide Complexes

A number of derivatives of the tartaric acid structure have been examined as substitutes for the tartrate ester in the asymmetric epoxidation catalyst. These have included a variety of tartramides, some of which are effective in catalyzing asymmetric epoxidation (although none displays the broad consistency of results typical of the esters). One notable example is the dibenzyltartramide, which in a 1:1 ratio (in reality, a 2:2 complex as shown by an X-ray crystallographic structure determination [138]) with Ti(O-i-Pr)$_4$ catalyzes the epoxidation of allylic alcohols with the same enantiofacial selectivity as does the titanium–tartrate ester complex [18]. Remarkably, when the ratio of dibenzyltartramide to titanium is changed to 1:2, epoxidation is catalyzed with *reversed* enantiofacial selectivity. These results are illustrated for the epoxidation of α-phenylcinnamyl alcohol (eq. 12a).

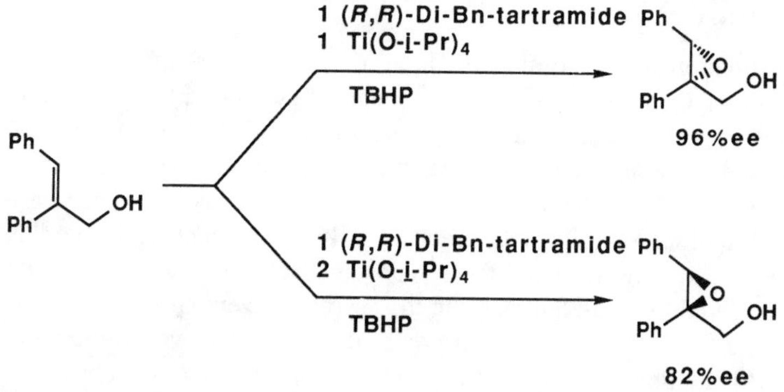

$$(12a)$$

[α-Phenylcinnamyl alcohol is a particularly felicitous substrate for asymmetric epoxidation; epoxidation of other allylic alcohols with the 1:2 dibenzyltartramide–titanium complex does not give as high enantioselectivities, but the reversed selectivity is consistent throughout [18].] An extensive listing of tartramides used in the epoxidation of α-phenylcinnamyl alcohol with both 1:1 and 1:2 catalysts has been tabulated elsewhere [6].

4.1.7.4. Ti(O-*i*-Pr)$_2$Cl$_2$-tartrate Complexes

As described in earlier sections of this chapter, certain epoxy alcohols (e.g., the 2-monosubstituted epoxy alcohols) are particularly susceptible to ring-opening processes. With the intent of controlling the ring-opening reaction, the epoxidation catalyst was modified by the use of Ti(O-*i*-Pr)$_2$Cl$_2$ in place of Ti(O-*i*-Pr)$_4$, the idea being to open the ring with chloride to produce a chlorodiol [18]. This modification was successful, with 3-chloro-1,2-diols being formed in yields of 60–80% with good regioselectivity. Epoxy alcohols were assumed to be intermediates in these reactions and can be regenerated from the chlorodiols by base-promoted ring closure. Unfortunately, the enantioselectivity of the process is variable, with enantiomeric purities generally in the range of 20–70% ee. A point of interest concerning the chlorohydroxylation process is that the reversal of enantiofacial selectivity from that of the normal asymmetric epoxidation process is not altered by changing the ratio of Ti(O-*i*-Pr)$_2$Cl$_2$ to tartrate from 1:1 to 2:1. Chlorohydroxylation of 2-(6-chloropyridin-2-yl)-2-propen-1-ol (shown in eq. 12b) followed by closure of the epoxide ring has provided a useful route to the optically active epoxy alcohol in 50% yield and with 90% ee [145b].

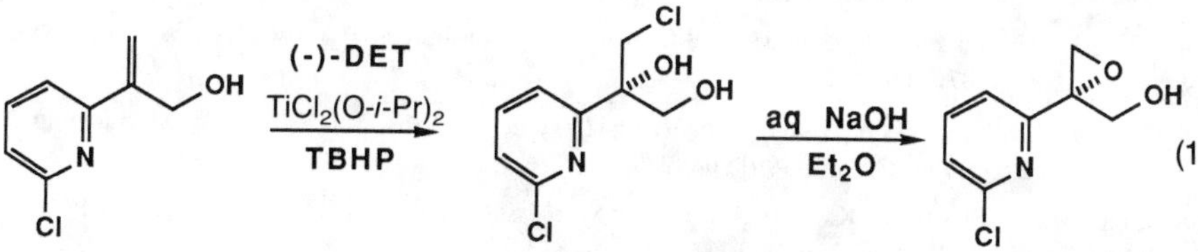

$$(12b)$$

4.1.8. Conclusion

Asymmetric epoxidation of allylic alcohols is a very reliable chemical reaction. More than a decade of experience has confirmed that the titanium–tartrate catalyst is extremely tolerant of structural diversity in the allylic alcohol substrate for epoxidation yet is highly selective in its ability to discriminate between the enantiofaces of the prochiral olefin. Today the practitioner of organic chemistry need provide only the allylic alcohol in order to perform the reaction. All other reagents and materials required for the reaction are available from supply houses and usually are sufficiently pure as received to be used directly in the asymmetric epoxidation process. [When purchasing *t*-butyl hydroperoxide in prepared solutions, however, the more concentrated 5.5 M solution in isooctane (2,2,4-trimethylpentane) should always be chosen over the 3.0 M solution.] If the considerations presented in this chapter are observed, with attention to the moderately stringent technique outlined, no difficulty should be encountered in performing this reaction.

Before 1986, asymmetric epoxidations frequently were performed by using a stoichiometric amount of the titanium–tartrate catalyst relative to the amount of allylic alcohol. This was necessary to obtain a reasonable reaction rate for the epoxidation as well as to drive the reaction to completion. The report in 1986 recommending the addition of activated molecular sieves to the reaction milieu makes it possible to use only catalytic amounts of the titanium–tartrate complex for nearly all asymmetric epoxidations.

The essence of the asymmetric epoxidation process, including correlation of enantiofacial selectivity with tartrate ester stereochemistry, is outlined in Figure 1. No exceptions to the face-selectivity rules shown in Figure 1 have been reported to date. Consequently, one can use this scheme with considerable confidence to predict and assign absolute configuration to the epoxides obtained from *prochiral allylic alcohols*. When allylic alcohols having chiral substituents at C-1, C-2, and/or C-3 are used in the reaction, the assignment of stereochemistry to the newly introduced epoxide group must be done with considerably more care.

A remaining goal related to asymmetric epoxidation is to obtain additional structural information about the titanium–tartrate catalyst as well as about the catalyst loaded with substrate and oxidant.

References

1. (a) Henbest, H. B. *Chem. Soc., Spec. Publ.* **1965,** *19,* 83. (b) Ewins, R. C.; Henbest, H. B.; McKervey, M. A. *J. Chem. Soc., Commun.* **1967,** 1085.

2. Katsuki, T.; Sharpless, K. B. *J. Am. Chem. Soc.* **1980,** *102,* 5974.

3. Hanson, R. M.; Sharpless, K. B. *J. Org. Chem.* **1986,** *51,* 1922.

4. Gao, Y.; Hanson, R. M.; Klunder, J. M.; Ko, S. Y.; Masamune, H.; Sharpless, K. B. *J. Am. Chem. Soc.* **1987,** *109,* 5765.

5. Reprinted with permission from *Comprehensive Organic Synthesis*, Vol. 7; Trost, B. M.; Fleming, I. (Eds.); Pergamon: Oxford, 1991; pp. 389–436.

6. Finn, M. G.; Sharpless, K. B. In *Asymmetric Synthesis*, Morrison, J. D. (Ed.); Academic Press: Orlando, FL, 1985; Vol. 5, p. 247.

7. (a) Rossiter, B. In *Asymmetric Synthesis;* Morrison, J. D. (Ed.); Academic Press: Orlando, FL, 1985; Vol. 5, p. 193. (b) Pfenniger, A. *Synthesis,* **1986,** 89.

8. Zeller, K. P. In *Houben-Weyl, Methoden der organische Chemie,* Vol. E13, Part 2; *Organische Peroxo-verbindungen;* Kropf, H. (Ed.); Thieme: Stuttgart, 1988; pp. 1210–1250

9. Hanson, R. M. *Chem. Rev.* **1991,** *91,* 437.

10. McGarvey, G. J.; Kimura, M.; Oh, T.; Williams, J. M. *Carbohydr. Chem.* **1984,** *3,* 125.

11. Behrens, C. H.; Sharpless, K. B. *Aldrichim. Acta,* **1983,** *16,* 67.

12. (a) Sharpless, K. B. *Proc. Robert A. Welch Found. Conf. Chem. Res.* **1984,** *27,* 59. (b) Sharpless, K. B. *ChemTech,* **1985,** *15,* 692. (c) Sharpless, K. B. *Chem. Br.* **1986,** *22,* 38.

13. Katsuki, T.; Martin, V. S. In preparation.

14. Martin, V. S.; Woodard, S. S.; Katsuki, T.; Yamada, Y.; Ikeda, M.; Sharpless, K. B. *J. Am. Chem. Soc.* **1981,** *103,* 6237.

15. (a) Klunder, J. M.; Ko, S. Y.; Sharpless, K. B. *J. Org. Chem.* **1986,** *51,* 3710. (b) Ko, S. Y.; Sharpless, K. B. *J. Org. Chem.* **1986,** *51,* 5413.

16. (a) Klunder, J. M.; Onami, T.; Sharpless, K. B. *J. Org. Chem.* **1989,** *54,* 1295. (b) Burgos, C. E.; Ayer, D. E.; Johnson, R. A. *J. Org. Chem.* **1988,** *53,* 4973. (c) Guivisdalsky, P. N.; Bittman, R. *Tetrahedron Lett.* **1988,** *29,* 4393. (d) Johnson, R. A.; Burgos, C. E.; Nidy, E. G. *Chem. Phys. Lipids,* **1989,** *50,* 119. (e) Burgos, C. E.; Nidy, E. G.; Johnson, R. A. *Tetrahedron Lett.* **1989,** *30,* 5081. (f) Guivisdalsky, P. N.; Bittman, R. *J. Am. Chem. Soc.* **1989,** *111,* 3077. (g) Byun, H.-S.; Bittman, R. *Tetrahedron Lett.* **1989,** *30,* 2751. (h) Guivisdalsky, P. N.; Bittman, R. *J. Org. Chem.* **1989,** *54,* 4637. (h) Guivisdalsky, P. N.; Bittman, R. *J. Org. Chem.* **1989,** *54,* 4643. (i) (*R*)- and (*S*)-glycidol are available in commercial quantity from ARCO Chemical Company, Newton Square, PA.

17. Ko, S. Y.; Masamune, H.; Sharpless, K. B. *J. Org. Chem.* **1987,** *52,* 667.

18. Lu, L. D. L.; Johnson, R. A.; Finn, M. G.; Sharpless, K. B. *J. Org. Chem.* **1984,** *49,* 728.

19. Hill, J. G.; Rossiter, B. E.; Sharpless, K. B. *J. Org. Chem.* **1983,** *48,* 3607.

20. (a) Carlier, P. R.; Sharpless, K. B. *J. Org. Chem.* **1989,** *54,* 4016. (b) Burns, C. J.; Martin, C. A.; Sharpless, K. B. *J. Org. Chem.* **1989,** *54,* 2826.

21. Farrall, M. J.; Alexis, M.; Trecarten, M. *Nouv. J. Chim.* **1983,** *7,* 449.

22. (a) Wadsworth, W. S. *Org. React,* **1977,** *25,* 73. (b) Walker, B. J. In *Organophosphorus Reagents in Organic Synthesis;* Cadogan, J. I. G. (Ed.); Academic Press: Orlando, FL, 1979; pp. 155–205.

23. Still, W. C.; Gennari, C. *Tetrahedron Lett.* **1983,** *24,* 4405.

24. Mancuso, A. J.; Huang, S.-L.; Swern, D. *J. Org. Chem.* **1978,** *43,* 2480.

25. Nicolaou, K. C.; Daines, R. A.; Uenishi, J.; Li, W. S.; Papahatjia, D. P.; Chakraborty, T. K. *J. Am. Chem. Soc.* **1987,** *109,* 2205; **1988,** *110,* 4672.

26. Schreiber, S. L.; Schreiber, T. S.; Smith, D. B. *J. Am. Chem. Soc.* **1987,** *109,* 1525.

27. Denmark, S. E.; Jones, T. K. *J. Org. Chem.* **1982,** *47,* 4595.

28. Marvell, E. N.; Li, T. *Synthesis,* **1973,** 457.

29. Jurczak, J.; Pikul, S.; Bauer, T. *Tetrahedron,* **1986,** *42,* 447.

30. Breitgoff, D.; Laumen, K.; Schneider, M. P. *J. Chem. Soc., Chem. Commun.* **1986,** 1523.

31. Kleemann, A.; Wagner, R. *Glycidol;* Huthig: New York, 1981.

32. Gao, Y.; Sharpless, K. B. *J. Org. Chem.* **1988,** *53,* 4081.

33. Dung, J. S.; Armstrong, R. W.; Anderson, O. P.; Williams, R. M. *J. Org. Chem.* **1983,** *48,* 3592.

34. Meister, C.; Scharf, H. D. *Justus Liebigs Ann. Chem.* **1983,** 913.

35. Tanner, D.; Somfai, P. *Tetrahedron,* **1986,** *42,* 5985.

36. Giese, B.; Rupaner, R. *Justus Liebigs Ann. Chem.* **1987,** 231.

37. Mori, K.; Ebata, T.; Takechi, S. *Tetrahedron,* **1984,** *40,* 1761.

38. (a) Schweiter, M. J.; Sharpless, K. B. *Tetrahedron Lett.* **1985,** *26,* 2543. (b) Adam, W.; Griesbeck, A.; Staab, E. *Tetrahedron Lett.* **1986,** *27,* 2839. (c) Adam, W.; Braun, M.; Griesbeck, A.; Lucchini, V.; Staab, E.; Will, B. *J. Am. Chem. Soc.* **1989,** *111,* 203.

39. (a) White, J. D.; Jayasinghe, L. R. *Tetrahedron Lett.* **1988,** *29,* 2138. (b) White, J. D.; Amedio, J. C., Jr.; Gut, S.; Jayasinghe, L. *J. Org. Chem.* **1989,** *54,* 4268.

40. (a) Masamune, S.; Choy, W.; Petersen, J. S.; Sita, L. R. *Angew. Chem., Int. Ed. Engl.* **1985,** *24,* 1. (b) Nishikimi, Y.; Iimori, T.; Sodeoka, M.; Shibasaki, M. *J. Org. Chem.* **1989,** *54,* 3354.

41. Rossiter, B. E.; Katsuki, T.; Sharpless K. B. *J. Am. Chem. Soc.* **1981,** *103,* 464.

42. Baker, R.; Cummings, W. J.; Hayes, J. F.; Kumar, A. *J. Chem. Soc., Chem. Commun.* **1986,** 1237.

43. Kuroda, C.; Theramongkol, P.; Engebrecht, J. R.; White, J. D. *J. Org. Chem.* **1986,** *51,* 956.

44. Honda, M.; Katsuki, T.; Yamaguchi, M. *Tetrahedron Lett.* **1984,** *25,* 3857.

45. Gorthy, L. A.; Vairamani, M.; Djerassi, C. *J. Org. Chem.* **1984,** *49,* 1511.

46. Hanessian, S.; Ugolini, A.; Dube, D.; Hodges, P. J.; Andre, C. *J. Am. Chem. Soc.* **1986,** *108,* 2776.

47. Wershofen, S.; Scharf, H. D. *Synthesis,* **1988,** 854.

48. Tung, R. D.; Rich, D. H. *Tetrahedron Lett.* **1987,** *28,* 1139.

49. Molander, G. A.; Hahn, G. *J. Org. Chem.* **1986,** *51,* 2596.

50. Corey, E. J., Pyne, S. G.; Su, W. G. *Tetrahedron Lett.* **1983,** *24,* 4883.

51. Furukawa, J.; Iwasaki, S.; Okuda, S. *Tetrahedron Lett.* **1983,** *24,* 5257.

52. Kitano, Y.; Kobayashi, Y.; Sato, F. *J. Chem. Soc., Chem. Commun.* **1985,** 498.

53. Roush, W. R.; Adam, M. A. *J. Org. Chem.* **1985,** *50,* 3752.

54. Oehlschlager, A. C.; Johnston, B. D. *J. Org. Chem.* **1987,** *52,* 940.

55. Lai, C. K., Gut, M. *J. Org. Chem.* **1987,** *52,* 685.

56. Dolle, R. E.; Nicolaou, K. C. *J. Am. Chem. Soc.* **1985,** *107,* 1691.

57. Pridgen, L. N.; Shilcrat, S. C.; Lantos, I. *Tetrahedron Lett.* **1984,** *25,* 2835.

58. Baker, S. R.; Boot, J. R.; Morgan, S. E.; Osborne, D. J.; Ross, W. J.; Shrubsall, P. R. *Tetrahedron Lett.* **1983,** *24,* 4469.

59. Katsuki, T.; Lee, A. W. M.; Ma, P.; Martin, V. S.; Masamune, S.; Sharpless, K. B.; Tuddenham, D.; Walker, F. J. *J. Org. Chem.* **1982,** *47,* 1373.

60. Ma, P.; Martin, V. S.; Masamune, S.; Sharpless, K. B.; Viti, S. M. *J. Org. Chem.* **1982,** *47,* 1378.

61. (a) Brunner, H.; Sicheneder, A. *Angew Chem., Int. Ed. Engl.* **1988,** *27,* 718. (b) Evans, D. A.; Williams, J. M. *Tetrahedron Lett.* **1988,** *29,* 5065. (c) Hughes, P.; Clardy, J. *J. Org. Chem.* **1989,**

54, 3260. (d) Nicolaou, K. C.; Prasad, C. V. C.; Somers, P. K.; Hwang, C.-K. *J. Am. Chem. Soc.* **1989**, *111*, 5330.

62. (a) Minami, N.; Ko, S. S.; Kishi, Y. *J. Am. Chem. Soc.* **1982**, *104*, 1109. (b) Roush, W. R.; Adam, M. A.; Walts, A. E.; Harris, D. J. *J. Am. Chem. Soc.* **1986**, *108*, 3422.

63. (a) Ko, S. Y.; Lee, A. W. M.; Masamune, S.; Reed, L. A., III; Sharpless, K. B.; Walker, F. J. *Science*, **1983**, *220*, 949. (b) Ko, S. Y.; Lee, A. W. M.; Masamune, S.; Reed, L. A., III; Sharpless, K. B.; Walker, F. J. *Tetrahedron*, **1990**, *46*, 245. (c) Reed, L. A., III; Ito, Y.; Masamune, S.; Sharpless, K. B. *J. Am. Chem. Soc.* **1982**, *104*, 6468.

64. Wrobel, J. E.; Ganem, B. *J. Org. Chem.* **1983**, *48*, 3761.

65. Behrens, C. H.; Ko, S. Y.; Sharpless, K. B.; Walker, F. J. *J. Org. Chem.* **1985**, *50*, 5687.

66. (a) Hoye, T. R.; Suhadolnik, J. C. *J. Am. Chem. Soc.* **1985**, *107*, 5312. (b) Hoye, T. R.; Suhadolnik, J. C. *Tetrahedron*, **1986**, *42*, 2855.

67. Doherty, A. M.; Ley, S. V. *Tetrahedron Lett.* **1986**, *27*, 105.

68. Baldwin, J. E.; Flinn, A. *Tetrahedron Lett.* **1987**, *28*, 3605.

69. Caron, M.; Sharpless, K. B. *J. Org. Chem.* **1985**, *50*, 1557.

70. Chong, J. M.; Sharpless, K. B. *J. Org. Chem.* **1985**, *50*, 1560.

71. Baker, R.; Swain, C. J.; Head, J. C. *J. Chem. Soc., Chem. Commun.* **1986**, 874.

72. Wood, R. D.; Ganem, B. *Tetrahedron Lett.* **1982**, *23*, 707.

73. Takahashi, T.; Miyazawa, M.; Veno, H.; Tsuji, J. *Tetrahedron Lett.* **1986**, *27*, 3881.

74. Mills, L. S.; North, P. C. *Tetrahedron Lett.* **1983**, *24*, 409.

75. (a) Suzuki, M.; Morita, Y.; Yanagisawa, A.; Noyori, R.; Baker, B. J.; Scheur, B. J. *J. Am. Chem. Soc.* **1986**, *108*, 5021. (b) Suzuki, M.; Morita, Y.; Yanagisawa, A.; Baker, B. J.; Scheuer, P. J.; Noyori, R. *J. Org. Chem.* **1988**, *53*, 286.

76. Denis, J.-N.; Greene, A. E.; Serra, A. A.; Luche, M.-J. *J. Org. Chem.* **1986**, *51*, 46.

77. Sharpless, K. B.; Behrens, C. H.; Katsuki, T.; Lee, A. W. M.; Martin, V. S.; Takatani, M.; Viti, S. M.; Walker, F. J.; Woodard, S. S. *Pure Appl. Chem.* **1983**, *55*, 589.

78. (a) Nagaoka, H.; Kishi, Y. *Tetrahedron*, **1981**, *37*, 3873. (b) Hirai, Y.; Chintani, M.; Yamazaki, T.; Momose, T. *Chem. Lett.* **1989**, 1449. (c) Ebata, T.; Mori, K. *Agric. Biol. Chem.* **1989**, *53*, 801. (d) Mori, K.; Takeuchi, T. *Justus Liebigs Ann. Chem.* **1989**, 453. (e) Boeckman, R. K., Jr.; Pruitt, J. R. *J. Am. Chem. Soc.* **1989**, *111*, 8286.

79. Still, W. C.; Ohmizu, H. *J. Org. Chem.* **1981**, *46*, 5242.

80. Garner, P.; Park, J. M.; Rotello, V. *Tetrahedron Lett.* **1985**, *26*, 3299.

81. Meyers, A. I.; Hudspeth, J. P. *Tetrahedron Lett.* **1981**, *22*, 3925.

82. Niwa, N.; Miyachi, Y.; Uosaki, Y.; Yamada, K. *Tetrahedron Lett.* **1986**, *27*, 4601.

83. Takabe, K.; Okisaki, K.; Uchiyama, Y.; Katagiri, T.; Yoda, H. *Chem. Lett.* **1985**, 561.

84. (a) Mori, K.; Ueda, H. *Tetrahedron*, **1981**, *37*, 2581. (b) Nicolaou, K. C.; Prasad, C. V. C.; Hwang, C.-K.; Duggan, M. E.; Veale, C. A. *J. Am. Chem. Soc.* **1989**, *111*, 5321.

85. Medina, J. C.; Kyler, K. S. *J. Am. Chem. Soc.* **1988**, *110*, 4818.

86. Aziz, M.; Rouessac, F. *Tetrahedron*, **1988**, *44*, 101.

87. Morimoto, Y.; Oda, K.; Shirahama, H.; Matsumoto, T.; Omura, S. *Chem. Lett.* **1988**, 909.

88. (a) Takahashi, K.; Ogata, M. *J. Org. Chem.* **1987**, *52*, 1877. (b) Naruta, Y.; Nishigaichi, Y.;

Maruyama, K. *Tetrahedron Lett.* **1989**, *30*, 3319. (c) Nicolaou, K. C.; Duggan, M. E.; Hwang, C.-K. *J. Am. Chem. Soc.* **1989**, *111*, 6676.

89. Rastetter, W. H.; Adams, J. *Tetrahedron Lett.* **1982**, *23*, 1319.

90. Evans, D. A.; Bender, S. L.; Morris, J. *J. Am. Chem. Soc.* **1988**, *110*, 2506.

91. Reddy, K. S.; Ko, O. H.; Ho, D.; Persons, P. E.; Cassidy, J. M. *Tetrahedron Lett.* **1987**, *28*, 3075.

92. Yamada, S.; Shiraishi, M.; Ohmori, M.; Takayama, H. *Tetrahedron Lett.* **1984**, *25*, 3347.

93. Mori, K.; Okada, K. *Tetrahedron*, **1985**, *41*, 557.

94. Roush, W. R.; Blizzard, T. A. *J. Org. Chem.* **1984**, *49*, 4332.

95. Bonadies, F.; Rossi, G.; Bonini, C. *Tetrahedron Lett.* **1984**, *25*, 5431.

96. Sodeoka, M.; Iimori, T.; Shibasaki, M. *Tetrahedron Lett.* **1985**, *26*, 6497.

97. (a) Meyers, A. G.; Porteau, P. J.; Handel, T. M. *J. Am. Chem. Soc.* **1988**, *110*, 7212. (b) Williams, D. R.; Brown, D. L.; Benbow, J. W. *J. Am. Chem. Soc.* **1989**, *111*, 1923.

98. Erickson, T. J. *J. Org. Chem.* **1986**, *51*, 934.

99. Marshall, J. A.; Jenson, T. M. *J. Org. Chem.* **1984**, *49*, 1707.

100. Hamon, D. P. G.; Shirley, N. J. *J. Chem. Soc., Chem. Commun.* **1988**, 425.

101. Pettersson, L.; Frejd, T.; Magnusson, G. *Tetrahedron Lett.* **1987**, *28*, 2753.

102. Rizzi, J. P.; Kende, A. S. *Tetrahedron*, **1984**, *40*, 4693.

103. Acemoglu, M.; Uebelhart, P.; Rey, M.; Eugster, C. H. *Helv. Chim. Acta*, **1988**, *71*, 931.

104. Carlier, P. R.; Mungall, W. S.; Schroder, G.; Sharpless, K. B. *J. Am. Chem. Soc.* **1988**, *110*, 2978.

105. (a) Kitano, Y.; Matsumoto, T.; Takeda, Y.; Sato, F. *J. Chem. Soc., Chem. Commun.* **1986**, 1323. (b) Kitano, Y.; Matsumoto, T.; Sato, F. *Tetrahedron*, **1988**, *44*, 4073. (c) Kitano, Y.; Okamoto, S.; Sato, F. *Chem. Lett.* **1989**, 2163.

106. (a) Russell, A. T.; Procter, G. *Tetrahedron Lett.* **1987**, *28*, 2041. (b) Procter, G.; Russell, A. T.; Murphy, P. J.; Tan, T. S.; Mather, A. N. *Tetrahedron*, **1988**, *44*, 3953.

107. (a) Katsuki, T. *Tetrahedron Lett.* **1984**, *25*, 2821. (b) Schwab, J. M.; Ho, C.-K. *J. Chem. Soc., Chem. Commun.* **1986**, 872. (c) Schwab, J. M.; Ray, T.; Ho, C.-K. *J. Am. Chem. Soc.* **1989**, *111*, 1057.

108. Kitano, Y.; Matsumoto, T.; Wakasa, T.; Okamoto, S.; Shimazaki, T.; Kobayashi, Y.; Sato, F.; Miyaji, K.; Arai, K. *Tetrahedron Lett.* **1987**, *28*, 6351.

109. Kitano, Y.; Matsumoto, T.; Okamoto, S.; Shimazaki, T.; Kobayashi, Y.; Sato, F. *Chem. Lett.* **1987**, 1523.

110. Roush, W. R.; Spada, A. P. *Tetrahedron Lett.* **1983**, *24*, 3693.

111. Aristoff, P. A.; Johnson, P. D.; Harrison, A. W. *J. Am. Chem. Soc.* **1985**, *107*, 7967.

112. Overman, L. E.; Lin, N.-H. *J. Org. Chem.* **1985**, *50*, 3670.

113. Aggarwal, S. K.; Bradshaw, J. S.; Eguchi, M.; Parry, S.; Rossiter, B. E.; Markides, K. E.; Lee, M. L. *Tetrahedron*, **1987**, *43*, 451.

114. Ogata, M.; Matsumoto, H.; Takahashi, K.; Shimizu, S.; Kida, S.; Murahayashi, A.; Shiro, M.; Tawara, K. *J. Med. Chem.* **1987**, *30*, 1054.

115. (a) Sugiyama, S.; Honda, M.; Komori, T. *Justus Liebigs Ann. Chem.* **1988**, 619. (b) Tanaka, A.;

Suzuki, H.; Yamashita, K. *Agric. Biol. Chem.* **1989**, *53*, 2253. (c) Chamberlin, A. R.; Dezube, M.; Reich, S. H.; Sall, D. J. *J. Am. Chem. Soc.* **1989**, *111*, 6247. (d) Rama Rao, A. V.; Khrimian, A. P.; Radha Krishna, P.; Yagadiri, P.; Yadav, J. S. *Synth. Commun.* **1989**, *18*, 2325. (e) Discordia, R. P.; Dittmer, D. C. *J. Org. Chem.* **1990**, *55*, 1414.

116. Marshall, J. A.; Flynn, K. E. *J. Am. Chem. Soc.* **1982**, *104*, 7430.

117. Mihelich, E. D. Unpublished results.

118. Alvarez, E.; Manta, E.; Martin, J. D.; Rodriquez, M. L.; Ruiz-Perez, C. *Tetrahedron Lett.* **1988**, *29*, 2093.

119. Ronald, R. C.; Ruder, S. M.; Lillie, T. S. *Tetrahedron Lett.* **1987**, *28*, 131.

120. Ekhato, I. V.; Silverton, J. V.; Robinson, C. H. *J. Org. Chem.* **1988**, *53*, 2180.

121. (a) Marshall, J. A.; Flynn, K. E. *J. Am. Chem. Soc.* **1984**, *106*, 723. (b) Marshall, J. A.; Audia, V. H. *J. Org. Chem.* **1987**, *52*, 1106.

122. (a) Hatakeyama, S.; Sakurai, K.; Takano, S. *J. Chem. Soc., Chem. Commun.* **1985**, 1759. (b) Hafele, B.; Schroter, D.; Jager, V. *Angew. Chem., Int. Ed. Engl.* **1986**, *25*, 87. (c) Babine, R. E. *Tetrahedron Lett.* **1986**, *27*, 5791. (d) Schreiber, S. L.; Schreiber, T. S.; Smith, D. B. *J. Am. Chem. Soc.* **1987**, *109*, 1525. (e) Askin, D.; Volante, R. P.; Reamer, R. A.; Ryan, K. M.; Shinkai, I. *Tetrahedron Lett.* **1988**, *29*, 277.

123. Bergens, S.; Bosnich, B. *Comments Inorg. Chem.* **1987**, *6*, 85.

124. Schreiber, S. L.; Goulet, M. T.; Schule, G. *J. Am. Chem. Soc.* **1987**, *109*, 4718.

125. Mori, K.; Seu, Y. B. *Tetrahedron*, **1985**, *41*, 3429.

126. Mori, K.; Otsuka, T. *Tetrahedron*, **1985**, *41*, 553.

127. Lewis, M. D.; Duffy, J. P.; Blough, B. E.; Crute, T. D. *Tetrahedron Lett.* **1988**, *29*, 2279.

128. Corey, E. J.; Tramontano, A. *J. Am. Chem. Soc.* **1984**, *106*, 462.

129. (a) Roush, W. R.; Brown, R. J. *J. Org. Chem.*. **1982**, *47*, 1373. (b) Roush, W. R.; Brown, R. J. *J. Org. Chem.* **1983**, *48*, 5093. (c) Bulman-Page, P. C.; Carefull, J. F.; Powell, L. H.; Sutherland, I. O. *J. Chem. Soc., Chem. Commun.* **1985**, 822.

130. (a) Kufner, U.; Schmidt, R. R. *Angew Chem., Int. Ed. Engl.* **1986**, *25*, 89. (b) Frater, G.; Muller, J. *Helv. Chim. Acta.* **1989**, *72*, 653. (c) Ibuka, T.; Tanaka, M.; Yamamoto, Y. *J. Chem. Soc., Chem. Commun.* **1989**, 967.

131. Katsuki, T.; Sharpless, K. B. Unpublished results.

132. Rossiter, B. E.; Sharpless, K. B. *J. Org. Chem.* **1984**, *49*, 3707.

133. Ikeqami, S.; Katsuki, T.; Yamaguchi, M. *Chem. Lett.* **1987**, 83.

134. (a) Corey, E. J.; Ha, D.-C. *Tetrahedron Lett.* **1988**, *29*, 3171. (b) Hosokawa, T.; Kono, T.; Shinohara, T.; Murahashi, S.-I. *J. Organometal. Chem.* **1989**, *370*, C13.

135. Sharpless, K. B.; Woodard, S. S.; Finn, M. G. *Pure Appl. Chem.* **1983**, *55*, 1823.

136. (a) Woodard, S. S.; Finn, M. G.; Sharpless, K. B. *J. Am. Chem. Soc.* **1991**, *113*, 106. (b) Finn, M. G.; Sharpless, K. B. *J. Am. Chem. Soc.* **1991**, *113*, 113.

137. Corey, E. J. *J. Org. Chem.* **1990**, *55*, 1693.

138. Williams, I. D.; Pedersen, S. F.; Sharpless, K. B.; Lippard, S. J. *J. Am. Chem. Soc.* **1984**, *106*, 6430.

139. Mimoun, H.; Chaumette, P.; Mignard, M.; Saussine, L.; Fischer, J.; Weiss, R. *Nouv. J. Chim.* **1983**, *7*, 467.

140. (a) Jorgensen, K. A.; Wheeler, R. A.; Hoffmann, R. *J. Am. Chem. Soc.* **1987,** *109,* 3240. (b) Jorgensen, K. A. *Chem. Rev.* **1989,** *89,* 431.

141. (a) Katsuki, T.; Sharpless, K. B. U.S. Patent 4,471,130, Sept. 11, 1984; *Chem. Abstr.* **1985,** *102,* 24872m. (b) Miyano, S.; Lu, L. D.-L.; Viti, S. M.; Sharpless, K. B. *J. Org. Chem.* **1983,** *48,* 3611. (c) Miyano, S.; Lu, L. D.-L.; Viti, S. M.; Sharpless, K. B. *J. Org. Chem.* **1985,** *50,* 4350.

142. (a) Kobayashi, Y.; Kusakabe, M.; Kitano, Y.; Sato, F. *J. Org. Chem.* **1988,** *53,* 1586. (b) Kamatani, T.; Tsubuki, M.; Tatsuzaki, Y.; Honda, T. *Heterocycles,* **1988,** *27,* 2107. (c) Kusakabe, M.; Kitano, Y.; Kobayashi, Y.; Sato, F. *J. Org. Chem.* **1989,** *54,* 2085. (d) Kusakabe, M.; Sato, F. *J. Org. Chem.* **1989,** *54,* 3486. (e) Kamatani, T.; Tatsuzaki, Y.; Tsubuki, M.; Honda, T. *Heterocycles,* **1989,** *29,* 1247.

143. Kitano, Y.; Kusakabe, M.; Kobayashi, Y.; Sato, F. *J. Org. Chem.* **1989,** *54,* 994.

144. Chan, T. H.; Chen, L. M.; Wang, D. *J. Chem. Soc., Chem. Commun* **1988,** 1280.

145. (a) Kihara, M.; Ohnishi, K.; Kobayashi, S. *J. Heterocyclic Chem.* **1988,** *25,* 161. (b) Jikihara, T.; Katsurada, M.; Ikeda, O.; Yoneyama, K.; Takematsu, T. Presentation ORGN 220; 198th National Meeting of the American Chemical Society; Sept. 15–20, 1989.

4.2

Asymmetric Catalytic Epoxidation of Unfunctionalized Olefins

Eric N. Jacobsen

Department of Chemistry Roger Adams Laboratory
University of Illinois
Urbana, Illinois

4.2.1. Introduction

Enantioselective alkene epoxidation constitutes an extremely appealing strategy for the synthesis of optically active organic compounds, as is well illustrated by the dramatic emergence of the titanium tartrate catalyzed epoxidation of allylic alcohols as one of the most widely applied reactions in asymmetric synthesis. Katsuki and Sharpless entitled their initial 1980 report on this reaction "The First Practical Method for Asymmetric Epoxidation" [1], clearly indicating that other synthetically useful procedures were anticipated. New methods for asymmetric epoxidation emerged only very gradually after 1980, however, particularly compared with the rapid advances made in other areas of catalytic asymmetric synthesis. Fortunately, thanks to concerted research efforts by several laboratories throughout the world, a number of highly promising new strategies for enantioselective epoxidation catalysis have been developed successfully in the past few years [2].

A foremost aim of research in this area has been the achievement of high enantioselectivity in the epoxidation of alkenes bearing no functionality to precoordinate to the catalyst, where selectivity is determined solely through nonbonded interactions. When only the steric and electronic properties of the double bond undergoing epoxidation are relevant to the enantioselection of the reaction, the pool of potential substrates could in principle be extremely broad. This chapter presents an overview of the various existing strategies for asymmetric epoxidation of substrates other than allylic alcohols, with particular emphasis on catalytic epoxidation of unfunctionalized olefins.

4.2.2. Transition Metal Based Catalysts

The reactive oxidizing species in metal-catalyzed epoxidations is naturally a central consideration in the classification of different oxidants, but in very few cases has it been possible in fact to obtain physical characterization of such species in catalytic systems. Instead, epoxidation catalysts may best be classified by the nature of the stoichiometric oxidants that they can employ. According to this criterion, we distinguish simply two classes: oxo transfer catalysts that employ oxygen atom sources such as iodosylarenes, hypochlorite, and persulfate; and peroxo transfer catalysts that make use of hydrogen peroxide or alkyl hydroperoxides as oxidants. No synthetic catalysts have yet been developed that accomplish highly enantioselective epoxidations with molecular oxygen as the stoichiometric oxidant, although such systems are likely to emerge. It is likely that these catalysts will be oxo transfer agents analogous to the cytochrome P-450 class of heme-containing proteins.

4.2.2.1. Porphyrin-Based Oxo Transfer Catalysts

The impetus for the study of transition metal porphyrin complexes in oxo transfer catalysis arose from Groves' landmark discovery that Fe(III) porphyrin complexes are models for cytochrome P-450 [3]. Indeed, chiral porphyrin monooxygenase model systems have received the greatest attention thus far in the context of potentially viable asymmetric epoxidation catalysts. The first example of asymmetric epoxidation of simple olefins catalyzed by chiral porphyrin complexes was described in 1983 by Groves and Meyers [4]. Oxidation reactions were carried out with catalysts **1** of **2** employing iodosylmesitylene as the stoichiometric oxidant (Scheme 1). Styrene derivatives afforded the highest selectivities with these catalysts, and the best result was obtained in the epoxidation of 4-chlorostyrene (51% ee). To account for the observed selectivities and for the general observation that cis olefins are more reactive than trans olefins in porphyrin-catalyzed reactions, Groves proposed a transition state model for oxygen atom transfer involving side-on ap-

Scheme 1

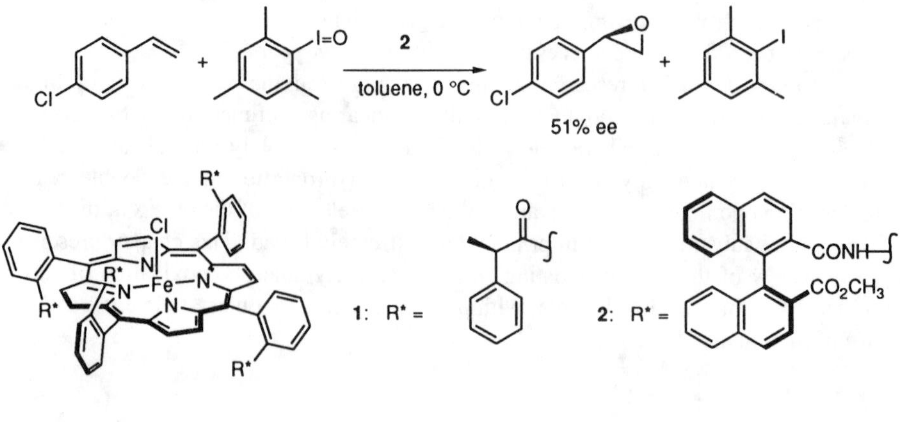

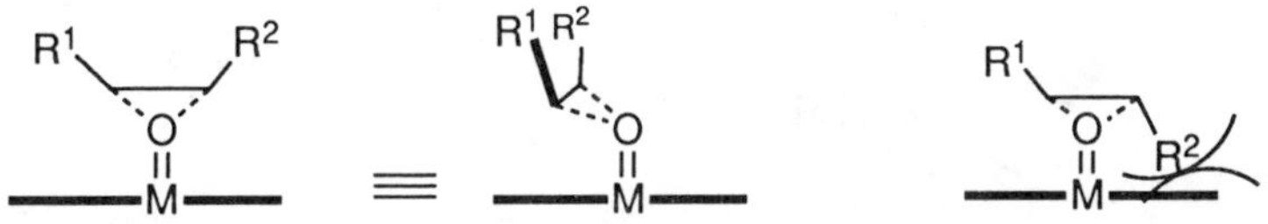

Figure 1 Side-on approach model for oxygen transfer illustrating the less-hindered approach of *cis*-alkenes to the metal oxo moiety. The porphyrin ligand is symbolized by the heavy line.

proach of olefin to the putative iron–oxo intermediate (Fig. 1). Although the precise angle of alkene approach to the oxo intermediate remains open to some controversy [5], the side-on approach postulate and variants thereof have gained wide acceptance and have provided an extremely useful model for the design of other chiral porphyrin derivatives. The X-ray crystal structure of the Ru(II) porphyrin/styrene oxide complex **3** has been interpreted as lending indirect physical support to the side-on approach model (Fig. 2) [6].

More recently, Groves has investigated the chemistry of the vaulted binaphthyl derivatives **4** and **5** (Chart 1). These catalysts also catalyzed the epoxidation of simple olefins with modest enantioselectivities, with the iron derivative **4** generally affording better results [7]. Epoxidation of 4-chlorostyrene occurred with 39% ee at room temperature, while *cis*-β-methylstyrene was epoxidized with the highest level of asymmetric induction (58% ee) under the same conditions. The enantioselectivity with *cis*-β-methylstyrene increased to 72% at −15°C, although the yield also dropped dramatically under these conditions (9% based on iodosylbenzene). Complex **4** also catalyzed the oxidation of sulfides and hydroxylation at the benzylic position of alkyl benzene derivatives. The selectivity in each of these reactions was rationalized according to a steric model in which a wedgelike pocket in the catalyst permits the smaller group on the substrate to fit into the wedge (Fig. 3). This model succeeds in predicting the sense of asymmetric induction in the oxidation of several substrates (Table 1), although the epoxidation of *tert*-butylethylene provided a notable exception to the steric convention. This latter result suggests that epoxidation of aryl- and alkyl-substituted olefins by oxo transfer catalysts may proceed via

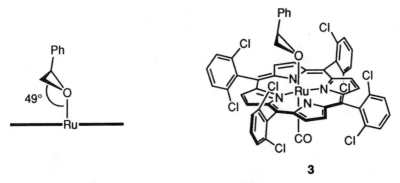

3

Figure 2 Schematic representation of the coordination of epoxide to a Ru(II)–porphyrin complex, as exhibited in the X-ray crystal structure of **3**.

Chart 1

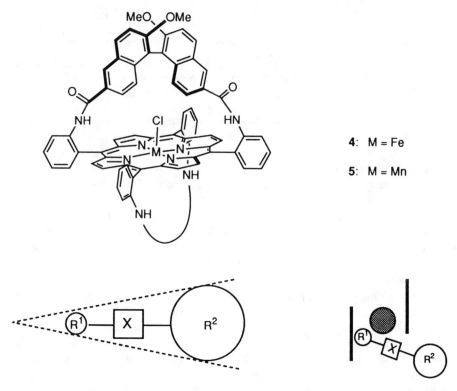

4: M = Fe

5: M = Mn

Figure 3 Schematic steric model for asymmetric oxidation reactions catalyzed by **4** and **5**.

Table 1 Asymmetric Oxidation Reactions Catalyzed by **3**

Substrate	Major Product	% ee	Yield (%)[a]
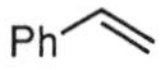 Ph∕═	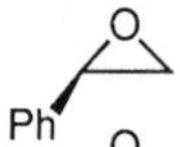 Ph∕△═O	30	23
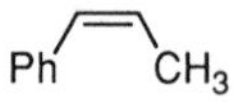 Ph∕═CH₃	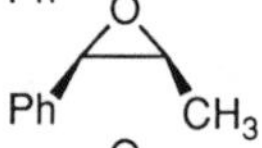 Ph∕△═O∕CH₃	58/20°C 72/-15°C	64 9
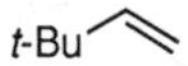 t-Bu∕═	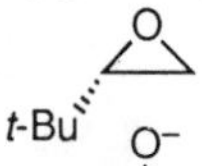 t-Bu∕△═O	40	14
Ph-S-CH₃	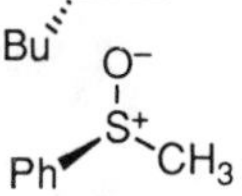 Ph∕S⁺(O⁻)∕CH₃	24	84
Ph-CH₂CH₃	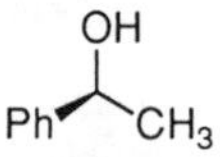 Ph∕(OH)∕CH₃	40	40

[a] Yields based on iodosylbenzene.

completely different mechanisms, an idea that is considered in detail later in this chapter.

Several other groups have pursued the design and synthesis of chiral porphyrin derivatives bearing conformationally restricted bridging ligands. The resulting complexes include Mansuy's "basket handle" porphyrin **6** [8], Naruta's "twin coronet" porphyrins **7** and **8** [9], Collman's "picnic basket" derivative **9** [10], and Inoue's "strapped" porphyrin catalyst **10** [11] (Chart 2). Epoxidation of styrene derivatives with these catalysts takes place with low to moderate enantioselectivity, and representative examples are listed in Table 2. In the case of catalysts **7** and **8**, Naruta has drawn a correlation between substrate electronic properties and observed enantioselectivities in epoxidations. Thus, electron-rich derivatives such as 2-methoxystyrene were epoxidized with very low selectivity, while electron-deficient derivatives were epoxidized with much higher selectivities (up to 89% ee for 2-nitrostyrene). A secondary attractive stacking interaction between the aromatic ring of the substrate and the electron-rich naphthyl group on the ligand was proposed to account for this remarkable effect of the electronic properties of substrate on enantioselectivity (Fig. 4). In principle, such π-π* secondary interactions may provide an effective strategy for limiting degrees of freedom in bimolecular oxo transfer reactions.

Unlike most chiral porphyrin based systems, both Collman's catalyst **9** and Inoue's catalyst **10** have chemically nonequivalent faces as a result of a bridge spanning only one of the sides of the porphyrin. Although this configuration can lead to competitive pathways as a result of oxo transfer from both faces of the metal complex, in each case the open face of the porphyrin is blocked by coordination of a large ligand, which cannot fit into the cavity created by the strap. Epoxidation of

Table 2 Asymmetric Epoxidations with Catalysts 6–10

Substrate	Catalyst	% ee	Turnover #
p-ClC$_6$H$_4$	6	50	a
Ph	7	22	50
o-MeOC$_6$H$_4$	7	0	32
o-NO$_2$C$_6$H$_4$	7 8	80 89	26 46
C$_6$F$_5$	7	74	36
Ph	9	13	a
Ph	10	50[b]	a

[a] Not reported. [b] Reaction carried out in the presence of 10 mol% added 1-ethylimidazole.

Chart 2

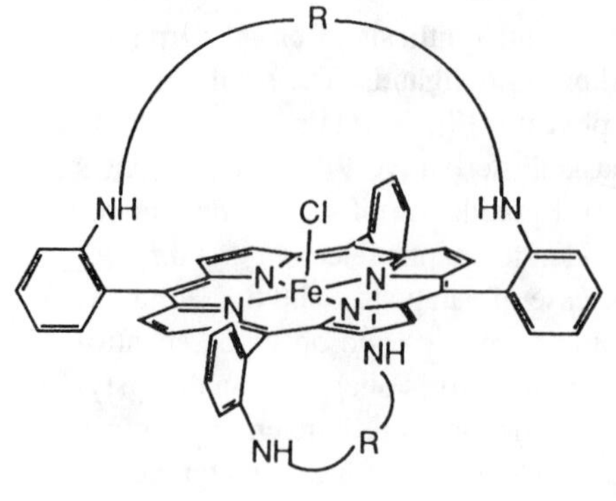

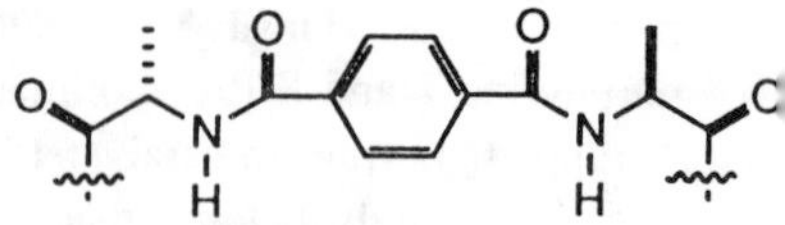

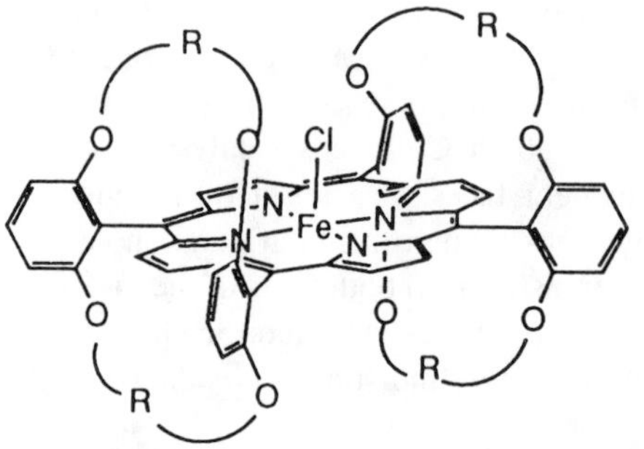

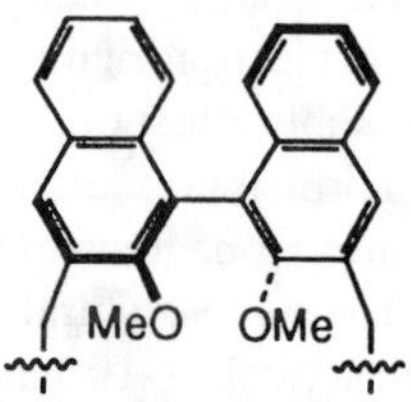

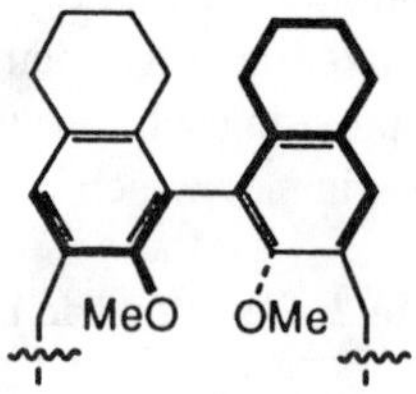

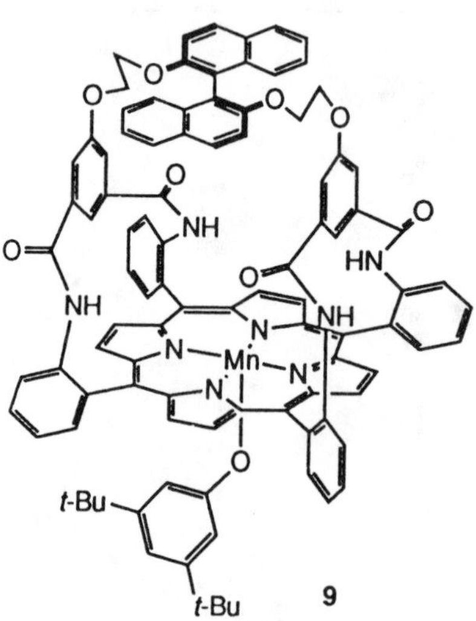

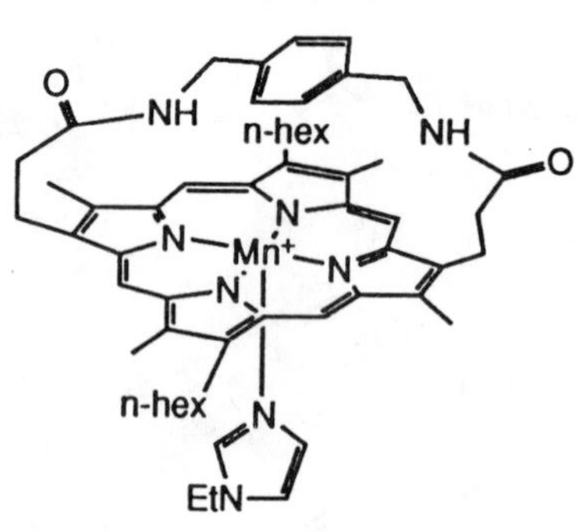

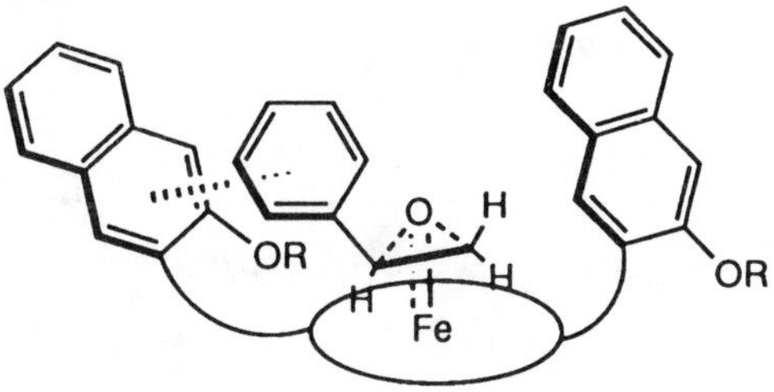

Figure 4 Proposed π-interaction in the epoxidation of styrene derivatives catalyzed by **7**.

simple olefins is thus believed to occur within the asymmetric cavity, leading in principle to optimal steric communication. However, epoxidation of styrene afforded only 13% ee with catalyst **9**, a result that was attributed by the authors to the considerable gap between the basket strap and the metal center [10]. Reactions with catalyst **10** were somewhat more enantioselective, with up to 58% ee in the epoxidation of indene [11].

Several practical problems associated with chiral porphyrin based catalysts limit their potential applicability in organic synthesis. In addition to the generally low enantioselectivities that have been obtained to date in olefin epoxidation, the catalysts described above tend to be unstable under the reaction conditions of epoxidation. Since these chiral complexes are obtained by multistep syntheses in extremely low overall yields, high catalyst turnovers are essential if such systems are to be synthetically viable. To limit catalyst degradation, substrates have typically been employed in large excess relative to the iodosylarene derivative, and conversion to product has been limited to below 30%.

The presence of electron-donating substituents on the *meso* phenyl groups in catalysts such as **6–9** markedly sensitizes the ligands toward oxidative degradation. An important step toward overcoming this very serious practical concern was achieved by O'Malley and Kodadek in the development of the "chiral wall" porphyrin **11** [12]. With its electron-withdrawing meso substituents, catalyst **11** displays remarkable stability and activity under the epoxidation conditions first developed by Meunier for achiral systems involving sodium hypochlorite as the stoichiometric oxidant (Scheme 2) [13]. The high turnover numbers observed in this reaction (> 3000) help compensate for the generally limited access to this and related chiral porphyrin complexes, and the use of bleach as a stoichiometric oxidant with complete conversion of starting alkene constitutes an enormous improvement over the iodosylarene-based systems used formerly. Unfortunately, enantioselectivities with the chiral wall porphyrin **11** are low (up to 40% ee for *cis*-β-methylstyrene).

More recently, Halterman and Jan have reported the first D_4 symmetric metalloporphyrin **12** (Chart 3) [14]. Similar to Kodadek's system, the absence of electron-withdrawing groups at the meso positions of **11** render the catalyst very robust under Meunier's conditions (> 2000 catalyst turnovers in the epoxidation of olefins with aromatic substituents. Enantioselectivities with Halterman's catalyst system are significant, with up to 76% ee obtained for *cis*-β-methylstyrene.

Scheme 2

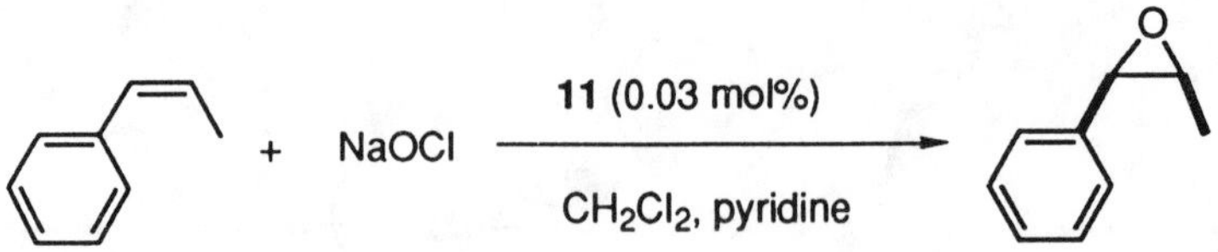

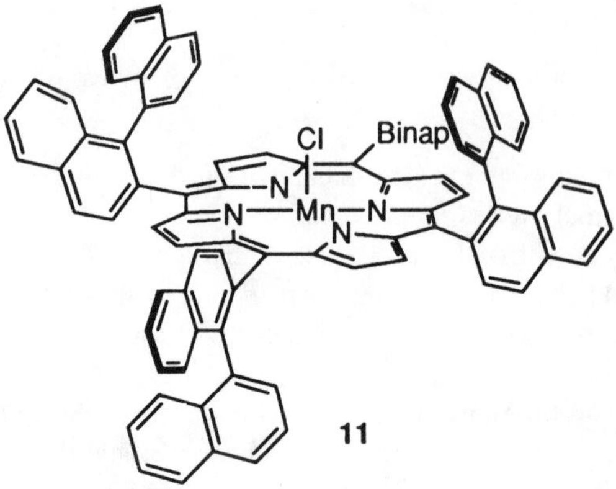

11

Chart 3

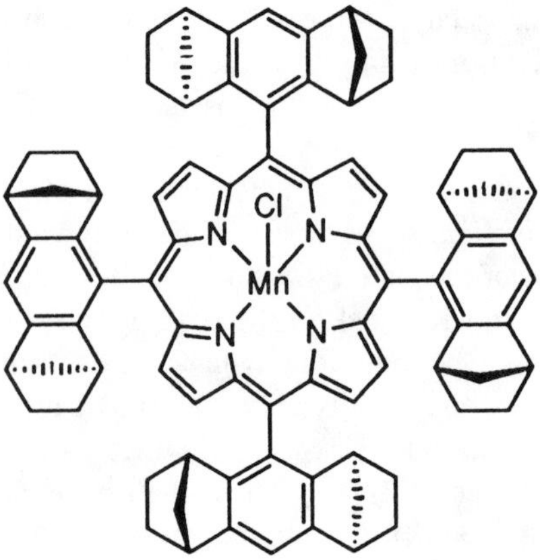

12

4.2.2.2. Salen-Based Oxo Transfer Catalysts

Chiral salen complexes with the general structure **B** shown in Figure 5 possess several structural and chemical features in common with porphyrins that render them appealing templates for chiral catalyst design. Both classes of coordination compounds are sterically well defined and kinetically nonlabile, and thus they

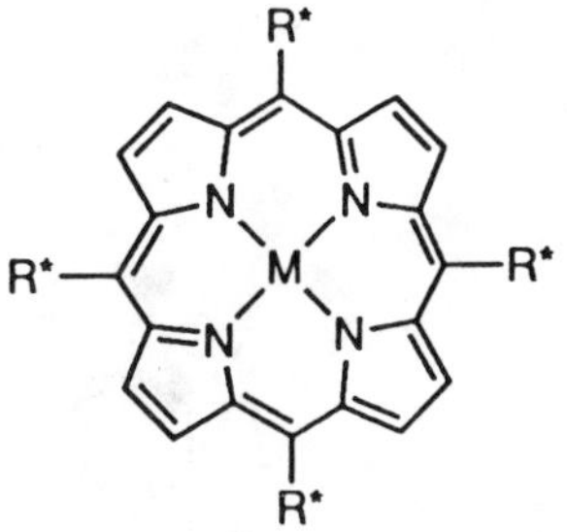

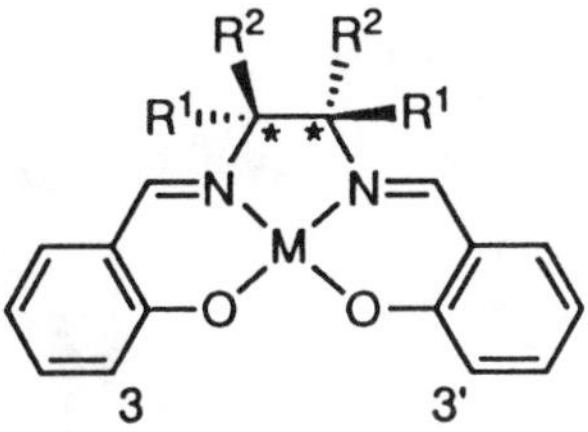

Figure 5   Generalized structures for chiral porphyrin (**A**) and chiral salen (**B**) complexes.

provide a sensible matrix for rational ligand design. Salen complexes also share with porphyrins the ability to catalyze the epoxidation of unfunctionalized alkenes, as demonstrated by Kochi [15] and by Burrows [16] in their research with achiral complexes of first-row transition metals. However, unlike porphyrin systems, salen complexes bear tetravalent and thus potentially stereogenic carbon centers in the vicinity of the metal binding site. Stereochemical communication in epoxidation can thus be enhanced, at least in principle, as a result of the proximity of the reaction site to the ligand dissymmetry [17]. Also, the synthesis of salen complexes from chiral diamines is generally highly efficient and extremely straightforward, and extensive screening of ligand structural types can be quickly and readily accomplished (see below). Indeed, the ability to carry out such screening can represent the most critical aspect to the development of a successful chiral catalyst system.

Chiral manganese salen complexes that are effective and practical catalysts for the asymmetric epoxidation of a variety of alkenes have been developed recently in our laboratories (Scheme 3) [18]. Systematic variation of the steric and electronic environment of the complexes has led to the discovery of catalysts that are particularly effective for the epoxidation of various important classes of olefins [19].

Catalyst Preparation

The preparation of C_2 symmetric salen ligands is achieved in a general sense and in excellent yield by the condensation of appropriately substituted 1,2-diamines with 2 equiv. of a salicylaldehyde derivative in ethanol (Scheme 4). Chiral salen complexes of all the first-row transition metals have been prepared in our laboratories, with Mn(III) derivatives displaying superior selectivity and the highest turnovers in epoxidation of most alkenes [20]. The generation of (salen)Mn(III) complexes is readily accomplished by refluxing an ethanolic solution of a salen ligand with 2 equiv. of $Mn(OAc)_2 \cdot 4H_2O$ in air. When the ligand is sparingly soluble in hot ethanol, generation of the soluble dipotassium salt by treatment with 2 equiv. of KOH facilitates subsequent formation of the Mn complex. The intermediate (salen)Mn(III)OAc complex is treated in situ with LiCl, and precipitation of the resulting (salen)Mn(III)Cl complex as a dark brown air- and moisture-stable powder

Scheme 3

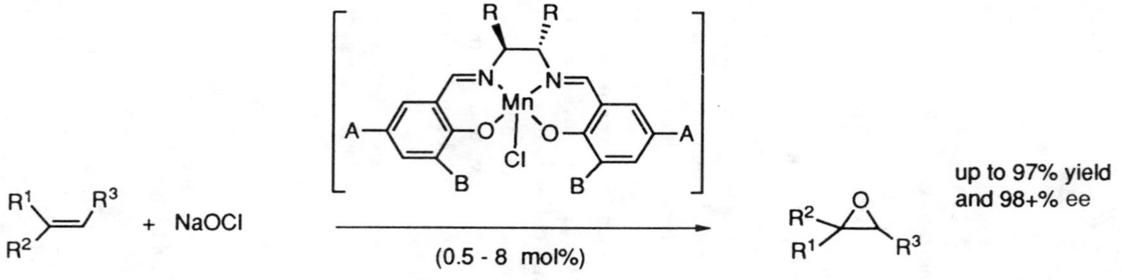

up to 97% yield
and 98+% ee

(0.5 - 8 mol%)

+ NaOCl

Scheme 4

1. EtOH/H₂O, reflux
2. Mn(OAc)₄•4H₂O (2 equiv.)
3. LiCl, air

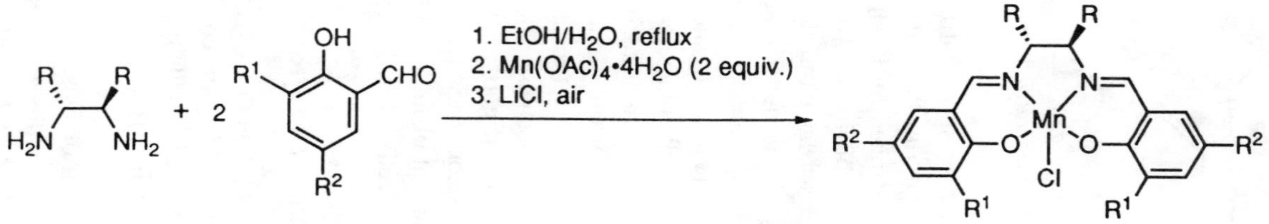

is then accomplished by addition of water [21]. This procedure is general for the preparation of Schiff base complexes derived from 1,2- 1,3- and 1,4-diamines, and our group has applied it to the synthesis of more than 50 optically active complexes (Chart 4) [22]. Subsequent to our initial report of enantioselective epoxidation by

Chart 4

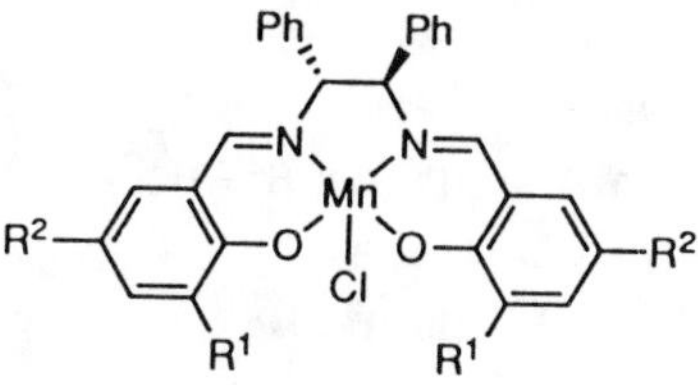

13 R^1 = H, R^2 = H
14 R^1 = t-Bu, R^2 = H
15 R^1 = t-Bu, R^2 = Cl
16 R^1 = t-Bu, R^2 = OMe
17 R^1 = 1-methylcyclohexyl, R^2 = Me
18 R^1 = C(Ph)$_2$Me, R^2 = H
19 R^1 = C(Et)$_2$Me, R^2 = H
20 R^1 = Me$_3$Si, R^2 = H
21 R^1 = t-Bu, R^2 = Me
22 R^1 = 9-methyl-9-fluorenyl, R^2 =Me
23 R^1 = 1-adamantyl, R^2 =Me
24 R^1 = R^2 = Br
25 R^1 = t-Bu, R^2 = NO$_2$

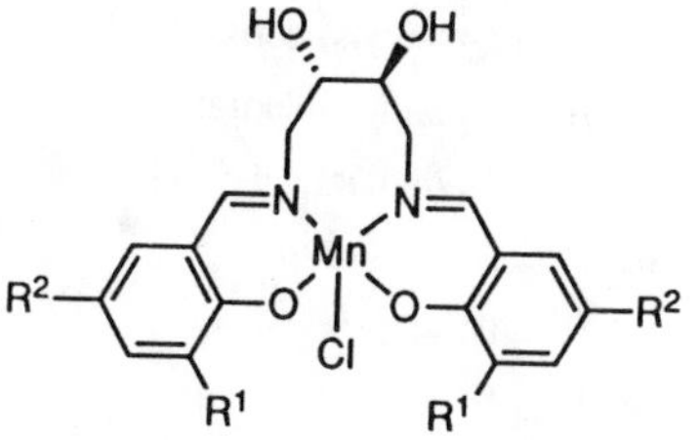

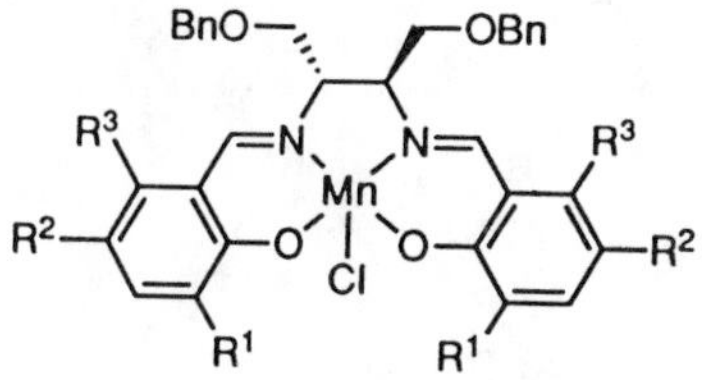

29 R^1 = R^2 = R^3 = H
30 R^1 = R^3 = H, R^2 = t-Bu
31 R^1 = R^3 = H, R^2 = NO$_2$
32 R^1 = R^2 = Br, R^3 = H
33 R^1 = H, R^2, R^3 = -CH=CH-CH=CH-
34 R^1 = t-Bu, R^2 =Me, R^3 = H

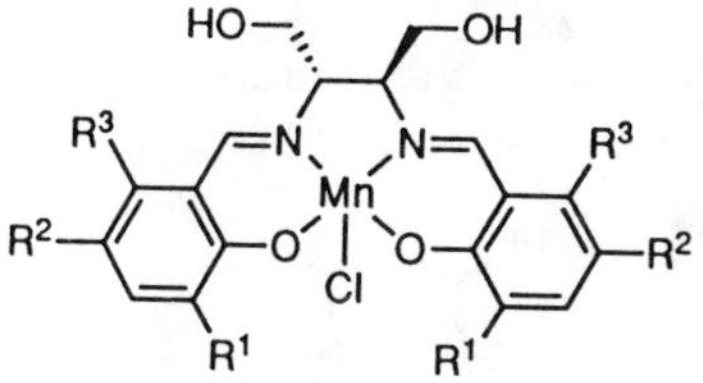

35 R^1 = R^2 = R^3 = H
36 R^1 = R^3 = H, R^2 = t-Bu
37 R^1 = R^3 = H, R^2 = NO$_2$
38 R^1 = R^2 = Br, R^3 = H
39 R^1 = H, R^2, R^3 = -CH=CH-CH=CH-
40 R^1 = t-Bu, R^2 = Me, R^3 = H

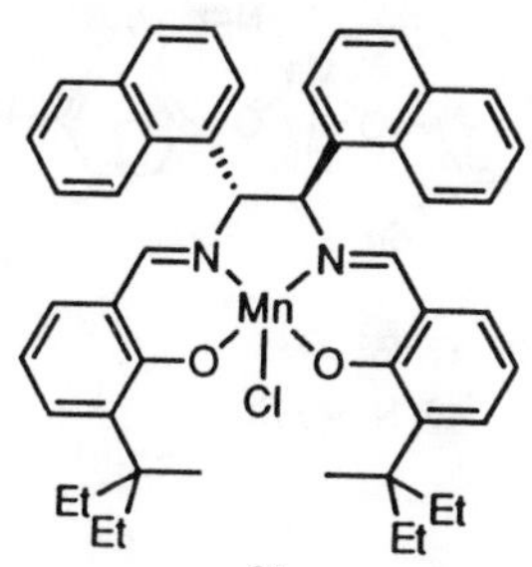

26 R^1 = R^2 = H
27 R^1 = H, R^2 = NO$_2$
28 R^1 = t-Bu. R^2 = Me

41

Chart 4 (*continued*)

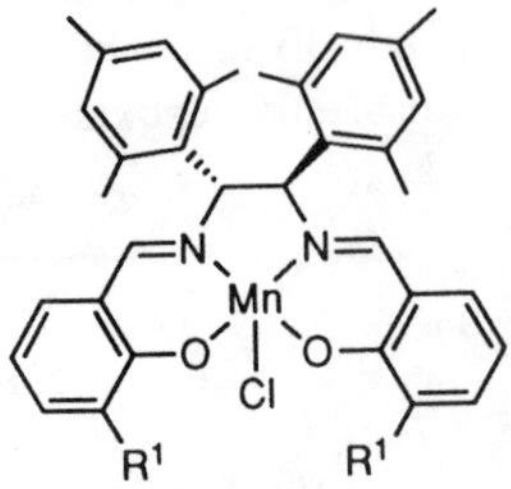

42 $R^1 = t$-Bu
43 $R^1 = C(Ph)_2Me$

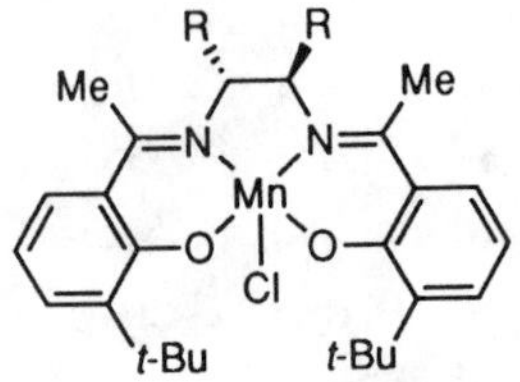

44 R = H
45 R = Ph (unstable)

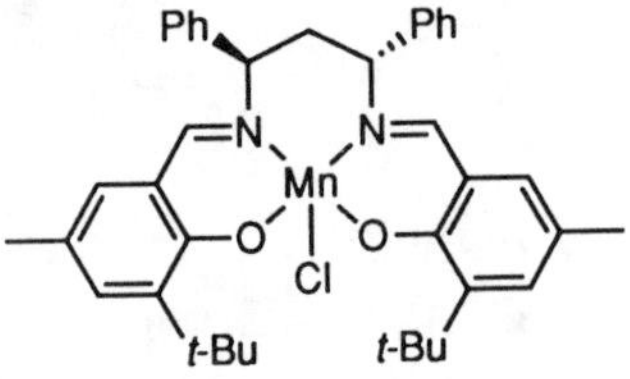

46

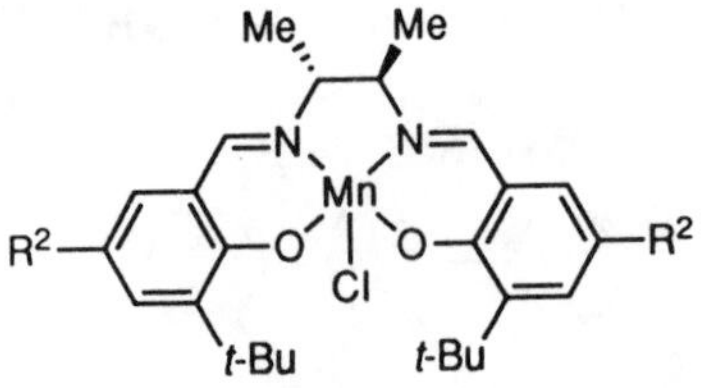

47 $R^2 = Me$
48 $R^2 = t$-Bu

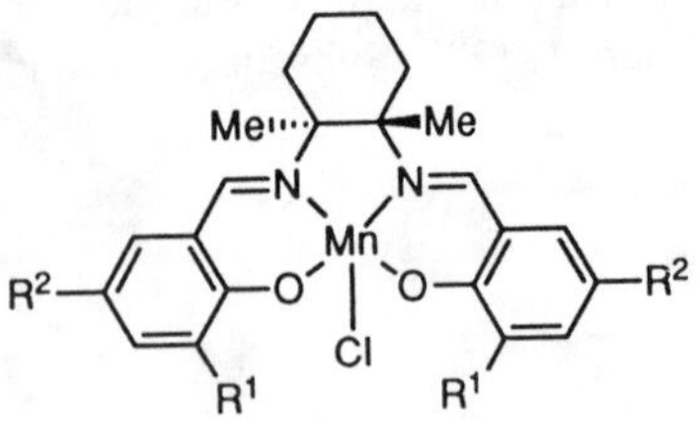

49 $R^1 = R^2 = H$
50 $R^1 = C(Ph)_2Me, R^2 = H$
51 $R^1 = R^2 = t$-Bu
52 $R^1 = t$-Bu, $R^2 = Me$
53 $R^1 = $ 1-methylcyclohexyl, $R^2 = Me$

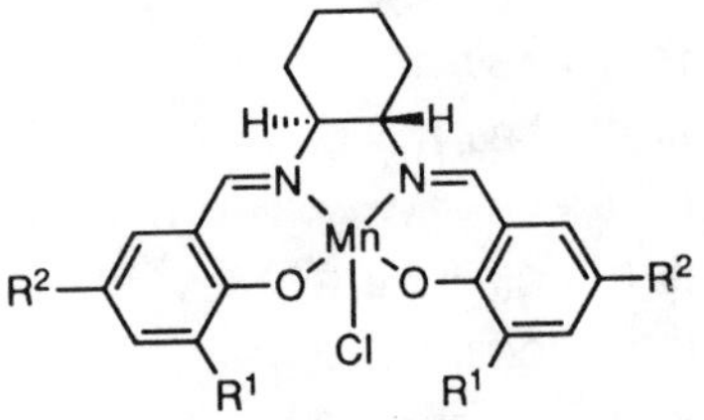

54 $R^1 = H, R^2 = H$
55 $R^1 = t$-Bu, $R^2 = Me$
56 $R^1 = t$-Bu, $R^2 = Cl$
57 $R^1 = t$-Bu, $R^2 = H$
58 $R^1 = t$-Bu, $R^2 = OMe$
59 $R^1 = t$-Bu, $R^2 = NO_2$
60 $R^1 = $ 1-methylcyclohexyl, $R^2 = Me$
61 $R^1 = $ 9-methyl-9-fluorenyl, $R^2 = Me$
62 $R^1 = $ 1-adamantyl, $R^2 = Me$
63 $R^1 = R^2 = t$-Am
64 $R^1 = R^2 = Br$
65 $R^1 = R^2 = t$-Bu

Chart 5

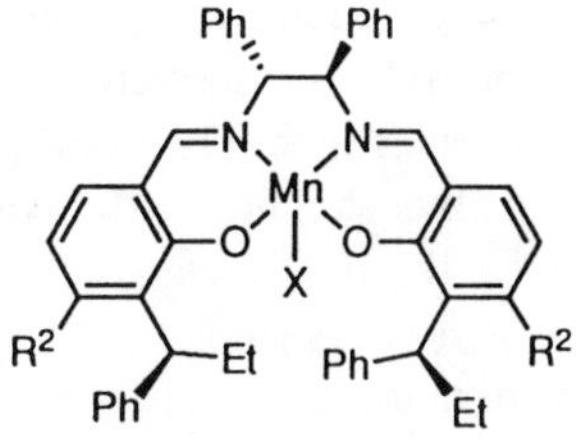

66 R^2 = H, X = OAc
67 R^2 = Me, X = PF$_6$

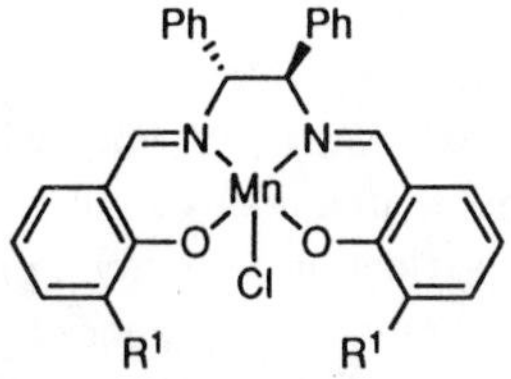

73 R^1 = TMS
74 R^1 = TBDMS

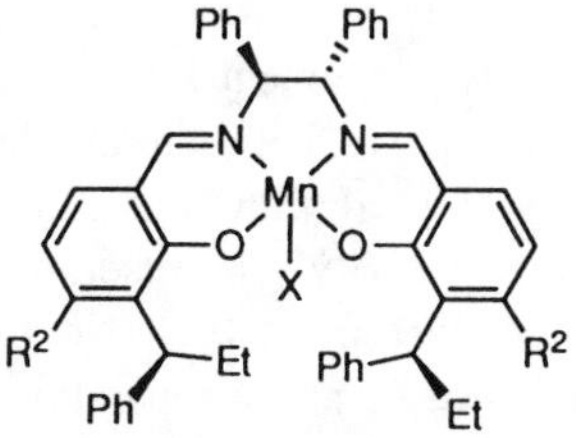

68 R^2 = H, X = OAc
69 R^2 = Me, X = PF$_6$

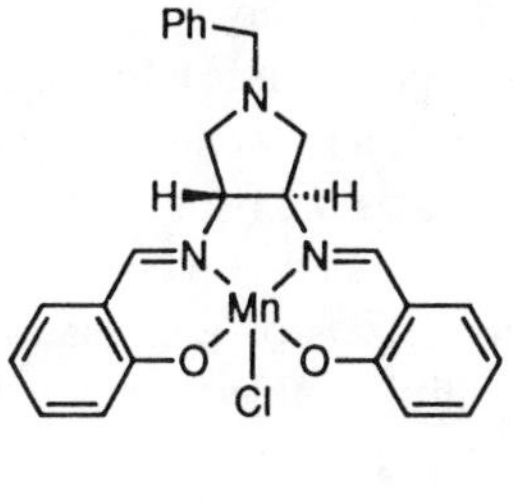

75

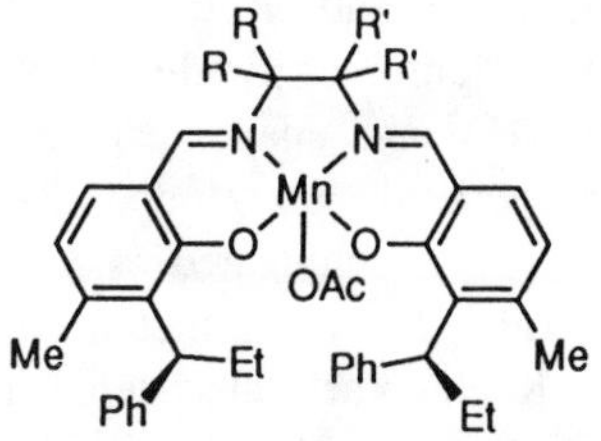

70 R^1 = R^2 = H
71 R^1 = R^2 = Me
72 R^1 = H, R^2 = Me

chiral (salen)Mn(III) complexes, Katsuki [23], Burrows [24], and Thornton [25] have disclosed interesting related systems. The catalysts studied by these groups are listed in Chart 5. Racemic diamines often can be derived synthetically from tartaric acid or resolved in a straightforward manner with tartaric acid, mandelic acid, or camphorsulfonic acid—each of which is available at low cost in optically pure form as either optical antipode. Thus, both enantiomers of a variety of chiral salen derivatives are readily accessible.

Epoxidation Method

The selection of stoichiometric reagents is a crucial consideration, which is closely tied to the practical utility of any catalytic process. Iodosylarene derivatives were first examined with both achiral and chiral (salen)Mn(III) catalysts [18], but it was found that aqueous sodium hypochlorite is a more effective stoichiometric oxidant in epoxidations employing these catalysts [21]. In general, enantioselectivities obtained with iodosylarenes and with bleach are identical, strongly suggesting a common Mn(V) oxo intermediate as the active species with both oxidants [21]. Bleach is of course an appealing oxidant because of its extremely low cost, low corrosiveness, lack of explosiveness, and relatively innocuous stoichiometric by-product (aqueous NaCl). Conditions for epoxidation with (salen)Mn(III) catalysts have been developed involving a two-phase system, with an aqueous buffered commercial bleach phase and an organic phase composed of a solution of substrate and catalyst in a suitable solvent [21]. Dichloromethane is especially convenient for laboratory procedures on a small to moderate scale, but solvents such as 1,2-dichloroethane, *tert*-butyl methyl ether, ethyl acetate, or toluene may be used with generally only minor effects on epoxidation rate, yield, and enantioselectivity. The reactions are typically complete with a few hours at room temperature or at 0°C. No phase transfer catalyst is necessary, and workup is accomplished by phase separation and epoxide isolation by recrystallization, distillation, or chromatography.

Many of the catalysts in Chart 4 successfully effect epoxidation of olefins with widely varying electronic properties, and in general complete substrate conversion can be achieved using 0.5–8 mol % of catalyst. Substrates include electron-rich olefins such as simple alkyl- and aryl-substituted alkenes, and electron-deficient olefins such as conjugated esters. In certain cases, the addition of pyridine *N*-oxide derivatives serves to improve both catalyst turnovers and enantioselectivities [23c, 26]. This is particularly true for substrates that tend to undergo epoxidation sluggishly with the (salen)Mn(III)-based systems, such as conjugated esters and simple alkyl-substituted olefins. The function of *N*-oxide additives in metallosalen-catalyzed epoxidations, which has been examined by Kochi in achiral systems [15], appears to involve direct complexation to the high valent oxo intermediate with rate acceleration in the subsequent epoxidation. In epoxidations employing bleach as oxidant, the catalyst resides entirely in the organic phase of the two-phase reaction medium, so water-insoluble pyridine *N*-oxide derivatives bearing hydrophobic substituents, such as the commercially available 4-phenylpyridine *N*-oxide, are most effective.

Ligand Design

The 1,2-diphenylethylenediamine derivative **13** was one of the first epoxidation catalysts examined in our laboratories. It displayed negligible enantioselectivity in the epoxidation of most substrates, with only *trans*-stilbene affording promising results (33% ee). The low enantioselectivities observed with this and related catalysts were suggestive of a side-on approach transition state similar to the one

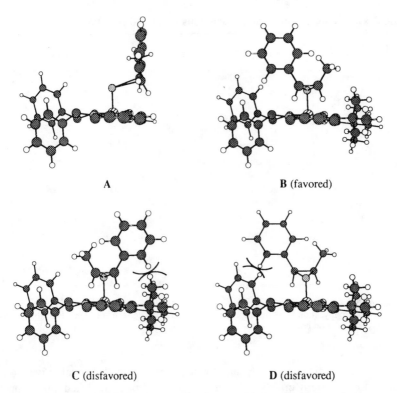

A

B (favored)

C (disfavored)   D (disfavored)

Figure 6 Proposed transition structures for the epoxidation of *cis*-β-methylstyrene. (**A**) epoxidation with catalyst **13** and (**B–D**) epoxidation with catalyst **14**. [Structures created with the program Chem 3D based on the coordinates from the X-ray crystal structure of **13** (PF$_6^-$ salt).

proposed by Groves and discussed above [4]. With (salen)Mn(V) oxo intermediates in which the ligand was lacking 3 and 3′ substituents, alkene side-on approaches would be accessible with minimal stereochemical communication with the dissymmetric ligand (Fig. 6A). This possibility was supported by the observation that bulky *tert*-butyl substituents introduced at the 3 and 3′ positions of salen complex **14** resulted in a dramatic increase in selectivity in the epoxidation of a variety of olefins, particularly with cis-disubstituted alkenes (Table 3) [18]. In accordance with the side-on approach model, the preferred transition state in the epoxidation of *cis*-β-methylstyrene with **14** was proposed to be the one in Figure 6B. Two important features of the chiral salen ligand were suggested to contribute to the good enantioselectivities observed. The bulky *tert*-butyl substituents probably enforce olefin approach near the dissymmetric diimine bridge and thus improve stereochemical communication in the reaction. The approach in Figure 6C is presumably disfavored because of a repulsive steric interaction between the larger substituent on the substrate and the bulky *tert*-butyl groups on the ligand. The dissymmetry of the diimine bridge presumably disfavors attack syn to the phenyl group (approach D),

Table 3 Asymmetric Epoxidation Reactions Catalyzed by **14**

Entry	Olefin	Method[a]	Isolated Yield(%)	% ee	Configuration[b]
1	Ph (styrene)	A	75	57	R-(+)
2	(bicyclic dioxolane-cyclohexene)	B	65	94	(−)[c]
3	Ph (cis-β-methylstyrene)	A	73	81	1R,2S-(−)
4	(dihydronaphthalene)	B	72	78	1R,2S-(+)
5	Ph (2-phenylpropene / isomer)	A	36	30	R-(+)

[a]Method A: NaOCl (pH 11.3), CH_2Cl_2 solvent, 0 °C. Method B: Same as method A but with 0.2 equiv 4-phenylpyridine N-oxide employed as additive. [b]The sign corresponds to that of $[\alpha]_D$. [c]Absolute configuration not ascertained.

but leaves accessible approach B, anti to the phenyl group. This qualitative model is based entirely on steric effects and is almost certainly oversimplified, but it nonetheless predicts the sense of absolute stereoinduction of each of the reactions listed in Table 3.

A logical sequence of ligand modifications, first involving two unsuccessful strategies, eventually led to the development of the highly enantioselective epoxidation catalyst **65**. The initial approach involved the screening of bulky complexes such as **41–43**, with the idea that greater steric hindrance could lead to better differentiation of diastereomeric approaches of the substrate. However, not only did these complexes afford poorer enantioselectivities than **14**, but congestion of the reaction site also tended to result in extremely sluggish epoxidation and increased by-product formation. The second strategy involved synthesis of relatively unhindered salen complexes derived from C_2 symmetric α-tertiary diamines, since axially locked substituents were anticipated to enhance differentiation of approaches of the substrate analogous to those shown in Figure 6 (B and D). Thus, catalysts **49–53** were prepared from *trans*-1,2-diamino-1,2-dimethylcyclohexane [27], but these again exhibited only moderate enantioselectivities (49–55% ee) in the epoxidation of *cis*-β-methylstyrene [19]. The most significant observation made in this study was that the sense of asymmetric induction was opposite from that anticipated according to the simple steric models illustrated in Figure 6. This led us to consider new mechanistic alternatives, and in particular whether catalysts derived from unhindered 1,2-diamines might undergo competitive side-on alkene approach from over the diimine bridge (Fig. 7).

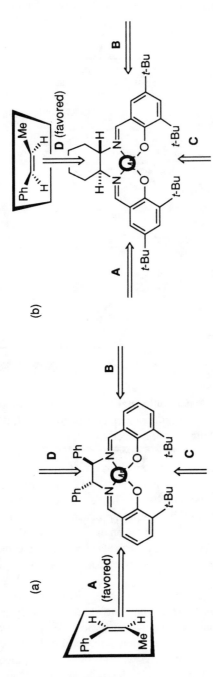

Figure 7 Side-on approaches of *cis*-β-methylstyrene to (a) **14** and (b) **65**. In both cases the oxo ligand is oriented out of the plane of the page.

175

Table 4 Asymmetric Epoxidation of Representative Olefins by Catalyst **65**

Entry	Olefin	Method[a]	Isolated Yield (%)	% ee of major epoxide	Equiv. 65 Required for Complete Reaction
1	Ph—Me (cis)	A	84[b]	92	0.04
2	(1,2-dihydronaphthalene)	B	67	86	0.04
3	(2,2-dimethyl-2H-chromene)	A	87	98	0.02
4	(indene)	A	80	88	0.01
5	Me₃Si—=—cyclohexyl	A	65[c]	98	0.04
6	(cyclohexene dioxolane spiro)	B	63	94	0.15
7	Ph—CO₂Et (cis)	B	67[e]	97	0.08

[a]See Table 3. [b]Isolated yield of epoxide mixture (cis:trans = 11.5:1). [c]Isolated yield of epoxide mixture (cis:trans = 1:5.2). [d]Isolated yield of epoxide mixture (cis:trans = 5:1).

This possibility was evaluated with catalyst **65**, which was designed to disfavor all side-on olefin approaches except approach D (Fig. 7b). As illustrated in Table 4, catalyst **65** in fact displays high enantioselectivity with a variety of cis-disubstituted alkenes [19]. Indeed, this is the most selective catalyst developed to date for the epoxidation of a wide range of unfunctionalized olefins.

It is remarkable that of the dozens of catalysts in Chart 4, **65** clearly has one of the *least* dissymmetric ligand structures. The key feature of this catalyst thus appears to be its ability to limit competing approaches of the substrate, such that substrate interaction with the dissymmetric environment is maximized. Access to **65** from 1,2-diaminocyclohexane is also a particularly appealing feature of this catalyst system, since this diamine is one of the most readily available C_2 symmetric chiral building blocks [28].

In a series of elegant papers, Katsuki and coworkers have studied catalytic epoxidations with catalysts **66–72**, which are similar to catalysts **14–25** but bear additional stereogenic centers on the 3- and 3'-positions of the salen ligand [23]. The highest level of asymmetric induction achieved by these systems was in the epoxidation of 1,2-dihydronaphthalene with **69** (83% ee). Although higher enantioselectivities in the epoxidation of cis olefins have been obtained with catalysts that simply bear *tert*-butyl groups at the 3- and 3'-positions (such as **14** and **65**), Katsuki's systems display moderate enantioselectivity with certain trans olefins (*trans*-β-methylstyrene: 50% ee with **66** and added 2-methylimidazole; *trans*-stilbene: 48% ee with **67**). By comparison, catalysts **14** and **65** afford very low enantioselectivities (< 20% ee) with trans-disubstituted substrates. In an interesting experiment, Katsuki demonstrated that epoxidation of *trans*-stilbene takes place with 48% ee using catalyst **72**, which bears no stereogenic centers on the diimine bridge. He concluded that the asymmetry at the 3- and 3'-positions of catalysts **66–72** has little influence on the epoxidation of cis olefins but is the dominant factor in the epoxidation of trans olefins [23e].

Mechanistic Considerations

As discussed earlier, (salen)Mn(V) oxo complexes are strongly implicated as intermediates in catalytic epoxidations involving (salen)Mn complexes by the fact that identical enantioselectivities obtained with either iodosylmesitylene or sodium hypochlorite are indistinguishable within experimental error. As the first highly enantioselective nonenzymatic epoxidation reaction that involves transfer of an oxo ligand, the (salen)Mn(III)-catalyzed epoxidation reaction constitutes a potentially powerful and previously unavailable mechanistic tool for studying oxygen atom transfer from transition metals to alkenes. At the very least, it is reasonable to assume that substrates that undergo epoxidation with high enantioselectivity (e.g., > 90% ee) afford epoxide via a single pathway, since the chances are very small that two different mechanisms would afford selectivity exceeding 20:1 with the same sense of asymmetric induction. At best, the stereochemical information that is provided by these catalysts may provide a detailed picture of the transition state structure of the first irreversible step in epoxidation.

The side-on approach mechanism for olefin epoxidation discussed above has provided a useful model for making rational modifications and improvements to the catalyst system, but a more subtle issue in these epoxidation reactions is the identity, if any, of reactive intermediates along the oxidation pathway. Indeed, even within the context of the side-on approach, several different initial events can be envisioned between alkene and metal oxo species. The simplest possible mechanism would involve concerted, although not necessarily synchronous, formation of both oxygen carbon bonds (Fig. 8a). Alternatively, stepwise bond formation may take place, through either polar or nonpolar intermediates (Fig. 8b). Finally, evidence for rate-

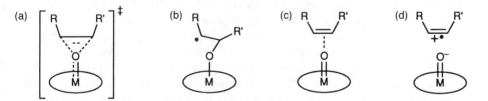

Figure 8 Proposed mechanisms for oxygen atom transfer: (a) transition state for concerted mechanism, (b) nonpolar intermediate in stepwise mechanism, (c) charge transfer complex formation, and (d) electron transfer.

limiting electron transfer (Fig. 8c) [29] or charge transfer complex formation (Fig. 8d) has been presented in porphyrin systems [30].

The latter two mechanisms appear to be unlikely in asymmetric epoxidations catalyzed by complexes such as **65**. The high levels of enantioselectivity are consistent with a highly ordered stereo-determining transition state, and this is very difficult to explain if the stereo-determining step involves only noncovalent interactions. Of course, charge transfer complex formation or electron transfer *preceding* the first irreversible, stereo-determining step cannot be ruled out by this argument, and indeed these pathways could constitute initial, reversible steps in the concerted and stepwise mechanisms depicted in Figures 8a and 8b.

The most straightforward mechanistic clue concerning the events involved in oxygen atom transfer in these systems derives from the formation of both cis and trans epoxides as primary products from cis olefins. As illustrated in Table 5, the extent of trans epoxide formation with catalyst **65** depends strongly on the nature of the substrate. Simple alkyl-substituted cis olefins are epoxidized stereospecifically, while aryl-substituted cis olefins afford cis epoxides as the predominant products with varying amounts of trans epoxides as by-products. In contrast, conjugated dienes and enynes generally afford trans epoxides as the major products. These observations are readily interpreted according to a stepwise mechanism in which a discrete intermediate (**A**, Scheme 5) undergoes competitive collapse to cis epoxide and rotation/collapse to trans epoxide. The propensity to form trans product thus correlates well with the predicted stabilization of intermediate **A** through π-overlap, with the longer-lived intermediates expected to have more time to reorient prior to collapse. As illustrated by entries 4 and 5 of Table 5, both the corresponding cis and trans enynes afford epoxide mixtures of similar diastereomeric composition, indicating that in these cases the intermediates have ample time to rotate prior to collapse. These arguments hold regardless of whether the intermediate is polar, although the remarkable absence of solvent effects on this reaction would suggest that a radical species is more likely.

In the case of cis-disubstituted olefins, the trans pathway results in formation of diastereomeric epoxides. In contrast, for terminal olefins such as styrene, the trans pathway results instead in partitioning to enantiomers. Indeed, the epoxidation of *cis*-β-deuteriostyrene with catalyst **14** was shown to generate *cis*- and *trans*-2-deuteriostyrene oxide in an 8:1 ratio. Thus, the diminished enantioselectivity observed with styrene (60–70% ee) relative to sterically similar cis-disubstituted olefins (> 80% ee for *cis*-β-methylstyrene) can be attributed simply to "enantiomeric leakage" via a trans pathway. Suppression of this pathway has not been accomplished successfully, and synthetically useful enantioselectivities with terminal olefins are yet to be achieved using the chiral (salen)Mn(III) systems.

Although discrete intermediates are strongly implicated in the epoxidation of conjugated olefins, it appears likely that isolated olefins proceed instead through a concerted mechanism. The existence of such a mechanistic divergence is supported by a variety of phenomena. The most significant evidence from a synthetic perspective is the very low enantioselectivity generally observed in the epoxidation of isolated olefins (0–60% ee with all substrates examined). These substrates also react

Table 5 Nonstereospecific Epoxidation Reactions Catalyzed by **65**

Entry	Olefin	cis epoxide/ trans epoxide	cis epoxide % ee	trans epoxide % ee
1	 t-Bu, Et	>99:1	34	--
2	 Ph, Me	92:8	92	83
3	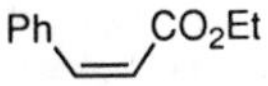 Ph, CO$_2$Et	78:22	97	78
4	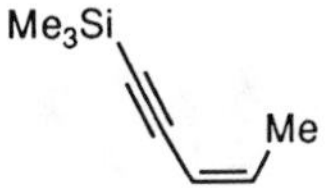 Me$_3$Si, Me	29:71	48	90
5	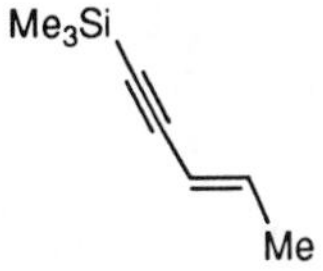 Me$_3$Si, Me	33:67	28	46
6	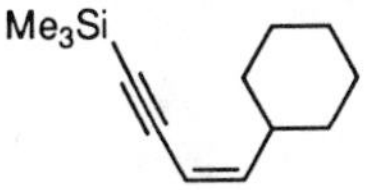 Me$_3$Si, (cyclohexyl)	16:84	64	98
7	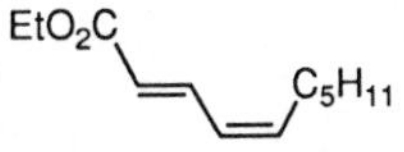 EtO$_2$C, C$_5$H$_{11}$	10:90	60	87

one to two orders of magnitude more slowly than sterically similar conjugated olefins. A dramatic difference in enantioselectivity with alkyl- and aryl-substituted olefins was also noted in the porphyrin system **3** developed by Groves (see Table 1).

A concerted mechanism in the epoxidation of the hypersensitive radical clock **76** [31] by catalyst **14** was implied by the observation of epoxide as the only oxidation product (Scheme 6) [32]. Since the radical **78** has been predicted to rearrange irreversibly at a rate of at $\geq 10^{10}$ s^{-1} on the basis of model experiments [31], the absence of any products resulting from such rearrangement renders the intermediacy of **78** extremely unlikely in this reaction.

The dramatic difference in enantioselectivities observed in epoxidations of isolated and of conjugated olefins with (salen)Mn(III) catalysts may be understood in terms of this mechanistic divergence. In a stepwise process, the intermediacy of a

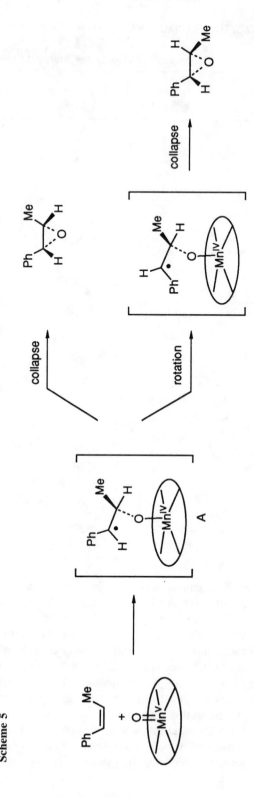

Scheme 5

Scheme 6

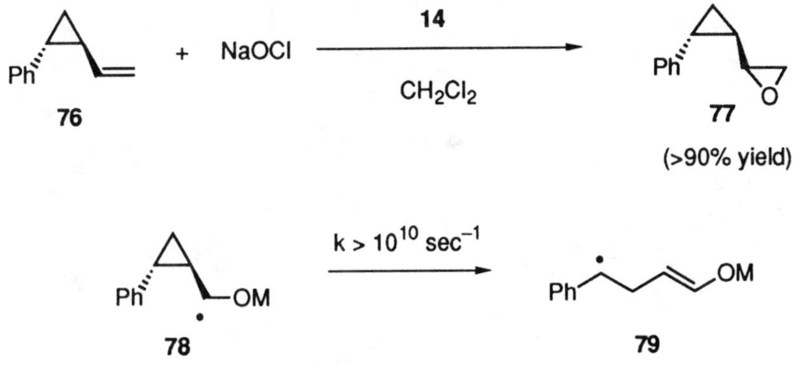

(>90% yield)

high energy species would require an endothermic first step (Fig. 9b). Since olefin isomerization has never been detected with these systems, it is clear that this first step is irreversible, and therefore rate- and stereo-determining. According to Hammond postulate arguments, the transition state in such a reaction should be productlike, with a relatively tight interaction between substrate and catalyst. In contrast, a concerted process would predict a transition state that is earlier along the reaction coordinate (Fig. 9a). Given that the reactants consist of two noninteracting species, an early transition state would be expected to result in poor stereochemical communication between substrate and the dissymmetric environment of the catalyst.

It is also important to note that the stereoelectronic requirements in concerted and in stepwise mechanisms may be completely different. In the stepwise process, formation of a single bond in the stereo-determining step requires overlap between one orbital on alkene with one orbital on the metal oxo species. In a concerted process, simultaneous formation of two C—O bonds requires overlap between two orbitals (one filled, the other empty) on each reactive partner. This latter type of interaction very likely has geometrical constraints different from those of the former, and indeed side-on approach may be stereoelectronically untenable in a concerted mechanism.

Electronic Effects on Catalyst Selectivity

The (salen)Mn(III) epoxidation catalysts bear a ligand template wherein not only the steric environment but also the electronic properties of the metal center may be tuned in a relatively straightforward manner. Since the electronic structure of the substrates has such an important effect on the enantioselectivities in asymmetric epoxidation, there naturally arises the intriguing possibility that enantioselectivity with a given substrate could be tuned through variation of the electronic properties of the catalysts. Indeed, variation of the 5- and 5'-substituents of salen catalysts had a measurable effect on the selectivity of epoxidation of different types of substrates. The change in enantioselectivity was small for isolated olefins such as **80**, which

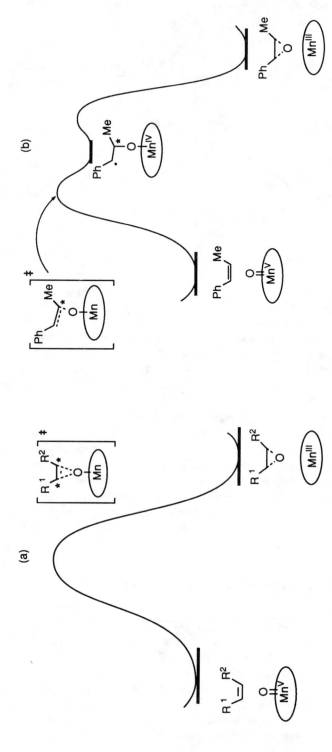

Figure 9 Energy diagrams for oxo transfer to alkenes: (a) concerted mechanism and (b) stepwise nonpolar mechanism.

were epoxidized with only marginal selectivity, but it was strikingly large for 2,2-dimethylchromene (**82**), where enantioselectivities obtained in reactions at room temperature ranged from 22% ee with the NO_2-substituted catalyst **25** up to 96% ee for the MeO-substituted analog **16**. This corresponds to a difference of the $\Delta\Delta G^{\ddagger}$ for the two reactions of 2.0 kcal/mol. The data for the three substrates **80–82** are presented in Hammett plots in Figure 10. The linear correlations in each of these plots indicate that these effects are indeed due to electronic rather than steric factors.

Electronic effects on enantioselectivity may be attributed to changes imparted by the substituents on the reactivity of the metal oxo intermediates. Electron-withdrawing groups such as nitro destabilize the Mn(V)oxo intermediate relative to oxidation, whereas electron-donating groups attenuate the reactivity of the oxidant. In accordance with the Hammond postulate arguments presented in the preceding sections, more reactive oxidants should effect the stereo-determining initial C—O bond forming event via a more reactantlike transition structure, with greater separation between substrate and catalyst and concomitantly poorer steric differentiation of diastereomeric transition structures. A milder oxidant would be expected in turn to effect epoxidation via a more productlike transition state.

A study of secondary isotope effects in the epoxidation of styrene was undertaken to test this hypothesis. The relative rates of epoxidation of styrene and β-deuteriostyrene were examined in competition experiments using the catalysts in Figure 11, and isotope effects were determined by proton NMR analysis of substrate and product distributions [33]. As illustrated in Figure 11, a trend is observed with the more enantioselective, electron-rich catalysts exhibiting more pronounced isotope effects. Since k_H/k_D reflects the change in hybridization at the β-carbon in the transition structure, it provides an indirect estimation of the position of the transition state along the reaction coordinate between starting materials and products. Thus, all other factors being equal, epoxidations with more productlike transition states should exhibit more pronounced inverse isotope effects. The correlation between the electron-donating abilities of the 5 and 5′ substituents with both enantioselectivity and k_H/k_D provides strong evidence that a late transition state is indeed a crucial factor in attaining high enantioselectivity.

Synthetic Applications

At this stage the success of the (salen)Mn(III)-catalyzed asymmetric epoxidation method remains largely limited in its synthetic utility to conjugated olefins, with the notable exception of cyclic ketal derivatives of conjugated enones, which are epoxidized with excellent enantioselectivities. Nonetheless, the conjugated alkenes that may be epoxidized successfully include a variety of interesting and important substrates.

Remarkably high enantioselectivities are observed in the epoxidation of 2,2-dimethylchromene derivatives by **65** (Table 6) [34]. The epoxidation of 6-cyano derivative **83** was applied to highly efficient syntheses of the antihypertensive agents cromakalim and the related compound EMD-52,692 (Scheme 7) [34]. The oxidation of *cis*-methylcinnamate with **65** constituted the first direct epoxida-

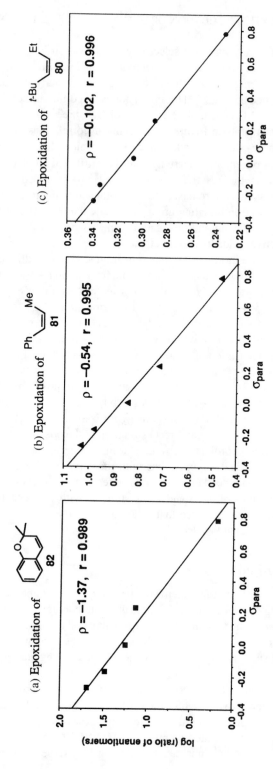

Figure 10 Hammett plots depicting enantiomeric composition of epoxides generated by oxidation of the indicated alkenes by catalysts **16, 21, 14, 15,** and **25**. Enantiomeric excess ranges: (a) 22–96, (b) 49–83, and (c) 26–37% ee.

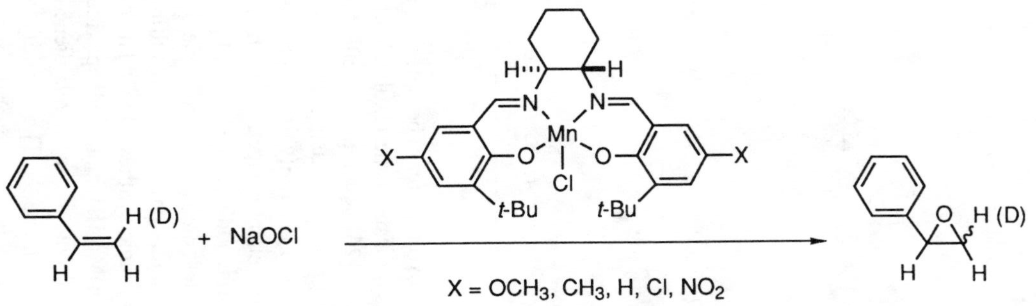

X	k_H / k_D
NO_2	0.95
Cl	0.89
H	0.86
CH_3	0.84
OCH_3	0.82

Figure 11 Electronic effects of catalyst on k_H/k_D in the epoxidation of styrene.

Table 6 Epoxidation of 2,2-Dimethylchromene Derivatives Catalyzed by **65**

Entry	Olefin	ee (%)	Isolated Yield (%)	Absolute Configuration
1	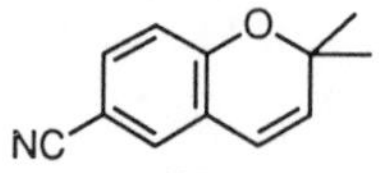 NC— **83**	97	96	(3R, 4R)-(+)
2	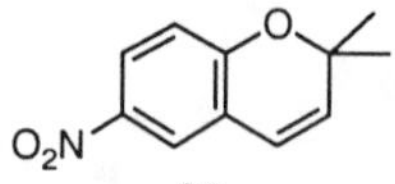 O$_2$N— **84**	94	76	(3R, 4R)-(+)
3	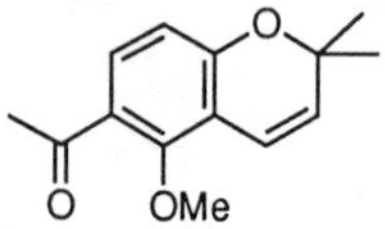 O OMe **85**	98	75	(3R, 4R)-(+)
4	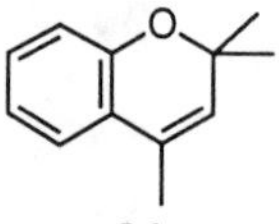 **86**	97	51	(3R, 4R)-(+)
5	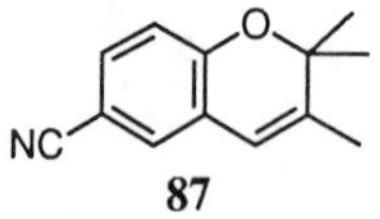 NC— **87**	>98	82	(3R, 4R)-(+)

tion of an α,β-unsaturated ester via oxo transfer. The potential utility of this procedure is illustrated by the practical synthesis of the taxol side chain **88** via epoxidation of *cis*-ethyl cinnamate (Scheme 8) [26]. An important effect of 4-phenylpyridine-*N*-oxide was noted, not only with regard to increasing catalyst turnovers, but also in terms of substantially increased enantioselectivity. It ‡as proposed that the *N*-oxide additive suppresses a competitive low-enantioselectivity pathway to epoxide involving Lewis acid catalyzed hypochlorite addition to the conjugated ester.

Among conjugated olefin substrates, dienes constitute one of the most interesting substrate classes for asymmetric epoxidation, since the resulting vinyl epoxides are extremely useful intermediates in organic synthesis [35]. The dominance of the so-called trans pathway in the (salen)Mn(III)-catalyzed epoxidation of conjugated enynes and dienes suggests a general route to *trans,trans*-diene monoepoxides from the corresponding *cis,trans*-dienes (Scheme 9). The viability of such a strategy is supported by the observation that cis olefins are generally more reactive than elec-

Scheme 7

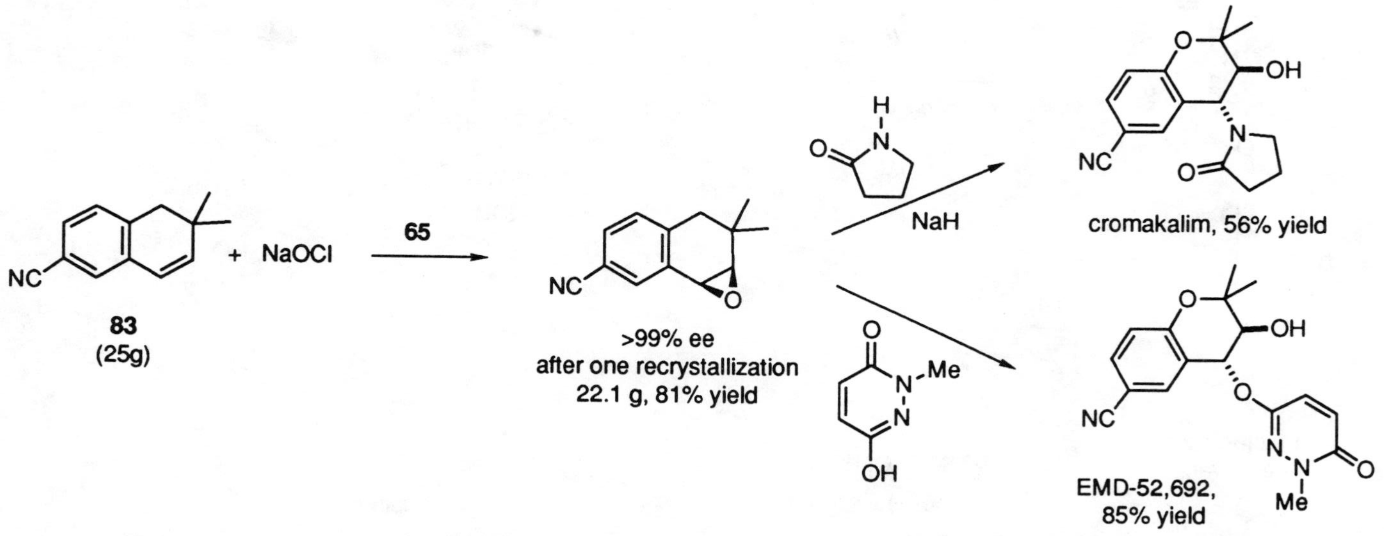

187

Scheme 8

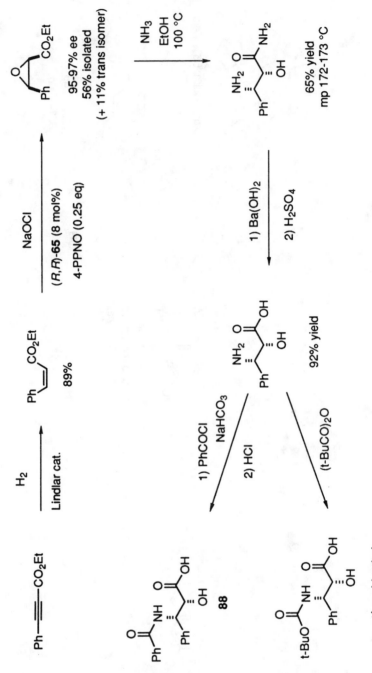

taxotère side chain

Scheme 9

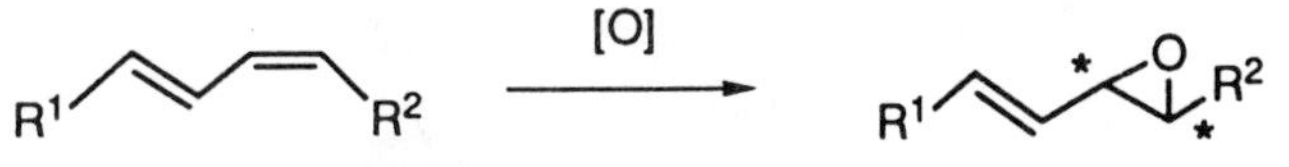

tronically similar trans olefins, and that conjugated olefins are more reactive then isolated olefins in epoxidations catalyzed by (salen)Mn complexes.

The successful application of a (salen)Mn(III) catalyzed regio- and enantioselective polyene monoepoxidation is illustrated in the formal synthesis of LTA_4, in which the key step is the highly regioselective and moderately enantioselective monoepoxidation of the *cis,trans,trans*-triene **89** (Scheme 10) [36].

4.2.2.3. Other Oxo Transfer Catalysts

In addition to porphyrin and salen complexes, a variety of metal complexes bearing cyclic and acyclic multidentate ligands have been studied as asymmetric epoxidation catalysts. An iron(III) complex of ligand **90** has been reported to catalyze olefin epoxidation with iodosylbenzene, but racemic product was formed with styrene as substrate [37]. Similar results were obtained with Ni(II) complexes of ligands **91** and **92** [38]. The complete absence of asymmetric induction in these reactions may be due to the involvement of free-radical mechanisms involving active oxidants that are not coordinated to the chiral catalyst [39].

Epoxidation of simple olefins with metal complexes of chiral tetradentate bis-amide ligands also has met with limited success. For example, epoxidation of styrene by iodosylbenzene catalyzed by chiral anionic cobalt(III) complexes **93–99** (Chart 7) afforded the corresponding epoxide with low enantioselectivity (0–17% ee) [40]. High valent Co(V) oxo intermediates were suggested as the active oxygen transfer species in these reactions.

Both Fe(II) and Fe(III) complexes of the antitumor antibiotic bleomycin (BLM) and its synthetic analog PYML-6 are effective catalysts for the epoxidation of olefins [41]. Molecular oxygen in the presence of a reducing agent was shown to be an effective stoichiometric oxidant with the Fe(II) system, while H_2O_2 was em-

Chart 6

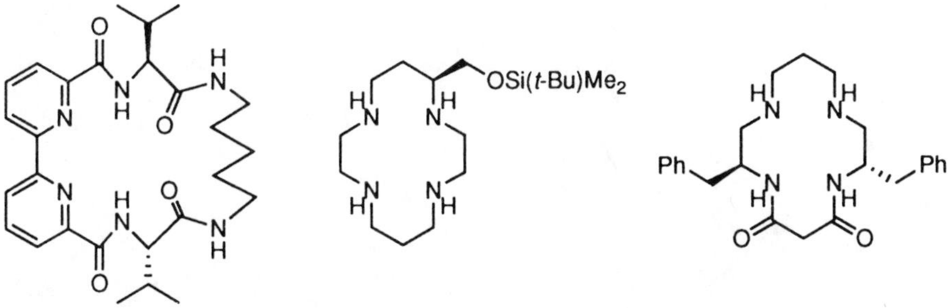

90   91   92

Scheme 10

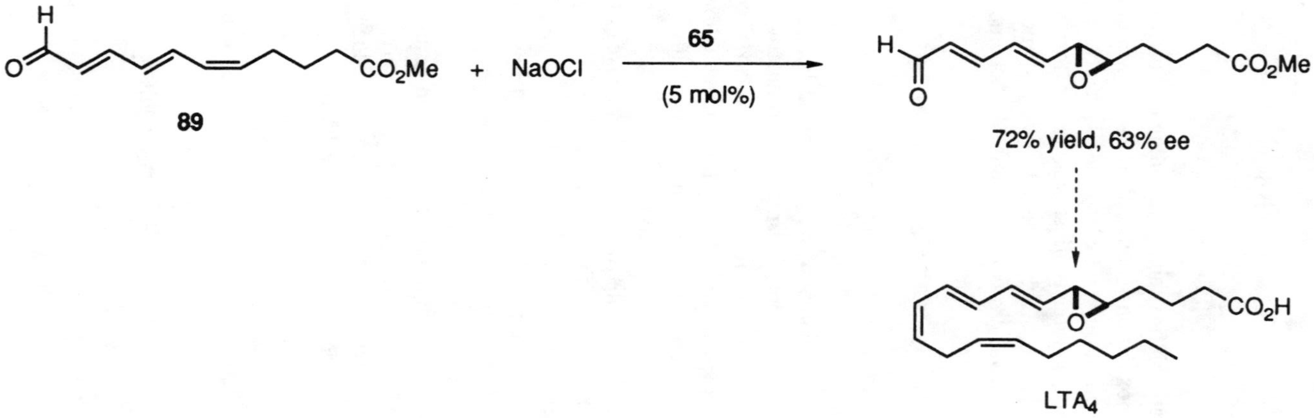

72% yield, 63% ee

65
(5 mol%)

89

LTA₄

Chart 7

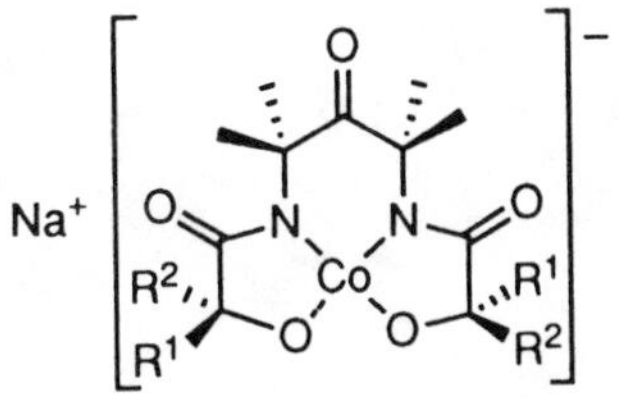

93 R^1 = Me, R^2 = Et
94 R^1 = Et, R^2 = Ph

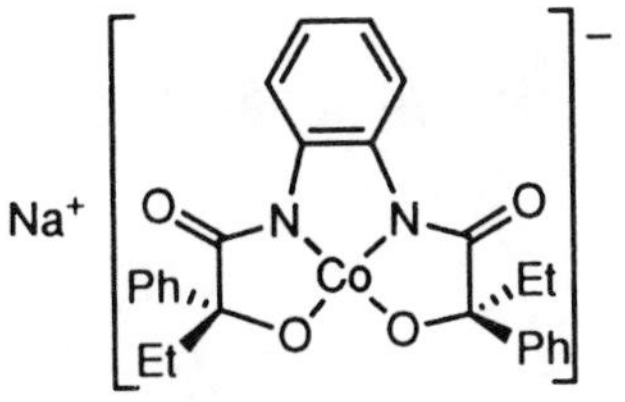

95

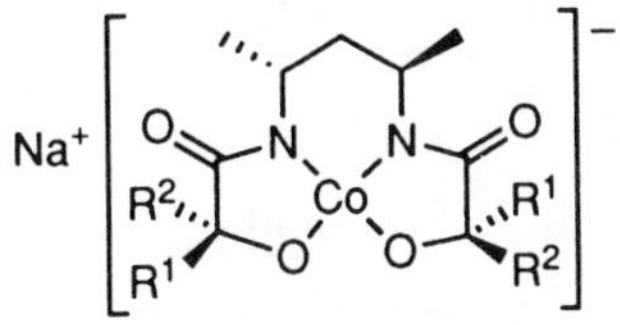

96 R^1 = Me, R^2 = Et
97 R^1 = Et, R^2 = Ph

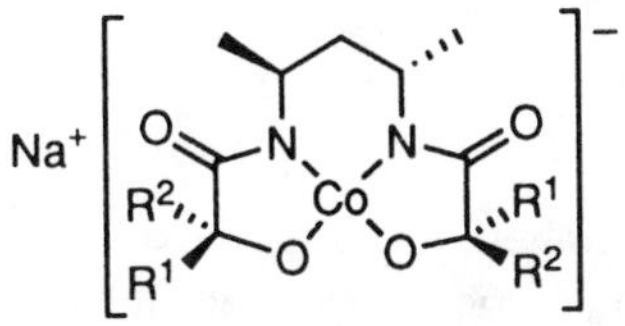

98 R^1 = Me, R^2 = Et
99 R^1 = Et, R^2 = Ph

ployed with the Fe(III) catalysts. The structures proposed by the authors for the Fe(II)-O_2 complexes of BLM (**100**) and PYML (**101**) are shown in Chart 8. The highest selectivity (51% ee) was obtained for the epoxidation of *cis*-β-methylstyrene with the Fe(II)-PYML-6 catalyst. As has generally been observed with porphyrin- and salen-based systems, the enantioselectivity for the epoxidation of *trans*-β-methylstyrene was found to be very low (< 5% ee).

4.2.2.4. Peroxo-Based Catalysts

Inorganic peroxide-based systems constitute the most important class of homogeneous catalysts for the racemic epoxidation of alkyl-substituted olefins. The enormous success of the Halcon process for the epoxidation of propylene naturally

Chart 8

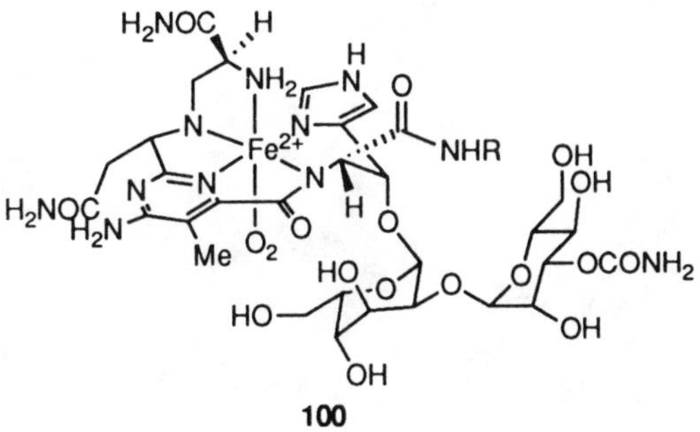

100

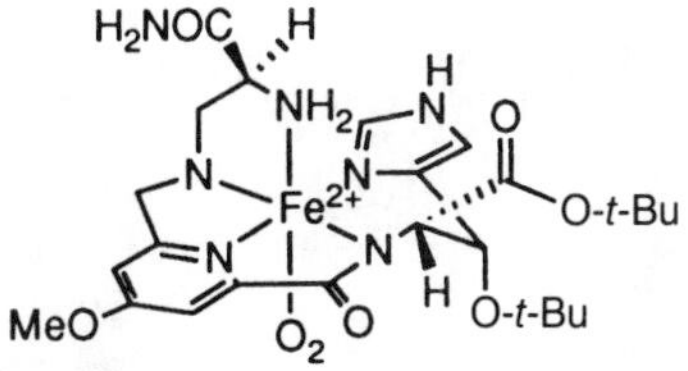

101

Scheme 11

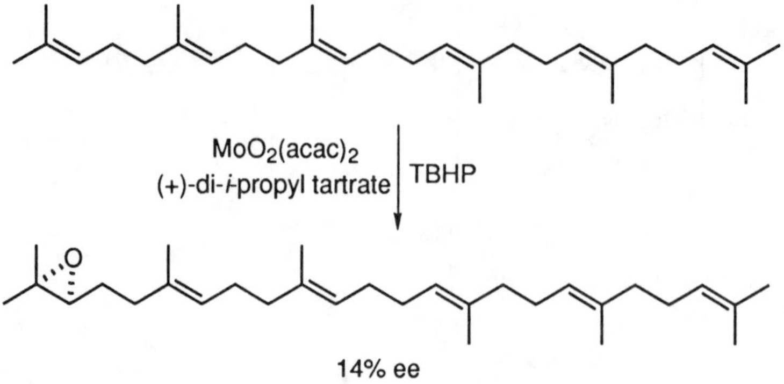

14% ee

attracted attention to the possible use of chiral peroxides or alkylperoxy complexes for asymmetric epoxidation. Indeed, the first catalytic enantioselective epoxidation of unfunctionalized olefins was effected by Otsuka in 1979 utilizing a straightforward chiral modification of catalyst systems employed in the Halcon process [42]. Thus, catalysts prepared in situ from $MoO_2(acac)_2$ and dialkyl tartrates or other carbohydrate derivatives effectively catalyzed oxygen transfer from t-butylhydroperoxide (TBHP) to simple olefins, although enantioselectivities were reported to be very low. The best enantioselectivity reported by Otsuka was 14% ee in the epoxidation of squalene (Scheme 11).

In 1979 Kagan reported that the molybdenum(VI) peroxo complex **102** (Chart 9) effected the epoxidation of simple olefins with enantioselectivities in the range of 5.1–34.8% ee [43]. This stoichiometric reaction has been refined through a ligand variation approach, with enantioselectivities as high as 53% ee reported for the epoxidation of a tetracyclic olefin [44]. It was subsequently found that addition of excess chiral diol to the reaction system induced a kinetic resolution process mediated by a chiral molybdenum–diol complex. Under these conditions the enantiomeric composition of the epoxide was observed to be time dependent, with extended reaction time leading to very high enantioselectivities (up to 95% ee) at the expense of epoxide yield [45]. Chiral bidentate ligands were found to be substantially more effective than corresponding monodentate ligands, which led only to low enantioselectivities (0.7–8.5% ee) in olefin epoxidation [46].

Chart 9

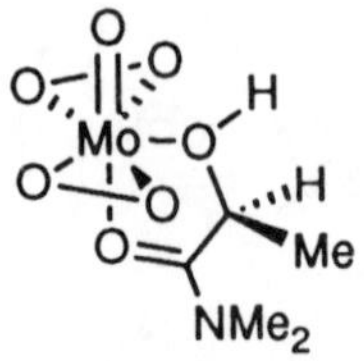

102

Chart 10

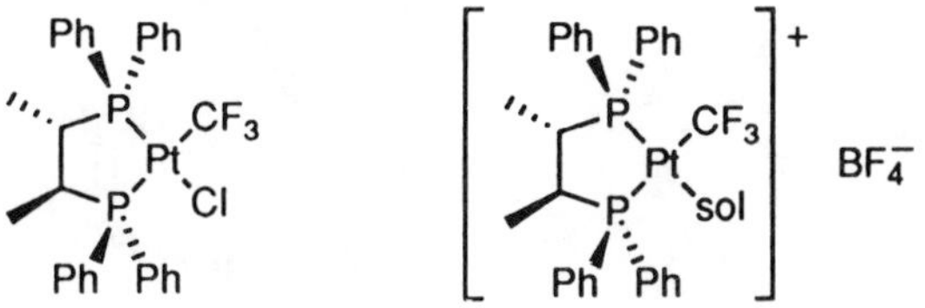

103   **104**

In 1987 Struckul reported that platinum(II) complexes of chiral diphosphines also catalyze asymmetric epoxidation of simple olefins in the presence of H_2O_2 [47]. Two different reaction mechanisms were proposed. Hydroperoxo–platinum(II) intermediates were implicated in the reaction with neutral catalyst **103**, while a Wacker-like mechanism involving direct attack on coordinated alkenes by H_2O_2 was proposed with cationic catalyst **104** (Chart 10). Catalysts of the latter type afforded higher enantioselectivities (up to 41% ee) in the epoxidation of propene and 1-octene. Although subsequent studies with catalysts of this type have not appeared, catalyst **104** remains the most effective nonenzymatic catalyst reported to date for the epoxidation of terminal olefins with aliphatic substituents.

Very recently, Halterman has described the first application of catalysts based on chiral cyclopentadienyl (Cp) to the asymmetric epoxidation of olefins (Scheme 12) [48]. Although selectivities were very moderate, this first report opens yet another promising avenue for ligand design in asymmetric epoxidation.

Scheme 12

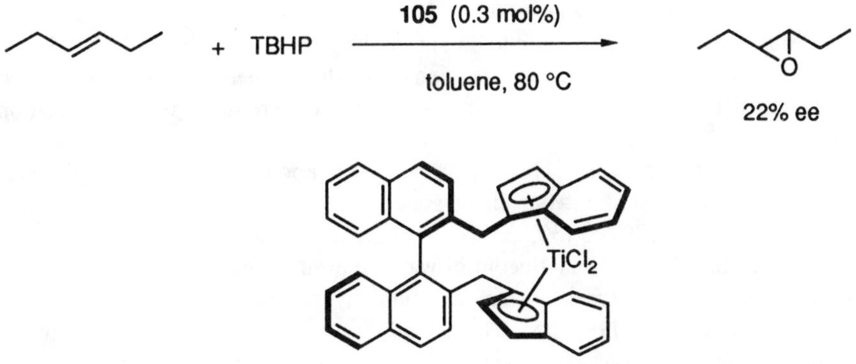

105

4.2.3. Asymmetric Epoxidation with Organic Oxidants

4.2.3.1. Electrophilic Oxidants

Although asymmetric epoxidation with organic electrophiles has typically been restricted to stoichiometric processes, an overview of known reagents is provided

Chart 11

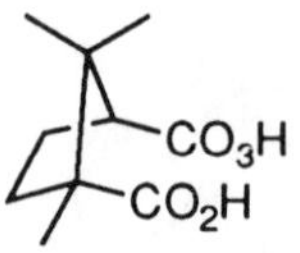

106

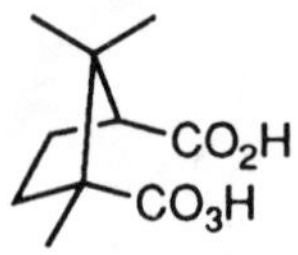

107

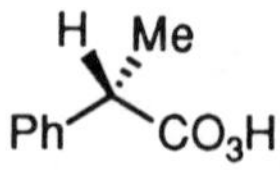

108

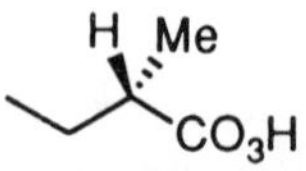

109

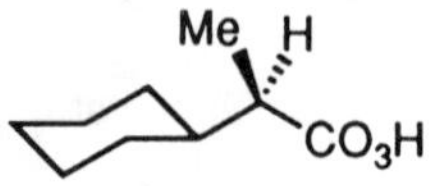

110

111

here, since their modification into catalytic systems may be achievable, at least in principle.

In 1965 Henbest reported the first asymmetric epoxidation reaction of olefins by a chiral peroxy acid, (+)-peroxycamphoric acid (106) [49]. This discovery is primarily of historical, rather than synthetic, significance since the enantioselectivities obtained with this reagent were only 1.0–2.4% ee for a variety of substrates [50]. Pirkle's reevaluation of the peroxycamphoric acid that was used in early reactions led to the discovery that the oxidant actually contained two diastereomers (106 and 107) that effected epoxidation with opposite facial selectivities [51]. However, the use of purified (+)-peroxycamphoric acid afforded epoxides with enantioselectivities only 1.5–2 times greater than those obtained with the isomeric mixture of oxidants. For instance, the epoxidation of styrene by 106 afforded less than 10% ee. Various other chiral peroxy acids (e.g., 108–111: Chart 11) have been tested with very limited success [52]. Failure to obtain significant enantioselectivities with such systems may be attributable to the location of the dissymmetry of the peracid—too far removed from the site of oxygen transfer in the stereo-determining transition state.

Several different classes of chiral hydroperoxy compounds (e.g., 112–115: Chart 12) have also been studied, but enantioselectivities obtained with these reagents have not exceeded 10% ee [50,53,54]. A single exception to this trend lies in the 1986 report that absolute enantiocontrol was achieved for the epoxidation of *trans*-stilbene and α-methylstyrene using the chiral hydroperoxyimine generated from 2-cyanoheptahelicene (116) [55]. The helical dissymmetric environment surrounding the hydroperoxy moiety may be responsible for the extremely high enantioselectivities, but it should be noted that the chemistry of this reagent has not been elaborated since the original report.

Chiral dioxiranes 117 and 118 (Chart 13) [56], generated by the reaction of Oxone with enantiomerically pure ketones, have also been studied as reagents for the epoxidation of simple olefins. The enantioselectivities obtained with these oxidants for the epoxidation of 1-methylcyclohexene and *trans*-β-methylstyrene were only ≈10% ee, which is only marginally better than those obtained with chiral

Chart 12

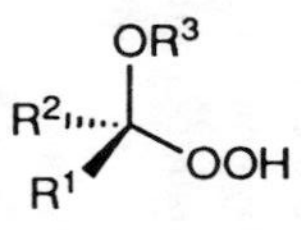

112 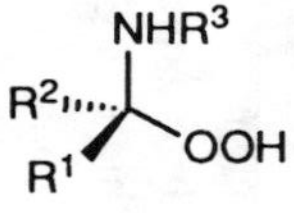 **113** 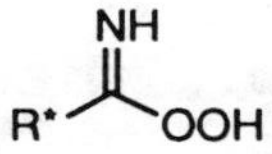 **114** 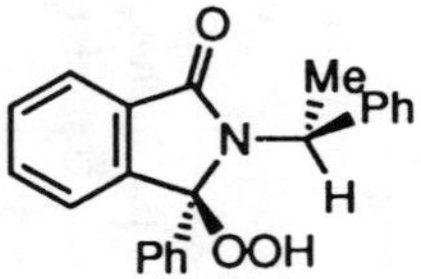 **115** 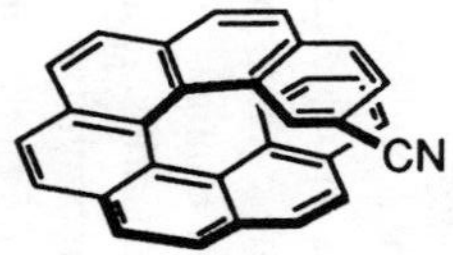 **116**

Chart 13

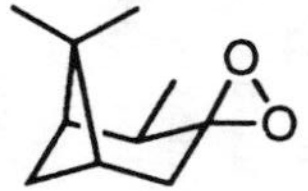

117 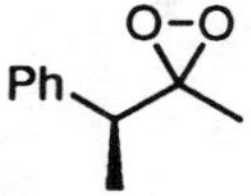 **118**

Chart 14

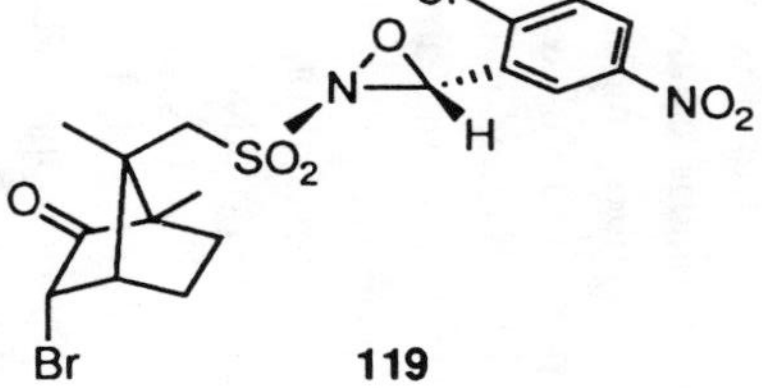

119 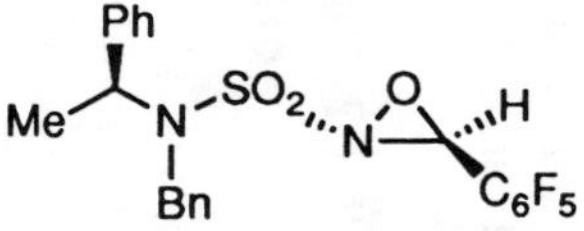 **120** 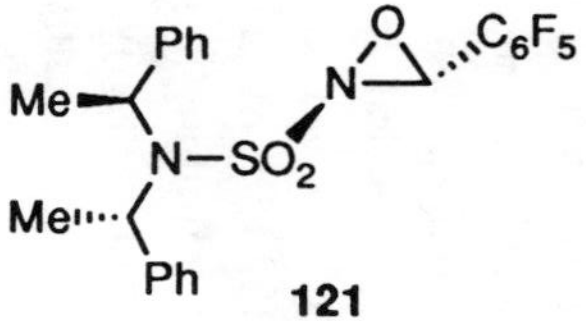 **121**

peroxy acids. However, the use of ketones in substoichiometric amounts (30 mol %) with no appreciable loss in selectivity is significant and suggests that further research in this area is certainly well justified.

Without question the most successful approach to stoichiometric asymmetric epoxidation employing electrophilic organic oxidants was developed by Davis and his coworkers through the use of chiral oxaziridine derivatives. Epoxidation of simple alkenes with chiral oxaziridine **119** (Chart 14) afforded the corresponding epoxides with enantiomeric purities up to 40% ee [57]. Structural changes of the reagent led to further improvements, the highest published enantioselectivity (64.7% ee) having been obtained in the epoxidation of *trans*-β-methylstyrene by **120** [58]. A planar orientation of the oxaziridine with reacting olefin in the stereo-determining transition state was proposed based on the observed sense and degree of enantioselectivity.

Very recently, Davis and Przeslawski disclosed that **121** effected epoxidation of styrene and *trans*-β-methylstyrene with better than 90% ee [59]. Although these latter reactions were very slow (2 weeks at 60°C) and chiral oxaziridine based systems have not succumbed in general to catalytic modification, the high selectivities obtained thus far in stoichiometric oxidations clearly bear considerable significance.

4.2.3.2 Nucleophilic Oxidants: Asymmetric Epoxidation of Conjugated Ketones

In contrast to the limited success attained thus far in catalytic epoxidations by electrophilic non-metal-based oxidants, asymmetric catalytic epoxidation of certain α,β-unsaturated carbonyl compounds with nucleophilic oxidants has been developed very successfully. These reactions typically employ hydrogen peroxide or an alkyl hydroperoxide under basic conditions as the stoichiometric oxidant, and an asymmetric catalyst consisting of a chiral organic phase transfer agent.

Wynberg pioneered the use of chiral ammonium salts to catalyze asymmetric epoxidation of α,β-unsaturated ketones. Enantioselectivities as high as 55% ee were obtained in the epoxidation of *trans*-chalcone derivatives with basic hydrogen peroxide using quinine-derived **122** as catalyst (Scheme 13) [60,61]. Other stoichiometric oxidants were employed with varying degrees of success. Epoxidation of *trans*-chalcone derivatives with sodium hypochlorite in the presence of **122** afforded product with about 25% ee, comparable to that obtained by employing the H_2O_2/NaOH system. However, the absolute configuration of the major enantiomer obtained with bleach was opposite to that obtained by using H_2O_2 as oxidant [62]. Compound **122** also catalyzed the epoxidation of quinone derivative **123** with 78% ee in the presence of anhydrous TBHP as oxidant and powdered NaOH (Scheme 14) [63].

Cyclodextrins have also been applied as phase transfer catalysts in the asymmetric epoxidation of α,β-unsaturated ketones. Low enantioselectivities have been reported for the epoxidation of *trans*-chalcone using bleach and α- or β-cyclodextrin [64], and attempts to improve enantioselectivity by modifications at the hydroxy

Scheme 13

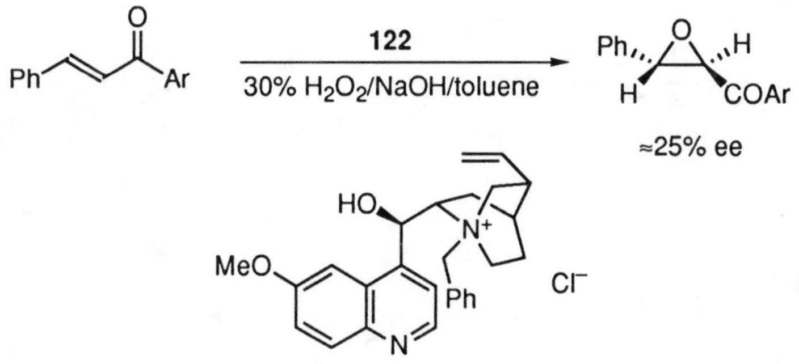

Scheme 14

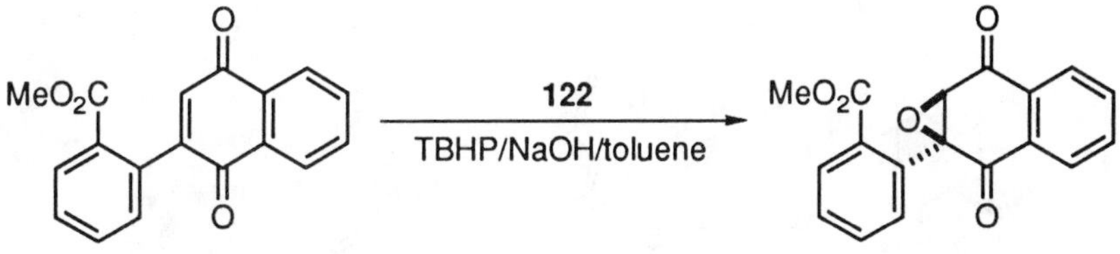

Scheme 15

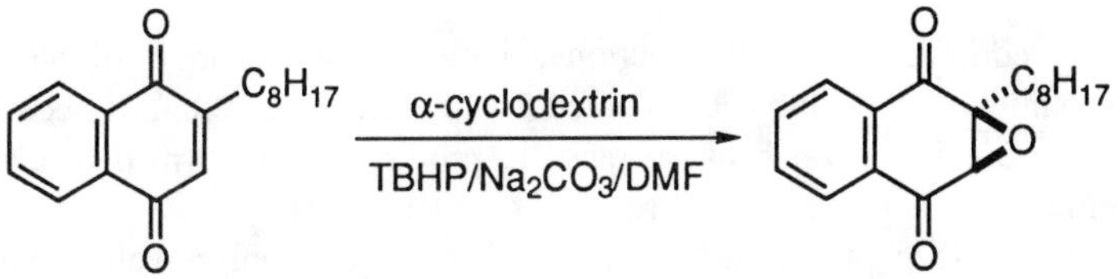

groups of cyclodextrins have thus far been unsuccessful [65]. Epoxidation of *trans*-cinnamaldehyde with H_2O_2 has also been reported to be catalyzed by α-cyclodextrin with low enantioselectivity (up to 8% ee) [66]. More promising results were obtained in the epoxidation of substituted benzoquinones (up to 48% ee) using TBHP as oxidant (Scheme 15) [67].

The application of synthetic peptides in asymmetric epoxidations of α,β-unsaturated ketones has led to the development of highly enantioselective catalytic systems [68]. In 1980 Juliá introduced a poly[(S)-alanine] catalyst (**124**) for the epoxidation of *trans*-chalcone in the presence of H_2O_2/NaOH under phase transfer conditions, with up to 97% ee [69]. Polymer **124** was prepared by treatment of the N-carboxyanhydride of alanine with butylamine (Scheme 16). A three-phase mix-

Scheme 16

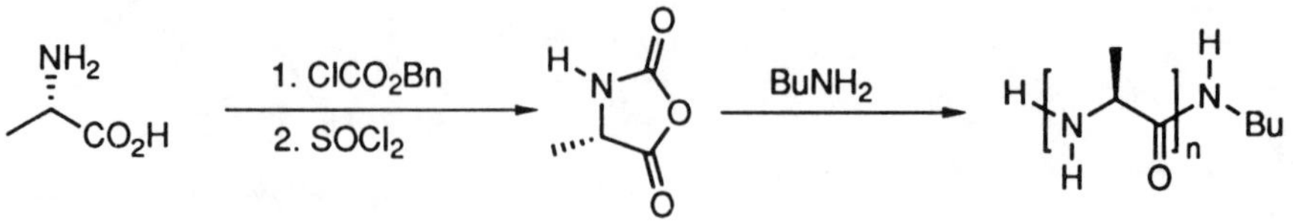

124

Scheme 17

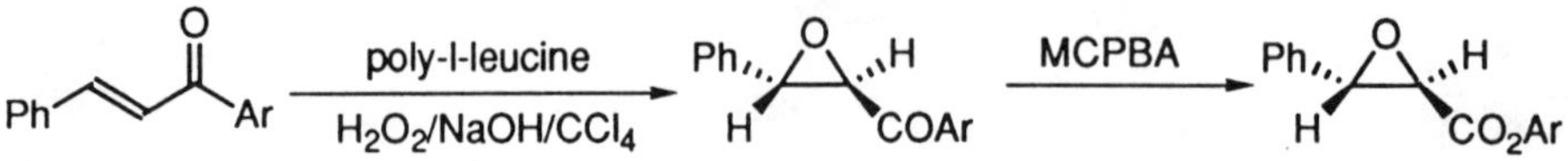

Scheme 18

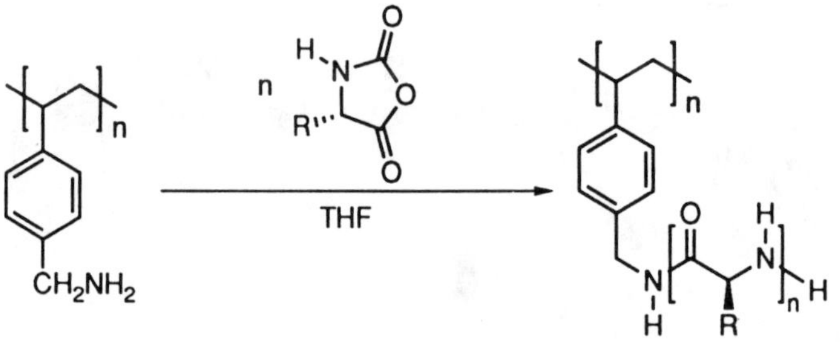

ture is involved in these catalytic reactions, since the peptides are insoluble in either H_2O or organic solvents. It has been noted that the triphasic system is necessary for high asymmetric induction because racemic products were formed in a biphasic system employing organic hydroperoxide as oxidant [61].

Other synthetic peptides such as poly-L-leucine and poly-L-valine have been reported to effect the highly enantioselective epoxidation of a variety of *trans*-chalcone derivatives [70]. The enantioselectivities are highly dependent on the size of the polymer, and it has been reported that asymmetric induction drops significantly if the polymer chain contains fewer than 10 amino acid residues.

These peptide-catalyzed reactions have been applied to a variety of α,β-unsaturated ketones and other electron-deficient olefins such as nitroalkenes and α,β-unsaturated nitriles, although asymmetric induction reported for such substrates has generally been very low [71]. Indeed, high enantioselectivities appear to be limited to *trans*-chalcone derivatives. However, a certain degree of flexibility in this chemistry was introduced through the combination of asymmetric epoxidation with Baeyer–Villiger oxidation to afford the corresponding glycidic esters (Scheme 17) [72].

Polymer-supported polyamino acids have more recently been studied as catalysts in asymmetric epoxidation (Scheme 18) [73]. High enantioselectivity was again achieved in the epoxidation of *trans*-chalcone (up to 99% ee), and the separation of the catalyst from the reaction system was greatly simplified. The catalysts could be reused without a significant loss of reactivity. Catalyst immobilization is likely to bring similar practical benefits to other known asymmetric epoxidation systems.

4.2.4. Conclusion

A wide variety of approaches toward enantioselective catalytic epoxidation catalysts have been developed over the past few years, with systems based on porphyrin, salen, and polyamino acid showing the most promise at this stage. Epoxidation catalysts based on salen ligands have already been applied to practical syntheses of important optically active compounds and appear to provide one of the more viable new directions for research in this field. Nonetheless, enormous challenges remain in asymmetric epoxidation, with catalysts yet to be discovered for highly enantioselective epoxidation of most trans olefins and simple alkyl-substituted olefins. It is likely that the scope of substrates for the systems developed thus far will be greatly expanded as the mechanistic subtleties of olefin oxidation continue to be elucidated.

Acknowledgments

I am extremely grateful to Dr. Wei Zhang, who initiated my group's research on epoxidation with chiral salen complexes and whose thesis served as the basis for much of the background in this chapter, and to Dr. John Young for assistance in the preparation of this chapter. I thank Professors Thomas Kodadek, Ronald Halterman, Franklin Davis, Tsutomu Katsuki, and James Collman for open discussions and correspondence about their research on asymmetric epoxidation. I am also indebted to Professor K. Barry Sharpless for numerous enlightening discussions. The writing of this chapter was made possible by a Beckman Fellowship in the Center for Advanced Study, and I also thank the National Institutes of Health, the National Science Foundation, and the David and Lucille Packard Foundation for financial support of our work.

References

1. Katsuki, T.; Sharpless, K. B. *J. Am. Chem. Soc.* **1980**, *102*, 5974.

2. For a brief review on recent progress in asymmetric epoxidation, see: Bolm, C. *Angew. Chem., Int. Ed. Engl.* **1991**, *30*, 403.

3. (a) Groves, J. T.; Nemo, T. E.; Myers, R. S. *J. Am. Chem. Soc.* **1979**, *101*, 1032. (b) McMurry, T. J.; Groves, J. T. In *Cytochrome P-450;* Ortiz de Montellano, P. R. (Ed.); Plenum: New York, 1986; Chapter 1.

4. Groves, J. T.; Myers, R. S. *J. Am. Chem. Soc.* **1983**, *105*, 5791.

5. (a) Ostovic, D.; Bruice, T. C. *J. Am. Chem. Soc.* **1988**, *110*, 6906. (b) Ostovic, D.; Bruice, T. C. *J. Am. Chem. Soc.* **1989**, *111*, 6511. (c) He, G.-X.; Mei, H.-Y.; Bruice, T. C. *J. Am. Chem. Soc.* **1991**, *113*, 5644.

6. Groves, J. T.; Han, Y.; Van Engen, D. V. *J. Chem. Soc., Chem. Commun.* **1990**, 436.

7. Groves, J. T.; Viski, P. *J. Org. Chem.* **1990**, *55*, 3628.

8. Mansuy, D.; Battoni, P.; Renaud, J.-P.; Guerin, P. *J. Chem. Soc., Chem. Commun.* **1985**, 155.

9. (a) Naruta, Y.; Tani, F.; Maruyama, K. *Chem. Lett.* **1989,** 1269. (b) Naruta, Y.; Tani, F.; Ishihara, N.; Maruyama, K. *J. Am. Chem. Soc.* **1991,** *113,* 6865.

10. Collman, J. P.; Zhang, X.; Hembre, R. T.; Brauman, J. I. *J. Am. Chem. Soc.* **1990,** *112,* 5356.

11. Konishi, K.; Oda, K.-I.; Nishida, K.; Aida, T.; Inoue, S. *J. Am. Chem. Soc.* **1992,** *114,* 1313.

12. O'Malley, S.; Kodadek, T. *J. Am. Chem. Soc.* **1989,** *111,* 9176.

13. Guilmet, E.; Meunier, B. *Nouv. J. Chim.* **1982,** *6,* 511.

14. Halterman, R. L.; Jan, S.-T.; *J. Org. Chem.* **1991,** *56,* 5253.

15. (a) Srinivasan, K.; Michaud, P.; Kochi, J. K. *J. Am. Chem. Soc.* **1986,** *108,* 2309. (b) Samsel, E. G.; Srinivasan, K.; Kochi, J. K. *J. Am. Chem. Soc.* **1985,** *107,* 7606.

16. Yoon, H.; Burrows, C. J. *J. Am. Chem. Soc.* **1988,** *110,* 4087.

17. For early examples of the preparation and use of chiral salen complexes, see: (a) Cesarotti, E.; Pasini, A.; Ugo, R. *J. Chem. Soc., Dalton Trans.* **1981,** 2147, and references therein. (b) Nakajima, K.; Kojima, M.; Fujita, J. *Chem. Lett.* **1986,** 1483.

18. Zhang, W.; Loebach, J. L.; Wilson, S. R.; Jacobsen, E. N. *J. Am. Chem. Soc.* **1990,** *112,* 2801.

19. Jacobsen, E. N.; Zhang, W.; Muci, A. R.; Ecker, J. R.; Deng, L. *J. Am. Chem. Soc.* **1991,** *113,* 7063.

20. Palucki, M.; Jacobsen, E. N. Unpublished results.

21. Zhang, W.; Jacobsen, E. N. *J. Org. Chem.* **1991,** *56,* 2296.

22. Zhang, W. Ph.D. thesis, University of Illinois; September 1991.

23. (a) Irie, R.; Noda, K.; Ito, Y.; Matsumoto, N.; Katsuki, T. *Tetrahedron Lett.* **1990,** *31,* 7345. (b) Irie, R.; Noda, K.; Ito, Y.; Katuski, T. *Tetrahedron Lett.* **1991,** *32,* 1055. (c) Irie, R.; Ito, Y.; Katsuki, T. *Synlett,* **1991,** *2,* 265. (d) Irie, R.; Noda, K.; Ito, Y.; Matsumoto, N.; Katsuki, T. *Tetrahedron Asymmetry,* **1991,** *2,* 481. (e) Hosoya, N.; Irie, R.; Ito, Y.; Katsuki, T. *Synlett,* **1991,** 691.

24. O'Connor, K. J.; Wey, S. J.; Burrows, C. J. *Tetrahedron Lett.* **1992,** *33,* 1001.

25. Reddy, D. R.; Thornton, E. R. *J. Chem. Soc., Chem. Commun.* **1992,** 172.

26. Deng, L.; Jacobsen, E. N. *J. Org. Chem.* **1992,** *57,* 4320.

27. Zhang, W.; Jacobsen, E. N. *Tetrahedron Lett.* **1991,** *32,* 1711.

28. Szmant, H. H. *Organic Building Blocks of the Chemical Industry;* Wiley: New York, 1989; p. 423.

29. Traylor, T. G.; Miksztal, A. R. *J. Am. Chem. Soc.* **1989,** *111,* 7443.

30. He, C.-X.; Arasasingham, R. D.; Zhang, G.-H.; Bruice, T. C. *J. Am. Chem. Soc.* **1991,** *113,* 9828.

31. He, G.-X.; Bruice, T. C. *J. Am. Chem. Soc.* **1991,** *113,* 2747, and referenced cited therein.

32. Fu, H.; Look, G. C.; Zhang, W.; Jacobsen, E. N.; Wong, C.-H. *J. Org. Chem.* **1991,** *56,* 6497.

33. Güler, M. L.; Jacobsen, E. N. Manuscript in preparation.

34. Lee, N. H.; Muci, A. R.; Jacobsen, E. N. *Tetrahedron Lett.* **1991,** *32,* 5055.

35. Lee, N. H.; Jacobsen, E. N. *Tetrahedron Lett.* **1991,** *32,* 6533.

36. Chang, S. B.; Lee, N. H.; Jacobsen, E. N. Manuscript in preparation.

37. Hopkins, R. B.; Hamilton, A. D. *J. Chem. Soc., Chem. Commun.* **1987,** 171.

38. (a) Kinneary, J. F.; Wagler, T. R.; Burrows, C. J. *Tetrahedron Lett.* **1988,** *29,* 877. (b) Wagler, T. R.; Burrows, C. J. *Tetrahedron Lett.* **1988,** *29,* 5091.

39. Yoon, H.; Thomas, R.; Wagler, T. R.; O'Connor, K. J.; Burrows, C. J. *J. Am. Chem. Soc.* **1990,** *112,* 4568.

40. Ozaki, S.; Mimura, H.; Yasuhara, N.; Masui, M.; Yamagata, Y.; Tomita, K. *J. Chem. Soc., Perkin Trans. 1,* **1990,** 353.

41. Kaku, Y.; Otsuka, M.; Ohno, M. *Chem. Lett.* **1989,** 611.

42. Tani, K.; Hanafusa, M.; Otsuka, S. *Tetrahedron Lett.* **1979,** 3017.

43. Kagan, H. B.; Mimoun, H.; Mark, C.; Schurig, V. *Angew. Chem., Int. Ed. Engl.* **1979,** *18,* 485.

44. (a) Broser, E.; Krohn, K.; Hintzer, K.; Schurig, V. *Tetrahedron Lett.* **1984,** *25,* 2463. (b) Winter, W.; Mark, C.; Schurig, V. *Inorg. Chem.* **1980,** *19,* 2045.

45. Schurig, V.; Hintzer, K.; Leyrer, U.; Mark, C.; Pitchen, P.; Kagan, H. B. *J. Organomet. Chem.* **1989,** *370,* 81.

46. Bortoloni, O.; Di Furia, F.; Modena, G.; Schionato, A. *J. Mol. Catal.* **1986,** *35,* 47.

47. Sinigalia, R.; Michelin, R. A.; Pinna, F.; Strukul, G. *Organometallics,* **1987,** *6,* 728.

48. Colletti, S. L.; Halterman, R. L. *Tetrahedron Lett.* **1992,** *33,* 1005.

49. Henbest, H. B. *Chem. Soc., Spec. Publ.* **1965,** *19,* 83.

50. (a) Bowman, R. M.; Grundon, M. F. *J. Chem. Socl, C,* **1967,** 2368. (b) Bowman, R. M.; Collins, J. F.; Grundon, M. F. *J. Chem. Soc., Chem. Commun.* **1967,** 1131. (c) Morrison, J. D.; Mosher, H. S. *Asymmetric Organic Reactions,* American Chemical Society: Washington, DC, 1971; p. 336.

51. Pirkle, W. H.; Rinaldi, P. L. *J. Org. Chem.* **1977,** *42,* 2080.

52. (a) Montanari, F. *J. Chem. Soc., Chem. Commun.* **1969,** 135. (b) Bowman, R. M.; Collins, J. F.; Grundon, M. F. *J. Chem. Soc., Perkin Trans. 1,* **1973,** 626.

53. (a) Rebek, J., Jr.; McCready, R. *J. Am. Chem. Soc.* **1980,** *102,* 5602. (b) Nanjo, K.; Suzuki, K.; Sekiya, M. *Chem. Lett.* **1978,** 1143.

54. Rebek, J., Jr.; Wolf, S.; Mossman, A. *J. Org. Chem.* **1978,** *43,* 180.

55. (a) Hassine, B.; Gorsane, M.; Geerts-Evrard, F.; Pecher, J.; Martin, R. H.; Castelet, D. *Bull. Soc. Chim. Belg.* **1986,** *95,* 547. (b) Hassine, B.; Gorsane, M.; Pecher, J.; Martin, R. H. *Bull. Soc. Chim. Belg.* **1986,** *95,* 557.

56. Curci, R.; Fiorentino, M.; Serio, M. R. *J. Chem. Soc., Chem. Commun.* **1984,** 155.

57. Davis, F. A.; Harakal, M. E.; Awad, S. B. *J. Am. Chem. Soc.* **1983,** *105,* 3123.

58. Davis, F. A.; Chattopadhyay, S. *Tetrahedron Lett.* **1986,** *27,* 5079.

59. Davis, F. A.; Przeslawski, R. M. *Abstracts of Papers,* 201st National Meeting of the American Chemical Society, Atlanta; American Chemical Society: Washington, DC, 1991; ORGN 0105.

60. Helder, R.; Hummelen, J. C.; Laane, R. W. P. M.; Wiering, J. S.; Wynberg, H. *Tetrahedron Lett.* **1976,** 1831.

61. (a) Wynberg, H.; Gerijdanus, B. *J. Chem. Soc., Chem. Commun.* **1978,** 427. (b) Marsman, B.; Wynberg, H. *J. Org. Chem.* **1979,** *44,* 2312.

62. Hummelen, J. C.; Wynberg, H. *Tetrahedron Lett.* **1978,** 1089.

63. (a) Wynberg, H.; Marsman, B. *J. Org. Chem.* **1980,** *45,* 158. (b) Harigaya, Y.; Yamaguchi, H.; Onda, M. *Heterocycles,* **1981,** *15,* 183.

64. Banfi, S.; Colonna, S.; Juliá, S. *Synth. Commun.* **1983,** *13,* 1049.

65. Colonna, S.; Banfi, S.; Papagni, A. *Gazz. Chim. Ital.* **1985,** *115,* 81.

66. Hu, Y.; Harada, A.; Takahashi, S. *Synth. Commun.* **1988,** *18,* 1607.

67. Colonna, S.; Manfredi, A.; Annunziata, R.; Gaggero, N. *J. Org. Chem.* **1990,** *55,* 5862.

68. For a review, see: Colonna, S.; Manfredi, A.; Spadoni, M. *Org. Synth.; Mod. Trends, Proc. IUPAC Symp. 6th* **1986,** 275.

69. Juliá, S.; Masana, J.; Vega, C. *Angew. Chem., Int. Ed. Engl.* **1980,** *19,* 929.

70. (a) Colonna, S.; Molinari, H.; Banfi, S. *Tetrahedron,* **1983,** *39,* 1635. (b) Banfi, S.; Colonna, S.; Molinari, H. *Tetrahedron,* **1984,** *40,* 5207. (c) Bezuidenhoudt, C. B.; Swanepoel, A.; Augustyn, A. N. *Tetrahedron Lett.* **1987,** *28,* 4857.

71. Juliá, S.; Gulxer, J.; Masana, J.; Rocas, J. *J. Chem. Soc., Perkin Trans. 1,* **1982,** 1317.

72. Baures, P. W.; Eggleston, D. S.; Flisak, J. R.; Gombatz, K.; Lantos, I.; Mendelson, W.; Remich, J. J. *Tetrahedron Lett.* **1990,** *31,* 6501.

73. Itsuno, S.; Sakakura, M.; Ito, K. *J. Org. Chem.* **1990,** *55,* 6047.

4.3

Asymmetric Oxidation of Sulfides

Henri B. Kagan

Laboratoire de Synthèse Asymétrique
Institut de Chimie Moléculaire d'Orsay
Université Paris-Sud, Orsay, France

4.3.1. Introduction

Chiral sulfoxides are an important class of compounds that are finding increasing use as chiral auxiliaries in asymmetric synthesis; for reviews, see References 1–6. Current interest in chiral sulfoxides also reflects the existence of products with biological properties that need a sulfinyl group with a defined configuration. Some materials can also be based on a chiral sulfoxide structure (e.g., liquid crystals). For all these reasons then, it is very important to develop efficient methods to prepare chiral sulfoxides with high enantiometic purity; reviews of this area can be found in References 3 and 5–7. Among the various approaches to enantiomerically pure sulfoxides, the most practical one is the Andersen method, which uses a diastereomerically and enantiomerically pure sulfinate [7]. The main difficulty associated with this method, however, lies in achieving high diastereoselectivity, apart from suitable cases such as menthyl p-tolylsulfinate, where crystallization can be combined with in situ epimerization [1,7]. Asymmetric oxidation of achiral R^1-S-R^2 sulfides is in principle a very straightforward route to chiral sulfoxides, with much flexibility in the choice of R^1 and R^2 groups. Unfortunately, for a long time the enantioselectivity of such reactions remained very low, and this route was devoid of synthetic interest [8]. Renewed interest came in the early 1980s with progress obtained in various approaches. Among these methods, oxidation by hydroperoxides in the presence of chiral complexes, the use of chiral oxaziridines [9], electrochemical oxidation with chiral electrodes [10], and enzymatic or microbial reactions seem to be the most attractive for the synthesis. This chapter reviews the progress made since 1983 by selecting reactions that are catalytic with respect to the source of chirality. Surveys of the asymmetric oxidation of sulfides (catalytic and stoichiometric) can be found in articles devoted to various aspects of sulfoxide chemistry [5–8,11].

The oxidations of sulfides with hydroperoxides mediated or catalyzed by chiral titanium alcoholates are discussed first. Then, results obtained with various chiral metal–Schiff bases or metal–porphyrin combinations as catalysts are presented. Heterogeneous chiral catalysts are briefly discussed. Biomimetic oxidations using a chiral catalyst or complexing agents such as flavins, cyclodextrin, and bovin serum albumin (BSA), also are included in this chapter, which closes by summarizing the most efficient enzymatic oxidations of thio ethers.

4.3.2. Oxidations in the Presence of Chiral Titanium Alcoholates

The Sharpless asymmetric epoxidation of allylic alcohols with hydroperoxides in the presence of a chiral titanium complex consisting of $Ti(O-i-Pr)_4$ and diethyl tartrate (DET) or diisopropyl tartrate (1:1), discovered in 1980, soon evolved as a major synthetic methodology for enantiomerically pure compounds [12,13]. The success of this reaction attracted much attention to the titanium–chiral alcoholate combinations as potential mediators or catalysts in various reactions. In 1984 we attempted in Orsay to oxidize simple thio ethers with t-butyl hydroperoxide in the presence of the Sharpless reagent. We isolated racemic mixtures first but then discovered that addition of one mole equivalent of water provided sulfoxides with quite high enantiomeric purity [14,15a]. Under these conditions, the epoxidation of allylic alcohols is completely blocked. Independently that same year, Modena et al. in Padua reported the beneficial effect of using a large excess of diethyl tartrate (4 mol equiv.) with respect to $Ti(O-i-Pr)_4$ [16]. This stoichiometry also blocks epoxidation of allylic alcohols [14]. The Orsay and Padua procedures seem to involve closely related complexes, since the results are very similar for comparable examples. These systems are described in Sections 4.3.2.1 and 4.3.2.2. Originally titanium complexes were used in stoichiometric amounts [14–16]. Later we devised a procedure that allows the use of the titanium complex in catalytic amounts. For that reason, the results obtained for stoichiometric reactions are presented, since they are closely related to the catalytic reaction.

4.3.2.1. Complexes Based on Titanium Tartrate–Water Combination

We found serendipitously that methyl p-tolyl sulfide is oxidized to methyl p-tolyl sulfoxide with high enantiomeric purity (in the range of 80–90% ee) when the Sharpless reagent is modified by addition of one mole equivalent of water [14,15]. The story of this discovery is related in a recent review [17]. Sharpless conditions gave racemic sulfoxide and sulfone. Careful optimization of the stoichiometry of the titanium complex in the oxidation of methyl p-tolyl sulfide allowed us to select the

Table 1 Asymmetric Oxidation of Sulfide Ar-S-R by t-BuOOH in the Presence of Ti(O-i-Pr)$_4$/(+)-DET/H$_2$O in a 1:2:1 Ratio[a]

Entry	Ar	R	Isolated Yield (%)	% ee[b]	Ref.
1	p-Tolyl	Methyl	90	89	15a
2	p-Tolyl	Ethyl	71	74	15a
			60	83	15b
3	p-Tolyl	n-Butyl	75	75	15a
4	1-Naphthyl	Methyl	98	89	18
5	2-Naphthyl	Methyl	88	90	15a
6	2-Naphthyl	n-Propyl	78	24	15a
7	9-Anthracenyl	Methyl	33	86	18
8	o-Tolyl	Methyl	77	89	18
9	p-OMeC$_6$H$_4$	Methyl	72	86	15a
10	o-OMeC$_6$H$_4$	Methyl	70	84	15a
11	Phenyl	Cyclopropyl	73	95	18
12	Phenyl	CH$_2$Cl	60	47	18
13	Phenyl	CH$_2$CN	85	34	18
14	2-Pyridyl	Methyl	63	77	15a

[a] Reaction performed at 5 mmol scale. (Sulfide) = (reagent) = 2×10^{-1}M in CH$_2$Cl$_2$ at $-20°$C.

[b] Measured by proton NMR spectroscopy with Eu(hfc)$_3$ or (R)-(3,5-dinitrobenzoyl)-1-phenylethylamine [24]. All sulfoxides have (R) configuration.

combination Ti(O-i-Pr)$_4$/(R,R)-DET/H$_2$O (1:2:1) as our standard system [15a]. In the beginning of our investigations, the standard conditions implied a stoichio metric amount of the chiral titanium complex with respect to the prochiral sulfide [14,15a,18–21]. Later, we found conditions that served to decrease the amount of the titanium complex without too much alteration of enantioselectivity [22,23].

Table 1 lists representative results from asymmetric oxidation of thio ethers with t-butyl hydroperoxide under our standard conditions (in dichloromethane at $-20°$C). Enantioselectivities are especially good (80–95% ee) for oxidation of aryl methyl sulfoxides (Table 1). Phenyl cyclopropyl sulfide is also an excellent substrate (95% ee). A significant decrease in enantioselectivity is seen for oxidation of sulfides of type aryl-S-alkyl in which alkyl is larger than methyl, such as n-propyl and n-butyl.

In the case of the oxidation of methyl alkyl sulfides, enantioselectivity remains in the range of 50–60% ee (Table 2). Disufides R-S-S-R′, sulfenamides R-S-NR′$_2$, and sulfenates R-S-O-R′ were oxidized into chiral thiosulfinates, sulfinamides, and sulfinates respectively (< 52% ee, Table 3) [19].

Enantioselectivity is very much dependent on the solvent employed. A screening of appropriate solvents for the oxidation of methyl p-tolyl sulfoxide showed a dramatic solvent effect (Table 4) [20]. The best solvents were dichloromethane and 1,2-dichloroethane, which have similar dielectric constants: 1.6 and 1.44, respectively.

The nature of the hydroperoxides employed is of importance in the enantiose-lective oxidation of sulfides [21–23]. The best hydroperoxide is cumene hydro-

Table 2 Asymmetric Oxidation of Dialkyl Sulfides (R^1-S-R^2) by t-BuOOH and the Reagent Ti(O-i-Pr)$_4$/(+)-DET/H$_2$O in a Ratio of 1 : 2 : 1[a]

Entry	R^1	R^2	Isolated Yield (%)	% ee[b]	Ref.
1	t-Butyl	Methyl	72	53	15a
2	n-Octyl	Methyl	77	53	22
3	Cyclohexyl	Methyl	67	54	15a
4	PhCH$_2$	Methyl	88	35	22
5	PhCH$_2$CH$_2$CH$_2$	Methyl	84	50	15a
6	Methyl	CH$_2$CO$_2$Et	84	63	22
7	Methyl	CH$_2$CH$_2$CO$_2$Me	85	64	22

[a] Reaction performed at 5 mmol scale. [Sulfide] = [reagent] = 2×10^{-1}M in CH$_2$Cl$_2$ at -20°C.
[b] Measured by proton NMR spectroscopy with Eu(hfc)$_3$ or (R)-(3,5-dinitrobenzoyl)-1-phenylethylamine [25]. All sulfoxides have (R) configuration.

Table 3 Asymmetric Oxidation of R-S-X by t-BuOOH and the Reagent Ti(O-i-Pr)$_4$/(+)-DET/H$_2$O in a Ratio of 1 : 2 : 1[a]

Entry	R	X	Isolated Yield (%)	% ee[b]	Config.
1	Methyl	S-Methyl	60	41	(S)
2	i-Propyl	S-i-Propyl	43	52	(S)
3	t-Butyl	S-t-Butyl	34	41	(S)
4	Phenyl	O-Methyl	86	29	(R)
5	p-Tolyl	O-Methyl	88	36	(R)
6	p-Tolyl	NH-i-Propyl	28	24	(S)
7	p-Tolyl	N-Diethyl	60	35	(S)

[a] Reaction performed at 5 mmol scale. [Sulfide] = [reagent] = 2×10^{-1}M in CH$_2$Cl$_2$ at -20°C.
[b] Measured by proton H NMR spectroscopy with Eu(hfc)$_3$ or (R)-(3,5-dinitrobenzoyl)-1-phenylethylamine [25].
Source: Reference 19.

Table 4 Enantioselectivity in Asymmetric Oxidation of Methyl p-Tolyl sulfide by t-BuOOH in the Presence of (TiO-i-Pr)$_4$/(R,R)-DET/H$_2$O in a Ratio of 1 : 2 : 1 in Various Solvents

	Solvent					
	CCl$_4$	CHCl$_3$	CH$_2$Cl$_2$	ClCH$_2$CH$_2$Cl	Toluene	Acetone
% ee	4.5	70	85	86	26	62
Config.	(S)	(R)	(R)	(R)	(R)	(R)

Source: Reference 20.

Table 5 Enantiometric Excess in Asymmetric Oxidation by Various Hydroperoxides in the Presence of Ti(O-i-Pr)$_4$/(R)(R)-DET/H$_2$O in a Ratio 1 : 2 : 1[a]

Sulfide	Cumene Hydroperoxide	t-BuOOH	Ph$_3$COOH
Me-S-(p-tolyl)	96	89	16
Me-S-(o-anisyl)	93	74	
Me-S-phenyl	93	88	
Me-S-(n-octyl)	80	53	32
Me-S-benzyl	61	35	

[a] Reactions performed at $-20°$C in CH$_2$Cl$_2$.

Source: Reference 22.

peroxide, a readily available compound. Table 5 shows comparisons of t-BuOOH, PhC(Me)$_2$OOH, and Ph$_3$COOH in the oxidation of typical sulfides.

As often found in asymmetric synthesis reaction, *temperature* is an experimental parameter that can increase enantioselectivity. Here, however, a decrease in the reaction temperature does not always increase the enantioselectivity. An optimum temperature was found to be in the range of -20 to $-25°$C for oxidation of methyl p-tolylsulfide [15a], and this temperature range was retained for the standard oxidations. In the case of monooxidation of dithianes, the maximum enantioselectivity is obtained at ca. $-40°$C [26] (see Scheme 1).

Mechanistic studies gave some information on the titanium–tartrate combination, although many aspects of the reaction remained obscure [15a]. Infrared spectroscopy in dichloromethane showed the carbonyl stretching bands of tartrate at 1745 cm^{-1} (free) and 1675 cm^{-1} (chelated). The latter is more intense than that in the Sharpless reagent (which had adsorptions at 1735 and 1635 cm^{-1}) [27]. The Hammett plot (on p-R-C$_6$H$_4$-S(O)-Me) gave very good linear correlation with a ρ value of -1.02, indicative of an electrophilic attack on sulfur [15a]. There is no evidence (IR or polarimetric) of precoordination of the sulfides to the titanium complex. We advanced the hypothesis that water addition is beneficial for building Ti—O—Ti units (as is known in controlled hydrolysis of titanium alcoholates [83]), although other explanations are possible. The selective hydrolysis of one ester function of diethyl tartrate or dimethyl tartrate could be excluded. A very good correlation exists between abolute configuration of tartrate and sulfoxides. Scheme 2 provides excellent predictions, by taking L and S groups (on steric grounds), as large and small, respectively.

Scheme 1 (From Ref. 26.)

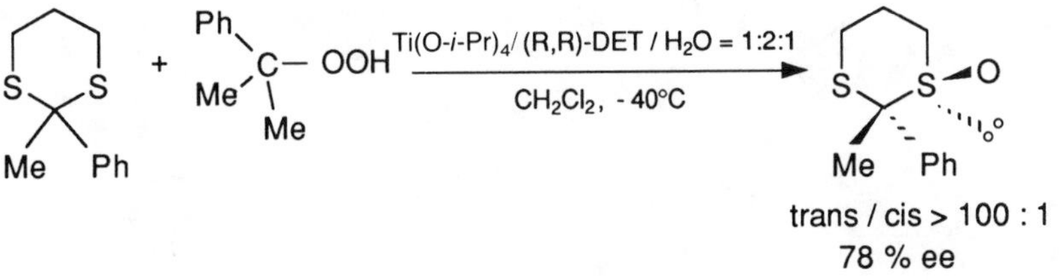

trans / cis > 100 : 1
78 % ee

Scheme 2   (From Refs. 15a, 22.)

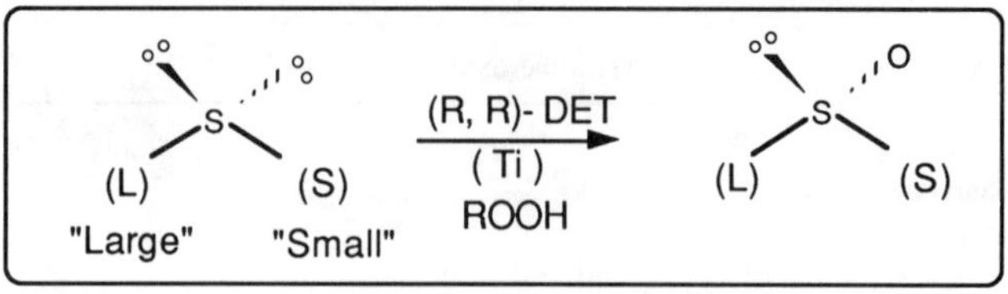

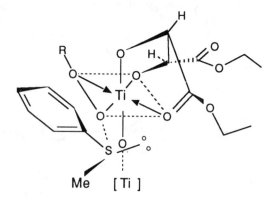

$$L = Ar \quad S = alkyl$$
$$L = t\text{-}Bu \quad S = n\text{-}alkyl$$
$$L = C{\equiv}C \quad S = Me$$

The aromatic rings have a directing effect, which is a combination of steric and polar effects. Polar effects were seen in the oxidation of (p-R)-C_6H_4-S-Me: enantioselectivity decreases with σ_R values (e.g., R = p-NO$_2$, 17% ee; R = p-OMe, 90% ee). A triple bond behaves like a phenyl group in Scheme 2; for example, n-Bu—C≡C—S—Me gave the corresponding (R)-sulfoxide with 75% ee [22]. If one assumes a tridentate tartrate around a titanium peroxide moiety in a Ti—O—Ti binuclear complex, one can propose the Figure 1 as the preferred transition state. This mechanism is based on the hypothesis that the nucleophilic attack of sulfide takes place along the O—O bond of the coordinated peroxide.

Diethyl tartrate is the best ligand for enantioselective oxidation of thio ethers used to date. This was established for the asymmetric oxidation of methyl p-tolylsulfide with cumene hydroperoxide: 96% ee (DET); 87% ee (diisopropyl tartrate); 62% ee (dimethyl tartrate) [22], 1.5% ee (bis N,N-dimethyltartramide, t-BuOOH as the oxidant) [15]. Recently, Yamamoto et al. found that methyl p-tolyl sulfide gave the corresponding sulfoxide with 84% ee using 1,2-bisaryl 1,2-ethanediol as the chiral ligand [28]. Curiously there is an inversion of configuration going from the bis(o-anisyl)ligand **A** to the bis(p-anisyl)ligand **B** of the same absolute configuration (Scheme 3). The same inversion was observed in the oxidation of methyl benzyl sulfide with the ligands **A** and **B**: that is, 66% ee (S) and 43% ee (R), respectively.

Figure 1

Scheme 3 (From Ref. 28.)

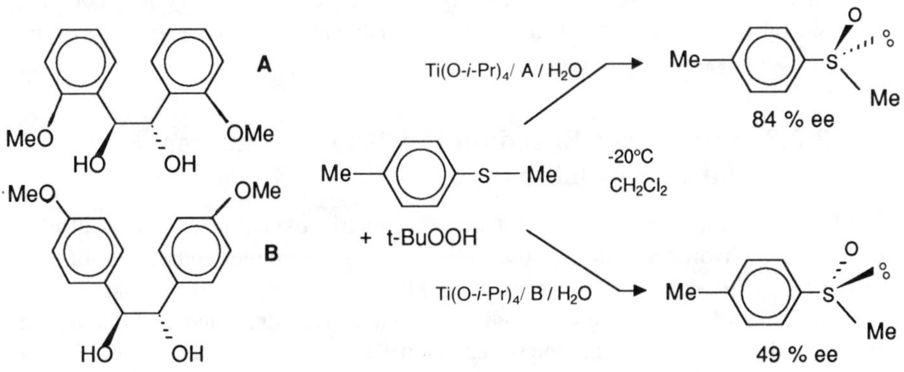

Catalytic reactions in Sharpless epoxidation were achieved in 1986 by addition of molecular sieves, which suppress the formation of nonenantioselective complexes by moisture already present in the medium or produced during the reaction [29]. We faced similar problems because a decrease in the concentration of titanium complexes parallels a decrease in the enantiomeric purity of sulfoxides. The first difficulty might be due to an increase in the uncatalyzed pathway. It was estimated that the latter reaction was almost 200 times slower than the titanium-mediated oxidation with t-BuOOH [15a]. The sluggishness of direct oxidation under the standard conditions left room to set up a catalytic process. Because of the various equilibria involved, many titanium species are potential catalysts. After many variations in experimental conditions, we found that cumene hydroperoxide allows for catalytic conditions [22]. Table 5 shows typical results from the enantioselective oxidation of methyl p-tolyl sulfide. The enantioselectivity remains very good (85% ee) until the concentration of the titanium complex is 0.2 mol equiv. A significant decrease in enantioselectivity starts for 0.1 mol equiv. or below, although the chemical yield remains good. Obviously undesired catalytic species are quite active at low catalyst concentrations. Curiously, addition of molecular sieves (pellets) helps to maintain good enantioselectivity, perhaps by efficient regulation of the amount of water. For preparative work at the 5–20 mmol scale, it is very convenient to use 0.5 mol equiv. of the titanium complex. A detailed procedure has been published [23]. It has been applied for example, to the oxidation of 2,2-disubstituted 1,3-dithianes [30]. Uemura et al. [30] found that (R)-1,1'-binaphthol can replace (R, R)-diethyl tartrate in our water-modified catalyst, giving very good results (up to 73% ee) in the oxidation of methyl p-tolyl sulfide with t-BuOOH (at −20°C, in toluene). The chemical yield is close to 90% with the use of a catalytic amount (10 mol %) of the titanium complex [Ti(O-i-Pr)$_4$/(R)-binaphthol/H$_2$O = 1:2:20]. Uemura et al. studied the effect of added water and found that high enantioselectivity was obtained when using 0.5–3.0 equiv. of water with respect to the sulfide. In the absence of water enantioselectivity was very low. The beneficial effect of water is clearly established here, but the amount of water needed is much higher than in the case the catalyst with diethyl tartrate. They assumed that a mononuclear titanium complex with two binaphthyl ligands was involved in which water affects the structure of the titanium complex and its rate of formation.

4.3.2.2. Complexes Based on the Titanium–Excess Tartrate Combination

In 1984 the Padua group described the asymmetric oxidation of some sulfides with t-butyl hydroperoxide in the presence of one mole equivalent of the $Ti(O-i-Pr)_4/(R,R)$-DET (1:4) combination [16]. The reactions were mainly performed at $-20°C$ in toluene or 1,2-dichloroethane. Results, listed in Table 6, are similar to those given by using the water-modified reagent developed by us (cf. data in Tables 1, 2, and 6). For example, (R)-methyl p-tolyl sulfoxide with virtually the same enantiomeric purity was obtained in both cases (i.e., the Orsay complex, 89% ee; the Padua complex, 88% ee). Also, there is a substantial decrease in enantioselectivity upon replacing dichloromethane or 1,2-dichloroethane with toluene in both systems.

It was hypothesized that identical species could be involved in both systems if an excess of tartrate were to bring an uncontrolled amount of water into the system, since in the Padua system addition of molecular sieves (4A, powder) produced racemic sulfoxide [22]. The procedure with excess diethyl tartrate was applied to the asymmetric oxidation of 1,3-dithiolanes (Table 7) [31,32].

The oxidation of the corresponding 1,3-dithianes or 1,3-oxathiolanes gave much lower enantioselectivity. An interesting application is the resolution of racemic ketones through their transformation into 1,3-dithiolanes by asymmetric S-monooxidation followed by separation of diastereomers and finally by regeneration of the carbonyl group. This procedure has been applied to dl-menthone; it gave $(-)$-menthone with 93% ee. β-Hydroxysulfides of structure $PhCH(OH)CH_2$-S-R (racemic mixture) gave some kinetic resolution by partial conversion to the correspond-

Table 6 Asymmetric Oxidation of R^1-S-R^2 by t-BuOOH in the Presence of $Ti(O-i-Pr)_4/(+)$-DET in a Ratio of 1:4[a]

Entry	R^1	R^2	Isolated Yield (%)	% ee[b]	Config.
1	p-Tolyl	Methyl	46	64	R
2	p-Tolyl	Methyl	60[c]	88	R
3	p-Tolyl	t-Butyl	99	34	$(+)$
4	p-ClC$_6$H$_4$	CH$_2$CH$_2$OH	41	14	$(-)$
5	Benzyl	Methyl	70[c]	46	$(+)$

[a] 1 mol equiv. of Ti complex, at $-20°C$, in toluene unless stated.
[b] In 1,2-dichloroethane.
[c] In dichloromethane at $-77°C$.

Source: Reference 16.

Table 7 Asymmetric Monooxidation of 1,3-Dithiolanes by t-BuOOH in the Presence of Ti(O-i-Pr)$_4$/(R,R)-DET in a Ratio of 1 : 4[a]

Substrate	Yield (%)[a]	Diastereomer Ratio	% ee
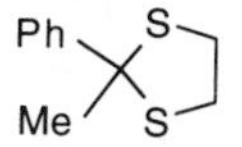 Ph, Me (S-S)	66	97:3	83
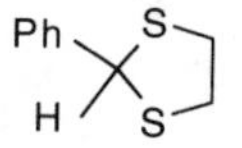 Ph, H (S-S)	76	94.6	76
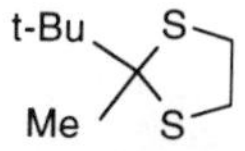 t-Bu, Me (S-S)	61	99:1	68
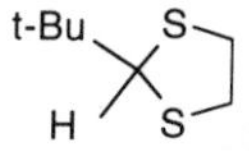 t-Bu, H (S-S)	82	99:1	70
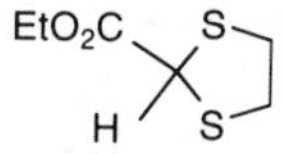 EtO$_2$C, H (S-S)	62	85:15	85

[a] Reaction at $-20°C$, in 1,2-dichloroethane. The relative stereochemistry of the S-monooxide is preferentially trans (between oxygen and the bulky group); the absolute configuration has not been established.

Source: References 31, and 32.

ing sulfoxide [33]. Stereoselectivity was much improved by protection of the hydroxyl group via silylation or acylation [34].

4.3.2.3. Some Applications

Most of the applications described in the literature deal with the water-modified titanium system. Some examples are described here in Scheme 4 [35,36]. Beckwith et al., during their investigations of homolytic substitution at sulfur, prepared sulfoxides **1** and **2** by oxidation with t-BuOOH in the presence of Ti(O-i-Pr)$_4$/(R,R)-DET/H$_2$O. Absolute configurations were assigned on the basis of the rule shown in Scheme 2 [15a,22] and were later confirmed by X-ray crystallography of sulfoxide **1**. Total synthesis of itomanindole A (**3**), an indolic compound isolated from red algae, was effected by oxidation with cumene hydroperoxide in the presence of the Ti(O-i-Pr)$_4$/(R,R)-DET/H$_2$O combination, and the product assigned an (R) configuration based on the rule of Scheme 2. A Syntex team prepared both enantiomers of p-anisyl methyl sulfoxide (> 95 % ee) by oxidation with cumene hydroperoxide in the presence of the titanium complex and used it as a chiral auxiliary and a building block for the synthesis of a cardiovascular drug [24]. Oxidation of 2-aryl-1,3-dithiolane or 2-aryl-1,3-oxathiolane with t-BuOOH in the presence of the same chiral titanium complex gave a superior level of diastereoselectivity and enantioselectivity by respect to oxidation catalyzed by flavin-containing monooxygenase

Scheme 4

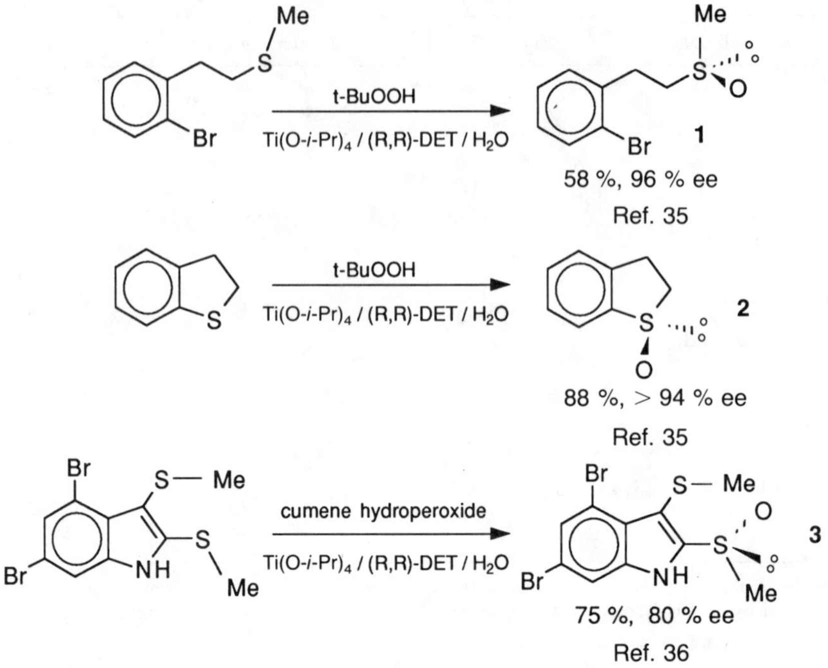

58 %, 96 % ee
Ref. 35

88 %, > 94 % ee
Ref. 35

75 %, 80 % ee
Ref. 36

[41]. Finally the high enantioselectivity (90% ee) obtained in the oxidation of methyl 1-naphthyl sulfide (Table 1, entry 3) should be even higher if *t*-BuOOH is replaced by cumene hydroperoxide [22,23]. This is of special interest since the report of Sakuraba and Ushiki [42], which showed that anion of methyl 1-naphthyl sulfoxide reacts with several alkyl phenyl ketones with diastereomeric excess close to 100%. Desulfurization with Raney nickel provided enantiomerically pure tertiary alcohols. The Ti(O-*i*Pr)₄/(R,R)-DET (1:4) combination has been useful for the resolution of 2,2′-dimethylthio-[1,1′-binaphthalene]. Through demethylation it is then possible to recover the both enantiomers of 2,2′-dithiol-[1,1′-binaphthalene], a useful chiral auxiliary [43,44].

4.3.3 Chiral Titanium–Schiff Base Catalysts

The high enantioselectivity achieved by the chiral titanium alcoholates described in the preceding section is counterbalanced by low catalytic efficiency. Pasini et al. have developed chiral oxotitanium(IV)–Schiff base complexes **4** (Scheme 5) which are good catalysts (catalyst:substrate ratio = 1:1000 to 1:1500) for the oxidation of methyl phenyl sulfide with 35 % H₂O₂ in aqueous ethanol or dichloromethane [45]. Unfortunately enantioselectivity is low (< 20% ee) and some sulfone is formed. The authors proposed that sulfide would coordinate to titanium (with asymmetric induction at sulfur) prior to the external attack of hydrogen peroxide (instead of a

Scheme 5

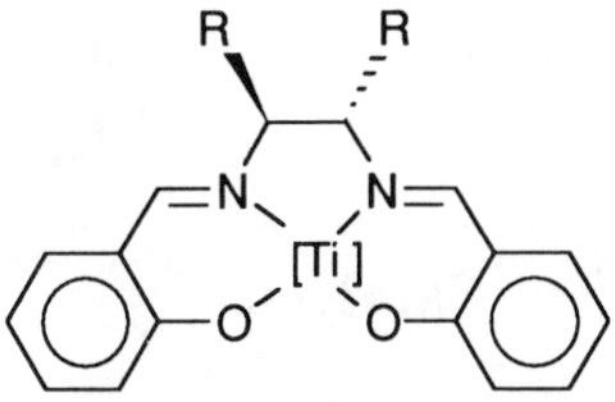

4   R = Me or Ph
Ref. 45

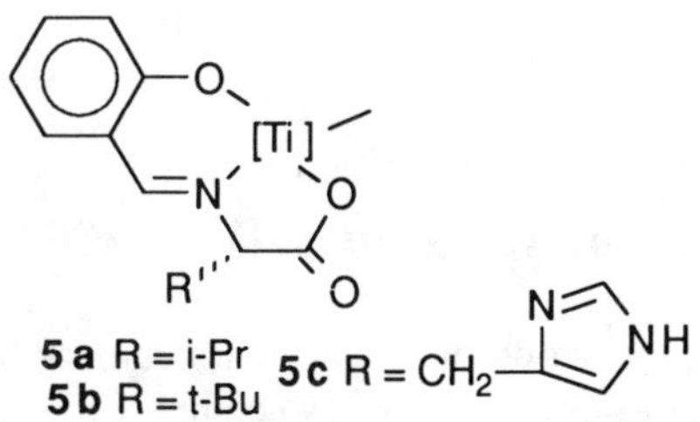

5 a R = i-Pr   **5 c** R = CH₂—
5 b R = t-Bu
Ref. 46

(Ti) = Ti=O (presumably as polymeric complex with Ti-O-Ti chains and pseudo octahedral structure).

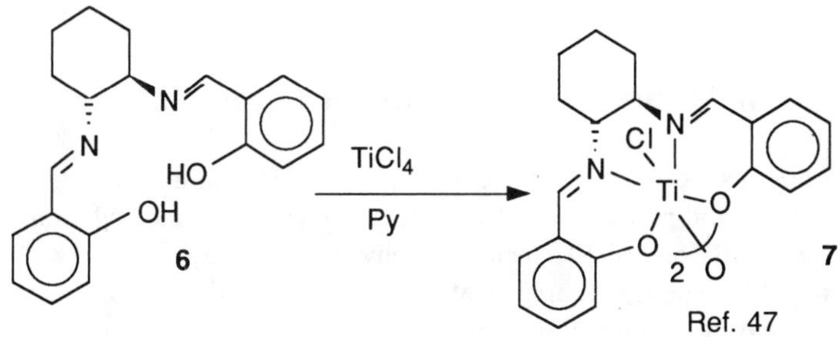

7
Ref. 47

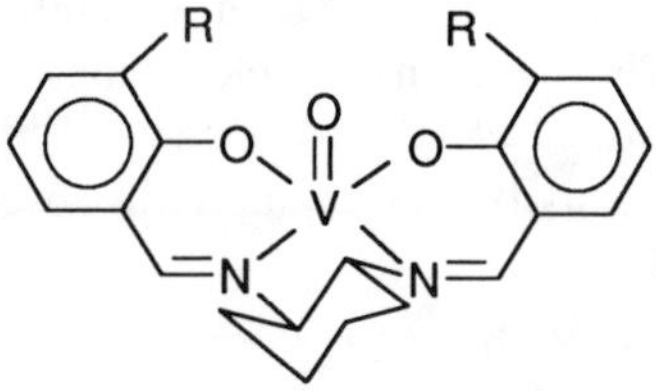

8   R = OMe, OEt or t-Bu
Ref. 48

peroxotitanium species). A low level of asymmetric induction was also observed by Colonna et al. with chiral titanium complexes of *N*-salicylidene-L-amino acids **5** (Scheme 5) [46]. These catalysts (0.1 mol equiv.) gave enantioselectivities below 25% ee in the oxidation of methyl *p*-tolyl sulfide and various sulfides with *t*-BuOOH in benzene at room temperature.

A more promising approach is related to the work of Fujita et al., where active catalysts are prepared by the reaction of a Schiff base of (*R,R*)-1,2-cyclohexa-nediamine **6** with TiCl₄ in pyridine. The isolated complex is a catalyst (4 mol % equiv.) for the asymmetric oxidation of methyl phenyl sulfide by trityl hydroperoxide in methanol at 0°C [47]. The (*R*)-sulfoxide with 53% ee is formed in good yield. Other hydroperoxides (TBHP and cumene hydroperoxide) gave inferior results. The X-ray crystal structure of an isolated complex surprisingly revealed an addition of oxygen (see **7**, Scheme 5) acting as a bridge between two titanium metals (Ti—O—Ti unit). Each Ti metal is coordinated octahedrally; the planes of the Ti and the Schiff base are almost parallel to each other. The authors believe that the structure of the complex in solution is different from the structure in the solid state.

4.3.4. Chiral Vanadium(IV)–Schiff Base Catalysts

Asymmetric oxidation of sulfides with cumene hydroperoxide is also catalyzed by the Schiff base–oxovanadium(IV) complex formulated as **8** (Scheme 5) [48]. Many aryl methyl sulfides were investigated in this reaction (room temperature in dichloromethane with 0.1 mol equiv. catalyst). Chemical yields are excellent, but enantioselectivities are not higher than 40% ee (methyl phenyl sulfoxide). Complex **5a**, in which [Ti] is replaced by VO, has also been employed in the oxidation of sulfides to give racemic sulfoxides [46].

4.3.5. Iron– or Manganese–Porphyrin Catalysts

Oxometalloporphyrins were taken as models of intermediates in the catalytic cycle of cytochrome P-450 and peroxidases. The oxygen transfer from iodosyl aromatics to sulfides with Fe(III) or Mn(III) metalloporphyrins as catalysts is very clean, giving sulfoxides [49]. The first examples of the asymmetric oxidation of sulfides to sulfoxides with significant enantioselectivity were published only very recently. Naruta et al. prepared chiral "twin coronet" iron porphyrin **9** (Fig. 2) [37]. This C_2 symmetric complex efficiently catalyzes the oxidation of sulfides with iodosylbenzene (turnover number up to 290) with enantioselectivities of up to 73% ee (Table 8). Addition of 1-methylimidazole, which acts as an axial ligand of iron, is necessary to achieve good enantioselectivity. For example, in absence of 1-meth-

Table 8 Asymmetric Oxidation of Ar-S-Me by PhIO
         Catalyzed by Iron(II)–Porphyrin **9**

Entry	Ar	Temp. (°C)	Turnover Number[b]	% ee[b]	Config.
1	C_6H_5	−15	139	46	S
2	$2\text{-}NO_2C_6H_4$	−5	88	24	S
3	$3\text{-}NO_2C_6H_4$	−15	128	45	S
4	$4\text{-}NO_2C_6H_4$	0	120	53	S
5	C_6F_5	−15	55	73	S
6	$4\text{-}MeC_6H_4$	−15	144	54	S
7	$2\text{-}N_{ap}hthyl$	−15	168	34	R

[a] Reaction performed in CH_2Cl_2, PhIO = 260 μmol, **9** = 1 μmol, sulfide = 500 μmol, 1-methylimidazole = 100 μmol.
[b] Based on the amount of isolated sulfoxides.

Source: Reference 37.

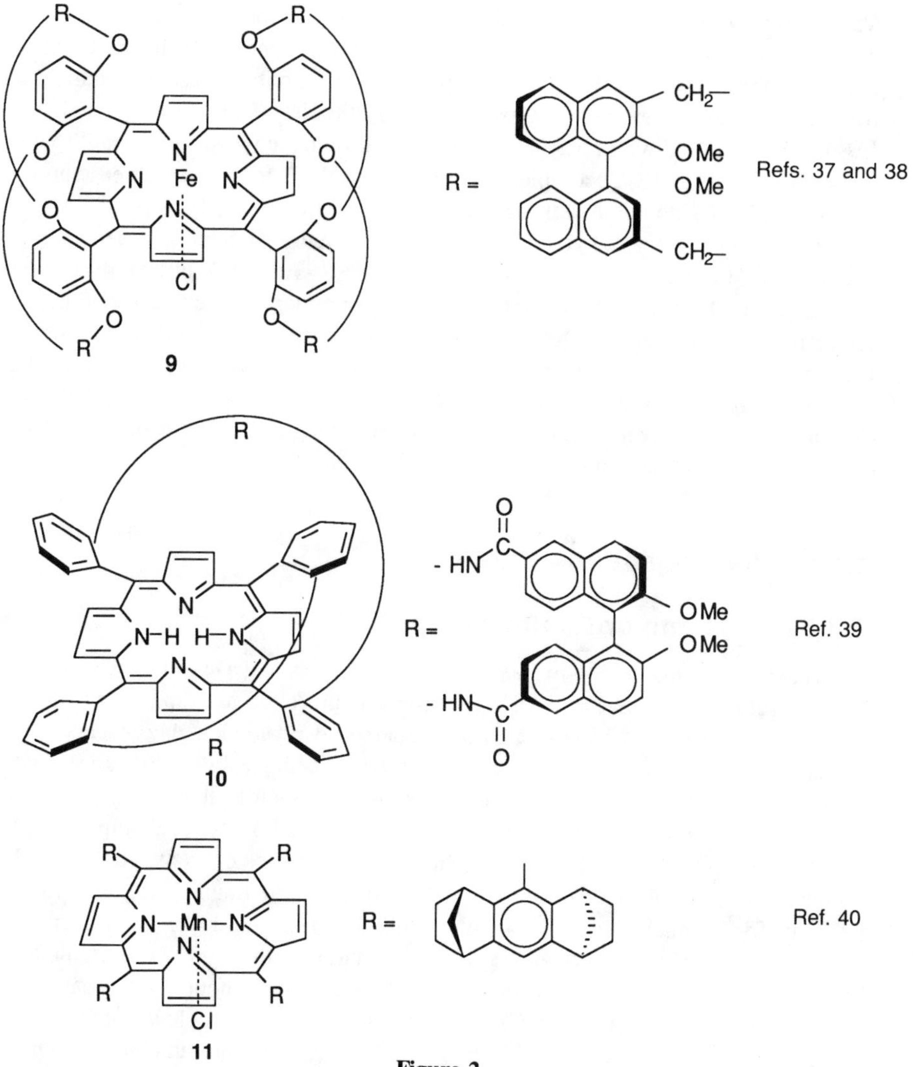

Figure 2

ylimidazole pentafluoro phenyl methyl sulfoxide is formed with only 31% ee (compare this result with entry 4, Table 8). It was proposed that asymmetric induction is mainly controlled by the steric hindrance around sulfur rather than by electronic effects. In 1991 the same authors disclosed full details of their catalytic system and proposed a mechanism for explaining asymmetric induction [38]. This mechanism is based on the steric approach control of a sulfide to the oxo-iron center in the molecular cleft.

Groves and Viski reported similar results with binaphthyl iron(III)–tetraphenyl-porphyrin **10** as the catalyst (0.1% mol equiv.) in the asymmetric oxidation of sulfides with iodosylbenzene [39]. Enantioselectivities up to 48% ee were achieved.

Very recently Halterman et al. provided a new example of the asymmetric oxidation of sulfides by iodosylbenzene, catalyzed by a D_4 symmetric manganese–tetraphenylporphyrin complex **11** (Fig. 2) [40]. Catalytic activity is excellent in dichloromethane at 20°C (catalyst/PhIO/sulfides = 1:200:400), and methyl phenyl sulfoxide was produced with 55% ee and methyl *o*-bromophenyl sulfoxide at 68% ee.

Chiral (salen)Mn(III)Cl complexes are useful catalysts for the asymmetric epoxidation of isolated double bonds. Jacobsen et al. used these catalysts for the asymmetric oxidation of aryl alkyl sulfides with unbuffered 30% hydrogen peroxide in acetonitrile [50]. The catalytic activity is excellent (2–3 mol %), but the maximum enantioselectivity achieved (68% ee for methyl *o*-bromophenyl sulfoxide) remains rather modest. The chiral salen ligands used in the catalysts are derived from **6** (Scheme 5) with subtituents at the ortho and meta positions with respect to the phenol moiety. The structure of these ligands can easily be modified, and it is likely that substantial improvements will be obtained by changing the steric and electronic properties of the substituents.

4.3.6. Heterogeneous Catalysts

4.3.6.1. Montmorillonite Support

It has been shown that an ion-exchanged adduct of a clay and a chiral metal complex can be useful in the resolution of a racemic mixture or in asymmetric synthesis. Usually when one enantiomer of a chiral chelate is adsorbed on a clay, it leaves half the surface unoccupied (while the racemic mixture occupies all active sites). The empty sites can be occupied by prochiral sulfides, which enables an asymmetric photooxidation to occur [51]. In this process photoexcited Λ or Δ-(2,2'-bipy)$_3$RuCl$_2$ first reacts with O$_2$ to provide O$_2^-$ and [Ru(bipy)$_3$]$^{3+}$. Attack of an oxygen molecule on the cation radical of a sulfide yields the corresponding sulfoxide. Reaction is performed in methanol/water (1/4) by stirring the clay–chiral chelate adduct (0.5 mol equiv.) under bubbling oxygen gas and irradiation with a 500 W tungsten lamp. There is complete conversion of sulfides to sulfoxides after 2 hours, and no sulfones are detected. Unfortunately, the sulfoxides thus formed have low enantiomeric purities (i.e., 15–20% ee for a variety of alkyl groups R). The ruthenium complex itself under homogeneous conditions led to racemic sulfoxides, in agreement with the hypothesis of an asymmetric control during the oxidation of adsorbed sulfides. Clay–chiral chelate adducts were also used as templates in the presence of an oxidant such as sodium metaperiodate, MCPBA, or K$_2$O$_8$S$_2$ in water [52]. Δ-Ni (phen)$_3$$^{2+}$–montmorillonite clay showed an appreciable enantioselectivity. Cyclohexyl-S-Ph at room temperature gave the corresponding sulfoxide in 78% ee (90% yield) using NaIO$_4$ and 62% ee (90% yield) using MCPBA. Similar results were obtained in the oxidation of *n*-Bu-S-Ph and Bn-S-Ph. Unfortunately, the sulfide must be preadsorbed on montmorillonite–clay, with the latter in large excess, making the transition into a practical catalytic process difficult.

4.3.6.2. Chiral Electrodes

Electrooxidation of sulfides to sulfoxides on electrodes chemically modified with optically active compounds is an attractive approach. The first attempts were disappointing ($<$ 2% ee) [53]. High enantioselectivity was reported by Komori and Nonaka in 1984 [54]. The authors prepared various types of poly amino acid coated electrode. The electrodes were platinum or graphite plates, and in some cases polypyrrole films were coated or covalently bound to the base electrode surface. Oxidations were carried out by means of a controlled-potential method in acetonitrile containing (n-Bu$_4$)NBF$_4$ and water. Alkyl aryl sulfides were investigated in detail with several modified electrodes: Ph-S-Me gave very low enantioselectivity. The best results were observed for sulfides bearing bulky alkyl groups. The most appropriate electrode is the one prepared by dip-coating of a platinum electrode modified chemically with polypyrrole and then poly(L-valine). With the use of this electrode, the following results were obtained for Ph-S-R:

R = i-Bu	44% ee
R = cyclohexyl	54% ee
R = i-Pr	73% ee
R = t-Bu	93% ee

Sulfoxides yielded have the (S)-configuration. A detailed picture of the origin of asymmetric induction is unknown. These chiral electrodes are reusable without loss of enantioselectivity. This approach is synthetically promising, but preparation of coated electrodes is a delicate undertaking, making the method difficult in practice unless robust chiral electrodes become commercially available.

4.3.7. Chiral Flavins as the Catalysts

Biological oxidation of sulfides involves cytochromes P-450 or flavin-dependent oxygenases. A chiral flavin model was recently prepared by Shinkai et al. and used as the catalyst in the asymmetric oxidation of aryl methyl sulfides [55]. Flavinophane 12 (Scheme 6) is a compound with planar chirality. It catalyzes the oxidation of sulfides with 35% H$_2$O$_2$ in aqueous methanol at $-20°$C in the dark.

Flavinophane 12 acts as a true catalyst with turnover number up to 800. The results in the oxidation of p-R-C$_4$H$_4$-S-Me were:

R = H	47% ee
R = Me	65% ee
R = t-Bu	42% ee
R = CN	25% ee

The reaction is much slower for R being CN compared to R being H, indicative of an electrophilic character in the oxygen transfer at sulfur, presumably through the intermediate depicted in Scheme 6.

Scheme 6

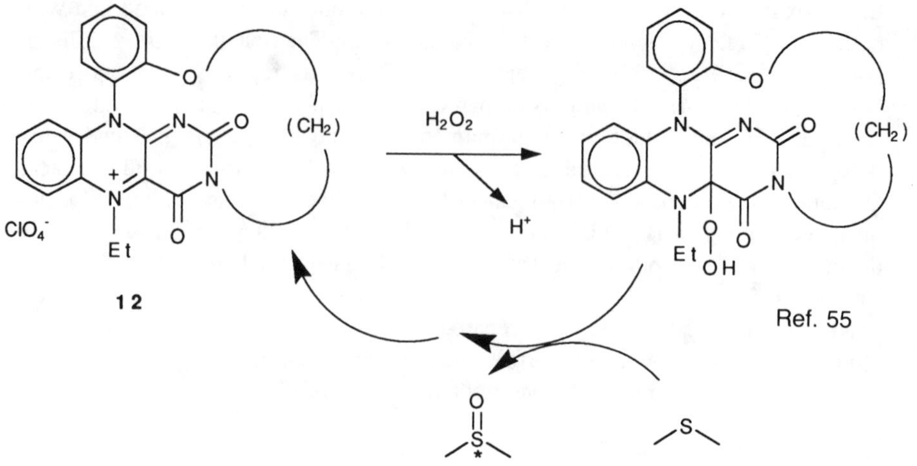

12

Ref. 55

4.3.8. Template Effects

4.3.8.1. Sulfoxidations in the Presence of Cyclodextrins

It is known that cyclodextrins have a hydrophobic cavity (a binding site for aromatics) and a hydrophilic external surface. A "template-directed" asymmetric sulfoxidation has been attempted with various aryl alkyl sulfides [56]. Oxidations were performed by using metachloroperbenzoic acid in water in the presence of an excess of β-cyclodextrin. The best result (33% ee) was attained for *meta*-(*t*-Bu)phenyl ethyl sulfoxide. The decrease in the amount of β-cyclodextrin below 1 mol equiv. causes a sharp decrease in enantioselectivity because of competition with oxidation of free substrate by the oxidant. Similarly Drabowicz and Mikolajczyk observed modest asymmetric induction (27% ee) in oxidation of Ph-S-*n*-Bu with H_2O_2 the presence of β-cyclodextrin [57].

4.3.8.2. Sulfoxidations in the Presence of Bovin Serum Albumin (BSA)

Sugimoto et al. found that BSA, a carrier protein in biological systems, is a host for aromatic sulfides. Based on this observation, the oxidation of sulfides with $NaIO_4$ was attempted in aqueous solution (pH 9.2) in the presence of BSA (0.3–2.0 mM, e.g., 0.06 to 0.5 mol equiv. with respect to sulfide) [58,59]. The best results were obtained by using 0.3 mol equiv. of BSA. Results are shown in Table 9.

There is a strong dependence of enantioselectivity on the structure of the aromatic moiety (reminiscent of enzymatic reactions); for example, Ph-S(O)-*i*-Pr and *p*-Tol-S(O)-*i*-Pr were isolated in about 80% yields with 81% ee (*R*) and 34% ee (*S*) respectively. The enantioselectivity is highly pH-dependent (i.e., it vanishes at pH

Table 9 Asymmetric Oxidation of Sulfides Ar-S-R by NaIO$_4$
Containing BSA (0.3 mol equiv.)

Entry	Ar	R	Isolated Yield (%)	% ee[b]	Config.
1	Phenyl	Methyl	47	7	R
2	Phenyl	Ethyl	58	29	R
2	Phenyl	i-Propyl	78	81	R
3	Phenyl	n-Butyl	87	36	R
4	Phenyl	i-Butyl	86	22	S
5	Phenyl	t-Butyl	86	75	R
6	Phenyl	Benzyl	52	49	R
7	p-Tolyl	i-Propyl	82	34	S

Source: Reference 58.

< 5). This property could be related to conformational changes in the gross protein structure. With MPCBA or H$_2$O$_2$ as oxidant, lower enantiomeric excess or over-oxidation to sulfones was unavoidable. The authors found that 30% hydrogen peroxide at pH 9.2 in the presence of isobutyl phenyl sulfide and BSA (0.33 mol equiv.) gave the sulfoxide with enantiomeric purity that increased with conversion. This is indicative of a kinetic resolution process [60]. Indeed, kinetic resolution of racemic sulfoxides was realized; for example, enantioselectivities at 50% conversion were 18, 33, 21, and 6% ee for i-Pr-S(O)-Ph, t-Bu-S(O)-Ph, PhCH$_2$-S(O)-Ph, and i-Pr-S(O)-p-Tol, respectively. By the simultaneous combination of asymmetric oxidation and kinetic resolution, it was possible to increase substantially the enantiomeric purities of phenyl alkyl sulfoxides. Thus, oxidation of i-Pr-S-Ph gave the (R)-sulfoxide with 62% ee in 78% yield, while by overoxidation (R)-sulfoxide with 93% ee was recovered in 47% yield. This was interpreted as a two-stage process in the binding domain of BSA, with preferential formation of the (R)-sulfoxide and subsequent preferential destruction of the (S)-sulfoxide. However no such effect is observed with p-tolylsulfides, where overoxidation tends to decrease the enantiomeric purity of the sulfoxides.

Asymmetric oxidation of formaldehyde dithioacetals with aqueous NaIO$_4$ was realized by Ogura et al. in the presence of a catalytic amount of BSA (0.005–0.02 mol equiv.) [61]. In the conditions the authors employed, the starting sulfide is virtually insoluble in water (pH 9.2), and the best results were obtained at low concentrations of BSA. It is noteworthy that the concentration of BSA (i.e., not the BSA/sulfide ratio) is the controlling factor in enantioselectivity. With protocol, this p-Tol-S-CH$_2$-S-p-Tol was transformed into monosulfoxide with 60% ee. The same protocol gave isopropyl phenyl sulfoxide with 60% ee.

Colonna et al. later investigated a wide range of sulfides to study periodate oxidation catalyzed by BSA [62,63]. The reactions were performed by stirring a heterogeneous mixture of sulfides, NaIO$_4$, and BSA (0.05 mol equiv.). In this study, lowering or increasing the amount of BSA had a detrimental effect on enantioselectivity. The enantiomeric purities of the sulfoxides thus obtained are as follows:

Table 10 Asymmetric Oxidation of Sulfides Ar-S-R by Dioxiranes Catalyzed by BSA at 4°C

Entry	Ar	Me	Dioxirane Precursor	Yield (%)	% ee	Config.
1	Ph	Me	Acetone[a]	98	7	S
2	Ph	i-Pr	Acetone[a]	56	79	R
3	Ph	t-Bu	Acetone[a]	70	73	R
4	p-Tolyl	i-Pr	Acetone[a]	50	29	S
5	Ph	i-Pr	CH_3COCF_3	67	89	R
6	Ph	CH_2Ph	CH_3COCF_3	56	67	R

[a] Sulfide/$KHSO_5$/ketone/BSA = 1:2:13:0.05.
[b] Sulfide/$KHSO_5$/ketone/BSA = 1:2:0.44:0.05.
[c] Sulfide/$KHSO_5$/ketone/BSA = 1:0.5:0.11:0.0125.

Source: References 64 and 65.

n-butyl t-butyl sulfoxide, 33% ee; mesityl phenyl sulfoxide, 40% ee; and p-Tol-S-CH_2-S(O)-p-Tol, 40% ee. Oxidation to sulfur of racemic sulfides with a stereocenter α or β gave two diastereomeric sulfoxides, each with some enantiomeric excess. Spectral data obtained electronically and by circular dichroism on a mixture of BSA and some sulfides indicated that the latter are not tightly bound to BSA under the reaction conditions.

Recently dioxiranes were generated in situ from ketones and caroate ($KHSO_5$) in the presence of BSA and sulfides at pH 7.5–8.0 [64,65]. Reactions were performed at 4°C with reaction times ranging from 15 to 180 minutes, depending on the substrates and dioxiranes. The involvement of dioxiranes as the actual oxidant is well supported by the significant differences in enantioselectivity and absolute configurations that were found by changing the ketones, which were the precursors of the dioxiranes. Some representative examples are listed in Table 10. Yields are satisfactory, and enantioselectivities are up to 89% ee. Moreover, the catalyst amount (1.25–5 mol %) is noteworthy, although the high molecular weight of BSA (170,000) necessitates a decrease in the quantity of BSA if the method is to be synthetically valuable.

4.3.9. Enzymatic Reactions

Biooxidation of chiral sulfides was initially investigated in the 1960s, especially through the pioneering works of Henbest et al. [66]. Since then many developments have been reported in the literature, and these are summarized in reviews [67,68]. It would be helpful to reveal some structural or mechanistic details of the enzymes involved in the oxidative processes. Biotransformations are also of great current interest for the preparation of chiral sulfoxides, which are useful as synthetic intermediates or chiral auxiliaries. Since extensive review of these transformations is beyond the scope of this chapter, only highlights are discussed, for comparison with the abiotic enantioselective oxidations described earlier. Biooxidations by microorganisms and by isolated enzymes are treated in Sections 4.3.9.1 and 4.3.9.2.

4.3.9.1. Microbiological Oxidations

Microbiological oxidation is the easiest procedure, since it uses the intact cells. Scheme 7 shows results obtained by using *Aspergillus niger* [66]. Enantioselectivity can be very high, but experiments are performed on a small scale with a low yield of sulfoxides. Both enantiomers of methyl *p*-tolyl sulfoxides were prepared by Sih et al. with *Mortierella isabellina* NRRL 1757, giving (*R*)-sulfoxides with 100% ee in 60% yield or with *Helminthosporium* sp. NRRL 4671, giving (*S*)-sulfoxides with 100% ee in 50% yield [69]. A similar result was obtained for ethyl *p*-tolyl sulfide.

Corynebacterium equi IFO 3730 gave high enantioselectivity in the oxidation of aryl alkyl sulfides [70]. The results listed in Scheme 7 arise from experiments with no formation of sulfone, which occurs quite easily in several cases. Thio ketals and

Scheme 7 Reactions catalyzed by *A. niger* [66], *C. equi* [70], and *M. isabellina* [75].

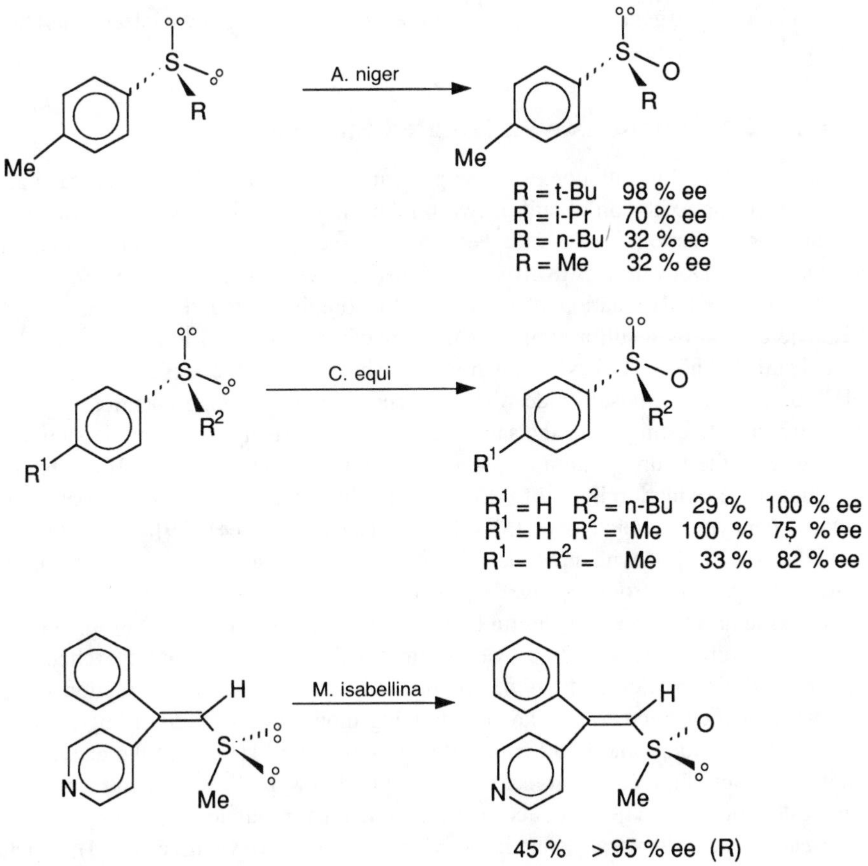
A. niger
Me
R
Me
R
O
R = t-Bu 98 % ee
R = i-Pr 70 % ee
R = n-Bu 32 % ee
R = Me 32 % ee
C. equi
R1
R2
R1
R2
O
R1 = H R2 = n-Bu 29 % 100 % ee
R1 = H R2 = Me 100 % 75 % ee
R1 = R2 = Me 33 % 82 % ee
M. isabellina
H
N
Me
H
O
N
Me
45 % >95 % ee (R)

thio acetals were oxidized to mono S-oxides by various fungal species with enantioselectivities up to 70% ee [71,72]. *Corynebacterium equi* was very successfully used in the oxidation of formaldehyde dithioacetals to mono S-oxide or sulfone sulfoxide depending on the substrate. Thus, n-Bu-S-CH-S-n-Bu was transformed into n-Bu-SO$_2$-CH$_2$-S(O)-n-Bu with more than 95% ee in 70% yield. *Saccharomyces cerevisiae* (baker's yeast) has a desaturase that is able to oxidize some sulfides such as methyl 9-thiasterate [73]. To obtain a less symmetrical sulfoxide, oxidation of Bn-S-(CH$_2$)$_7$-CO$_2$Me has been carried out with *S. cerevisiae* to give (*S*)-sulfoxide with 70% ee. The stereochemical course of the oxidation of 9-thiostereate with cultures of baker's yeast has been recently reinvestigated, and it is demonstrated that the biosulfoxidation on this quasi-symmetrical substrate is highly enantioselective (> 95% ee) [74]. *Mortierella isabellina* is able to oxidize a vinylic sulfide to the corresponding sulfoxide with very high enantiomeric purity, as shown in Scheme 7 [75]. Oxidation of various methyl β-arylvinyl sulfides with *Helminthosporium* sp. and by *Fusarium oxysporum* were also performed [76]. The corresponding sulfoxides were sometimes obtained with enantioselectivities greater than 98% ee. Comparisons were made between various chemical oxidations, including the catalysis of the water-modified titanium complex [22] and microbiological oxidations. In general, the latter method gives higher enantioselectivities but lower chemical yields.

4.3.9.2. Oxidations with Isolated Enzymes

Cytochrome P-450 is a monooxygenase present in mammalian tissues. Takata et al. investigated the oxidation of sulfides with rabbit liver microsomes, and substantial asymmetric induction (up to 54% ee) was observed with some simple prochiral sulfides [77]. Dopamine β-hydroxylase (DBH), a copper monooxygenase, catalyzes the benzylic oxidation of dopamine to norepinephrine. Replacement of a benzylic carbon by a sulfur atom was investigated by May and Philipps [78]. The model sulfide PhSCH$_2$CH$_2$NH$_2$ was treated with oxygen in the presence of DBH at pH 5.0. The (*S*)-aminosulfoxide was produced with a very high enantiomeric excess, which interestingly has the same stereochemistry as the phenylethanolamines arising from the hydroxylation of phenylethylamines catalyzed by DBH. Walsh et al. studied the oxidation in air of *p*-tolyl ethyl sulfide catalyzed by two cytochrome P-450 isoenzymes, which gave (*S*)-sulfoxide (up to 80% ee) [79]. Oxidation of *p*-tolyl methyl sulfide with a monooxygenase containing flavin adenine dinucleotide (FAD) and purified from pig liver microsomes gave the (*R*)-sulfoxide (95 % ee) [80]. Cashman et al. made a careful comparison of the oxidation of 2-(*p*-methoxyphenyl)-1,3-dithiolane and 2-(*p*-cyanophenyl)-1,3-oxathiolane with the chemical system (NaIO$_4$/BSA, chiral titanium complex/t-BuOOH) described in Section 4.3.2 and with purified microsomal flavin-containing monooxygenase from hog liver as well as from cytochrome P450 from rat or mouse liver [41]. In all the enzymatic oxidations, very high enantioselectivities (> 90% ee) were observed, with the trans sulfoxide being the major product, as in the chemical oxidation.

Pseudomonas oleovorans contains *P. oleovorans* monooxygenase (POM), which is a typical "ω-hydroxylase" for hydroxylation of terminal methyl of alkanes as well as epoxidation of terminal olefins. The ω-hydroxylation system of *P. oleovorans* was reconstituted from purified components, POM, rubridoxin, and a flavoprotein reductase. In the presence of NADH and oxygen, it oxidizes a wide range of aliphatic methyl alkyl sulfides. Enantioselectivities are very much dependent on the length of the alkyl chain of Me-S(O)-R, as exemplified by the following results:

R = *n*-Pr   80% ee
R = *n*-Bu   80% ee
R = *n*-pentyl   60% ee
R = *n*-hexyl   30% ee
R = *n*-heptyl   70% ee

Chloroperoxidase, a heme protein, was used by Colonna et al. as a catalyst for the oxidation of thio ethers into sulfoxides [82]. *t*-Butyl hydroperoxide is the best oxidant (i.e., H_2O_2, PhIO, and other hydroperoxides are less effective). Reactions were performed in water at pH 5 at 4°C with 1.6×10^{-5} mol equiv. of catalyst. The following (*R*)-sulfoxides, among others, have been prepared in good to excellent yields: PhCH$_2$S(O)Me (91% ee), PhS(O)Me (76% ee), *p*-TolS(O)Me (80% ee), *p*-ClC$_6$H$_4$S(O)Me (85% ee), and *p*-MeOC$_6$H$_4$S(O)Me (92% ee).

4.3.10. Conclusion

Asymmetric oxidation of sulfoxides has been widely expanded in the last decade, both by chemical and by biochemical methods. Chemical methods are now able to give rise to a wide range of chiral sulfoxides with enantioselectivity often higher than 90% ee by using, for example, hydroperoxides and chiral titanium complexes. However, the catalyst efficiency of these systems is low. On the other hand, highly active catalyst systems (e.g., chiral porphyrin or Schiff base–metal complexes) have not reached a very high level of enantioselectivity. One can expect, in both cases, significant improvements in the near future. Biooxidations of sulfides are now possible, with many microorganisms, giving very satisfactory results.

References

1. Solladié, G. *Synthesis,* **1981,** 185–196.

2. Barbachyn, M. R.; Johnson, C. R. In *Asymmetric Synthesis;* Morrison, J. D. (Ed.); Academic Press: Orlando, FL, 1984; Vol. 4, pp. 227–261.

3. Mikolajczyk, M.; Drabowicz, J. *Top Stereochem.* **1982,** *13,* 333–468.

4. Posner, G. *Acc. Chem. Res.* **1987,** *20,* 72–78.

5. Posner, G. H. In *The Chemistry of Sulphones and Sulphoxides;* Patai, S.; Rappoport, Z; Stirling, C. J. M. (Eds.); John Wiley & Sons, Chichester, UK, 1988; Chapter 16, pp. 823–848.

6. Drabowicz, J.; Kielbasinski, P.; Mikolajczyk, M. In *The Chemistry of Sulphones and Sulphoxides;* Patai, S.; Rappoport, Z.; Stirling, C. J. M. (Eds.); John Wiley & Sons, Chichester, UK, 1988; Chapter 8, pp. 233–253.

7. Andersen, K. K. In *The Chemistry of Sulphones and Sulphoxides;* Patai, S.; Rappoport, Z.; Stirling, C. J. M. (Eds.); John Wiley & Sons, Chichester, UK, 1988; Chapter 3, pp. 55–92.

8. Early attempts are reported in Morrison, J. D.; Mosher, H. S. *Asymmetric Organic Reactions;* Prentice Hall: Englewood Cliffs, NJ, 1971; pp. 336–351.

9. Davis, F. A.; Mc Cauley, J. P., Jr.; Harakal, M. E. *J. Org. Chem.* **1984,** *49,* 1465–1467.

10. Komori, T.; Nonaka, T. *J. Am. Chem. Soc.* **1983,** *105,* 5690–5691.

11. Madesclaire, M. *Tetrahedron,* **1986,** *42,* 5459–5495.

12. Katsuki, T.; Sharpless, K. B. *J. Am. Chem. Soc.* **1980,** *102,* 5974–5976.

13. Rossiter, B. E.; Katsuki, T.; Sharpless, K. B. *J. Am. Chem. Soc.* **1981,** *103,* 464–465.

14. Pitchen, P.; Kagan, H. B. *Tetrahedron Lett.* **1984,** *25,* 1049–1952.

15. (a) Pitchen, P.; Deshmukh, M.; Dunach, E.; Kagan, H. B. *J. Am. Chem. Soc.* **1984,** *106,* 8188–8193. (b) Glahsl, G.; Herrmann, R. *J. Chem. Soc., Perkin Trans,* **1988,** 1753–1757.

16. Di Furia, F.; Modena, G.; Seraglia, R. *Synthesis,* **1984,** 325–326.

17. Kagan, H. B.; Rebiere, F. *Synlett,* **1990,** 643–650.

18. Dunach, E.; Kagan, H. B. *New J. Chem.* **1985,** *9,* 1–3.

19. Nemecek, C.; Dunach, E.; Kagan, H. B. *New J. Chem.* **1986,** *10,* 761–764.

20. Kagan, H. B.; Dunach, E.; Nemecek, C.; Pitchen, O.; Samuel, O.; Zhao, S. H. *Pure Appl. Chem.* **1985,** *57,* 1911–1916.

21. Zhao, S.; Samuel, O.; Kagan, H. B. *C. R. Acad. Sci. Paris, Ser. B,* **1987,** 304, 273–275.

22. Zhao, S.; Samuel, O.; Kagan, H. B. *Tetrahedron,* **1987,** *43,* 5135–5144.

23. Zhao, S.; Samuel, O.; Kagan, H. B. *Org. Synth.* **1989,** *68,* 49–56.

24. Davis, F.; Kern, J. R.; Kurtz, L. J.; Pfister, J. R. *J. Am. Chem. Soc.* **1988,** *110,* 7873–7874.

25. Deshmukh, M.; Dunach, E.; M.; Jugé, S.; Kagan, H. B. *Tetrahedron Lett.* **1984,** *25,* 3467–3470. Corrigendum, *ibid.* **1985,** *26,* 402.

26. Samuel, O.; Ronan, B.; Kagan, H. B. *J. Organomet. Chem.* **1989,** *370,* 43–50.

27. Finn, M. G.; Sharpless, K. B. *J. Am. Chem. Soc.* **1991,** *113,* 113–126.

28. Yamamoto, K.; Ando, H.; Shuetake, T.; Chikamatsu, H. *J. Chem. Soc., Chem. Commun.* **1989,** 754–755.

29. Hanson, R. M.; Sharpless, K. B. *J. Org. Chem.* **1986,** *51,* 1922–1925.

30. Komatsu, K.; Nishibayashi, Y.; Sugita, T.; Uemura, S. *Tetrahedron Lett.* **1992,** *33,* 5391.

31. Bortolini, O.; Di Furia, F.; Licini, G.; Modena, G.; Rossi, M. *Tetrahedron Lett.* **1986,** *27,* 6257–6260.

32. (a) Bortolini, O.; Di Furia, F.; Licini, G.; Modena, G. In *The Role of Oxygen in Chemistry and Biochemistry,* Ando, W.; Moro-oka, Y. (Eds.); *Studies in Organic Chemistry;* Elsevier: Amsterdam, 1988; Vol. 33, pp. 193–200. (b) Bortolini, O.; Di Furia, F.; Licini, G.; Modena, G. *Rev. Heteroatom Chem.* **1988,** *1,* 66–79.

33. Bortolini, O.; Di Furia, F.; Licini, G.; Modena, G. *Phosphorus Sulfur,* **1988,** *37,* 171–174.

34. Conte, V.; Di Furia F.; Licini, G.; Modena, G. *Tetrahedron Lett.* **1989,** *30,* 4859–4862.

35. Beckwith, A. L. J.; Boate, D. R. *J. Chem. Soc., Chem. Commun.* **1986,** 189–190.

36. Tanaka, J.; Higa, T.; Bernardinelli, G.; Jefford, C. W. *Tetrahedron*, **1989**, *45*, 7301–7310.

37. Naruta, Y.; Tani, F.; Maruyama, K. *J. Chem. Soc., Chem. Commun.* **1990**, 1378–1380.

38. Naruta, Y.; Tani, F.; Maruyama, K. *Tetrahedron: Asymmetry*, **1991**, *2*, 533–542.

39. Groves, J. T.; Viski, P. *J. Org. Chem.* **1990**, *55*, 3628–3634.

40. Halterman, R. L.; Jan, S. T.; Nimmens, H. L. *Synlett*, **1991**, 791–792.

41. Cashman, J. R.; Olsen, L. D.; Bornheim, L. M. *J. Am. Chem. Soc.* **1990**, *112*, 3191–3195.

42. Sakuraba, H.; Ushiki, S. *Tetrahedron Lett.* **1990**, *31*, 5349–5352.

43. Di Furia, F.; Licini, G.; Modena, G.; Valle, G. *Bull. Soc. Chim. Fr.* **1990**, 734–744.

44. Di Furia, F.; Licini, G.; Modena, G. *Tetrahedron Lett.* **1989**, *30*, 2575–2576.

45. Colombo, A.; Marturano, G.; Pasini, A. *Gazz. Chim. Ital.* **1986**, *116*, 35–40.

46. Colonna, S.; Manfredi, A.; Spadoni, M.; Casella, L.; Gullotti, M. *J. Chem. Soc., Perkin Trans. 1*, **1987**, 71–73.

47. Nakajima, K.; Sasaki, C.; Kojima, M.; Aoyama, T.; Ohba, S.; Saito, Y.; Fujita, J. *Chem. Lett.* **1987**, 2189–2192.

48. Nakajima, K.; Kojima, M.; Fujita, J. *Chem. Lett.* **1986**, 1483–1486.

49. Ando, W.; Tajima, R.; Takata, T. *Tetrahedron Lett.* **1982**, *23*, 1685–1688.

50. Palucki, M.; Hanson, P.; Jacobsen, E. N. *Tetrahedron Lett.* **1992**, *33*, 7111.

51. Hikita, T.; Tamaru, K.; Yamagishi, A.; Iwamoto, T. *Inorg. Chem.* **1989**, *28*, 2221–2223.

52. Yamagishi, A. *J. Chem. Soc., Chem. Commun.* **1986**, 290–291.

53. Firth, B. E.; Miller, L. L. *J. Am. Chem. Soc.* **1976**, *98*, 8272–8273.

54. Komori, T.; Nonaka, T. *J. Am. Chem. Soc.* **1984**, 2656–2659.

55. Shinkai, S.; Yamaguchi, T.; Manabe, O.; Toda, F. *J. Chem. Soc., Chem. Commun.* **1988**, 1399–1401.

56. Czarnik, A. W. *J. Org. Chem.* **1984**, *49*, 924–927.

57. Drabowicz, J.; Mikolajczyk, M. *Phosphorus Sulfur*, **1984**, *21*, 245.

58. (a) Sugimoto, T.; Kokubo, T.; Miyazaki, J.; Tanimoto, S.; Okano, M. *J. Chem. Soc., Chem. Commun.* **1989**, 402–404. (b) Sugimoto, T.; Kokubo, T.; Miyazaki, J.; Tanimoto, S.; Okano, M. *J. Chem. Soc., Chem. Commun.* **1989**, 1052–1053.

59. Sugimoto, T.; Kokubo, T.; Miyazaki, J.; Tanimoto, S.; Okano, M. *Bioorg. Chem.*, **1981**, *10*, 311–323.

60. Kagan, H. B.; Fiaud, J. C. *Top. Stereochem.* **1988**, *18*, 249–339.

61. Ogura, K.; Fujita, M.; Iida, H. *Tetrahedron Lett.* **1980**, *21*, 2233–2236.

62. Colonna, S.; Banfi, S.; Sommaruga, M. *J. Org. Chem.* **1985**, *50*, 769–771.

63. Colonna, S.; Banfi, S.; Annunziata, R.; Casella, L. *J. Org. Chem.* **1986**, *51*, 891–895.

64. Colonna, S.; Gaggero, N. *Tetrahedron Lett.* **1989**, *30*, 6233–6236.

65. Colonna, S.; Gaggero, N.; Leone, M.; Pasta, P. *Tetrahedron*, **1991**, *47*, 8385–8398.

66. Auret, B. J.; Boyd, D. R.; Henbest, H. B.; Ross, S. *J. Chem. Soc. C*, **1968**, 2371–2376.

67. Madesclaire, M. *Tetrahedron*, **1986**, *42*, 5459–5495.

68. Holland, H. L. *Chem. Rev.* **1988,** *88,* 473–485.

69. Abushanab, E.; Reed, D.; Suzuki, F.; Sih, C. J. *Tetrahedron Lett.* **1978,** *37,* 3415–3418.

70. Ohta, H.; Okamoto, Y.; Tsuchihashi, G. *Chem. Lett.* **1984,** 205–208.

71. Auret, B. J.; Boyd, D. R.; Cassidy, E. S.; Hamilton, R.; Turley, F.; Drake, A. F. *J. Chem. Soc., Perkin Trans. 1,* **1985,** 1547–1552.

72. Auret, B. J.; Boyd, D. R.; Dunlop, R.; Drake, A. F. *J. Chem. Soc., Perkin Trans. 1,* **1988,** 2827–2829.

73. Buist, P. H.; Marecak, D. M.; Partington, E. T.; Skala, P. *J. Org. Chem.* **1990,** *55,* 5667–5699.

74. Buist, P. H.; Marecak, D. M. *J. Am. Chem. Soc.* **1991,** *113,* 5877.

75. Madesclaire, M.; Fauve, A.; Metin, J.; Carpy, A. *Tetrahedron: Asymmetry,* **1990,** *1,* 311–314.

76. Rossi, C.; Fauve, A.; Madesclair, M.; Roche, D.; Davis, F. A.; Thimma Reddy, R. *Tetrahedron: Asymmetry,* **1992,** *3,* 629.

77. Takata, T.; Yamazaki, M.; Fujimori, K.; Kim, Y. H.; Oae, S.; Iyanagi, T. *Chem. Lett.* **1980,** 1441–1444.

78. May, S. W.; Phillips, R. S. *J. Am. Chem. Soc.* **1980,** *102,* 5981–5983.

79. Light, D. R.; Waxman, D. J.; Walsh, C. *Biochemistry,* **1982,** *21,* 2490–2493.

80. Waxman, D. J.; Light, D. R.; Walsh, C. *Biochemistry,* **1982,** *21,* 2499–2502.

81. Katopodis, A. G.; Smith, H. A., Jr.; May, S. W. *J. Am. Chem. Soc.,* **1988,** *110,* 897–899.

82. Colonna, S.; Gaggero, N.; Manfredi, A.; Casella, L.; Gullotti, M. *J. Chem. Soc., Chem. Commun.* **1988,** 1451–1452.

83. Bradley, D. C.; Mehrota, R. C.; Gaur, D. P. In *Metal Alkoxide*; Academic Press, London, UK, 1978.

4.4

Catalytic Asymmetric Dihydroxylation

Roy A. Johnson

The Upjohn Company
Kalamazoo, Michigan

K. Barry Sharpless

The Scripps Research Institute
La Jolla, California

4.4.1. Introduction

The cis dihydroxylation of olefins mediated by osmium tetroxide represents an important general method for olefin functionalization [1,2]. For the purpose of introducing the subject of this chapter, it is useful to divide osmium tetroxide mediated cis dihydroxylations into four categories: (1) the stoichiometric dihydroxylation of olefins, in which a stoichiometric equivalent of osmium tetroxide is used for an equivalent of olefin; (2) the catalytic dihydroxylation of olefins, in which only a catalytic amount of osmium tetroxide is used relative to the amount of olefin in the reaction; (3) the stoichiometric, asymmetric dihydroxylation of olefins, in which osmium tetroxide, an olefinic compound, and a chiral auxiliary are all used in equivalent or stoichiometric amounts; and (4) the catalytic, asymmetric dihydroxylation of olefins. The last category is the focus of this chapter. Many features of the reaction are common to all four categories, and are outlined briefly in this introductory section.

In the reaction with osmium tetroxide (1), the olefin (2) is first osmylated to form an osmium(VI) monoglycolate ester for which Criegee proposed structure 3 [1] (see Scheme 1) in analogy to the cyclic ester postulated by Böeseken and van Giffen for permanganate glycolates [3]. The osmylation process has been proposed to proceed by either [3 + 2] cycloaddition leading directly to the monoglycolate ester or a reversible [2 + 2] cycloaddition leading to a metallaoxetane intermediate (4) [4], which undergoes irreversible rearrangement to the monoglycolate ester (3). Arguments favoring each of these pathways have been put forward and have been reviewed recently [5]. In the absence of hydrolytic conditions, the monoglycolate ester 3 may form a dimeric complex (5) [6], while under hydrolytic conditions, hydrolysis of the glycolate ester(s) occurs with breaking of the osmium–oxygen bonds, releasing the diol 6 and an osmium(VI) oxo complex.

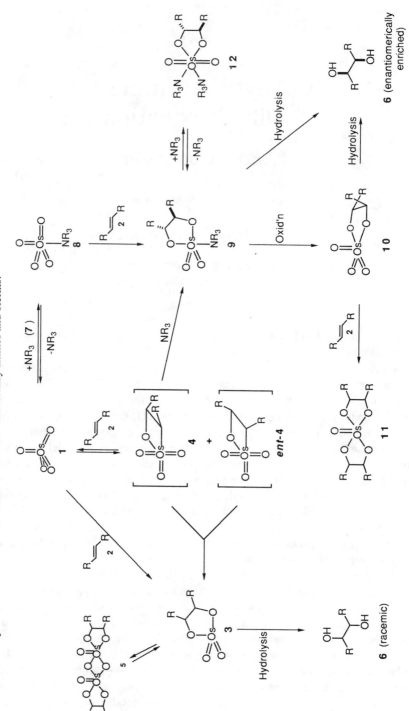

Scheme 1 Complexation and reaction of osmium tetroxide with tertiary amines and olefins.

Inclusion in the reaction of a cooxidant serves to return the osmium to the osmium tetroxide level of oxidation and allows for the use of osmium in catalytic amounts. Various cooxidants have been used for this purpose; historically, the application of sodium or potassium chlorate in this regard was first reported by Hofmann [7]. Milas and coworkers [8,9] introduced the use of hydrogen peroxide in *t*-butyl alcohol as an alternative to the metal chlorates. Although catalytic cis dihydroxylation using perchlorates or hydrogen peroxide usually gives good yields of diols, it is difficult to avoid overoxidation, which with some types of olefins becomes a serious limitation to the method. Superior cooxidants that minimize overoxidation are alkaline *t*-butylhydroperoxide, introduced by Sharpless and Akashi [10], and tertiary amine oxides such as *N*-methylmorpholine-*N*-oxide (NMO), introduced by Van-Rheenen, Kelly, and Cha (the Upjohn process) [11]. A new, important addition to this list of cooxidants is potassium ferricyanide, introduced by Minato, Yamamoto, and Tsuji in 1990 [12].

Acceleration of osmylation by the addition of pyridine to the reaction system was first observed by Criegee et al. [13] and is a property shared by other tertiary amines as well. The reactions and equilibria included in Scheme 1 can be postulated on the basis of spectral and crystallographic characterization of various reaction components. Osmium tetroxide (**1**) and tertiary amines (**7**) have been observed to form monoamine complexes such as **8** [14]. Osmium monoglycolate–amine complexes such as **9** have also been isolated and characterized [15]. However, the mechanism by which complex **9** forms is not yet clear. Whether **9** arises from osmylation of the olefin (**2**) by complex **8** or via the metallaoxetane **4** remains to be determined. Oxidation of complex **9** gives the putative trioxoglycolate **10** [1], which can react with a second olefin and form the osmium(VI) bisglycolate ester **11**, an event with important consequences to the catalytic asymmetric dihydroxylation process (see Section 4.4.2.2.) [16]. Alternatively, complex **9** can add a second amine ligand giving complex **12**, which is found to be unreactive to further oxidation or hydrolysis [1,17]. High binding constants for complex **12** are observed when the amine is pyridine or a chelating diamine, a somewhat lower binding constant is seen for quinuclidine, and weak binding constants are seen for derivatives of the cinchona alkaloids dihydroquinidine (**13**) and dihydroquinine (**14**) (see Chart 1), which are used as chiral ligands in asymmetric dihydroxylation. Hydrolysis of the glycolate esters **9** and **10** releases the diols.

Amine complexation by osmium tetroxide opened the door to asymmetric dihydroxylation of olefins when Hentges and Sharpless found that osmylation of olefins in the presence of dihydroquinidine acetate (**15**) or dihydroquinine acetate (**16**) under stoichiometric conditions gave, after hydrolysis, optically active diols with 25–94% ee [18]. The crystallographically determined structure shown in Figure 1 for the complex of dioxo[(3*S*,4*S*)-2,2,5,5-tetramethyl-3,4-hexanediolato]osmium(VI) with dihydroquinine *p*-chlorobenzoate clearly shows the osmium(VI) monoglycolate complexed to the quinuclidine nitrogen of the alkaloid [19]. Following the report of Hentges and Sharpless, several other groups have described stoichiometric asymmetric dihydroxylation systems. Generally, these systems use chiral, chelating diamines as the chiral auxiliary with osmium tetroxide for induction of

Chart 1 Structures of chiral ligands.

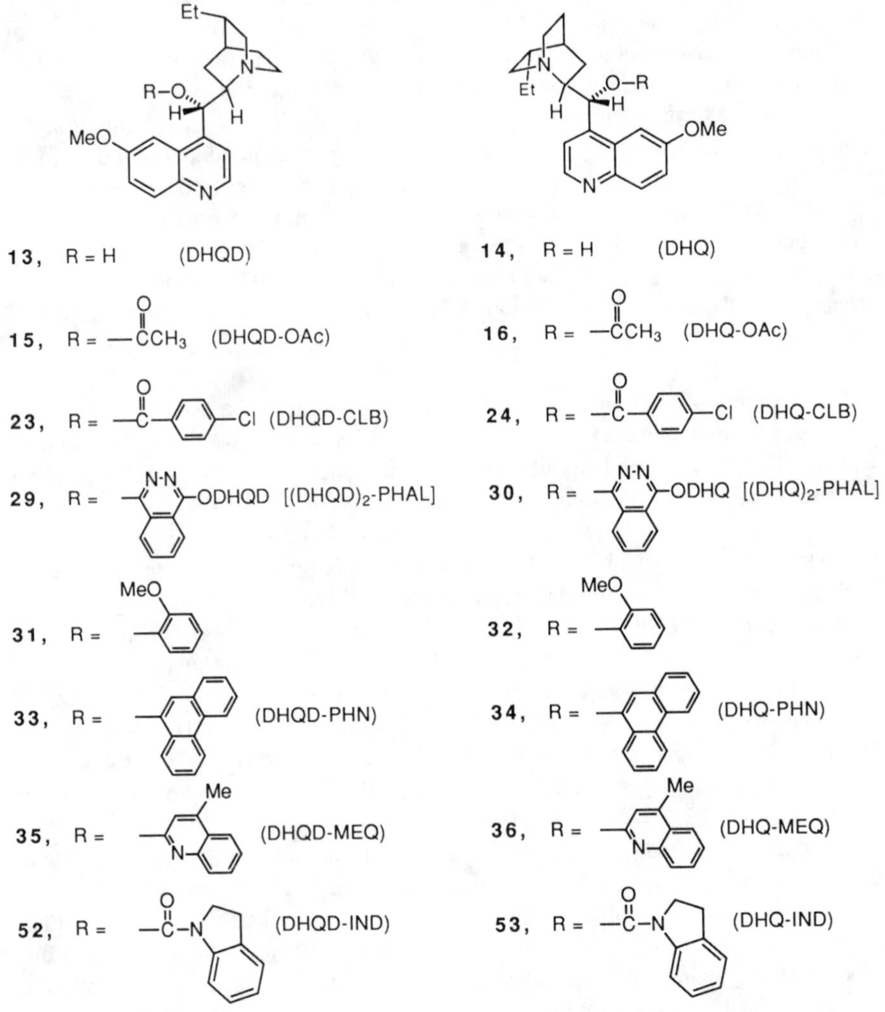

13, R = H   (DHQD)

14, R = H   (DHQ)

15, R = —$\overset{O}{\overset{\|}{C}}$CH₃   (DHQD-OAc)

16, R = —$\overset{O}{\overset{\|}{C}}$CH₃   (DHQ-OAc)

23, R = —C(=O)—⟨C₆H₄⟩—Cl (DHQD-CLB)

24, R = —C(=O)—⟨C₆H₄⟩—Cl (DHQ-CLB)

29, R = ⟨phthalazine⟩—ODHQD [(DHQD)₂-PHAL]

30, R = ⟨phthalazine⟩—ODHQ [(DHQ)₂-PHAL]

31, R = ⟨2-MeO-phenyl⟩

32, R = ⟨2-MeO-phenyl⟩

33, R = ⟨phenanthryl⟩ (DHQD-PHN)

34, R = ⟨phenanthryl⟩ (DHQ-PHN)

35, R = ⟨4-Me-quinolinyl⟩ (DHQD-MEQ)

36, R = ⟨4-Me-quinolinyl⟩ (DHQ-MEQ)

52, R = —C(=O)—N⟨indolinyl⟩ (DHQD-IND)

53, R = —C(=O)—N⟨indolinyl⟩ (DHQ-IND)

asymmetry in the diol products. Diamines used as chiral auxiliaries include 1,4-piperidinylbutan-2,3-diol ketals and acetals (**17**) [20], (−)-(*R*,*R*)-*N*,*N*,*N*′*N*′-tetramethylcyclohexane-1,2-*trans*-diamine (**18**) [21], 1,2-(*trans*-3,4-diaryl)pyrrolidinylethanes (**19**) [22], *N*,*N*′-dialkyl-2,2′-bipyrrolidines (**20**) [23], and 1,2-diphenyl-*N*,*N*′-bis(2,4,6-trimethylbenzylidene)-1,2-diaminoethane (**21**) [24].

Asymmetric induction also occurs during osmium tetroxide mediated dihydroxylation of olefinic molecules containing a stereogenic center, especially if this center is near the double bond. In these reactions, the chiral framework of the molecule serves to induce the diastereoselectivity of the oxidation. These diastereoselective reactions are achieved with either stoichiometric or catalytic quantities of osmium

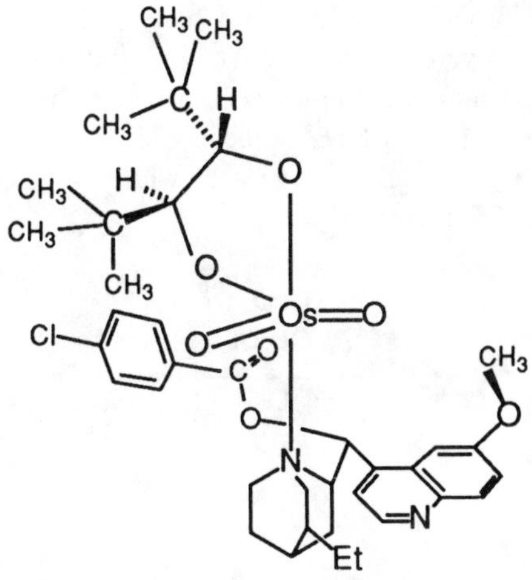

Figure 1 Structure of the dioxo[(3S,4S)-2,2,5,5-tetramethyl-3,4-hexanediolato]osmium(VI) complex with dihydroquinine p-chlorobenzoate (DHQ-CLB).

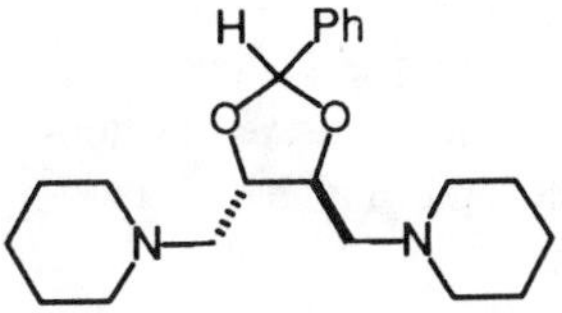

tetroxide. The possibility exists for pairing or "matching" this diastereoselectivity with the face selectivity of asymmetric dihydroxylation to achieve enhanced or "double diastereoselectivity" [25], as discussed further later in the chapter.

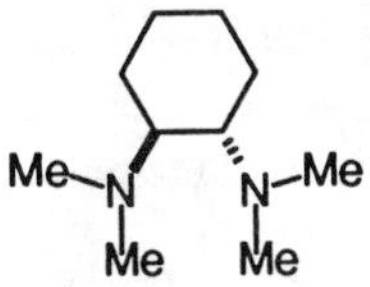

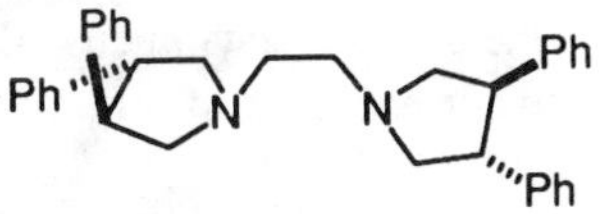

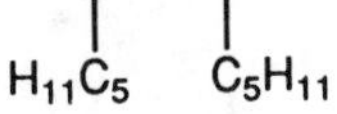

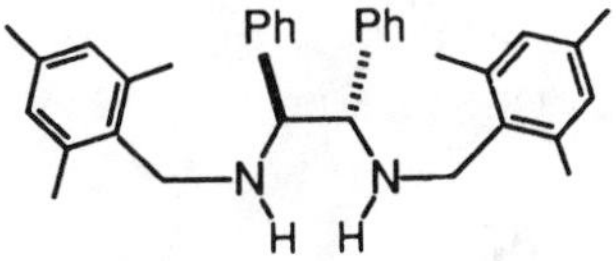

This brief outline of historical developments in osmium tetroxide mediated olefin hydroxylation brings us to our main subject, catalytic asymmetric dihydroxylation. The transition from *stoichiometric* to *catalytic* asymmetric dihydroxylation was made in 1987 with the discovery by Sharpless and coworkers that the stoichiometric process became catalytic when N-methylmorpholine-N-oxide (**22**, NMO) was used as the cooxidant in the reaction [26]. Dihydroquinidine p-chlorobenzoate (**23**) and dihydroquinine *p*-chlorobenzoate (**24**) (see Chart 1) also were introduced as new chiral ligands with improved enantioselective properties. Using 0.002 mol of osmium tetroxide and 0.134 mol of a chiral auxiliary (0.033 mol when the reaction was performed at 0°C), one mole of (*E*)-stilbene (**25**) was converted to (*R,R*)-(+)-dihydrobenzoin (**26**) in 80% yield with 88% ee while using 1.2 mol of NMO as the oxidant for this catalytic process.

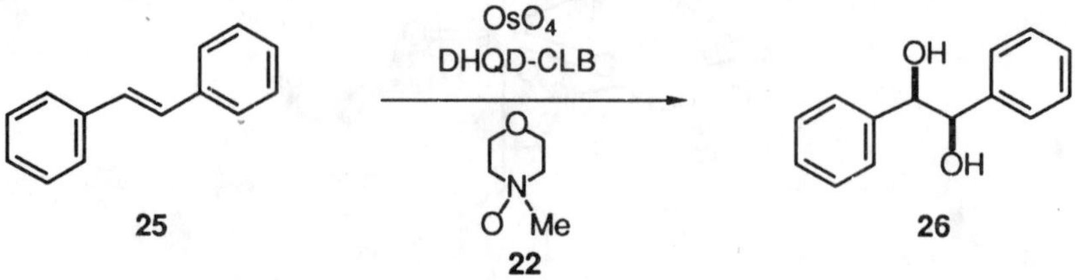

The original communication also described the asymmetric dihydroxylation of seven other olefins, among which were representatives from three of the six possible substitution patterns (see Chart 2). The observation of a catalytic asymmetric dihydroxylation of methyl (*E*)-*p*-methoxycinnamate (**27**) with NMO as oxidant to give diol **28** (70% ee) was described independently by Gredley in a patent application published in 1989 [27] and reported in a communication in 1990 [28]. Since 1987, the catalytic asymmetric dihydroxylation process has undergone intensive development resulting in a number of improvements, including several very recent mod-

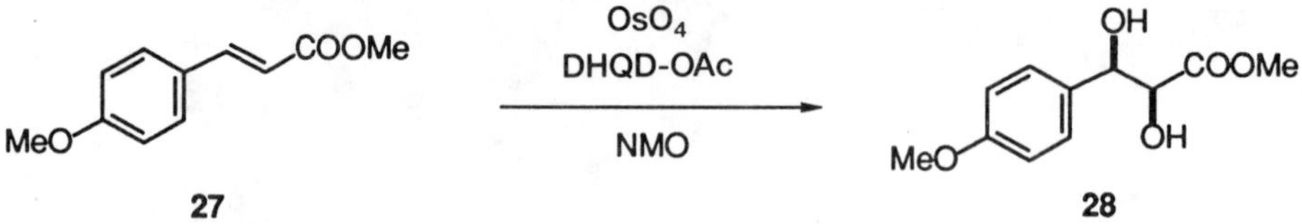

Chart 2   Olefin substitution patterns.

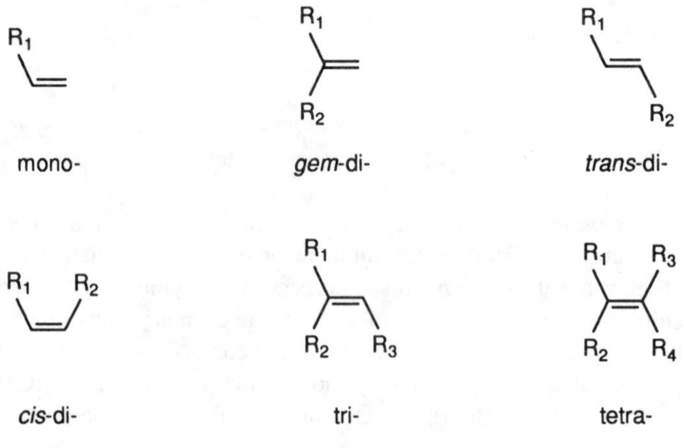

ifications of reaction conditions and of the chiral auxiliary, which significantly improve enantioselectivity over a broad range of olefin structures. Development activity continues and will lead to additional improvements.

The focus of this chapter is to acquaint the reader with details of catalytic asymmetric dihydroxylation with osmium tetroxide and the scope of results that one can expect to achieve with current optimum conditions. The literature through mid-1992 has been reviewed in compiling this chapter. Osmium tetroxide catalyzed hydroxylations of olefins and acetylenes are the subject of an extensive review by Schröder published in 1980 [2a]. A comprehensive review of research and industrial applications of asymmetric dihydroxylations is in preparation [2b].

4.4.2. General Features of Osmium-Catalyzed Asymmetric Dihydroxylation

Catalytic asymmetric dihydroxylations (ADs) are easy reactions to perform. The reaction actually requires water and is insensitive to oxygen and so can be carried out without fear of exposure to the atmosphere. The reaction uses a multicomponent reagent system, which allows for considerable variation of each component. In the years since the discovery of the catalytic asymmetric process, an evolution of improvements in each of the variables has occurred. These improvements led to the publication of the following "recipe" [29] for the catalytic asymmetric dihydroxylation of one millimole of an olefin: potassium ferricyanide (0.94 g, 3.0 mmols), potassium carbonate (0.41 g, 3.0 mmols), either 1,4-bis(9-O-dihydroquinidine)phthalazine, (DHQD)$_2$-PHAL (**29**), or 1,4-bis(9-O-dihydroquinine)phthalazine, (DHQ)$_2$-PHAL (**30**) (see Chart 1), (0.0078 g. 0.010 mmol), and potassium osmate(VI) dihydrate (0.00074 g, 0.0020 mmol) in a 1:1 mixture of t-butyl alcohol and water (5 mL of each). [*However, see the next paragraph for an improved version of this recipe.*] The recommended temperature for the reaction is 0°C [29]. The addition of a sulfonamide to the reaction should also be considered according to the guidelines outlined in Section 4.4.2.5. One requirement for this set of reaction conditions is that the reaction mixture be stirred vigorously, since addition of the inorganic salts causes separation of the solvent into two phases. This separation of phases is itself thought to be important to the reaction (see Section 4.4.2.2.).

The total weight of the four solid ingredients used for the one-millimole scale reaction is 1.36 g, of which less than a milligram is potassium osmate(VI) dihydrate. Ready-made mixtures of the four solid components, for which the names AD-mix-β [containing (DHQD)$_2$-PHAL, see Section 4.4.2.6 for derivation of names] and AD-mix-α [containing (DHQ)$_2$-PHAL] have been coined, are available from the Aldrich Chemical Company. AD-mix-α and AD-mix-β were used for many of the entries in the tables of results discussed below [29]. The AD-mixes offer a convenient alternative to the need to weigh out each component for one or more small-scale reactions. The stability of the mixture shows promise for a practical shelf life [29]. The original formula [29] for one kilogram of AD-mix is: K$_3$Fe(CN)$_6$, 699.96 g; K$_2$CO$_3$, 294.00 g; (DHQD)$_2$- or (DHQ)$_2$-PHAL, 5.52 g; and K$_2$OsO$_2$(OH)$_4$, 0.52 g, and the ratio of ligand to Os is 5:1. *This formula is improved* *for general use by increasing the amount of $K_2OsO_2(OH)_4$ fivefold, to 2.60 g.* The advantage of this formula is that olefins such as α,β-unsaturated amides, which are hydroxylated only sluggishly with the original formula, are now hydroxylated at a practical rate.

In the following sections, the reaction components and variables for catalytic AD are discussed in greater detail.

4.4.2.1. Osmium

Descriptions in the literature of osmium tetroxide frequently include several qualifying statements which express the essential emotions accompanying the use of this reagent. These are: (1) the *cis* dihydroxylation of olefins with osmium tetroxide is a reliable and powerful synthetic method, (2) osmium tetroxide is an expensive reagent, and (3) osmium tetroxide is regarded as a toxic material. Judging from its frequent use, the utility of osmium tetroxide clearly outweighs the drawbacks of cost and danger. At the same time, the problems of cost and toxicity exert pressure for the development of efficient methods for the use of osmium tetroxide at the catalytic level. Fortunately, in this regard, the discovery early in the history of osmium tetroxide that the molecule is very easily reduced and reoxidized has led to extensive use of the reagent in a catalytic mode.

Osmium tetroxide is the traditional osmium species used in the dihydroxylation of olefins. For large-scale reactions, osmium tetroxide may be weighed and transferred as the solid. For many catalytic applications on a laboratory scale, the amount of osmium tetroxide required is too small to be weighed conveniently. In these cases, advantage can be taken of the solubility of osmium tetroxide in organic solvents by the preparation of a stock solution of known concentration and the use of an aliquot for the small scale reaction.

Alternate sources of osmium can be used and should be considered, as they offer advantages in ease of handling. Both osmium(III) chloride hydrate [OsCl·xH$_2$O] [30] and potassium osmate(VI) dihydrate [K$_2$OsO$_2$(OH)$_4$] [31] have been used as replacements for osmium tetroxide, and the latter reagent is specified in the preceding standard set of conditions for catalytic AD.

Curiously, osmium tetroxide's reputation for chronic toxicity is not supported in the toxicology literature [32]. In fact, an aqueous solution of osmium tetroxide is used to treat refractory rheumatoid arthritis in humans by direct injection into the knee joint [33,34].

Osmium tetroxide, like the skunk, carries its own warning system—a strong odor described variously as resembling chlorine, bromine, or ozone. Clearly, a commonsense approach to the use of this reagent in an efficient hood is called for, and under these conditions osmium tetroxide should be regarded as no more dangerous than many other reagents found in daily use in the laboratory.

4.4.2.2. Oxidants

Either amine oxides (usually NMO) [11,26] or potassium ferricyanide/potassium carbonate [12,35] are used as cooxidants for catalytic AD. The choice of oxidant carries with it the choice of solvent for the reaction, and the details of the catalytic cycle appear to be quite different depending on which oxidant–solvent combination is used. When potassium ferricyanide/potassium carbonate is used as the oxidant, the solvent used for the reaction is a 1:1 mixture of *t*-butyl alcohol and water [35,36]. This solvent mixture, normally miscible, separates into two liquid phases upon addition of the inorganic reagents. The sequence of reactions summarized below in equations 1–5 has been postulated as occurring under these conditions. This reaction sequence is further illustrated in the reaction cycle shown in Scheme 2, which also emphasizes the role of the two-phase solvent system.

Scheme 2 Catalytic cycle for asymmetric dihydroxylation using potassium ferricyanide as cooxidant.

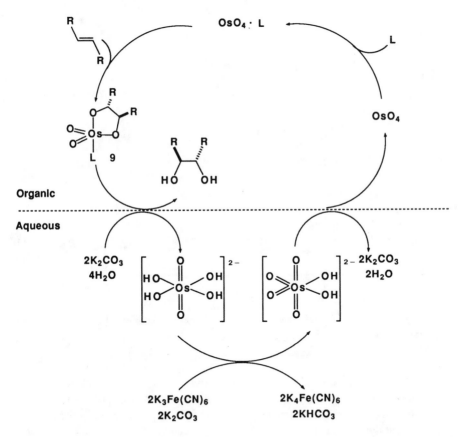

First, osmylation of the olefin proceeds to form the osmium(VI) monoglycolate–amine complex (**9**) as shown in equation 1. Formation of **9** is presumed to occur in the organic phase in which all the involved species are soluble. Next, at the organic–aqueous interface, hydrolysis of the glycolate ester releases the diol into the organic phase and the reduced osmium into the aqueous phase as the hydrated osmate(VI) dianion (shown as the potassium salt in eq. 2). Oxidation of osmate(VI) by potassium ferricyanide regenerates osmium tetroxide via an intermediate perosmate(VIII) dianion (eq. 3). Loss of two hydroxide groups from the perosmate(VIII) ion gives osmium tetroxide, which then migrates back into the organic phase to restart the cycle (eq. 4) [36]. Summation of equations 1–4 gives equation 5, showing that in this catalytic cycle both oxygens of the diol are provided by water.

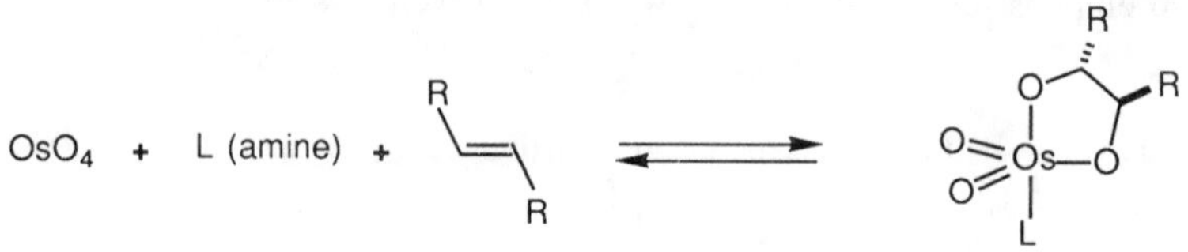

$$OsO_4 \; + \; L \text{ (amine)} \; + \; \text{(olefin)} \; \rightleftharpoons \; \text{(osmate ester)} \tag{1}$$

$$\text{(osmate ester)} \; + \; 4H_2O \; + \; 2K_2CO_3 \; \longrightarrow \; \text{(diol)} \; + \; K_2OsO_2(OH)_4 \; + \; 2KHCO_3 \; + \; L \tag{2}$$

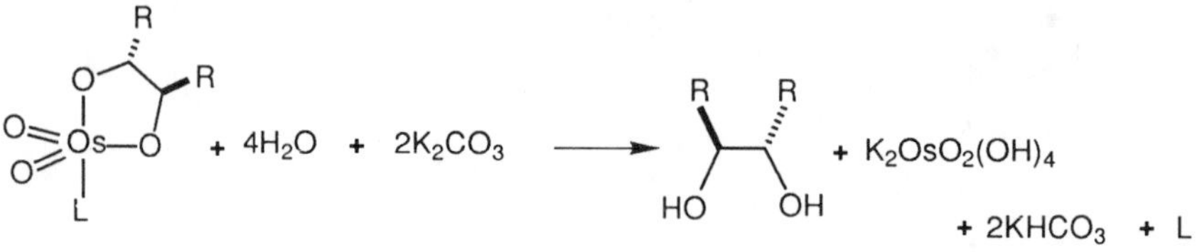

$$K_2OsO_2(OH)_4 \; + \; K_3Fe(CN)_6 \; + \; 2K_2CO_3 \; \longrightarrow \; K_2OsO_4(OH)_2 \; + \; 2K_4Fe(CN)_6 \; + \; 2KHCO_3 \tag{3}$$

$$K_2OsO_4(OH)_2 \; + \; 2KHCO_3 \; \rightleftharpoons \; OsO_4 \; + \; 2K_2CO_3 \; + \; 2H_2O \tag{4}$$

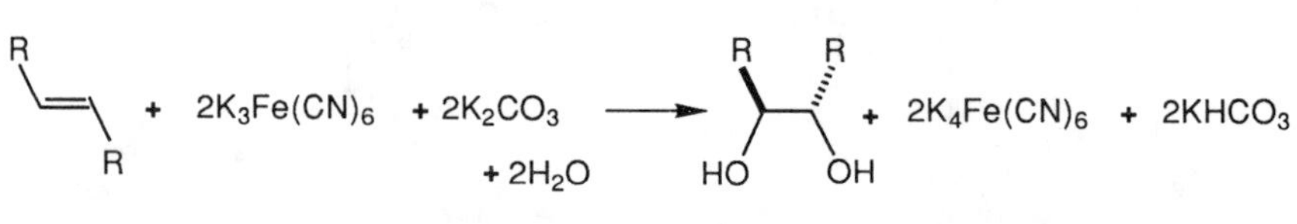

$$\text{(olefin)} \; + \; 2K_3Fe(CN)_6 \; + \; 2K_2CO_3 \; + \; 2H_2O \; \longrightarrow \; \text{(diol)} \; + \; 2K_4Fe(CN)_6 \; + \; 2KHCO_3 \tag{5}$$

   A new development is that electrochemical oxidation of ferrocyanide to ferricyanide can be coupled with AD to give a very efficient electrocatalytic process [37]. Under these conditions, the amount of potassium ferricyanide needed for the reaction becomes catalytic and equations 6 and 7 can be added following equation 4. Summation of equation 1–4, 6, and 7 gives equation 8, showing that only water in addition to electricity is needed for the conversion of olefins to asymmetric diols and that hydrogen gas, released at the cathode, is the only by-product of this process. In practice, sodium ferrocyanide is used in the reaction and the amount of this reagent used in comparison to the potassium ferricyanide method mentioned above has been reduced from 3.0 equiv. to 0.15 equiv. (relative to an equivalent of olefin).

$$2K_4Fe(CN)_6 \; - \; 2e^- \longrightarrow 2K_3Fe(CN)_6 \; + \; 2K^+ \qquad (6)$$

$$2H_2O \; + \; 2e^- \longrightarrow 2OH^- \; + \; H_2 \qquad (7)$$

$$\begin{array}{c} R \\ \diagdown \hspace{-0.5em} = \\ R \end{array} \; + \; 2H_2O \longrightarrow \begin{array}{c} R \quad R \\ \diagup \hspace{-0.3em} \diagdown \\ HO \quad OH \end{array} \; + \; H_2 \qquad (8)$$

For laboratory scale reactions, this electrocatalytic AD generally is performed in a glass H-type cell in which the anode and cathode compartments are separated by a semipermeable Nafion cation-exchange membrane and platinum electrodes are used. A 5% aqueous solution of phosphoric acid is used in the cathode compartment, and the reaction in the anode compartment is stirred vigorously. Under a controlled anode potential of 0.4 V (vs. Ag/AgCl) and with $(DHQD)_2$-PHAL as chiral ligand, α-methylstyrene was converted to R-2-phenyl-1,2-propanediol in 15 hours with the electrical consumption of 2.1 F/mol. The product was isolated in 100% yield with 92% ee [37].

When NMO is used as the oxidant, typically with acetone–water (10:1) as a solvent, the reaction mixture is essentially homogeneous and a different catalytic cycle is observed [16,36]. The cycle can be broken down into equations 9–13 and can be illustrated as shown in Scheme 3. The first step (eq. 9) is the same as above, that is, formation of the osmium(VI) monoglycolate–amine complex **9**. Next the organic soluble oxidant NMO is postulated to bind reversibly to the osmium(VI) glycolate ester (**9**), as shown in equation 10. Transfer of oxygen to osmium leads to formation of an osmium(VIII) trioxoglycolate (**10**) as shown in equation 11, and hydrolysis of this trioxoglycolate (eq. 12) gives the diol and regenerated osmium tetroxide. Equation 13 simply indicates the equilibrium between complexed and free osmium tetroxide to complete the cycle. Summation of equations 9–13 gives equation 14 and shows that as this catalytic system recycles, NMO and water provide one oxygen each to the newly formed diol.

In addition to participating in the primary catalytic cycle just outlined, the osmium(VIII) trioxoglycolate (**10**) shown in equation 11 has access to a secondary reaction cycle, which has a devastating effect on the overall enantioselectivity of the dihydroxylation process [16,36]. The trioxoglycolate (**10**) can add to (osmylate) a second olefin molecule, forming a bisglycolate (**11**) as shown for the secondary cycle in Scheme 3. Completion of this secondary cycle involves hydrolysis of the bisglycolate (**11**) to the monoglycolate (**9**) followed by reoxidation leading back to the osmium (VIII) trioxoglycolate (**10**) intermediate [36]. By modification of experimental conditions, the enantioselectivity of the secondary cycle can be determined and is found to be very low [16]. In fact, in the specific case examined the diol produced in this way has a slight preponderance of the configuration opposite to that of the primary cycle. Clearly, avoidance of this secondary cycle is necessary to attain the highest degree of enantioselectivity possible for a given catalytic AD. The

Scheme 3 Catalytic cycle for asymmetric dihydroxylation using *N*-methylmorpholine-*N*-oxide as cooxidant.

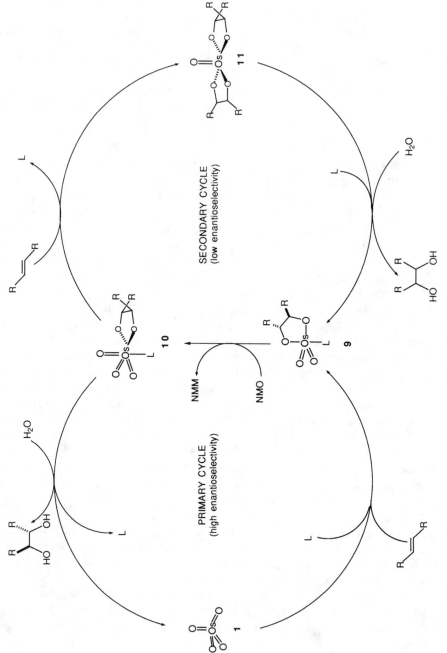

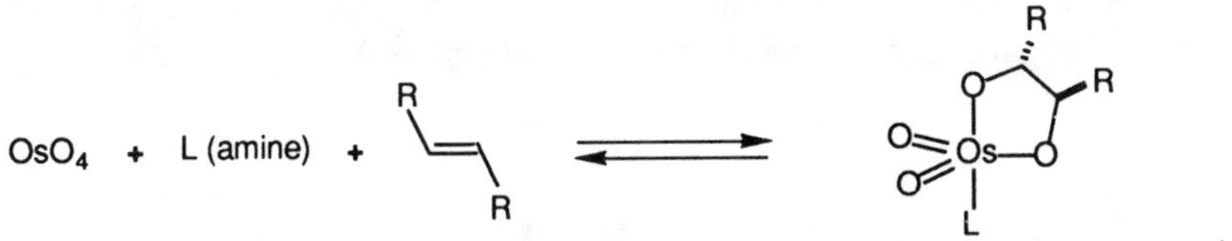

$$OsO_4 \ + \ L \ (amine) \ + \ \text{olefin} \ \rightleftharpoons \ \text{osmate ester} \tag{9}$$

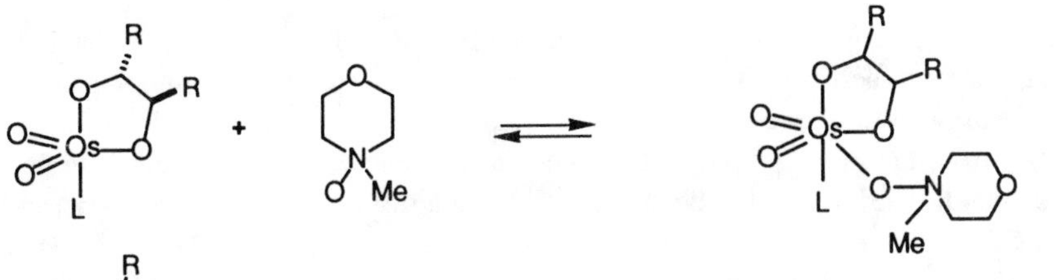

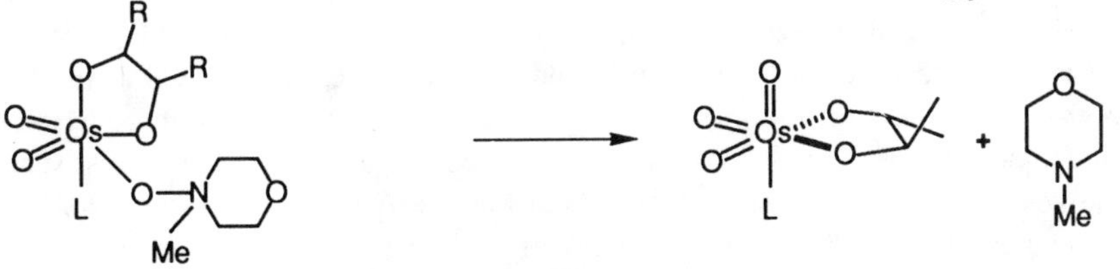

$$\text{osmate ester} \ + \ \text{morpholine-N-oxide} \ \rightleftharpoons \ \text{adduct} \tag{10}$$

$$\text{adduct} \ \rightarrow \ \text{trioxoglycolate} \ + \ \text{morpholine} \tag{11}$$

$$\text{trioxoglycolate} \ + \ H_2O \ \rightarrow \ OsO_4 \cdot L \ + \ \text{diol} \tag{12}$$

$$OsO_4 \cdot L \ \rightleftharpoons \ OsO_4 \ + \ L \tag{13}$$

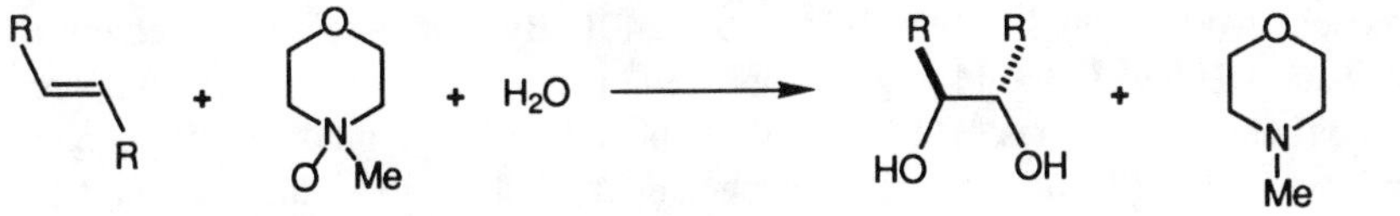

$$\text{olefin} \ + \ \text{morpholine-N-oxide} \ + \ H_2O \ \rightarrow \ \text{diol} \ + \ \text{morpholine} \tag{14}$$

extent to which the secondary cycle will be followed depends on the rate of hydrolysis of the trioxoglycolate ester, the concentration of olefin available for reaction to form the bisglycolate, and the rate of the latter process. Consistent with this hypothesis is the finding that slow addition of olefin to the reaction is accompanied by an increase in the enantiomeric purity of the diol product [16]. These findings with NMO have added much insight into the processes by which osmium tetroxide catalyzes the dihydroxylation of olefins and have led to practical improvements in the enantiomeric purity of products obtained under this set of conditions. However, with the advent of the potassium ferricyanide/potassium carbonate cooxidant modification, which appears to allow for hydrolysis of the osmium(VI) monoglycolate ester (9) before further oxidation can occur [36], a considerably easier method is in hand for avoiding the secondary reaction cycle and obtaining optimum enantioselectivities.

4.4.2.3. Chiral Ligand (Auxiliary)

The chiral ligands for catalytic AD are derived from dihydroquinidine (13, Chart 1) and dihydroquinine (14). Dihydroquinidine and dihydroquinine are minor components of the naturally occurring cinchona alkaloids and are separated in sufficient quantity from the more abundant quinidine and quinine to satisfy current commercial demand. They may also be obtained by catalytic reduction of quinidine and quinine. The cinchona alkaloids have an illustrious history in the fields of chemistry and medicine, areas that have been frequently reviewed [38]. Dihydroquinidine and dihydroquinine become very effective ligands for AD when derivatized at the C_9 hydroxyl group (new results indicate that derivatives of the parent quinidine and quinine alkaloids are almost as effective when used as ligands [29]). The best such derivatives (to date) are discussed further in the next paragraph. In their role as chiral ligands, these pairs of derivatives function almost as if they are enantiomers, although they are actually diastereoisomers because of differences in the attachment of the ethyl group (cf. 13 and 14 in Chart 1). Since they operate like enantiomers in AD, the term "pseudoenantiomers" has been applied to these pairs [39]. In the preparation of enantiomeric diols, the enantiomeric purity of the diol obtained using the dihydroquinidine derivative usually exceeds by 2–10% that of the diol obtained using the dihydroquinine derivative. This difference is generally seen regardless of the derivative used.

More than 250 derivatives, mainly of the cinchona alkaloids, have been made and tested as chiral ligands in the catalytic AD process. (Recall that the acetate was used in the original stoichiometric AD work.) The initial catalytic AD used the p-chlorobenzoate derivative, and this pair of ligands (15 and 16) was used in much subsequent work through 1990 [16,17,26,35,40]. By that time several ethers of DHQD and DHQ were found to give improved enantioselectivities when compared with the p-chlorobenzoates [41]. Particularly notable in this regard was the o-methoxyphenyl ethers (31 and 32), the 9′-phenanthryl ethers (33 and 34), and the 4′-methyl-2′-quinolyl ethers (35 and 36). Most recently, a new pair of ligands, the bis-DHQD and bis-DHQ ethers of phthalazine-1,4-diol (29 and 30), show significant improvement over the earlier ethers in the AD of most olefins [29].

Attachment of the alkaloid derivatives to a polymer support has been examined as a way in which to recycle the AD catalyst [42]. Polymers 37–40 were prepared by copolymerization of the appropriate alkaloid olefin monomer with acrylonitrile. After complexation with osmium tetroxide, the polymers were used as heterogeneous catalysts in the AD of (E)-stilbene. The reaction time with polymer 37, in which the quinuclidine ring of the alkaloid is attached directly to the polymer chain, was unacceptably long (7 days). The other three polymers (38–40) with spacer groups between the alkaloid and polymer chain were effective catalysts for AD, although reaction times are still longer than those with homogeneous catalysts. At room temperature with 25 mol % alkaloid and 1 mol % osmium tetroxide, reaction

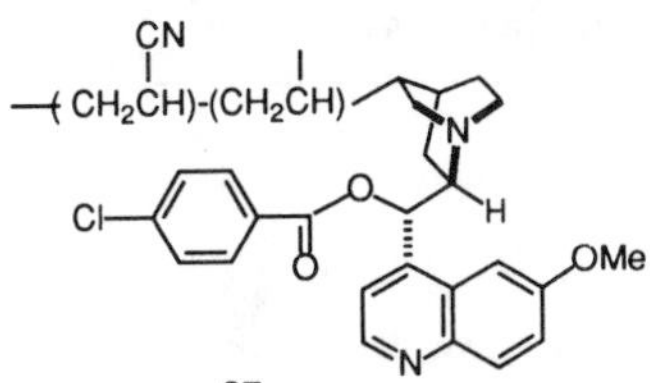

37

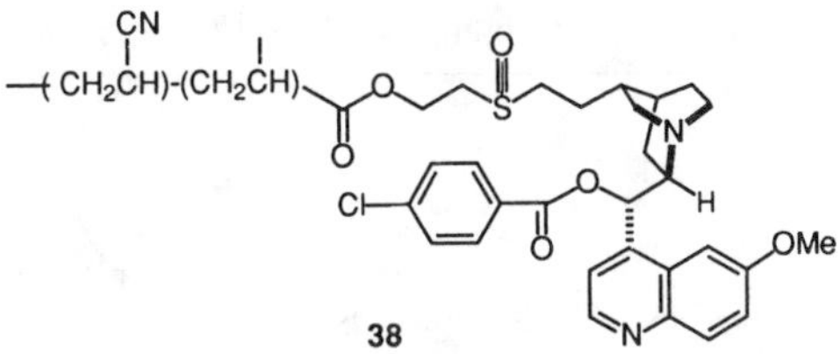

38

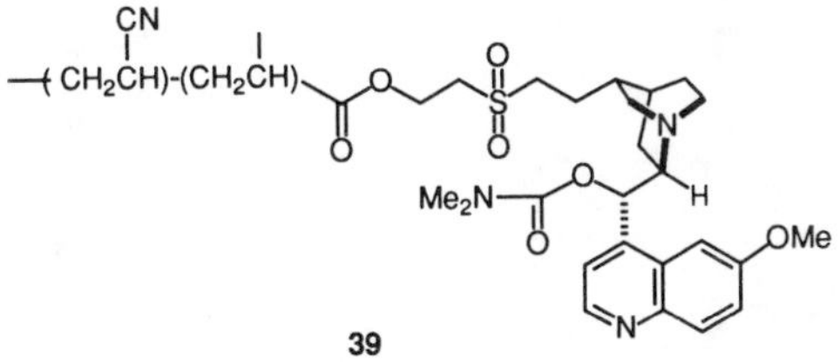

39

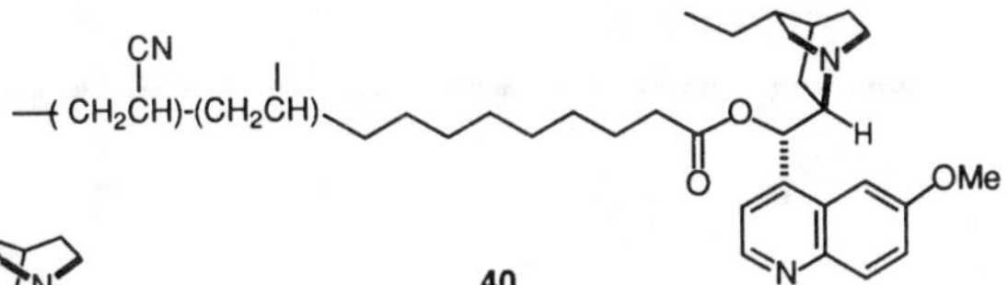

40

times of 18–48 hours gave stilbene diol in yields of 75–96%. The enantiomeric purities of the diols ranged from 85–93% ee with polymer **38** and were 80–82% ee with polymers **39** and **40** when NMO was used as cooxidant. With polymers **39** and **40** an improvement in enantioselectivity to 86–87% ee was observed when $K_3Fe(CN)_6/K_2CO_3$ was used as cooxidant. Recovery and reuse of polymer **38** in a second hydroxylation gave diol with very slightly reduced enantiomeric purity, indicating a potential for continuous use [42].

4.4.2.4. Stoichiometry

From the practical perspective of laboratory use, the catalytic AD process requires a minimum of concern over stoichiometry. The new osmium tetroxide–chiral ligand complexes are so efficient that for most olefins, 0.2 mol % of osmium will provide a satisfactory rate of reaction at 0°C [29]. In the occasional case where hydroxylation is slow under these conditions, the quantity of osmium in the catalyst should be increased to 1 mol % and the reaction temperature kept at 0°C. In the rare case where hydroxylation is still slow under these conditions, the temperature may be raised to 25°C and, to ensure no loss in enantioselectivity, the ligand concentration may be increased from 1 mol % to 2 mol %.

There are two main hindrances to the achievement of 100% ee in the catalytic AD reaction, namely, the nonenantioselective catalysis of dihydroxylation by species other than the desired complex (such as osmium tetroxide itself or the trioxoglycolate **10**) and less than a 100% exclusive fit of one face of the prochiral olefin to the chiral catalyst at the transition state. The dependence of enantioselectivity on the chiral ligand:osmium tetroxide ratio has been examined carefully in reaching the prescribed 5:1 stoichiometry of these reagents. First, the effect on enantioselectivity by changing this ratio was examined using *trans*-5-decene as the substrate for

Table 1   Relationship of Ligand Concentration to Enantiometric Excess

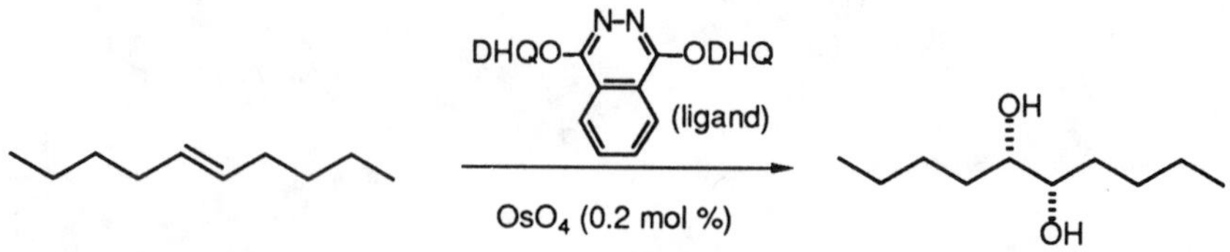

Ligand (mol %)	% ee	
	rt	0°C
10	93	
8	92	
6	92	
4	92	
2	90	94
1	89	93
0.5		93
0.25		91

catalytic AD. The results are shown in Table 1 (note that the DHQ derived PHAL is used in this experiment) and show a near-maximum 93% ee for the diol when the reaction is run at 0°C with a 5:1 ratio of ligand to OsO_4. The catalytic efficiency of the PHAL ligand–osmium tetroxide system is dramatically portrayed in the AD of (E)-stilbene [29]. Under the original conditions (a 5:1 ligand/OsO_4 ratio) and at 0°C, (E)-stilbene is dihydroxylated with better than 99.5% ee. The effect on enantioselectivity of lowering this ratio was examined, and it was found that even when the ratio of ligand to OsO_4 was only 1:20, diol with 96% ee was obtained from AD of stilbene. In other words, under the conditions of this experiment, there were 10,000 molecules of stilbene and 20 molecules of OsO_4 for every molecule of chiral ligand, yet the ligand–OsO_4 complex imparted optical activity to 9600 molecules of olefin, leaving only 400 molecules to be hydroxylated by OsO_4 alone (Table 1).

The chiral ligand is the crucial component controlling fit of the olefin on the catalyst and, clearly, for E-stilbene the fit must be nearly perfect. Efforts at modeling the chiral ligand–OsO_2–olefin complex with the intent to understand olefin fit have suggested some possible features of importance [43], but improvements in the chiral ligand have thus far depended largely on a trial-and-error approach to design.

4.4.2.5. Additives

When the secondary reaction cycle shown in Scheme 3 was discovered, it became clear that an increase in the rate of hydrolysis of trioxoglycolate 10 should reduce the role played by this cycle. The addition of nucleophiles such as acetate (tetraethylammonium acetate is used) to osmylations is known to facilitate hydrolysis of osmate esters. Addition of acetate ion to catalytic ADs using NMO as cooxidant was found to improve the enantiomeric purity for some diols, presumably as a result of

accelerated osmate ester hydrolysis [16]. The subsequent change to potassium ferricyanide as cooxidant appears to result in nearly complete avoidance of the secondary cycle (see Section 4.4.2.2.), but the turnover rate of the new catalytic cycle may
still depend on the rate of hydrolysis of the osmate ester **9**. The addition of a
sulfonamide (usually methanesulfonamide) has been found to enhance the rate of
hydrolysis for osmate esters derived from 1,2-disubstituted and trisubstituted olefins
[29]. On the other hand, for reasons that are not yet understood, addition of a sulfonamide to the catalytic AD of terminal olefins (i.e., monosubstituted and 1,1-disubsti-
tuted olefins) actually slows the overall rate of the reaction. Therefore, when called
for, the sulfonamide is added to the reaction at the rate of one equivalent per equivalent of olefin. This enhancement in rate of osmate hydrolysis allows most sluggish
dihydroxylation reactions to be run at 0°C rather than at room temperature [29].

4.4.2.6. Enantioselectivity Mnemonic

The asymmetric complex formed by osmium tetroxide, chiral ligand, and the
olefinic substrate delivers two oxygen atoms to one face or the other of the olefin
depending on which antipode of the chiral auxiliary is used in the reaction (see
Section 4.4.2.3 for a discussion of the pseudoenantiomeric character of DHQ and
DHQD). Figure 2 presents a mnemonic showing olefin orientation and face selectivity [26]. In the mnemonic, the olefin is oriented to fit the size constraints, where
R_L = largest substituent, R_M = medium-sized substituent, and R_S = smallest
substituent other than hydrogen. The oxygens will then be delivered from the upper
or β-face if a dihydroquinidine (DHQD) derived chiral auxiliary is used (this is the
source of the β designation used in the name AD-mix-β) and from the lower or
α-face if a dihydroquinine (DHQ) derived auxiliary is used (AD-mix-α). Any olefin
can be oriented four ways in the plane of this mnemonic (the number of orientations
is reduced for olefins that are symmetrically substituted). Two orientations are
possible for each face of an olefin, and they are related by a 180° rotation in
the plane of the paper. Both orientations of one face yield the same diol upon
cis dihydroxylation. Consider the two orientations (**41** and **41′**) shown for one face
of *trans*-3-decene, which are related by a simple 180° rotation. Hydroxylation of
either with osmium tetroxide and (DHQD)$_2$-PHAL (**29**) as chiral ligand will produce (*R,R*)-decane-3,4-diol (**42**).

The mnemonic shown in Figure 2 is empirical and is based primarily on the
enantioselectivities observed in the synthesis of the diols derived from the olefins
listed in Tables 2–5 in Section 4.4.3. Conversely, the absolute configurations of the
diols obtained from these olefins can be assigned from the mnemonic, and from
there it is a small step to the use of Figure 2 in the prediction of enantioselectivities
for new dihydroxylations. However, this is a step which must be taken with caution.
Unlike the case of asymmetric epoxidation (Chapter 4.1), where a similarly styled
mnemonic can be safely used to predict absolute configurations of epoxides, the
mnemonic for AD is only suggestive of new diol configurations. The reasons for
this lie in the much greater structural diversity (all olefins vs. only allylic alcohols)
available for AD as well as in the nature of the interactions between substrate and
catalyst (allylic alcohols coordinate as the alkoxides to the metal of the asymmetric

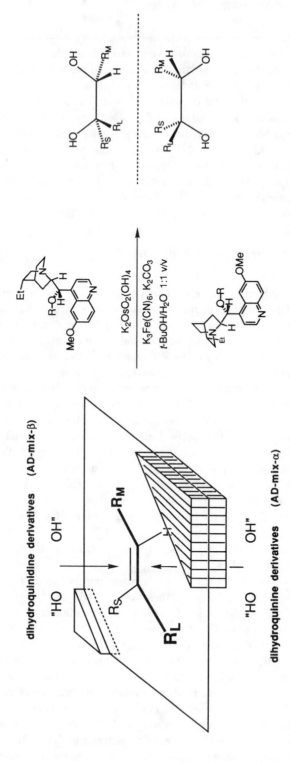

Figure 2 Enantioselectivity mnemonic scheme.

244

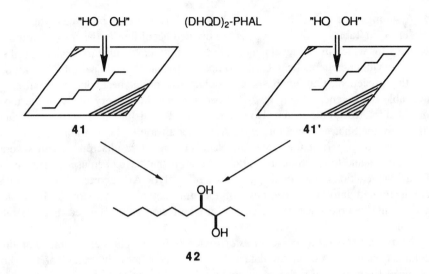

epoxidation catalyst). For some classes of olefins, such as the trans-disubstituted olefins, Figure 2 should correctly predict the enantioselectivity of dihydroxylation for all prochiral substrates for reasons such as those given in the preceding discussion of **41**. The prediction of enantioselectivities in the dihydroxylation of monosubstituted olefins should also be quite reliable for similar reasons. In other cases, particularly the 1,1-disubstituted olefins, the difficulty in applying the mnemonic when the two substituents are very similar in nature quickly becomes apparent. Figure 2, therefore, should be regarded as a useful tool in the analysis of AD reactions.

4.4.3. Catalytic Asymmetric Dihydroxylations by Olefin Substitution Pattern

The challenges facing the development of any catalytic process for the asymmetric functionalization of olefins are large when one considers, first, the number and variety of olefins that are potential substrates for the reaction and, second, a useful asymmetric synthesis should yield an enantioselectivity of at least 90% ee, with the ultimate goal being 100% ee. When the catalytic version of AD was announced in 1988, the hydroxylation of only one olefin, E-stilbene, approached this standard for enantioselectivity. At the start of 1992, examples from four of the six substitution classes of olefins can be hydroxylated with 94% ee or higher, and the number of examples meeting this standard is limited primarily by the number of olefins that have been subjected to the catalytic process. Nevertheless, it is clear that this catalytic system will not be effective for every olefin even within these four substitution classes. It is important therefore to examine carefully the structural nature of the olefin when considering application of AD. Examination of an olefin with the mnemonic (Fig. 2) in mind should be of help in this regard.

A summary of results from experiments exploring the scope of catalytic AD is presented in Tables 2–6. The results are divided according to the five substitution patterns for which useful levels of enantioselectivity have thus far been achieved so that the synthetic chemist can easily compare a potential candidate for AD with the existing precedent. Our emphasis in this chapter is on the best conditions currently available for a general catalytic AD process, since this is very likely of the most practical interest to the reader. However, Tables 2–6 include most of the published data from the Sharpless laboratory for AD under a variety of conditions. These data are included to permit the reader to place in context results reported in several publications and to illustrate the variety of olefins that have been used at least once for an AD reaction. Additional applications of catalytic AD reported in the literature are integrated into the discussions of the different olefin categories. Examples illustrating "double diastereoselectivity" are collected and discussed in a separate section.

Please note the following features of Tables 2–5. The tables are arranged in nine columns each; the numerical data refer to the enantiomeric excess of the diol obtained from the olefin shown in column 2. Yields are generally in the range of 75–95%. The absolute configurations of the diols are known and can be found in the references cited or deduced with the aid of the enantioselectivity mnemonic of Figure 2. References are found in the footnotes to the tables. The absence of entries for an olefin in a column means simply that the experiment has not been performed. Several olefins have a nearly complete series of entries across the table (see Table 2, entry 6; Table 3, entries 16 and 17; and Table 5, entry 3), which serve to illustrate the degree of improvement attained as beneficial modifications of the reaction were discovered. Note that numbers *without parentheses* refer to experiments using DHQD ligands and numbers *within parentheses* are for experiments using DHQ ligands. *Columns 1 and 2* list entry number and show olefin structure, respectively. *Column 3* lists the enantiomeric purity (% ee) obtained under *stoichiometric* AD conditions using *p*-chlorobenzoate (CLB) derivatives of the alkaloid ligands. *Column 4* lists results obtained under initial catalytic AD conditions with NMO as cooxidant. *Column 5* lists results obtained following discovery of the secondary cycle shown in Scheme 2 and introduction of the method of slow addition of olefin to suppress this cycle. Results in column 5 are divided between slow addition without/with acetate added to the reaction. *Column 6* lists results obtained using potassium ferricyanide as cooxidant and data are divided between reactions run at 0°C/25°C. *Column 7* lists results obtained using the 4'-methyl-2'-quinolyl ether (MEQ) derivative of DHQD as the chiral ligand. *Column 8* lists results obtained using the 9'-phenanthryl ether (PHN) derivative of DHQD as the chiral ligand. *Column 9* lists results obtained with the bis-DHQD and bis-DHQ ethers of phthalazine-1,4-diol (PHAL) as the chiral ligands.

4.4.3.1. Monosubstituted Olefins

The enantiomeric excesses obtained to this point for the catalytic AD of monosubstituted olefins (see Table 2 [16,26,29,31,40,44–46,49]) are lower than for *trans-*

Table 2 Asymmetric Dihydroxylation of Monosubstituted Olefins (% ee)

1 Entry	2 Olefin	3[a] CLB stoich.; 0°C	4[b] CLB cat.; NMO 0°C/25°C	5[c] CLB cat.; NMO slo/OAc⁻	6[d] CLB cat.; Fe 0°C/25°C	7[e] MEQ cat.; Fe 0°C	8[f] PHN cat.; Fe 0°C	9[g] PHAL cat.; Fe 0°C
1	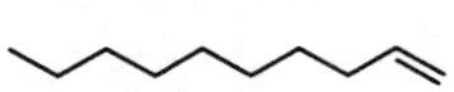				–/45	65	74	84(80)
2	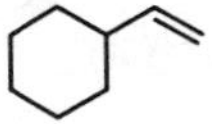		46/--				80	
3	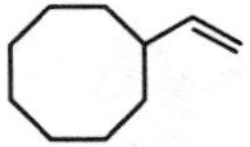				–/64	85	93	
4	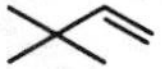				–/44	79	79	
5	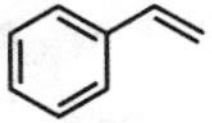	61	62(60)/--	60/--	73/74	87	78	97(97)
6	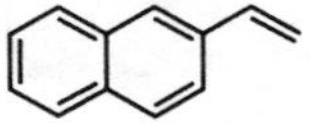				–/88	93	83	
7	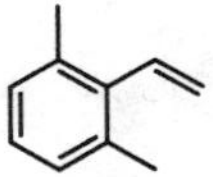		65/--					88(77)
8	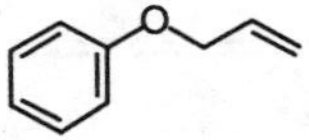					66		
9	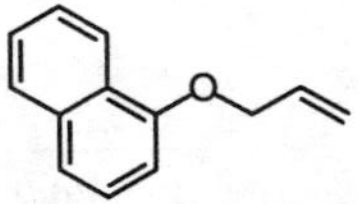			40/--	–/60	44		91(88)

(Continued)

247

Table 2 (*Continued*)

Entry	Olefin	3[a] CLB stoich.; 0°C	4[b] CLB cat.; NMO 0°C/25°C	5[c] CLB cat.; NMO slo/OAc⁻	6[d] CLB cat.; Fe 0°C/25°C	7[e] MEQ cat.; Fe 0°C	8[f] PHN cat.; Fe 0°C	9[g] PHAL cat.; Fe 0°C
10	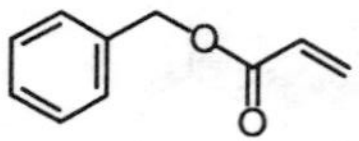						50(49)	77(70)
11	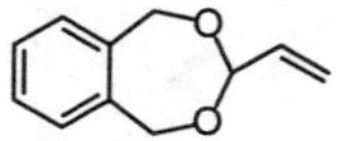						84	67
12	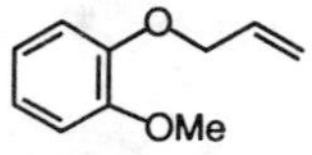							
13							38	63(56)
14							44	54
15							53	72
	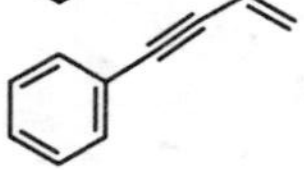							73

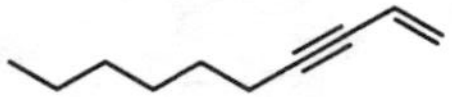

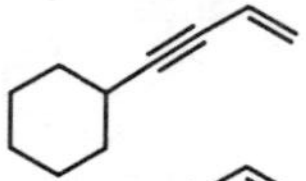

[a]Reference for column 3: entry 5 [40]. [b]Ref's for column 4: entries 2,5,7 [26]. [c]Ref's for column 5: entry 5 [16], entry 9 [49]. [d]Ref's for column 6: entries 1,3,4,6 [31]; entry 5 [31,35]; entry 9 [49]. [e]Ref's for column 7: entries 1, 3-6 [31]; entries 8,9 [44]. [f]Ref's for column 8: entries 1,3,4,5,6 [31]; entries 2,10 [44]; entry 11 [45]; entries 13-15 [46]. [g]Ref's for column 9: entries 1,5,10 [29]; entry 4,8,9,11,12 [44]; entries 13-15 [46].

248

Table 6 ... Asymmetric Dihydroxylation of C=C Substrates ... cont. (continued)

Entry	Olefin	CLB stoich.; 0°C	CLB cat.; NMO 0°C	CLB cat.; NMO slo/OAc⁻	CLB cat.; Fe 0°C/25°C	MEQ cat.; Fe 0°C	PHN cat.; Fe 0°C	PHAL cat.; Fe 0°C
		3	4	5	6	7	8	9
1	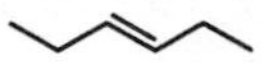	69-71	20	70/--				
2		71-73	20	--/69	74-79/--	90	95	97(93)
3	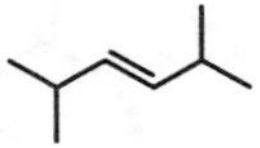	79-80	12	46/76				
4	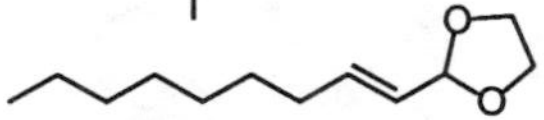	83		56/66				
5	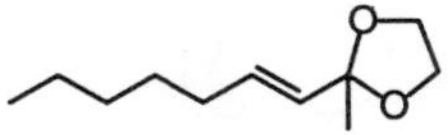	50		46/50				
6	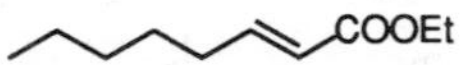 COOEt	66-67				85	94	99(96)
7	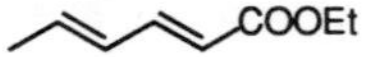 COOEt							92
8	COOEt							93
9	COO-i-Pr			76de/--				
10	COOMe	77						
11	COOMe	54		52de/--				97de

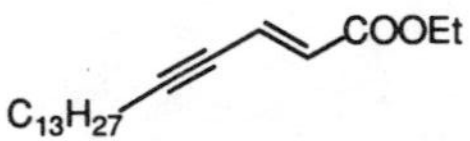

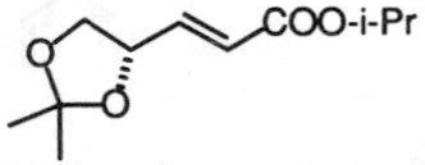

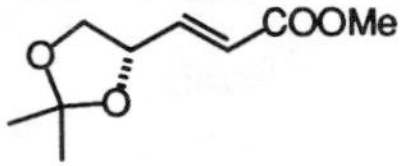

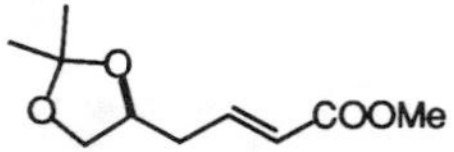

(Continued)

Table 3 (*Continued*)

Entry	Olefin	CLB stoich.; 0°C [a]	CLB cat.; NMO 0°C [b]	CLB cat.; NMO slo/OAc⁻ [c]	CLB cat.; Fe 0°C/25°C [d]	MEQ cat.; Fe 0°C [e]	PHN cat.; Fe 0°C [f]	PHAL cat.; Fe 0°C [g]
		3	4	5	6	7	8	9
12	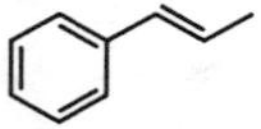 (styrene, Me)	87	66(55)	86/--	--/91			(97)
13	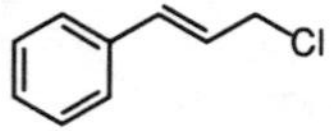 (cinnamyl Cl)	79		78/--				
14	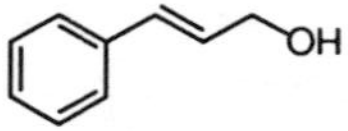 (cinnamyl OH)	66		66/--				
15	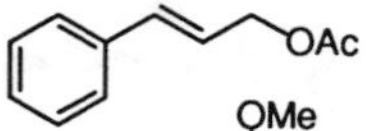 (cinnamyl OAc)	82	76	79/---				
16	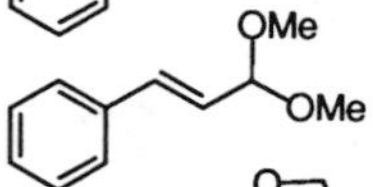 (OMe OMe)	82		75/83				
17	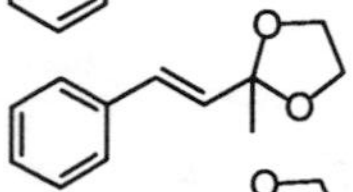 (dioxolane Me)	85		60/89				
18	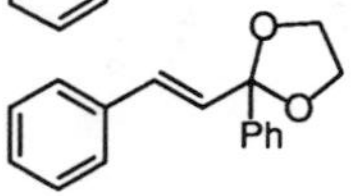 (dioxolane Ph)	86		84/87				
19	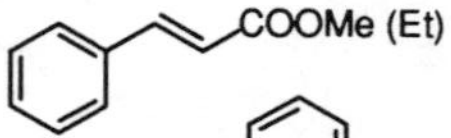 (COOMe (Et))	89	60	86-88/--	91/95	98	98	97(95)
20	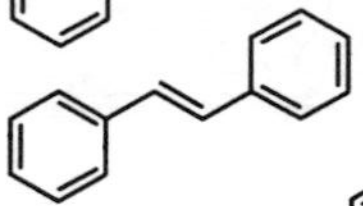 (stilbene)	99	88(75)	95/--	--/99	98	99	>99.5(>99.5)
21	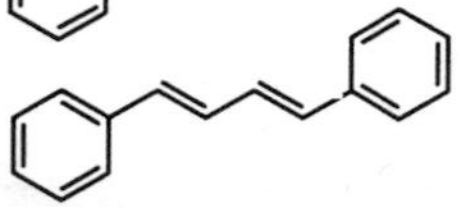 (diphenylbutadiene)							97

>99(>99)

93
94
73
90
97

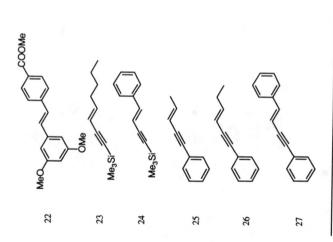

22

23

24

25

26

27

[a]References for column 3: entries 1-3,6 [40,41]; entries 4,5,10-20 [40]. [b]Ref's for column 4: entries 1,12,15, [26]; entries 2, 19 [35]; entry 3 [16]; entry 20 [47]. [c]Ref's for column 5: entries 1-5,10-19 [40]; entry 20 [35,40]. [d]Ref's for column 6: entries 2,19 [31,35]; entries 12,15,20 [35]. [e]Ref's for column 7: entries 2,6,19,20 [31]. [f] Ref's for column 8: entries 2,6,19,20 [31]; entries 23-27 [46]. [g]Ref's for column 9: entries 2,6-9,19-21 [29]; entry 22 [48].

251

Table 4 Asymmetric Dihydroxylation of 1,1-Disubstituted Olefins (% ee)

1	2	3	4[a]	5	6[b]	7[c]	8[d]	9[e]
Entry	Olefin	CLB stoich.; 0°C	CLB cat.; NMO 0°C	CLB cat.; NMO slo/OAc⁻	CLB cat.; Fe 0°C/25°C	MEQ cat.; Fe 0°C	PHN cat.; Fe 0°C	PHAL cat.; Fe 0°C
1	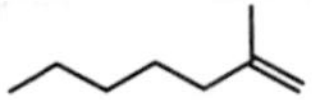							78(76)
2	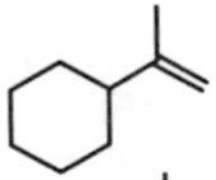		33		–/37	73	82	94(93)
3								
4	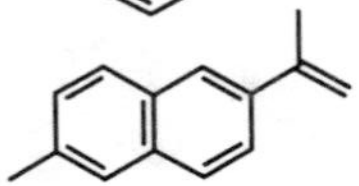				–/74	88	69	
5	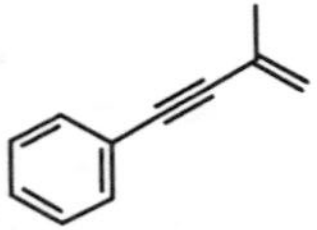						48	79

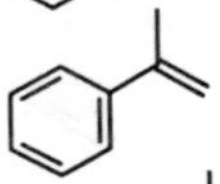

[a]Reference for column 4: entry 3 [26]. [b]Ref. for column 6: entries 2,4 [31]. Ref. for column 7: entries 2,4 [31]. [c]Ref's for column 8: entries 2,4 [31]; entry 5 [46]. Refs for column 5 [46]. [d]Ref for column 9: entries 1,3 [29]; entry 5 [46]. [e]entries 2,4 [31].

Table 5 Asymmetric Dihydroxylation of Trisubstituted Olefins (% ee)

1	2	3[a]	4	5[b]	6[c]	7[d]	8[e]	9[f]
Entry	Olefin	CLB stoich.; 0°C	CLB cat.; NMO 0°C	CLB cat.; NMO slo/OAc⁻	CLB cat.; Fe 0°C/25°C	MEQ cat.; Fe 0°C	PHN cat.; Fe 0°C	PHAL cat.; Fe 0°C
1	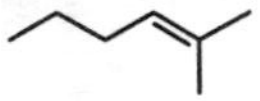	55		--/54				
2	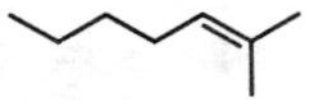							98(95)
3	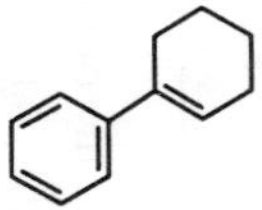	79		78/81	91/--	92	93	99(97)
4	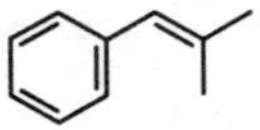	55		53/--	74/--	81	84	
5	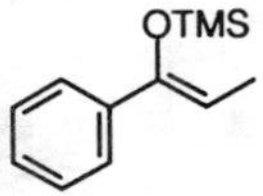	85		72/78				
6	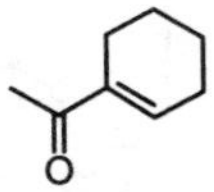	56		--/54				

[a]Reference for column 3: entries 1,3-6 [40]. [b] Ref. for column 5: entries 1,3-6 [40]. [c]Ref. for column 6: entries 3,4 [31]. [d]Ref. for column 7: entries 3,4 [31]. [e]Ref. for column 8: entries 3,4 [31]. [f]Ref. for column 9: entries 2,3 [29].

Table 6 Asymmetric Dihydroxylation of *cis*-
Disubstituted Olefins

entry	olefin	% ee, rt	% ee, 0°C
1	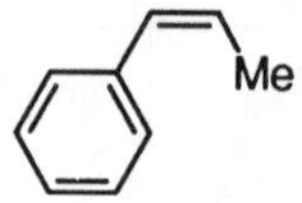 Me	65	72
2	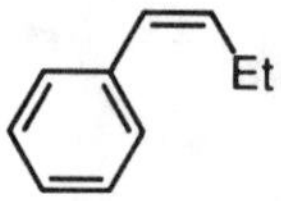 Et	67	
3	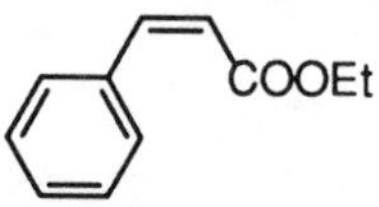 COOEt	75	78
4	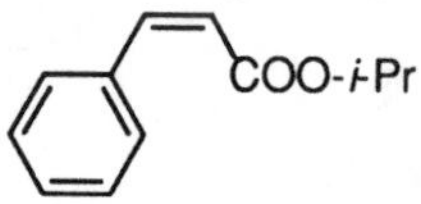 COO-*i*-Pr	74	80
5	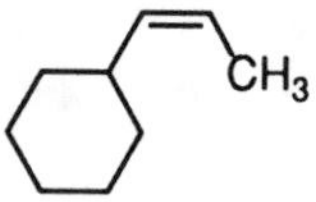 CH$_3$	49	
6	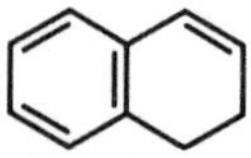	14	

disubstituted olefins (Table 3). The entries in column 9 show enantiomeric purities ranging from 54 ee to 97% ee for dihydroxylations with the (DHQD)$_2$-PHAL and (DHQ)$_2$-PHAL pair of chiral ligands. Several monosubstituted olefins having branching at the α-position (e.g., entries 2–4 and 11) are dihydroxylated with higher enantioselectivities when DHQD-PHN is used as the chiral ligand instead of (DHQD)$_2$-PHAL. Recently, a new ligand for terminal olefins has been discovered [48b].

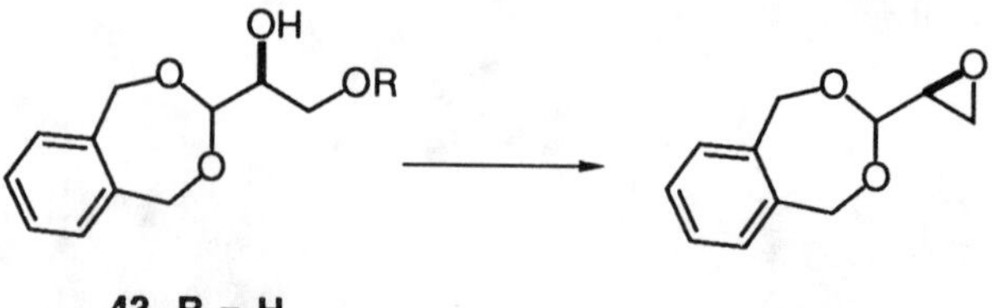

43, R = H
44, R = Tos

45

The diol (**43**) obtained from dihydroxylation of acrolein benzene-1,2-dimethanol acetal (entry 11) is a masked glyceraldehyde and has the potential to be a very useful synthon. Although the enantiomeric purity of the crude diol formed in this reaction is 84% ee, one recrystallization from ethyl acetate improves it to 97% ee in 55% recovery yield. The masked glyceraldehyde **43** is converted via the tosylate **44** to the masked glycidaldehyde **45** in an overall yield of 85%. Both these masked aldehydes are superior to the free aldehydes in terms of handling ease, stability, and safety. The aldehydes can be released from the acetal under the mild conditions of catalytic hydrogenolysis [45].

Diols obtained from the catalytic AD of aryloxyallyl ethers such as those of entries 8, 9, and 12 are useful as precursors for the synthesis of the pharmacologically important β-blockers [49].

4.4.3.2. trans-Disubstituted Olefins

Olefins of the trans-disubstituted type have given diols with excellent enantiomeric purities when dihydroxylated using the $(DHQD)_2$-PHAL/(DHQ-PHAL pair of chiral ligands with osmium tetroxide (see Table 3 [16,26,29,31,35,40,41,46–48]). All the entries but one in column 9 for diols obtained with these ligands exceed 90% ee (or 90% de). Judging from other entries in Table 3, particularly those of column 3 for earlier stoichiometric ADs with the DHQD-CLB ligand, good enantioselectivities are anticipated for the dihydroxylation of most trans-disubstituted olefins when the PHAL ligands are used.

Included in this class of olefins is (*E*)-stilbene (entry 20), which throughout studies of AD has usually been the olefin dihydroxylated with the highest degree of enantioselectivity. Availability of (*R,R*) or (*S,S*)-1,2-diphenyl-1,2-ethanediol (also referred to as stilbenediol or dihydrobenzoin) with high enantiomeric purities has led to reports of a number of applications, including incorporation into chiral dioxaphospholanes [50], chiral boronates [51], chiral ketene acetals [52], chiral crown ethers [53], and conversion into 1,2-diphenylethane-1,2-diamines [54]. Dihydroxylation of the substituted *trans*-stilbene **46** with OsO_4/NMO and DHQD-CLB gives the *R,R*-diol **47** with 82% ee in 88% yield [55].

46   **47**

Details for the large scale synthesis of (*R,R*)-1,2-diphenyl-1,2-ethanediol using the DHQD-CLB/NMO variation of catalytic AD have been published [47]. Under these conditions the crude diol is produced with 90% ee and upon crystallization, essentially enantiomerically pure diol is obtained in 75% yield. Subsequent improvements in the catalytic AD process now allow this dihydroxylation to be achieved with $> 99.8\%$ ee (entry 20, column 9); however, the *Organic Synthesis* procedure [47] is still an excellent choice for preparing large amounts of the enantiomerically pure stilbene diols since it is fast and needs little solvent (the reaction is run at 2 molar in stilbene).

The asymmetric dihydroxylation of dienes has been examined, originally with the use of *N*-methylmorpholine oxide (NMO) as the cooxidant for osmium [56a] and, more recently, with potassium ferricyanide as the cooxidant [56b]. Tetraols are the main product of the reaction when NMO is used, but with $K_3Fe(CN)_6$, ene-diols are produced with excellent regioselectivity. The example of dihydroxylation of *trans,trans*-1,4-diphenyl-1,3-butadiene is included in Table 3 (entry 21). One double bond of this diene is hydroxylated in 84% yield with 99% ee when the amounts of $K_3Fe(CN)_6$ and K_2CO_3 are limited to 1.5 equiv. each. Unsymmetrical dienes are also dihydroxylated with excellent regioselectivity. In these dienes, preference is shown for (a) a trans over a cis olefin, (b) the terminal olefin in $\alpha,\beta,\gamma,\delta$-unsaturated esters, and (c) the more highly substituted olefin [56b].

A class of *trans*-disubstituted olefin that is encountered frequently in organic chemistry is the α,β-unsaturated ester. Catalytic AD of several examples may be found in Table 3 (entries 6–11, 19) and in these cases, diols with high enantiomeric purities are produced. The olefin of entry 7 is an interesting variation of this class, being an $\alpha,\beta,\gamma,\delta$-bis-unsaturated ester. Catalytic AD of this diene under controlled conditions (so that only enough oxidant is present to allow dihydroxylation of one double bond) gives the γ,δ-dihydroxy-α,β-unsaturated ester with 88% ee (using $(DHQ)_2$-PHAL) in 76% yield. Dihydroxylation of methyl *p*-methoxycinnamate has been reported, and the $2R,3S$-diol was taken on in a synthesis of (+)-diltiazem, a vasodilating agent [28]. The catalytic AD of other α,β-unsaturated esters found in chiral molecules is described in Section 4.4.3.7.

4.4.3.3. 1,1-Disubstituted Olefins

A limited number of 1,1-disubstituted olefins have been subjected to catalytic AD and these are listed in Table 4 [26,29,31,46]. The results shown there indicate that selected members of this class of olefins can be dihydroxylated with high enantioselectivity but it should be recognized that these olefins all have relatively dissimilar substituents. As the two substituents becomes more alike, the discriminatory capability of the catalyst is expected to lessen and lower enantioselectivities to be observed. Such a trend may be found in the comparison of entries 1 and 5 with entry 3. The difference in size of substituents clearly is greater in the case of entry 3 and the diol derived from this olefin has the highest enantiomeric purity.

4.4.3.4. Trisubstituted Olefins

A listing of trisubstituted olefins that have been subjected to catalytic AD is given in Table 5 [29,31,40]. Only three compounds of this type have been dihydroxylated using the newer, more efficient chiral ligands, and all three are converted to diols with high enantiomeric purities under these conditions.

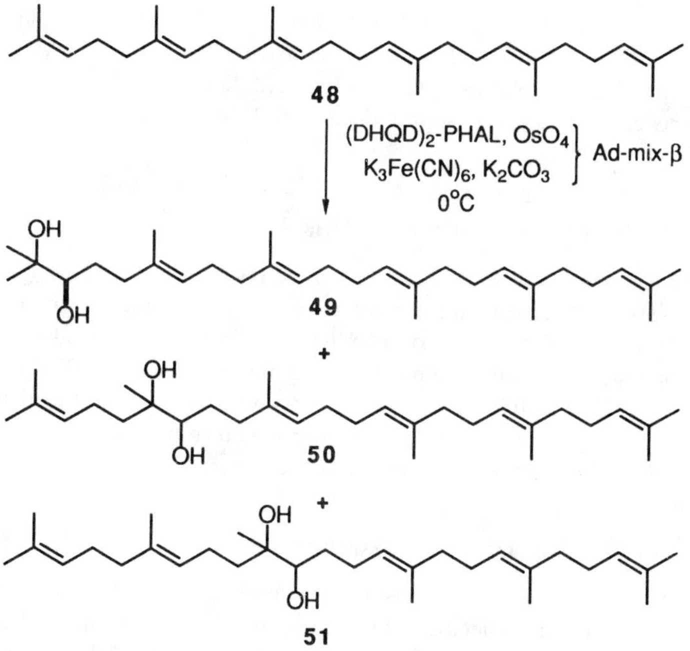

Squalene (**48**) presents an interesting array of trisubstituted double bonds with which to test the selectivity of catalytic AD. Osmium tetroxide catalyzed dihydroxylation of squalene in the *absence* of a chiral ligand generates a mixture of the 2,3-diol (**49**), 6,7-diol (**50**), and 10,11-diol (**51**) in a ratio of 1:1:1. (Squalene is used in excess in these experiments to minimize multiple-dihydroxylations of the squalene molecule.) Dihydroxylation with (DHQD)$_2$-PHAL as the chiral ligand produces the diols **49**, **50**, and **51** in a ratio of 46:35:19. Stereochemical analysis of the 2,3-diol (**49**) indicates formation of the 2,3R-diol with 96% ee [57a]. Perhydroxylation of squalene has also been achieved with 98% ee or de for each of the six dihydroxylation events required in the process [57b].

4.4.3.5. cis-Disubstituted Olefins

cis-Disubstituted olefins have proven to be the most difficult class of substrates from which to obtain diols with high enantiomeric purities. With any of the previously discussed chiral ligands, enantioselectivity in the AD of these olefins has been low. Using (Z)-1-phenylpropene (*cis*-β-methylstyrene) as an example, AD with DHQD-CLB gave 1-phenylpropan-1,2-diol with 35% ee, DHQD-MEQ gave 26% ee, DHQD-PHN gave 22% ee, and (DHQD)$_2$-PHAL gave 29% ee. Continued searching for better chiral ligands for this class of olefins has recently led to the discovery of a new pair of ligands that imparts significantly higher enantiomeric purity to the diols [57c]. The new ligands are 9-*O*-indolinylcarbamoyldihydroquinidine (**52**, Chart 1) and the pseudoenantiomeric dihydroquinine derivative (**53**) (DHQD-IND and DHQ-

IND, respectively). Although work with these new ligands is still in progress, current results for AD using DHQD-IND are listed in Table 6. *cis*-β-Methylstyrene now is converted to (1*R*,2*S*)-1-phenylpropan-1,2-diol with 72% ee (entry 1) and *i*-propyl *cis*-cinnamate gives isopropyl-(2*R*,3*R*)-2,3-dihydroxy-3-phenylpropionate with 80% ee (entry 4).

4.4.3.6. Tetrasubstituted Olefins

Using the AD recipe given in Section 4.4.2, tetrasubstituted silyl enol ethers give good yields of α-hydroxyketones after workup with enantioselectivities ranging from 60–97% ee [57d]. The AD of other tetrasubstituted olefins is also possible by running the reaction at room temperature with a more powerful AD recipe consisting of 1 mol % $K_2OsO_2(OH)_4$, 5 mol % ligand, and 3 equivalents of methanesulfonamide [57e]. The use of this more active recipe is recommended whenever a very unreactive (in the turnover sense) olefin is encountered.

4.4.3.7. Double Diastereoselectivity

A diastereoselective osmylation must have occurred in the first reaction of osmium tetroxide with a chiral molecule, which may have been of pinene by Hofmann in 1912 [7]. However, at that time even the cis nature of the dihydroxylation was unknown and was not determined until the careful work of Criegee in the 1930s. By the late 1930s, diastereoselectivity was clearly recognized in the osmium tetroxide catalyzed dihydroxylations of natural products and, for example, played an important role in structural assignments and synthetic elaborations of the corticosteroid side chain. A discussion of these diastereoselective reactions and references to the original literature can be found in *Steroids,* the classical account by Fieser and Fieser [58] of the early decades of steroid chemistry. Increased efforts at exercising stereocontrol in synthesis, particularly in the last 10–15 years, have led to frequent observations of diastereoselectivity in olefin osmylation and to efforts at developing models for use in predictions of this selectivity.

With the discovery of asymmetric dihydroxylation (AD) and subsequent catalytic modifications came the possibility for "matching and mismatching" [25] diastereoselectivity with the enantiofacial selectivity of the asymmetric process. Knowledge of the diastereoselectivity of a given olefin hydroxylation is needed to match substrate with the correct choice of chiral catalyst. The best way to determine the diastereoselectivity for dihydroxylation of a chiral olefin is to perform the osmylation in the absence of a chiral ligand and then determine the ratio (de) of the resulting diols. A much less rigorous alternative is the inspection of molecular models, which may suggest that there is a diastereoselective preference for the dihydroxylation of a particular chiral olefin. Such predictions of diastereoselectivity may be reliable for some rigid cyclic molecules but will be difficult for olefins found in acyclic environments. The most frequently studied acyclic olefins are those in which the double bond is attached to a chiral carbon carrying a heteroatom substituent. Several analyses of the diastereoselectivity of osmylation of these allylic systems, including the closely related γ-substituted α,β-unsaturated esters, have been put forward [59].

From consideration of the rationale behind these results, it appears that as a first approximation, the osmylation of such olefins will occur from the sterically more accessible face rather than in response to the electronic nature of the allylic system.

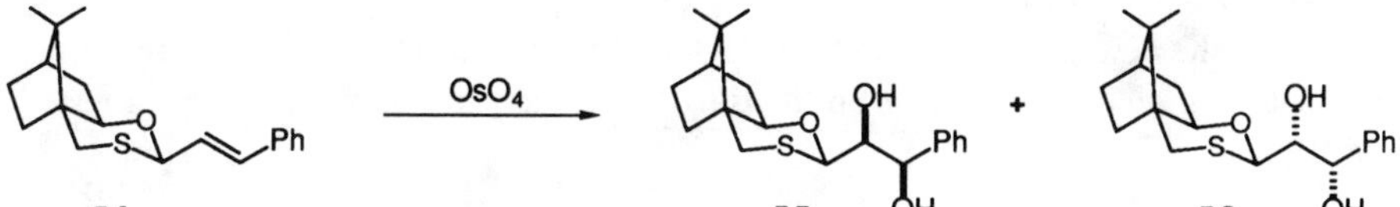

quinuclidine	2.5 : 1	
DHQD-OAc	40 : 1	
DHQ-OAc	1 : 16	

The matching and mismatching of chiral olefin **54** and catalyst was examined briefly using *stoichiometric* quantities of osmium tetroxide with achiral and chiral ligands [60]. The monothio acetal derived from camphor (**54**) was dihydroxylated with osmium tetroxide in the presence of quinuclidine, DHQD-OAc, or DHQ-OAc. With the achiral quinuclidine as ligand, the ratio of (2S,3R) to (2R,3S) diasteriomers **55** and **56** was 2.5:1. With DHQD-OAc as the chiral ligand, catalyst and substrate are matched and the ratio is enhanced to 40:1 while with DHQ-OAc catalyst and substrate are mismatched and a reversed selectivity of 1:16 is observed.

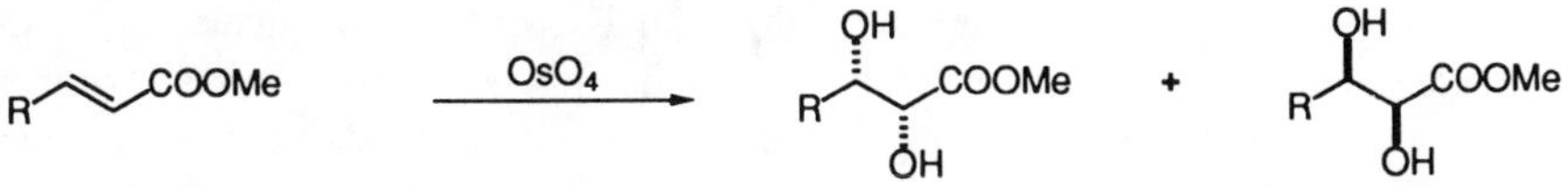

R	Reagent	Ratio 58 : 59
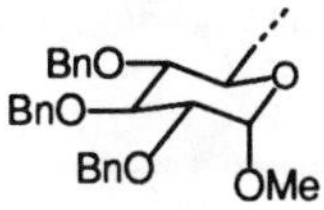 **57a**	OsO$_4$ only	10.3 : 1
	DHQD-CLB	1.3 : 1
	DHQ-CLB	20.5 : 1
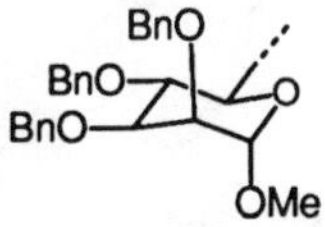 **57b**	OsO$_4$ only	7.4 : 1
	DHQD-CLB	3.4 : 1
	DHQ-CLB	15.9 : 1
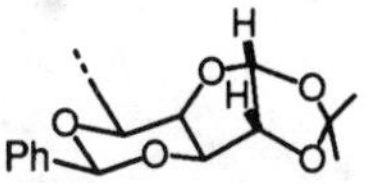 **57c**	OsO$_4$ only	1 : 2.2
	DHQD-CLB	1 : 5.3
	DHQ-CLB	1 : 1.6

A series of α,β-unsaturated esters (**57a–c**) has been matched and mismatched with catalyst in a study of the catalytic AD of octuronic acid derivatives [61]. In this study, hydroxylations with osmium tetroxide alone were compared with hydroxylations with DHQD-CLB and DHQ-CLB as the chiral ligands. Olefins **57a** and **57b** are matched with DHQ-CLB and olefin **57c** is matched with DHQD-CLB. In all three cases, matching of olefin diastereoselectivity with catalyst enantioselectivity enhances the overall selectivity of the dihydroxylation. The authors of this work observed greater enhancement of selectivity with stoichiometric reagents in comparison with catalytic amounts of reagents, a discrepancy that may be reduced with application of the ferricyanide procedure for the catalytic process, if indeed it does not disappear.

Using the newest chiral ligands, (DHQD)$_2$-PHAL and (DHQ)$_2$-PHAL, the effect of matching and mismatching on the dihydroxylation of olefin **60** has been tested [62].

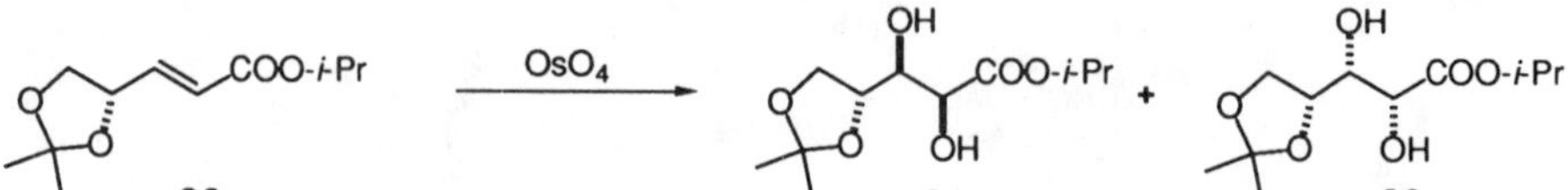

Reagent	Ratio 61 : 62
no ligand	2.8 : 1
(DHQD)$_2$-PHAL (Ad-mix-β)	39 : 1
(DHQ)$_2$-PHAL (Ad-mix-α)	1 : 1.3

The ratio of diols **61:62** from dihydroxylation with osmium tetroxide alone is 2.8:1, while with (DHQD)$_2$-PHAL the ratio is 39:1, and with (DHQ)$_2$-PHAL the ratio is 1:1.3, showing the cumulative effect that can be achieved by matching of diastereoselectivities and also revealing that AD may not be as reliable as is asymmetric epoxidation (Chapter 4.1) for attaining good results when diastereoselectivities are mismatched.

Other comparisons of diastereoselective osmylations in the absence and in the presence of chiral ligands have been reported. A study of the dihydroxylation of unsaturated side chains attached to various steroid nuclei allows some comparisons of matching and mismatching of diastereoselectivities [63].

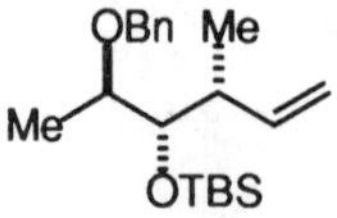

63

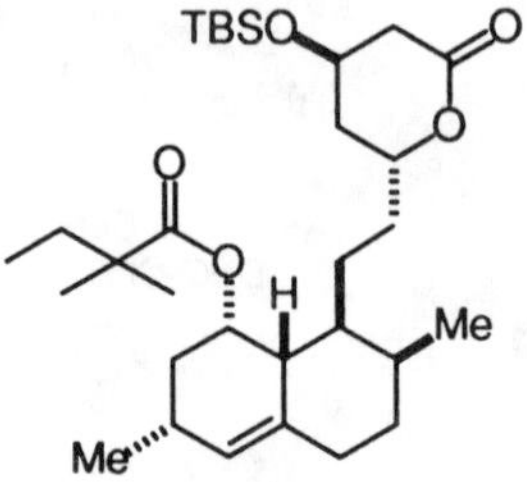

64

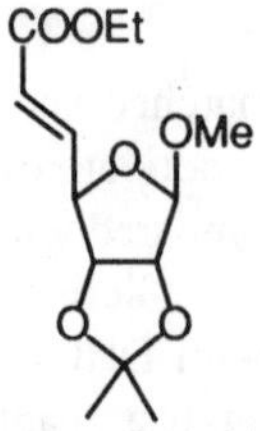

65

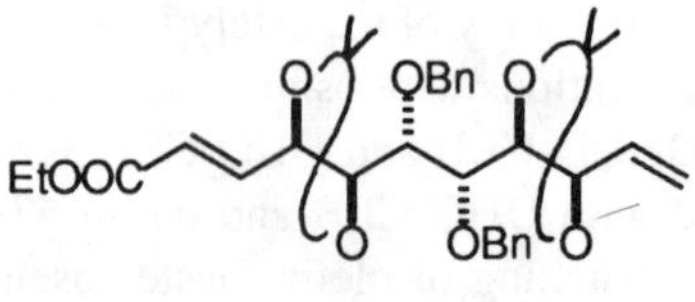

66

Dihydroxylation of the double bond in the synthetic intermediate **63** gives a 3.5:1 ratio of diastereoisomers in the absence of a chiral catalyst and a 5:1 ratio when matched with DHQ-CLB [64]. Synthetic intermediate **64** is dihydroxylated with a 12:1 ratio of diastereoselectivity without a chiral catalyst and, under otherwise identical conditions, in a 20:1 ratio with DHQ-CLB [65]. It should be noted that dihydroxylation of **64**, the subject of extensive process development, is very sensitive to solvent composition, and an improved diastereoselective ratio (39:1) was obtained by altering this variable. Dihydroxylation of **65** is an example of the stereochemistry required for a synthetic goal necessitating the use of a mismatched system. The ratio of diastereomeric products in the absence of chiral ligand was not given but was stated to be "reversed" from the 2.5:1 ratio observed with DHQD-CLB [66]. Finally, both olefins in the synthetic intermediate **66** were dihydroxylated at the same time using DHQ-CLB; the diastereofacial selectivity at each olefin was determined to be 12.9:1 [67]. The use of the newer chiral ligands, which are capable of inducing higher levels of enantioselectivity, is expected to give improved results in future applications of AD for achieving double diastereoselectivity.

4.4.4. Diol Activation

This section outlines three chemical transformations designed to allow further synthetic elaboration of the diols obtained from AD. The first and most broadly applicable method is the conversion of the diols into cyclic sulfates, a functionality that has reactive properties like an epoxide but is even more electrophilic than an epoxide [68]. The second approach to diol activation is the regioselective conversion of one of the hydroxyl groups into a sulfonate ester [69]. This approach requires that the diol be substituted in a way that leads to regioselective derivatization of one of the two hydroxyl groups, diol esters being a prime example of such an arrangement. The third very convenient approach currently under development is the conversion of diols into acetoxy halides from which epoxides are easily obtained.

4.4.4.1. Cyclic Sulfates, Sulfites, and Sulfamidates

The use of cyclic sulfates in synthetic applications has been limited in the past because, although cyclic sulfites are easily prepared from diols, a convenient method for oxidation of the cyclic sulfites to cyclic sulfates had not been developed. The experiments of Denmark [70] and of Lowe and coworkers [71] with stoichiometric ruthenium tetroxide oxidations and of Brandes and Katzenellenbogen [72a] and Gao and Sharpless [68] with catalytic ruthenium tetroxide and sodium periodate as cooxidant have led to an efficient method for this oxidation step. Examples of the conversion of several diols (**67**) to cyclic sulfites (**68**) followed by oxidation to cyclic sulfates (**69**) are listed in Table 7. The cyclic sulfite/cyclic sulfate sequence has been applied to 1,2-, 1,3-, and 1,4-diols with equal success. Cyclic sulfates, like epoxides, are excellent electrophiles and, as a consequence of their stereoelectronic makeup, are less susceptible to the elimination reactions that usually accompany

Table 7 Cyclic Sulfates (**69**) from Diols (**67**)

Entry	R_1	R_2	Yield (%)
1	COO-i-Pr	COO-i-Pr	90
2	COOEt	COOEt	69
3	COOMe	COOMe	63
4	n-C$_8$H$_{17}$	H	92
5	c-C$_6$H$_{11}$	H	97
6	n-C$_4$H$_9$	n-C$_4$H$_9$	89
7	n-C$_{15}$H$_{31}$	COOMe	90
8	c-C$_6$H$_{11}$	COOEt	95
9	H	COO-c-C$_6$H$_{11}$	88
10	H	CONHCH$_2$Ph	64
11	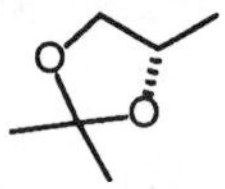	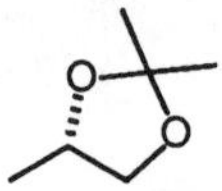	94
12	H	CH$_2$OSiMe$_2$-t-Bu	87
13	COOMe	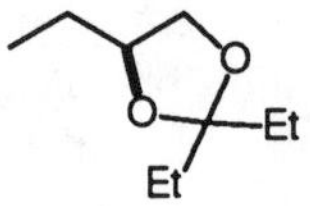	95

attack by nucleophiles at a secondary carbon. With the development of convenient methods for their syntheses, the reactions of cyclic sulfates have been explored. Most of the reactions have been nucleophilic displacements with opening of the cyclic sulfate ring. The variety of nucleophiles used in this way is already extensive and includes H$^-$ [68], N$_3^-$ [68,73–76], F$^-$ [68,72,74], PhCOO$^-$ [68,73,74], NO$_3^-$ [68], SCN$^-$ [68], PhCH$_2^-$ [68], (ROOC)$_2$CH$^-$ [68], RNH$_2$ [76], (RS)$_2$CH$^-$ and (RS)$_3$C$^-$ [77], RC≡C$^-$ [78], PhC(NH$_2$)=NH [79], and RPH$_2$ [80], a list that is sure to grow. Upon opening of the cyclic sulfate (**69**), one obtains a sulfate ester (**70**), itself a versatile functional group. Hydrolysis of the sulfate ester (**70**) with aqueous acid regenerates the hydroxyl group, giving **71**; conditions suitable for both the generation of the cyclic sulfate and the hydrolysis of the sulfate ester in the presence of other acid sensitive groups have been developed [73].

Alternatively, the remaining sulfate ester of **70** may serve as a leaving group for a second nucleophilic displacement reaction. When this displacement is by an intramolecular nucleophile, a new ring is formed, as was first shown in the synthesis of a cyclopropane using malonate as the nucleophile [68] and of aziridines using amines as the nucleophiles [76]. The concept is further illustrated in the double displacement on (*R,R*)-stilbenediol cyclic sulfate (**72**) by benzamidine (**73**) to produce the chiral imidazoline **74** [79]. Conversion of the imidazoline (**74**) to (*S,S*)-

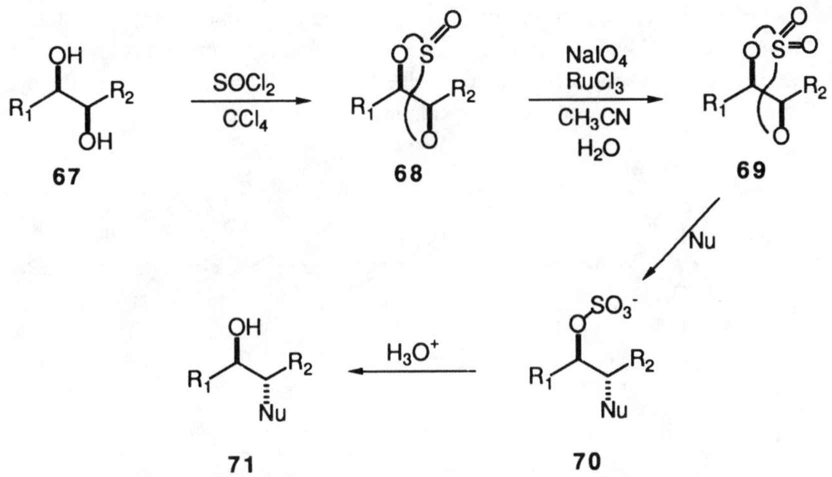

stilbenediamine **75** demonstrates an alternative route to optically active 1,2-diamines. Acylation of **75** with chloroacetyl chloride forms a bisamide, which after reduction with diborane is cyclized to the enantiomerically pure *trans*-2,3-diphenyl-1,4-diazabicyclo[2.2.2]octane (**76**) [81].

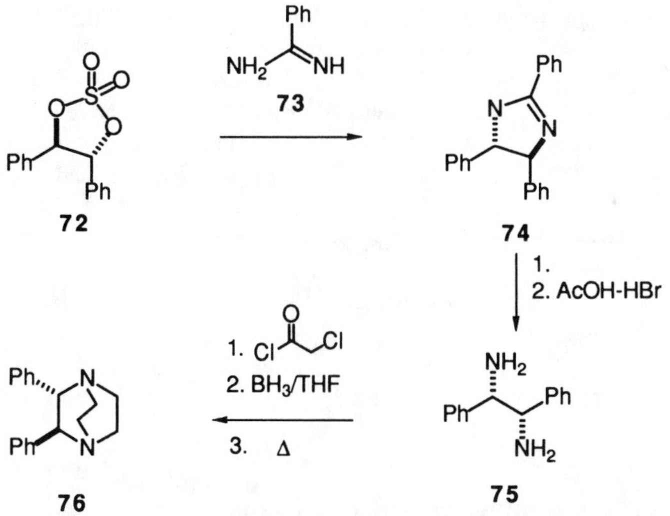

Cyclic sulfites (**68**) also are opened by nucleophiles, although they are less reactive than cyclic sulfates and require higher reaction temperatures for the opening reaction. Cyclic sulfite **77**, in which the hydroxamic ester is too labile to withstand ruthenium tetroxide oxidation of the sulfite, is opened to **78** in 76% yield by reaction with lithium azide in hot DMF [82]. Cyclic sulfite **79** is opened with nucleophiles such as azide ion [83] or bromide ion [84], using elevated temperatures in polar aprotic solvents. Structures such as **80** generally are not isolated but as in the case of **80** are carried on (when X = N$_3$) to amino alcohols [83] or (when X = Br) to maleates [84] by reduction. Yields are good and for compounds unaffected by the harsher conditions needed to achieve the displacement reaction, use of the cyclic sulfite eliminates the added step of oxidation to the sulfate.

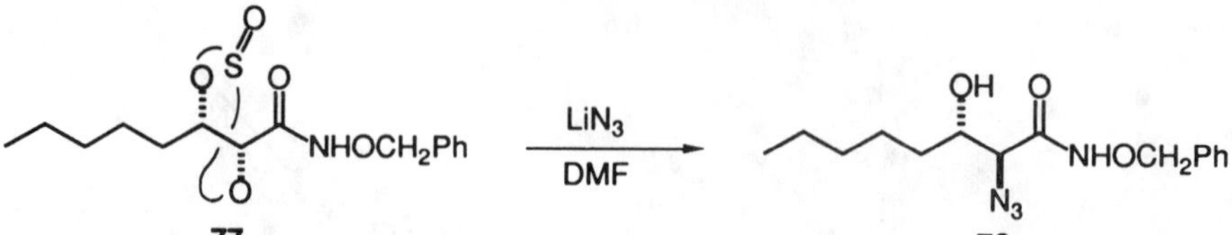

In the same vein as the cyclic sulfate activation of diols, cyclic sulfamidates have been prepared from 1,2- and 1,3-amino alcohols for the purpose of activating the carbinol toward nucleophilic attack [85–88]. (S)-Prolinol [85], N-benzyl serine t-butyl ester [86], and 2-(2-hydroxyethyl)piperidine [87] react with thionyl chloride to form the corresponding cyclic sulfamidite, and the latter are oxidized with RuO$_4$/NaIO$_4$ to the cyclic sulfamidates **81**, **82** and **83**, respectively. Nucleophilic displacements on the oxy carbon of the cyclic sulfamidate have been achieved with RNH$_2$ and ROH [85]; with H$_2$O, N$_3^-$, SCN$^-$, and CN$^-$ [86]; and with F$^-$ [88].

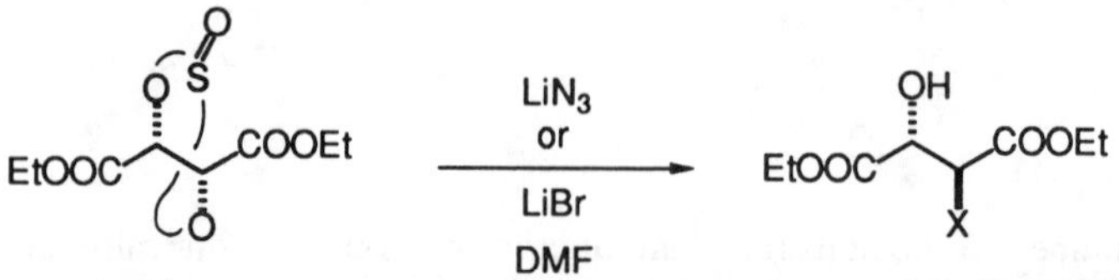

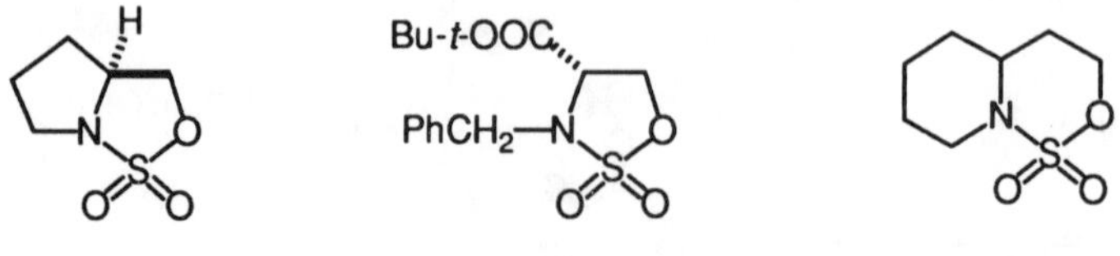

4.4.4.2. Regioselective Sulfonylation

2,3-Dihydroxy esters, such as **84**, are monosulfonylated in good yield and with high regioselectivity for the 2-hydroxyl group with either p-toluenesulfonyl chloride or p-nitrosulfonyl chloride [69]. Minor side products formed in the reaction are the bissulfonate ester and the α-sulfonyloxy-α,β-unsaturated ester, the formation in both cases being kept to a minimum by carrying out the reaction under relatively dilute conditions. Under mild alkaline conditions, the monosulfonates (**85**) are converted to glycidic esters (**86**) in high yield. The nosylates, but not the tosylates,

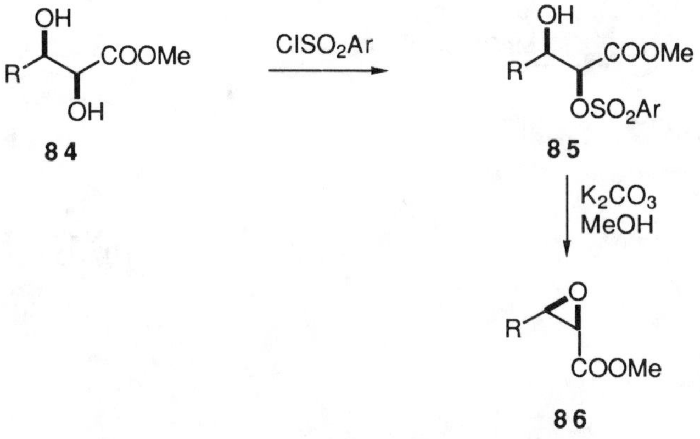

84 → **85** → **86**

undergo displacement by azide ion to give α-azido-β-hydroxy esters, which are envisioned as precursors to β-hydroxyamino acids [69].

4.4.4.3. Regioselective Acetoxyhalogenation and Conversion into Epoxides

1,2-Diols such as **87** are converted in excellent yields [89] into acetoxy chlorides (**88**) by treatment with trimethyl orthoacetate and trimethylsilyl chloride [90] or into acetoxybromides (**89**) with trimethyl orthoacetate and acetyl bromide [91]. These reactions proceed through nucleophilic attack on an intermediate 1,3-dioxolan-2-ylium cation [91] with inversion of configuration. In the presence of an aryl substituent as in **87**, displacement occurs exclusively at the benzylic position. With aliphatic diols such as **90**, the halide is introduced mainly at the less hindered position and acetoxybromides **91** and **92** are formed in a ratio of 7:1. Treatment of the acetoxy halides **88** or **89** under mildly alkaline conditions affords epoxide **93** in 84–87% yield, while the mixture of **91** and **92** is converted to epoxide **94** in 94% yield. Since both steps can be performed in the same reaction vessel, this reaction sequence constitutes an extremely efficient method for the direct conversion of 1,2-diols into epoxides with overall retention of configuration [89].

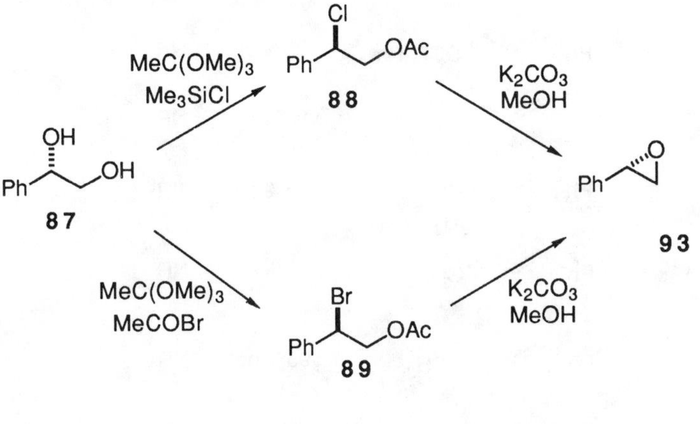

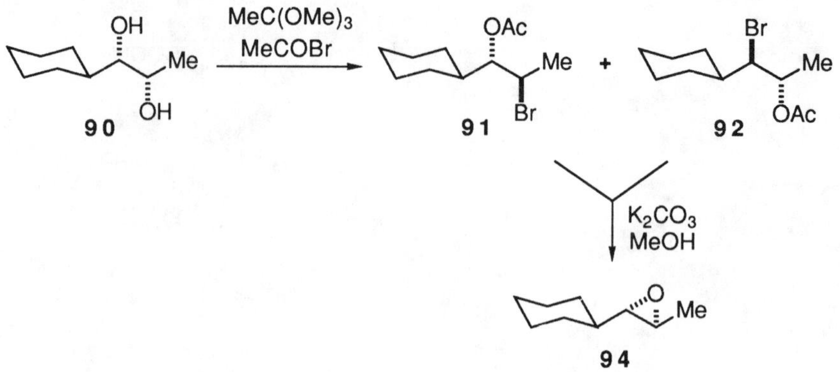

4.4.5. Conclusion

With the very recent discoveries of the phthalazine class of ligands (**29** and **30**) and the acceleration of osmate ester hydrolysis in the presence of organic sulfonamides, the osmium-catalyzed AD process has reached a level of effectiveness and simplicity unique among asymmetric catalytic transformations. A striking characteristic of this system, which is shared with Mn(III) salen catalyzed asymmetric epoxidation (see Chapter 4.2), is the ability to deliver very high enantioselectivities without

Chart 3 Comparison of asymmetric dihydroxylation with asymmetric epoxidation of a series of olefins.

	AD	AE
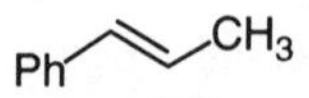 Ph~~~CH₃	>95% ee	NR
Ph~~~CH₂OH	>95% ee	>95% ee
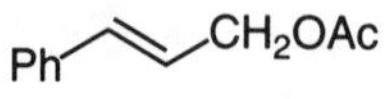 Ph~~~CH₂OAc	>95% ee	NR
Ph~~~OCH₂Ph	>95% ee	NR
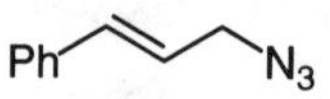 Ph~~~N₃	>95% ee	NR
Ph~~~Cl	>95% ee	NR
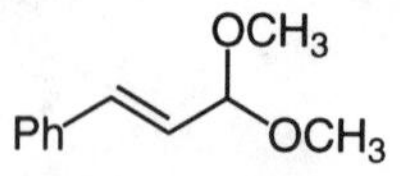 Ph~~~CH(OCH₃)OCH₃	>95% ee	NR
Ph~~~C(O)OCH₃	>95% ee	NR

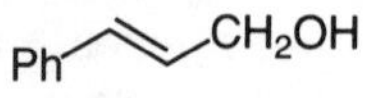

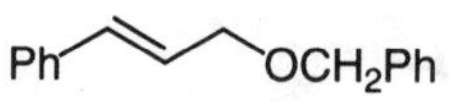

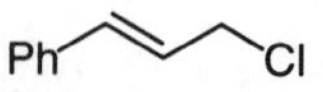

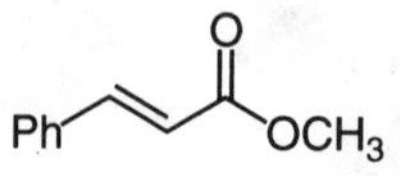

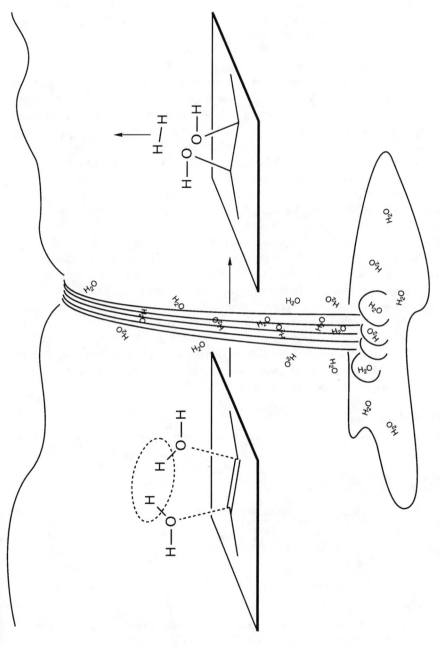

Figure 3 Schematic representation of electrocatalytic asymmetric dihydroxylation of olefins.

the need for an ancillary functional group to act as binding tether. This latter requirement is essential for titanium tartrate catalyzed asymmetric epoxidation (AE, see Chapter 4.1) and catalytic asymmetric hydrogenation (see Chapter 1) and has serious consequences for the scope of these two otherwise excellent systems. This point stands out clearly in Chart 3, which compares the outcome of the catalytic AD and AE processes for a closely related family of olefins.

Broad scope and high enantioselectivity are important for any catalytic asymmetric transformation, but they are not sufficient to ensure that a process will be widely used, especially on a commercial scale. To achieve this latter goal the process must also be economical and easy to perform. From this point of view it is hard to imagine a selective catalytic process that could have been more responsive to improvement than has catalytic AD. The adjustable parameters of the catalytic AD reaction were amenable to improvement at virtually every turn and have resulted in a procedure requiring only trace amounts of the expensive alkaloid ligand and the osmium catalyst. The key responsible variables are the large ligand acceleration effect and the binding constant $(L + OsO_4 \rightleftarrows L \cdot OsO_4)$. Both these variables are dramatically temperature dependent, such that everything related to the enantioselectivity of the process favors operating at lower temperature. The slow rate of osmate ester hydrolysis prevented use of lower temperature with the more substituted olefins, but now the sulfonamide effect has overcome this limitation.

Last but not least, from the process improvement point of view, a research group at Sepracor led by Yun Gao has developed an extremely effective electrocatalytic version of the catalytic AD process (see Section 4.4.2.2) [37]. As symbolized in Figure 3, this electrocatalytic approach to AD appears to be ideal. Enantiomerically pure diols arise from electricity, water, and olefins compounds, with hydrogen gas and a little water over the dam as the only by-products.

Acknowledgment

We thank David J. Berrisford for the use of his excellent bibliography of the asymmetric dihydroxylation literature.

References

1. Criegee, R. *Justus Liebigs Ann. Chem.* **1936,** 75.

2. (a) Schröder, M. *Chem. Rev.* **1980,** *80,* 187. (b) VanNieuwenhze, M.; Gao, Y.; Sharpless, K. B. *Aldrichemica Acta,* Dec. 1993, in preparation.

3. Böeseken, J.; van Giffen, J. *Rec. Trav. Chim. Pays-Bas,* **1921,** *39,* 183.

4. Sharpless, K. B.; Teranishi, A. Y.; Bäckvall, J.-E. *J. Am. Chem. Soc.* **1977,** *99,* 3120.

5. Jorgensen, K. A.; Schiott, B. *Chem. Rev.* **1990,** *90,* 1483.

6. For an X-ray crystallographic structure determination of such a dimeric complex, see: Collin, R.; Griffith, W. P.; Phillips, F. L.; Skapski, A. C. *Biochim. Biophys. Acta,* **1973,** *320,* 745.

7. Hofmann, K. A. *Chem. Ber.* **1912,** *45,* 3329.

8. Milas, N. A.; Sussman, S. *J. Am. Chem. Soc.* **1936,** *58,* 1302.

9. Milas, N. A.; Trepagnier, J. H.; Nolan, J. T., Jr.; Iliopulos, M. I. *J. Am. Chem. Soc.* **1959,** *81,* 4730.

10. Sharpless, K. B.; Akashi, K. *J. Am. Chem. Soc.* **1976,** *98,* 1986.

11. VanRheenen, V.; Kelly, R. C.; Cha, D. Y. *Tetrahedron Lett.* **1976,** 1973, wherein the invention of this method is attributed to: Schneider, W. P.; McIntosh, A. V. U.S. Patent 2,769,824 Nov. 6, 1956.

12. Minato, M.; Yamamoto, K.; Tsuji, J. *J. Org. Chem.* **1990,** *55,* 766.

13. Criegee, R.; Marchand, B.; Wannowius, H. *Justus Liebigs Ann. Chem.* **1942,** *550,* 99.

14. For an X-ray crystallographic structure determination of one such monoamine complex, see: Svendsen, J. S.; Markó, I.; Jacobsen, E. N.; Pulla Rao, C.; Bott, S.; Sharpless, K. B. *J. Org. Chem.* **1989,** *54,* 2263.

15. For X-ray crystallographic structure determinations of complexes of this type, see Ref. 19 and: Cartwright, B. A.; Griffith, W. P.; Schroder, M.; Skapski, A. C. *J. Chem. Soc., Chem. Commun.* **1978,** 853.

16. Wai, J. S. M.; Markó, I.; Svendsen, J. S.; Finn, M. G.; Jacobsen, E. N.; Sharpless, K. B. *J. Am. Chem. Soc.* **1989,** *111,* 1123.

17. Jacobsen, E. N.; Markó, I.; France, M. B.; Svendsen, J. S.; Sharpless, K. B. *J. Am. Chem. Soc.* **1989,** *111,* 737.

18. Hentges, S. G.; Sharpless, K. B. *J. Am. Chem. Soc.* **1980,** 4263.

19. Pearlstein, R. M.; Blackburn, B. K.; Davis, W. M.; Sharpless, K. B. *Angew, Chem., Int. Ed. Engl.* **1990,** *29,* 639.

20. Yamada, T.; Narasaka, K. *Chem. Lett.* **1986,** 131.

21. Tokles, M.; Snyder, J. K. *Tetrahedron Lett.* **1986,** *34,* 3951.

22. Tomioka, K.; Nakajima, M.; Koga, K. *J. Am. Chem. Soc.* **1987,** *109,* 6213.

23. Hirama, M.; Oishi, T.; Ito, S. *J. Chem. Soc., Chem. Commun.* **1989,** 665.

24. Corey, E. J.; Jardine, P. D.; Virgil, S.; Yuen, P.-W.; Connell, R. D. *J. Am. Chem. Soc.* **1989,** *111,* 9243.

25. Masamune, S.; Choy, W.; Petersen, J. S.; Sita, L. R. *Angew. Chem., Int. Ed. Engl.* **1985,** *24,* 1.

26. Jacobsen, E. N.; Markó, I.; Mungall, W. S.; Schröder, G.; Sharpless, K. B. *J. Am. Chem. Soc.* **1988,** *110,* 1968.

27. Gredley, M. PCT Int. Appl. WO 89 02,428; *Chem. Abstr.* **1989,** *111,* 173782v.

28. Watson, K. G.; Fung, Y. M.; Gredley, M.; Bird, G. J.; Jackson, W. R.; Gountzos, H.; Matthews, B. R. *J. Chem. Soc., Chem. Commun.* **1990,** 1018.

29. Sharpless, K. B.; Åmberg, W.; Bennani, Y. L.; Crispino, G. A.; Hartung, J.; Jeong, K.-S.; Kwong, H.-L.; Morikawa, K.; Wang, Z.-M.; Xu, D.; Zhang, X.-L. *J. Org. Chem.* **1992,** *57,* 2768.

30. See footnote 6 of Ref. 26.

31. Sharpless, K. B.; Amberg, W.; Beller, M.; Chen, H.; Hartung, J.; Kawanami, Y.; Lübben, D.; Manoury, E.; Ogino, Y.; Shibata, T.; Ukita, T. *J. Org. Chem.* **1991,** *56,* 4585.

32. McLaughlin, A. I. G.; Milton, R.; Perry, K. M. A. *Brit. J. Ind. Med.,* **1946,** *3,* 183; *Chem. Abstr.,* **1946,** *40,* 5841⁴.

33. Nissila, M. *Scand. J. Rheumatol.,* **1978,** *Suppl. 29,* 1–44.

34. Hinckley, C. C.; Bemiller, J. N.; Strack, L. E.; Russell, L. D. *"Platinum, Gold, and Other Metal Chemotherapeutic Agents,"* Lippard, S. J. (Ed.), ACS Symp. Series, No. 209, **1983**, 421–437.

35. Kwong, H.-L.; Sorato, C.; Ogino, Y.; Chen, H.; Sharpless, K. B. *Tetrahedron Lett.* **1990**, *31*, 2999.

36. Ogino, Y.; Chen, H.; Kwong, H.-L.; Sharpless, K. B. *Tetrahedron Lett.* **1991**, *32*, 3965.

37. Gao, Y.; Zepp, C. M. Sepracor, Inc. Unpublished results.

38. Verpoorte, R.; Schripsema, J.; van der Leer, T. In *The Alkaloids;* Brossi, A. (Ed.); Academic Press: Orlando, FL; **1988**; Vol. 34, pp. 331–398.

39. See Ref. 30, footnote 20.

40. (a) Lohray, B. B.; Kalantar, T. H.; Kim, B. M.; Park, C. Y.; Shibata, T.; Wai, J. S.; Sharpless, K. B. *Tetrahedron Lett.* **1989**, *30*, 2041. (b) Hashiyama, T.; Morikawa, K.; Sharpless, K. B. *J. Org. Chem.*, **1992**, *57*, 5069.

41. Shibata, T.; Gilheany, D. G.; Blackburn, B. K.; Sharpless, K. B. *Tetrahedron Lett.* **1990**, *31*, 3817.

42. Kim, B. M.; Sharpless, K. B. *Tetrahedron Lett.* **1990**, *31*, 3003.

43. Ogino, Y.; Chen, H.; Manoury, E.; Shibata, T.; Beller, M.; Lübben, D.; Sharpless, K. B. *Tetrahedron Lett.* **1991**, *32*, 5761.

44. Wang, Z.-M.; Zhang, X.-L.; Sharpless, K. B. *Tetrahedron Lett.* **1993**, *34*, 2267.

45. Oi, R.; Sharpless, K. B. *Tetrahedron Lett.* **1992**, *33*, 2095.

46. Jeong, K.-S.; Sjö, P.; Sharpless, K. B. *Tetrahedron Lett.* **1992**, *33*, 3833.

47. McKee, B. H.; Gilheany, D. G.; Sharpless, K. B. *Org. Synth.* **1991**, *70*, 47.

48. (a) Keinan, E.; Sinha, S. C.; Sharpless, K. B. Unpublished results. (b) Crispino, G. A.; Jeong, K. S.; Kolb, H.; Wang, Z.-M.; Xu, D.; Sharpless, K. B. *J. Org. Chem.* in press.

49. Rama Rao, A. V.; Gurjar, M. K.; Joshi, S. V. *Tetrahedron: Asymmetry,* **1990**, *1*, 697.

50. Wink, D. J.; Kwok, T. J.; Yee, A. *Inorg. Chem.* **1990**, *29*, 5006.

51. (a) Hoffmann, R. W.; Ditrich, K.; Köster, G.; Stürmer, R. *Chem. Ber.* **1989**, *122*, 1783. (b) Stürmer, R. *Angew. Chem., Int. Ed. Engl.* **1990**, *29*, 59. (c) Stürmer, R.; Hoffman, R. W. *Synlett,* **1990**, 759.

52. (a) Konopelski, J. P.; Boehler, M. A.; Tarasow, T. M. *J. Org. Chem.* **1989**, *54*, 4966. (b) Eid, C. N., Jr.; Konopelski, J. P. *Tetrahedron Lett.* **1991**, *32*, 461.

53. Crosby, J.; Fakley, M. E.; Gemmell, C.; Martin, K.; Quick, A.; Slawin, A. M. Z.; Shahriai-Zavareh, H.; Stoddart, J. F.; Williams, D. J. *Tetrahedron Lett.* **1989**, *30*, 3849.

54. Pini, D.; Iuliano, A.; Rosini, C.; Salvadori, P. *Synthesis,* **1990**, 1023.

55. Hirsenkorn, R.; *Tetrahedron Lett.* **1990**, *31*, 7591; **1991**, *32*, 1775.

56. (a) Xu, D.; Crispino, G. A.; Sharpless, K. B. *J. Am. Chem. Soc.* **1992**, *114*, 7570. (b) Park, C. Y.; Kim, B. M.; Sharpless, K. B. *Tetrahedron Lett.* **1991**, *32*, 1003.

57. (a) Crispino, G. A.; Sharpless, K. B. *Tetrahedron Lett.* **1992**, *33*, 4273. (b) Crispino, G. A.; Ho, P.-T.; Sharpless, K. B. *Science,* **1993**, *259*, 64. (c) Wang, L.; Sharpless, K. B. *J. Am. Chem. Soc.* **1992**, *114*, 7568. (d) Morikawa, K.; Hashiyama, T.; Sharpless, K. B. Unpublished results (1992). (e) Morikawa, K.; Andersson, P.; Park, J.; Hashiyama, T.; Sharpless, K. B. *J. Org. Chem.* in press.

58. Fieser, L. F.; Fieser, M. *Steroids;* Reinhold: New York, 1959; pp. 612–632.

59. *Cf.* (a) Cha, J. K.; Christ, W. J.; Kishi, Y. *Tetrahedron,* **1984,** *40,* 2247. (b) Dent, W. H., III; Vedejs, E. *J. Am. Chem. Soc.* **1989,** *111,* 6861. (c) Houk, K. N.; Duh, H.-Y.; Wu, Y.-D.; Moses, S. *J. Am. Chem. Soc.* **1986,** *108,* 2754.

60. Annuziata, R.; Cinquini, M.; Cozzi, F.; Raimondi, L.; Stefanelli, S. *Tetrahedron Lett.* **1987,** *28,* 3139.

61. Brimacombe, J. S.; McDonald, G.; Rahman, M. A. *Carbohydr. Res.* **1990,** *205,* 422.

62. Sharpless, K. B.; Morikawa, K.; Kim, B.-M. Unpublished results.

63. (a) Sun, L.-Q.; Zhou, W.-S.; Pan, X.-F. *Tetrahedron: Asymmetry,* **1991,** *2,* 973. (b) Zhou, W.-S.; Huang, L.-F.; Sun, L.-Q.; Pan, X.-F. *Tetrahedron Lett.* **1991,** *32,* 6745.

64. Ireland, R. E.; Wipf, P.; Roper, T. D. *J. Org. Chem.* **1990,** *55,* 2284.

65. DeCamp, A. E.; Mills, S. G.; Kawaguchi, A. T.; Desmond, R.; Reamer, R. A.; DiMichele, L.; Volante, R. P. *J. Org. Chem.* **1991,** *56,* 3564.

66. Cooper, A. J.; Salomon, R. G. *Tetrahedron Lett.* **1990,** *31,* 3813.

67. Ikemoto, N.; Schreiber, S. L. *J. Am. Chem. Soc.* **1990,** *112,* 9657.

68. Gao, Y.; Sharpless, K. B. *J. Am. Chem. Soc.* **1988,** *110,* 7538.

69. Fleming, P. R.; Sharpless, K. B. *J. Org. Chem.* **1991,** *56,* 2869.

70. Denmark, S. E. *J. Org. Chem.* **1981,** *48,* 3144.

71. (a) Lowe, G.; Salamone, S. J. *J. Chem. Soc., Chem. Commun.* **1983,** 1392. (b) Lowe, G.; Parratt, M. J. *Bioorg. Chem.* **1988,** *16,* 283.

72. (a) Brandes, S. J.; Katzenellenbogen, J. A. *Mol. Pharmacol.* **1987,** *32,* 391. (b) Liu, A.; Katzenellenbogen, J. A.; VanBrocklin, H. F.; Mathias, C. J.; Welch, M. J. *J. Nucl. Med.* **1991,** *32,* 81.

73. Kim, B. M.; Sharpless, K. B. *Tetrahedron Lett.* **1989,** *30,* 655.

74. Vanhessche, K.; Van der Eycken, E.; Vandewalle, M. *Tetrahedron Lett.* **1990,** *31,* 2337.

75. (a) Machinaga, N.; Kibayashi, C. *Tetrahedron Lett.* **1990,** *31,* 3637. (b) Machinaga, N.; Kibayashi, C. *J. Chem. Soc., Chem. Commun.* **1991,** 405. (c) Machinaga, N.; Kibayashi, C. *J. Org. Chem.* **1991,** *56,* 1386.

76. Lohray, B. B.; Gao, Y.; Sharpless, K. B. *Tetrahedron Lett.* **1989,** *30,* 2623.

77. (a) van der Klein, P.; Boons, G. J. P. H.; Veeneman, G. H.; van der Marel, G. A.; van Boom, J. H. *Synthesis,* **1990,** 311. (b) van der Klein, P.; de Nooy, A. E. J.; van der Marel, G. A.; van Boom, J. H. *Synthesis,* **1991,** 347.

78. Bates, R. W.; Fernández-Moro, R.; Ley, S. V. *Tetrahedron Lett.* **1991,** *32,* 2651.

79. Oi, R.; Sharpless, K. B. *Tetrahedron Lett.* **1991,** *32,* 999.

80. Burk, M. J. *J. Am. Chem. Soc.* **1991,** *113,* 8518.

81. Oi, R.; Sharpless, K. B. *Tetrahedron Lett.* **1991,** *32,* 4853.

82. Kim, B. M.; Sharpless, K. B. *Tetrahedron Lett.* **1990,** *31,* 4317.

83. Lohray, B. B.; Ahuja, J. R. *J. Chem. Soc., Chem. Commun.* **1991,** 95.

84. Gao, Y.; Zepp, C. M. *Tetrahedron Lett.* **1991,** *32,* 3155.

85. Alker, D.; Doyle, K. J.; Harwood, L. M.; McGregor, A. *Tetrahedron: Asymmetry,* **1990,** *1,* 877.

86. Baldwin, J. E.; Spivey, A. C.; Schofield, C. J. *Tetrahedron: Asymmetry,* **1990,** *1,* 881.

87. Lowe, G.; Reed, M. A. *Tetrahedron: Asymmetry,* **1990,** *1,* 885.

88. White, G. J.; Garst, M. E. *J. Org. Chem.* **1991,** *56,* 3177.

89. Kolb, H.; Sharpless, K. B. *Tetrahedron,* **1992,** *48,* 10515.

90. (a) Newman, M. S.; Olson, D. R. *J. Org. Chem.* **1973,** *38,* 4203. (b) Dansette, P.; Jerina, D. M. *J. Am. Chem. Soc.* **1974,** *94,* 1224.

91. Harmann, W.; Heine, H.-G.; Wendisch, D. *Tetrahedron Lett.* **1977,** 2263.

Asymmetric
Carbonylation

Giambattista Consiglio

Swiss Federal Institute of Technology
Department of Industrial and Engineering Chemistry
ETH Zentrum
Zürich, Switzerland

5.1. Introduction

A survey of the reviews published to date on asymmetric carbonylation reactions—
that is, hydroformylation (Scheme 1) and hydrocarbalkoxylation (Scheme 2)—
shows that quite often the conclusion on the scope of these reactions is not very
optimistic [1–9]. The rather pessimistic impression of the results so far obtained in
these reactions is more or less inevitable when this type of reaction is discussed in
the broadest context of catalytic asymmetric reactions with chiral transition metal
complexes in comparison with other more successful reaction types [10–31]. Nevertheless, asymmetric carbonylations have high potential as useful synthetic methods in research laboratories and as commercial processes in industry.

Scheme 1

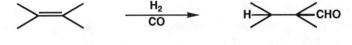

Scheme 2

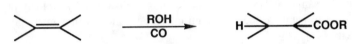

A very important aspect of hydroformylation and hydrocarbalkoxylation is related to the control of the regiochemistry when two different groups are added to the
different termini of a carbon–carbon double bond. Of course, through asymmetric

catalysis we are interested in the synthesis of chiral non-racemic products. More than 50 years after the discovery of hydroformylation [32,33], there is no system known that allows the synthesis of the 2-methyl-substituted aldehyde starting from aliphatic 1-substituted ethylenes with regioselectivity better than 50% [34,35].

From the view point of steric control, asymmetric catalysis of transition metal complexes is based, in almost all cases reported, on the modification of the metallic component by a chiral ligand [36]. To be effective, the chiral ligand must remain coordinated to the metal. In the cases in which displacement reactions of the chiral ligand are possible, an effective asymmetric bias can be obtained only if coordination of the chiral ligand brings about an increase of the catalytic activity relative to the species not containing the chiral ligand [37]. Let us analyze the possible hydroformylation catalysts from this point of view. The first attempts to achieve asymmetric hydroformylation were carried out by using compounds that were better known at the time, namely cobalt carbonyl modified with chiral monophosphines [38]; some years later diphosphines were used as modifiers, but in both cases the results were disappointing [39]. The reasons for the failure of the cobalt catalysts to give even a fair asymmetric induction are illustrated in Scheme 3.

Scheme 3

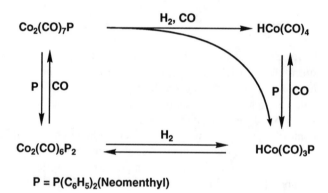

$$P = P(C_6H_5)_2(\text{Neomenthyl})$$

In the presence of hydrogen and carbon monoxide, cobalt carbonyl systems modified by phosphines such as $Co_2(CO)_6P_2$ (where P is, e.g., neomenthyldiphenylphosphine [40]) are shifted toward the formation of hydridocobalt tetracarbonyl even if a very low partial pressure of carbon monoxide is applied. The unmodified cobalt hydride is able to promote a much faster hydroformylation than the phosphine-containing species. It should be noted that in the absence of carbon monoxide, the same catalytic system containing the neomenthyldiphenylphosphine ligand is able to catalyze the asymmetric hydrogenation with a fairly good enantioselectivity [41].

Reduction of catalytic activity by phosphine ligands appears to be less dramatic in the case of rhodium systems [42]; the displacement of these ligands by carbon monoxide appears to be much less effective even if different species having different ratios of phosphine ligand to carbon monoxide can coexist under hydroformylation conditions [43–45]. In fact, appreciable levels of asymmetric induction have been achieved by using these catalysts, but the results have no practical significance. Investigation of platinum complexes began somewhat later. If these complexes are to have good catalytic activity, they must be modified either with counteranions having low coordinating ability [46,47] or with tin dichloride [48,49]. Homoleptic carbonyl platinum systems show a much lower catalytic activity than those modified by phosphine ligands [50]. The best results reported to date in asymmetric hydroformylation have been obtained with these chiral platinum catalyst systems. Catalyst systems containing other metals [39,51] have been investigated hardly at all, probably because of their low catalytic activity. Although enantioselective hydroformylation has been extensively investigated, the related hydrocarbalkoxylation has received much less attention. The activity of catalyst systems known to carbonylate olefins to esters or acids is generally not high [52]. Accordingly, high temperatures are required to promote the reaction, which presents an obstacle, at least from a psychological point of view, for research on asymmetric catalysis of these systems.

Since the last exhaustive review on asymmetric carbonylation in 1985 [9], very interesting and promising results have appeared in the literature: (1) very high enantioselectivity for the chiral platinum complex catalyzed hydroformylation of aromatic olefins, (2) a highly effective enantioface selection in the synthesis of syndiotactic poly(1-oxo-2-phenyltrimethylene) through the palladium catalyzed carbonylation of styrene, (3) the first examples of enantiomer differentiating carbonylations of organic halides and aziridines, (4) the regioselective and enantioselective hydrocarboxylation of aromatic olefins under mild conditions, and (5) the enantiotopic group discriminating carbonylation of vinyl halides. Therefore, a reexamination of the complete subject now appears necessary and appropriate.

5.2. Enantioface Discriminating Carbonylation Reactions of Olefinic Substrates

Carbonylation reactions of olefinic substrates can give rise to different types of product. The reactions that have been most extensively investigated in relation to enantioface discrimination are hydroformylation for the synthesis of aldehydes [5] and the hydrocarbalkoxylation/hydrocarbohydroxylation for the synthesis of esters or related compounds such as carboxylic acids and lactones) [6]. This is probably because modification of the known catalytic systems (e.g., RhH(CO)(PPh$_3$)$_3$ [53], PtH(SnCl$_3$)(CO)(PPh$_3$)$_2$ [49], and PdCl$_2$(PPh$_3$)$_2$ [54]) with chiral phosphine or diphosphine ligands looked straightforward. However, no attempts have been reported for the catalytic asymmetric synthesis of α-alkoxy esters (Scheme 4) and succinic acid derivatives (Scheme 5), which are known to be catalyzed by PdCl$_2$ in the presence of an oxidant such as CuCl$_2$ [55,56].

Scheme 4

Scheme 5

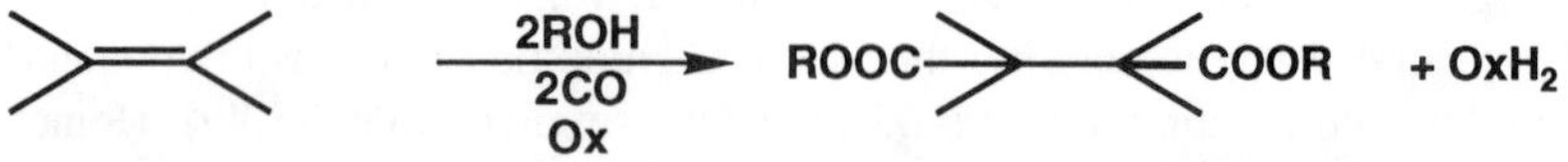

$$\text{>=<} \quad \xrightarrow[\substack{2CO \\ Ox}]{2ROH} \quad \text{ROOC>—<COOR} \quad + OxH_2$$

The synthesis of saturated or unsaturated ketones (Scheme 6), which is catalyzed by palladium complexes modified with phosphorus or nitrogen ligands and weakly coordinating or noncoordinating counteranions, seems to be regioselective only with styrene as the substrate [57]. Furthermore, with phosphine or diphosphine ligands these catalytic systems allows the production of a linear isomer with virtually complete regioselectivity [58].

Scheme 6

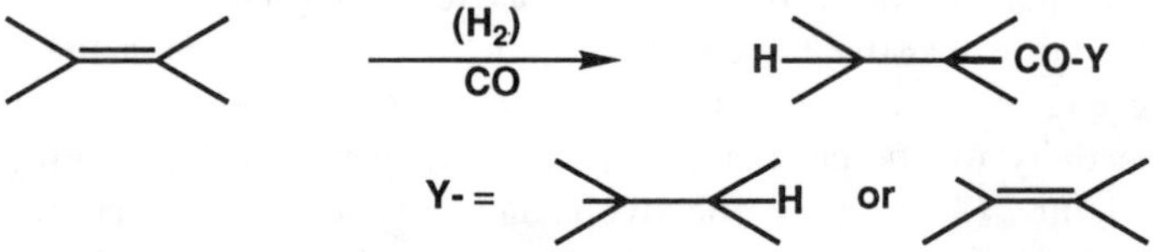

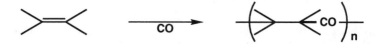

A new direction in this field is the asymmetric copolymerization of carbon monoxide with olefins (Scheme 7). Enantioface discrimination phenomena are responsible for the formation of stereoregular polymers, which can be formed through an enantiomorphic site control, a chain end control or both: See the related and thoroughly investigated examples of insertion polymerization, Ziegler–Natta catalysis [59].

Scheme 7

$$\text{>=<} \quad \xrightarrow{CO} \quad \left(\text{>—<}CO\right)_n$$

One of the most important issues in the investigation of stereodifferentiating carbonylation reactions is the choice of chiral modifying ligands. At first chiral monophosphine ligands were employed [51,60–63], but later chiral diphosphine ligands [64–96] began to be used almost exclusively in asymmetric hydroformylation with rhodium catalyst systems. For platinum, except for one example with a chiral monophosphine [97], only chiral diphosphines [98–118] have been used. Chiral diphosphines are preferentially used because of the stabilization of the bonding due to chelation. This stabilization appears to be favorable in view of the aforementioned possibility of competition with carbon monoxide for coordination at the metal. In rhodium-catalyzed asymmetric hydroformylation, other chiral ligands such as chiral β-diketonate [119], chiral substituted cyclopentadienyl [120] or phosphites [71,121,122] have been examined without success.

As previously discussed, chiral phosphine ligands do not cause efficient enantioselection in the case of cobalt-catalyzed hydroformylation [39,40,123]. A few results were reported on the use of chiral Schiff bases in large excess toward the metal component [124,125]. However, an asymmetric induction higher than 10% ee was observed only in the case of styrene [1]. Chiral clusters gave no asymmetric induction, either [123,126].

In the hydrocarbalkoxylation of olefins, palladium catalysts modified with chiral diphosphines have mainly been used [127–137]; only recently chiral ligands of other types have been reported in conjunction with catalyst systems working under mild conditions [138,139].

5.2.1. Enantioselective Hydroformylation

Although different metal catalyst systems have been found to be active for hydroformylation, only rhodium and platinum have been extensively investigated. Very few results, with only low enantioselectivities, were reported for cobalt, ruthenium, and iridium. Therefore, our discussion is limited to the first two catalyst systems.

Although the details can be rather different, the general mechanistic pictures for the two catalyst systems are quite alike [140,141]; for example, both catalyst systems promote *syn* addition to olefinic double bonds [142,143]. Scheme 8 illustrates the generally accepted mechanism for the hydroformylation of terminal olefins [144].

Scheme 8

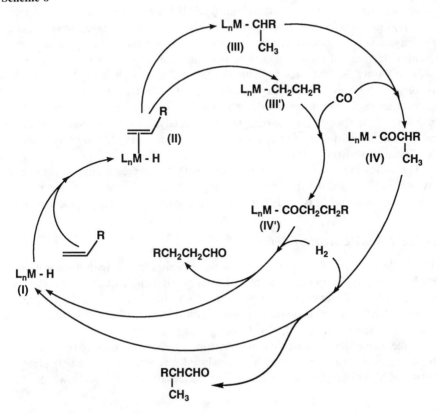

Important intermediates in this mechanistic scheme are the π-olefin–metal complex **II**, the alkyl–metal complex **III**, and the acyl–metal complex **IV**. In the presence of enantiomerically pure ligands (and if we neglect the metal's being a stereogenic center), the number of intermediates **II**, **III**, and **IV** should be doubled as a result of the existence of pairs of diastereomeric intermediates. The ability of the intermediate **III** to arise from different olefinic isomers allows us to verify in a very simple way whether the enantioselectivity measured in the aldehydes produced was determined before or after the formation of the intermediate **III**; that is the asymmetric hydroformylation of isomeric substrates such as 1-butene and 2-butene, which can yield the same chiral aldehyde, should give us a clear answer [1,5]. It has been shown that asymmetric induction is determined, at least predominantly, at the formation of the alkyl–metal intermediate **III** [5,115]. Although there is an interesting question about whether the enantioselectivity is associated with the kinetic or thermodynamic enantioface selection at the formation of the π-olefin–metal complex **II**, no information clarifying this point is found in the literature at present. For substrates giving rise to the formation of two isomeric aldehydes, the enantioselectivity observed in the products can be different from the extent of the initial enantioface selection of the prochiral olefin even if the formation of π-olefin–metal complex **II** is irreversible. In fact, this result may be due to a possible different regioselectivity for the two enantiofaces of the substrate to give the intermediates **III** and **III′**, respectively [145].

As Scheme 8 shows, once the intermediate **III** has been formed, the asymmetric induction is virtually (if not completely) determined. This scheme, however, does not explain why the concentrations of carbon monoxide and hydrogen influence the enantioselectivity of the reaction. If it is assumed that the mechanism depicted in Scheme 8 truly illustrates the formation of aldehydes through the hydroformylation of olefins, the most simple explanation for this effect is that a change in the concentration of carbon monoxide and hydrogen can change the nature (and the number) of the species (**I**) responsible for the catalysis [115]. Actually, the possibility of the coexistence of different catalytic species during hydroformylation has long been considered [146].

The following discussions, which deal with the representative enantioselectivities obtained for the different types of substrate, independent of the reaction conditions used, will give a better idea of the synthetic aspects of the reactions.

5.2.1.1. Aliphatic Olefins

Very little work has been performed on the enantioselective hydroformylation of aliphatic olefins. From the viewpoint of asymmetric bias, aliphatic olefins are not very attractive substrates, since asymmetric discrimination is based only on repulsive van der Waals interactions. The best enantioselectivity for each substrate in the literature is summarized in Tables 1 and 2.

In the case of terminal olefins, asymmetric induction in the range of 20% ee has been obtained for the chiral rhodium catalyzed reaction. For the chiral platinum catalyzed hydroformylation of 1-butene, the best result is 67.1% ee. However, if the

Table 1 Asymmetric Hydroformylation of Aliphatic Olefins
with Chiral Rhodium Catalysts

Substrate	Ligand	Isomeric Ratio (L:B)[a]	% ee (config.)	Ref.
1-Butene	(R,R)-DIOP	93:7	18.8(R)	65[b]
1-Pentene	(R,R)-DIOP	93:7	19.7(R)	65
1-Octene	(R,R)-DIOP	90:10	15.2(R)	65
3-Methyl-1-butene	(R,R)-DIOP	93:7	15.2(R)	65
3,3-Dimethyl-1-butene	DIOCOL	93:7	1.0(S)	82
2,3-Dimethyl-1-butene	(S,S)-CHIRAPHOS	~100:0	21.8(R)	86[c]
2,3,3-Trimethyl-1-butene	DIOCOL	~100:0	1.1(R)	82
2-Ethyl-1-hexene	(R,R)-DIOP	~100:0	1.1(R)	1
(Z)-2-Butene	(S,S)-CHIRAPHOS	0:100	18.4(S)	86[d]
(E)-2-Butene	(S,S)-CHIRAPHOS	1:99	18.5(S)	86[e]
(Z)-2-Hexene	(R,R)-DIOP	40:60[f]	1.4(S)[g] 5.8(R)[h]	65
(E)-2-Hexene	(R,R)-DIOP	42:58[f]	4.6(S)[g] 2.9(R)[h]	65
Bicyclo[2.2.2]oct-2-ene	(R,R)-DIOP	—	4.2(R)	74[i]
Norbornene	(S,S)-CHIRAPHOS	exo	16.4(1R,2R,4S)	86[c]

[a] L = Linear (straight-chain or less branched) aldehyde; B = branched (or more branched) aldehyde. The chiral product (or the one to which the enantiomeric purity refers) is underlined.
[b] Other data in References 70, 75, 82, 86, and 95.
[c] Other data in Reference 95.
[d] Other data in References 39, 65, 70, 74, 75, 77, 82, 86, and 95.
[e] Other data in References 65, 77, and 95.
[f] Molar ratio of 2-ethylpentanal/2-methylhexanal.
[g] Refers to 2-ethylpentanal.
[h] Refers to 2-methylhexanal.
[i] See also Reference 39.

Table 2 Asymmetric Hydroformylation of Aliphatic Olefins
with Chiral Platinum Catalysts

Substrate	Ligand	Isomeric Ratio (L:B)[a]	% ee (config.)	Ref.
1-Butene	(R,R)-BCO-DPP	86:14	67.1(S)	115[b]
1-Pentene	(R,R)-DIOP	97:3	29.1(R)	5
1-Hexene	(R,R)-DIOP	87:13	10.9(R)	5
1-Octene	(R,R)-DIOP	96:4	18.6(R)	5
2-Methyl-1-butene	NMDPP	n.r.[c]	9.4(R)	97
2,3-Dimethyl-1-butene	(R,R)-BCO-DBP	~100:0	46.1(R)	115[d]
(Z)-2-Butene	(R,R)-BCO-DBP	13:87	30.4(R)	115[e]
(E)-2-Butene	(R,R)-BCO-DBP	13:87	28.9(R)	115[e]
Norbornene	(R,R)-BCO-DPP	exo	29.8(1R,2R,4S)	115[d]

[a] See footnote a in Table 1.
[b] Other data in References 86 and 105.
[c] n.r. = not reported.
[d] Other data in References 95 and 105.
[e] Other data in References 75 and 105.

rather low regioselectivity normally observed for the formation of the branched isomer is taken into account, the best yield for the pure enantiomer in this case seldom exceeds 20%. Stereoselectivities of 60 and 16% ee were reported as the best results for the reaction of norbornene catalyzed by chiral platinum and rhodium catalyst systems, respectively.

The problem of regioselectivity does not exist for substrates having C_{2v} or C_{2h} symmetry. Rhodium catalysts in the presence of excess chiral ligands do not normally give products arising from migration of the double bond in the substrate; the best results are, however, around 20% ee. A modest enantioselectivity (28% ee) was reported for the hydroformylation of (Z)-2-butene by a polymer supported chiral rhodium catalyst [77]. With chiral platinum catalysts, the enantioselectivity is somewhat better (30% ee) than that for the chiral rhodium catalyst systems, but the reactions are less selective as a result of the isomerization of the substrate and also the consequent formation of isomeric straight-chain aldehydes.

Vinylidene olefins represent very interesting substrates from the standpoint of possible synthetic applications [2]. In fact, regioselectivity in most cases favors almost exclusively the less branched product. Enantioselectivity is, however, normally low with chiral rhodium catalyst; the best result is 22% ee for 2,3-dimethyl-1-butene with the CHIRAPHOS ligand. The best result (46.1% ee) for the same substrate was realized with chiral platinum systems. Internal C_s olefins are not promising substrates for the chiral platinum catalysts because of their strong tendency to isomerize; probably this is why no results have been reported. With a chiral rhodium system, only low regioselectivity and enantioselectivity were reported for (Z)- and (E)-2-hexene.

5.2.1.2. Aromatic Olefins

The hydroformylation of aromatic olefins, particularly that of styrene, has received much more attention than that of aliphatic olefins. In fact, styrene represents a suitable model for vinyl aromatics. The hydroformylation of vinyl aromatics can give immediate precursors for 2-arylpropionic acids, a very important class of anti-inflammatory agents [147]. Some of the best results obtained for the hydroformyla-

Table 3 Asymmetric Hydroformylation of Styrene with Chiral Rhodium Catalysts

Ligand	Isomeric Ratio (L:B)[a]	% ee (config.)	Ref.
(R)-Me(n-Pr)PhP	n.r.[b]	21.0(S)	60
(R,R)-DIOP	32:68	25.2(R)	6
(R,R)-DIOP-DBP	11:89	27.6(R)	70
(S,S)-CHIRAPHOS	6:94	24.2(R)	86
(S)-ValNOP	6:94	30.6(R)	90
(R,R)-DIOP[c]	10:90	20.0(R)	91

[a] See a in Table 1.
[b] Not reported.
[c] Hydroformylation of tricarbonyl(η^6-styrene)chromium.

Table 4   Asymmetric Hydroformylation of Aromatic Olefins
with Chiral Rhodium Catalysts

Substrate	Ligand	Isomeric Ratio (L:B)[a]	% ee (config.)[b]	Ref.
Indene	(S)-Me(n-Pr)PhP	95:5	n.r.	60
Isoeugenol	(S)-Me(n-Pr)PhP	85:15	n.r.	60
Cinnamic alcohol	(S)-Me(n-Pr)PhP	~100:0[c]	n.r.	60
Cinnamic aldehyde propylene glycol acetal	(S)-Me(n-Pr)PhP	55:45	n.r.	60
Phenyl vinyl ether	NMDPP	n.r.	0.3(R)	63
2-Phenyl-1-propene	(S,S)-CHIRAPHOS	99:1	21.4(R)	86
2-Phenyl-1-butene	(R,R)-DIOP	~100:0	1.8(R)	64
3-Phenyl-1-propene	(R,R)-DIOP	88:12	15.5(R)	64
3-Phenyl-2-propene	(R,R)-DIOP	88:12	14.4(R)	64

[a] See note a in Table 1; alternatively, the ratio of α-formylation/β-formylation (with respect to the aromatic substituent) is given.
[b] n.r. = not reported.
[c] Some isomerization (~25%) of the substrate and formation of 3-phenylpropanal takes place.

tion of styrene using chiral rhodium catalysts are summarized in Table 3. The regioselectivity is excellent in some cases, but the enantioselectivity is low (20–30% ee). A few other aromatic substrates have also been employed, but asymmetric induction remains low (Table 4).

Much better enantioselectivities were obtained with chiral platinum catalysts (Table 5). For the chiral diphosphine–platinum catalyst systems, substitution of the diphenylphosphino groups on the diphosphine ligand by dibenzophosphole groups improves regioselectivity as well as enantioselectivity in general. For styrene and ring-substituted styrenes, enantioselectivities approaching 90% ee have been obtained under standard hydroformylation conditions [108,110,111]. Almost complete enantioselectivities ($\geq$ 96% ee) have been reported [108,117] when the reaction using the (2S,4S)-BPPM or the (2S,4S)-BPPM-DBP ligands is carried out in triethylorthoformate, conditions that transform in situ the stereochemically labile chiral aldehyde into the corresponding acetal (Scheme 9). Under these conditions, however, the reactions are very slow and thus require a long period of time to obtain reasonable conversion. [Note: All the attempts to reproduce these results on a preparative scale (a few grams) in the author's and other laboratories have not been successful [115]. *Editor's comment: The validity of the reported results was confirmed by the editor. It seems that the preparation of the chiral catalyst needs expertise, and the reaction is awfully slow. The most important message here is that the chiral platinum catalysts can indeed bring about excellent asymmetric induction, but the rapid racemization of the nearly enantiomerically pure aldehyde thus formed hampers the achievement of excellent enantioselectivity. If a good method is developed for the in situ conversion of the resulting aldehyde to a stereochemically stable derivative without seriously affecting the reaction rate, excellent results will emerge. The use of orthoformate by Stille provides an important hint to solve this problem.*]

Table 5 Asymmetric Hydroformylation of Styrene and Substituted Styrenes with Chiral Platinum Catalysts

Substrate	Ligand	Isomeric Ratio (L:B)[a]	% ee (config.)	Ref.
Styrene	(R,R)-DIOP	43:57	12.2(S)	100
Styrene	(R,R)-DIOP-DBP	18:82	79.8(S)	100,104
Styrene	(2S,4S)-BPPM	69:31	70(S) ≥96(S)[b]	103,108
Styrene	(2S,4S)-BPPM-DBP	24:76	40(S) ≥96(S)[b]	117
Styrene	Poly-BPPM[c]	69:31	72(S)	103
Styrene	(S,S)-CHIRAPHOS	38:62	45.0(R)	86
Styrene	(S)-PRONOP	59:41	48.1(S)	109
Styrene	(S,S)-BDPP	28:72	88.8(S)	110,111
Styrene	(R,R)-BCO-DPP	57:43	25(S)	113,115
Styrene	(R,R)-BCO-DBP	14:86	85(S)	113,115
Styrene	(S)-BINAP	67:33	68.8(S)	118
Styrene	(R,R)-DIOP[d]	73:27	46(R)	91
Styrene	(2S,4S)-BPPM[d]	24:76	40(S)	91
Styrene	(S)-BINAP[d]	32:68	0	91
Styrene	(S,S)-CHIRAPHOS[d]	65:35	6(R)	91
p-Methoxystyrene	(2S,4S)-BPPM	63:37	70(S)	108
p-Methylstyrene	(2S,4S)-BPPM	64:36	73(S)	108
p-Bromostyrene	(2S,4S)-BPPM	63:37	75(S)	108
p-Acetylstyrene	(2S,4S)-BPPM	53:47	85(S)	108
p-Nitrostyrene	(2S,4S)-BPPM	42:58	38(S)	108
p-Isobutylstyrene	(2S,4S)-BPPM-DBP	33:67	39(S) ≥96(S)[b]	117
m-Benzoylstyrene	(2S,4S)-BPPM-DBP	23:77	27(S) ≥96(S)[b]	117
p-Benzoylstyrene	(2S,4S)-BPPM-DBP	24:76	37(S) ≥96(S)[b]	117
m-Fluoro-p-phenylstyrene	(2S,4S)-BPPM-DBP	21:79	19(S) ≥96(S)[b]	117
p-Thiophenoylstyrene	(2S,4S)-BPPM-DBP	~20:80	9(S) ≥96(S)[b]	117
p-Ind-styrene[e]	(2S,4S)-BPPM-DBP	~0:100	0 ≥60(S)[b]	117

[a] See note a in Table 1.

[b] The reaction was carried out in triethylorthoformate to give the corresponding diethylacetal.

[c] Polymer-bound ligand.

[d] Hydroformylation of tricarbonyl(η^6-styrene)chromium.

[e] 4-(1,3-Dihydro-1-oxo-2H-isoindol-2-yl)styrene.

The effect of *para* substituents of styrenes on enantioselectivity was examined by using (2S,4S)-BPPM as the chiral ligand [108]. Higher regioselectivity for the branched isomer was observed both for electron-withdrawing and electron-donating substituents. For enantioselectivity, the effect of the *para* substituent turned out to be insignificant with the exception of the *p*-nitro- (lower enantioselectivity) and *p*-acetyl derivatives (higher enantioselectivity).

Some other aromatic olefinic substrates have been investigated (Table 6). For

Scheme 9

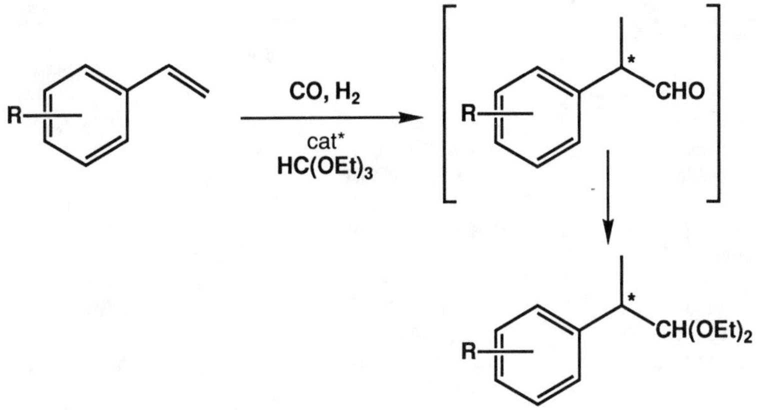

α-substituted vinyl aromatics, the asymmetric induction is found to be much lower than that for styrenes. *Para* substitution has a small effect (if any) on the regio- and enantioselectivity of 2-phenylpropenes for an (R,R)-DIOP–platinum catalyst system. For 2-vinylnaphthalene and 6-methoxy-2-methylnaphthalene, extremely high enantioselectivities (≥ 96% ee) have been claimed when the reactions are carried out in triethylorthoformate [117].

5.2.1.3. Functionalized Olefins

The hydroformylation of functionalized olefins can be a valuable tool for the synthesis of polyfunctionalized synthetic intermediates [148]. Only a few results have

Table 6 Asymmetric Hydroformylation of Aromatic Olefins with Chiral Platinum Catalysts

Substrate	Ligand	Isomeric Ratio (L:B)[a]	% ee (config.)	Ref.
2-Phenylpropene	(R,R)-DIOP	>99:1	15(S)	116[b]
2-(p-Methoxyphenyl)propene	(R,R)-DIOP	>99:1	17	116
2-(p-Chlorophenyl)propene	(R,R)-DIOP	>99:1	12	116
2-(p-Trifluoromethylphenyl)propene	(R,R)-DIOP	>99:1	13	116
2-Phenyl-1-butene	(R,R)-DIOP	~100:0	20.7(S)	99[b]
Indene	(R,R)-BCO-DBP	>95:5	45	113
Acenaphthylene	(R,R)-BCO-DBP	—	48	113
2-Vinylnaphthalene	(2S,4S)-BPPM-DBP	9:91	39(S) ≥96(S)[c]	117
6-Methoxy-2-vinylnaphthalene	(2S,4S)-BPPM-DBP	23:77	37(S) ≥96(S)[c]	117

[a] See note *a* in Table 1; alternatively, the ratio of α-formylation to β-formylation with respect to the aromatic substituent is given.

[b] See also References 98 and 101.

[c] Reaction was carried out in triethylorthoformate.

Table 7 Asymmetric Hydroformylation of Functionalized Olefins
with Chiral Rhodium Catalysts

Substrate	Ligand	Isomeric Ratio (L:B)[a]	% ee (config.)	Ref.
Methallyl alcohol	(R,R)-DIOP	~100:0	14.2(R)	69
2,5-Dihydrofuran	(R,R)-DIOP	—	7(R)	74[b]
N-Vinylphthalimide	(R,R)-DIOP-DBP	~0:100	38.3(R)	80
N-Vinylsuccinimide	(R,R)-DIOP-DBP	~0:100	41(R)	80
Vinyl acetate	(R,R)-DIOP-DBP	<25:75	51(R)	81[c]
Vinyl propionate	(R,R)-DIOP-2-NP	<25:75	36(S)	81
Vinyl benzoate	(S,S)-DIOP	<25:75	30(R)	81
Vinylferrocene	(R,R)-DIOP	29:71	~0	85
Dimethyl itaconate	(R,R)-DIOP	95:5	9.1(S)	88
Methyl methacrylate	(S,S)-CHIRAPHOS	31:69	3(R)	93
Methyl 2-(N-acetamido)acrylate	(R,R)-DIOP	~0:100	59(R)	94
Benzyl 2-(N-acetamido)acrylate	(R,R)-DIOP	~0:100	46(R)	96
Methyl 2-(N-BOC)alaninate[d]	(R,R)-DIOP	~0:100	46(R)	96

[a] See note a in Table 1; alternatively, the ratio α-formylation/β-formylation (with respect to the substituent) is given.
[b] See also Reference 82.
[c] Other data in References 76 and 89.
[d] Methyl N-[(tert-butyloxy)carbonyl]dehydroalaninate.

been reported for chiral rhodium catalyzed reactions (Table 7). The best enantioselectivity (59% ee) coupled with a complete regioselectivity was obtained with methyl 2-(N-acetamido)acrylate by using (R,R)-DIOP as the chiral ligand [93]. Enantioselectivities of 30–40% ee were obtained for vinyl esters and ~40% ee for vinyl imides [80]. Better results have been reported for the chiral platinum catalyzed reactions (Table 8). Enantioselectivities of 70–73% ee were obtained for vinyl

Table 8 Asymmetric Hydroformylation of Functionalized Olefins
with Chiral Platinum Catalysts

Substrate	Ligand	Isomeric Ratio (L:B)[a]	% ee (config.)	Ref.
Methyl methacrylate	(2S,4S)-BPPM	~100:0	60(R)	108[b]
t-Butyl methacrylate	(R,R)-DIOP	~100:0	35(S)	93
Phenyl methacrylate	(R,R)-DIOP	~100:0	49.5(R)	93
Dimethyl itaconate	(R,R)-DIOP	~100:0	83.6(R)	107[c]
Diphenyl itaconate	(R,R)-DIOP	~100:0	42.5(S)	107
N-Vinylphthalimide	(2S,4S)-BPPM	67:33	73(R)[e]	108
Vinyl acetate	(2S,4S)-BPPM	—	70(R)[e]	108[d]
Vinylferrocene	(R,R)-DIOP	79:21	11.9	85
2-Vinyl-5-benzoylthiophene	(2S,4S)-BPPM-DBP	17:83	≥96(S)	117
		4:96	≥96(S)[e]	

[a] See note a in Table 1.
[b] Other data in References 93, 107, and 110.
[c] See also Reference 110.
[d] See also Reference 106.
[e] Reaction was carried out in triethylorthoformate.

acetate and N-vinylphthalimide [108]. Nearly complete enantioselectivities ($\geq 96\%$ ee) were claimed for 2-vinylnaphthalene and 6-methoxy-2-vinylnaphthalene, and 2-vinyl-5-benzoylthiophene when the reaction was carried out in triethylorthoformate [117] (see Editor's comment, above). An enantioselectivity of 83.6% ee was achieved for dimethyl itaconate, but much lower selectivity (42.5% ee) was obtained for the corresponding diphenyl ester.

5.2.1.4. Dienes and Acetylenes

The enantioselective hydroformylation of acetylenes and dienes has no obvious synthetic significance, since the same product can be obtained from the corresponding much less expensive olefinic substrates. Only rhodium-based catalytic systems modified with DIOP were used for these substrates. Phenylacetylene, 1-octyne, and 2-butyne were hydroformylated with enantioselectivities lower than 1% ee [68%]. There are some indications that the triple bond is hydroformylated, at least in part, giving the corresponding unsaturated aldehyde, which is then hydrogenated under the reaction conditions to give the saturated product. 1,3-Butadiene, 2-methyl-1,3-butadiene, and 2,3-dimethyl-1,3-butadiene were hydroformylated with low enantioselectivity [79]. 2-Methylbutanal, 3-methylpentanal, and 3,4-dimethylpentanal were formed with 0.1, 5.7, and 32.3% ee, respectively [79]. The latter two aldehydes were identified through oxidation to the corresponding acids. Comparison of the results for the corresponding olefins indicates that the observed enantioselectivity may be ascribed to the enantioselective hydrogenation of the initially formed unsaturated aldehyde.

5.2.1.5. Comments on Enantioselective Hydroformylation

The preceding sections on asymmetric hydroformylation have been summarized from a synthetic point of view and should give an idea of the state of the art. High enantioselectivities coupled with high regioselectivities have been achieved only in the case of the chiral platinum catalyzed hydroformylation of vinyl aromatics. However, these catalysts are normally much less chemoselective than the corresponding rhodium catalysts because of competing hydrogenation and isomerization (where applicable). Chiral phosphine and diphosphine ligands, containing the diphenylphosphino substituent, have given comparable enantioselectivities in the rhodium-catalyzed reaction. On the contrary, only chiral diphosphines have been used for platinum catalysts. Fully alkylated chiral diphosphines depress the catalytic activity, particularly in the platinum systems [95], and a few such chiral ligands were employed to give very low enantioselectivity. Furthermore, the platinum complexes with such chiral peralkyldiphosphines should be very basic and therefore should racemize, in situ, the aldehydes formed [149].

The influence on enantioselectivity of the carbon monoxide partial pressure as well as the molar ratio of chiral ligand to metal has been investigated for chiral rhodium catalyzed reactions [63]. From these results it seems evident that there is competition between the ligands and carbon monoxide for the coordination sites on the metal. In the case of chiral platinum catalyzed reactions, the carbon monoxide partial pressure also influences the extent of asymmetric induction. However, in this case an increase in the chiral ligand/metal molar ratio causes only a decrease in activity and does not affect the asymmetric induction [101]. The existence of different catalytic species has been proposed to explain the effect of carbon monoxide [105]. The effect of the hydrogen partial pressure on the enantioselectivity of hydroformylation is much less clearly understood.

The effect of the reaction temperature has been investigated for both rhodium and platinum systems. In general, the extent of the asymmetric induction decreases with increasing temperature, but sometimes a change in the prevailing enantiomer is observed [110]. This effect might be associated with the existence of different catalytic species: that is, the relative concentration of such species may well be influenced by the reaction temperature. Another explanation for the BINAP–platinum catalyst species was proposed in which an influence of temperature on the population of conformers arising from the rotation of the phenyl rings was invoked [118]. Attempts have been made to support the chiral rhodium catalyst systems on alumina [150]. Rhodium and platinum catalysts containing polymeric chiral diphosphine ligands gave results comparable to those obtained with similar homogeneous catalyst systems [106,151].

The broad scope of hydroformylation with respect to the structure of the substrate permits chiral centers to arise in the formation of C—C bonds (e.g., in the case of styrene) or in the formation of C—H bonds (e.g., in the case of 2-phenylpropene). Attempts have been made to rationalize the results on enantioselectivity and regioselectivity with respect to different substrates for a given catalytic system. As a consequence, a model has been developed for the transition state involved in the transformation of the π-complex to the alkyl complex (Scheme 8, **II** and **III**) [5]. Despite its simplicity, the model does allow a good level of predictability, probably because the crucial catalytic species are not substantially influenced by the structure of the substrate.

5.2.2. Enantioselective Hydrocarbonylations

The *syn* addition to the double bond found for the hydroformylation reaction holds also for the hydrocarbalkoxylation of olefinic substrates [152]. For this reaction, a mechanism similar to that of hydroformylation starting with a hydrido–metal complex was originally proposed [51]; that is, the only difference is related to the fate of the acyl–metal intermediates (cf. Scheme 8, **IV** and **IV′**), which undergo alcoholysis to give the ester products in the case of hydrocarbalkoxylation. Recently, another mechanism in which the active species is a carbalkoxy–metal complex was proposed. This mechanism seems to be operative depending on the reaction conditions (Scheme 10) [153]. In any event, the results obtained by using linear isomeric butenes as the substrates show that the extent of asymmetric induction is determined at the insertion of the olefin either into the metal–hydrogen bond (Scheme 8) or into the carbalkoxy–metal bond (Scheme 10). In fact, it was found for some substrates that the olefin complex intermediates do not undergo olefin dissociation [128]. This observation would suggest that asymmetric induction is determined, at least partially, at the olefin complexation step.

Scheme 10

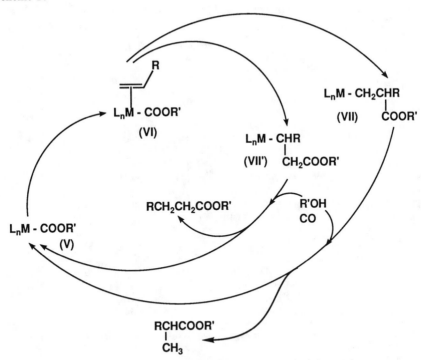

5.2.2.1. Enantioselective Hydrocarbalkoxylation and Hydrocarboxylation

Enantioselective hydrocarbalkoxylation has been investigated less than hydroformylation, probably because much more severe conditions (e.g., temperatures $>$ 100°C) were required for the reaction when the classical $PdCl_2$ catalysts modified with phosphines and diphosphines were used. Curiously, a report related to the apparently much more active cationic palladium systems has remained isolated [133]. Only recently, an enantioselective version of the milder carboxylation procedure was reported using a $PdCl_2/CuCl_2$-based catalyst with crotyl alcohol and some vinyl aromatics. Representative results for the hydrocarbalkoxylation of different olefinic hydrocarbons and *para*-substituted 2-phenylpropenes, using $PdCl_2$/phosphine systems as the catalyst precursors, are summarized in Table 9. Table 10 shows the results concerning some functionalized substrates. The extent of asymmetric induction achieved is low to moderate in most cases. The best results (up to ~60% ee) were obtained with 2-aryl-1-alkenes.

Modification of bis(dibenzylideneacetone)palladium(0) in methanol with trifluoroacetic acid and neomenthyldiphenylphosphine promotes the carbonylation of styrene at 50°C and ambient pressure to give methyl 2-phenylpropanoate with high regioselectivity (96%) and fairly good enantioselectivity (52% ee) [133].

Table 9 Asymmetric Hydrocarbalkoxylation of Olefinic Hydrocarbons
with Chiral Phosphine Modified PdCl$_2$ Catalysts

Substrate	Ligand	ROH:R	Isomeric Ratio (L:B)[a]	% ee (config.)	Ref.
1-Butene	(R,R)-DIOP	t-C$_4$H$_9$	78:22	20.2(S)	6
3-Methyl-1-butene	(R,R)-DIOP	CH$_3$	90:10	10.3(S)	6
3,3-Dimethyl-1-butene	(R,R)-DIOP	CH$_3$	95:5	2.0(S)	6
2-Methyl-1-butene	(R,R)-DIOP	t-C$_4$H$_9$	~100:0	4.3(R)	6
2,3-Dimethyl-1-butene	(R,R)-DIOP	t-C$_4$H$_9$	~100:0	4.6(S)	6
2,3,3-Trimethyl-1-butene	(R,R)-DIOP	t-C$_4$H$_9$	~100:0	19.5(S)	6
(Z)-2-Butene	(R,R)-DIOP-DBP	C$_2$H$_5$	~1:99	43.0(R)	135
(E)-2-Butene	(R,R)-DIOP	t-C$_4$H$_9$	2:98	23.2(S)	6
Styrene	(R,R)-DIOP-DBP	CH$_3$	29:71	22.0(S)	135
2-Phenylpropene	(R,R)-DIOP-DBP	t-C$_4$H$_9$	n.r.	59.0(S)	132
2-Phenyl-1-butene	(R,R)-DIOP	t-C$_4$H$_9$	~100:0	54.7(S)	6
1-Phenyl-1-propene	(R,R)-DIOP-DBP	CH$_3$	44:50[c]	12.0[d] 9.0[e]	135
2-(p-Methoxyphenyl)propene	(R,R)-DIOP-DBP	i-C$_3$H$_7$	87:13	30	116
2-(p-Chlorophenyl)propene	(R,R)-DIOP-DBP	i-C$_3$H$_7$	80:20	36	116
2-[(p-Trifluoromethyl) phenyl]propene	(R,R)-DIOP-DBP	i-C$_3$H$_7$	96:4	30	116
Indene	(R,R)-DIOP-DBP	CH$_3$	n.r.	5	135
Norbornene	(R,R)-DIOP	CH$_3$	exo	4.5(1R,2R,4S)	6

[a] See note a in Table 1; alternatively the ratio of α-ester to β-ester (with respect to the substituent) is given.
[b] n.r. = not reported.
[c] Molar ratio between 2-methyl-3-phenylpropanoate and 2-phenylbutanoate.
[d] Refers to 2-methyl-3-phenylpropanoate.
[e] Refers to 2-phenylbutanoate.

Under aerobic conditions, systems based on PdCl$_2$/CuCl$_2$ appear to be even more active for the regioselective hydrocarboxylation of alkenes to branched products at ambient temperature and pressure. Recently, it has been reported that the chiral modification of this catalytic system by poly-L-leucine promotes the carbonylation of but-2-en-1-ol to α-methyl-γ-butyrolactone with 61% ee [138]. The same catalyst system modified with enantiomerically pure 1,1'-binaphthyl-2,2'-diyl hydrogen

Table 10 Asymmetric Hydrocarbalkoxylation of Functionalized Olefins
with Chiral Phosphine Modified PdCl$_2$ Catalysts

Substrate	Ligand	ROH:R	Isomeric Ratio (L:B)[a]	% ee (config.)	Ref.
N-Vinyl succinimide	(R,R)-DIOP	C$_2$H$_5$	91:9	17.1(S)	131
Methyl methacrylate	(R,R)-DIOP	CH$_3$	~100:0	49.0(S)	6
(E)-N-Propenylphthalimide	(R,R)-DIOCOL	C$_2$H$_5$	88:12	3.0(R)	134
N-Allylphthalimide	(R,R)-DIOCOL	C$_2$H$_5$	71:29	4.0(R)	134
Methacrylamide	(R,R)-DIOP	CH$_3$	~100:0	37.3(R)	93

[a] See note a in Table 1; alternatively the ratio of α-ester to β-ester (with respect to the substituent) is given.

phosphate promotes the regioselective hydrocarboxylation of *p*-isobutylstyrene and 2-vinyl-6-methoxynaphthalene to the corresponding 2-arylpropionic acids with 84 and 91% ee, respectively (Scheme 11) [139]. These results are very promising, even though the catalyst systems do not give high turnover numbers.

Scheme 11

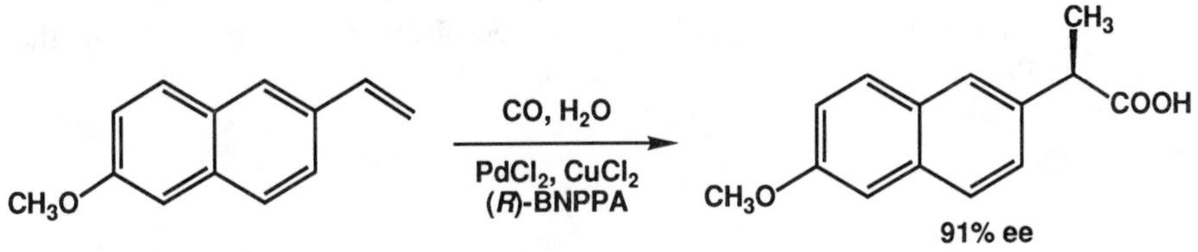

5.2.2.2. *Comments on Enantioselective Hydrocarbonylations*

The scope and the mechanistic implications of the active catalytic systems based on the chiral phosphate/PdCl$_2$/CuCl$_2$ system remain to be clarified. As far as the catalytic systems of the type L$_2$PdCl$_2$ are concerned, the extent of asymmetric induction achieved is normally rather low—indeed, lower than that obtained in the hydroformylation of the same substrates. Better enantioselectivities were obtained only in the case of 2-aryl-1-alkenes. In a manner similar to the case of hydroformylation, asymmetric induction is influenced by the partial pressure of carbon monoxide [6] and also by the presence of ligands in addition to the chiral one [130]. These results can be interpreted by assuming the existence of different catalytic species; that is, the relative concentration of key catalyst species is influenced by the concentration of the ligands. Furthermore, the extent of asymmetric induction is affected by the solvent, the alcohol used as the hydrogen donor and, to a small extent, by the halide present in the catalyst precursor. A model similar to that mentioned for hydroformylation provides some explanation for the results obtained using a single catalytic system for different substrates [6]. The level of predictability is good albeit not perfect.

5.2.3. Stereoselective Copolymerization of Olefins with Carbon Monoxide

Although catalysts capable of copolymerizing olefins with carbon monoxide were identified soon after the discovery of the hydroformylation reaction [154], the catalytic activity of these systems was rather low and no extrapolation to C_s olefins was investigated until the very recent work of Drent and coworkers [155]. Unfortunately, most of the work is covered in the patent literature, where the reported data are seldom complete. Some features, however, show very interesting aspects of the problem.

In principle, in a copolymer of carbon monoxide with a monosubstituted ethylene, the same types of polymer structure found in vinyl polymers (i.e., isotactic, syndiotactic, and atactic) are possible [156]. The only difference is that from the viewpoint of the microstructure, the number of the possible stereosequences (*n*-ads) is larger than those for vinyl polymers [157].

The copolymerization of styrene and carbon monoxide using (PHEN)Pd(*p*-CH$_3$C$_6$H$_4$SO$_3$)$_2$ (PHEN=1,10-phenanthroline) as the catalyst precursor was found to be completely regioselective and to produce poly(1-oxo-2-phenyltrimethylene), which was the strictly alternating copolymer of carbon monoxide and styrene (Scheme 12) [158]. Moreover, a high diastereoselectivity was observed in the process [159].

Scheme 12

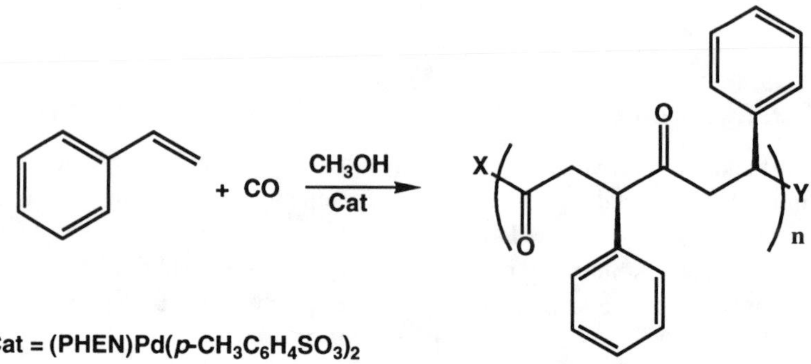

Cat = (PHEN)Pd(*p*-CH$_3$C$_6$H$_4$SO$_3$)$_2$

X = OCH$_3$; Y = COOCH$_3$ or -CH=CH-C$_6$H$_5$

In fact, the recovered copolymer was overwhelmingly syndiotactic, the content of racemic dyads being higher than 90%. The nature of the true catalytic species is largely unknown. The high enantioface discrimination that takes place during the process could be due to elements of chirality present in the catalytic species. This is the case in the polymerization of terminal olefins with Ziegler–Natta catalysts in which the chirality of the catalytic site is the crucial factor for determining enantioface selection [157]. The ^{13}C NMR spectrum of the copolymer, however, shows only three of the four possible triads. This result infers a chain end control of the stereochemistry [159,160]. Labeling experiments have shown that the chain begins through insertion of a styrene unit into a carbomethoxypalladium species and that the insertions of the olefin monomers are secondary [160]. Assuming a chain growth mechanism based on successive insertions of styrene into acyl–metal bond as well as carbon monoxide into an alkyl–metal bond [161], the very effective chain end controlled enantioface selection in the formation of the macromolecular chain is rather surprising. In fact, in such a mechanism the spacer carbon monoxide is present between the chirality inducing center and the newly formed chiral center.

The use of chelate thio ether ligands such as 1,2-bis(thioethyl)ethane brings about a remarkable change in the microstructure of the copolymers produced [162]. The results disclosed that the alternating nature of the copolymerization and the regiospecificity of the enchainment remain, but stereochemical control is completely lost. In the product poly(1-oxo-2-phenyltrimethylene), the distribution of all possible triads is statistical and is close to the thermodynamic equilibrium. The reasons for the loss of stereoselectivity are not clear at the moment. Through coordination to the metal center, the sulfur atoms of the ligand become stereogenic; furthermore, the five-membered chelate ring adopts a puckered chiral conformation. Energy barriers for inversion for these stereogenic elements are known to be very low [163,164]. It appears possible that the production of a sterically irregular material is a consequence of the sterically unequivocal structure of the catalytic center.

In a manner similar to other carbonylation reactions, the control of regioselectivity in this copolymerization is much more difficult when aliphatic olefins are the substrates. The catalyst capable of catalyzing the copolymerization of these substrates, namely $(DPPP)Pd(CF_3COO)_2$, produces a regioirregular polymer, which appears to be also very stereoirregular. The regioregularity is substantially improved by using catalytic systems containing basic diphosphines [165]. The use of chiral ligands such as DIOP or BuDIOP causes a more stereoregular enchainment which, however, cannot be evaluated accurately at the moment [166]. In any case, these results do indicate the possibility of bringing about enantioface selection even in the copolymerization of aliphatic terminal olefins, which may open a new way to the synthesis of stereoregular poly(1-oxo-2-alkyltrimethylene).

5.3. Enantiomer Discriminating Carbonylation Reactions

Enantiomer discriminating reactions (kinetic resolutions) are based on the difference in rate of reaction between two enantiomers with a chiral reagent (or a chiral catalyst) [167]. The main disadvantage of this type of asymmetric reaction is that the maximum yield is limited to 50% unless a rapid racemization process of the starting material exists or there is an intermediate in which the chiral information of the substrate is lost. This does not appear to be the case for the examples of enantiomer discriminating carbonylation reactions that have been investigated. The data are mainly expressed as reactions rates between enantiomers, from which the highest obtainable enantiomeric excess can be evaluated in combination with the conversion [167].

5.3.1. Enantiomer Discrimination of Racemic Olefins by Hydrocarbonylation

The enantiomer discriminating hydroformylation of chiral olefins of the type $RCH(CH_3)CH{=}CH_2$ (where R = C_2H_5, n-C_3H_7, i-C_3H_7, t-C_4H_9, and C_6H_5) and of $C_2H_5CH(CH_3)C(CH_3){=}CH_2$ using rhodium catalysts modified with DIOP has been reported [5]. In the case of chiral terminal olefins, the two isomeric aldehydes can be formed, and therefore enantiomer discrimination should also reflect this possibility. The observed ratio of the rates of the two enantiomers is close to unity, as calculated on the basis of the conversion as well as the enantiomeric purity of the unreacted olefin. Discrimination seems to be mostly related to steric effects, the highest ratios being 1.326 and 1.107 for R = t-C_4H_9, and C_6H_5, respectively. The same catalyst precursor shows no difference in rate for the enantiomers of $C_2H_5CH(CH_3)C(CH_3)$=CH_2, for which formylation takes place only at the terminal position. In the hydrocarbomethoxylation of $RCH(CH_3)CH$=CH_2 (R = C_2H_5 and t-C_4H_9) catalyzed by [(R,R)-DIOP]PdCl$_2$, the ratios of 1.024 and 1.002, respectively, were observed [6,168]. A comparable value (1.017) was also found for $C_2H_5CH(CH_3)C(CH_3)$=CH_2. Discrimination of the two antipodes relative to regioselectivity has been observed in the hydroformylation of 3-phenyl-1-butene to 4-phenylpentanal and 2-methyl-2-phenylbutanal with a rhodium catalyst modified with DIOP [169].

5.3.2. Enantiomer Discrimination of Racemic Halides by Carbonylation

Better results have been obtained in the enantiomer discriminating carbonylation of 1-phenylethyl bromide to 2-phenylpropionic acid (Scheme 13) under phase transfer conditions using (DBA)$_2$ palladium (DBA-dibenzylideneacetone) modified with different chiral ligands under ambient conditions [170]. Monophosphines (e.g., neomenthyldiphenylphospine (NMDPP) and diphosphines (e.g., DIOP) were found to be inefficient for this reaction, whereas aminophosphines and 2-substituted-3,1,2-oxazaphospholanes gave significant enantiomer discrimination. The best ligand used was $(2R,5S)$-(2,6-dimethylphenoxy)-3,2,1-oxazaphospholane (Scheme 13).

Scheme 13

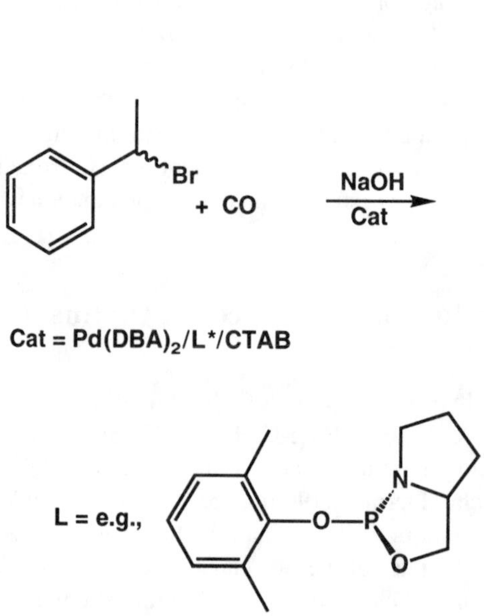

Cat = Pd(DBA)$_2$/L*/CTAB

L = e.g.,

At the ligand-to-metal ratio of 6, the ratio of the rates of the two enantiomers was found to be 5.1. This ratio decreases to 2.4 at the ligand-to-metal ratio of 4. The alternative diastereomeric ligand with the opposite absolute configuration at phosphorus gave essentially no discrimination. It is important to note that the presence of a phase transfer agent is necessary for the enantiomer discrimination to occur in this reaction.

5.3.3. Enantiomer Discrimination of Racemic Aziridines by Carbonylation

It has been shown that monocyclic β-lactams are synthesized through carbonylation of 2-arylaziridines in the presence of rhodium-based catalysts (e.g., [Rh(CO)$_2$Cl]$_2$ or [Rh(1,5-COD)Cl]$_2$ (COD=1,5-cyclooctadiene)) (Scheme 14) [171]. Insertion of carbon monoxide takes place regiospecifically into the aryl-bearing carbon–nitrogen bond with retention of configuration. Chiral diphosphines such as DIOP and BINAP or β-cyclodextrin do not cause appreciable enantiomer discrimination in this catalytic system. Some enantiomer discrimination is realized by diethyl tartrate. Impressive results were obtained using menthol as the chiral modifier which was used at a very high molar ratio with respect to the substrate (3:1). There is some influence of the rhodium precursor used, and the best results have been obtained with [Rh(1,5-COD)Cl]$_2$. Thus, 1-t-butyl-2-phenylaziridine was carbonylated by using l-menthol at 90°C and 20 atm of carbon monoxide to (S)-1-t-butyl-3-phenyl-azetidin-2-one with 99.5% ee in 25% yield. The enantiomeric purity of the recovered substrate (56%) was 85% ee. The N-adamantylaziridine gave comparable results. The ratio of the rates of carbonylation for the two enantiomers cannot be estimated exactly, since only isolated yields were given and the conversion was not reported. The ratio must be quite high, judging from the enantiomeric purity of the

Scheme 14

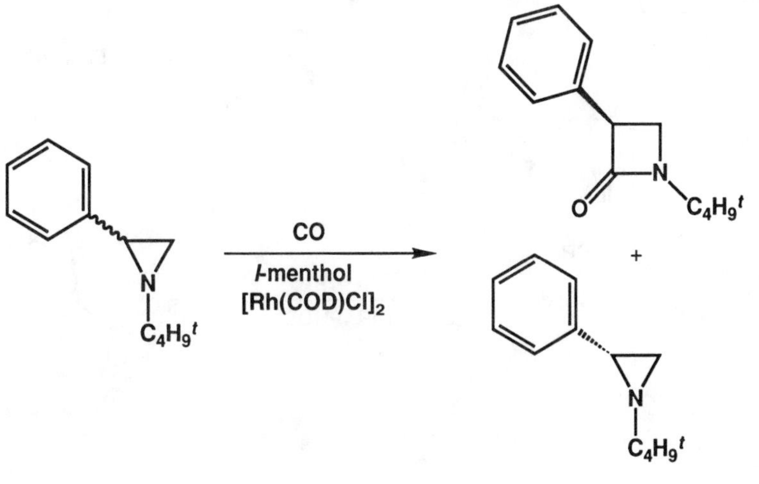

product and the recovered starting material. A further investigation of this reaction is worthwhile.

5.4. Enantiotopic Group Discriminating Synthesis of Lactones

The first example of asymmetric synthesis of lactones entailing enantiotopic group discrimination was reported only very recently [172]. Alkenyl halides containing a plane of symmetry (Schemes 15 and 16) were carbonylated to the corresponding lactones in the presence of a chiral diphosphine–palladium complex and a base. Enantioselectivity should arise from an enanatiotopic group (either OH or $OSiMe_3$) discriminating attack of an acyl–metal intermediate, which is generated through oxidative addition of the halide to the chiral palladium complex followed by carbonyl insertion [172]. In the reaction illustrated in Scheme 15, oxidation of the product to the corresponding ketolactone was observed. The ketolactone thus formed showed a different enantiomeric excess from that of the hydroxylactone. Several chiral ligands were examined for the reaction, and the best enantioselectivity was achieved with BINAP (57% ee). Somewhat lower enantioselectivity (39% ee) was obtained for the analogous reaction shown in Scheme 16. In this case, the corresponding diols gave only racemic product.

Scheme 15

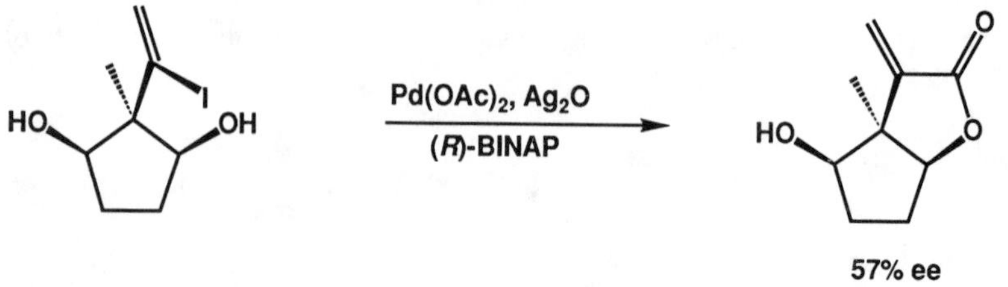

57% ee

Scheme 16

39% ee

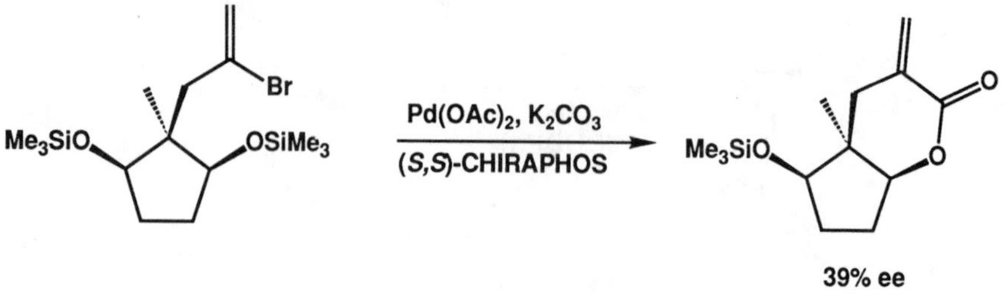

5.5. Conclusion and Outlook

Since the first report of an asymmetric carbonylation reaction nearly 25 years ago, the results achieved in this field are still not very practical. Because of the intrinsic interest of this field, however, further research is obviously worthwhile. Rhodium catalysts are very selective hydroformylation catalysts, but we have not been able to devise appropriate chiral ligands which give well-defined catalytic systems. A promising direction in this field probably is the development of chiral phosphole ligands. For example, 1,2,5-triphenylphosphole exhibited unique behavior with respect to selectivity and seems to give rather unequivocal catalytic species [173,174].

Chiral platinum catalysts are much more enantioselective, but their use is accompanied by problems in chemoselectivity. Nevertheless, very good results have been obtained in the hydroformylation of vinyl aromatics [175], and theoretical work on these systems has recently appeared [176].

The classical palladium dichloride based catalysts for hydrocarbalkoxylation do not appear to be very active and, therefore, improvements in enantioselectivity for these systems may be difficult. The related cationic systems should be investigated further, since they are apparently active under much milder reaction conditions. Similarly, the scope of the $PdCl_2/CuCl_2$ catalytic systems modified with chiral ligands is worth exploring.

The enantiomer discriminating reactions do not seem to be promising at present. The only system that has given excellent results in enantioselectivity to date is the carbonylation of aziridines.

An interesting development can be expected for the asymmetric and catalytic versions of the Pauson–Kand reaction for the syntheses of cyclopentenone derivatives. Pioneering works on both asymmetric [177–179] and catalytic versions [180] of this reaction have been reported. Another possibility for the synthesis of cyclopentane derivatives is offered by hydroacylation reaction [181–184]. In this reaction, an enantioselectivity of 73% ee was obtained in the cyclization of 4-butyl-4-penten-1-al to 4-butylcyclopentanone [183] with a chiral rhodium catalyst. An enantioselective hydrocarbonylation–cyclization of 1,4-dienes to substituted cyclopentanones is possible, as well [185]. Furthermore, the investigation of the asymmetric catalytic synthesis of 2-substituted butanedioic acid derivatives through the double carbonylation of olefins is worthwhile [55,56,160].

Another very interesting aspect of asymmetric carbonylation is represented by the diastereoselective synthesis of copolymers of olefins with carbon monoxide. In the case of styrene, either a syndiotactic or an atactic copolymer is available through suitable modification of a palladium catalyst system. The next question is how to synthesize the corresponding isotactic structure. As far as the copolymerization of aliphatic terminal olefins is concerned, the first problem to be solved appears to be the complete control of regioselectivity in enchainment [186]. Regioregularity is a necessary premise for the control of enantioface discrimination for syndio- and isotactic growth of the macromolecular chain. Furthermore, the possibility of obtaining copolymers with cyclic olefins [187] renders this research very attractive for the investigation of the enantiomer discrimination during the polymerization process, and the syntheses of new polymers that might have interesting properties.

References

1. Pino, P.; Consiglio, G.; Botteghi, C.; Salomon, C. *Adv. Chem. Ser.* **1974**, *132*, 295–323.

2. Siegel, H.; Himmele, W. *Angew. Chem.* **1980**, *92*, 182–187.

3. Cornils, B. In *New Syntheses with Carbon Monoxide;* Falbe, J. (Ed.); Springer-Verlag: Berlin, 1980, pp. 133–137.

4. Botteghi, C.; Gladiali, S.; Bellagamba, V.; Ercoli, R.; Gamba, A. *Chim. Ind. (Milan)* **1980**, *62*, 757–768.

5. Consiglio, G.; Pino, P. *Top. Curr. Chem.* **1982**, *105*, 77–123.

6. Consiglio, G.; Pino, P. *Adv. Chem. Ser.* **1982**, *196*, 371–388.

7. Markó, L. In *Aldehyde (Methoden der Organischen Chemie, Houben-Weyl);* Thieme: Stuttgart, 1983; Vol. E3, pp. 224–230.

8. Stille, J. K. In *Catalysis of Organic Reactions;* Augustine, R. L. (Ed.); Dekker: New York, 1985; pp. 23–36.

9. Ojima, I.: Hirai, K. In *Asymmetric Synthesis:* Morrison, J. D. (Ed.); Academic Press: Orlando, FL, 1985; Vol. 5, pp. 103–146.

10. Markó, L.; Heil, B. *Catal. Rev.* **1973**, *8*, 269–293.

11. Bayer, E.; Schurig, V. *ChemTech.* **1976**, 212–214.

12. Pino, P.; Consiglio, G. In *Fundamental Research in Homogeneous Catalysis;* Tsutsui, M.; Ugo, R. (Eds.); Plenum: New York; Vol. 2, 1977, pp. 147–167.

13. Pittman, C. U., Jr. *Polym. News,* **1978**, *4*, 226–227.

14. Valentine, D., Jr.; Scott, J. W. *Synthesis,* **1978**, 329–356.

15. Pino, P.; Consiglio, G. in *Fundamental Research in Homogeneous Catalysis;* Tsutsui, M. (Ed.); Plenum: New York; Vol. 3, 1979, pp. 519–536.

16. ApSimon, J. W.; Seguin, R. P. *Tetrahedron,* **1979**, *35*, 2797–2842.

17. Kagan, H. B. *Ann. N.Y. Acad. Sci.* **1980**, *333*, 1–15.

18. Stille, J. K.; Fritschel, S. J.; Takaishi, N.; Masuda, T.; Imai, H.; Bertelo, C. A. *Ann. N.Y. Acad. Sci.* **1980**, *333*, 35–44.

19. Bosnich, B.; Fryzuk, M. D. In *Topics in Inorganic and Organometallic Stereochemistry (Topics in Stereochemistry,* Vol. 12); Geoffroy, G. (Ed.); Wiley: New York, 1981; pp. 119–154.

20. Matteoli, U.; Frediani, P.; Bianchi, M.; Botteghi, C.; Gladiali, S. *J. Mol. Catal.* **1981**, *12*, 265–319.

21. Merril, R. E. *Chem. Tech.* **1981**, 118–127.

22. Pino, P.; Consiglio, G. *Pure Appl. Chem.* **1983**, *55*, 1781–1790.

23. Kagan, H. B. In *Comprehensive Organometallic Chemistry;* Wilkinson, G.; Stone, F. G. A.; Abel E. W. (Eds.); Pergamon Press: Oxford, 1983; Vol. 8, pp. 463–498.

24. ApSimon, J. W.; Collier, T. L. *Tetrahedron,* **1986,** *42,* 5157–5254.

25. Chaloner, P. A.; Parker, D. In *Reactions of Coordinated Ligands;* Braterman, P. S. (Ed.); Plenum. New York; 1986; Vol. 1, pp. 789–802.

26. Brunner, H. In *Topics in Stereochemistry;* Eliel, E. L.; Wilen, S. H. (Eds.); Wiley: New York, 1988; Vol. 18, pp. 129–247.

27. Brunner, H. *Synthesis,* **1988,** 645–654.

28. Kagan, H. B. *Bull. Soc. Chim. Fr.* **1988,** 846–853.

29. Ojima, I.; Clos, N.; Bastos, C. *Tetrahedron,* **1989,** *45,* 6901–6939.

30. Noyori, R.; Kitamura, M. In *Modern Synthetic Methods 1989;* Scheffold, R. (Ed.); Springer-Verlag: Berlin, 1989; Vol. 5, pp. 115–198.

31. Blystone, S. L. *Chem. Rev.* **1989,** *89,* 1663–1679.

32. Roelen, O. German Patent 849,548 (1938); *Chem. Zentr.* **1953,** 927.

33. Roelen, O. *Chem. Exp. Didakt.* **1977,** *3,* 119–124.

34. Pittman, C. U., Jr.; Hirao, A. *J. Org. Chem.* **1978,** *43,* 640–646.

35. Hughes, O. R.; Unruh, J. D. *J. Mol. Catal.* **1981,** *12,* 71–83.

36. Exceptions are known: for example, Pino P.; Cioni, P.; Wei, J. *J. Am. Chem. Soc.* **1987,** *109,* 6189–6191.

37. For the concept of ligand promoted activity see, for example: Woodard, S. S.; Finn, M. G.; Sharpless, K. B. *J. Am. Chem. Soc.* **1991,** *113,* 106–113.

38. Pino, P.; Botteghi, C.; Consiglio, G.; Pucci, S. In *Proceedings of the Symposium on Chemistry of Hydroformylation and Related Reactions,* Veszprém, Hungary, 1972, pp. 1–5; *Chem. Abstr.* **1972,** *77,* 139149.

39. Piacenti, F.; Menchi, G.; Frediani, P.; Matteoli, U.; Botteghi, C. *Chim. Ind. (Milan)* **1978,** *60,* 808–809.

40. Major, A. J. *Dissertation,* ETH, Zürich, **1987,** No. 8372.

41. Le Maux, P.; Massonneau, V.; Simmoneaux, G. *J. Organomet. Chem.* **1985,** *284,* 101–108.

42. Garland, M. Private communication.

43. Brown, J. M.; Kent, A. G. *J. Chem. Soc., Perkin Trans. 2,* **1987,** 1597–1607.

44. Hughes, O. R.; Young, D. A. *J. Am. Chem. Soc.* **1981,** *103,* 6636–6642.

45. Oswald, A. A.; Merola, J. S.; Mozeleski, E. J.; Kastrup, R. W.; Reisch, J. C. *Adv. Chem. Ser.* **1981,** *171,* 503–509.

46. Mrowca, J. J.; U.S. Patent 3,876,672 (1972); *Chem. Abstr.* **1976,** *84,* 30432; cf. also Belgian Patent 825,835 (1975).

47. Botteghi, C.; Paganelli, S.; Matteoli, U.; Scrivanti, A.; Ciorciaro, R.; Venanzi, L. M. *Helv. Chim. Acta,* **1990,** *73,* 284–287.

48. Schwager, I.; Knifton, J. F. *J. Catal.* **1976,** *45,* 256–267.

49. Hsu, C.-Y.; Orchin, M. *J. Am. Chem. Soc.* **1975,** *97,* 3553.

50. Longoni, G.; Chini, P. *J. Am. Chem. Soc.* **1976,** *98,* 7225–7231.

51. Tinker, H. B.; Solodar, A. J. Canadian Patent 1,027,141 (1973); *Chem. Abstr.* **1978,** *89,* 42440.

52. Pino, P.; Piacenti, F.; Bianchi, M. In *Organic Syntheses via Metal Carbonyls,* Pino, P.; Wender, I., Eds.; Wiley: New York, 1977; Vol. 2, pp. 233–296.

53. Brown, C. K.; Wilkinson, G. *J. Chem. Soc. A,* **1970,** 2753–2764.

54. Bittler, K.; von Kutepow, N.; Neubauer, D.; Reis, H. *Angew. Chem.* **1968,** *80,* 352–359.

55. James, D. E.; Hines, L. F.; Stille, J. K. *J. Am. Chem. Soc.* **1976,** *98,* 1806–1809.

56. James, D. E.; Stille, J. K. *J. Am. Chem. Soc.* **1976,** *98,* 1810–1823.

57. Pisano, C.; Consiglio, G.; Sironi, A.; Moret, M. *J. Chem. Soc., Chem. Commun.* **1991,** 421–423.

58. Pisano, C.; Mezzetti, A.; Consiglio, G. *Organometallics,* **1992,** *11,* 20–22.

59. Pino, P.; Giannini, U.; Porri, L. In *Encyclopedia of Polymer Science and Engineering,* 2nd ed.; Wiley: New York, 1987; Vol. 8, pp. 147–220.

60. Himmele, W.; Siegel, H.; Aquila, W.; Mueller, F. J. H. German Patent 2,132,414 (1971); *Chem. Abstr.* **1973,** *78,* 97328.

61. Ogata, I.; Ikeda, Y. *Chem. Lett.* **1972,** 487–488.

62. Tanaka, M.; Watanabe, Y.; Mitsudo, T.; Yamamoto, K.; Takegami, Y. *Chem. Lett.* **1972,** 483–485.

63. Tanaka, M.; Watanabe, Y.; Mitsudo, T.; Takegami, Y. *Bull. Chem. Soc. Jpn.* **1974,** *47,* 1698–1703.

64. Salomon, C.; Consiglio, G.; Botteghi, C.; Pino, P. *Chimia,* **1973,** *27,* 215–217.

65. Consiglio, G.; Botteghi, C.; Salomon, C.; Pino, P. *Angew. Chem.* **1973,** *85,* 663–665.

66. Stern, R.; Hirschauer, A.; Sajus, L. *Tetrahedron Lett.* **1973,** 3247–3250.

67. Botteghi, C.; Consiglio, G.; Pino, P. *Justus Liebigs Ann. Chem.* **1974,** 864–869.

68. Botteghi, C.; Salomon, C. *Tetrahedron Lett.* **1974,** 4285–4288.

69. Botteghi, C. *Gazz. Chim. Ital.* **1975,** *105,* 233–245.

70. Tanaka, M.; Ikeda, Y.; Ogata, I. *Chem. Lett.* **1975,** 1115–1118.

71. Tanaka, M.; Ogata, I. *J. Chem. Soc., Chem. Commun.* **1975,** 735.

72. Stefani, A.; Pino, P. *Helv. Chim. Acta,* **1977,** *60,* 518–520.

73. Lednor, P. W.; Beck, W.; Fick, H. G.; Zippel, H. *Chem. Ber.* **1978,** *111,* 615–628.

74. Piacenti, F.; Menchi, G.; Frediani, P.; Matteoli, U.; Micera, G. *Chim. Ind. (Milan)* **1978,** *60,* 16–17.

75. Hayashi, T.; Tanaka, M.; Ikeda, Y.; Ogata, I. *Bull. Chem. Soc. Jpn.* **1979,** *52,* 2605–2608.

76. Watanabe, Y.; Mitsudo, T.; Yasunori, Y.; Kikuchi, J.; Takegami, Y. *Bull. Chem. Soc. Jpn.* **1979,** *52,* 2735–2736.

77. Fritschel, S. J.; Ackerman, J. J. H.; Keyser, T.; Stille, J. K. *J. Org. Chem.* **1979,** *44,* 3152–3157.

78. Stefani, A.; Tatone, D.; Pino, P. *Helv. Chim. Acta* **1979,** *62,* 1098–1102.

79. Botteghi, C.; Branca, M.; Saba, A. *J. Organomet. Chem.* **1980,** *184,* C17–C19.

80. Beker, Y.; Eisenstadt, A.; Stille, J. K. *J. Org. Chem.* **1980,** *45,* 2145–2151.

81. Hobbs, C. F.; Knowles, W. S. *J. Org. Chem.* **1981,** *46,* 4422–4427.

82. Gladiali, S.; Faedda, G.; Marchetti, M.; Botteghi, C. *J. Organomet. Chem.* **1983,** *244,* 289–302.

83. Hayashi, T.; Tanaka, M.; Ogata, I.; Kodama, T.; Takahashi, T.; Uchida, Y.; Uchida, T. *Bull. Chem. Soc. Jpn.* **1983**, *56*, 1780–1785.

84. Gladiali, S.; Chelucci, G.; Marchetti, M.; Azzena, U. *Chim. Ind. (Milan)* **1985**, *67*, 387–391.

85. Kollár, L.; Floris, B.; Pino, P. *Chimia*, **1986**, *40*, 201–202.

86. Consiglio, G.; Morandini, F.; Scalone, M.; Pino, P. *J. Organomet. Chem.* **1985**, *279*, 193–202.

87. Brown, J. M.; Cook, S. J. *Tetrahedron*, **1986**, *42*, 5105–5109.

88. Kollár, L.; Consiglio, G.; Pino, P. *Chimia*, **1986**, *40*, 428–429.

89. Thomson, R. J.; Jackson, W. R.; Haarburger, D.; Klabunovsky, E. I.; Pavlov, V. A. *Aust. J. Chem.* **1987**, *40*, 1083–1106.

90. Pottier, Y.; Mortreux, A.; Petit, F. *J. Organomet. Chem.* **1989**, *370*, 333–342.

91. Doyle, M. M.; Jackson, W. R.; Perlmutter, P. *Tetrahedron Lett.* **1989**, *30*, 5357–5360.

92. Kollár, L.; Consiglio, G.; Pino, P. *J. Organomet. Chem.* **1990**, *386*, 389–394.

93. Consiglio, G.; Kollár, L.; Koelliker, R. *J. Organomet. Chem.* **1990**, *396*, 375–383.

94. Gladiali, S.; Pinna, L. *Tetrahedron: Asymmetry*, **1990**, *1*, 693–696.

95. Consiglio, G.; Rama, F. *J. Mol. Catal.* **1991**, *66*, 1–5.

96. Gladiali, S.; Pinna, L. *Tetrahedron: Asymmetry*, **1991**, *2*, 623–632.

97. Hsu, C.-Y. Dissertation, Cincinnati, OH, 1975; *Chem. Abstr.* **1975**, *82*, 154899.

98. Consiglio, G.; Pino, P. *Helv. Chim. Acta*, **1976**, *59*, 642–645.

99. Consiglio, G.; Pino, P. *Isr. J. Chem.* **1976/1977**, *15*, 221–222.

100. Kawabata, Y.; Suzuki, T. M.; Ogata, I. *Chem. Lett.* **1978**, 361–362.

101. Consiglio, G.; Arber, W.; Pino, P. *Chim. Ind. (Milan)* **1978**, *60*, 396–400.

102. Pittman, C. U., Jr.; Kawabata, Y.; Flowers, L. I. *J. Chem. Soc., Chem. Commun.* **1982**, 473–474.

103. Stille, J. K.; Parrinello, G. *J. Mol. Catal.* **1983**, *21*, 203–210.

104. Consiglio, G.; Pino, P.; Flowers, L. I.; Pitmann, C. U., Jr. *J. Chem. Soc., Chem. Commun.* **1983**, 612–613.

105. Haelg, P.; Consiglio, G.; Pino, P. *J. Organomet. Chem.* **1985**, *296*, 281–290.

106. Parrinello, G.; Deschenaux, R.; Stille, J. K. *J. Org. Chem.* **1986**, *51*, 4189–4195.

107. Kollár, L.; Consiglio, G.; Pino, P. *J. Organomet. Chem.* **1987**, *330*, 305–314.

108. Parrinello, G.; Stille, J. K. *J. Am. Chem. Soc.* **1987**, *109*, 7122–7127.

109. Mutez, S.; Mortreux, A.; Petit, F. *Tetrahedron Lett.* **1988**, *29*, 1911–1914.

110. Kollár, L.; Bakos, J.; Tóth, I.; Heil, B. *J. Organomet. Chem.* **1988**, *350*, 227–284.

111. Kollár, L.; Bakos, J.; Tóth, I.; Heil, B. *J. Organomet. Chem.* **1989**, *370*, 257–261.

112. Kollár, L.; Bakos, J.; Heil, B.; Sándor, P.; Szalontai, G. *J. Organomet. Chem.* **1990**, *385*, 147–152.

113. Consiglio, G.; Nefkens, S. C. A. *Tetrahedron: Asymmetry*, **1990**, *1*, 417–421.

114. Paganelli, S.; Matteoli, U.; Scrivanti, A. *J. Organomet. Chem.* **1990**, *397*, 119–125.

115. Consiglio, G.; Nefkens, S. C. A.; Borer, A. *Organometallics*, **1991**, *10*, 2046–2051.

116. Consiglio, G.; Roncetti, L. *Chirality,* **1991,** *3,* 341–344.

117. Stille, J. K.; Su, H.; Brechot, P.; Parrinello, G.; Hegedus, L. S. *Organometallics,* **1991,** *10,* 1183–1189.

118. Kollár, L.; Sádor, P.; Szalontai, G. *J. Mol. Catal.* **1991,** *67,* 191–198.

119. Schurig, V. *J. Mol. Catal.* **1979,** *6,* 75–77.

120. Bailey, N. A.; Jassal, V. S.; Vefghi, R.; White, C. *J. Chem. Soc., Dalton Trans.* **1989,** 2815–2822.

121. Vaccher, C.; Mortreux, A.; Petit, F.; Picavet, J.-P.; Sliwa, H.; Murrall, N. W.; Welch, A. J. *Inorg. Chem.* **1984,** *23,* 3613–3617.

122. Wink, D. J.; Kwok, T. J.; Yee, A. *Inorg. Chem.* **1990,** *29,* 5006–5008; for recent successful developments cf.: Sakai, N.; Nozaki, K.; Mashima, K.; Takaya, H. *Tetrahedron: Asymmetry,* **1992,** *3,* 583–7.

123. Collin, J.; Jossart, C.; Balavoine, G. *Organometallics,* **1986,** *5,* 203–208.

124. Botteghi, C.; Consiglio, G.; Pino, P. *Chimia,* **1972,** *26,* 141–142.

125. Pino, P.; Salomon, C.; Botteghi, C.; Consiglio, G. *Chimia,* **1972,** *26,* 655.

126. Mignani, G.; Patin, H.; Dabard, R. *J. Organomet. Chem.* **1979,** *169,* C19–C21.

127. Botteghi, C.; Consiglio, G.; Pino, P. *Chimia,* **1973,** *27,* 477–478.

128. Consiglio, G. *Helv. Chim. Acta,* **1976,** *59,* 124–126.

129. Consiglio, G.; Pino, P. *Chimia,* **1976,** *30,* 193–194.

130. Consiglio, G. *J. Organomet. Chem.* **1977,** *132,* C26–C28.

131. Consiglio, G.; Haelg, P.; Pino, P. *Atti Accad. Naz. Lincei, Cl. Sci. Fis., Mat. Nat., Rend.* **1980,** *68,* 533–538.

132. Hayashi, T.; Tanaka, M.; Ogata, I. *Tetrahedron Lett.* **1978,** 3925–3926.

133. Cometti, G.; Chiusoli, G. P. *J. Organomet. Chem.* **1982,** *236,* C21–C22.

134. Delogu, G.; Faedda, G.; Gladiali, S. *J. Organomet. Chem.* **1984,** *268,* 167–174.

135. Hayashi, T.; Tanaka, M.; Ogata, I. *J. Mol. Catal.* **1984,** *26,* 17–30.

136. Japanese Patent Tokkyo Koho 58,167,541, to the Agency of Industrial Sciences and Technology; *Chem. Abstr.* **1984,** *100,* 68012.

137. Hiyama, T.; Wakasa, N.; Kusumoto, T. *Synlett,* **1991,** 569–570.

138. Alper, H.; Hamel, N. *J. Chem. Soc., Chem. Commun.* **1990,** 135–136.

139. Alper, H.; Hamel, N. *J. Am. Chem. Soc.* **1990,** *112,* 2803–2804.

140. Pino, P. *Ann. N.Y. Acad. Sci.* **1983,** *415,* 111–128.

141. Consiglio, G.; Morandini, F.; Haelg, P.; Pino, P. *J. Mol. Catal.* **1990,** *60,* 363–374.

142. Stefani, A.; Consiglio, G.; Botteghi, C.; Pino, P. *J. Am. Chem. Soc.* **1977,** *99,* 1058–1063.

143. Haelg, P.; Consiglio, G.; Pino, P. *Helv. Chim. Acta,* **1981,** *64,* 1865–1869.

144. Tolman, C. A.; Faller, J. W. In *Homogeneous Catalysis with Metal Phosphines Complexes;* Pignolet, L. H. (Ed.); Plenum: New York, 1983; p. 13–109.

145. Pino, P.; Stefani, A.; Consiglio, G. In *Catalysis in Chemistry and Biochemistry, Theory and Experiments,* Pullman B. (Ed.); Reidel: Dordrecht, 1979; pp. 347–354.

146. Pino, P.; Piacenti, F.; Bianchi, M. In *Organic Syntheses via Metal Carbonyls;* Pino, P.; Wender, I. (Eds.); Wiley: New York, 1977; pp. 43–215.

147. Rieu, J.-P.; Boucherle, A.; Cousse, H.; Mouzin, G. *Tetrahedon*, **1986**, *42*, 4095–4131.

148. Botteghi, C.; Ganzerla, R.; Lenarda, M.; Moretti, G. *J. Mol. Catal.* **1987**, *40*, 129–182.

149. Slough, G. A.; Bergman, R. G.; Heathcock, C. H. *J. Am. Chem. Soc.* **1989**, *111*, 938–949.

150. Kijenski, J.; Glinski, M.; Bielawski, K. In *Heterogeneous Catalysis and Fine Chemicals;* Guisnet, M.; Barrault, J.; Bouchoule, C.; Duprez, D.; Montassier, C.; Perót, G. (Eds.); Elsevier: Amsterdam, 1988; pp. 379–386.

151. Stille, J. K. In *Macromolecules;* Benoit, H.; Rempp, P. (Eds.); Pergamon: Oxford, 1982; p. 99–111.

152. Consiglio, G.; Pino, P. *Gazz. Chim. It.* **1975**, *105*, 1133–1135.

153. Milstein, D. *Acc. Chem. Res.* **1988**, *21*, 428–434.

154. Ballauf, F.; Bayer, O.; Teichmann, L. German Patent 863,711 (1941); see also Hoyer, H.; Fitzky, H.-G. *Makromol. Chem.* **1972**, *161*, 49–56.

155. Drent, E. European Patent Application 181,014 (1985); *Chem. Abstr.* **1985**, *105*, 98172.

156. Pino, P.; Muelhaupt, R. *Angew. Chem.* **1980**, *92*, 869–887.

157. Farina, M. In *Topics in Stereochemistry;* Eliel, E. L.; Wilen, S. H. (Eds.); Wiley: New York, 1987; Vol. 17, pp. 1–111.

158. Drent, E. European Patent Application 229,408 (1986); *Chem. Abstr.* **1988**, *108*, 6617.

159. Corradini, P.; De Rosa, C.; Panunzi, A.; Petrucci, G.; Pino, P. *Chimia*, **1990**, *44*, 52–54.

160. Barsacchi, M.; Consiglio, G.; Medici, L.; Petrucci, G.; Suter, U. W. *Angew. Chem.* **1991**, *103*, 992–994.

161. Lai, T.-W.; Sen, A. *Organometallics*, **1984**, *3*, 866–870.

162. VanDorn, J. A.; Drent, E. European Patent Application 345,847 (1986); *Chem. Abstr.* **1990**, *112*, 199339.

163. Cross, R. J.; Rycroft, D. S.; Sharp, D. W. A. *J. Chem. Soc., Dalton Trans.* **1980**, 2434–2441.

164. Abel, E. W.; Orrell, K. G.; Bhargava, S. K. In *Progress in Inorganic Chemistry;* Lippard, S. J. (Ed.); Wiley: New York, 1984; Vol. 32, pp. 1–118.

165. Van Doorn, J. A.; Wong, P. K.; Sudmeier, O. European Patent Application 376,364 (1989); *Chem. Abstr.* **1991**, *114*, 24797.

166. Wong, P. K. European Patent Application 384,517 (1989); *Chem. Abstr.* **1991**, *114*, 103079.

167. Kagan, H. B.; Fiaud, J. C. In *Topics in Stereochemistry;* Eliel, E. L.; Wilen, S. H. (Eds.); Wiley: New York, 1988; Vol. 18, pp. 249–330.

168. Consiglio, G. Unpublished results.

169. Stefani, A.; Tatone, D.; Pino, P. *Helv. Chim. Acta*, **1976**, *59*, 1639–1641.

170. Arzoumanian, H.; Buono, G.; Choukrad, M.; Petrignani, J.-F. *Organometallics*, **1988**, *7*, 59–62.

171. Calet, S.; Urso, F.; Alper, H. *J. Am. Chem. Soc.* **1989**, *111*, 931–934.

172. Suzuki, T.; Uozumi, Y.; Shibasaki, M. *J. Chem. Soc., Chem. Commun.*, **1991**, 1593–1595.

173. Neibecker, D.; Reau, R. *J. Mol. Catal.* **1989**, *57*, 153–163.

174. Bergiunhou, C.; Neibecker, D.; Reau, R. *J. Chem. Soc., Chem. Commun.*, **1988**, 1370–1371.

175. Botteghi, C.; Paganelli, S.; Schionato, A.; Marchetti, M. *Chirality,* **1991**, *3,* 355–369.

176. Castonguay, L. A.; Rappé, A. K.; Casewit, C. J. *J. Am. Chem. Soc.* **1991**, *113,* 7177–7183.

177. Brunner, H. *Pure Appl. Chem.,* **1990**, *62,* 589–594.

178. Brunner, H. In *Organometallics in Organic Synthesis;* Werner, H.; Erker, G. (Eds.); Springer-Verlag: Berlin, 1989; Vol. 2, pp. 277–289.

179. Poch, M.; Valento, E.; Moyano, A.; Pericas, M. A.; Castro, J.; DeNicola, A.; Greene, A. E. *Tetrahedron Lett.* **1990**, *31,* 7505.

180. Rautenstrauch, V.; Mégard, P.; Conesa, J.; Kuester, W. *Angew. Chem.* **1990**, *102,* 1441–1443.

181. James, B. R.; Joung, C. G. *J. Chem. Soc., Chem. Commun.,* **1983**, 1215–1216.

182. James, B. R.; Joung, C. G. *J. Organomet. Chem.* **1985**, *285,* 321–332.

183. Taura, Y.; Tanaka, M.; Funakoshi, K.; Sakai, K. *Tetrahedron Lett.* **1989**, *30,* 6349–6353.

184. Taura, Y.; Tanaka, M.; Wu, X.-M.; Funakoshi, K.; Sakai, K. *Tetrahedron,* **1991**, *47,* 4879–4888.

185. Eilbracht, P.; Acker, M.; Haedrich, I. *Chem. Ber.* **1988**, *121,* 519–524.

186. Batistini, A.; Consiglio, G.; Suter, U. W. *Angew. Chem.* **1992**, *104,* 306–7; for recent developments cf.: Batistini, A.; Consiglio, G. *Organometallics,* **1992**, *11,* 1766–69.

187. Van Deursen, J. H.; Van Doorn, J. A.; Drent, E.; Wong, P. K. European Patent Application 390,237 (1990); *Chem. Abstr.* **1991**, *114,* 103010.

CHAPTER

6

Asymmetric Hydrosilylation

Henri Brunner

Institut für Anorganische Chemie
Universität Regensburg
Regensburg, Germany

Hisao Nishiyama and Kenji Itoh

School of Materials Science
Toyohashi University of Technology
Toyohashi, Japan

6.1. Introduction

Asymmetric hydrosilylation of ketones, imines, and olefins with chiral transition metal catalysts provides an effective route to optically active alcohols, amines, and alkanes [1]. In the hydrosilylation of ketones and imines, silyl ethers and silyl amines are formed, which can be hydrolyzed to give the corresponding alcohols and amines, respectively. The adducts of hydrosilanes with olefins can be utilized for the preparation not only of organosilanes with chiral alkyl groups but also of optically active alcohols via oxidative cleavage of the silicon–carbon bonds.

The asymmetric hydrosilylation of ketones with rhodium catalysts bearing optically active phosphines originated in the early 1970s [2], after the catalytic activity of the Wilkinson catalyst $RhCl(PPh_3)_3$ for the hydrosilylation had been reported [3]. The enantioselectivities attained in the 1970s were moderate: for example, 50–85% ee with DIOP–Rh and AMPHOS–Rh catalysts.

In the early 1980s, the asymmetric hydrosilylation of ketones using chiral nitrogen ligands began. These chiral nitrogen ligands containing a pyridine skeleton increased the asymmetric inductions year by year. Recently, some of them have attained an extremely high level of enantioselectivity: for example, 95–99% ee with Pythia–(Et,H)–Rh and Pybox–*i*-Pr–catalysts. In the 1970s, asymmetric induction in the hydrosilylation of olefins, catalyzed by palladium and platinum catalysts, was relatively low (i.e., around 50% ee). About 10 years ago, the discovery of the oxidative cleavage of carbon–silicon bonds with hydrogen peroxide or *m*-chloroper-benzoic acid opened a valuable route to optically active secondary alcohols and other chiral functionalized compounds from olefins via asymmetric hydrosilylation as a synthetic method [4]. Therefore, efforts in the asymmetric hydrosilylation of olefins have continued, but remarkable progress was not reported in the 1980s.

In 1990 the enantioselectivity of an intramolecular olefin hydrosilylation reached more than 90% ee, giving a chiral 1,3-diol [5]. Furthermore, a remarkable break-through in the asymmetric hydrosilylation of simple 1-alkenes emerged in 1991 using a palladium catalyst with a newly designed chiral monodentate phosphine ligand, MOP [6].

Another important aspect of asymmetric hydrosilylation is asymmetric induction on the silicon atom, as a way to supply optically active silicon derivatives. 1-Naphthylphenylsilane has frequently been used as a prochiral silane that is con-verted to 1-NpPhSiH(*R*) via hydrosilylation of ketones and subsequent alkylation. Although the enantioselectivity with diethyl ketone was below 50% ee, the di-astereoselective reaction with (−)-menthone catalyzed by a DIOP–Rh catalyst at-tained more than 80% ee [7]. Remarkable progress in the enantioselectivity of this reaction was recently made using a Cy–BINAP–Rh catalyst and diethyl ketone [8].

Accordingly, it is evident that with the design of highly efficient chiral ligands and with the wide range of substrates, asymmetric hydrosilylation has reached extremely high levels of enantioselectivity. Representative chiral ligands for asym-metric hydrosilylation are listed in Charts 1 and 2.

6.2. Asymmetric Hydrosilylation of Ketones

The enantioselective hydrosilylation of ketones was first reported using chiral plat-inum catalysts in 1972 [9a]. However, the "Wilkinson complex," the $RhCl(PPh_3)_3$, proved to be also an effective catalyst. Thus, most of the studies on the asymmetric hydrosilylation of ketones used chiral rhodium complexes with optically active phosphine ligands [1,3]. The reactions were frequently carried out with acetophe-none as the model ketone and dihydrosilanes, such as diphenylsilane and 1-naphthylphenylsilane, either in benzene or toluene solution or without solvent at 0°C ~ room temperature: after hydrolysis of the reaction mixture, optically active 1-phenylethanol was obtained. 1-Naphthylphenylsilane frequently induced higher enantioselectivities than diphenylsilane in the reaction catalyzed by chiral phos-phine–rhodium systems.

Until the early 1980s, most of the chiral phosphine–rhodium catalysts afforded only low to moderate enantioselectivities. Among them, 58% ee (*S*) with (+)-DIOP/[Rh(COD)Cl]₂ [10] and 65% ee (*S*) with [Glucophinite–Rh(COD)]BF₄ [11] in the reduction of acetophenone with 1-NpPhSiH₂ were the two best results (Scheme 1).

In 1982 it was reported that nitrogen-containing ligands such as pyridylimines **11–14** derived from 2-pyridinecarboxaldehyde and optically active alkylamines (1-phenethylamine and 3-aminomethylpinane) were superior to optically active phosphine ligands for the rhodium complex catalyzed reactions. Thus, the catalyst

Chart 1   Chiral phosphine ligands for asymmetric hydrosilylation.

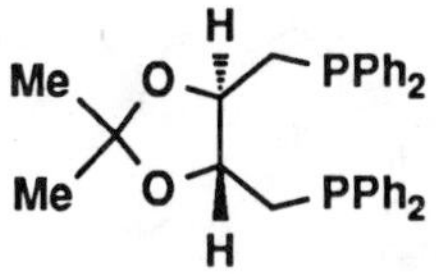

1 (*R,R*)-(-)-DIOP 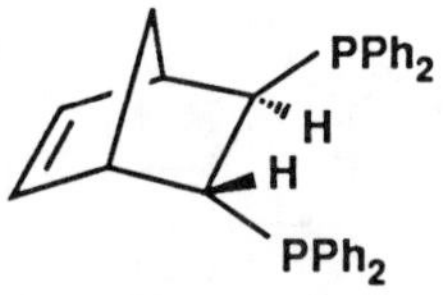   **2** (*R,R*)-(-)-NORPHOS 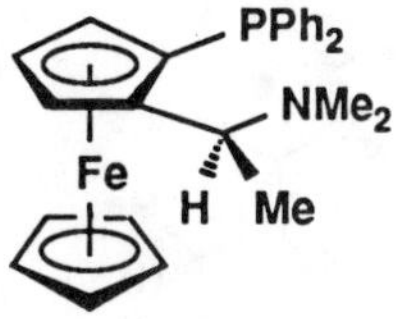   **3** (*S*)-(*R*)-(+)-PPFA

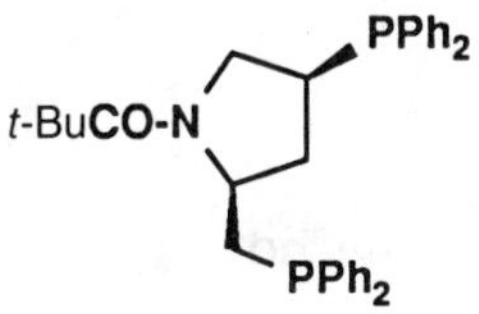

4 (*S,S*)-(-)-PPPM 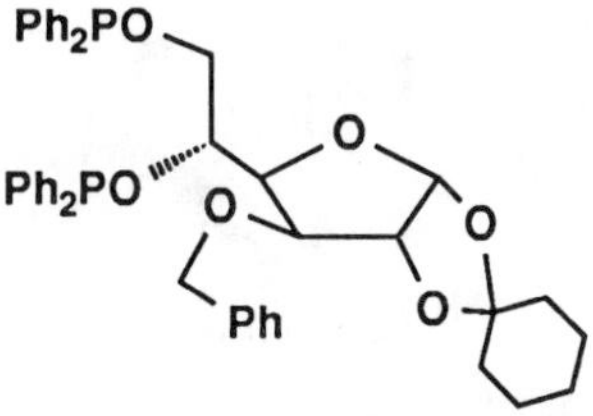   **5** Glucophinite 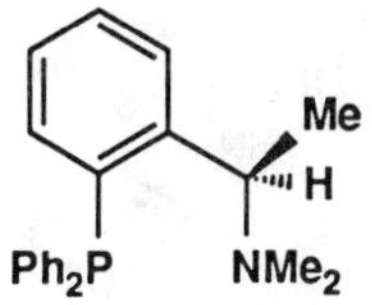   **6** (*R*)-(+)-AMPHOS

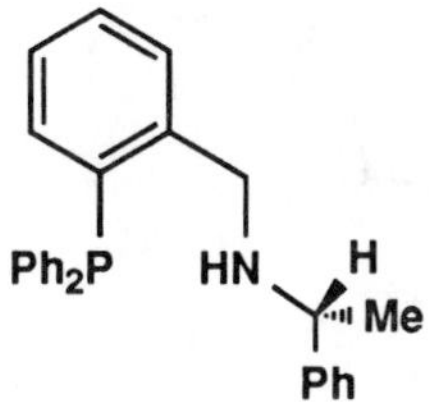

7 (*R*)-(+)-AMINPHOS 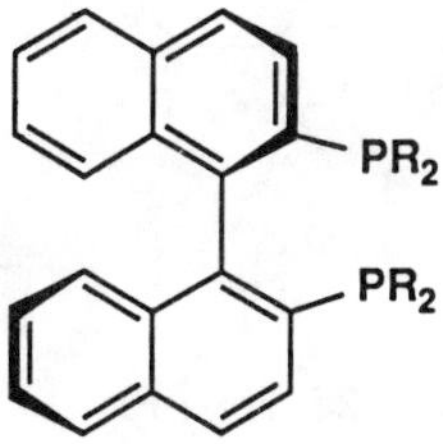   **8** (*R*)-(+)-BINAP   **10**   (*S*)-(-)-MOP
                                     R = Ph
                                9 (*R*)-(+)-Cy-BINAP
                                     R = cyclohexyl

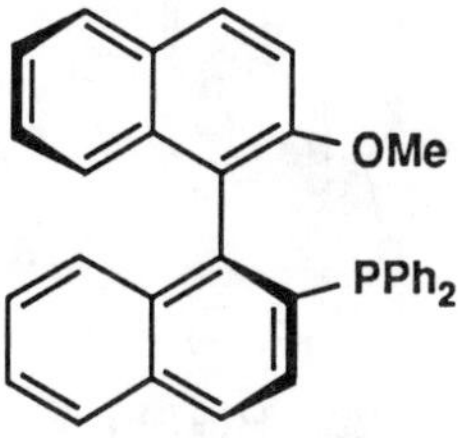

generated in situ from imine **12** and [Rh(COD)Cl]$_2$ gave 79% ee (*S*) in the reaction of acetophenone with diphenylsilane [12]. Excess of the ligand, up to tenfold with respect to rhodium, was necessary to achieve high enantioselectivity. In contrast to optically active phosphine ligands, the chiral nitrogen ligands are easily accessible from readily available optically active amines. Another example is the catalyst arising from menthyl-α,α'-dipyridine derivative **15** and [Rh(COD)Cl]$_2$, which attained 72% ee for acetophenone with the use of diphenylsilane [13].

Chart 2   Chiral nitrogen ligands for asymmetric hydrosilylation.

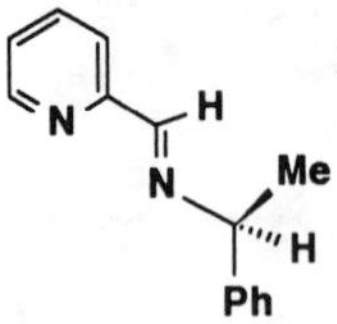

11

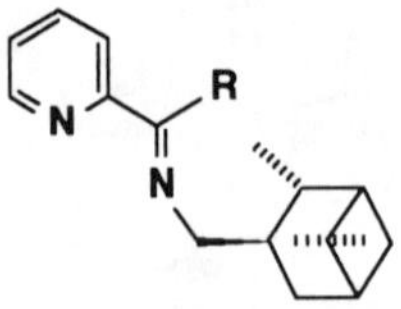

12 R = H
13 R = Me
14 R = Ph

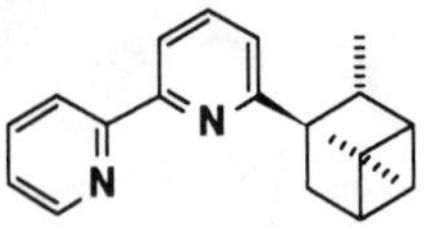

15

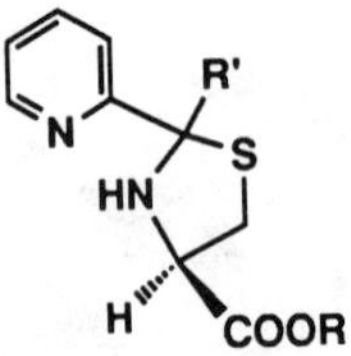

(*R*)-Pythia

16 R,R' = Me,H
17 R,R' = Me,Me
18 R,R' = Et,H

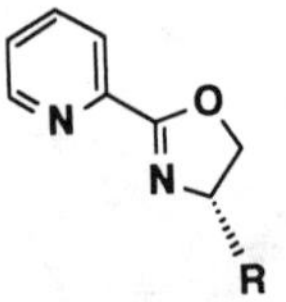

(*S*)-Pymox

19 R = *i*-Pr
20 R = *t*-Bu

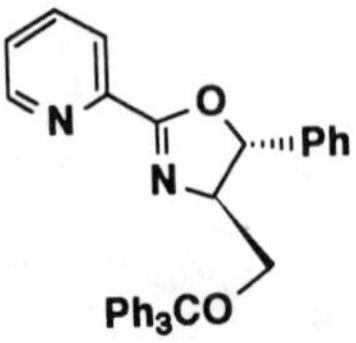

21

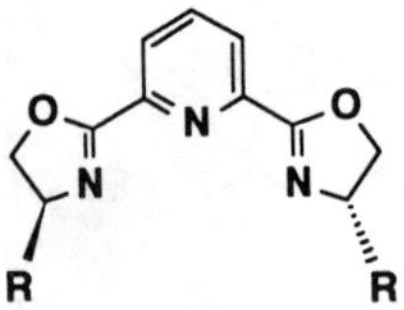

(*S,S*)-Pybox

22 R = *i*-Pr
23 R = *t*-Bu
24 R = CH₂Ph

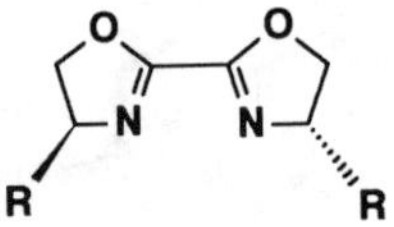

25 R = *i*-Pr
26 R = CH₂Ph

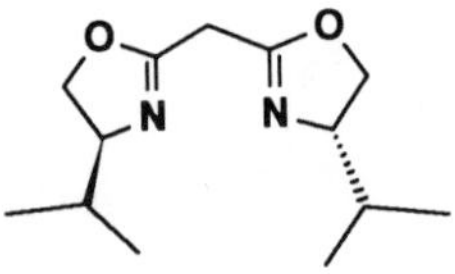

27

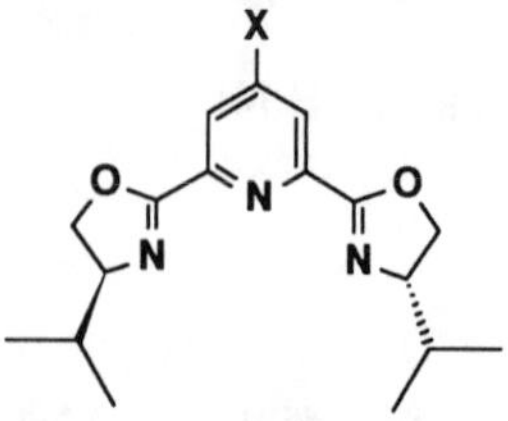

(*S,S*)-4-X-Pybox

28 X = Me₂N
29 X = MeO
30 X = Cl

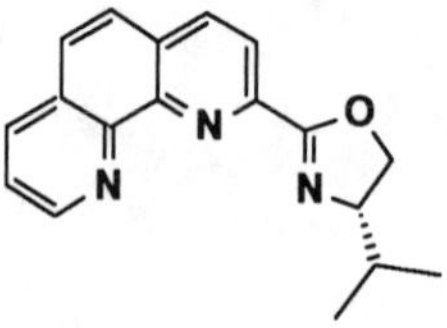

31

Scheme 1

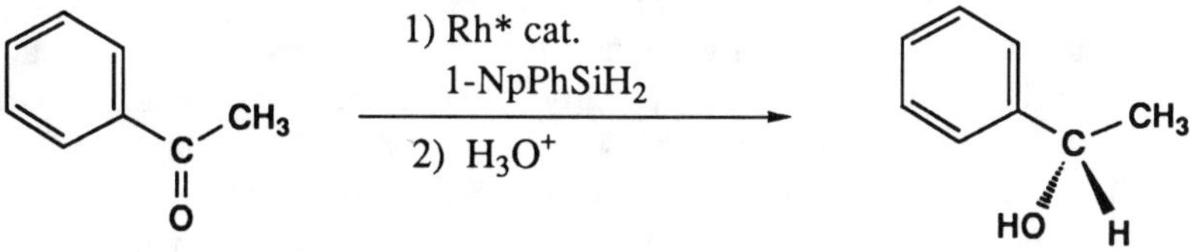

Rh* cat	yield (%)	%ee
[Rh(COD)Cl]$_2$, (+)-DIOP	100	58 (S)
[Rh(C$_2$H$_4$)$_2$Cl]$_2$, Glucophinite	65	65 -

An interesting new ligand system, 2-oxazolinylpyridines (*mono-oxazolinylpyri-dine*, Pymox), appeared for the Cu-catalyzed asymmetric monophenylation of diols with triphenylbismuthdiacetate in 1986 [14]. The chirality of the oxazoline ring is derived from optically active β-amino alcohols, which are readily available in large numbers and quantities from optically active amino acids. Acetophenone was reduced to 1-phenylethanol with 60–62% ee (*R*) by using (*S*)-Pymox–*i*-Pr (**19**) and 83–91% ee (*R*) by using (*S*)-Pymox–*t*-Bu **20** in the presence of [Rh(COD)Cl]$_2$ and diphenylsilane [15,16]. Another substituted oxazolinylpyridine (**21**) exhibited 80% ee (*R*) for the hydrosilylation of acetophenone with [Rh(C$_2$H$_4$)$_2$Cl]$_2$ and 1-naphthyl-phenylsilane, but the enantioselectivity decreased to below 70% ee with diphenyl-silane [17].

Normally, the hydrosilylation of ketones is carried out either without solvent or with solvents such as toluene or benzene. However, in the case of the Pymox–rhodium catalysts the use of tetrachloromethane leads to a remarkable increase of the enantioselectivity compared to toluene, ether, or dichloromethane ("CCl$_4$ effect"). This is probably due to a change in the catalytically active species caused by the oxidative addition of CCl$_4$ [16]. In fact, stoichiometric CCl$_4$ quantities with respect to Rh are sufficient to bring about the "CCl$_4$ effect." Similar oxidative addition reactions of alkyl chlorides to rhodium(I) species in the presence of another nitrogen ligand, Pybox, were recently reported to give stable alkyl–RhCl$_2$(Pybox) complexes together with RhCl$_3$(Pybox) [18].

In 1984 chiral *pyridinethiaz*olidines (Pythia) were introduced as the best nitrogen ligands to date for the asymmetric hydrosilylation of acetophenone with [Rh(COD)Cl]$_2$ [19]. In the presence of 0.66 mol % of [Rh(COD)Cl]$_2$, (*R*)-Pythia–(Et,H), **18**, reduced acetophenone to 1-phenylethanol with 97.6% ee (*R*) with a 13-fold excess of the ligand at −20°C for 120 hours or with 86.7% ee (*R*) under the standard conditions with a 8-fold excess of the ligand at 0°C for 18 hours [19]. Pythia–(Me,H) (**16**) and Pythia–(Me,Me) (**17**) also induced high enantioselec-tivities, 87.2% and 84.1% ee, respectively, in this reaction. In the Pythia systems, the use of 1-naphthylphenylsilane as a reducing agent gave relatively low enan-tioselectivities (i.e., ca. 50% ee) [20] (Scheme 2).

In 1989 the chiral *bis*(*oxazolinyl*)*pyridine*, Pybox, was introduced as a C_2-symmetric nitrogen ligand for the asymmetric hydrosilylation of ketones [15]. Pybox is a terdentate pyridine ligand readily synthesized by the condensation of pyridine-2,6-dicarboxylic acid and optically active β-amino alcohols. The combination of (*S*,*S*)-Pybox–*i*-Pr **22** (3 mol %) and [Rh(COD)Cl]$_2$ (1 mol %) with diphenylsilane reduced acetophenone at 0°C for 28 hours with 76% ee (*S*). Excellent enantioselectivity (83% ee, *S*) was found in the same hydrosilylation by using (Pybox–*i*-Pr)RhCl$_3$ (**32**) (1 mol %) and AgBF$_4$ (2 mol %), which is the best result to date with the use of an equimolar amount of a nitrogen ligand with respect to rhodium metal. Further improvement of the enantioselectivity up to 95% ee (*S*) was attained simply by using a four-fold excess of Pybox **22** in the same system [21] (Scheme 2).

Scheme 2

[Rh(COD)Cl]$_2$ (0.66 mol%)
Pythia-(Et,H) (**18**) (13 equiv. to Rh)   99 %, 97.6 %ee (*R*)
Ph$_2$SiH$_2$ (1.2 equiv.), -20°C, 120 h
then H$_3$O$^+$

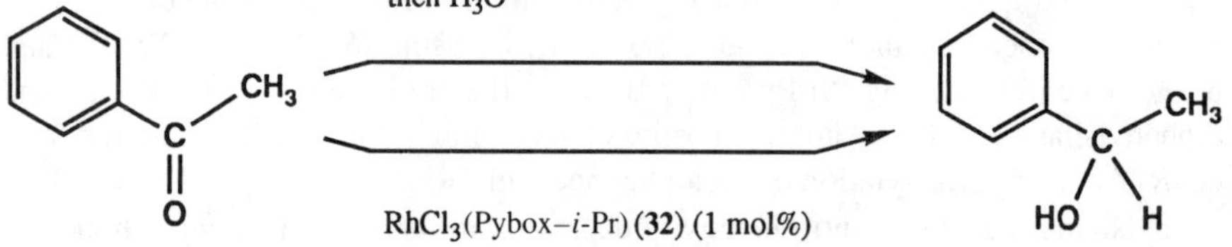

RhCl$_3$(Pybox–*i*-Pr) (**32**) (1 mol%)
Pybox–*i*-Pr (**22**) (0–4 mol%)
AgBF$_4$ (2 mol%), Ph$_2$SiH$_2$ (1.6 equiv.)
THF, 0°C, 2–6 h, then H$_3$O$^+$

Pybox–*i*-Pr(**22**) (no addn)	89 %,	83 %ee (*S*)
(4 mol%)	94 %,	95 %ee (*S*)

The rhodium trichloride complex **32** (Chart 3) itself does not react with diphenylsilane at ambient temperature. Its activation by a silver salt or a Lewis acid is crucial for formation of the corresponding cationic rhodium complex, which can smoothly catalyze the hydrosilylation of ketones. Enantioselectivity with the Pybox–Rh systems is also influenced by counteranions.

In the presence of excess ligand in the Pybox–Rh system, it was found that a facile and fast ligand exchange took place between the coordinating Pybox and the free Pybox during the hydrosilylation of ketones. For example, addition of a 1:1 mixture of (*S*,*S*)-Pybox–*i*-Pr **22** and (*R*,*R*)-Pybox–*i*-Pr (four-fold to **32**) decreased the enantioselectivity from 95% ee (*S*) to 39% ee (*S*) for acetophenone. Moreover, the experiments with carefully mixed ligands exhibited linearity between the composition of the chiral catalyst and the enantiomeric excess of the product. Thus, it is reasonable to postulate that the active catalyst has only one Pybox ligand attached to the rhodium atom [21].

Chart 3

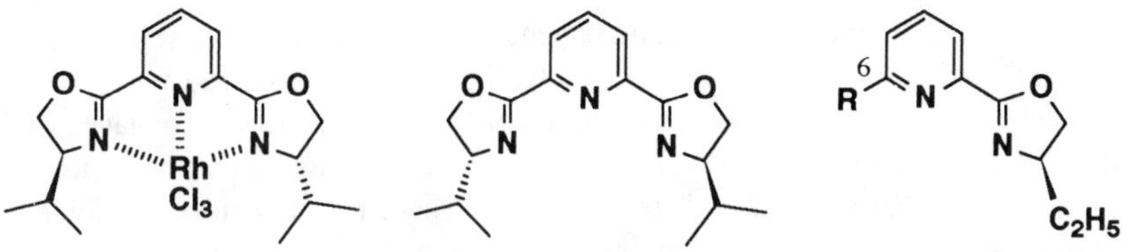

(R,R)-Pybox

32   **33**

34 R = H
35 R = Me

As far as the choice of the substituents on the oxazoline ring is concerned, a *tert*-butyl group in the Pymox ligand is superior to an isopropyl group in terms of enantioselectivity [15,16], while in the Pybox system an isopropyl group has proven to be the best substituent [21].

For the direction of the chirality induction, (R)-Pythia systems and (S)-Pymox systems give the same absolute configuration (R) in the secondary alcohols produced. In contrast, (S,S)-Pybox systems afford the secondary alcohols with (S) absolute configuration (Chart 4). In the Pymox systems, however, the introduction of a methyl group at the 6-position of the Pymox skeleton inverts the chirality of the product from (S) with (R)-Pymox–Et (**34**) to (R) with (R)-6-Me–Pymox–Et (**35**)

Chart 4 Chirality induction with Pythia, Pymox, and Pybox.

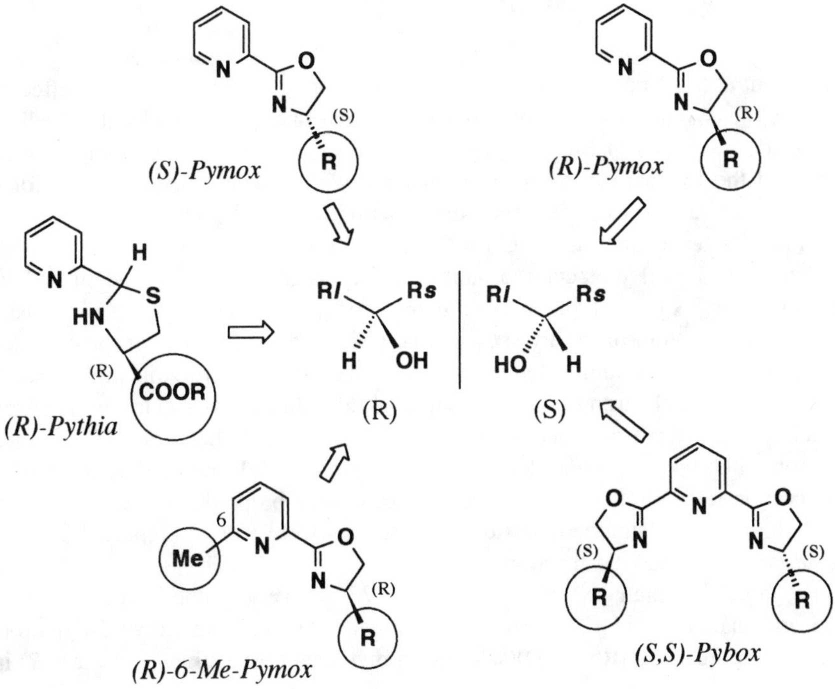

[22]. Another monooxazoline (**31**) proved to be a poor ligand for asymmetric hydrosilyation [23].

As mentioned above, excess ligand is beneficial for rhodium catalysts containing nitrogen ligands to achieve high enantioselectivity. This is also true for the Pymox ligands [16]. However, the rhodium complex 6-Me–Pymox–Et (**35**) behaved differently. The asymmetric inductions for the reduction of acetophenone were identical for the ratios of Rh to **35** of 1:1 and 1:5 [22]. These results can be explained by assuming an equilibrium between the species **I** and **II** in Scheme 3. Species **I** contains a bidentate ligand L—L′ (Pymox) and a vacant coordination site, whereas species **II** has a ligand L—L′ bonded in a monodentate fashion in addition to the bidentate ligand L—L′. Excess ligand should favor species **II**, to which a higher enantioselectivity is ascribed [24]. On the contrary, the 6-methyl substituent in the rhodium–Pymox systems should prevent the formation of species **II** and thus these ligands are not as effective.

Scheme 3

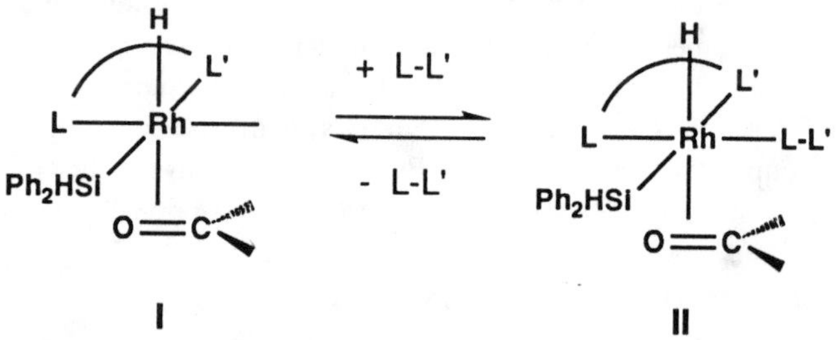

The 4-substituted Pybox–*i*-Pr derivatives **28–30** were synthesized [25]. The substituent at the 4-position on the pyridine skeleton of the Pybox systems affected not only the enantioselectivity but also the reaction rate. For example, the rhodium trichloride complex of 4-Me$_2$N–Pybox (**28**) and AgBF$_4$ with no extra addition of **28** catalyzed the hydrosilylation of acetophenone with diphenylsilane at 20°C for 6 hours to give 92% ee (*S*), while the same reaction with 4-Cl–Pybox (**30**) took place in a much faster manner at −5°C for 3 hours to give 83% ee (*S*).

Extended Hückel molecular orbital calculations of the Pybox system and its rhodium complex proved that 4-substituents strongly influence the electron density on the nitrogen atom of the Pybox pyridine and the rhodium atom of the complex [25]. If the rate-determining step is the reductive elimination from the higher valent rhodium species, electron-donating groups should stabilize this species or the corresponding transition state, decreasing the reaction rate (Scheme 4). In contrast, electron-withdrawing groups, by a corresponding destabilization, should increase the reaction rate. Similar manipulations of electronic properties by remote substituents in catalytic asymmetric oxidation reactions were recently reported as *electronic tuning* of catalytic reactions [26].

The chiral bidentate oxazoline derivatives **25–27** were synthesized from oxalic acid and malonic acid [27]. However, only the benzylic derivative **26** (tenfold excess with regard to [Rh(COD)Cl]$_2$ gave a high enantioselectivity (84% ee, *R*) in

Scheme 4

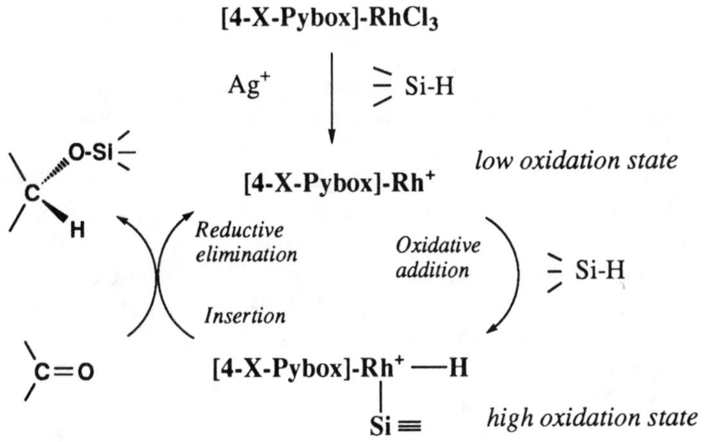

the hydrosilylation of acetophenone with diphenylsilane in CCl_4. The Rh–COD complexes of **25** and **27** exhibited no asymmetric induction.

The enantioselective hydrosilylation of ketones has been extended to alkyl aryl and dialkyl ketones other than acetophenone. The chiral nitrogen ligands Pythia and Pybox with rhodium complexes are much more successful catalysts in the hydrosilylation of alkylaryl ketones than chiral phosphine ligands [19–21].

The reduction of propiophenone with a rhodium catalyst containing Glucophinite (**5**) as a chiral phosphine system gave 61% ee as the highest value [11]. The rhodium catalyst with Pybox–*i*-Pr (**22**) as the chiral ligand increased the asymmetric induction to 91% ee (*S*) [21]. In general, enantioselectivities around 90% can be obtained for the substituted acetophenone and acetonaphthone derivatives with rhodium catalysts bearing Pythia (**18**) and Pybox (**22**). For example, 3-methoxyphenyl methyl ketone was reduced with the Pythia (**18**)–Rh system to give 93.3% ee (*R*) [20], and 2-carbomethoxyphenyl methyl ketone was reduced with the Pybox (**22**)–Rh system to give 96% ee (*S*) [21]. An almost complete enantioselection (99% ee, *S*) was observed in the asymmetric hydrosilylation of α-tetralone with the Pybox (**22**)–Rh catalyst [21] (Scheme 5).

Scheme 5

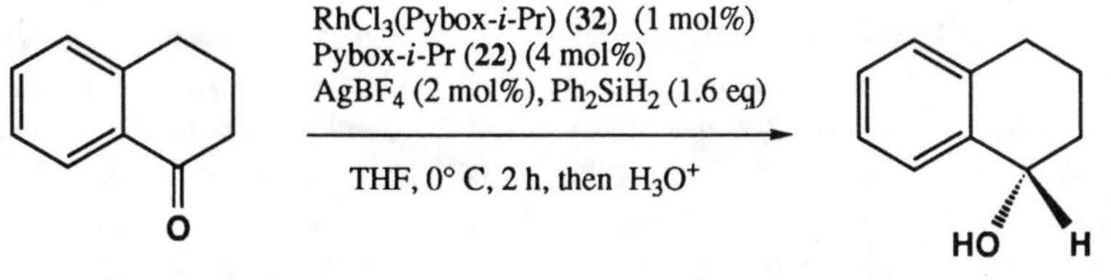

92 %, 99 %ee (*S*)

o-Methoxybenzyl methyl ketone was reduced using the same catalyst with 82% ee (*S*), whereas benzyl methyl ketone gave only 71% ee (*S*) under similar conditions [21]. The increase in enantioselectivity for the former case can be ascribed to steric and chelation effects.

In the hydrosilylation of dialkyl ketones using chiral phosphine ligands, enantioselectivity have not exceeded 50% ee [28,29]. The chiral aminophosphine ligand (*S*)-AMPHOS **6** improved the enantioselectivity up to 72% for the reaction of methyl *t*-butyl ketone with $[Rh(C_2H_4)_2Cl]_2$ and diphenylsilane [30]. The nitrogen ligand Pythia–(Me,Me) (**17**) gave 68.5% ee (*R*) for methyl *t*-butyl ketone [19].

A simple linear ketone 2-octanone was reduced by using the Pybox **22**–Rh system with a good enantioselectivity (63% ee, *S*) [21]. The 4-Cl–Pybox (**30**)–Rh system reduced 2-octanone with 80% ee (*S*) [25]. An optically active, naturally occurring secondary alcohol, (*S*)-sulcatol (**36**), was obtained with 70% ee (*S*) by the hydrosilylation of $CH_3-CO-(CH_2)_2CH=C(CH_3)_2$ with diphenylsilane catalyzed by the Pybox (**22**)–Rh system (Scheme 6) [21].

Scheme 6

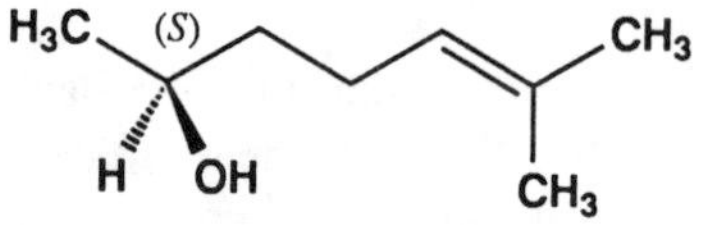

36   Sulcatol

Thus, enantioselectivity in the asymmetric hydrosilylation of ketones has been dramatically improved in the last decade.

6.3. Asymmetric Hydrosilylation of α,β-Unsaturated Ketones

The hydrosilylation of α,β-unsaturated ketones with rhodium–phosphine catalysts is known to bring about 1,2-addition or 1,4-addition, depending markedly on the nature of the hydrosilanes [31]. Monohydrosilanes such as triethyl and dimethylethylsilane predominantly undergo 1,4-addition to give silyl enol ethers, which yield saturated ketones after hydrolysis. In contrast, dihydrosilanes such as diphenylsilane give 1,2-addition products yielding silyl ethers, which are converted into allylic alcohols.

The enantioselective hydrosilylation of α,β-unsaturated ketones was reported with (+)-DIOP–rhodium catalysts [32,33]. In most cases, the enantioselectivities in these reactions are lower than 50% ee. 2-Methyl-2-cyclohexenone was reduced by using the (+)-DIOP-rhodium catalyst and 1-naphthylphenylsilane to give the corresponding allylic alcohol with 52% ee (*S*) [32]. The rhodium catalysts with nitrogen ligands Pythia **18** and Pybox **22** gave relatively low enantioselectivity, except for a 71% ee in the reaction of chalcone with the Pybox (**22**)–rhodium system [21]. In the hydrosilylation of (R)-carvone with diethylsilane and a (−)-DIOP–rhodium catalyst, a good diastereoselectivity (60% ee) was observed [31].

6.4. Asymmetric Hydrosilylation of Keto Esters

The enantioselective hydrosilylation of α-keto esters giving α-hydroxy esters after hydrolysis can be carried out efficiently with a DIOP–[Rh(COD)Cl]$_2$ catalyst system [34]. Pyruvates gave better enantioselectivies than phenylglyoxylates [34]. (R)-n-Propyl lactate was obtained with 85.4% ee by using 1-naphthylphenylsilane and the (−)-DIOP–rhodium catalyst (Scheme 7).

Scheme 7

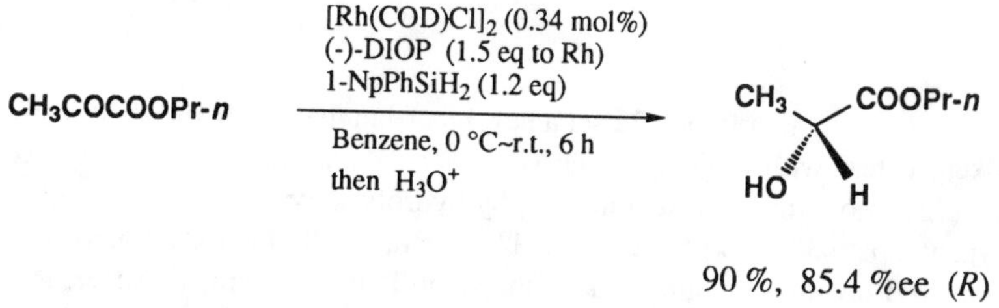

CH$_3$COCOOPr-**n** → [reaction conditions: [Rh(COD)Cl]$_2$ (0.34 mol%), (−)-DIOP (1.5 eq to Rh), 1-NpPhSiH$_2$ (1.2 eq), Benzene, 0 °C~r.t., 6 h, then H$_3$O$^+$] → CH$_3$—CH(OH)—COOPr-**n** (product with HO and H stereochemistry)

90 %, 85.4 %ee (R)

Diastereoselective hydrosilylations of menthyl pyruvate with 1-naphthylphenylsilane catalyzed by (+)- and (−)-DIOP–rhodium complexes were shown to give menthyl lactate with 85.6% ee (S) and 82.8% ee (R), respectively [34]. Thus, asymmetric induction was only slightly influenced by the chirality of the ester group and was dominated mainly by the chirality of the catalyst (catalyst control). The high enantioselectivity was ascribed to the coordination of the ester group to the metal center of the catalyst.

N-(α-Ketoacyl)-α-amino esters turned out to be suitable substrates for asymmetric hydrosilylation [34]. The reduction of the keto function provides α-hydroxyacylamino esters, which serve as building blocks for the synthesis of depsipeptides. The formation of the new chiral center in the product is mainly governed by the chirality of the catalyst in a manner similar to that observed for methyl pyruvate.

β-Keto esters such as acetoacetate and benzoylacetate did not give asymmetric inductions above 70% ee in the hydrosilylation with DIOP–rhodium catalysts [34]. Use of Pybox **22** resulted in 27% ee (S) for ethyl acetoacetate [21].

In contrast, high asymmetric inductions were attained in the hydrosilylation of γ-keto esters (levulinate derivatives) by the combination of 1-naphthylphenylsilane and DIOP–rhodium catalysts [34]. The hydrosilylation catalyzed by a (+)-DIOP–rhodium complex and subsequent lactonization converted isobutyl levulinate to (S)-4-methyl-γ-butyrolactone **37** with 84.4% ee. The results clearly show that bulkier ester groups give higher enantioselectivities, which strongly suggests the existence of chelation control (Scheme 8).

Scheme 8

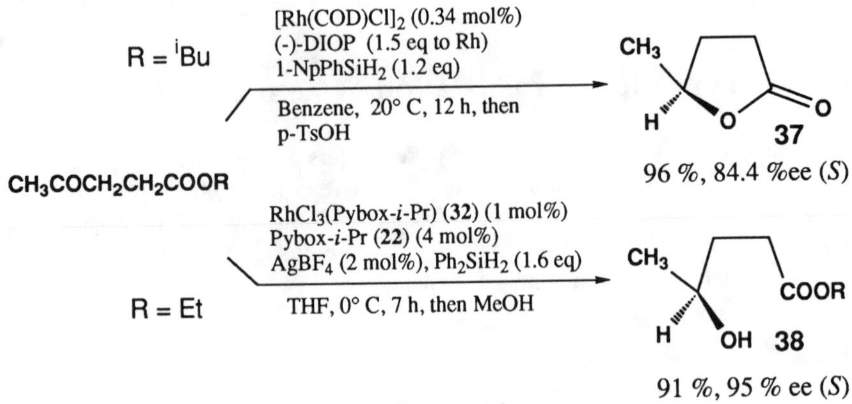

R = ^{i}Bu

[Rh(COD)Cl]$_2$ (0.34 mol%)
(-)-DIOP (1.5 eq to Rh)
1-NpPhSiH$_2$ (1.2 eq)

Benzene, 20° C, 12 h, then
p-TsOH

CH$_3$COCH$_2$CH$_2$COOR

37

96 %, 84.4 %ee (S)

R = Et

RhCl$_3$(Pybox-i-Pr) (**32**) (1 mol%)
Pybox-i-Pr (**22**) (4 mol%)
AgBF$_4$ (2 mol%), Ph$_2$SiH$_2$ (1.6 eq)

THF, 0° C, 7 h, then MeOH

38

91 %, 95 % ee (S)

It is noteworthy that Pybox **22** set a new record in the asymmetric hydrosilylation of γ-keto esters with diphenylsilane [21]. Ethyl levulinate was reduced with the Pybox (**22**)–rhodium catalyst, followed by hydrolysis with methanol, to give 4-hydroxypentanoate **38** with 95% ee (S) [21] (Scheme 8). In the reduction of isopropyl and menthyl levulinates catalyzed by the same Pybox system, however, the enantioselectivity decreased slightly compared to that for the ethyl ester [35].

6.5. Asymmetric Hydrosilylation of Imines

In the early 1970s, the Wilkinson catalyst RhCl(PPh$_3$)$_3$ was applied to the asymmetric hydrosilylation of prochiral ketimines to produce the corresponding silylated amines [36,37]. The products were converted to optically active secondary amines by hydrolysis. For example, the ketimines **39** and **40**, derived from acetophenone, were reduced to the corresponding optically active secondary amines **41** (47% ee, S) and **42** (65% ee, S), respectively, with a (+)-DIOP–rhodium catalyst and diphenylsilane [37] (Scheme 9). Thus, the enantioselectivity was similar to that for the asymmetric hydrosilylation of ketones with chiral phosphine–rhodium catalysts.

A similar procedure was applied in the synthesis of alkaloid precursors. The hydrosilylation of the cyclic ketimines **43** and **44** followed by treatment of the product silylamines with acid anhydride gave cyclic amides **45** (65.1% ee) and **46** (61% ee), respectively [38] (Scheme 10).

Several chiral aminophosphine–phosphinite ligands were used in the asymmetric hydrosilylation of ketimines. However, the enantioselectivities obtained were below 30% ee [39].

Oxime derivatives can be reduced with DIOP–rhodium catalysts and diphenylsilane (> 3 equiv. relative to the oxime) to give the corresponding primary amines.

Scheme 9

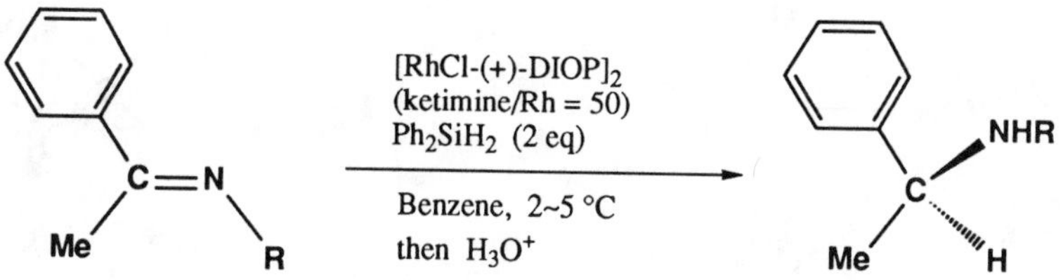

39	R = Ph	**41**	R = Ph	90 %, 47 % ee (S)
40	R = CH$_2$Ph	**42**	R = CH$_2$Ph	97 %, 65 % ee (S)

Scheme 10

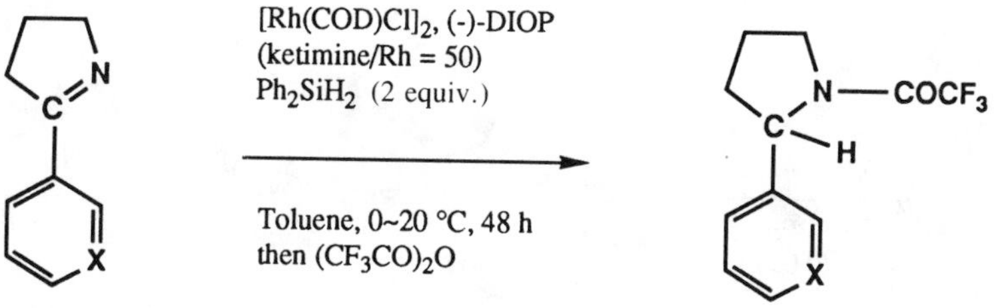

43	X = CH	**45**	X = CH	90 %, 65.1 % ee (R)
44	X = N	**46**	X = N	28 %, 66.1 % ee (S)

Thus, a prochiral ketone, PhCO-t-Bu, was converted to PhCH(NH$_2$)-t-Bu with 36% ee via its oxime [40].

6.6. Asymmetric Hydrosilylation of Olefins

Hydrosilylation of olefins was developed into versatile routes to functionalized organosilanes. In 1971 the asymmetric hydrosilylation of olefins by using platinum(II) complexes with chiral phosphines to give chiral organosilanes was first reported by Yamamoto et al. [41a] (Scheme 11).

In 1978 an efficient oxidative cleavage of silicon–carbon bonds was developed to extend the potential utility of organosilicon compounds for organic synthesis [4a]. The resulting alkylsilanes can readily be converted to the corresponding alcohols and organic halides. For this transformation to occur, it is important that the hydrosilanes used for the hydrosilylation of olefins have at least one electronegative group (Cl or MeO), an allyl, or an aryl group on the silicon atom.

A moderate enantioselectivity (53% ee) was reported for the reaction of norbor-

Scheme 11

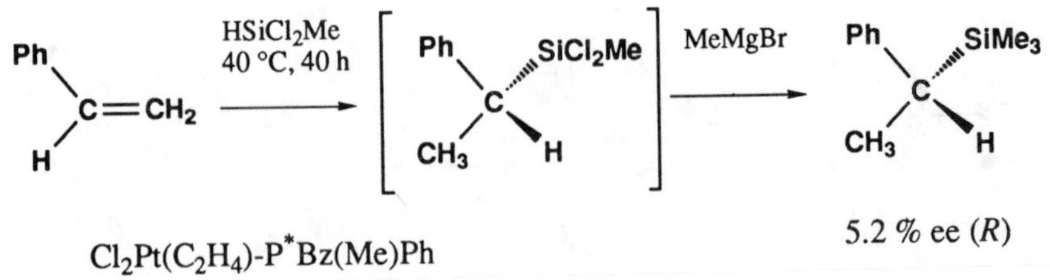
Ph
C=CH2
H
HSiCl2Me
40 °C, 40 h
Ph
SiCl2Me
C
CH3
H
MeMgBr
Ph
SiMe3
C
CH3
H
5.2 % ee (R)
Cl2Pt(C2H4)-P*Bz(Me)Ph

nene with trichlorosilane using a PPFA (**3**)–Pd catalyst [42] (Scheme 12). The hydrosilylation product was converted to norbornyl bromide with KF and *N*-bromosuccinimide (NBS) via a pentafluorosilicate derivative. Hydrosilylation of styrene with the same catalyst followed by oxidation with *m*-chloroperbenzoic acid (MCPBA) in DMF gave 1-phenylethanol with 52% ee (*S*).

Scheme 12

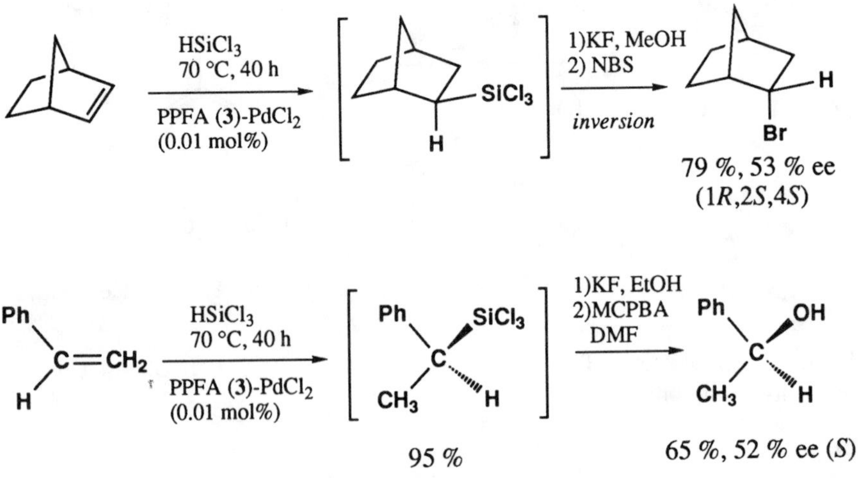
HSiCl3
70 °C, 40 h
PPFA (3)-PdCl2
(0.01 mol%)
SiCl3
H
1)KF, MeOH
2) NBS
inversion
H
Br
79 %, 53 % ee
(1R,2S,4S)
Ph
C=CH2
H
HSiCl3
70 °C, 40 h
PPFA (3)-PdCl2
(0.01 mol%)
Ph
SiCl3
C
CH3
H
95 %
1)KF, EtOH
2)MCPBA
DMF
Ph
OH
C
CH3
H
65 %, 52 % ee (S)

Asymmetric hydrosilylation of 1,3-dienes was carried out using palladium complexes of PPFA (**3**) and modified PPFA ligands, **47** and **48**, as catalysts [43,44]. Hydrosilylation of 1-phenylbutadiene with trichlorosilane catalyzed by PPFA (**47**)–$PdCl_2$ followed by oxidation with H_2O_2 gave allylic alcohol **49** with 66% ee (*R*) [44] (Scheme 13). Through the same procedure, cyclopentadiene was converted to 1-penten-3-ol with 57% ee by using PPFA (**48**)–$PdCl_2$ as the catalyst [44].

Intramolecular hydrosilylation–oxidation provides an effective method for the stereoselective synthesis of polyol skeletons from allylic alcohols [4,45]. In 1990 the enantioselective version of intramolecular hydrosilylation was successfully applied to the synthesis of chiral 1,3-diol derivatives from a 1,4-pentadien-3-ol skeleton [5].

The 1,4-pentadien-3-ol **50** was converted to the corresponding diarylsilyl ethers

Scheme 13

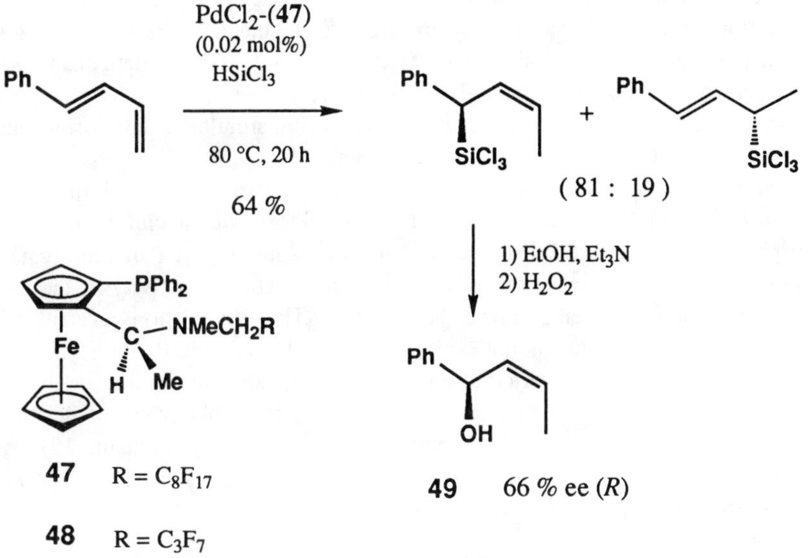

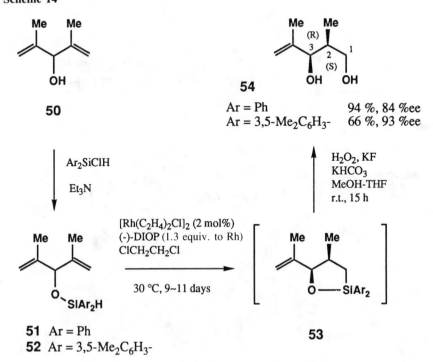

Scheme 14

51 and **52**, which were subjected to intramolecular hydrosilylation–cyclization with a (−)-DIOP–[Rh(C$_2$H$_4$)$_2$Cl]$_2$ catalyst (2 mol %) in ClCH$_2$CH$_2$Cl (Scheme 14). The cyclization product **53** was readily transformed to the (2S,3R)-1,3-diol **54** with virtually complete diastereoselectivity (*syn/anti* > 98:2) and excellent enantioselectivity (up to 93% ee, R), especially with the use of the 3,5-dimethylphenylsilane derivative **52** [5]. Thus, two vicinal chiral centers are simultaneously formed at one of the two double bonds of the meso compound **50**.

Remarkable improvement in the asymmetric hydrosilylation of olefins was accomplished using a palladium catalyst (0.1 mol %) with a chiral monodentate phosphine ligand (S)-MOP (**10**) derived from binaphthol [6]. Palladium catalysts with chelating chiral diphosphines did not exhibit catalytic activity in the hydrosilylation of olefins at temperatures below 80°C. This observation gave rise to the design of the novel chiral monodentate phosphine ligand, MOP. With this ligand, simple 1-alkenes were converted to optically active 2-alkanols with more than 94% ee (e.g., 95% ee (R) for 1-octene and 97% ee (S) for 4-phenyl-1-butene through hydrosilylation at 40°C and subsequent oxidation with H$_2$O$_2$) (Scheme 15).

Scheme 15

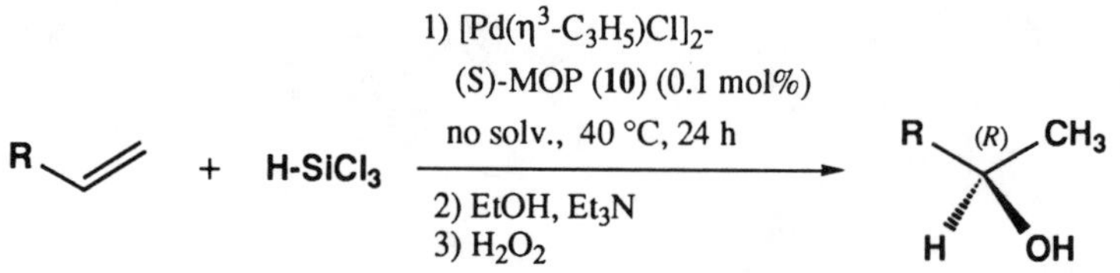

$$R = {}^nC_6H_{13}- \qquad 71\%, \ 95 \ \%ee \ (R)$$
$$R = PhCH_2CH_2- \qquad 68\%, \ 97 \ \%ee \ (S)$$

Scheme 16

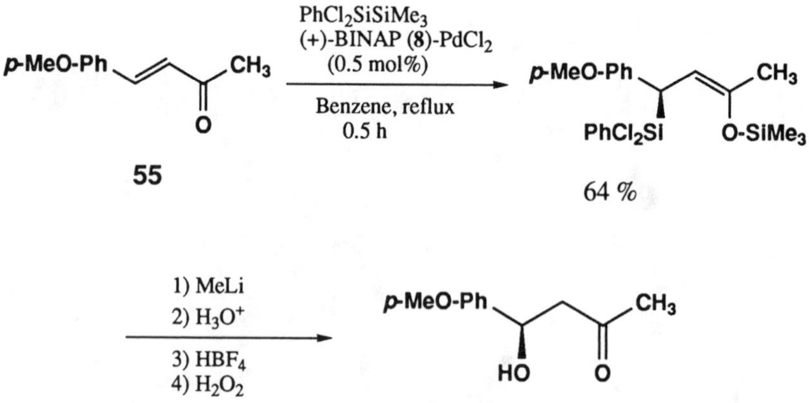

56   81 %, 92 % ee

Alternatively, an excellent asymmetric transformation of styrene to 1-phenyl-ethanol was accomplished by catalytic hydroboration with BINAP–rhodium catalysts [46]. The reaction of styrene with catecholborane (1.1 equiv.), (+)-BINAP **8**–[Rh(COD)$_2$]BF$_4$ (2 mol %) at −78°C for 2 hours, followed by oxidation, gave 1-phenylethanol with 96% ee (R) in 91% yield.

Asymmetric double silylation of α,β-unsaturated ketones with PhCl$_2$SiSiMe$_3$ is promoted by BINAP (**8**)–palladium catalysts to give optically active β-hydroxy ketones [47] (Scheme 16). In the reaction with the α,β-unsaturated ketone **55**, an enantioselectivity of 92% ee was obtained.

6.7. Asymmetric Synthesis of Chiral Silicon Compounds

Hydrosilylation of prochiral dihydrosilanes in the presence of chiral rhodium catalysts provides optically active monohydrosilanes. Most of the studies were carried out in the 1970s. The highest enantioselectivity attained was 46% ee for 1-NpPhSiHMe **57** through the reaction of 1-NpPhSiH$_2$ with diethyl ketone using a (+)-DIOP–rhodium catalyst followed by methylation [28] (Scheme 17).

Scheme 17

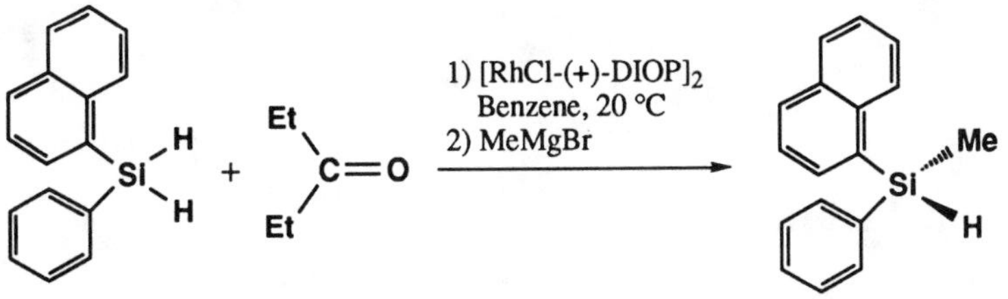

57 92 %, 46 % ee (R)

Scheme 18

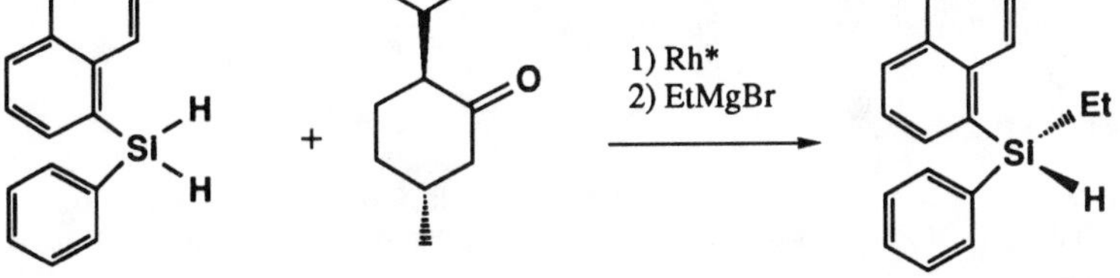

(-)-Menthone

Rh*	% ee
RhCl(PPh$_3$)$_3$	67 % (R)
[RhCl-(+)-DIOP]$_2$	82 % (R)
[RhCl-(-)-DIOP]$_2$	46 % (R)

The diastereoselective reaction of 1-NpPhSiH$_2$ with a chiral ketone, (−)-menthone, afforded a good asymmetric induction at the silicon center. A chirality transfer of 67% from (−)-menthone to the silicon atom was found with the use of a nonchiral catalyst, RhCl(PPh$_3$)$_3$ [29] (Scheme 18). The highest diastereoselection (82% de) was realized by the combination of the (+)-DIOP–rhodium catalyst and (−)-menthone, which was a better match than the (−)-DIOP–rhodium catalyst and (−)-menthone combination [29].

Recently, the enantioselectivity was improved to 85% ee in the reaction of 1-NpPhSiH$_2$ with diethyl ketone catalyzed by the rhodium complex with Cy–BINAP **9** [8] (Scheme 19). However, the enantioselectivity varied drastically depending on the structure of the ketones.

Scheme 19

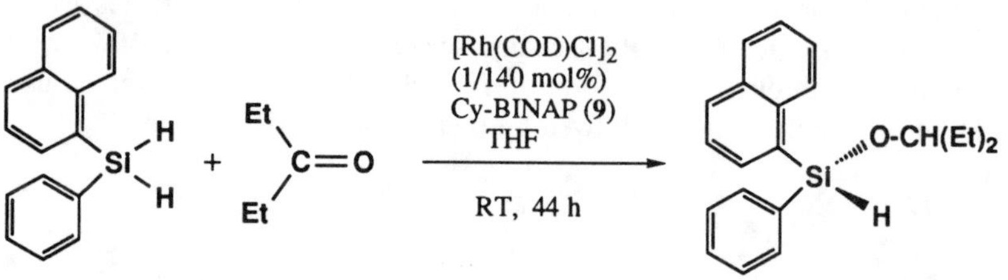

64 %, 85 % ee (*R*)

6.8. Conclusion

Metal-catalyzed asymmetric hydrosilylations have recently made spectacular progress, especially with respect to enantioselectivity. This success has been largely dependent on the development of new and efficient chiral ligands. However, in spite of intensive efforts, our comprehension of the mechanistic details of the hydrosilylation reactions is limited. Further studies are necessary to elucidate the catalytic cycles. Recent advances in computer-assisted molecular modeling may help us to devise new catalyst systems and ligands.

References

1. Reviews: (a) Ojima, I.; Yamamoto, K.; Kumada, M. In *Aspects of Homogeneous Catalysis;* Ugo, R. (Ed.); Reidel Dordrecht, 1977; Vol. 3, pp. 185–228. (b) Ojima, I.; Hirai, K. In *Asymmetric Synthesis;* Morrison, J. D. (Ed.); Academic Press: Orlando, FL, 1985; Vol. 5, pp. 103–146. (c) Nogradi, M. In *Stereoselective Synthesis;* VCH: Weinheim, 1987; pp. 90–95. (d) Brunner, H. *Synthesis,* **1988,** 645. (e) Ojima, I.; Clos, N.; Bastos, C. *Tetrahedron,* **1989,** *45,* 6901. (f) Ojima, I. In *The Chemistry of Organic Silicon Compounds,* Part 2; Patai, S.; Rappoport, Z. (Eds.); Wiley: New York, 1989; pp. 1479–1526.

2. Yamamoto, K.; Uramoto, Y.; Kumada, M. *J. Organomet. Chem.* **1971,** *31,* C9.

3. (a) Ojima, I.; Nihonyanagi, M.; Nagai, Y. *J. Chem. Soc. Chem. Commun.* **1972,** 938. (b) Ojima, I,; Kogure, T.; Nihonyanagi, M.; Nagai, Y. *Bull. Chem. Soc. Jpn.* **1972,** *45,* 3506.

4. (a) Tamao, K.; Kakui, T.; Kumada, M. *J. Am. Chem. Soc.* **1978**, *100*, 2268. (b) Tamao, K.; Ishida, N.; Tanaka, T.; Kumada, M. *Organometallics*, **1983**, *2*, 1694. (c) Tamao, K. *Yuki Gousei Kagaku Kyoukai Shi*, **1988**, *46*, 861.

5. Tamao, K.; Tohma, T.; Inui, N.; Nakayama, O.; Itoh, Y. *Tetrahedron Lett.* **1990**, *31*, 7333.

6. Hayashi, T.; Uozumi, Y. *J. Am. Soc. Chem.* **1991**, *113*, 9887.

7. (a) Corriu, R. J. P.; Moreau, J. J. E. *J. Organomet. Chem.* **1975**, *91*, C27. (b) Corriu, R. J. P.; Moreau, J. J. E. *Nouv. J. Chem.* **1977**, *1*, 71.

8. Tsuneto, A.; Ohta, T.; Takaya, H. Spring meeting of the Chemical Society of Japan, **119**, entry 2P17.

9. (a) Yamamoto, K.; Hayashi, T.; Kumada, M. *J. Organomet. Chem.* **1972**, *46*, C65. (b) Hayashi, T.; Yamamoto, K.; Kumda, M. *J. Organomet. Chem.* **1976**, *112*, 253.

10. Dumont, W.; Poulin, J. C.; Dang, T.-P.; Kagan, H. B. *J. Am. Chem. Soc.* **1973**, *95*, 8295.

11. Johnson, T.; Klein, K.; Thomen, S. *J. Mol. Catal.* **1981**, *12*, 37.

12. (a) Brunner, H.; Riepl, G. *Angew. Chem., Int. Ed. Engl.* **1982**, *21*, 377; *Angew. Chem. Suppl.* **1982**, 769. (b) Brunner, H.; Reiter, B.; Riepl, G. *Chem. Ber.* **1984**, *117*, 1330.

13. Botteghi, C.; Schionato, A.; Chelucci, G.; Brunner, H.; Kürzinger, A.; Obermann, U. *J. Organomet. Chem.* **1989**, *370*, 17.

14. Brunner, H.; Obermann, U.; Wimmer, P. *J. Organomet. Chem.* **1986**, *316*, C1.

15. Nishiyama, H.; Sakaguchi, H.; Nakamura, T.; Horihata, M.; Kondo, M.; Itoh, K. *Organometallics*, **1989**, *8*, 846.

16. Brunner, H.; Obermann, U. *Chem. Ber.* **1989**, *122*, 499.

17. Balavoine, G.; Client, J. C.; Lellouche, I. *Tetrahedron Lett.* **1989**, *30*, 5141.

18. Nishiyama, H.; Horihata, M.; Hirai, T.; Wakamatsu, S.; Itoh, K. *Organometallics*, **1991**, *10*, 2706.

19. (a) Brunner, H.; Riepl, G.; Weitzer, H. *Angew. Chem., Int. Ed. Engl.* **1983**, *22*, 331; *Angew. Chem. Suppl.* **1983**, 445. (b) Brunner, H.; Becker, R.; Riepl, G. *Organometallics*, **1984**, *3*, 1354.

20. Brunner, H.; Kürzinger, A. *J. Organomet. Chem.* **1988**, *346*, 413.

21. Nishiyama, H.; Kondo, M.; Nakamura, T.; Itoh, K. *Organometallics*, **1991**, *10*, 500.

22. Brunner, H.; Brandl, P. *J. Organomet. Chem.* **1990**, *390*, C81.

23. Gladiali, S.; Pinna, L.; Delogu, G.; Graf, E.; Brunner, H. *Tetrahedron: Asymmetry*, **1990**, *1*, 937.

24. Brunner, H.; Brandl, P. *Tetrahedron: Asymmetry*, **1991**, *2*, 919.

25. (a) Nishiyama, H.; Kondo, M.; Wakamatsu, S.; Yamaguchi, S.; Itoh, K. In *Sixth IUPAC Symposium on Organometallic Chemistry Directed Towards Organic Synthesis;* Utrecht: Aug. 25–29, 1991; No. S-14. (b) Nishiyama, H.; Yamaguchi, S.; Kondo, M.; Itoh, K. *J. Org. Chem.* **1992**, *57*, 4306.

26. Jacobsen, E. N.; Zhang, W.; Güler, M. L. *J. Am. Chem. Soc.* **1991**, *113*, 6703.

27. Helmchen, G.; Krotz, A.; Ganz, K.-T.; Hansen, D. *Syn. Lett.* **1991**, 257.

28. Kagan, H. B.; Peyronel, J. F.; Yamagishi, T. *Adv. Chem. Ser.* **1979**, *173*, 50–66.

29. (a) Corriu, R. J. P.; Moreau, J. J. E. *J. Organomet. Chem.* **1975**, *85*, 19. (b) Corriu, R. J. P.; Moreau, J. J. E. *J. Organomet. Chem.* **1974**, *64*, C51. (c) Hayashi, T.; Yamamoto, K.; Kumada, M. *Tetrahedron Lett.* **1974**, 331.

30. Payne, N. C.; Stephan, D. W. *Inorg. Chem.* **1982**, *21*, 182.

31. Ojima, I.; Kogure, T. *Organometallics*, **1982**, *1*, 1390.

32. (a) Ojima, I.; Kogure, T.; Nagai, Y. *Chem. Lett.* **1975,** 985. (b) Ojima, I.; Kogure, T. *J. Organomet. Chem.* **1982,** *234,* 239.

33. Hayashi, T.; Yamamoto, K.; Kumada, M. *Tetrahedron Lett.* **1975,** 3.

34. (a) Ojima, I,; Kogure, T.; Kumagai, M. *J. Org. Chem.* **1977,** *42,* 1671. (b) Ojima, I.; Tanaka, T.; Kogure, T. *Chem. Lett.* **1981,** 823.

35. Nishiyama, H.; Yamaguchi, S.; Nakamura, T.; Itoh, K. Unpublished work.

36. Ojima, I.; Kogure, T.; Nagai, Y. *Tetrahedron Lett.* **1973,** 2475.

37. (a) Langlois, N.; Dang, T.-P.; Kagan, H. B. *Tetrahedron Lett.* **1973,** 4865. (b) Kagan, H. B.; Langlois, N.; Dang, T.-P. *J. Organomet. Chem.* **1975,** *90,* 353.

38. (a) Becker, R.; Brunner, H.; Mahboobi, S.; Wiegrebe, W. *Angew. Chem., Int. Ed. Engl.* **1985,** *24,* 995. (b) Brunner, H.; Kürzinger, A.; Mahboobi, S.; Wiegrebe, W. *Arch. Pharm.* **1988,** *321,* 73.

39. Kokel, N.; Mortreux, A.; Petit, F. *J. Mol. Catal.* **1989,** *57,* L5.

40. (a) Brunner, H.; Becker, R. *Angew. Chem., Int. Ed. Engl.* **1984,** *23,* 222. (b) Brunner, H.; Becker, R.; Gauder, S. *Organometallics,* **1986,** *5,* 739.

41. (a) Yamamoto, K.; Hayashi, T.; Kumada, M. *J. Am. Chem. Soc.* **1971,** *93,* 5301. (b) Yamamoto, K.; Hayashi, T.; Zembayashi, M.; Kumada, M. *J. Organomet. Chem.* **1976,** *118,* 161.

42. Hayashi, T.; Tamao, K.; Katsuro, Y.; Nakae, I.; Kumada, M. *Tetrahedron Lett.* **1980,** *21,* 1871.

43. (a) Hayashi, T.; Kabeta, K.; Yamamoto, T.; Tamao, K.; Kumada, M. *Tetrahedron Lett.* **1983,** *24,* 5661. (b) Hayashi, T.; Kabeta, K. *Tetrahedron Lett.* **1985,** *26,* 3023.

44. Hayashi, T.; Matsumoto, Y.; Morikawa, I.; Ito, Y. *Tetrahedron: Asymmetry,* **1990,** *1,* 151.

45. (a) Tamao, K.; Nakajima, T.; Sumiya, R.; Arai, H.; Higuchi, N.; Ito, Y. *J. Am. Chem. Soc.* **1986,** *108,* 6090. (b) Tamao, K.; Nakagawa, Y.; Arai, H.; Higuchi, N.; Ito, Y. *J. Am. Chem. Soc.* **1988,** *110,* 3712.

46. (a) Hayashi, T.; Matsumoto, Y.; Ito, Y. *J. Am. Chem. Soc.* **1989,** *111,* 3426. (b) Hayashi, T.; Matsumoto, Y.; Ito, Y. *Tetrahedron: Asymmetry,* **1991,** *2,* 601.

47. Hayashi, T.; Matsumoto, Y.; Ito, Y. *J. Am. Chem. Soc.* **1988,** *110,* 5579.

7

Asymmetric Carbon–Carbon Bond Forming Reactions

7.1. Asymmetric Allylic Substitution and Grignard Cross-Coupling
Tamio Hayashi

7.2. Asymmetric Aldol Reactions
Masaya Sawamura and Yoshihiko Ito

7.1

Asymmetric Allylic Substitution and Grignard Cross-Coupling

Tamio Hayashi

Catalysis Research Center and Graduate School of Pharmaceutical Sciences
Hokkaido University, Sapporo, Japan

7.1.1. Asymmetric Allylic Substitution Reactions Catalyzed by Palladium and Nickel Complexes

Allylic substitution reactions have been shown to be catalyzed by certain palladium or nickel complexes [1,2]. The catalytic cycle of the reactions involves a π-allyl–metal complex as a key intermediate (Scheme 1). Oxidative addition of an allylic substrate to palladium(0) or nickel(0) species forms a π-allyl–metal(II) complex, which undergoes attack of a nucleophile on the π-allyl moiety to give an allylic substitution product [1,2]. The substitution reactions proceed in an S_N or $S_{N'}$ manner depending on the catalysts, nucleophiles, and substituents on substrates.

Scheme 1

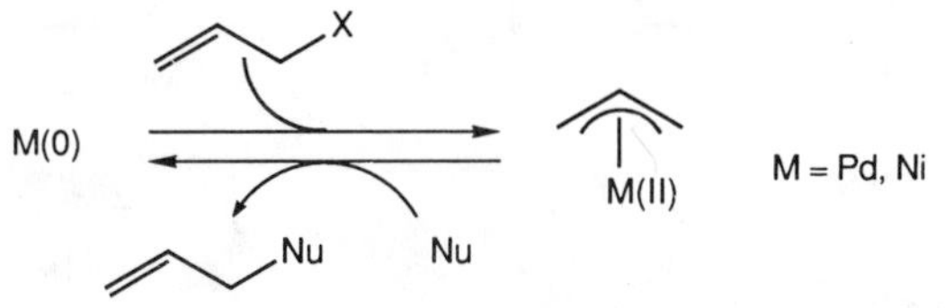

Stereochemical studies of oxidative addition and nucleophilic attack in stoichiometric reactions of π-allylpalladium complexes and catalytic allylic substitution reactions have demonstrated that the oxidative addition step forming the π-allyl-palladium species usually proceeds with inversion of configuration with respect to the stereogenic carbon center on allylic substrates and that soft nucleophiles attack a carbon atom of the π-allyl group from the side opposite to the palladium while hard nucleophiles, after formation of the nucleophile–palladium bond, attack the π-allyl carbon from the same side as the metal (Scheme 2).

Scheme 2

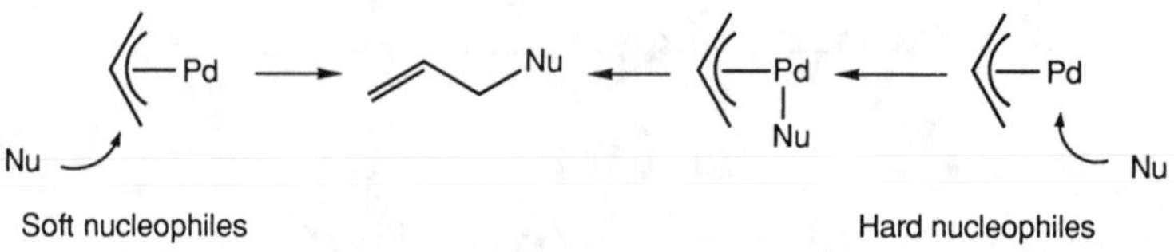

Soft nucleophiles   Hard nucleophiles

Incorporation of chiral ligands on the catalyst metal can, in principle, make the allylic substitution reaction result in the formation of chiral products. A new chiral carbon center can be created either in nucleophiles (type I) or in allylic substrates

Scheme 3

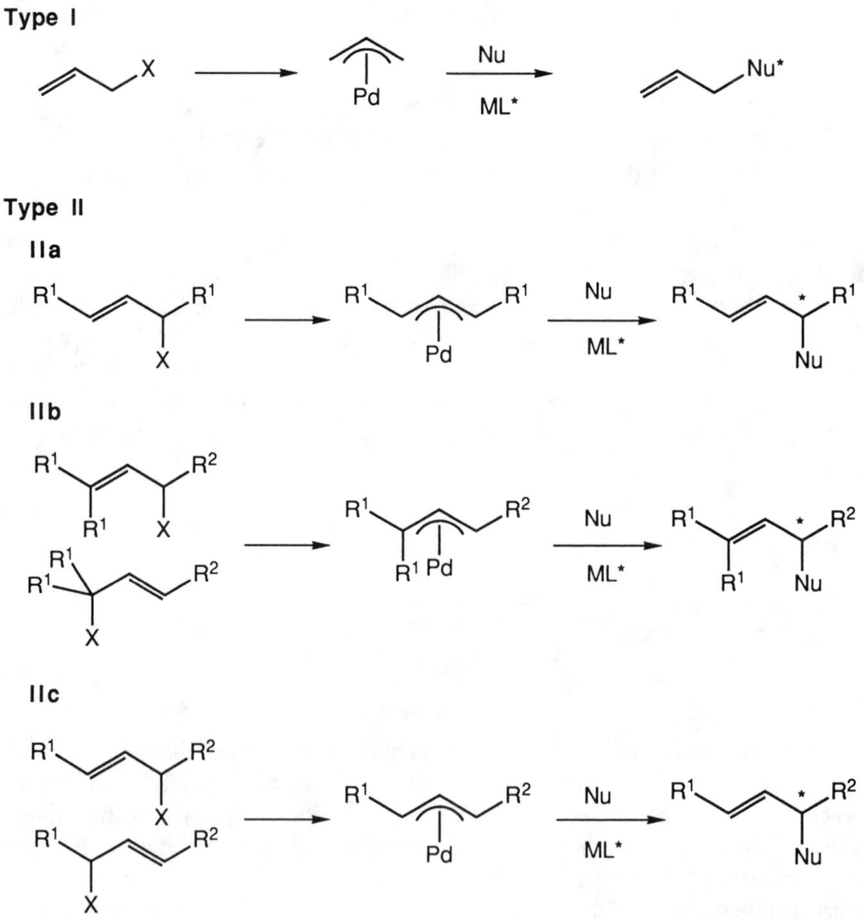

Type I

Type II

IIa

IIb

IIc

(type II) (Scheme 3). The type II reaction is classified into three categories, IIa, IIb, and IIc, according to the nature of the π-allyl intermediate. In type IIa, the π-allyl moiety in the intermediate is achiral, bearing the same substituents at the 1- and 3-positions, while the type IIb and IIc reactions proceed through diastereomeric π-allyl intermediates where the π-allyl moiety is chiral due to the different substituents at 1- and 3-positions. The chiral π-allyl group in type IIb can undergo an epimerization via the well-known σ-π-σ mechanism [2] (Scheme 4), whereas the π-allyl group in type IIc cannot. The nature of the π-allyl intermediate is significantly related to the mechanism of asymmetric induction in the catalytic asymmetric substitution reactions.

A substituent at the 1- or 3-position of a π-allyl group coordinated to a metal undergoes syn-anti interconversion with respect to the hydrogen at the 2-position by the σ-π-σ mechanism (Scheme 4). The syn-anti interconversion takes place through rotation of the alkenyl group by 180° in a σ-allyl intermediate. The isomerization of the substituent from syn to anti leads to the shift of palladium atom from the β face to the α face or vice versa. Thus, π-allyl complexes bearing different substituents at the 1- and 3-positions do not undergo racemization (or epimerization when palladium is coordinated with a chiral ligand) of the π-allyl–metal moiety, though they undergo syn-anti isomerization. They keep their chiral integrity during the syn-anti interconversion. On the other hand, racemization occurs in π-allyl complexes that have the same two groups on the π-allyl carbon at the 1- or 3-position through the σ-π-σ mechanism.

Scheme 4

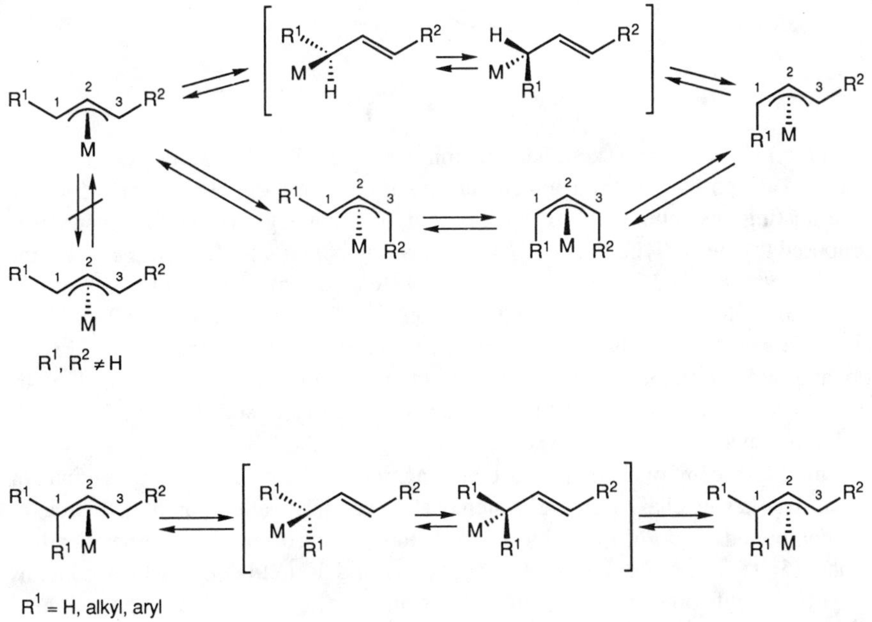

$R^1, R^2 \neq H$

$R^1 = H$, alkyl, aryl

7.1.1.1. Asymmetric Allylation Forming a Chiral Carbon Center in Nucleophiles (Type I)

The catalytic asymmetric allylation that introduced asymmetry in soft carbanions involved the reaction of active methine compounds with allyl phenyl ether in the presence of a palladium catalyst coordinated with (+)-DIOP (1) as a chiral bis-phosphine ligand [3] (Scheme 5). The stereoselectivity is dependent on the solvent and base: enantiomeric excesses of up to 10% are observed in the reaction of 2-acetylcyclohexanone (2a), 2-acetyl-1-tetralone (2b), and 2-phenylpropanal (2c), giving the corresponding allylation products 3a (7% ee), 3b (10% ee), and 3c (8% ee), respectively, which contain a chiral quaternary carbon center.

Scheme 5

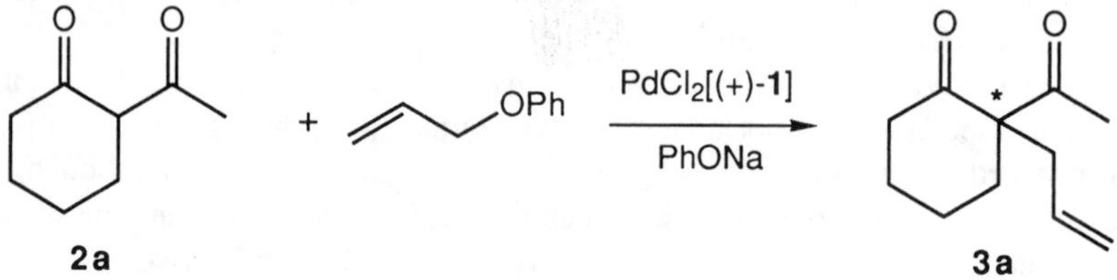

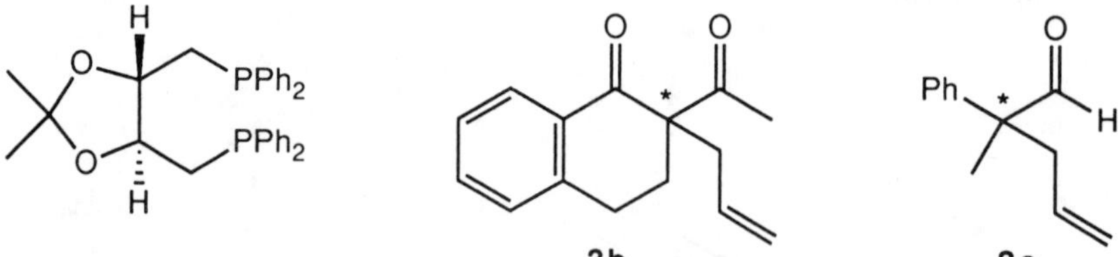

For type I asymmetric allylation, chiral ligands 4 and 5 were designed and prepared which have a chiral functional group at an appropriate distance from the coordinating phosphino groups. The chiral functional group on the ligand was proposed to interact with the nucleophile to bring about high selectivity [4] (Scheme 6). Ligands 4a and 5 were used for the reaction of the sodium enolate of 2-acetylcyclohexanone (2a) with allyl acetate at −30 or −50°C to give 3a with 52 and 31% ee, respectively. The importance of the distance between the chiral functional group and the phosphino groups coordinated to palladium is indicated by the lower selectivity (5% ee) that was observed on using ligand 4b, which has one additional methylene unit between the amide groups.

Chiral ferrocenylphosphines 6a–c bearing hydroxyl group(s) on the terminus of the ferrocene side chain were more stereoselective for the allylation of active methine compounds, giving optically active ketones with a chiral quaternary carbon center [5] (Scheme 7). The most effective ligand is (R)-(S)-6a, which contains 2-hydroxyethylamino group on the side chain. The reaction of 2-acetylcyclohex-

Scheme 6

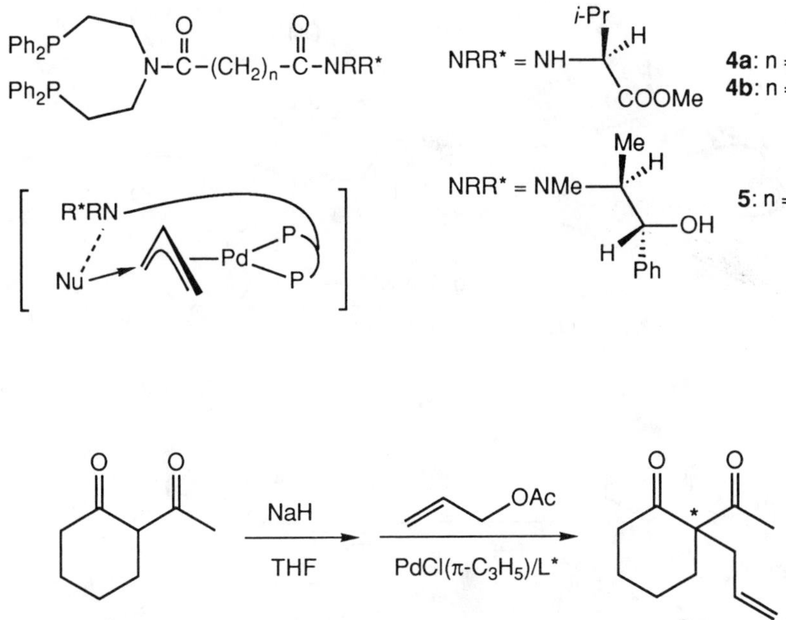

2a → **3a**

anone (**2a**), 2-acetyl-1-tetralone (**2b**), and 1-phenyl-2-methylbutane-1,3-dione (**2d**) gave the corresponding allylated products in 81% ee (*S*), 82% ee, and 60% ee, respectively. The high enantioselectivity associated with the ligand **6a** is ascribed to the stereocontrol effected by attractive interactions between the terminal hydroxyl group on the ligand and the prochiral enolate of a β-diketone, namely an attack on the π-allyl carbon on the π-allylpalladium intermediate from the side opposite palladium. A comparable enantioselectivity was observed with ferrocenylphosphine ligand **6d**, which was modified by introduction of a diaza crown ether on the side chain [6]. The allylation products **3a**, **3d**, and 2-allyl-2-acetylcyclopentanone (**3e**) were obtained in 60, 72, and 75% ee, respectively.

The ferrocenylphosphine **6a** has also been used for the palladium-catalyzed allylation of (α-isocyano)phenylacetate **7** to give allylated product **8** with 39% ee, which can be converted into optically active α-allylphenylglycine [7] (Scheme 8). Asymmetric allylation of Schiff base **9**, which is derived from glycine methyl ester, with allyl acetate or 2,3-dichloropropene in the presence of palladium–DIOP (**1**) gave the optically active Schiff base of allylglycine, **10**. The stereoselectivity was dependent on the ratio of the phosphine ligand to palladium, the highest (62% ee) being obtained with the molar ratio of 2 [8].

Asymmetric [3 + 2] cycloaddition of 2-(sulfonylmethyl)-2-propenyl carbonate (**11**) with methyl acrylate or methyl vinyl ketone is also catalyzed by the palladium–ferrocenylphosphine (**6b**) complex to product methylenecyclopentane derivatives with up to 78% ee [9] (Scheme 9). The stereochemistry is determined in the 1,4-

Scheme 7

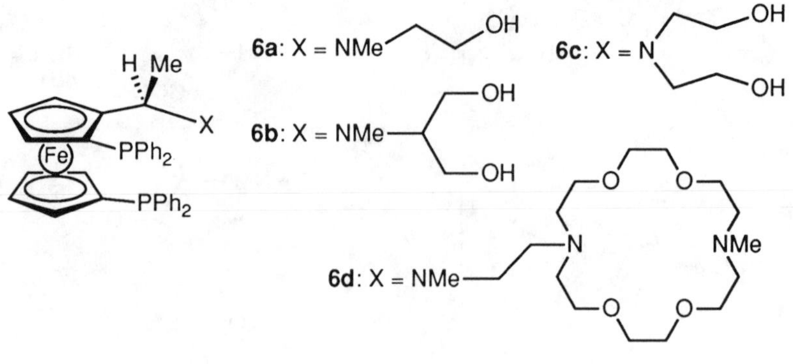

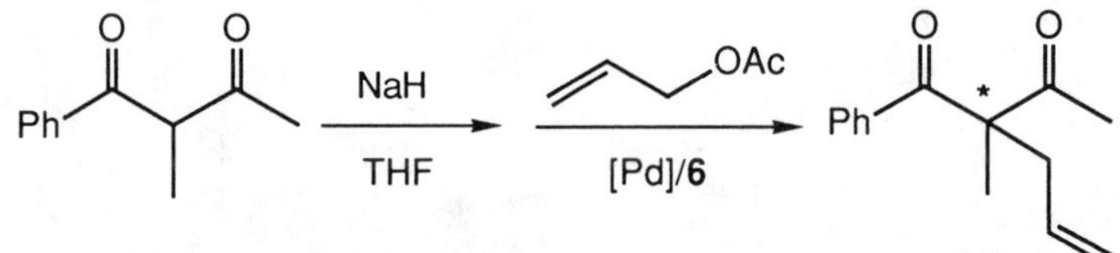

6a: X = NMe⟨CH₂CH₂OH⟩

6b: X = NMe⟨CH(CH₂OH)₂⟩

6c: X = N(CH₂CH₂OH)₂

6d: X = NMe⟨...⟩

2d → NaH / THF → ⟨OAc⟩ [Pd]/6 → **3d**

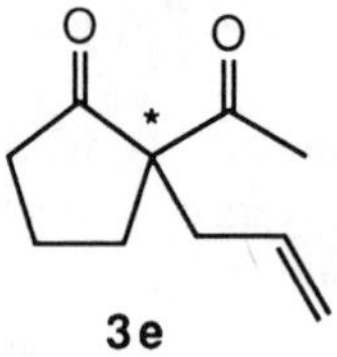

3e

Scheme 8

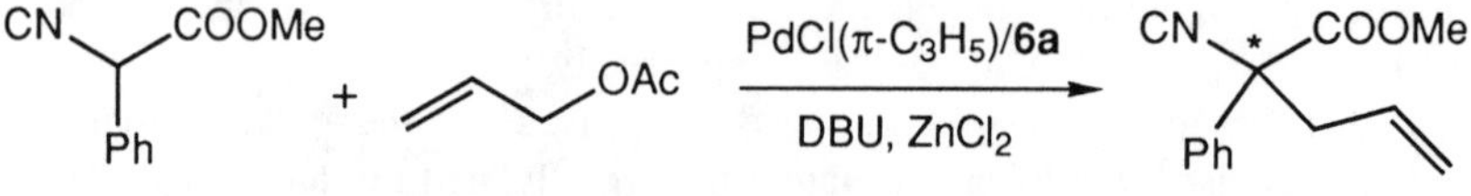

7 + ⟨OAc⟩ → PdCl(π-C₃H₅)/**6a** / DBU, ZnCl₂ → **8**

9 → LDA or NaH / THF → [Pd]/**1** → **10**

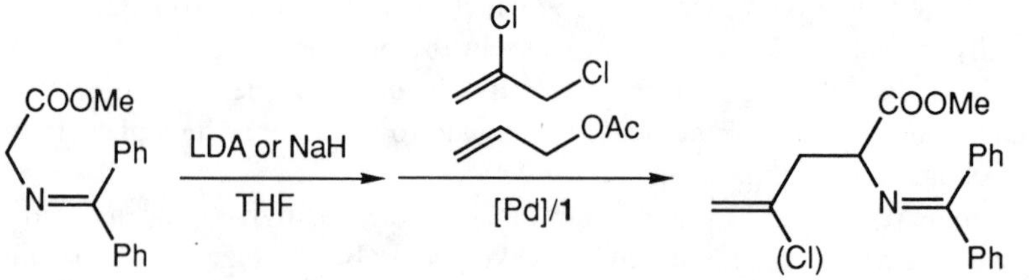

Scheme 9

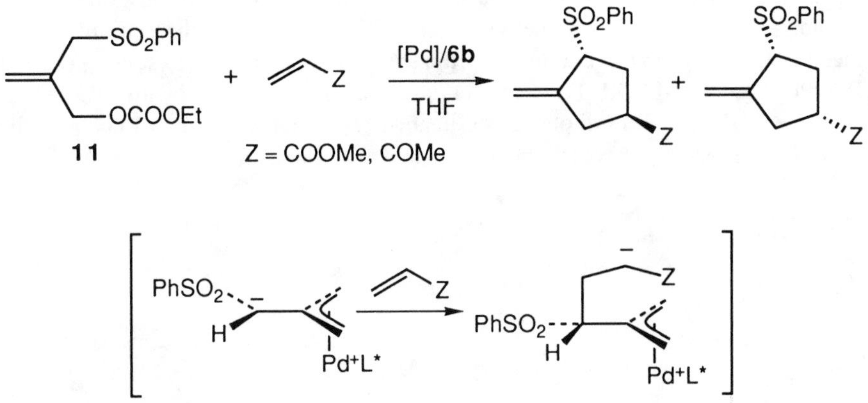

addition step of the zwitterionic π-allylpalladium intermediate to the electron-deficient olefin.

7.1.1.2. Asymmetric Allylic Substitution Forming a Chiral Carbon Center in Allylic Substrates (Type II)

Type IIa

The type IIa reaction catalyzed by a palladium complex proceeds via a π-allyl-palladium intermediate containing a meso-type π-allyl group. Both enantiomers of the racemic 2-propenyl carboxylate or related substrates bearing the same substituents at the 1- and 3-positions form, by oxidative addition to palladium(0), the meso-type π-allylpalladium intermediate. The two π-allyl carbons at the 1- and 3-positions are diastereotopic on coordination of a chiral phosphine ligand to palladium. Asymmetric induction arises from a preferential attack by the nucleophile on either of the two diastereotopic π-allyl carbon atoms (Scheme 10).

Scheme 10

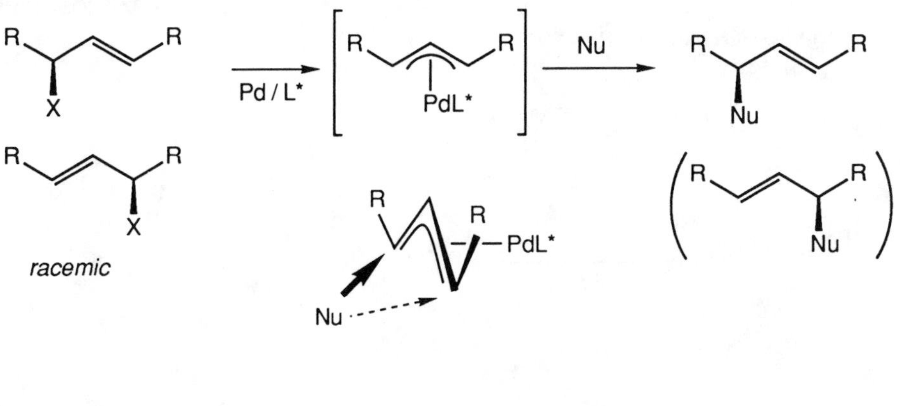

For allylic alkylation of type IIa, 3-acetoxycyclohexene derivatives (**12**) and 1,3-diphenyl-2-propenyl acetate (**13a**) have been often used as allylic substrates. Reaction of *cis*-3-acetoxy-5-carbomethoxycyclohexene (**12a**) with the sodium salt of methyl phenylsulfonylacetate in the presence of a palladium catalyst composed of Pd(PPh₃)₄ and (+)-DIOP (**1**) gave, after desulfonylation, (*R*,*R*)-**14** with 24% ee [10] (Scheme 11). Use of chiral phosphine ligands **15**, designed to control nucleophilic attack by introducing a chiral pocket in the π-allylpalladium intermediate, gave

Scheme 11

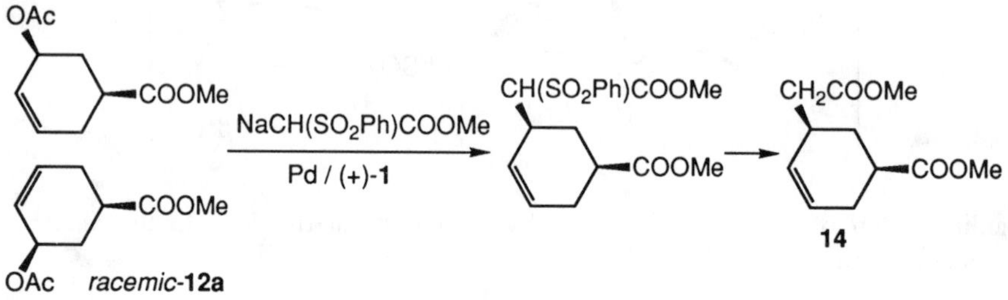

Scheme 12

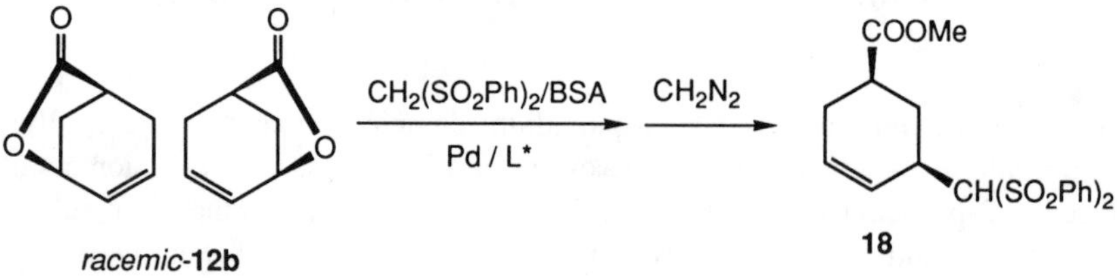

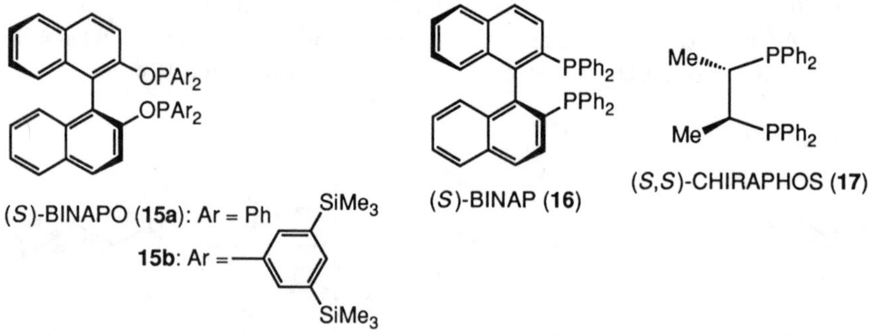

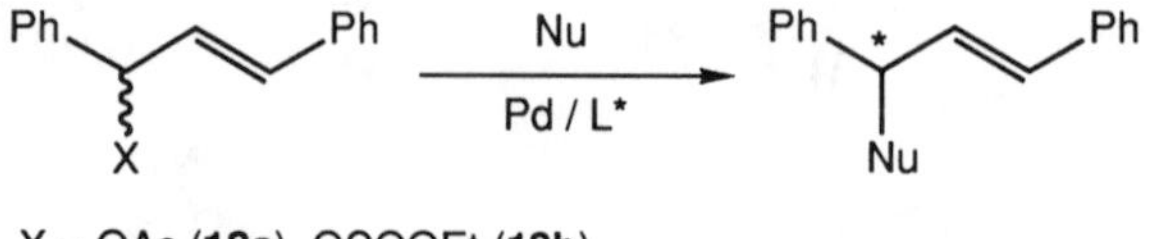

(*S*)-BINAPO (**15a**): Ar = Ph
15b: Ar =

(*S*)-BINAP (**16**)

(*S*,*S*)-CHIRAPHOS (**17**)

X = OAc (**13a**), OCOOEt (**13b**)

higher stereoselectivity [11] (Scheme 12). Thus, BINAPO (**15a**), a phosphinite analog of BINAP (**16**), and its 3,5-bis(trimethylsilyl)phenyl derivative **15b**, which are expected to possess a large bite angle on chelating to palladium, gave higher enantioselectivity than CHIRAPHOS (**17**), DIOP (**1**), or BINAP (**16**) for the reaction of lactone **12b** with an anion of bis(phenylsulfonyl)methane, **15b**, giving **18** with 69% ee. BINAPO (**15a**) is also effective for the alkylation of 1,3-diphenyl-2-propenyl acetate (**13a**). The results reported for the allylic substitution reactions of **13** are summarized in Table 1.

The chiral ferrocenylphosphines **6b** and **6c**, which contain hydroxyl groups on the pendant side chain, provide a different approach to obtaining higher stereoselectivity in type IIa reaction (entries 4–13 in Table 1). Reaction of 1,3-diaryl-2-propenyl acetate [Ar = Ph (**13a**), 1-Np (**19**), 3-MeOC$_6$H$_4$ (**20**)] with sodium acetylacetonate in the presence of a palladium catalyst coordinated with **6b** gave the corresponding alkylation products (*S*)-**21** with 90, 86, and 92% ee, respectively [12] (Scheme 13). The sodium salts of benzoylacetone and methyl acetoacetate can also be used for enantioselective alkylation. Use of **6e**, the ferrocenylphosphine bearing three hydroxyl groups, for the reaction of 1,3-diphenyl-2-propenyl acetate (**13a**) with sodium acetylacetonate increased the selectivity to 96% ee. The high enantioselectivity was not observed with ferrocenylphosphines lacking the hydroxyl functionality. It has been proposed that the hydroxyl group(s) on the ligand, which are located outside the π-allyl group of the π-allylpalladium intermediate, would

Scheme 13

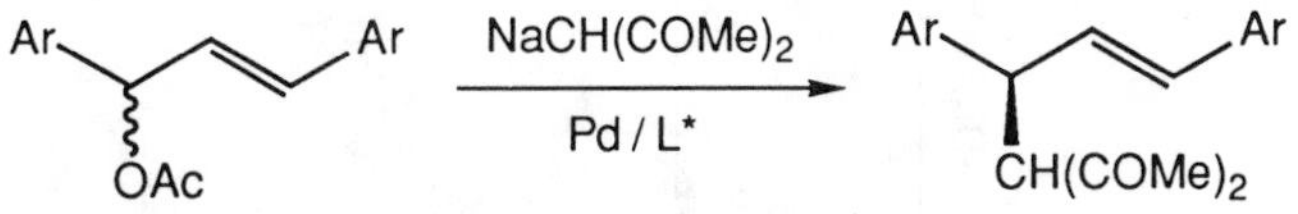

13a: Ar = Ph
19: Ar = 1-naphthyl
20: Ar = 3-MeOC$_6$H$_4$

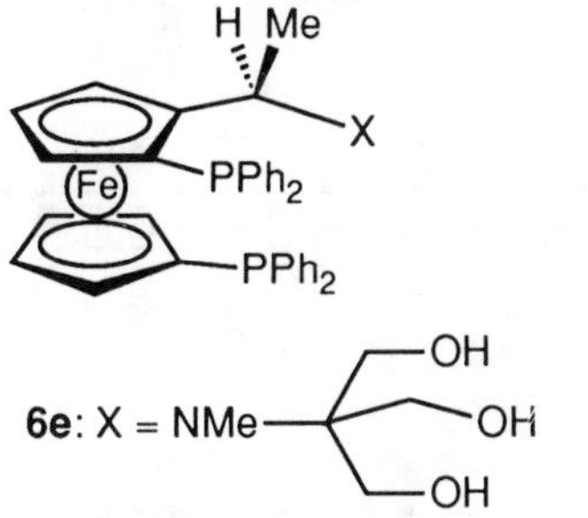

(*S*)-**21**

6e: X = NMe

(*R*)-(*S*)-BPPFA (**6f**): X = NMe$_2$
6g: X = Me

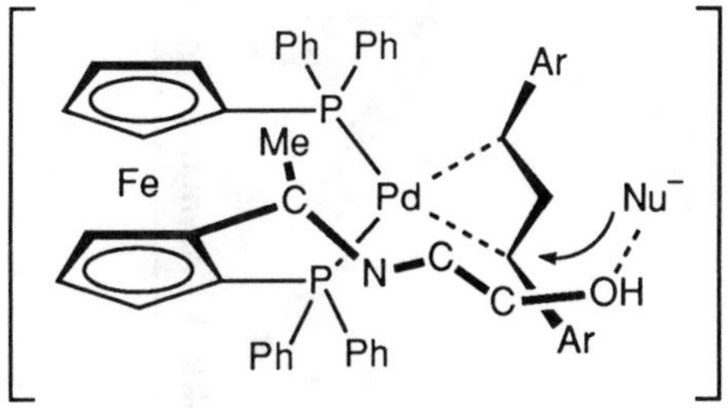

Table 1 Asymmetric Allylic Substitutions of 1,3-Diphenyl-2-propenyl Esters (13) with Nucleophiles in THF

Entry	X in Substrate 13	Nucleophile	Ligand	Reaction Conditions		Yield (%)	% ee (config.)	Ref.
				Temp. (°C)	Time (h)			
1	OAc (13a)	CHMe(COOMe)$_2$/BSA	(S)-BINAP (16)	66		81	50	11
2	OAc (13a)	CH$_2$(COOMe)$_2$/BSA	(S)-BINAPO (15a)	66		87	55	11
3	OAc (13a)	CHMe(COOMe)$_2$/BSA	(S)-BINAPO (15a)	25		75	68	11
4	OAc (13a)	NaCH(COMe)$_2$	(R)-(S)-FcP* (6e)	40	13–19	85	96 (S)	12b
5	OAc (13a)	NACH(COMe)$_2$	(R)-(S)-FcP* (6b)	40	13–19	97	90 (S)	12a
6	OAc (13a)	NaCH(COMe)$_2$	(R)-(S)-FcP* (6c)	40	13–19	86	81 (S)	12a
7	OAc (13a)	NaCH(COMe)$_2$	(R)-(S)-FcP* (6a)	40	13–19	86	71 (S)	12a
8	OAc (13a)	NaCH(COMe)$_2$	(R)-(S)-BPPFA (6f)	40	13–19	51	62 (S)	12a
9	OAc (13a)	NaCH(COMe)$_2$	(R)-(S)-FcP* (6g)	40	13–19	92	10 (R)	12a
10	OAc (13a)	NaCH(COMe)$_2$	(−)-DIOP (1)	40	13–19	88	0	12a
11	OAc (13a)	NaCH(COMe)COPh	(R)-(S)-FcP* (6b)	40	13–19	93	87 (S)	12a
12	OAc (13a)	NaCH(COMe)COOMe	(R)-(S)-FcP* (6b)	40	13–19	96	83 (S)	12a
13	OAc (13a)	NaCH(COOMe)$_2$	(R)-(S)-FcP* (6b)	40	13–19	98	48 (S)	12a
14	OAc (13a)	NaCH(COOMe)$_2$	(S)-BINAP (16)	25	44	80	30 (R)	14

15	OAc (13a)	NaCH(COOMe)$_2$	(S,S)-CHIRAPHOS (17)	25	44	86	90 (R)	14
16	OAc (13a)	NaCH(COOMe)$_2$	(S,S)-NORPHOS (24)	25	27	84	81 (R)	14
17	OAc (13a)	NaCMe(COOMe)$_2$	(S)-BINAP (16)	25	47	33	39 (R)	14
18	OAc (13a)	NaCMe(COOMe)$_2$	(S,S)-CHIRAPHOS (17)	25	48	75	80 (R)	14
19	OAc (13a)	NaCMe(COOMe)$_2$	(S,S)-NORPHOS (24)	25	44	86	78 (R)	14
20	OAc (13a)	NaC(NHCOMe)(COOMe)$_2$	(S)-BINAP (16)	25	120	92	94 (S)	14
21	OAc (13a)	NaC(NHCOMe)(COOMe)$_2$	(S,S)-CHIRAPHOS (17)	25	236	98	91 (S)	14
22	OAc (13a)	NaC(NHCOMe)(COOMe)$_2$	(S,S)-NORPHOS (24)	25	126	89	79 (S)	14
23	OAc (13a)	NaCH(COOMe)$_2$	(+)-26a	RT	39	79	77 (+)	15
24	OAc (13a)	NaCH(COOMe)$_2$	(+)-26b	Reflux	2	68	85 (+)	15
25	OAc (13a)	NaCH(COOEt)P(O)(OEt)$_2$	(+)-26a	RT	42	78	79 (+)	15
26	OAc (13a)	NaCH(COOEt)P(O)(OEt)$_2$	(+)-26b	Reflux	2	74	83 (+)	15
27	OAc (13a)	NaCH(COOMe)$_2$	30	50		86	77 (R)	16
28	OAc (13a)	NaCH(COOMe)$_2$	Sparteine (31)	RT	72	77	75 (R)	17
29	OCOOEt (13b)	H$_2$NCH$_2$Ph	(R)-(S)-FcP* (6b)	40	37	93	97 (R)	13
30	OCOOEt (13b)	H$_2$NCH$_2$Ph	(R)-(S)-FcP* (6b)	0	108	30	98 (S)	13
31	OCOOEt (13b)	H$_2$NCH$_2$Ph	(R)-(S)-FcP* (6a)	40	31	80	79 (S)	13
32	OCOOEt (13b)	H$_2$NCH$_2$Ph	(R)-(S)-BPPFA (6f)	40	24	79	31 (R)	13
33	OCOOEt (13b)	H$_2$NCH$_2$[3,4(MeO)$_2$C$_6$H$_3$]	(R)-(S)-FcP* (6b)	40	37	93	97 (R)	13

335

direct the incoming nucleophile preferentially to one of the two diastereotopic carbon atoms.

Ferrocenylphosphine **6b** was also effective for the reaction of *cis*-3-acetoxy-5-carbomethoxy cyclohexene (**12a**), the enantiomeric purity of the alkylation products being higher than 70% ee [12] (Scheme 14).

Scheme 14

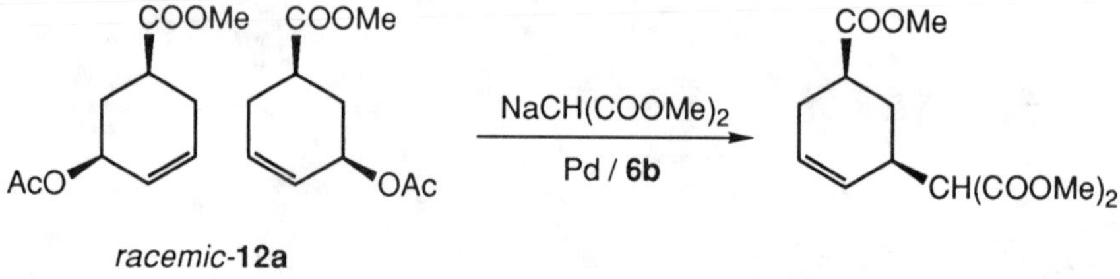

racemic-**12a**

The key role of the terminal hydroxyl group of ferrocenylphosphines **6a–c**, and **6e** in achieving high enantioselectivity was supported by an X-ray structure analysis and ^{31}P NMR data of π-allylpalladium complexes during studies on asymmetric allylic amination of **13** and related 1,3-disubstituted 2-propenyl esters [13]. Reaction of 1,3-diphenyl-2-propenyl ethyl carbonate (**13b**) with benzylamine in the presence of Pd/**6b** catalyst gave a quantitative yield of allylic amination product **22** with extremely high enantioselectivity ($\geq$ 97% ee) (entries 29–33). High enantioselectivity (82–97% ee) was also observed in the amination of allylic esters **23** substituted with *n*-Pr or *i*-Pr groups at 1- and 3-positions (Scheme 15). The allylic amines thus obtained were converted to optically active α-amino acids.

Scheme 15

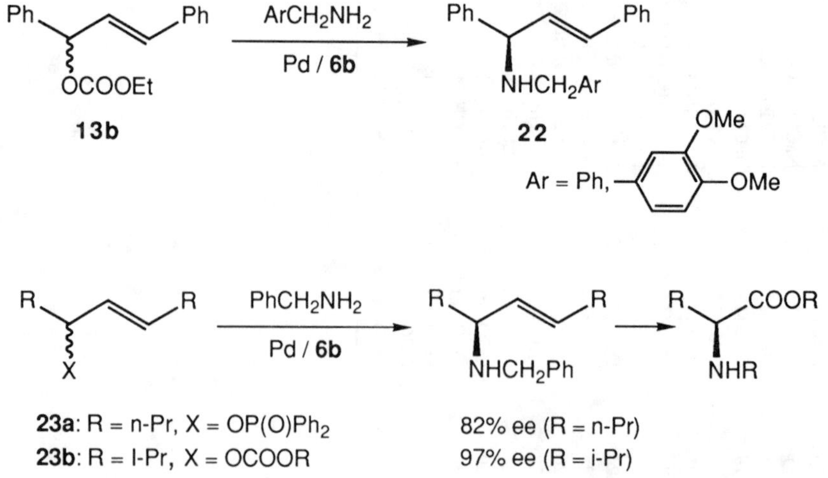

23a: R = n-Pr, X = OP(O)Ph$_2$ 82% ee (R = n-Pr)
23b: R = I-Pr, X = OCOOR   97% ee (R = i-Pr)

The stereoselectivity in the alkylation of **13a** was reported to be strongly dependent on the nucleophile, with the catalyst system consisting of palladium and

BINAP (**16**), CHIRAPHOS (**17**), or NORPHOS (**24**) [14] (Table 1: entries 14–22). For example, the reactions of **13a** with sodium salts of malonate, methylmalonate, and acetamidomalonate in the presence of BINAP (**16**) as the chiral ligand gave the corresponding alkylation products **25** with 30, 39, and 94% ee, respectively (Scheme 16).

Scheme 16

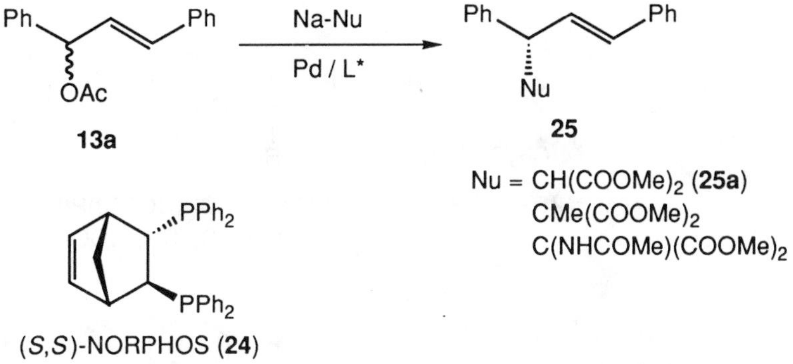

Nu = CH(COOMe)$_2$ (**25a**)
CMe(COOMe)$_2$
C(NHCOMe)(COOMe)$_2$

(*S,S*)-NORPHOS (**24**)

Chiral cycloalkylphosphines **26a,b**, bearing a carboxyl group at the β-position, have been used for the alkylation of **13a** (entries 23–26) and cyclohexenyl acetate (**27**) with dimethyl sodiomalonate and triethyl sodiophosphonoacetate [15] (Scheme 17). The phosphonoacetates **28a,b** resulting from the alkylation were converted to optically active acrylic acid derivatives **29**.

The asymmetric alkylation of **13a** with dimethyl sodiomalonate was also catalyzed by palladium coordinated with a chiral nitrogen ligand, tetrahydrobis(oxazole) **30** [16] or sparteine (**31**) [17], to give **25a** with 77 and 75% ee, respectively (entries 27 and 28).

High enantioselectivity has not been reported in palladium-catalyzed allylic alkylation with hard carbon nucleophiles such as organozinc and Grignard reagents. The chiral monophosphine ligands neomenthyldiphenylphosphine (**32**) and 1,2,5-triphenylpholane (**33**) gave phenylcyclohexene with around 10% ee in the phenylation of cyclohexenyl acetate (**27**) with phenylzinc chloride [18,19]. A little higher selectivity (30% ee) was observed [20] in the allylic vinylation of allyl ester **27** with Grignard reagent **34** catalyzed by a palladium complex coordinated with a chiral bisphosphine ligand (**35**) derived from an optically active amino alcohol (Scheme 18).

Nickel catalyst systems containing chelating chiral bisphosphine ligands exhibited much higher stereoselectivity than the palladium catalysts for the reaction of hard carbon nucleophiles [21,22]. Reaction of ethylmagnesium bromide with 3-phenoxycyclopentene (**36a**) and 3-phenoxycyclohexene (**36b**) in the presence of NiCl$_2$(CHIRAPHOS (**17**)) as the catalyst gave 3-ethylcyclopentene (**37a**: 90% ee) and 3-ethylcyclohexene (**37b**: 51% ee), respectively (Scheme 19). Use of (*R*)-6,6′-dimethyl-2,2′-bis(diphenylphosphino)-1,1′-biphenyl (BIPHEMP (**38**)) or 1,2-bis(diphenylphosphino)cyclopentane (**39**) as the chiral ligand increased the enan-

Scheme 17

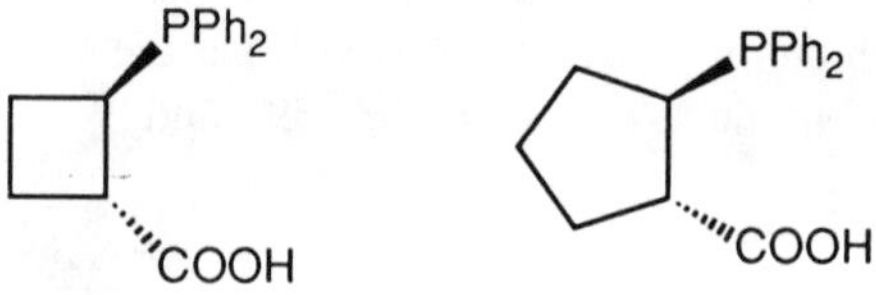

26a   26b

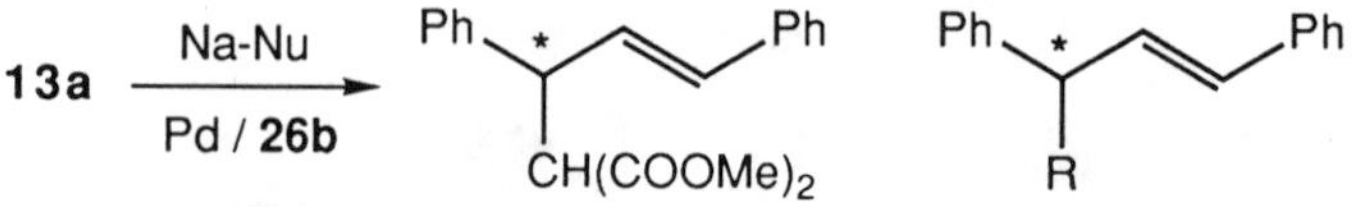

13a $\xrightarrow{\text{Na-Nu}}$ 25a

Pd / 26b

28a: R = CH(COOEt)P(O)(OEt)$_2$
29a: R = C(=CH$_2$)COOH

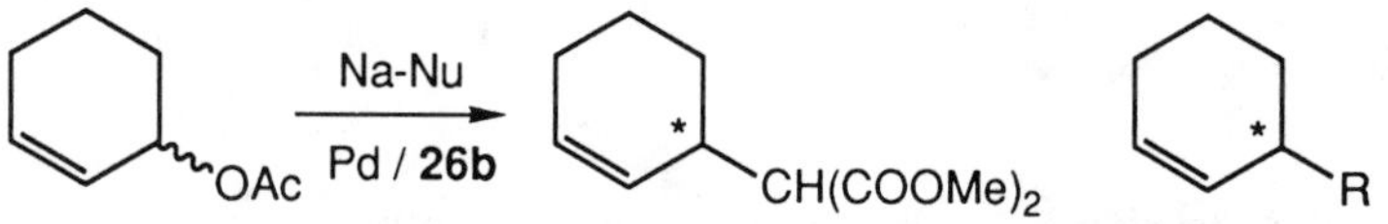

27   68% ee

28b: R = CH(COOEt)P(O)(OEt)$_2$
(55% ee)
29b: R = C(=CH$_2$)COOH

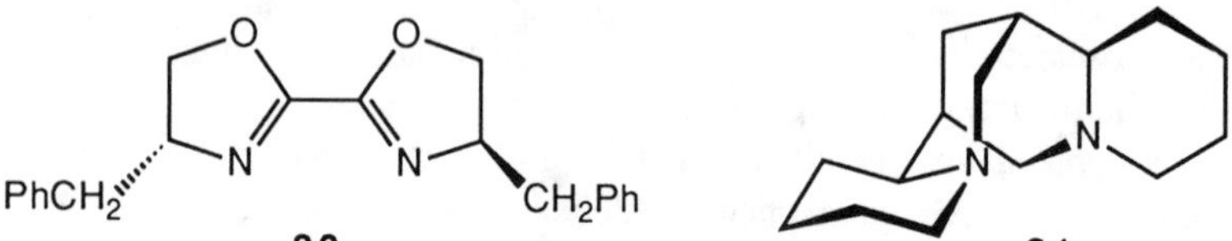

30   31

tioselectivity to 94% ee for **36a** and 83% ee for **36b**. Nickel-catalyzed allylic alkylation with hard carbon nucleophiles occurs through π-allyl(alkyl)nickel intermediate **40** coordinated to a chiral bisphosphine ligand. Reductive elimination of the two organic ligands from **40** leads to the allylic alkylation product. The stereochemistry of the product is probably determined by the preferential attack of the alkyl group on one of the diastereotopic π-allyl carbons from the same side as nickel.

Strong dependency of the stereoselectivity on the nucleophile was observed in the reaction of 3-phenoxycyclohexene (**37b**) catalyzed by NiCl$_2$(CHIRAPHOS (**17**)). Thus, the methyl, ethyl, vinyl, and phenyl Grignard reagents gave the corre-

Scheme 18

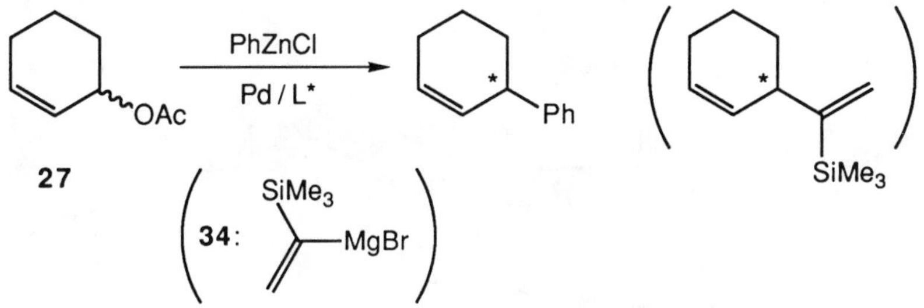

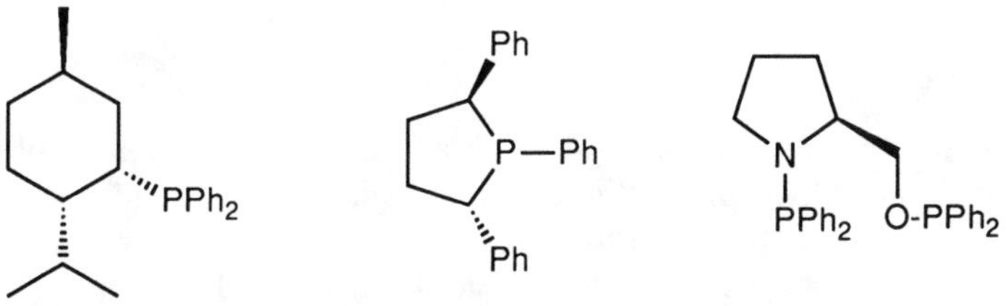

32 **33** **35**

sponding allylic alkylation products with 1, 51, 24, and 6% ee, respectively. A kinetic resolution of 2-cyclohexen-1-ol was reported in the methylation with methylmagnesium bromide catalyzed by $NiCl_2$(CHIRAPHOS (**17**)), the recovered alcohol having 14% ee at 70% conversion.

Scheme 19

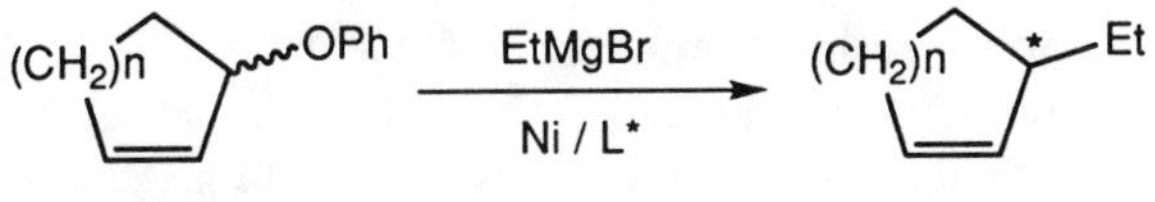

36a: n = 1
36b: n = 2

37a: n = 1
37b: n = 2

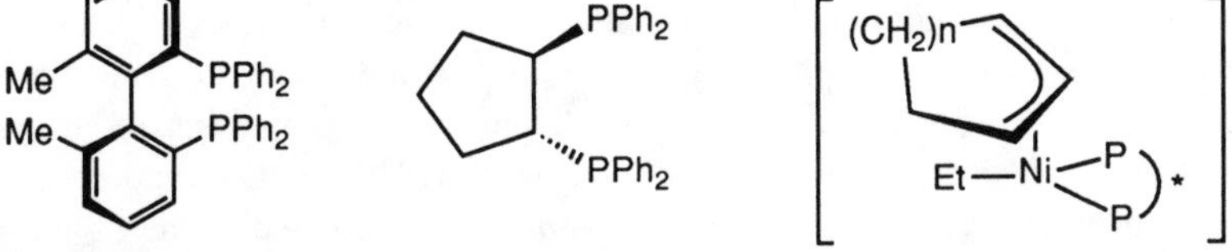

(*R*)-BIPHEMP (**38**) **39** **40**

The alkylation of allylic esters with the nickel catalyst $NiCl_2$(CHIRAPHOS (**17**)) has been used for the synthesis of optically active 1-arylpropionic acids [23] (Scheme 20). Asymmetric arylation of 3-penten-2-yl esters with 6-methoxy-2-

naphthylmagnesium bromide followed by oxidation of the carbon–carbon double bond gave naproxen with 64% ee.

Scheme 20

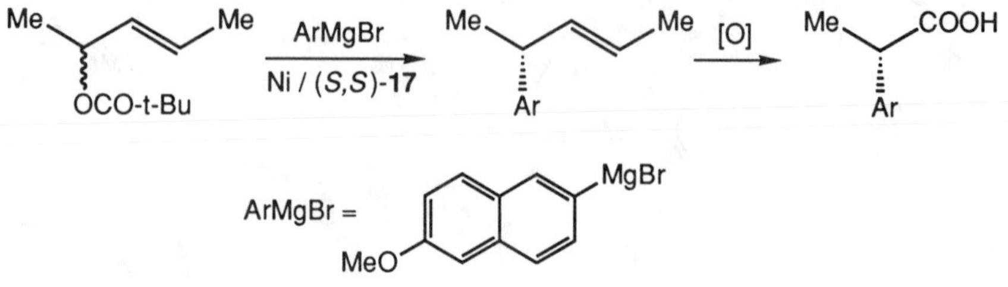

Type IIb

The most successfully designed system for asymmetric allylic substitutions of type IIb is the allylic alkylation that proceeds through a π-allylpalladium intermediate containing a 1,1-diaryl-substituted π-allyl group [24] (Scheme 21). Alkylation of racemic 1,3,3-triphenyl-2-propenyl acetate (**41a**) and 1,1,3-triphenyl-2-propenyl acetate (**42**) with dimethyl sodiomalonate in the presence of 5 mol % of [Pd(π-C_3H_5)((S,S)-CHIRAPHOS (**17**))]ClO$_4$ gave (R)-dimethyl(1,3,3-triphenyl-2-pro-penyl)malonate (**43a**) with 84–86% ee in high yields without its regioisomeric product. The π-allylpalladium intermediates **44a** and **44b** formed from **41a** or **42** have a chiral π-allyl group as a result of the unsymmetrical structure, and the intermediates **44a** and **44b** are epimeric isomers coordinated with a chiral phosphine ligand. The π-allyl complexes can undergo epimerization by inversion of the π-allyl group via the σ-π-σ process, which is accompanied by interchange of the syn and anti dispositions of the two phenyl groups substituted at the same carbon atom (see Scheme 4). The stereochemical results (i.e., acetates **41a** and **42** gave the alkylation product with essentially the same enantiomeric purity) indicate that the epimeriza-tion is fast compared with the nucleophilic attack, which forms the alkylation product. The high epimerization rate was attributed to the σ-π-σ process involving π-benzyl participation. The ratio of the diastereomeric palladium complexes **44**, after equilibration, was measured by ^{31}P NMR to be 6:1 under the reaction condi-tions. This ratio corresponds roughly to the enantiomeric purity of the product **43a**. The preferred configuration of the π-allylpalladium is proposed by taking into account the interaction of phenyl groups of the π-allyl with the chiral array of phenyl groups on the chiral ligand. An X-ray structure analysis of the related π-allylpalladium complex coordinated with (S,S)-CHIRAPHOS supports the pro-posed configuration of the intermediate **44b** [25]. Use of sparteine as the chiral ligand in place of CHIRAPHOS for the alkylation of **42** gave **43a** with similar enantiomeric purity (85% ee) [17].

Asymmetric allylic substitution reactions that proceed via π-allylpalladium inter-mediates with monosubstitution at the 1-position of the π-allyl are also classified as type IIb reactions.

Scheme 21

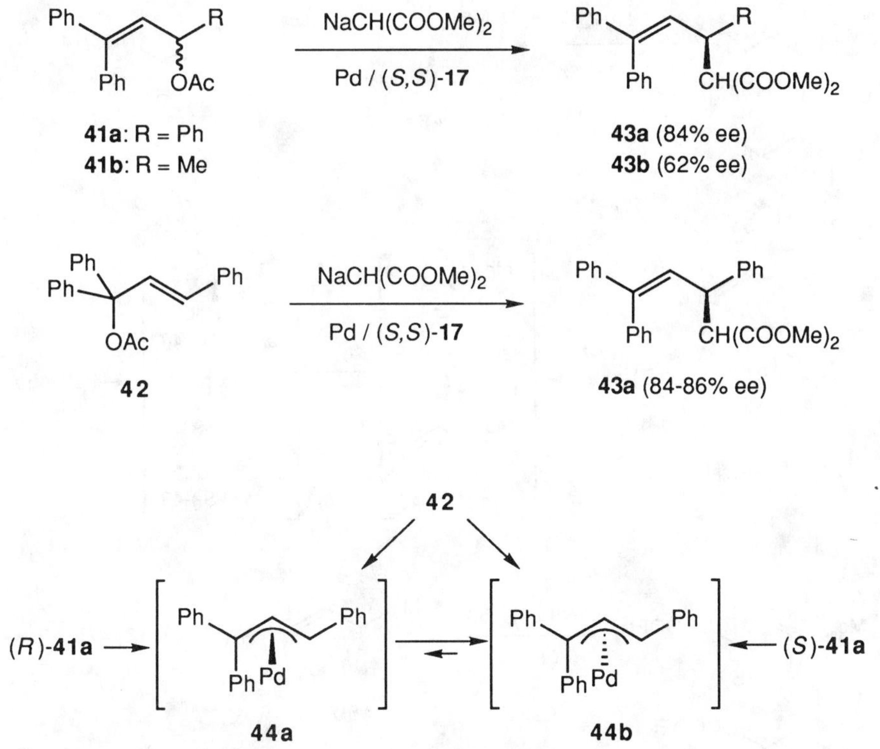

41a: R = Ph
41b: R = Me

43a (84% ee)
43b (62% ee)

42

43a (84-86% ee)

Allylic amination of (E)-2-butenyl acetate (**45a**) with benzylamine in the presence of 3 mol % of a palladium catalyst coordinated with chiral (hydroxyalkyl)ferrocenylphosphine ligand **6b** gave an amination product (S)-3-benzylamino-1-butene (**46a**) with 84% ee together with a small amount of an achiral regioisomer (E)-**47** in the ratio of 97:3 [26] (Scheme 22). The same amination with (Z)-**45a** resulted in the formation of (S)-**46a** with lower enantiomeric purity (53% ee) and achiral **47** with (Z) geometry, though the regioselectivity is still high (95:5). It follows that the nucleophilic attack of benzylamine takes place before an equilibration of the syn-anti isomerization of π-(1-methylallyl)palladium intermediates **48** has been reached. However, the epimerization of the π-allyl complexes **48** is fast compared with the nucleophilic attack. This is demonstrated by the reaction of the racemic and regioisomeric allyl acetate **49**, which gave optically active **46a** (64% ee). Use of di-*tert*-butyl iminodicarbonate **50** as a nucleophile gave allylic amination product **46b** with the same enantiomeric purity (61–63% ee (S)) from either (E)-2-butenyl carbonates **45b** or its (Z)-isomer, indicating that the nucleophilic attack of **50** is slow compared with the isomerizations of **48**.

Asymmetric allylic sulfonylation has been realized either by rearrangement of allylic sulfinates or by reaction of allylic acetates with sodium sulfinate [27] (Scheme 23). Treatment of (E)-2-butenyl p-toluenesulfinate (racemic) (**51**) with

Scheme 22

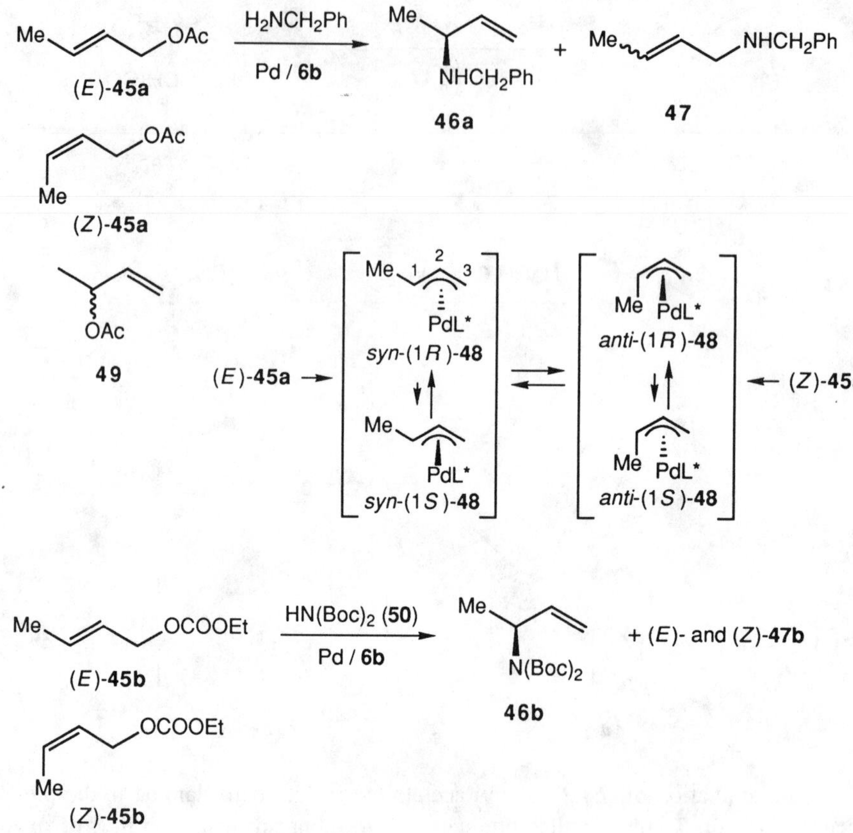

Scheme 23

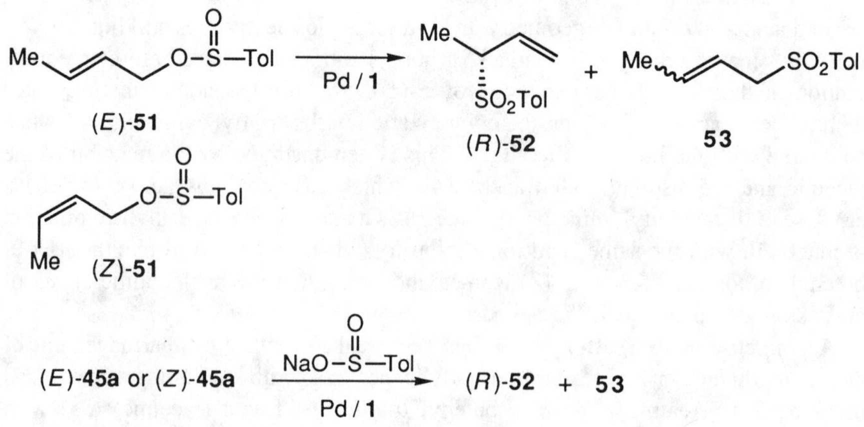

0.15 equiv. of Pd(PPh$_3$)$_4$ and 0.60 equiv. of (−)-DIOP (**1**) in THF at 0°C gave a 77% yield of (*R*)-sulfone **52** with 87% ee and a 15% yield of the achiral regioisomer **53**. The same stereochemical results (86% ee *R*) were obtained in the rearrangement of (*Z*)-sulfinate **51**. Intermolecular sulfonylation of (*E*)- or (*Z*)-2-butenyl acetate (**45a**) with sodium *p*-toluenesulfinate gave (*R*)-**52** with almost the same enantiomeric purity (88% ee). The small influence of the structure of the starting allylic substrates on the asymmetric induction indicates that the nucleophilic attack takes place after the complete equilibration of the π-allyl intermediates.

2-Butenyl phenyl carbonates, (*E*)-**54** and (*Z*)-**54**, and their regioisomeric carbonate **55** undergo decarboxylation in the presence of (*S,S*)-CHIRAPHOS (**17**)–palladium catalyst to give (*R*)-3-phenoxy-1-butene having an enantiomeric purity of 12–13% ee [28] (Scheme 24). The asymmetric decarboxylation is also catalyzed by nickel and rhodium complexes. The highest enantioselectivity (23% ee) was observed with the rhodium-catalyzed reaction of (*E*)-**54**.

Scheme 24

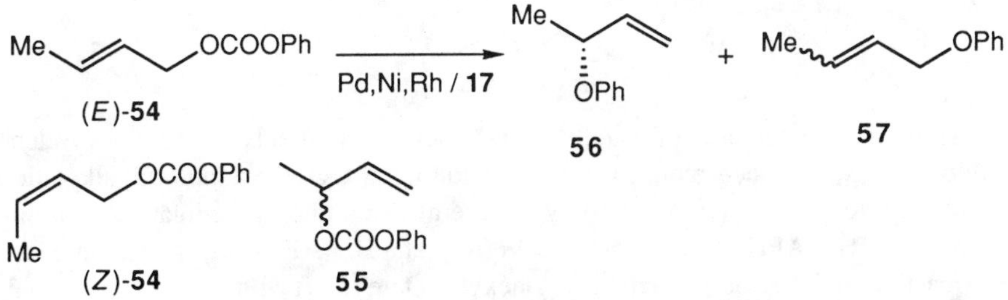

A platinum catalyst coordinated with DIOP (**1**) or its analog catalyzes the allylic alkylation of 2-butenyl acetate with dimethyl sodiomalonate giving optically active (*S*)-(buten-2-yl) malonate of 10–23% ee, though the catalytic activity is lower than the palladium system [29].

The asymmetric alkylation of 2-butenyl or 2-pentenyl alcohols and their esters with hard nucleophiles has been studied with nickel catalysts. Early work on the ethylation or methylation of the alcohols with Grignard reagents in the presence of a nickel catalyst coordinated with DIOP (**1**) or (*R*)-1,2-bis(diphenylphosphino)-1-phenylethane resulted in low enantioselectivity (< 15% ee) [21a,30]. Use of NiCl$_2$((*S,S*)-CHIRAPHOS (**17**)) for the phenylation of 2-butenyl phenyl ether (**57**) or its regioisomeric allyl ether **56** increased the enantioselectivity up to 60% ee [21b,c] (Scheme 25). The fact that (*R*)-3-phenyl-1-butene (**58**) was formed with almost the same enantiomeric purity (ranging between 48 and 60% ee) and with the same regioselectivity (65%) for all three isomeric ethers, (*E*)-**57**, (*Z*)-**57**, and **56**, indicates that the isomerization of π-(1-methylallyl)nickel intermediate **59** is fast compared with the reductive elimination forming **58**. The arylation of 2-butenyl pivalate with 6-methoxy-2-naphthylmagnesium bromide proceeded with higher enantioselectivity (89% ee), although the chemical yield is not as high (50%) [23] (Scheme 26).

Scheme 25

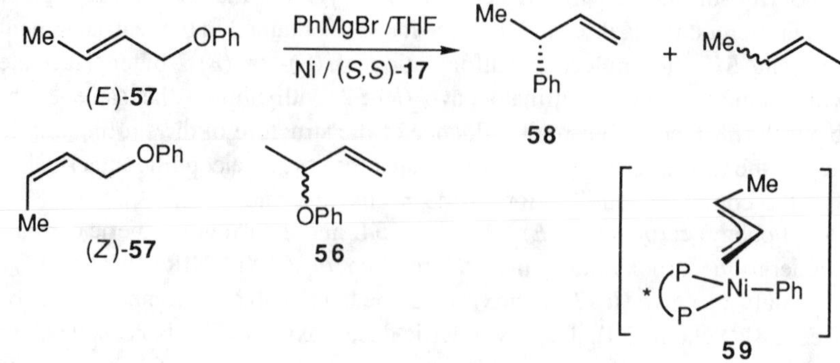

Scheme 26

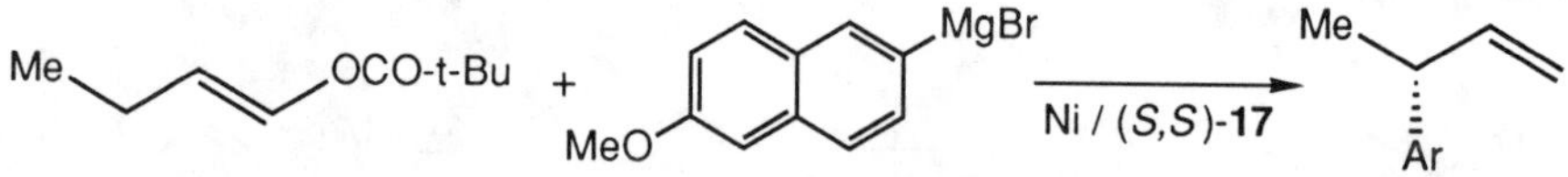

Asymmetric synthesis of axially chiral molecules, 4-substituted 1-alkylidene cyclohexanes, has been realized in the palladium-catalyzed asymmetric alkylation of allylic esters [31]. Stereoselectivity of several chiral phosphine ligands including DIOP (**1**), CHIRAPHOS (**17**), PROPHOS (**60**), and BINAP (**16**) were examined for the reaction of 4-*tert*-butyl-1-vinylcyclohexyl acetate (**61a**) with dimethyl sodiomalonate in THF giving axially chiral dimethyl 2-[4-*tert*-butylcyclohexylidene)methyl-malonate (**62**) (Scheme 27): BINAP (**16**) gave the best results, 40% ee for *trans*-**61a** and 25% ee for *cis*-**61a**. The enantiomeric purity of the product was dependent strongly on the nature of the leaving group [32]: methyl carbonate **61b**, benzoate **61c**, and 4-methoxybenzoate **61d** gave 27, 64, and 78% ee, respectively, with BINAP (**16**) as the chiral ligand. In dioxane, **62** was produced from **61d** with 90% ee. The strong influence of the leaving group on stereoselectivity indicates that the enantioselectivity-determining step is the oxidative addition of the allylic substrate to the chiral palladium(0) complex and that the isomerization between π-allyl intermediates *trans*-**63** (or *cis*-**66**) and *trans*-**64** (or *cis*-**64**) is slow compared with the subsequent nucleophilic attack of malonate.

Asymmetric elimination forming an optically active diene has been reported, which proceeds via π-allylpalladium intermediates **63** and **64** [33] (Scheme 28). Elimination of acetoacetate ester **61e** in the presence of a palladium catalyst coordinated with (*R*)-(*S*)-BPPFA (**6f**) gave (*R*)-4-*tert*-butyl-1-vinylcyclohexene (**65**) with up to 44% ee.

Intramolecular asymmetric allylic substitutions classified as type IIb have been employed for the synthesis of cyclic compounds. Treatment of allylic carbonate **66** with 5 mol % of palladium–(*S*)-(*R*)-BPPFA (**6f**) catalyst in benzene at 30°C for 0.7 hour gave 83% yield of cyclohexanone derivative **67**, which was converted by

Scheme 27

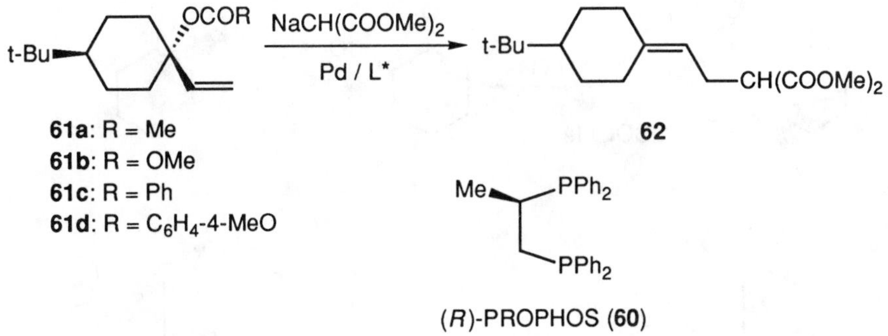

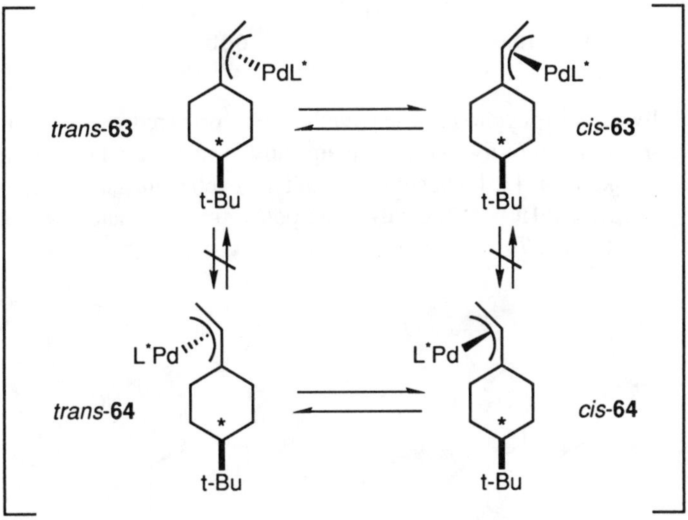

61a: R = Me
61b: R = OMe
61c: R = Ph
61d: R = C$_6$H$_4$-4-MeO

(R)-PROPHOS (60)

Scheme 28

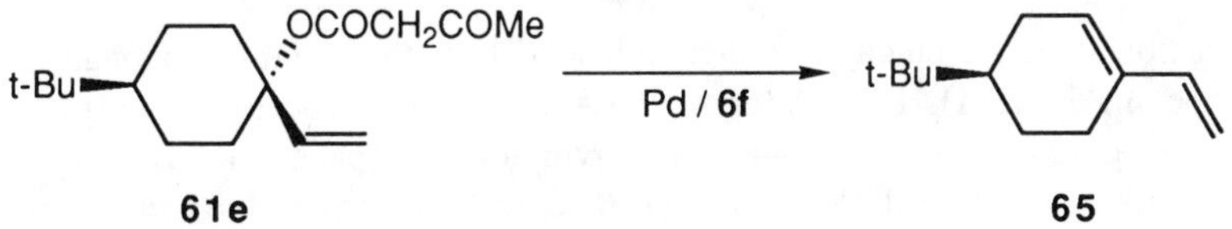

demethoxycarbonylation into (R)-3-vinylcyclohexanone with 48% ee [34] (Scheme 29). Stereochemical studies on the intramolecular alkylation of an optically active carbonate that is regioisomeric to **66** demonstrate that the enantioselectivity results from a diastereomeric equilibrium between the π-allyl intermediates **68a** and **68b**.

Scheme 29

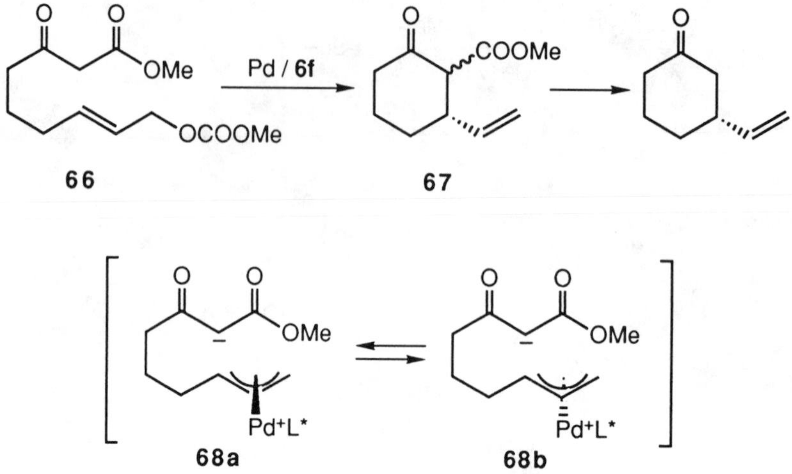

66   67

68a   68b

An optically active tricyclic ergoline synthon was prepared by the asymmetric intramolecular alkylation of **69**, which contains allylic acetate and nitroalkyl group on the indole skeleton [35] (Scheme 30). The cyclization takes place with palladium/(−)-CHIRAPHOS (**17**) catalyst and potassium carbonate or KF/alumina in THF or DME to give **70** with 70% ee.

Scheme 30

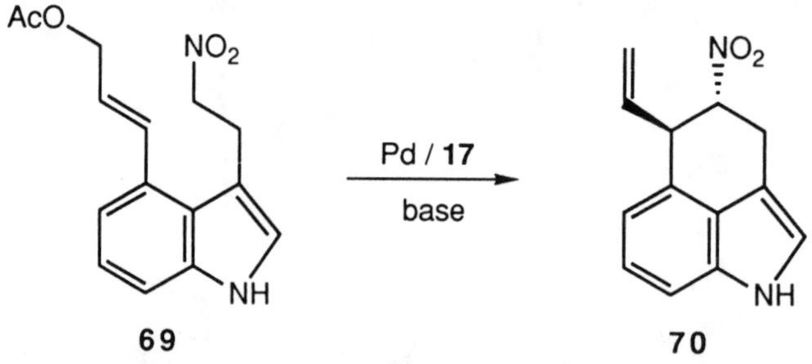

69   70

Reaction of (Z)-2-butenylene dicarbonate (**71**) with dimethyl malonate in the presence of $Pd_2(DBA)_3$/BPPFA (**6f**) (DBA = dibenzylideneacetone) catalyst at 0°C for 2 hours gave a low yield (20–40%) of 2-vinylcyclopropane-1,1-dicarboxylate **72** with up to 70% ee [36] (Scheme 31). The reaction proceeds through π-allyl-palladium complex **73** as the key intermediate, which is formed by oxidative addition of monoalkylation product **74** to palladium(0), and **73** undergoes intramolecular attack of the carbon nucleophile. A prolonged reaction time decreased the enantiomeric purity of cyclopropane **72**, indicating that the cyclization that forms the vinylcyclopropane is reversible and that the epimerization of the diastereomeric

π-allylpalladium intermediates **73a** and **73b** is fast relative to the cyclization. Reaction with methyl acetoacetate or acetylacetone took place in a different manner from that with malonate, giving dihydrofuran derivative **75** (50% ee), a product resulting from nucleophilic attack of enolate oxygen at the cyclization step.

Scheme 31

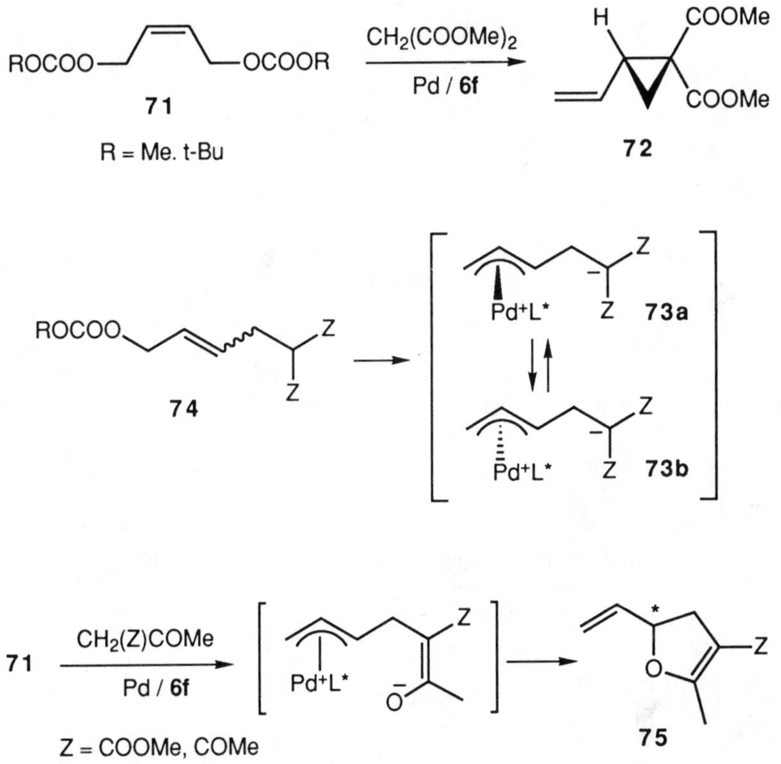

Treatment of 2-butenylene dicarbamates (**76**) with a palladium catalyst coordinated with ferrocenylphosphine **6b** brought about asymmetric cyclization to form optically active 4-vinyl-2-oxazolidones **77**, which were hydrolyzed to 2-amino-3-butenols [37] (Scheme 32). The enantiomeric purity of **77a** was 73 and 77% ee upon using (*E*)-butenylene *N,N*-diphenylcarbamate **76a** and its (*Z*)-isomer, respectively. Higher stereoselectivity was observed for the cyclization of 1,1-dimethyl-2-butenylene dicarbamate **78**, which gave oxazolidone **79** with 90% ee, although the regioselectivity is low (30%).

Type IIc

Asymmetric allylic substitution reactions of the racemic substrates bearing two different substituents at the 1- and 3-positions involve the chiral π-allylpalladium

Scheme 32

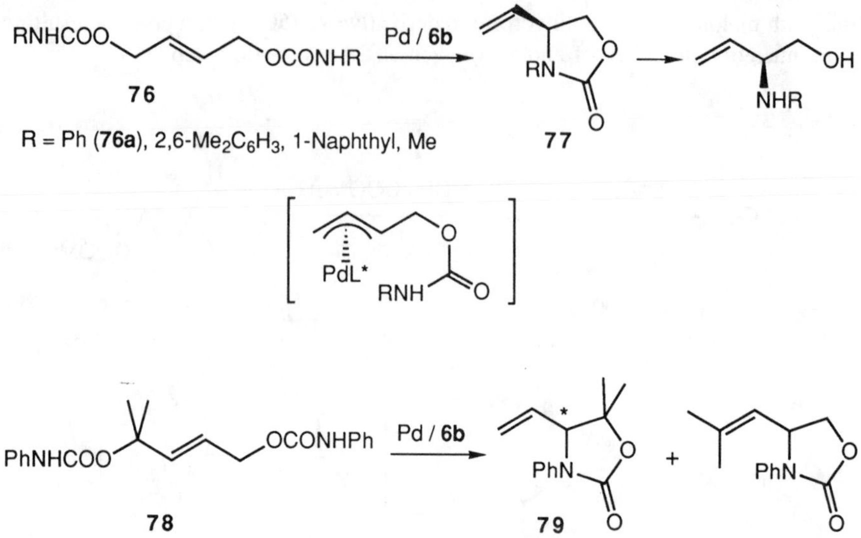
RNHCOO
OCONHR
Pd / 6b
76
R = Ph (76a), 2,6-Me2C6H3, 1-Naphthyl, Me
RN
77
OH
NHR
PdL*
RNH
PhNHCOO
OCONHPh
Pd / 6b
78
PhN
79
+

Scheme 33

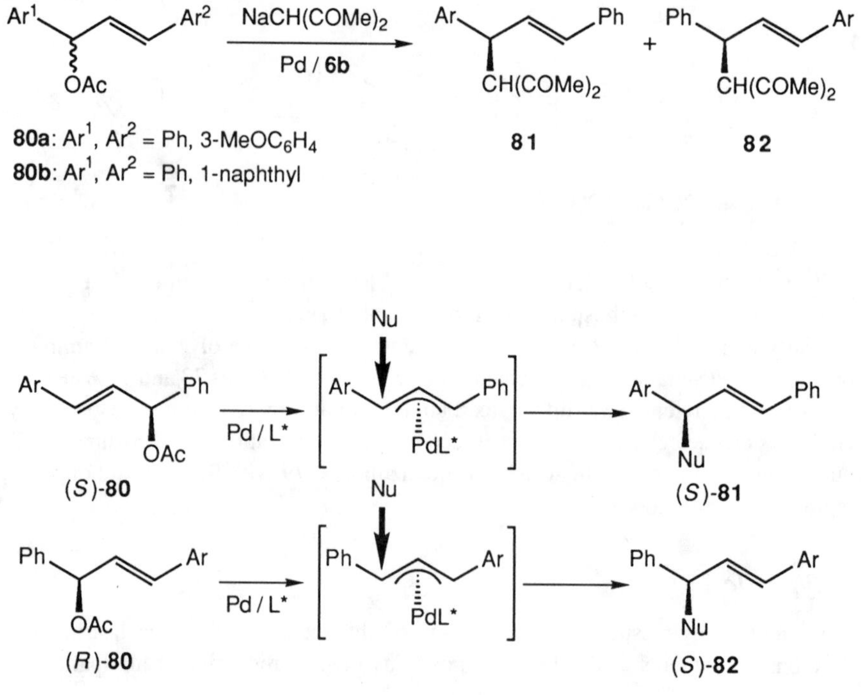
Ar1
Ar2
OAc
NaCH(COMe)2
Pd / 6b
Ar
Ph
CH(COMe)2
+
81
82
80a: Ar1, Ar2 = Ph, 3-MeOC6H4
80b: Ar1, Ar2 = Ph, 1-naphthyl
Nu
Pd / L*
PdL*
(S)-80
(S)-81
(R)-80
(S)-82

intermediates, which cannot undergo epimerization by the π-σ-π mechanism. Although the substitution reactions proceed with net retention of configuration, optically active products can be formed from racemic substrates by a regioselective nucleophilic attack on the π-allyl. Use of the chiral ferrocenylphosphine ligand **6b** for the alkylation of racemic allyl acetates **80**, which have slightly different aryl groups at the 1- and 3-positions, gave rise to high enantiomeric purity in the allylic alkylation products **81** and **82** [38] (Scheme 33). The high enantiomeric purity (up to 95% ee) of the products results from the regiocontrol by chiral ligand **6b**, which directs the nucleophilic attack on one of the π-allyl carbons (see also Scheme 13).

A kinetic resolution of a racemic allyl acetate has been reported in the reaction of 3-acetoxy-4-methyl-1-phenylpentene (**83**) with sodium acetylacetonate catalyzed by Pd/**6b** [39] (Scheme 34). The relative rate ratio for the enantiomers k_{fast}/k_{slow} is 14, practically enantiomerically pure (> 99% ee) (*R*)-**83** being recovered at 80% conversion. Allylic alkylation product (*S*)-**84** with high enantiomeric purity (> 98% ee) was also obtained at 40% conversion.

Scheme 34

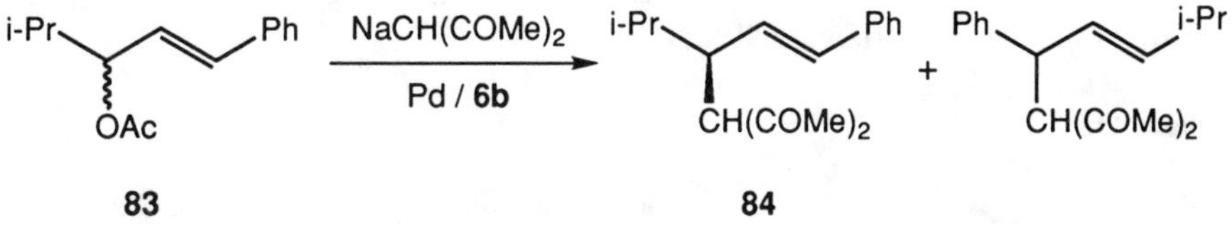

Scheme 35

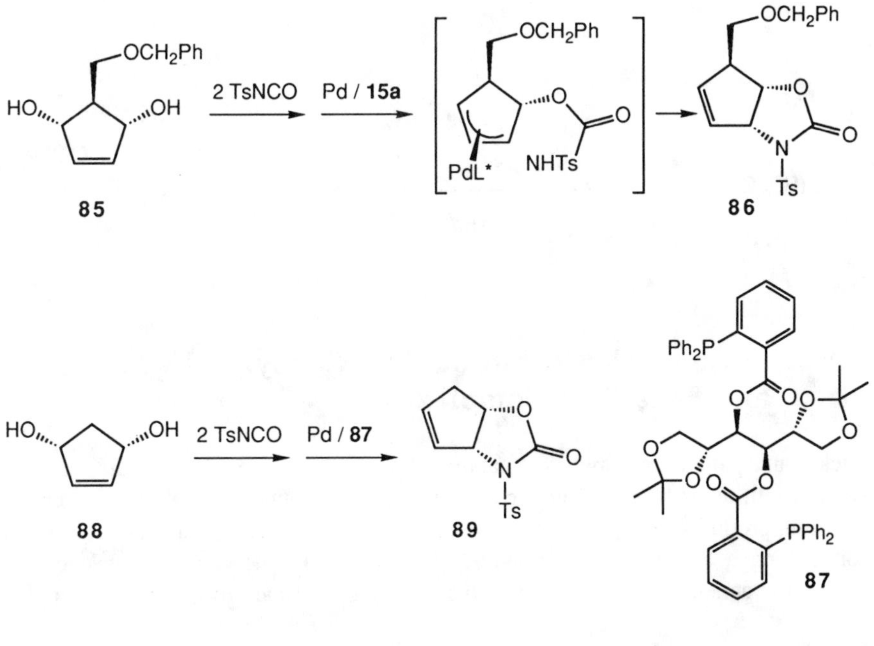

Treatment of cyclopentene diol derivative **85** with 2 equiv. of *p*-tolylsulfonyl isocyanate followed by 3 mol % of Pd$_2$(DBA)$_3$ and 6 mol % of (−)-BINAPO (**15a**) in THF gave oxazolidinone **86** with 59–65% ee. The cyclization proceeds via a dicarbamate of **85**, in a similar manner to the reaction shown in Scheme 32, and the observed enantiodifferentiation is the result of selective oxidative addition of the dicarbamate to the palladium(0) coordinated with the chiral ligand to form one of the chiral π-allyl intermediates [40] (Scheme 35). The use of 2-(diphenylphosphino)benzoic acid ester of optically active diol **87** as a chiral ligand for the cyclization of unsubstituted diol **88** led to the formation of oxazolidinone **89** with 65% ee, while BINAPO (**15a**) gave only 28% ee [41].

Asymmetric induction in allylic alkylation of cyclopentene diol esters has been reported by using (*S*)-BINAPO (**15a**) as the chiral ligand [42] (Scheme 36). Reaction of dibenzoate **90a** with dimethyl sodiomalonate in acetonitrile at 0°C gave a 38% yield of monoalkylated compound **91** with 57% ee, along with 32% of dialkylated product. The optically active bicyclic compound **92** was produced in the reaction with dimethyl sodio-3-ketoglutarate, where the allylic alkylation takes place twice and the enantioselectivity is determined at the first oxidative addition of the diester to palladium(0). The highest selectivity (55% ee) was obtained in the reaction of diphenyl carbonate **90b** at −40°C.

Scheme 36

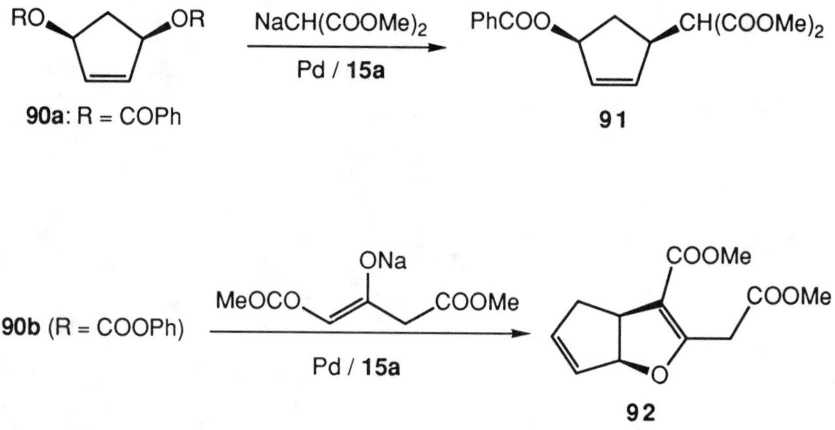

90a: R = COPh

91

90b (R = COOPh)

92

7.1.2. Asymmetric Cross-Coupling Catalyzed by Nickel and Palladium Complexes

Nickel and palladium complexes catalyze the reaction of organometallic reagents (R–m) with alkenyl or aryl halides and related compounds (R′–X) to give cross-coupling products (R–R′), which provides one of the most useful synthetic means for making a carbon–carbon bond [1] (Scheme 37). The catalytic cycle of the reaction is generally accepted to involve an unsymmetrical diorganometal complex

$L_nM(II)(R)R'$ as a key intermediate. From this intermediate the product R–R′ is released by reductive elimination to leave an $L_nM(0)$ species that undergoes oxidative addition to R′–X generating an intermediate $L_nM(II)(X)R'$. Transfer of an alkyl group from R–m to this intermediate regenerates the diorganometal complex.

Scheme 37

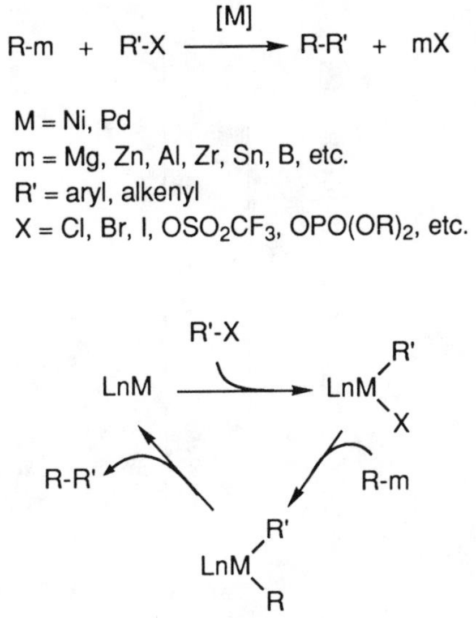
[M]
R-m + R'-X → R-R' + mX
M = Ni, Pd
m = Mg, Zn, Al, Zr, Sn, B, etc.
R' = aryl, alkenyl
X = Cl, Br, I, OSO2CF3, OPO(OR)2, etc.
R'-X
LnM
R'
LnM
X
R-R'
R-m
R'
LnM
R

Asymmetric synthesis by the catalytic cross-coupling reaction has been most extensively studied with secondary alkyl Grignard reagents. Asymmetric cross-coupling with chiral catalysts allows transformation of a racemic mixture of the secondary alkyl Grignard reagent into an optically active product by a kinetic resolution of the Grignard reagent. Since the secondary alkyl Grignard reagents usually undergo racemization at a rate comparable to the cross-coupling, an optically active coupling product is formed even if the conversion of the Grignard reagent is 100% (Scheme 38).

Scheme 38

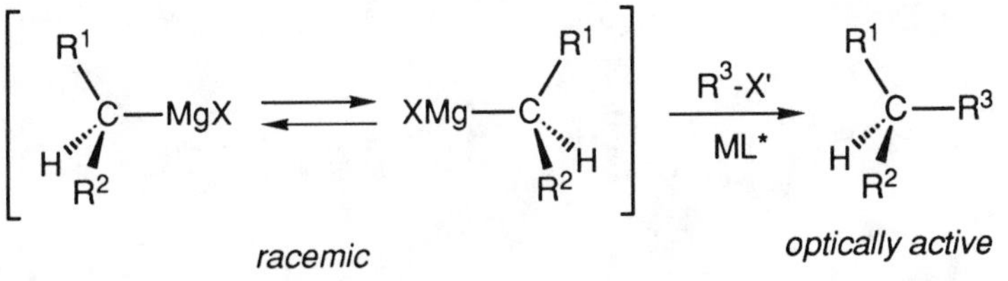
R1
C—MgX
H
R2
XMg—C
R1
H
R2
R3-X'
ML*
R1
C—R3
H
R2
racemic
optically active

The first reported examples of asymmetric Grignard cross-coupling entailed the use of a nickel catalyst coordinated with (−)-DIOP (**1**) as the catalyst [43,44].

Table 2 Asymmetric Cross-Coupling of 1-Phenylethylmagnesium (or Zinc) Reagent with Alkenyl Halides (**95**)

Entry	Halide (**95**)	PhCH(Me)MX (equiv. to **95**)	Catalyst (mol %)	Reaction Conditions			Yield (%)	% ee (Config.)	Ref.
				Solvent	Temp. (°C)	Time (h)			
1	**95a**	**93a**	(−)-DIOP (**1**)–Ni (0.4)	Ether	−80	20	74	7 (R)	43
2	**95a**	**93a** (3.9)	(−)-DIOP (**1**)–Ni (0.1)	Ether	0	1	81	13 (R)	44
3	**95b**	**93a** (4)	(S)-(R)-PPFA (**99a**)–Ni (0.5)	Ether	−20	24	>95	68 (R)	45a,b
4	**95b**	**93a** (3)	(S)-(R)-**99b**–Ni (0.5)	Ether	0	24	>95	62 (R)	45b
5	**95b**	**93a** (3)	(S)-(R)-**99c**–Ni (0.5)	Ether	0	24	43	42 (S)	45b
6	**95b**	**93a** (3)	(S)-(R)-PPFA (**99a**)–Pd (0.5)	Ether	25	70	82	61 (R)	45b
7	**95b**	**93a** (4)	(S)-(R)-BPPFA (**6f**)–Ni (0.5)	Ether	0	24	73	65 (R)	45b
8	**95b**	**93a** (4)	(S)-**100**–Ni (0.5)	Ether	0	24	>95	65 (S)	45a,b
9	**95b**	**93a** (3)	(R)-**101**–Ni (0.5)	Ether	0	24	86	5 (S)	45a,b
10	**95b**	**93a** (2)	(R)-(S)-**102**–Ni (0.5)	Ether	0	40	77	17 (R)	45c
11	**95c**	**93a** (2)	PPFA (**99a**)–Pd (2)	Ether	−25	~20	73		46
12	**95b**	**93a** (2)	**103a**–Ni (0.5)	Ether	0	40	>95	81 (S)	47
13	**95b**	**93a** (2)	**103b**–Ni (0.5)	Ether	0	40	>95	81 (S)	47
14	**95b**	**93a** (2)	**103c**–Ni (0.5)	Ether	0	40	77	83 (R)	47

15	95b	93a (2)	104–Ni (0.5)	Ether	0	40	73	49 (S)	48
16	95b	93a (2)	(R)-105a–Ni (0.8)	Ether	−5	16	>90	88 (R)	49a
17	95b	93a (2)	105b–Ni (0.5)	Ether	−40 → 0	14	>90	38 (S)	49b
18	95b	93a (2)	105c–Ni (0.5)	Ether	−40 → 0	14	50	65 (S)	49b
19	95b	93a (2)	106–Ni (0.5)	Ether	−10 → 0	17	95	46 (R)	50
20	95b	93a (2)	24–Ni (0.2)	Ether	−40 → RT	2	45	67 (S)	51
21	95a	93a	39–Ni (0.3)	Ether				47 (S)	52
22	95b	93a	107–Ni		0	24	65	11 (S)	53
23	95c	93a (1.5)	108–Pd (0.5)	Ether	−45	20	95	11 (R)	54
24	95c	93a (2.0)	109–Pd (0.5)		−45 → 0	20	46	40 (R)	55
25	95b	93a	110–Ni		−10		50	11 (R)	56
26	95b	93a (2.0)	111–Ni (0.7)	Ether	−78 → RT	12	50–52	12–17 (S)	57
27	95c	93a (2.0)	112–Pd (0.5)		−45 → 20	20	90	13 (S)	58
28	95b	93a	113–Ni	Ether	0	18	66	32 (R)	59
29	95b	93a (1.5–2.0)	114–Ni (1)	Ether	−78 → 25		<53	22 (S)	60
30	95b	93a + ZnI₂ (3)	(R)-(S)-PPFA (99a)–Pd (0.5)	THF	0	21	>95	86 (S)	62
31	95b	93a + ZnCl₂ (3)	120–Pd (0.5)	THF	0	20	>95	93 (R)	63
32	95b	93a + ZnCl₂ (3)	121–Pd (0.5)	THF	0	18	67	61 (S)	64
33	95b	93a (2)	(S)-105a–Ni (0.1)	Ether	0	21	95	61 (S)	65
34	95b	93a + ZnBr₂ (2)	(S)-105a–Ni (0.1)	Ether	−34	21	71	70 (R)	65

Scheme 39

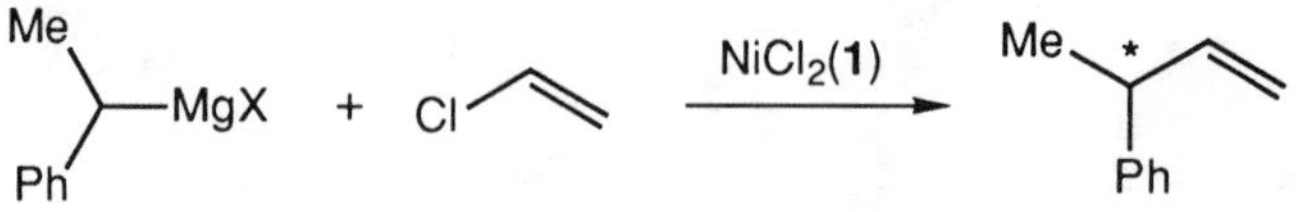

Me
$\searrow$
$\rangle$—MgX + Cl$\diagdown\diagup\diagdown$ $\xrightarrow{\text{NiCl}_2(1)}$ Me$\searrow$$_*$$\diagup\diagdown$
Ph$\diagup$ Ph

93   **95a**   **58**

Reaction of 1-phenylethyl (**93**) and 2-butyl (**94**) Grignard reagents with vinyl chloride (**95a**) and halobenzenes (**96**), respectively, gave the corresponding coupling products, (*R*)-3-phenyl-1-butene (**58**) (13% ee) and (*R*)-2-phenylbutane (**97**) (17% ee) (Schemes 39 and 40).

Scheme 40

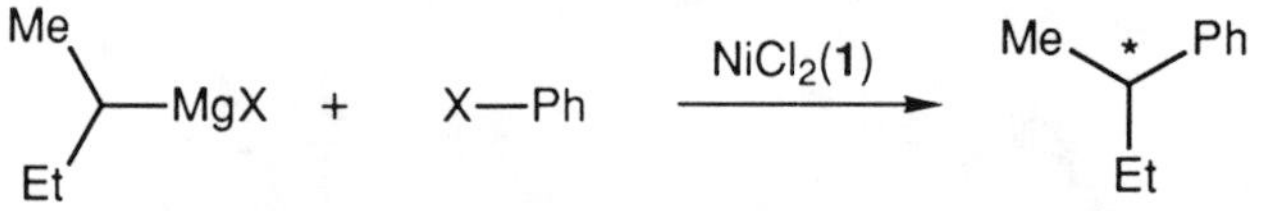

Me
$\searrow$
$\rangle$—MgX + X—Ph $\xrightarrow{\text{NiCl}_2(1)}$ Me$\searrow$$_*$$\diagup$Ph
Et$\diagup$ $|$
                                                        Et

94   **96**   **97**

After these findings, asymmetric cross-coupling was attempted using optically active phosphine ligands of various kinds. The reaction most extensively studied so far is that of 1-phenylethylmagnesium chloride (**93a**) with vinyl bromide (**95b**) or (*E*)-bromostyrene (**95c**), forming 3-phenyl-1-butene (**58**) or 1,3-diphenyl-1-butene (**98**), respectively (Scheme 41). Representative results are summarized in Table 2. The cross-coupling generally proceeds in high yields in ether at 0°C or at a lower temperature in the presence of not more than 1 mol % of a nickel–phosphine complex $NiCl_2P^*$ or a catalyst generated in situ from NiX_2 (X = Cl or Br) and a phosphine ligand L*. The preformed palladium complex $PdCl_2L^*$ also catalyzes asymmetric cross-coupling.

Table 2 shows that the ferrocenylphosphines containing a (dialkylamino)alkyl group on the side chain are effective for the cross-coupling of **93a** catalyzed by nickel or palladium complexes [45]: (*S*)-(*R*)-PPFA (**99a**) and (*S*)-(*R*)-BPPFA (**6f**) gave the coupling product **58** with 68 and 65% ee, respectively. The presence of the (dialkylamino)alkyl side chain is of primary importance for high enantioselectivity, and the selectivity is strongly affected by the structure of the dialkylamino group (entries 3–11). The amino group is proposed to coordinate with the magnesium atom in the Grignard reagent at the transmetallation step in the catalytic cycle, where the coordination occurs selectively with one of the enantiomers of the racemic Grignard reagent to bring about high enantioselectivity [45], although the coordination has not been supported by NMR studies of a palladium complex [46].

Based on the strong influence of the (dialkylamino)alkyl side chain on the ferrocenylphosphines **99**, a series of β-(dialkylamino)alkylphosphines (**103**) were prepared and used for cross-coupling. The aminoalkylphosphines substituted with sterically bulky alkyl group at the chiral carbon center are more effective than the ferrocenylphosphine ligands. VALPHOS (**103a**), ILEPHOS (**103b**), and *tert*-

LEUPHOS (**103c**), which were prepared from valine, isoleucine, and *tert*-leucine, respectively, gave **58** with $\geq 81\%$ ee [47] (entries 12–15). Use of polymer-supported β-(dialkylamino)alkylphosphine ligand **104** analogous to VALPHOS (**103a**) gave **58** with slightly lower enantiomeric purity [48]. Comparable enan-

Scheme 41

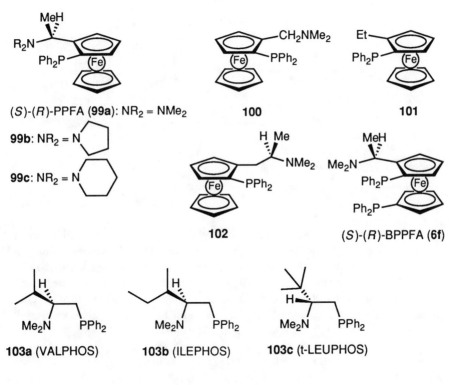

(*Continued*)

Scheme 41 *(Continued)*

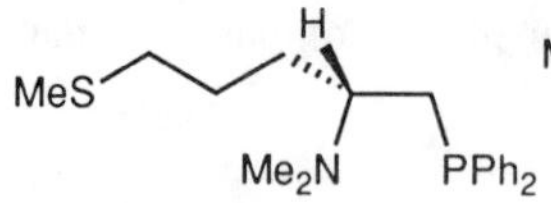

105a

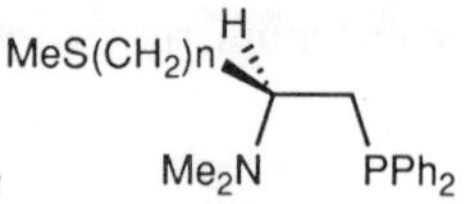

105b: n = 1
105c: n = 2

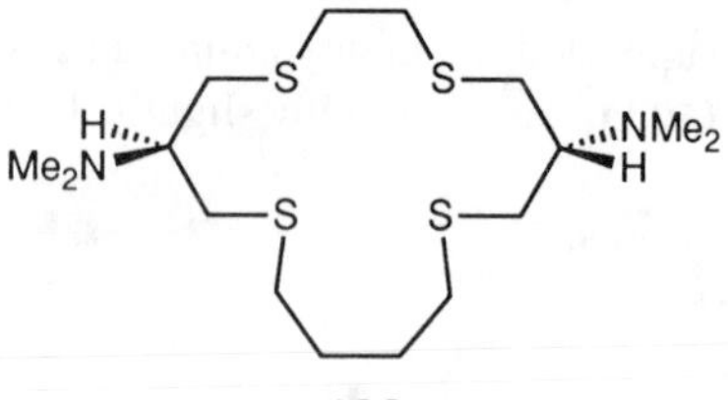

106

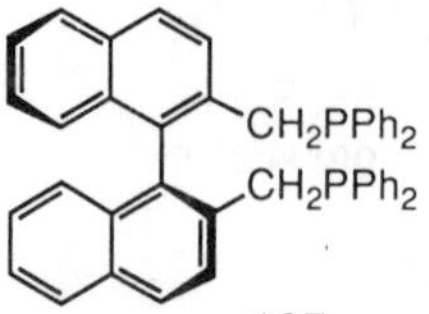

107

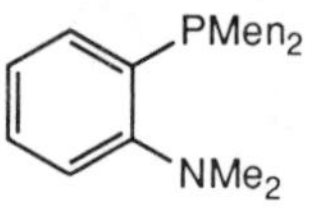

108

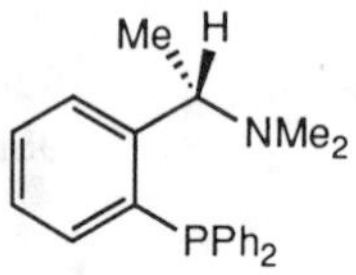

109

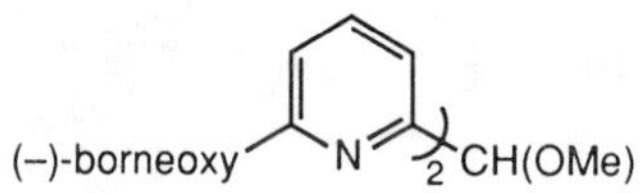

110

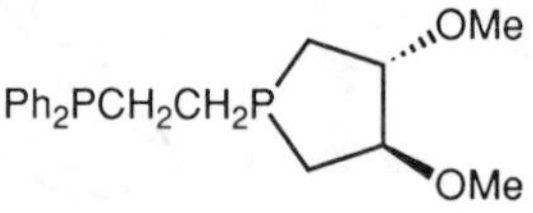

111

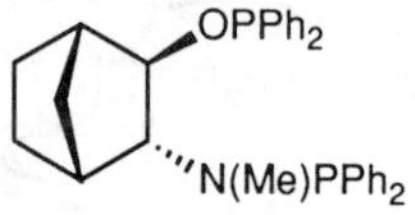

112

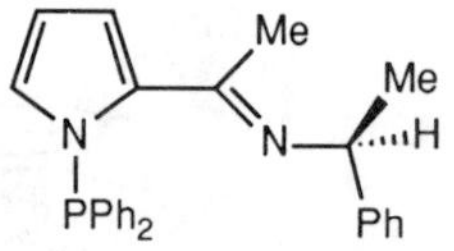

113

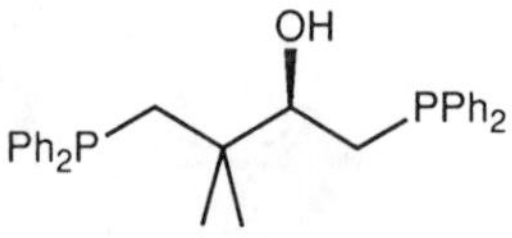

114

tioselectivity was observed with the β-(dialkylamino)alkylphosphines **105**, containing a sulfide group on the alkyl chain [49] (entries 16–18). The sulfur-bearing alkyl group is more effective than a simple alkyl side chain with the highest enantioselectivity (88% ee) being obtained with **105a**, a reagent derived from homomethionine. Several chiral macrocyclic sulfides have been prepared and examined as chiral ligands for the nickel-catalyzed coupling reaction, although the enantioselectivity was lower (46% ee with tetrasulfide ligand **106**) [50] (entry 19).

Nickel catalysts, complexed with unfunctionalized chelating bisphosphine ligands (*R*,*R*)-NORPHOS (**24**) [51] and **39** [52], also induced a high enantioselectivity in the reaction shown in Scheme 41 (see also Table 2, entries 20 and 21). The results reported with other chiral phosphine ligands (**107–114**) [53–60] are summarized in Table 2 (entries 22–29).

The chiral ferrocenylphosphines and β-(dialkylamino)alkylphosphines are used for the nickel-catalyzed asymmetric cross-coupling of 1-(4-substituted-phenyl)ethyl Grignard reagents **115a** and **115b** with vinyl bromide (**95b**) [47b, 61] (Scheme 42). The enantioselectivity is as high as that for the reaction of the 1-phenylethyl Grignard reagent (**93a**). The coupling products **116** were converted to optically active curcumene (**117**) and 2-(4-isobutylphenyl)propionic acid (ibuprofen) (**118**).

Scheme 42

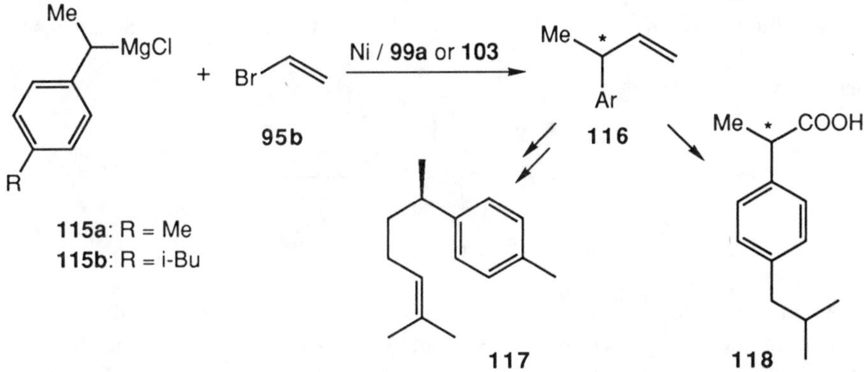

115a: R = Me
115b: R = i-Bu

Use of 1-arylethylzinc reagents in place of the corresponding Grignard reagents sometimes increases stereoselectivity (entries 30–32 in Table 2). The reaction of zinc reagents **119**, prepared from **93a** and **115a** with a zinc halide in THF in the presence of a palladium catalyst coordinated with a chiral ferrocenylphosphine [(R)-(S)-PPFA (**99a**)], proceeded with 85–86% enantioselectivity [62] (Scheme 43). A higher selectivity (93% ee) was obtained with C_2-symmetric ferrocenylphosphine ligand **120**, which has two phosphorus atoms and two aminoalkyl side chains on the ferrocene skeleton [63]. An aminoalkylphosphine ligand **121**, which is analogous to PPFA but has an (η^6-benzene)chromium structure in place of ferrocene, showed slightly lower selectivity (61% ee) in the reaction of 1-phenylethylzinc reagent [64]. A reversal of absolute configuration of the coupling product **58** by the addition of a zinc salt was observed in the reaction catalyzed by **105a**–Ni [65] (entries 33 and 34).

Scheme 43

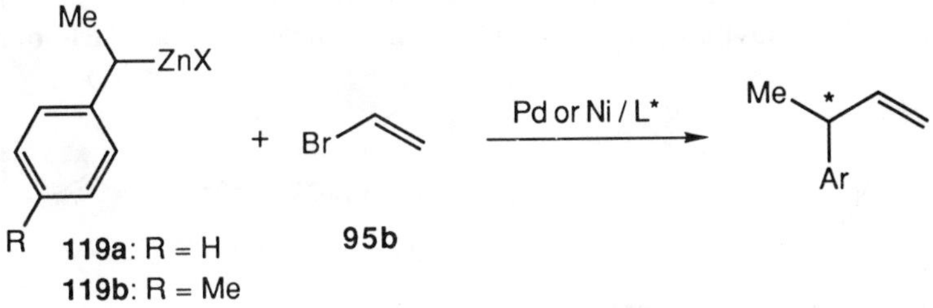

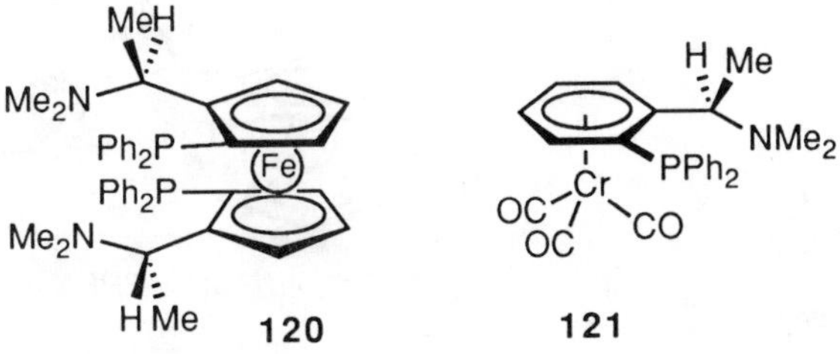

119a: R = H
119b: R = Me
95b

The 1-phenylethyl Grignard reagent **93a** was also used for coupling with allylic substrates such as allyl phenyl ether in the presence of NiCl$_2$[(S,S)-CHIRAPHOS (**17**)], which gave (R)-4-phenyl-1-butene (**122**) with 58% ee [21c] (Scheme 44).

Scheme 44

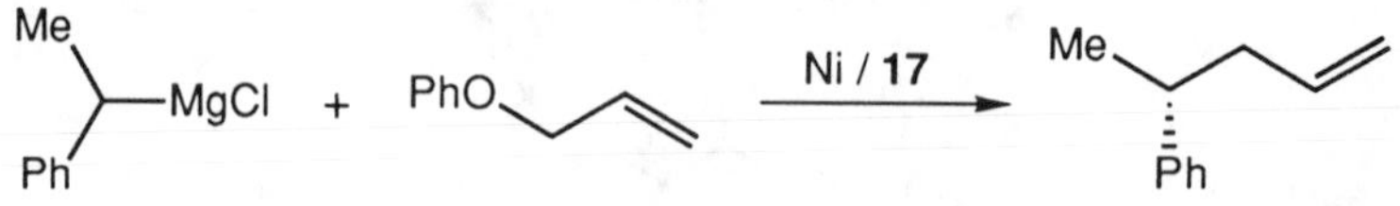

Asymmetric cross-coupling of secondary alkyl Grignard reagents that do not contain an aryl group such as phenyl on the chiral carbon center has not been as successful, in terms of enantioselectivity, as that of the 1-arylethyl Grignard reagent. The reaction of the 2-butyl Grignard reagents **94** with halobenzenes **96** was mainly studied with nickel catalysts complexed with chiral homologs of 1,2-bis(diphenylphosphino)ethane [52,66] (Scheme 40). The highest enantiomeric purity (55% ee) of the product, 2-phenylbutane, was obtained with 1,2-bis(diphenylphosphino)cyclopentane (**39**). Detailed studies on the reaction of **94** (X = Cl, Br, I) with **96** (X' = Cl, Br, I) in the presence of the nickel–(R)-PROPHOS (**60**) catalyst revealed that the absolute configuration of the coupling product as well as the enantioselectivity is dependent on the halogen atoms in both the Grignard reagent and halobenzenes. Reaction of **94** (X = Br) with **96** (X' = Br) gave (R)-2-phenylbutane (**97**) with 40% ee, whereas that of **94** (X = Cl) with **96** (X = I) gave (S)-**97** with 15% ee. In the reaction of substituted halobenzenes the enantioselectivity was found to be strongly influenced by steric factors but only slightly by electronic factors.

Asymmetric cross-coupling was successfully applied to the synthesis of optically active allylsilanes [67]. The reaction of α-(trimethylsilyl)benzylmagnesium bromide (**123**) with vinyl bromide (**95b**), (E)-bromostyrene (**95c**), and (E)-bromopro-

Scheme 45

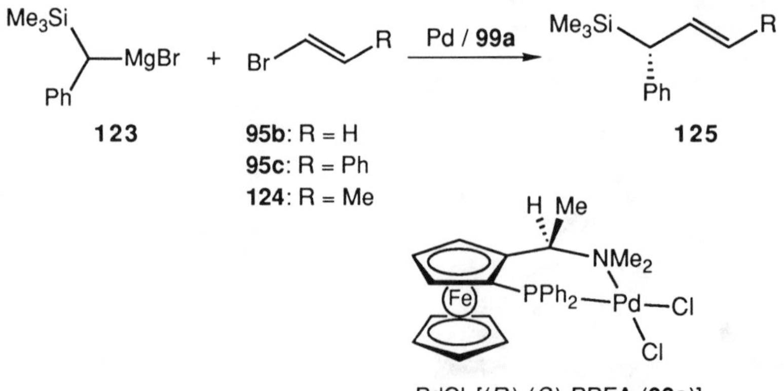

95b: R = H
95c: R = Ph
124: R = Me

PdCl$_2$[(R)-(S)-PPFA (**99a**)]

pene (**124**) in the presence of 0.5 mol % of a palladium complex coordinated with chiral ferrocenylphosphine, (*R*)-(*S*)-PPFA (**99a**), gave the corresponding (*R*)-allylsilanes (**125**) with 95, 85, and 95% ee, respectively, which were substituted with a phenyl group at the chiral carbon center bonded to the silicon atom (Scheme 45). These allylsilanes were used for the $S_{E'}$ reactions forming optically active homo-allyllic alcohols and π-allylpalladium complexes. Lower stereoselectivity was observed with (*Z*)-alkenyl bromides. The palladium–PPFA catalyst was also effective for the reaction of 1-(trialkylsilyl)ethylmagnesium chlorides with (*E*)-bromostyrene (**95c**). The enantioselectivity was dependent on the trialkylsilyl group, triethylsilyl being the best, producing 1-phenyl-3-silyl-1-butene with 93% ee (Scheme 46).

Scheme 46

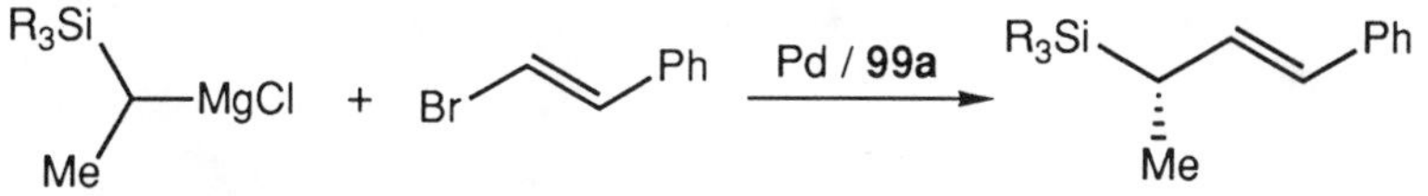

$$R_3Si = Me_3Si,\ PhMe_2Si,\ Et_3Si$$

The palladium-catalyzed asymmetric cross-coupling of α-(trimethylsilyl)benzyl-magnesium bromide (**123**) was also applied to the synthesis of optically active propargylsilane **128** (18% ee), using 1-bromo-2-phenylacetylene as a coupling partner [68] (Scheme 47).

Scheme 47

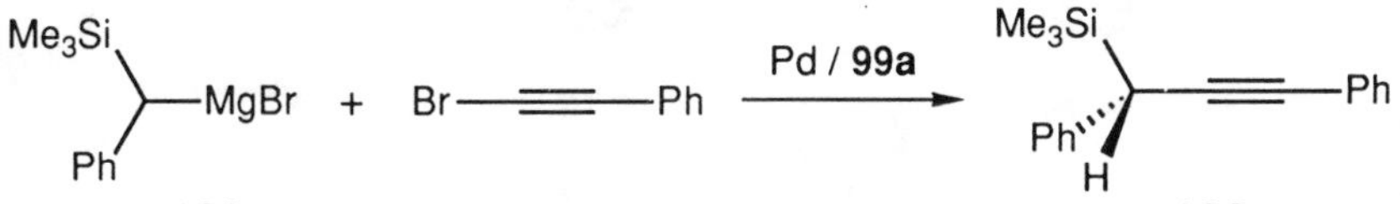

The Grignard reagents, 2-phenylpropylmagnesium chloride (**129**) and norbornyl Grignard reagent (**130**), which do not undergo racemization, have been kinetically resolved by asymmetric cross-coupling with less than 1 equiv. of vinyl bromide, although the efficiency of the resolution is not high [69] (Scheme 48). One of the enantiomers of racemic **130** underwent the coupling reaction in the presence of the VALPHOS (**103a**)–Ni catalyst 2.4 times faster than the other enantiomer. The enantiomeric purity of the coupling product **131** was 37% ee at 19% conversion.

Chiral ferrocenylsulfides **132** and **133**, which are analogous to PPFA (**99a**) or BPPFA (**6f**) but have a sulfide group instead of diphenylphosphino group, were used for the reaction of allylmagnesium chloride with 1-phenylethyl chloride in the presence of nickel or palladium catalysts to give 4-phenyl-1-pentene with up to 28% ee [70] (Schemer 49).

Scheme 48

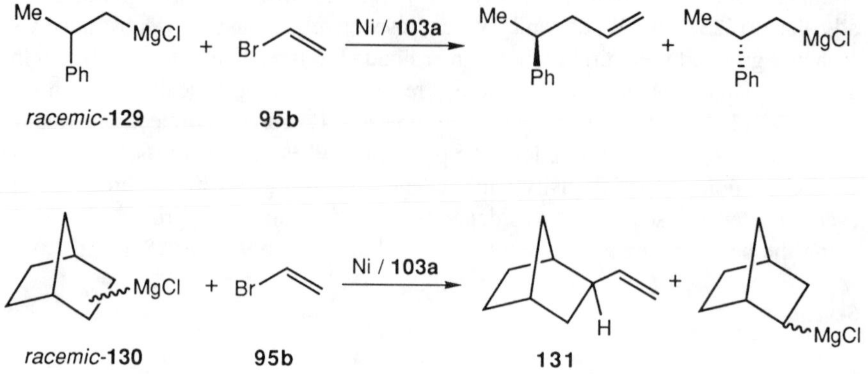

racemic-129   95b

racemic-130   95b   131

Scheme 49

132   133

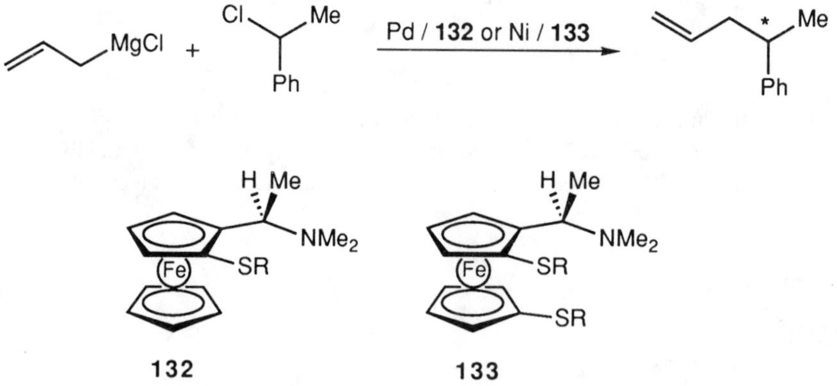

A chiral allene has been prepared by palladium-catalyzed cross-coupling of 4,4-dimethylpenta-1,2-dienylzinc chloride with iodobenzene or by that of 1-bromo-4,4-dimethylpenta-1,2-diene with phenylzinc chloride [71] (Scheme 50). The highest enantiomeric purity (25% ee) of the allene **134** was obtained in the former coupling using (R,R)-DIOP (**1**) as the chiral ligand. Interestingly, the enantiomeric purity was independent of the ratio of the reagents, although the reaction seems to involve a kinetic resolution of the racemic reagents.

Scheme 50

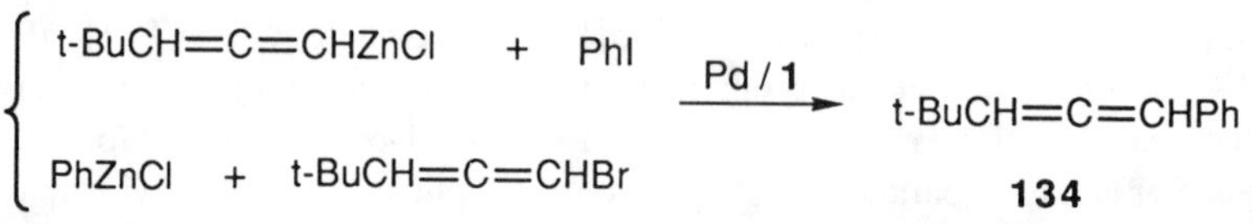

Preparation of axially chiral binaphthyls is one of the most exciting applications of the catalytic asymmetric cross-coupling reaction to organic synthesis. The reaction of 2-methyl-1-naphthylmagnesium bromide (**135**) with 1-bromo-2-methylnaphthalene (**136a**), forming 2,2'-dimethyl-1,1'-binaphthyl (**137a**), has been examined by using nickel catalysts coordinated with several chiral phosphine ligands (Scheme 51). Initial studies with (−)-DIOP (**1**), (*S*)-(*R*)-BPPFA (**6f**), and (*S*)-**107**, gave rather poor enantioselectivity (2, 5, and 13% ee, respectively) [53,72]. Use of the ferrocenylphosphine ligand (*S*)-(*R*)-**138**, which is a chiral monophosphine ligand containing a methoxy group on the side chain, dramatically increased the selectivity to produce a high yield of (*R*)-**137a** with 95% ee [73]. High enantioselectivity was also attained in the reaction with 1-bromonaphthalene (**136b**), which gave (*R*)-2-methyl-1,1'-binaphthyl (**137b**) with 83% ee.

Scheme 51

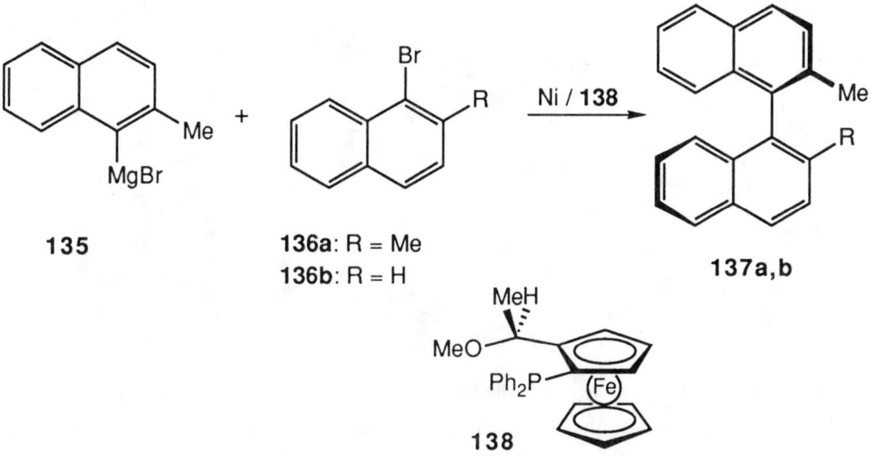

135   136a: R = Me   137a,b
             136b: R = H

138

The nickel-catalyzed cross-coupling of 2-methyl-1-naphthylmagnesium bromide (**135**) was extended to the asymmetric synthesis of ternaphthalenes [74] (Scheme 52). The reaction with 1,5- and 1,4-dibromonaphthalenes catalyzed by the (*S*)-(*R*)-**138**–Ni gave the corresponding optically active ternaphthalenes with 99% ee (**139**) and 95% ee (**140**), respectively, together with a small amount of meso compounds.

7.1.3. Conclusion

The data discussed in this chapter demonstrate that enantioselective carbon–carbon bond formation is catalyzed by chiral palladium and nickel complexes in reactions featuring asymmetric allylic substitution and cross-coupling. High enantioselectivities (> 90%) have been obtained in the allylic substitution reactions, where

Scheme 52

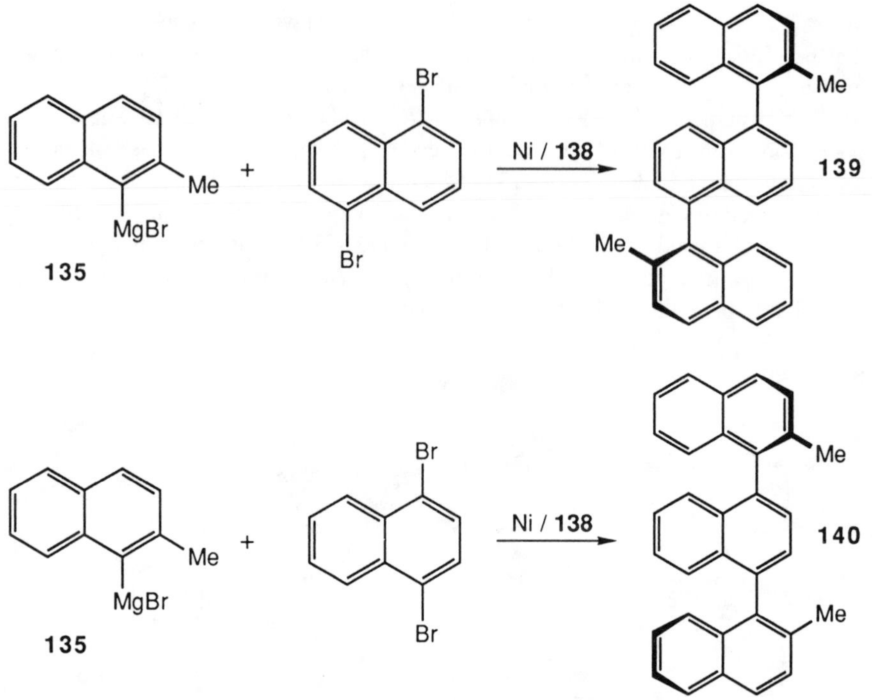

chiral phosphine ligands are well designed on the basis of the reaction mechanism. The chiral phosphine ligands that can control stereoselection by attractive interactions between functional groups of the ligand, and those in substrates at the nucleophilic attack on the π-allylpalladium intermediates have been successfully used, especially in type I and type IIa asymmetric reactions. In type IIb substitution reactions it is important to understand the enantio-determining step in the catalytic cycle. Depending on the enantio-determining step in the catalytic cycle (i.e., after the equilibration of diastereomeric π-allyl intermediates or prior to that), a proper chiral ligand must be designed for obtaining high stereoselectivity. In allylic substitution reactions, it would be very useful to develop chiral catalysts that are capable of inducing asymmetry to a great extent at the oxidative addition step of prochiral or meso-type allylic substrates to the catalysts. In asymmetric cross-coupling, the mechanism of enantioselection is not well understood, although high enantioselectivities have been achieved in the asymmetric synthesis of optically active allylsilanes and binaphthyls. More efficient catalyst systems will be developed when a detailed mechanistic understanding of the catalytic reactions in these areas is achieved. Applications of these reactions to asymmetric organic syntheses will continuously grow in the future.

References

1. (a) Trost, B. M.; Verhoeven, T. R. In *Comprehensive Organometallic Chemistry;* Wilkinson, G.; Stone, F. G. A.; Abel, E. W. (Eds.); Pergamon: Oxford, 1982; Vol. 8, p. 799. (b) Tsuji, J. *Tetrahedron,* **1986,** *42,* 4361. (c) Heck, R. F. *Palladium Reagents in Organic Synthesis,* Academic Press: Orlando, FL, 1985. (d) Jolly, P. W. In *Comprehensive Organometallic Chemistry;* Wilkinson, G.; Stone, F. G. A.; Abel, E. W. (Eds.); Pergamon: Oxford, 1982; Vol. 8, p. 713.

2. Consiglio, G.; Waymouth, R. M. *Chem. Rev.* **1989,** *89,* 257, and references cited therein.

3. Fiaud, J. C.; Hibon de Gournay, A.; Larcheveque, M.; Kagan, H. B. *J. Organomet. Chem.* **1978,** *154,* 175.

4. Hayashi, T.; Kanehira, K.; Tsuchiya, H.; Kumada, M. *J. Chem. Soc., Chem. Commun.* **1982,** 1162.

5. Hayashi, T.; Kanehira, K.; Hagihara, T.; Kumada, M. *J. Org. Chem.* **1988,** *53,* 113.

6. Sawamura, M.; Nagata, H.; Sakamoto, H.; Ito, Y. *J. Am. Chem. Soc.* **1992,** *114,* 2586.

7. Ito, Y.; Sawamura, M.; Matsuoka, M.; Matsumoto, M.; Hayashi, T. *Tetrahedron Lett.* **1987,** *28,* 4849.

8. (a) Genet, J. P.; Ferroud, D.; Juge, S.; Ruiz Montes, J. *Tetrahedron Lett.* **1986,** *27,* 4573. (b) Genet, J. P.; Juge, S.; Ruiz Montes, J.; Gaudin, J.-M. *J. Chem. Soc., Chem. Commun.* **1988,** 718.

9. Yamamoto, A.; Ito, Y.; Hayashi, T. *Tetrahedron Lett.* **1989,** *30,* 375.

10. Trost, B. M.; Strege, P. E. *J. Am. Chem. Soc.* **1977,** *99,* 1649.

11. Trost, B. M.; Murphy, D. J. *Organometallics,* **1985,** *4,* 1143.

12. (a) Hayashi, T.; Yamamoto, A.; Hagihara, T.; Ito, Y. *Tetrahedron Lett.* **1986,** *27,* 191. (b) Hayashi, T. *Pure Appl. Chem.* **1988,** *60,* 7.

13. Hayashi, T.; Yamamoto, A.; Ito, Y.; Nishioka, E.; Miura, H.; Yanagi, K. *J. Am. Chem. Soc.* **1989,** *111,* 6301.

14. (a) Yamaguchi, M.; Shima, T.; Yamagishi, T.; Hida, M. *Tetrahedron Lett.* **1990,** *31,* 5049. (b) Yamaguchi, M.; Shima, T.; Yamagishi, T.; Hida, M. *Tetrahedron: Asymmetry,* **1991,** *2,* 663.

15. Okada, Y.; Minami, T.; Umezu, Y.; Nishikawa, S.; Mori, R.; Nakayama, Y. *Tetrahedron: Asymmetry,* **1991,** *2,* 667.

16. Müller, D.; Umbricht, G.; Weber, B.; Pfaltz, A. *Helv. Chim. Acta,* **1991,** *74,* 232.

17. Togni, A. *Tetrahedron: Asymmetry,* **1991,** *2,* 683.

18. Fiaud, J. C.; Aribi-Zouioueche, L. *J. Organomet. Chem.* **1985,** *253,* 383.

19. Fiaud, J. C.; Legros, J. Y. *Tetrahedron Lett.* **1991,** *35,* 5089.

20. Fotiadu, F.; Cros, P.; Faure, B.; Buono, G. *Tetrahedron Lett.* **1990,** *31,* 77.

21. (a) Consiglio, G.; Morandini, F.; Piccolo, O. *Helv. Chim. Acta,* **1980,** *63,* 987. (b) Consiglio, G.; Morandini, F.; Piccolo, O. *J. Chem. Soc., Chem. Commun.* **1983,** 112. (c) Consiglio, G.; Piccolo, O.; Roncetti, L.; Morandini, F. *Tetrahedron,* **1986,** *42,* 2043.

22. Consiglio, G.; Indolese, A. *Organometallics,* **1991,** *10,* 3425.

23. Hiyama, T.; Wakasa, N. *Tetrahedron Lett.* **1985,** *26,* 3259.

24. (a) Auburn, P. R.; Mackenzie, P. B.; Bosnich, B. *J. Am. Chem. Soc.* **1985,** *107,* 2033. (b) Mackenzie, P. B.; Whelan, J.; Bosnich, B. *J. Am. Chem. Soc.* **1985,** *107,* 2046.

25. Farrar, D. H.; Payne, N. C. *J. Am. Chem. Soc.* **1985,** *107,* 2054.

26. Hayashi, T.; Kishi, K.; Yamamoto, A.; Ito, Y. *Tetrahedron Lett.* **1990,** *31,* 1743.

27. Hiroi, K.; Makino, K. *Chem. Lett.* **1986,** 617.

28. Consiglio, G.; Scalone, M.; Rama, F. *J. Mol. Catal.* **1989,** *50,* L11.

29. Brown, J. M.; MacIntyre, J. E. *J. Chem. Soc., Perkin Trans. 2,* **1985,** 961.

30. Cerest, M.; Felkin, H.; Umpleby, J. D.; Davis, S. G. *J. Chem. Soc., Chem. Commun.* **1981,** 681.

31. (a) Fiaud, J. C.; Legros, J. Y. *Tetrahedron Lett.* **1988,** *29,* 2959. (b) Fiaud, J. C.; Legros, J. Y. *J. Organomet. Chem.* **1989,** *370,* 383.

32. Fiaud, J. C.; Legros, J. Y. *J. Org. Chem.* **1990,** *55,* 4840.

33. Hayashi, T.; Kishi, K.; Uozumi, Y. *Tetrahedron: Asymmetry,* **1991,** *2,* 195.

34. (a) Yamamoto, K.; Tsuji, J. *Tetrahedron Lett.* **1982,** *23,* 3089. (b) Yamamoto, K.; Deguchi, R.; Ogimura, Y.; Tsuji, J. *Chem. Lett.* **1984,** 1657.

35. Genet, J. P.; Grisoni, S. *Tetrahedron Lett.* **1988,** *29,* 4543.

36. Hayashi, T.; Yamamoto, A.; Ito, Y. *Tetrahedron Lett.* **1988,** *29,* 669.

37. Hayashi, T.; Yamamoto, A.; Ito, Y. *Tetrahedron Lett.* **1988,** *29,* 99.

38. Hayashi, T.; Yamamoto, A.; Ito, Y. *Chem. Lett.* **1987,** 177.

39. Hayashi, T.; Yamamoto, A.; Ito, Y. *J. Chem. Soc., Chem. Commun.* **1986,** 1090.

40. Trost, B. M.; Van Vranken, D. L. *J. Am. Chem. Soc.* **1990,** *112,* 1261.

41. Trost, B. M.; Van Vranken, D. L. *J. Am. Chem. Soc.* **1991,** *113,* 6317.

42. Mori, M.; Nukui, S.; Shibasaki, M. *Chem. Lett.* **1991,** 1797.

43. Consiglio, G.; Botteghi, C. *Helv. Chim. Acta,* **1973,** *56,* 460.

44. Kiso, Y.; Tamao, K.; Miyake, N.; Yamamoto, K.; Kumada, M. *Tetrahedron Lett.* **1974,** 3.

45. (a) Hayashi, T.; Tajika, M.; Tamao, K.; Kumada, M. *J. Am. Chem. Soc.* **1976,** *98,* 3718. (b) Hayashi, T.; Konishi, M.; Fukushima, M.; Mise, T.; Kagotani, M.; Tajika, M.; Kumada, M. *J. Am. Chem. Soc.* **1982,** *104,* 180. (c) Hayashi, T.; Konishi, M.; Hioki, T.; Kumada, M.; Ratajczak, A.; Niedbala, H. *Bull. Chem. Soc. Jpn.* **1981,** *54,* 3615.

46. Baker, K. V.; Brown, J. M.; Cooley, N. A.; Hughes, G. D.; Taylor, R. J. *J. Organomet. Chem.* **1989,** *370,* 397.

47. (a) Hayashi, T.; Fukushima, M.; Konishi, M.; Kumada, M. *Tetrahedron Lett.* **1980,** *21,* 79. (b) Hayashi, T.; Konishi, M.; Fukushima, M.; Kanehira, K.; Hioki, T.; Kumada, M. *J. Org. Chem.* **1983,** *48,* 2195.

48. Hayashi, T.; Nagashima, N.; Kumada, M. *Tetrahedron Lett.* **1980,** *21,* 4623.

49. (a) Vriesema, B. K.; Kellogg, R. M. *Tetrahedron Lett.* **1986,** *27,* 2049. (b) Griffin, J. H.; Kellogg, R. M. *J. Org. Chem.* **1985,** *50,* 3261.

50. (a) Lemaire, M.; Vriesema, B. K.; Kellogg, R. M. *Tetrahedron Lett.* **1985,** *26,* 3499. (b) Vriesema, B. K.; Lemaire, M.; Butler, J.; Kellogg, R. M. *J. Org. Chem.* **1986,** *51,* 5169.

51. Brunner, H.; Pröbster, M. *J. Organomet. Chem.* **1981,** *209,* C1.

52. Consiglio, G.; Indolese, A. *J. Organomet. Chem.* **1991,** *417,* C36.

53. Tamao, K.; Yamamoto, H.; Matsumoto, H.; Miyake, N.; Hayashi, T.; Kumada, M. *Tetrahedron Lett.* **1977,** 1389.

54. Döbler, C.; Kinting, A.; *J. Organomet. Chem.* **1991**, *401*, C23.

55. Kreuzfeld, H.-J.; Döbler, C.; Abicht, H.-P. *J. Organomet. Chem.* **1987**, *336*, 287.

56. Wright, M. E.; Jin, M.-J. *J. Organomet. Chem.* **1990**, *387*, 373.

57. Brunner, H.; Limmer, S. *J. Organomet. Chem.* **1991**, *413*, 55.

58. Döbler, C.; Kreuzfeld, H.-J. *J. Organomet. Chem.* **1988**, *344*, 249.

59. Brunner, H.; Li, W.; Weber, H. *J. Organomet. Chem.* **1985**, *288*, 359.

60. Brunner, H.; Lautenschlager, H.-J.; König, W. A.; Krebber, R. *Chem. Ber.* **1990**, *123*, 847.

61. Tamao, K.; Hayashi, T.; Matsumoto, H.; Yamamoto, H.; Kumada, M.; *Tetrahedron Lett.* **1979**, *23*, 2155.

62. Hayashi, T.; Hagihara, T.; Katsuro, Y.; Kumada, M. *Bull. Chem. Soc. Jpn.* **1983**, *56*, 363.

63. Hayashi, T.; Yamamoto, A.; Hojo, M.; Ito, Y. *J. Chem. Soc., Chem. Commun.* **1989**, 495.

64. Uemura, M.; Miyake, R.; Nishimura, H.; Matsumoto, Y.; Hayashi, T. *Tetrahedron: Asymmetry,* **1992**, *3*, 213.

65. (a) Cross, G.; Kellogg, R. M. *J. Chem. Soc., Chem. Commun.* **1987**, 1746. (b) Cross, G.; Vriesema, B. K.; Boven, G.; Kellogg, R. M. *J. Organomet. Chem.* **1989**, *370*, 357.

66. (a) Consiglio, G.; Piccolo, O.; Morandini, F. *J. Organomet. Chem.* **1979**, *177*, C13. (b) Consiglio, G.; Morandini, F.; Piccolo, O. *Tetrahedron,* **1983**, *39*, 2699.

67. (a) Hayashi, T.; Konishi, M.; Ito, H.; Kumada, M. *J. Am. Chem. Soc.* **1982**, *104*, 4962. (b) Hayashi, T.; Konishi, M.; Okamoto, Y.; Kabeta, K.; Kumada, M. *J. Org. Chem.* **1986**, *51*, 3772.

68. Hayashi, T.; Okamoto, Y.; Kumada, M. *Tetrahedron Lett.* **1983**, *24*, 807.

69. Hayashi, T.; Kanehira, K.; Hioki, T.; Kumada, M. *Tetrahedron Lett.* **1981**, *21*, 137.

70. (a) Okoroafor, M. O.; Ward, D. L.; Brubaker, C. H., Jr. *Organometallics,* **1988**, *7*, 1504. (b) Naiini, A. A.; Lai, C.-K.; Ward, D. L.; Brubaker, C. H., Jr. *J. Organomet. Chem.* **1990**, *390*, 73.

71. de Graaf, W.; Boersma, J.; van Koten, G.; Elsevier, C. J. *J. Organomet. Chem.* **1989**, *378*, 115.

72. Tamao, K.; Minato, A.; Miyake, N.; Matsuda, T.; Kiso, Y.; Kumada, M. *Chem. Lett.* **1975**, 133.

73. Hayashi, T.; Hayashizaki, K.; Kiyoi, T.; Ito, Y. *J. Am. Chem. Soc.* **1988**, *110*, 8153.

74. Hayashi, T.; Hayashizaki, K.; Ito, Y. *Tetrahedron Lett.* **1989**, *30*, 215.

7.2

Asymmetric
Aldol Reactions

Masaya Sawamura and Yoshihiko Ito

Department of Synthetic Chemistry
Faculty of Engineering, Kyoto University
Kyoto, Japan

7.2.1. Introduction

The aldol reaction is one of the most important organic reactions, because it is a carbon–carbon bond forming reaction which produces highly functionalized compounds with a pair of newly generated chiral centers. Therefore, the pursuit of high stereoselectivity in the aldol reaction has been of much interest in recent years. In 1981 Masamune et al. [1] and Evans et al. [2], independently, reported extremely stereoselective asymmetric aldol reactions by means of chiral boron enolates. These discoveries are milestones in the history of asymmetric synthesis. Especially Evans's boron enolates, which can easily be prepared from optically active amino acids, have been widely applied to natural product syntheses. The next advance was made in 1982 by Iwasawa and Mukaiyama [3], who reported highly enantioselective aldol reactions of achiral tin(II) enolates with achiral aldehydes in the presence of a stoichiometric amount of chiral diamine ligands. This process does not require tedious procedures for attaching the auxiliary or removing it from the product. The next challenging goal in this field was the catalytic asymmetric aldol reaction.

7.2.2. Gold-Catalyzed Asymmetric Aldol Reaction of α-Isocyanocarboxylates

In 1986 Ito, Sawamura, and Hayashi [4] reported that gold(I) complexes prepared from cationic gold complex **1** and chiral ferrocenylphosphine ligands (**2**) bearing a tertiary amino group at the terminal position of a pendant chain are effective catalysts for asymmetric aldol reaction of isocyanoacetate **3a** with aldehydes, giving optically active 5-alkyl-2-oxazoline-4-carboxylates **4** (Scheme 1) [4–11]. The reaction shows high selectivity for the formation of trans isomers with enantiomeric excesses ranging as high as 97% (Table 1). Various substituents on aromatic aldehydes are acceptable for the highly stereoselective aldol reaction with the isocyanoacetate (Table 1: entries 5–10). Secondary and tertiary alkyl aldehydes give the corresponding *trans*-oxazolines almost exclusively with high enantioselectivity (entries 18, 19). The gold catalysts are effective for the reaction of α,β-unsaturated aldehydes as well (entries 20,21). The *trans*-oxazolines thus obtained can be readily hydrolyzed to β-hydroxy-α-amino acids and their derivatives. The turnover efficiency of the gold catalyst is shown in entry 6, where the [substrate]/[catalyst] ratio can be raised to 10,000:1 without significant loss of the enantiomeric purity of the *trans*-oxazoline, indicating that the gold-catalyzed aldol reaction may provide a practical process of great promise [12].

Scheme 1

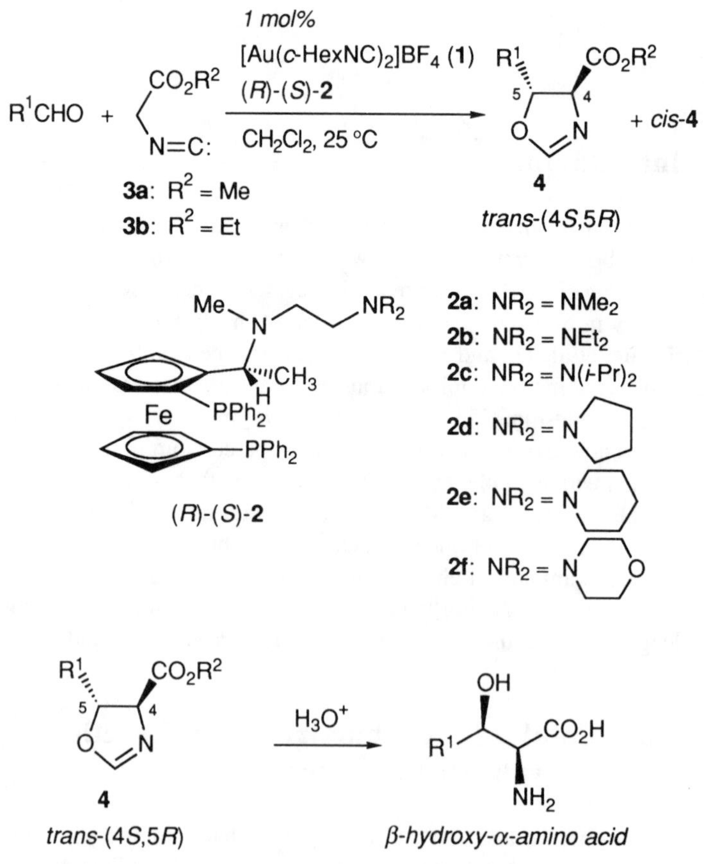

Both the enantio- and diastereoselectivity of this reaction are significantly dependent on the structure of the terminal amino group of the ligand (Table 1, entries 1–4,

Table 1 Gold-Catalyzed Asymmetric Aldol Reaction of Isocyanoacetate **3a** with Aldehydes (Scheme 1)[a]

Entry	Aldehyde	L*	Yield(%) of **4a**	trans/cis	% ee of trans-**4a**
1	PhCHO	**2a**	91	90/10	91
2	PhCHO	**2b**	98	89/11	93
3	PhCHO	**2e**	94	94/6	95
4	PhCHO	**2f**	93	95/5	95
5	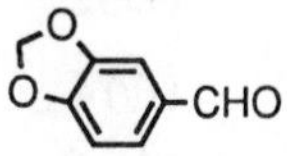 (benzodioxole)-CHO	**2f**	86	95/5	96
6[b]	(benzodioxole)-CHO	**2f**	c	91/9	91
7	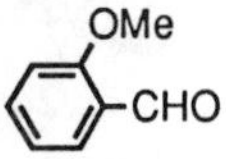 (2-OMe-phenyl)-CHO	**2f**	98	92/8	92
8	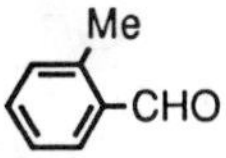 (2-Me-phenyl)-CHO	**2f**	98	96/4	95
9	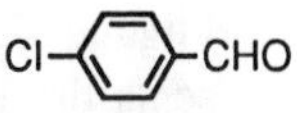 Cl-(phenyl)-CHO	**2f**	97	94/6	94
10	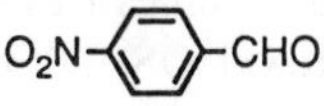 O$_2$N-(phenyl)-CHO	**2f**	80	83/17	86
11	MeCHO	**2a**	94	78/22	37
12	MeCHO	**2b**	100	84/16	72
13	MeCHO	**2c**	99	70/30	55
14	MeCHO	**2d**	83	87/13	74
15	MeCHO	**2e**	100	85/15	85
16	MeCHO	**2f**	99	89/11	89
17	i-BuCHO	**2f**	99	96/4	87
18	i-PrCHO	**2e**	99	99/1	94
19	t-BuCHO	**2f**	94	100/0	97
20	(E)-n-PrCH=CHCHO	**2f**	85	87/13	92
21	(E)-MeCH=CMeCHO	**2a**	89	91/9	93

[a]Reaction time: 20–40 h. [b]Substrates/catalyst = 1 × 10^4/L. Reaction at 40°C for 9 days. [c]Not isolated. Conversion: 100%.

11–16), indicating that the amino group plays a key role in the stereoselection [5,7]. In general, six-membered ring amino groups such as piperidino and morpholino groups are effective. Results listed in Table 2 [4] indicate that the existence of a terminal amino group in the proper position is crucial for high stereoselectivity.

Table 2 Gold-Catalyzed Asymmetric Aldol Reaction of Isocyanoacetate **3a** with PhCHO Using Various Chiral Ligands (Scheme 1)

Entry	L*	Time (h)	Yield of **4a**	trans/cis	% ee of trans-**4a**
1	**5**	40	99	89/11	23
2	**6**	22	91	69/31	37
3	**7**	51	80	68/32	racemic
4	**8**	90	87	86/14	racemic
5	(S,S)-CHIRAPHOS	100	90	75/25	racemic
6	(-)-DIOP	300	57	76/24	racemic
7	(+)-p-TolBINAP	200	69	69/31	8

Thus, the use of ligand **5** which is analogous to **2a** but with a longer tether between the terminal amino group and the ferrocene moiety, or ligand **6**, which has a hydroxyl group instead of the terminal amino group, causes a drastic decrease in stereoselectivity. Gold catalysts containing ferrocenylphosphines BPPFA (**7**) and **8**, which lack the terminal amino group, as well as CHIRAPHOS, DIOP, and p-Tol-BINAP, are catalytically much less active and give almost racemic oxazolines.

5: X = NMe(CH₂)₃NMe₂

6: X = NMeCH₂CH₂OH

7: X = NMe₂

8: X = OMe

The high efficiency of the gold catalysts has been explained by a hypothetical transition state as shown in Figure 1, where the chiral ligand chelates to gold with the two phosphorus atoms leaving the two nitrogen atoms in the pendant chain

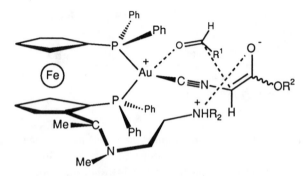

Figure 1 Proposed transition state model of the gold-catalyzed asymmetric aldol reaction.

uncoordinated, and the terminal amino group abstracts one of the α-methylene protons of the isocyanoacetate coordinated to the gold, forming an ion pair between enolate anion and ammonium cation. The attractive interaction (*secondary ligand–substrate interaction*) permits a favorable arrangement of the enolate and the aldehyde on the gold in the stereodifferentiating transition state.

The usefulness of the gold-catalyzed aldol reaction was demonstrated by application of the method to the asymmetric synthesis of the important membrane components D-*erythro* and *threo*-sphingosines, and their stereoisomers (Scheme 2) [13], and MeBmt, an unusual amino acid in the immunosuppressive undecapeptide cyclosporine (Scheme 3) [14].

Scheme 2

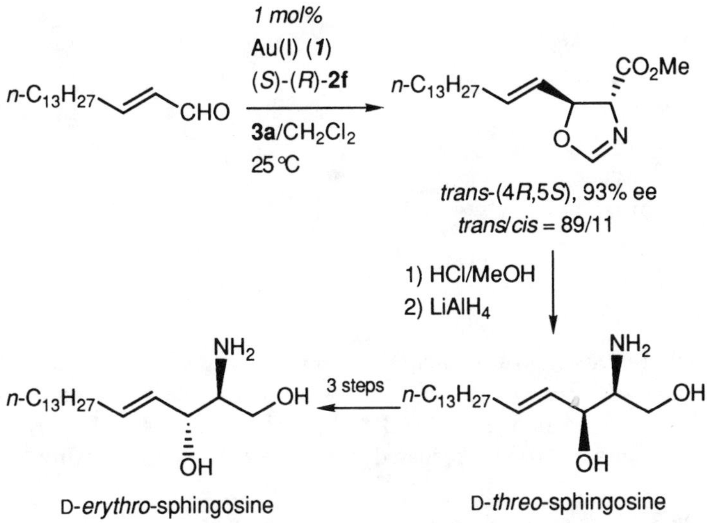

As pointed out by Togni and Pastor, enantioselectivities in the gold-catalyzed aldol reaction of aldehydes containing an α-heteroatom are significantly different from those of simple aldehydes (Table 3) [15,16]. Low enantioselectivities for *trans*-oxazolines are observed in the aldol reactions of 2-thiophene-, 2-furan-, and 2-pyridinecarboxaldehyde (entries 2, 4, 7). In the reactions of the 2-furan- and 2-pyridinecarboxaldehyde, *cis*-oxazolines with fairly high enantiomeric purities are formed as the minor product but in a rather low trans/cis ratio. A similar α-heteroatom effect is also observed in the aldol reaction of 2,3-O-isopropylidene-D-glyceraldehyde.

The catalytic asymmetric aldol reaction is tolerant of α-isocyanocarboxylates bearing an α-alkyl substituent (**9a–d**) as shown in Scheme 4 and Table 4 [17,18]. The enantioselectivities in the reaction with paraformadehyde are generally moderate, with the exception of the reaction of **9c** (81% ee), which has an α-isopropyl substituent (Table 4, entry 3). The oxazolines can be converted to optically active α-alkylserine derivative, a class of biologically interesting compounds.

Scheme 3

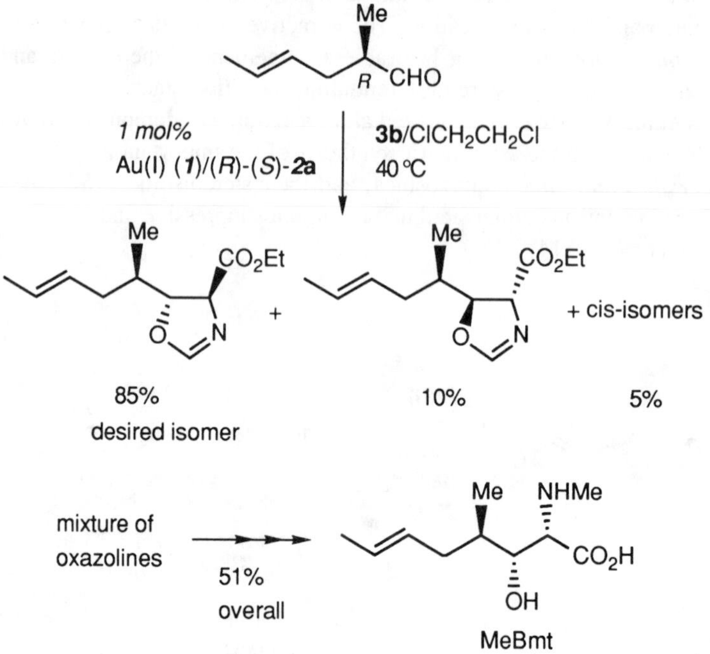

N,N-Dimethyl-α-isocyanoacetamide **10** is the substrate of choice in the reaction with acetaldehyde (98.6% ee) or primary aldehydes such as propionaldehyde (96.3% ee) or isovaleraldehyde (97.3% ee) (Scheme 5, Table 5) [19]. The oxazolinecarboxamides **11** thus prepared can be converted to β-hydroxy-α-amino

Table 3   Asymmetric Aldol Reaction of Isocyanoacetate **3b** with Functionalized Aldehydes in the Presence of Gold(I) Catalyst Containing Ligand **2a** (Scheme 1)[a]

Entry	Aldehyde R^1	Yield(%) of **4b**	trans/cis	% ee of trans-**4b**	% ee of cis-**4b**
1	3-thienyl	78	90/10	78	1 (4S,5S)
2	2-thienyl	90	95/5	33	17 (4S,5S)
3	3-furyl	80	86/14	87	7 (4S,5S)
4	2-furyl	62	68/32	32	83 (4S,5S)
5	4-pyridyl	38	88/12	75	78 (4S,5S)
6	3-pyridyl	55	88/12	79	62 (4S,5S)
7	2-pyridyl	45	75/25	6	84 (4S,5S)

[a] Reaction at 50°C in dichloromethane.

Scheme 4

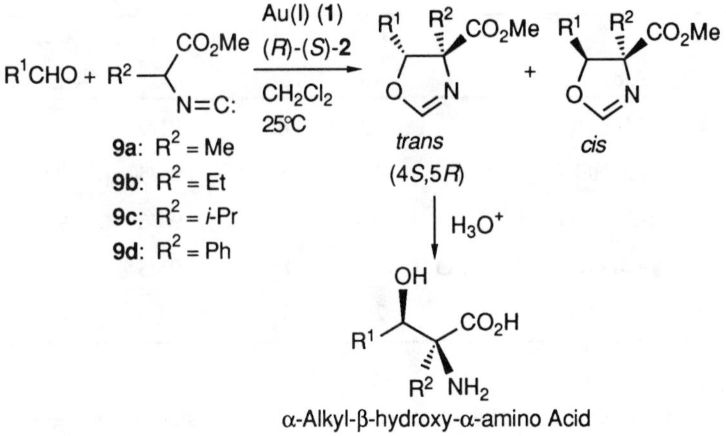

α-Alkyl-β-hydroxy-α-amino Acid

Scheme 5

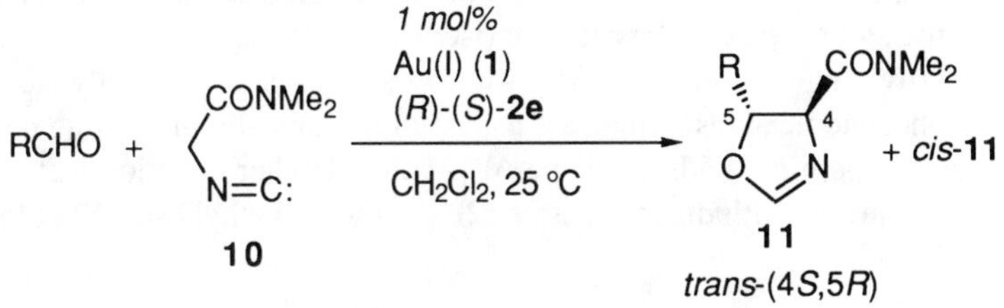

Table 4 Gold-Catalyzed Asymmetric Aldol Reaction of
α-Isocyanocarboxylates **9** with Aldehydes (Scheme 4)

Entry	Aldehyde	**9** R^2	Ligand	Yield (%)	trans/cis	% ee trans	% ee cis
1	(CH$_2$O)$_n$	Me	**2a**	100	-	64	-
2	(CH$_2$O)$_n$	Et	**2a**	89	-	70	-
3	(CH$_2$O)$_n$	i-Pr	**2e**	96	-	81	-
4	(CH$_2$O)$_n$	Ph	**2a**	75	-	67[a]	-
5	MeCHO	Me	**2f**	86	56/44	86	54 (4S,5S)
6	MeCHO	Et	**2f**	92	54/46	87	66 (4S,5S)
7	MeCHO	i-Pr	**2f**	100	24/76	26	51 (4S,5S)
8	PhCHO	Me	**2f**	97	93/7	94	53 (4S,5S)
9	PhCHO	i-Pr	**2e**	86	54/46	92	28 (4R,5R)

[a] The configuration has not been determined.

Table 5 Gold-Catalyzed Asymmetric Aldol Reaction of
Isocyanoacetamide **10** with Aldehydes (Scheme 5)

		Time	Yield(%)		% ee of
Entry	Aldehyde	(h)	of **11**	trans/cis	trans-**11**
1	MeCHO	40	85	91/9	98.6
2	EtCHO	40	84	95/5	96.3
3	i-BuCHO	6	92	94/6	97.3
4	p-BnOC$_6$H$_4$CH$_2$CHO	80	84	>95/5	94.5
5	PhCHO	25	74	94/6	94.1

acids by acidic hydrolysis. The aldol reaction of α-keto esters with isocyanoacetamide **10** proceeds with moderate to high enantioselectivity to give the corresponding
oxazolines with up to 90% ee [20].

The methodology of the catalytic asymmetric aldol reaction has been further
extended to the aldol-type condensation of (isocyanomethyl)phosphonates (**12**) with
aldehydes, providing a useful method for the synthesis of optically active (1-
aminoalkyl)phosphonic acids, which are a class of biologically interesting phospho-
rous analogs of α-amino acids (Scheme 6) [21,22]. Higher enantioselectivity and
reactivity are obtained with diphenyl ester **12b** than with diethyl ester **12a** (Table 6).

Scheme 6

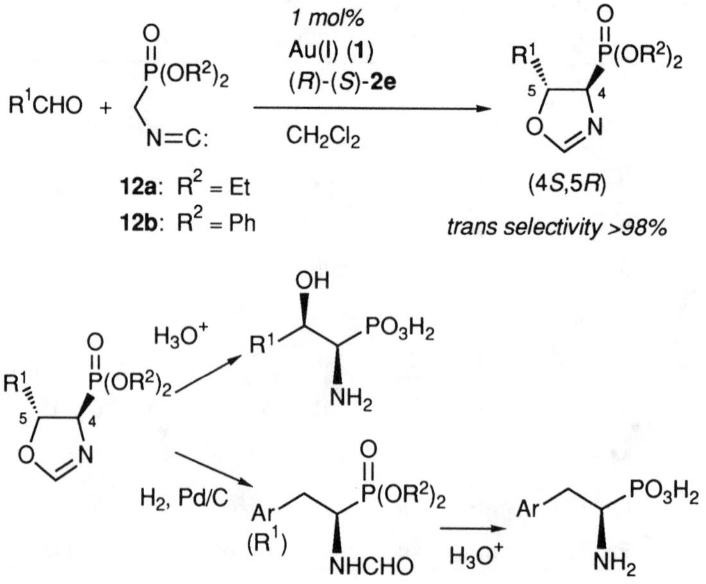

Table 6 Gold-Catalyzed Aldol-Type Condensation of (Isocyanomethyl)phosphonate **12** with Aldehydes (Scheme 6)

Entry	Isocyanide (R^2)	Aldehyde	Temp (°C)	Time (h)	Yield (%)	% ee
1	**12a** (Et)	PhCHO	40	90	78	92
2	**12a** (Et)	i-PrCHO	60	99	85	88
3	**12b** (Ph)	PhCHO	25	156	83	96
4	**12b** (Ph)	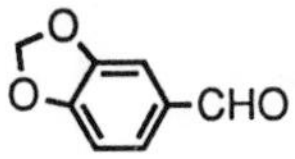 (see structure)—CHO	40	98	92	95
5	**12b** (Ph)	i-PrCHO	40	75	88	95

Pastor and Togni pointed out that the central chirality and the planar chirality in the ferrocenylphosphine ligand **2** are cooperative for stereoselection (the concept of *internal cooperativity of chirality*) [16,23,24]. As Table 7 shows, the change of chirality of the stereogenic carbon atom from *R* to *S* results in the formation of the other *trans*-oxazoline enantiomer with moderate enantiomeric excess.

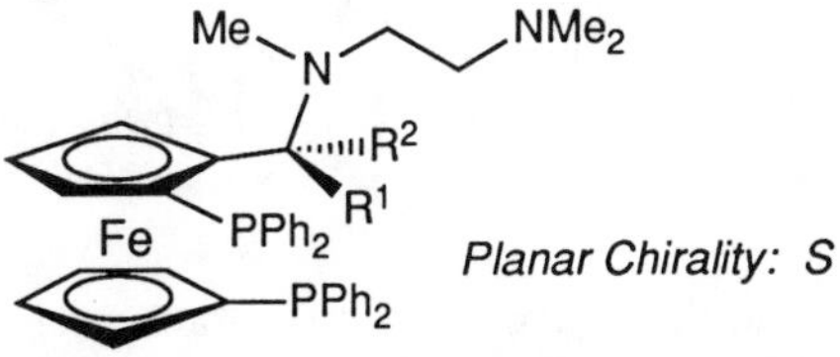

(*R*)-(*S*)-**2a**: R^1 = H, R^2 = Me
(*S*)-(*S*)-**2a**: R^1 = Me, R^2 = H

Recently it has been found that high stereoselectivity in the asymmetric aldol reaction of an isocyanoacetate is also obtainable with the silver(I) catalyst containing ferrocenylphosphine ligands **2e**, by keeping the isocyanoacetate concentration

Table 7 Gold-Catalyzed Asymmetric Aldol Reaction of Isocyanoacetate **3a** with PhCHO by the Use of (*R*)-(*S*)-**2a** or (*S*)-(*S*)-**2a** (Scheme 1)

Entry	L*	trans/cis	trans-**4a** % ee	cis-**4a** % ee
1	(*R*)-(*S*)-**2a**	90/10	91 (4*S*,5*R*)	7 (4*S*,5*S*)
2	(*S*)-(*S*)-**2a**	84/16	41 (4*R*,5*S*)	20 (4*S*,5*S*)

Table 8   Silver-Catalyzed Asymmetric Aldol Reaction of Isocyanoacetate **3a** with Aldehydes (Scheme 7)

Entry	Aldehyde	Yield(%) of **4a**	trans/cis	% ee of trans-**4a**
1	PhCHO	90	96/4	80
2	i-PrCHO	90	99/1	90
3	t-BuCHO	91	>99/1	88
4	CH$_2$=CMeCHO	90	97/3	87

low throughout the reaction by the slow addition of **3a** over a period of 1 hour (Scheme 7, Table 8) [25].

Scheme 7

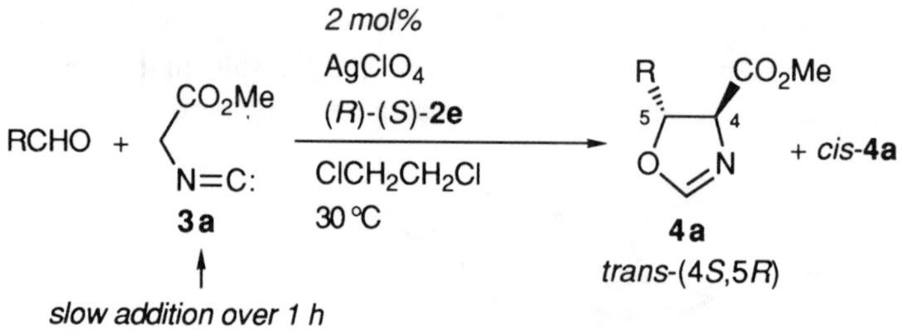

Mechanistic studies on the structure of gold(I) and silver(I) complexes coordinated with ferrocenylphosphine ligand **2** in the presence of an isocyanoacetate revealed that the most significant difference between these metal catalysts is in the number of the isocyanoacetates coordinating to the metal (Scheme 8). Thus, the gold complex adopts tricoordinated structure **13** with one isocyanoacetate even in the presence of a large excess of isocyanoacetate, while the silver complex is in equilibrium between tricoordinated complex **14** and tetracoordinated complex **15** bearing two isocyanoacetates [6,25]. The slow addition protocol in the silver-catalyzed aldol reaction should keep the isocyanoacetate concentration low in the reaction system, diminishing the unfavorable tetracoordinate species. The high stereoselectivity of the tricoordinated catalyst may result from the presence of one vacant coordination site on the metal, where an aldehyde can coordinate. The aldol reaction of an aldehyde in the chiral coordination site may well be more stereoselective (transition state as shown in Fig. 1) than that of the aldehyde without coordination to the metal catalyst.

Interestingly, aldol-type condensation of tosylmethyl isocyanide (**16**) with aldehydes is catalyzed by the silver catalyst more stereoselectively than that catalyzed by the gold catalyst under the standard reaction conditions (Scheme 9) [26]. Elucidation of the mechanistic differences between the gold and silver catalysts in the

Scheme 8

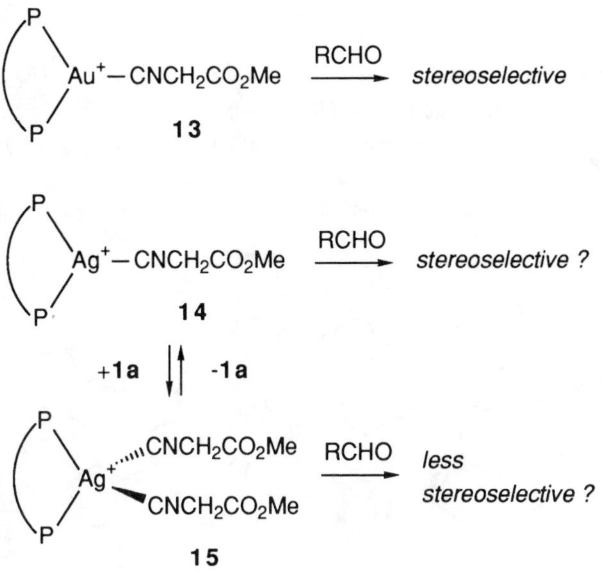

asymmetric aldol reaction of **16** needs further study. Oxazoline **17** can be converted to optically active α-alkyl-β-(*N*-methylamino)ethanols.

Scheme 9

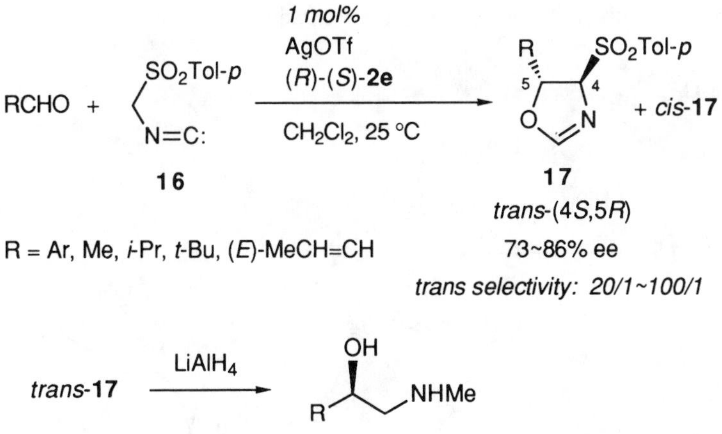

R = Ar, Me, *i*-Pr, *t*-Bu, (*E*)-MeCH=CH

7.2.3. Asymmetric Nitroaldol Reaction

Shibasaki et al. have reported an asymmetric nitroaldol reaction catalyzed by chiral lanthanum alkoxide **18** to produce an optically active 2-hydroxy-1-nitroalkane with

Scheme 10

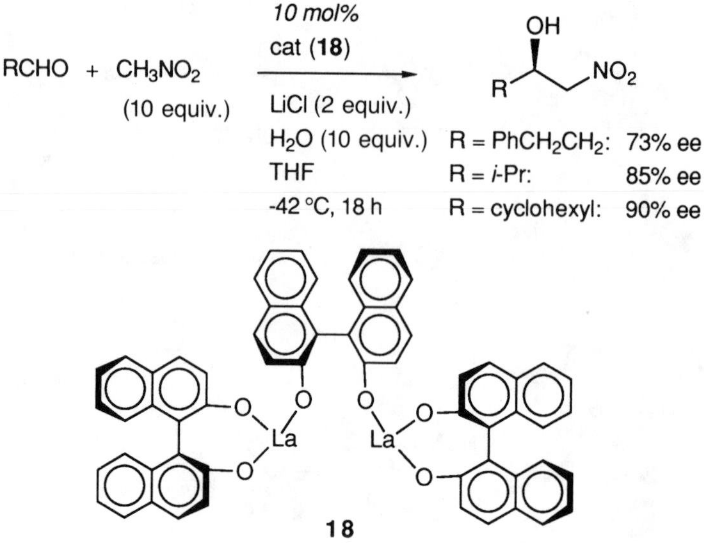

18

moderate to high enantiomeric excesses (Scheme 10) [27]. Apparently this novel catalyst acts as Lewis base. The proposed reaction mechanism is shown in Scheme 11, where the first step of the reaction is the ligand exchange between binaphthol and nitromethane. This reaction is probably the first successful example of the catalytic asymmetric reaction promoted by a Lewis base metal catalyst. Future application of this methodology is quite promising.

Scheme 11

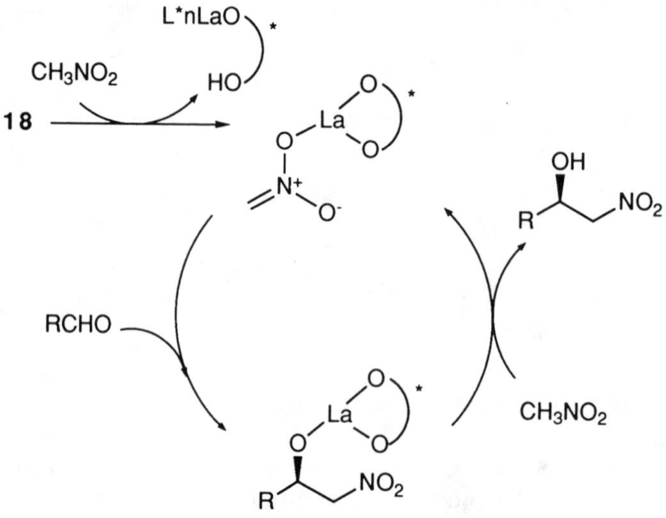

7.2.4. Lewis Acid Catalyzed Asymmetric Aldol Reaction

The aldol reaction of *p*-nitrobenzaldehyde with acetone is catalyzed by Zn^{2+} complexes of α-amino acid esters in MeOH, giving the optically active aldol adduct (Scheme 12) [28]. Although the enantiomeric excesses of the product have not been determined, the extent of asymmetric induction is dependent on the structure of α-substituents in the amino acids. The Zn^{2+} complexes of amino acid esters bearing an aromatic substituent such as esters of phenylalanine, tyrosine, and tryptophane (Trp) are more effective in terms of both catalytic activity and asymmetric induction. The highest asymmetric induction is observed with Trp–OEt ligand.

Scheme 12

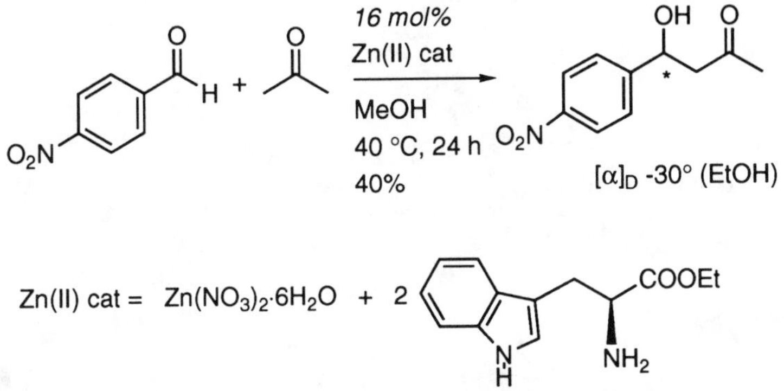

It has been reported that the chiral NMR shift reagent Eu(DPPM)$_3$, represented by structure **19**, catalyzes the Mukaiyama-type aldol condensation of a ketene silyl acetal with enantioselectivity of up to 48% ee (Scheme 13) [29–32]. The chiral alkoxyaluminum complex **20** [33] and the rhodium–phosphine complex **21** [34] under hydrogen atmosphere are also used in the asymmetric aldol reaction of ketene silyl

Scheme 13

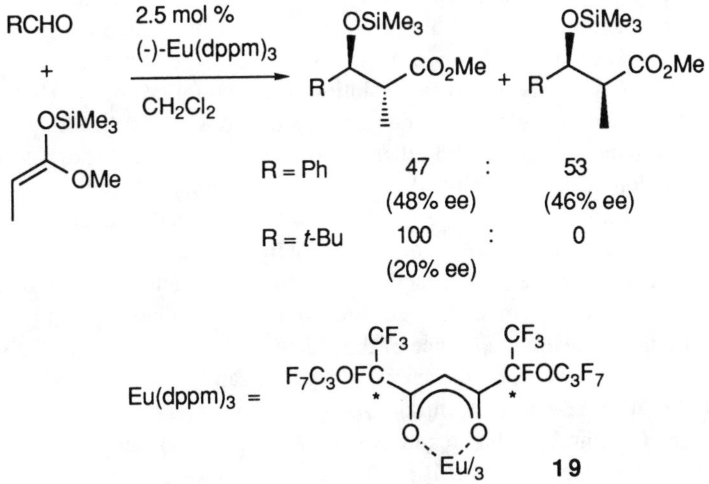

acetals (Scheme 14), although the catalyst turnover number is quite low for the former complex.

Scheme 14

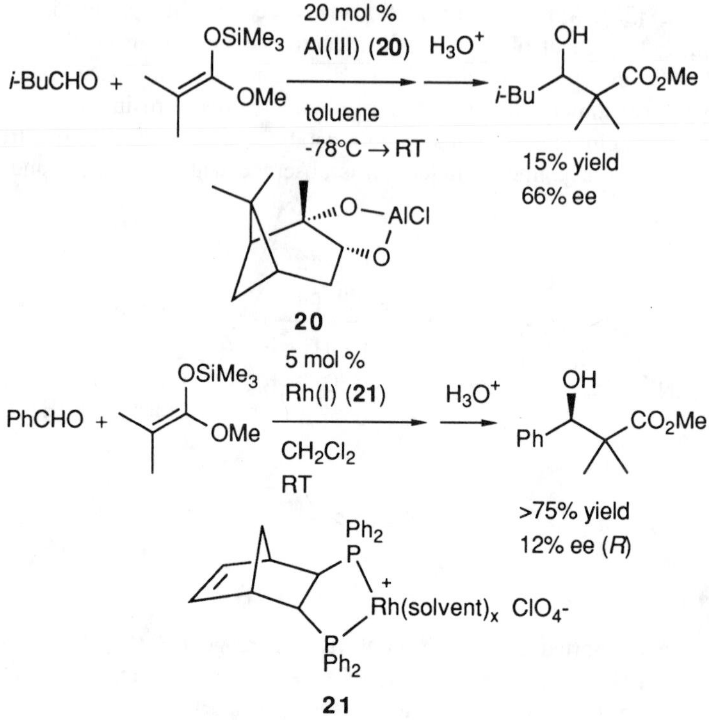

Recently, highly enantioselective Mukaiyama-type aldol reactions employing a substoichiometric amount of chiral Lewis acid have been reported from three research groups. Mukaiyama et al. have achieved a high enantioselectivity in the aldol reaction of silyl enol ethers of thioesters (**22**) with aldehydes in the presence of a substoichiometric amount of chiral promoters (20 mol %), which consist of tin(II) triflate and chiral diamine **23**, under reaction conditions in which a solution of an aldehyde and **22** is slowly added to a solution of the chiral promoters (Scheme 15) [35–40]. As shown in Table 9, various achiral aldehydes are applicable to the aldol reaction. The reaction of the silyl ether of propanethiolate (**22a**) shows selectivity for the formation of the *syn*-aldol adduct with high enantioselectivity (entries 1–7). α-Unsubstituted aldol adducts are produced with high enantioselectivities in the reactions of the silyl ether of ethanethiolate (**22b**) (entries 8–11) [38]. The slow addition protocol may keep the trimethylsilyl triflate concentration low to suppress the undesirable TMSOTf-mediated aldol reaction. The proposed transition state is shown in Figure 2, where the *re* face of the aldehyde is shielded by the naphthyl (or tetrahydronaphthyl) group of the chiral diamine ligand. The stereochemical outcome of the aldol reaction of chiral α-siloxyaldehydes **24** is almost completely controlled by the chiral catalyst regardless of the inherent diastereofacial preference of the chiral aldehydes (Scheme 16) [40].

Scheme 15

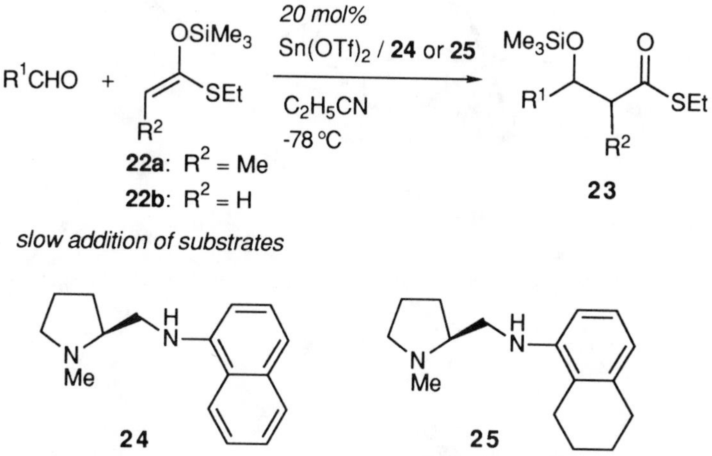

R¹CHO +
OSiMe3
SEt
R2
22a: R2 = Me
22b: R2 = H
20 mol%
Sn(OTf)2 / 24 or 25
C2H5CN
-78 °C
Me3SiO
O
R1
SEt
R2
23
slow addition of substrates
N
Me
H
N
24
25

The same research group has reported that the chiral oxotitanium(IV) complex **26** is an efficient catalyst for the asymmetric aldol reaction of the silyl ether of thioester **27** with aldehydes to produce the corresponding aldol adducts with moderate to high enantiomeric excess (Scheme 17) [41]. The bulkiness of the substituents on both silicon and sulfur atoms of **27** is important for enantioselection.

Mukaiyama's group also reported the catalytic asymmetric aldol reaction of ketene silyl acetals (**28**) promoted by chiral zinc complexes. These complexes are

Table 9 Asymmetric Aldol Reaction of **22** with Aldehydes Catalyzed by Chiral Tin(II) Catalysts (Scheme 15)

Entry	Aldehyde R^1	**22** R^2	L*	Addition Time(h)	Yield (%)	*syn*/*anti*	% ee of *syn*-**23**
1	Ph	Me	**24**	3	77	93/7	90
2	(*E*)-*n*-PrCH=CH	Me	**24**	3	93	97/3	93
3	*n*-C_7H_{15}	Me	**24**	3	80	100/0	>98
4	*c*-C_6H_{11}	Me	**24**	3	71	100/0	>98
5[a]	$Me_3SiC{\equiv}C$	Me	**24**	4.5	73	95/5	91
6[a]	PhC≡C	Me	**24**	4.5	82	90/10	86
7[a]	*n*-BuC≡C	Me	**24**	4.5	67	93/7	91
8[a]	*n*-C_7H_{15}	H	**25**	4	79	-	93
9	*c*-C_6H_{11}	H	**25**	4.5	81	-	92
10	(*E*)-*n*-PrC=C	H	**24**	20	65	-	72
11[a]	*n*-BuC≡C	H	**25**	5	68	-	88

[a] solvent: CH_2Cl_2

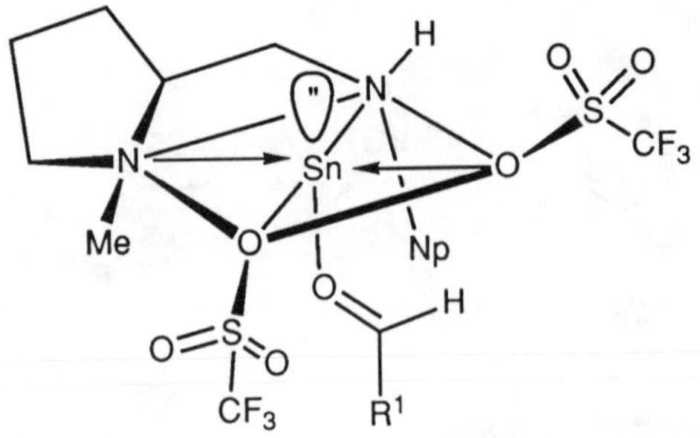

Figure 2 Proposed structure for a chiral Lewis acid–aldehyde complex.

Scheme 16

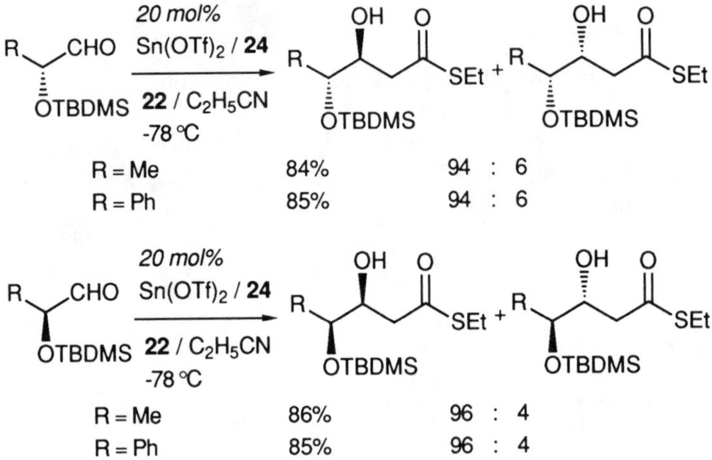

| R = Me | 84% | 94 : 6 |
| R = Ph | 85% | 94 : 6 |

| R = Me | 86% | 96 : 4 |
| R = Ph | 85% | 96 : 4 |

Scheme 17

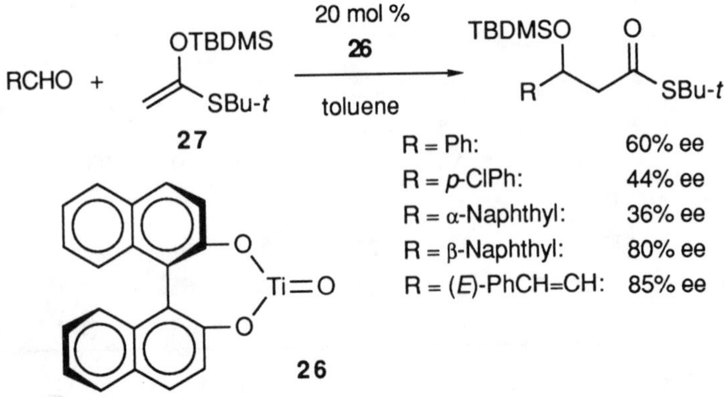

R = Ph:	60% ee
R = p-ClPh:	44% ee
R = α-Naphthyl:	36% ee
R = β-Naphthyl:	80% ee
R = (E)-PhCH=CH:	85% ee

Table 10 Asymmetric Aldol Reaction of Ketene Silyl Acetals **28** with Aldehydes Promoted by Chiral Zinc(II) Catalysts (Scheme 18)

	Aldehyde			Yield	
Entry	R^1	R^2	29	(%)	% ee
1	Ph	Et	29a	79	60
2	$c\text{-}C_6H_{11}$	Bn	29a	76	23
3	CCl_3	Bn	29a	70 [a]	72
4	CCl_3	Bn	29b	61 [a]	88
5	CBr_3	Bn	29b	66 [a]	93

[a] Isolated as alcohol form.

prepared in situ from diethylzinc and sulfonamide ligands **29**, derived from optically amino acids, in a ratio of 1:2 (Scheme 18) [42]. Although the enantioselectivities in the reaction with simple aldehydes such as benzaldehyde and cyclohexanecarboxaldehyde are low to moderate, high enantioselectivities of around 90% ee are obtainable with chloral and bromal (Table 10). A possible catalyst precursor generated in situ is presented as **30**.

Scheme 18

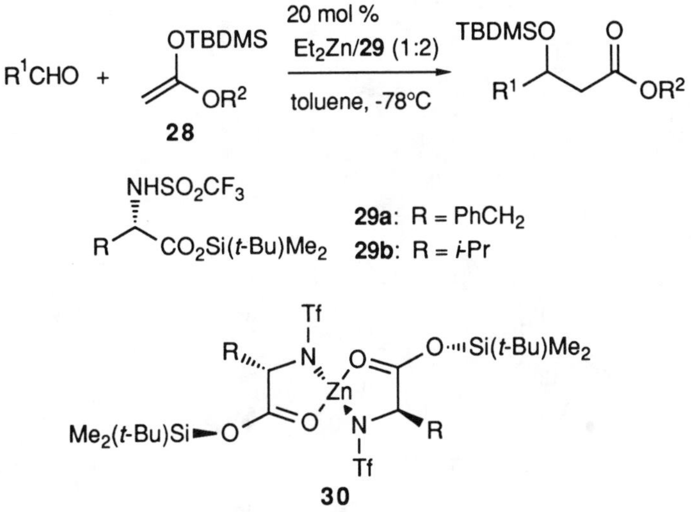

Recently, Yamamoto et al. have shown that the chiral acyloxyborane complex **31** is an excellent catalyst for the asymmetric Mukaiyama condensation of simple silyl enol ethers (Scheme 19; Table 11: entries 1–7) [43]. The *syn*-aldol adducts are formed preferentially with high enantiomeric excesses regardless of the stereochemistry (*E/Z*) of the silyl enol ethers, suggesting an extended transition state (entries 4, 7). This methodology has been extended to the aldol reaction of ketene silyl acetals (entries 8–14) [44]. Ketene silyl acetals derived from phenyl esters

Scheme 19

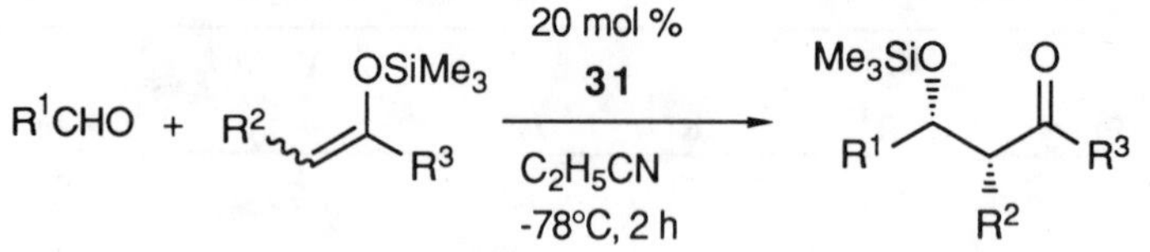

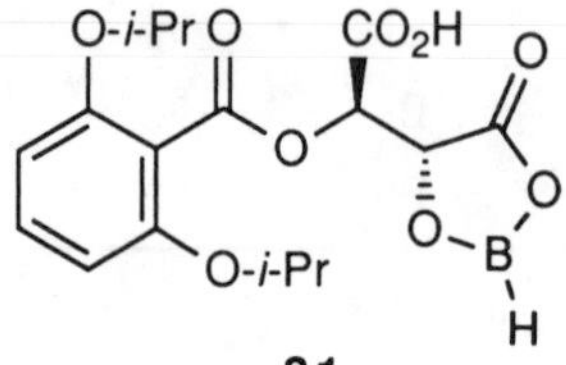

31

Table 11 Asymmetric Aldol Reaction Catalyzed by Chiral
Acyloxyborane Complex **31** (Scheme 19)

Entry		Aldehyde R^1	Yield (%)	syn/anti	syn % ee
1	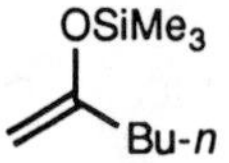 OSiMe₃	Ph	81	-	85
2	⤴ Bu-n	n-Bu	70	-	80
3	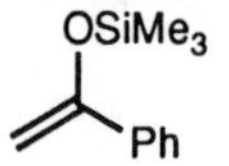 OSiMe₃ Ph	(E)-PhCH=CH	88	-	83
4	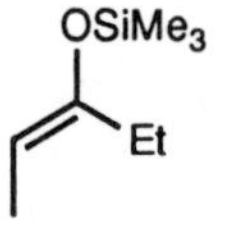 OSiMe₃ Et	Ph	96	94/6	96 (3R)
5		n-Pr	61	80/20	88 (3S)
6		(E)-MeCH=CH	79	>94/6	93 (3R)
7	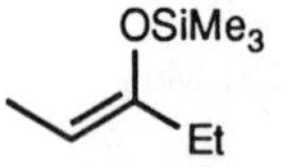 OSiMe₃ Et	Ph	97	93/7	94 (3R)
8	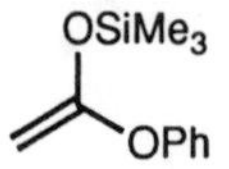 OSiMe₃ OPh	Ph	63	-	84 (3R)
9		n-Pr	49	-	76 (3S)
10	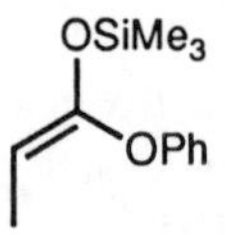 OSiMe₃ OPh	Ph	83	79/21	92 (3R)
11		n-Pr	57	65/35	88
12		i-Pr	45	64/36	79
13		(E)-n-PrCH=CH	97	96/4	97
14		(E)-MeCH=CMe	86	>95/5	94

reacted with higher diastereo- and enantioselectivities than those from the corresponding ethyl ester.

Another clever approach by Masamune et al. is an asymmetric aldol reaction of the ketene silyl acetal **32**. This reaction is effectively promoted by 20 mol % of chiral borane complexes **33**, prepared from BH3·THF and α,α-disubstituted glycine tosylamide **34**, under reaction conditions in which aldehydes are slowly added to the reaction mixture (Scheme 20, Table 12). The catalysts and the reaction conditions

Scheme 20

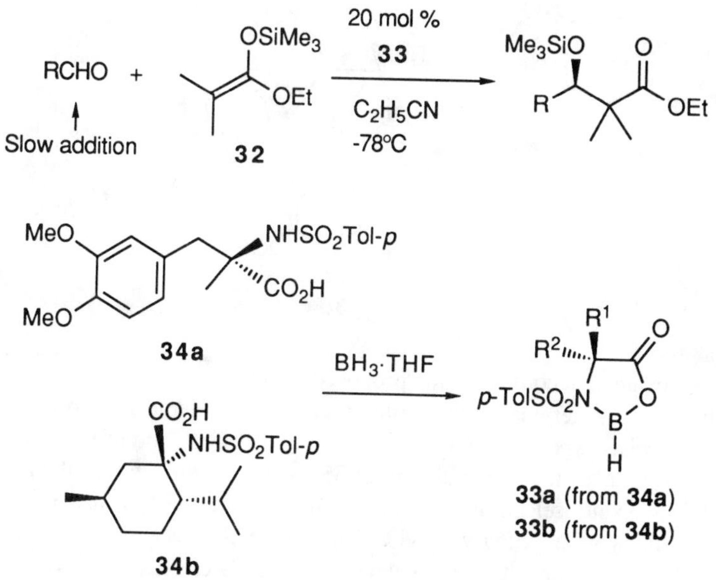

Table 12 Asymmetric Aldol Reaction of Ketene Silyl Acetal **32** with Aldehydes Catalyzed by Borane Complex **33** (Scheme 20)

Entry	Aldehyde R^1	Catalyst 33	Yield (%)	% ee
1	Ph	33a	80	84 (R)
2	Ph	33b	83	91 (R)
3	c-C6H11	33a	68	91 (R)
4	c-C6H11	33b	59	96 (R)
5	n-Pr	33a	81	>98
6	n-Pr	33b	82	>98
7	Ph(CH2)2	33a	83	>98
8	Ph(CH2)2	33b	83	>98
9	BnO(CH2)2	33a	86	99

Scheme 21

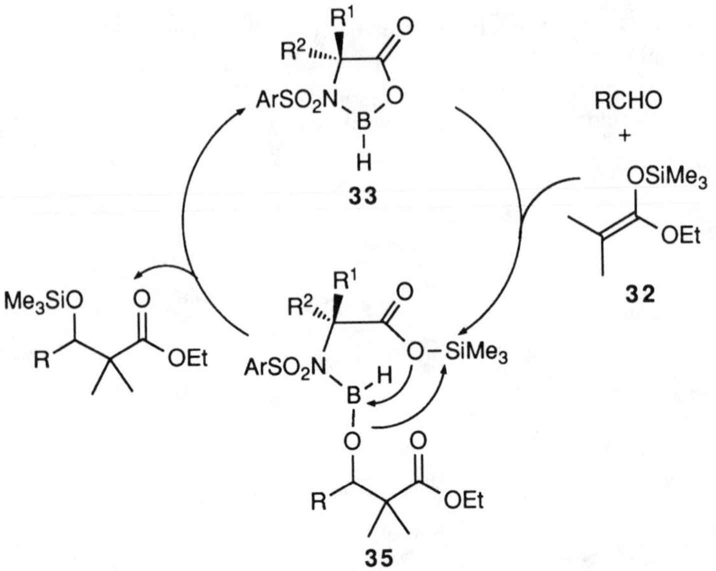

have been designed according to the proposed catalytic cycle shown in Scheme 21. Thus, the use of a geminally disubstituted catalyst accelerates the ring closure of intermediate **35** as expected from the Thorpe–Ingold effect, and the slow addition of the aldehyde reduces the accumulation of **35**, which might catalyze the aldol reaction with low enantioselectivity. Extremely high enantioselectivities are obtainable for the reactions with primary alkyl aldehydes. It has recently been shown that unsubstituted and monosubstituted ketene silyl acetals can also achieve high enantioselectivity [46].

Kiyooka et al. have shown that the asymmetric aldol reaction of ketene silyl acetals is promoted by 20 mol % of oxazaborolidine catalyst derived from (*S*)-valine with enantioselectivity employing nitromethane as the solvent [47].

7.2.5. Conclusion

Recent advancements of the catalytic asymmetric aldol reaction are reviewed. The gold-catalyzed asymmetric aldol reaction simultaneously provides high stereoselectivity and catalyst turnover efficiency, although the reaction is limited to α-isocyanocarboxylates or their analogs. Success in the gold-catalyzed aldol reaction has shown that a synthetic approach using multiple recognition of the reacting substrates is quite promising for the future development of catalytic asymmetric synthesis. Chiral Lewis base catalysts, one of which was first employed in the nitroaldol reaction, may provide a generally useful synthetic methodology, because most aldol reactions are catalyzed by base. Although it is now possible to achieve high enantioselectivities in some Mukaiyama-type aldol reactions using substoichiometric amounts of chiral Lewis acid catalysts, low catalyst turnover is still a major hurdle to be overcome. A major breakthrough in the catalytic asymmetric synthesis of aldols remains to be achieved.

References

1. Masamune, S.; Choy, W.; Kerdesky, F. A. J.; Imperiali, B. *J. Am. Chem. Soc.* **1981,** *103,* 1566.

2. Evans, D. A.; Bartroli, J.; Shih, T. L. *J. Am. Chem. Soc.* **1981,** *103,* 2127.

3. Iwasawa, N.; Mukaiyama, T. *Chem. Lett.* **1982,** 1441.

4. Ito, Y.; Sawamura, M.; Hayashi, T. *J. Am. Chem. Soc.* **1986,** *108,* 6405.

5. Ito, Y.; Sawamura, M.; Hayashi, T. *Tetrahedron Lett.* **1987,** *28,* 6215.

6. Sawamura, M.; Ito, Y.; Hayashi, T. *Tetrahedron Lett.* **1990,** *31,* 2723.

7. Hayashi, T.; Sawamura, M.; Ito, Y. *Tetrahedron,* **1992,** *48,* 1999.

8. Pastor, S. D. *Tetrahedron,* **1988,** *44,* 2883.

9. Togni, A.; Häusel, R. *Synlett,* **1990,** 633.

10. Pastor, S. D.; Togni, A. *Tetrahedron Lett.* **1990,** *31,* 839.

11. Togni, A.; Pastor, S. D. *J. Organomet. Chem.* **1990,** *381,* C21.

12. Ito, Y., et al. Unpublished result.

13. Ito, Y.; Sawamura, M.; Hayashi, T. *Tetrahedron Lett.* **1988,** *29,* 239.

14. Togni, A.; Pastor, S. D.; Rihs, G. *Helv. Chim. Acta,* **1989,** *72,* 1471.

15. Togni, A.; Pastor, S. D. *Helv. Chim. Acta,* **1989,** *72,* 1038.

16. Togni, A.; Pastor, S. D. *J. Org. Chem.* **1990,** *55,* 1649.

17. Ito, Y.; Sawamura, M.; Shirakawa, E.; Hayashizaki, K.; Hayashi, T. *Tetrahedron Lett.* **1988,** *29,* 235.

18. Ito, Y.; Sawamura, M.; Shirakawa, E.; Hayashizaki, K.; Hayashi, T. *Tetrahedron,* **1988,** *44,* 5253.

19. Ito, Y.; Sawamura, M.; Kobayashi, M.; Hayashi, T. *Tetrahedron Lett.* **1988,** *29,* 6321.

20. Ito, Y.; Sawamura, M.; Hamashima, H.; Emura, T.; Hayashi, T. *Tetrahedron Lett.* **1989,** *30,* 4681.

21. Sawamura, M.; Ito, Y.; Hayashi, T. *Tetrahedron Lett.* **1989,** *30,* 2247.

22. Togni, A.; Pastor, S. D. *Tetrahedron Lett.* **1989,** *30,* 1071.

23. Pastor, S. D.; Togni, A. *J. Am. Chem. Soc.* **1989,** *111,* 2333.

24. Pastor, S. D.; Togni, A. *Helv. Chim. Acta,* **1991,** *74,* 905.

25. Hayashi, T.; Uozumi, Y.; Yamazaki, A.; Sawamura, M.; Hamashima, H.; Ito, Y. *Tetrahedron Lett.* **1991,** *32,* 2799.

26. Sawamura, M.; Hamashima, H.; Ito, Y. *J. Org. Chem.* **1990,** *55,* 5935.

27. Sasai, H.; Suzuki, T.; Arai, S.; Arai, T.; Shibasaki, M. *J. Am. Chem. Soc.* **1992,** *114,* 4418.

28. Nakagawa, M.; Nakao, H.; Watanabe, K. *Chem. Lett.* **1985,** 391.

29. Mikami, K.; Terada, M.; Nakai, T. Paper 3Y29, presented at the 52th Annual Meeting of the Chemical Society of Japan, Kyoto, April 1–4, 1986.

30. Mikami, K.; Terada, M.; Nakai, T. *J. Org. Chem.* **1991,** *56,* 5456.

31. Mikami, K.; Terada, M.; Nakai, T. *Tetrahedron: Asymmetry,* **1991,** *2,* 993.

32. Terada, M.; Gu, J.-H.; Deka, D. C.; Mikami, K.; Nakai, T. *Chem. Lett.* **1992,** 29.

33. Reetz, M. T.; Kyung, S.-H.; Bolm, C.; Zierke, T. *Chem. Ind.* **1986,** 824.

34. Reetz, M. T.; Vougioukas, A. E. *Tetrahedron Lett.* **1987,** *28,* 793.

35. Mukaiyama, T.; Kobayashi, S.; Uchiro, H.; Shiina, I. *Chem. Lett.* **1990,** 129.

36. Kobayashi, S.; Fujishita, Y.; Mukaiyama, T. *Chem. Lett.* **1990,** 1455.

37. Mukaiyama, T.; Furuya, M.; Ohtsubo, A.; Kobayashi, S. *Chem. Lett.* **1991,** 989.

38. Kobayashi, S.; Furuya, M.; Ohtsubo, A.; Mukaiyama, T. *Tetrahedron: Asymmetry,* **1991,** *2,* 635.

39. Mukaiyama, T.; Uchiro, H.; Kobayashi, S. *Chem. Lett.* **1990,** 1147.

40. Kobayashi, S.; Ohtsubo, A.; Mukaiyama, T. *Chem. Lett.* **1991,** 831.

41. Mukaiyama, T.; Inubushi, A.; Suda, S.; Hara, R.; Kobayashi, S. *Chem. Lett.* **1990,** 1015.

42. Mukaiyama, T.; Takashima, T.; Kusaka, H.; Shimpuku, T. *Chem. Lett.* **1990,** 1777.

43. Furuta, K.; Maruyama, T.; Yamamoto, H. *J. Am. Chem. Soc.* **1991,** *113,* 1041.

44. Furuta, K.; Maruyama, T.; Yamamoto, H. *Synlett,* **1991,** 439.

45. Parmee, E. R.; Tempkin, O.; Masamune, S. *J. Am. Chem. Soc.* **1991,** *113,* 9365.

46. Parmee, E. R.; Hong, Y.; Tempkin, O.; Masamune, S. *Tetrahedron Lett.* **1992,** *33,* 1729.

47. Kiyooka, S.; Kaneko, Y.; Kume, K. *Tetrahedron Lett.* **1992,** *33,* 4927.

Asymmetric Phase Transfer Reactions

Martin J. O'Donnell

Department of Chemistry
Indiana University-Purdue University at Indianapolis
Indianapolis, Indiana

8.1. Introduction

Phase transfer catalysis (PTC) is a reaction method that typically involves a simple reaction procedure, mild conditions, inexpensive and safe reagents and solvents, and the ability to easily scale up the reaction [1–16]. The use of PTC for the preparation of chiral, nonracemic compounds from prochiral substrates using chiral catalysts has had some notable successes but, for the most part, it has been much less studied than achiral PTC [17–28]. This chapter reviews asymmetric phase transfer reactions and gives perspective to the field as a whole. Coverage has been limited to chiral quaternary ammonium compounds (quats), crown ethers [29–33], and associated species as catalysts. Related areas involving free-alkaloid catalysts [19,34,35], polymer-bound catalysts [36], asymmetric polymerizations [37,38], chiral stationary phases [39], resolutions [40–42], stoichiometric enzyme model studies [43], and micellar catalysis [44,45] are not covered.

8.2. The Phase Transfer System

8.2.1. Mechanism

A mechanistic scheme (Scheme 1) for the monoalkylation of active methylene compounds [46] is presented to illustrate the variables common to many of the systems that have been studied and to alert the reader to key problems that need to be addressed in such reactions. Three main steps are required in this process [14]: (1) deprotonation of the active methylene compound with base, which generally occurs at the interface between the two layers: liquid–liquid (L/L) or solid–liquid

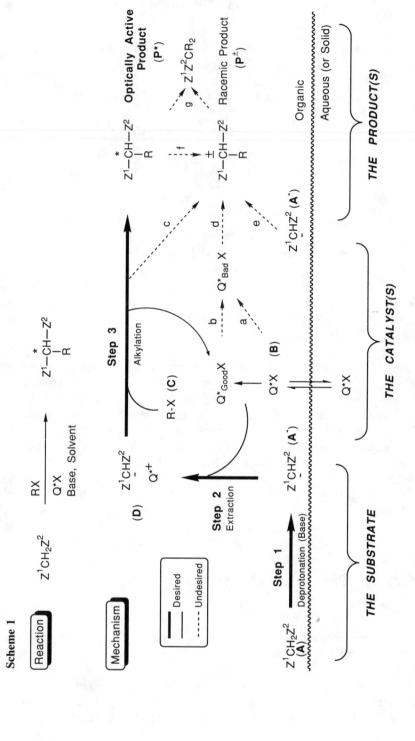

Scheme 1

(S/L) PTC; (2) extraction of the anion (A$^-$) into the bulk organic phase by ion exchange with the cation of the chiral quaternary ammonium compound to form a lipophilic ion-pair (**D**); and (3) creation of the new chiral center in product **P*** by alkylation of the ion-pair (**D**) with concomitant regeneration of the catalyst.

A number of undesirable processes can occur in competition with the formation of the optically active product: (1) alkylation of the "wrong" ion-pair, leading to the enantiomer of the desired product (step c); (2) side reactions of either the starting substrate [e.g., ester saponification or imine hydrolysis of starting materials such as **40** (Scheme 4, below) or the reaction product [racemization, step f, and/or dialkylation, step g) following product formation, as well as the hydrolysis reactions mentioned above for the starting material]; (3) interfacial alkylation (step e) of substrate anion (A$^-$) in the absence of the chiral quat cation, which necessarily yields racemic product; (4) reaction of the chiral quat (**B**) to form a new organic compound, which could function either as the reactive catalyst species ($Q^*_{Good}X$) or as a compound ($Q^*_{Bad}X$) (steps a or b), which is either an ineffective catalyst or one that leads to racemic product.

Key variables for controlling these processes include solvent, temperature, concentration of the various reactants, stirring rate, type of PTC process (L/L or S/L), structures and nature of various protecting groups in the nucleophilic and electrophilic reactants, structure of the catalyst and, finally, identity of counterions in the catalyst, alkyl halide, and base.

8.2.2. Chiral Nonracemic Catalysts

Catalysts derived from the *Cinchona* (Chart 1) [47] and *Ephedra* alkaloids (Chart 2) [47] have been utilized extensively in chiral PTC because the parent alkaloids are

Chart 1

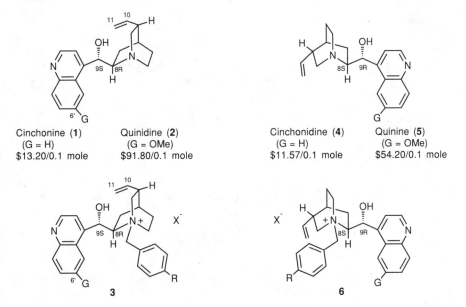

Cinchonine (1)	Quinidine (2)	Cinchonidine (4)	Quinine (5)
(G = H)	(G = OMe)	(G = H)	(G = OMe)
$13.20/0.1 mole	$91.80/0.1 mole	$11.57/0.1 mole	$54.20/0.1 mole

3   6

Chart 2

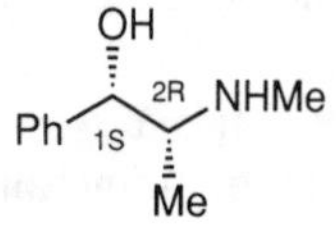

OH
2R
NHMe
Ph
1S
Me

(+)-Ephedrine (**7**)
\$9.12/0.1 mole

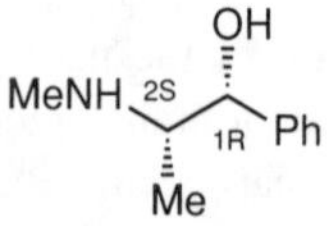

OH
MeNH
2S
1R
Ph
Me

(-)-Ephedrine (**9**)
\$6.47/0.1 mole

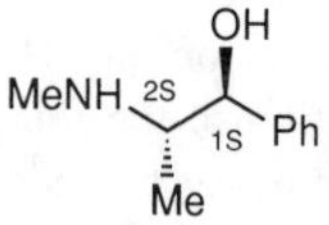

OH
MeNH
2S
1S
Ph
Me

(+)-Pseudoephedrine (**11**)
\$11.13/0.1 mole

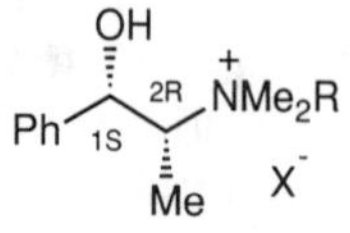

OH
+
2R
NMe2R
Ph
1S
Me
X-

8

OH
+
RMe2N
2S
1R
Ph
X-
Me

10

OH
+
RMe2N
2S
1S
Ph
X-
Me

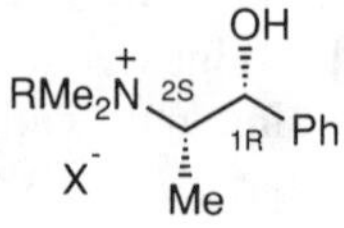

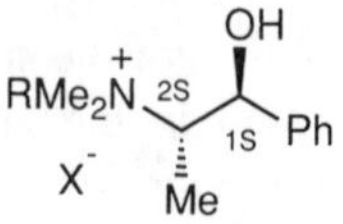

12

inexpensive and readily available in both enantiomeric forms and can easily be quaternarized to a variety of different salts. Typically one catalyst of the pair provides one enantiomer in excess while the other catalyst gives the opposite enantiomer as the major product [48]. Thus, by a simple choice of catalyst, it is possible to prepare either of the two enantiomeric products at will. Overall, the *Cinchona* quats have given more impressive enantioselectivities for a range of reactions than other classes of phase transfer catalyst (see Table 1, below), although recently *Ephedra* quats have been used effectively in 1,2-carbonyl and Michael additions.

Several other catalysts with structural features common to either the *Cinchona*- or *Ephedra*- derived catalysts have been prepared (Chart 3). The first of these catalysts (**13**) is of interest as a synthetic quat prepared as a rigid ammonium salt in

Chart 3

Me
N+
I-

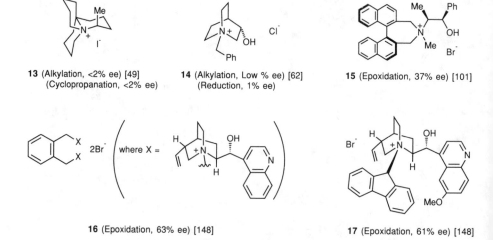

13 (Alkylation, <2% ee) [49]
(Cyclopropanation, <2% ee)

N+
OH
Cl-
Ph

14 (Alkylation, Low % ee) [62]
(Reduction, 1% ee)

Me
Ph
N+
OH
Me
Br-

15 (Epoxidation, 37% ee) [101]

X
X
2Br-
where X =
H
OH
+N
N
H

16 (Epoxidation, 63% ee) [148]

H
OH
Br-
+N
N
H
MeO

17 (Epoxidation, 61% ee) [148]

Chart 4

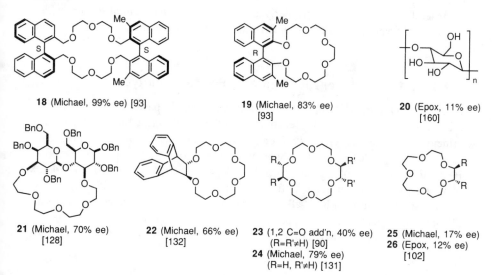

18 (Michael, 99% ee) [93]

19 (Michael, 83% ee) [93]

20 (Epox, 11% ee) [160]

21 (Michael, 70% ee) [128]

22 (Michael, 66% ee) [132]

23 (1,2 C=O add'n, 40% ee) (R=R'≠H) [90]
24 (Michael, 79% ee) (R=H, R'≠H) [131]

25 (Michael, 17% ee)
26 (Epox, 12% ee) [102]

which chirality is centered at the nitrogen and three of the four tetrahedral faces of the chiral nitrogen are blocked. Unfortunately, these systems are not effective chiral catalysts in either alkylation or cyclopropanation reactions [49–54]. Other catalysts in Chart 3 are discussed in subsequent sections.

Chiral crown enters (Chart 4) have given impressive enantioselectivities in Michael additions. Purely synthetic chiral crown ethers are of limited use on a large scale because of cost; although, in general, the crown ethers are less susceptible to catalyst degradation and, therefore, have higher catalyst turnover numbers than the chiral quaternary ammonium salts (see below). Other chelating agents such as cyclodextrins (**20**) and podands with chirality on sulfur (**27**, Chart 5) have also been reported. Recent studies in which starting materials from the chiral pool are used for preparation of chiral crown ether catalysts such as **21** are interesting.

A number of compounds of other types (Chart 5) have been employed as chiral catalysts in phase transfer reactions. Most of these embody the key structural component, a β-hydroxylammonium salt-type structure, which has been shown to be crucial to the success of the above-described *Cinchona*- and *Ephedra*-derived quats.

Chart 5

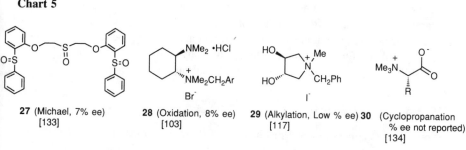

27 (Michael, 7% ee) [133]

28 (Oxidation, 8% ee) [103]

29 (Alkylation, Low % ee) [117]

30 (Cyclopropanation % ee not reported) [134]

Other catalysts such as the recently described betaines **30** have also been reported. In general, quats with chirality derived only from a single chiral center, which cannot participate in a multipoint interaction with other reactions species, have not been effective catalysts [55,56].

8.2.3. Catalyst Degradation and Product Analysis

β-Hydroxylammonium salts can react under the strongly basic reaction conditions present in many phase transfer reactions, and the newly formed products could, in principle, serve either as effective or ineffective catalysts (Scheme 1) [57]. Two routes of catalyst decomposition (Scheme 2, shown for *Cinchona* quat **3**, G = R = H) are deprotonation of the catalyst followed by fragmentation to form epoxide **32** or *O*-alkylation and subsequent fragmentation to the enol ether **35**. Both these tertiary amines can then be *N*-alkylated to form new chiral, nonracemic quat salts (**33** and **36**, respectively). The quaternary ammonium catalyst can also be dequaternarized by nucleophiles to a tertiary amine, which can undergo subsequent reactions [58–62].

Scheme 2

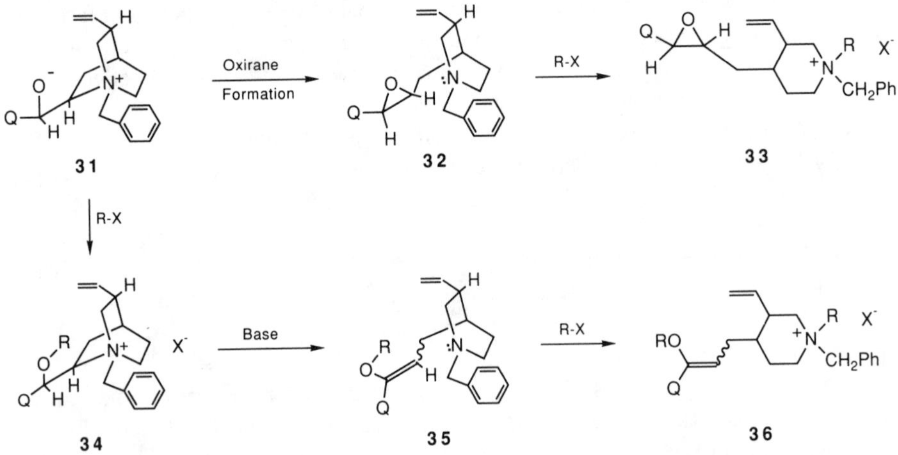

Decomposition products of the optically active catalyst may be carried through the reaction workup, where they contaminate the desired product, leading to false optical rotations. There are numerous reports of high inductions [55,63–68] which, on careful examination of the reaction products, have been questioned [56,58,69–73]. Because of this problem, methods for direct analysis of product mixtures such as chiral high performance liquid chromatography and NMR spectroscopy with chiral shift reagents are preferable to those based on optical rotations.

8.2.4. Other Studies

Various studies concerning the mechanism of the phase transfer process have appeared [1–5,7–9,11,14,15,74]. Recent catalyst [75,76] and structural studies by crystal structure [53,77–80], NMR [52,54,79–82], and calculations [78–80] are noted.

8.2.5. Isolation of Pure Enantiomers

Unless asymmetric inductions are perfect, it is necessary to remove the undesired enantiomer from the product mixture. While in conventional diastereoselective asymmetric syntheses this can typically be accomplished by crystallization or chromatography, the separation of enantiomeric products can be problematic. Often, though, with enantiomerically enriched samples it is possible to recrystallize either the racemate from the pure enantiomer or, preferably, one enantiomer from the other. Alternatively, chiral chromatography can be used for product purification. In the industrial scale synthesis of optically active compounds, these as well as other problems of economics are of prime importance [83,84].

8.3. Phase Transfer Reactions

Reactions that have been accomplished by enantioselective phase transfer catalysis are summarized in Table 1, according to type of catalyst and synthetic transformation. Highest reported enantioselectivities (% ee) or optical purities (% op) are listed to give perspective to the overall field [85]. Particular reaction classes are then discussed briefly [49,55,57,62,64,71,86–108]. The synthetic schemes are drawn using the following stereochemical convention: the cationic catalyst (for chelating catalysts such as crown ethers, this includes inorganic cations) is at the bottom, its associated anion (the nucleophile) is in the middle, and the electrophilic reaction partner comes in from the top.

8.3.1. Carbon–Carbon Bond Formation

8.3.1.1. Alkylation

Alkylation of the phenylindanone **37** by the Merck group demonstrates the reward that can accompany the careful and systematic study of a particular phase transfer reaction [20,22,57,77,109] (Scheme 3). The numerous reaction variables were optimized and the kinetics and mechanism of the reaction were studied in detail. Interestingly, the reaction systems 50% NaOH/toluene and 30% NaOH/toluene are different mechanistically.

Scheme 3

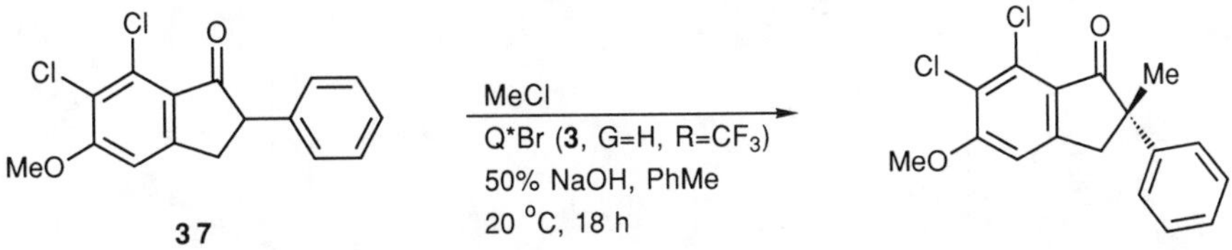

MeCl
——————————→
Q*Br (**3**, G=H, R=CF$_3$)
50% NaOH, PhMe
20 °C, 18 h

37

38 (95%, 92% ee) [77]

Table 1 Enantioselective PTC Reactions[a,b]

PTC Reaction	Cinchona Quat	Ephedra Quat	Crown Ether	Other
C—C bond formation				
Alkylation	94% ee [57]	7% op [86]	40% ee [90]	5% op [87]
1,2-Carbonyl addition	12% ee [88]	74% ee [89]		
Michael addition	87% ee [91]	68% ee [92]	99% ee [93]	0% [94]
Cyclopropanation		1% op ?? [64]		>2% ee [49]
Reduction				
Ketone	32% ee [71]	53% op [95]	8% ee [96]	1% ee [62]
Imine	11% op [97]	22% op [98]		
Other	25% ee [99]			
Oxidation				
Epoxidation	78% ee [100]	37% ee [101]	12% ee [102]	
α-Hydroxy ketone formation	79% ee [103]	7% ee [103]		8% ee [103]
Other	8% ee [104]	4% ee [104]		"Low" [105]
C—X bond formation				
C—N	19% op [106]	3% op [106]		
C—O	3% op [86]	0.03% op [86]		48% ee ?? [55]
C—S	36% op [107]	3% op [107]		
C—X	"Low" [108]	"Low" [108]		

[a] Enantiomeric excess (% ee) is the percentage of the major enantiomer minus that of the minor enantiomer; optical purity (% op) is the ratio, in percent, of the optical rotation of a mixture of enantiomers to that of the pure enantiomer.

[b] Disputed reactions indicated by ??; for others, see Refs. 55 and 63–68.

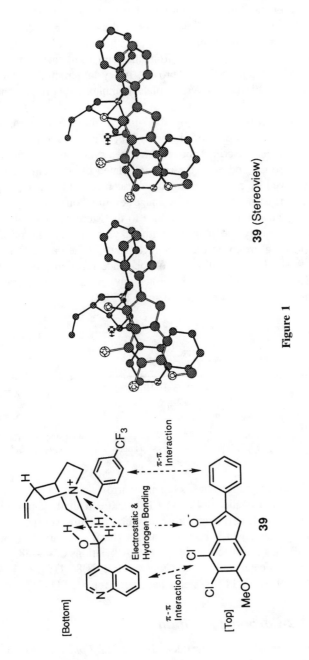

39 (Stereoview)

Figure 1

It has been proposed that the asymmetric induction step involves ion-pair **39** (Fig. 1), in which the enolate anion fits on top of the catalyst and is positioned by an electrostatic effect and hydrogen bonding as well as by π-π stacking interactions between the aromatic rings in the catalyst and the enolate. The electrophile then preferentially approaches the ion-pair from the top (front) face, since the catalyst effectively shields the bottom-face approach. A crystal structure of the catalyst as well as calculations of the catalyst–enolate complex support this interpretation [77,78].

α-Amino acids can be prepared by alkylation of the Schiff base ester **40** using chiral PTC with *Cinchona* quats (Scheme 4) [46]. The situation here is somewhat more complex than that described above, since the acyclic enolate can exist in either *E*- or *Z*-isomeric forms and also, the monoalkylated product is susceptible to deprotonation followed by either reprotonation (racemization) or a second alkylation (dialkylation). Furthermore, the imine and ester functionalities are, themselves, labile to strongly basic medium. In contrast to the Merck system, the best results were obtained in methylene chloride with alkyl bromides. Introduction of an electron-withdrawing group on the *N*-benzyl group of the catalyst did not have a major effect. A single recrystallization of the crude reaction product (64% ee) to remove racemic crystals followed by deprotection gave the product α-amino acid in high enantiomeric purity in a multigram scale [67,68].

Scheme 4

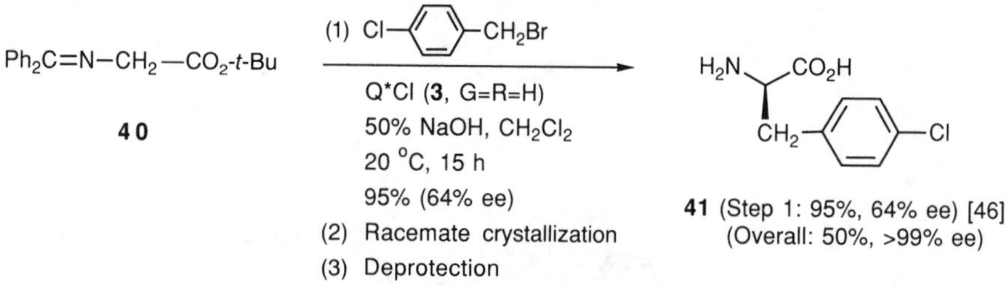

Ph$_2$C=N—CH$_2$—CO$_2$-*t*-Bu

40

(1) Cl—⟨ ⟩—CH$_2$Br

Q*Cl (**3**, G=R=H)
50% NaOH, CH$_2$Cl$_2$
20 °C, 15 h
95% (64% ee)
(2) Racemate crystallization
(3) Deprotection

H$_2$N   CO$_2$H

CH$_2$—⟨ ⟩—Cl

41 (Step 1: 95%, 64% ee) [46]
(Overall: 50%, >99% ee)

Other chiral PTC alkylation studies (Chart 6: arrows indicate point of alkylation) include preparations of α-alkyl- (**42**) [110,111] and α,α-dialkylamino acid derivatives **43** [110,111], **44** [112]) from either prochiral or chiral substrates. Alkylation of a protected glycinate, which itself contains a chiral auxiliary **45** (double asymmetric induction), has also been studied [113]. Alkylations of **46–48** [91,114,115] and other alkylations catalyzed by the *Cinchona* [62,68,86,116], *Ephedra* [65,86, 116], and other [49,62,87,117] systems have been reported [118].

8.3.1.2. 1,2-Carbonyl Addition

Diethylzinc has been added to benzaldehyde using an *Ephedra*-derived chiral quat to give optically active secondary alcohols (Scheme 5) [89]. This is an example in which the chiral catalyst affords a much higher enantioselectivity in the solid state than in solution. Other 1,2-carbonyl additions promoted by the *Cinchona* [108,119], *Ephedra* [108,119b], and crown ether [90] catalyst systems have been studied.

Chart 6

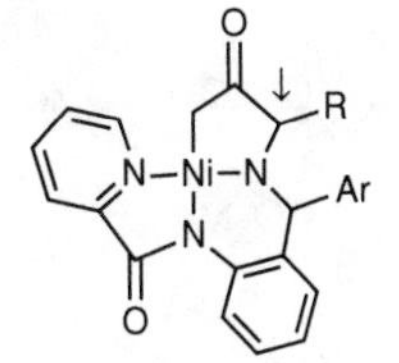
O
R
N–Ni–N
Ar
N
O

42 (Alkylation, 13% op) [110]
(R=H, Ar=Ph)
43 (Alkylation, 31% op)
(R=Me, Ar=H)

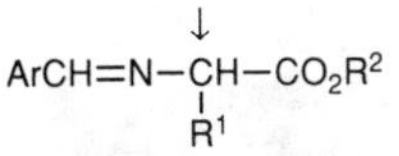
ArCH=N–CH–CO2R2
R1

44 (Alkylation, 50% ee) [112]

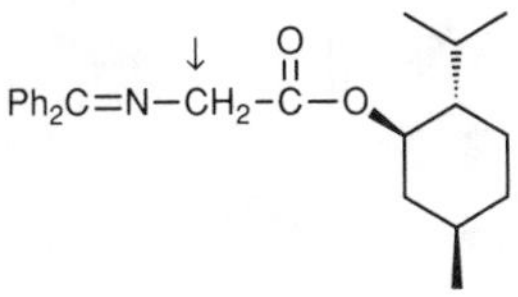
Ph2C=N–CH2–C–O

45 (Alkylation, 78% ee) [113]

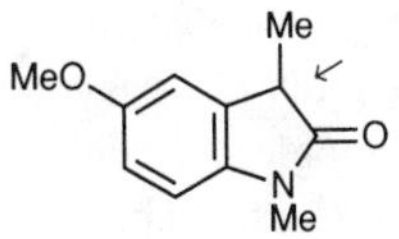
Me
MeO
O
N
Me

46 (Alkylation, 77% ee) [114]

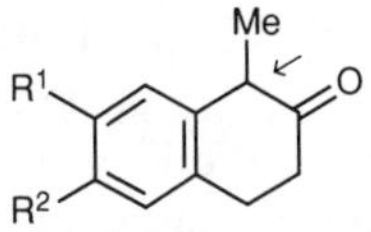
Me
R1
O
R2

47 (Alkylation, >70% ee) [91]
(Michael, 81% ee)

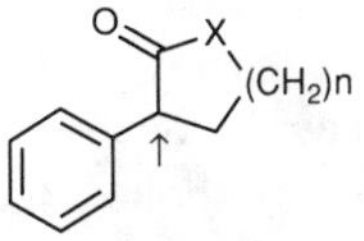
O
X
(CH2)n

48 (Alkylation, 36% ee) [91]
(Michael, 87% ee)

Scheme 5

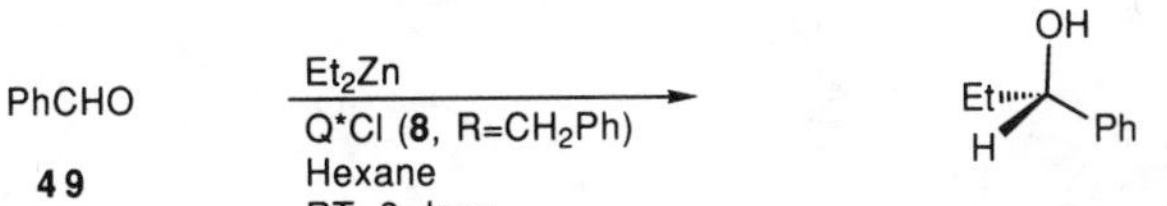
PhCHO
Et2Zn
Q*Cl (8, R=CH2Ph)
Hexane
RT, 3 days
OH
Et
H
Ph

49

50 (90%, 74% ee) [89]

Asymmetric PTC aldol condensation of protected glycinate **40** to give β-hydroxy-α-amino acid derivatives, **51** (*threo* or syn) and **52** (*erythro* or anti or *allo*), has been reported (Scheme 6) [88]. Unfortunately, the enantio- and diastereoselectivities in this reaction are not high.

Scheme 6

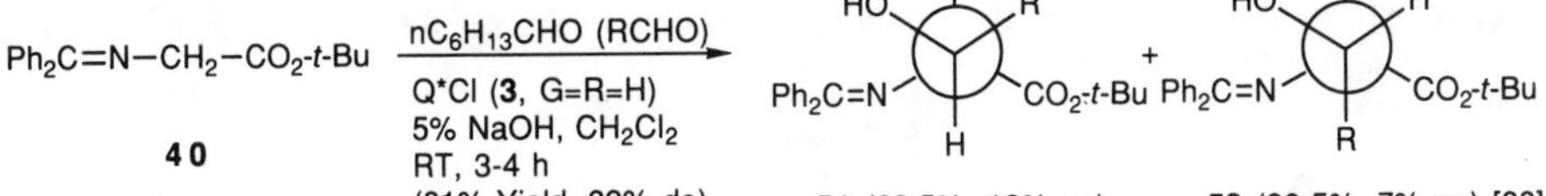
Ph2C=N–CH2–CO2-t-Bu
nC6H13CHO (RCHO)
Q*Cl (3, G=R=H)
5% NaOH, CH2Cl2
RT, 3-4 h
(81% Yield, 39% de)
H
HO
R
Ph2C=N
CO2-t-Bu
H
+
H
HO
H
Ph2C=N
CO2-t-Bu
R

40

51 (69.5%, 12% ee) **52** (30.5%, 7% ee) [88]

8.3.1.3. Michael Addition

A spectacular asymmetric induction (ca. 99% ee) has been achieved in the Michael addition of the cyclic substrate **53** to methyl vinyl ketone (MVK) in the presence of 4 mol % each of crown ether **18** and KO-*t*-Bu at −78°C (Scheme 7) [93]. As expected, higher reaction temperatures provide lower stereospecificity (e.g., when

Scheme 7

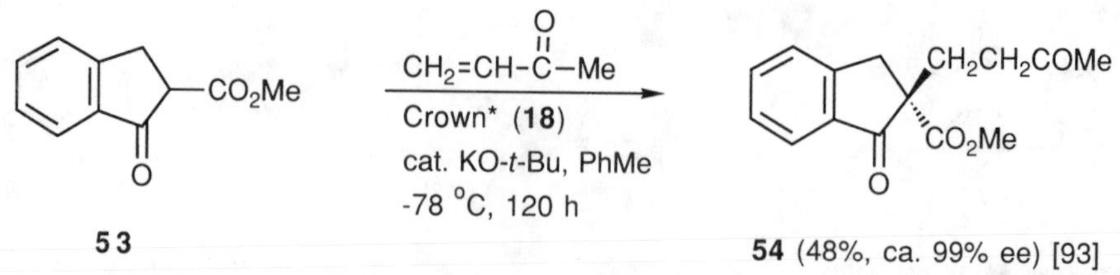

53

CH$_2$=CH-$\overset{\overset{\displaystyle O}{\|}}{C}$-Me

Crown* (**18**)

cat. KO-*t*-Bu, PhMe

-78 °C, 120 h

54 (48%, ca. 99% ee) [93]

the reaction was run at room temperature, the product with 67% ee was obtained). Using crown ether **19**, products with the opposite configurations were obtained. Acyclic substrates (PhCH$_2$CO$_2$Me and PhCH(Me)CO$_2$Me) gave Michael adducts with methyl acrylate at −78°C with up to 65 and 83% ee, respectively. The asymmetric induction was rationalized as shown in Figure 2. The rectangle containing K$^+$ symbolizes the plane of the ion-pair of the six-membered ring, which is approximately planar to the naphthalene rings in the crown ether. The C$^-$ occupies a position in the complex most favorable to electrophilic attack by MVK on one face (top) of its plane.

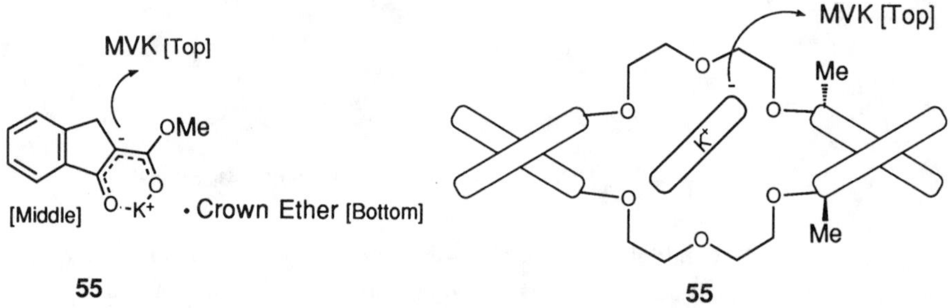

55   55

Figure 2

Several other high asymmetric inductions have been reported using chiral crown ethers as catalysts (see catalysts **21**, **22**, and **24** in Chart 4). Recently, an *Ephedra*-derived quat has been employed in the Michael addition of protected glycinate **56** to yield optically active α-amino acid derivatives (Scheme 8) [16,92,120]. Other

Scheme 8

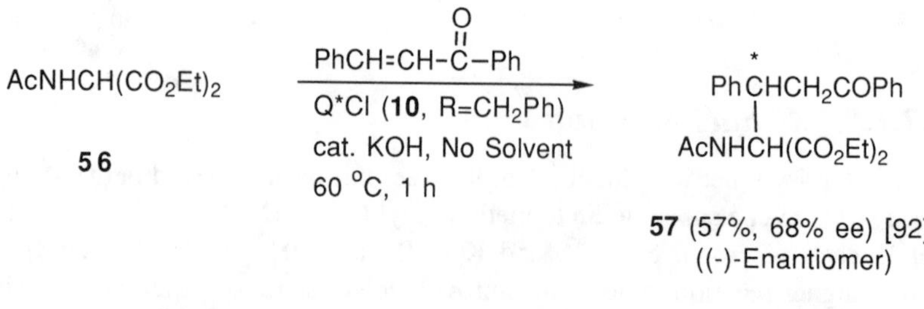

AcNHCH(CO$_2$Et)$_2$

56

PhCH=CH-$\overset{\overset{\displaystyle O}{\|}}{C}$-Ph

Q*Cl (**10**, R=CH$_2$Ph)

cat. KOH, No Solvent

60 °C, 1 h

PhĊHCH$_2$COPh
|
AcNHCH(CO$_2$Et)$_2$

57 (57%, 68% ee) [92]
((-)-Enantiomer)

Michael additions catalyzed by the *Cinchona* [91,107,121–126], *Ephedra* [107, 121,127], crown ether [102,128–132], and other [94,133] systems have been reported. Michael additions have been reviewed recently [23].

8.3.1.4. Cyclopropanation

Although asymmetric induction in the cyclopropanation of alkenes was reported early, this work has been disputed [64]. Other reports of cyclopropanations have yielded, at best, low asymmetric inductions [49,134].

8.3.2. Reduction

8.3.2.1. Ketone and Imine Reductions

The recent report of the asymmetric hydrosilylation of **58** by chiral PTC is interesting to note (Scheme 9) [95].

Scheme 9

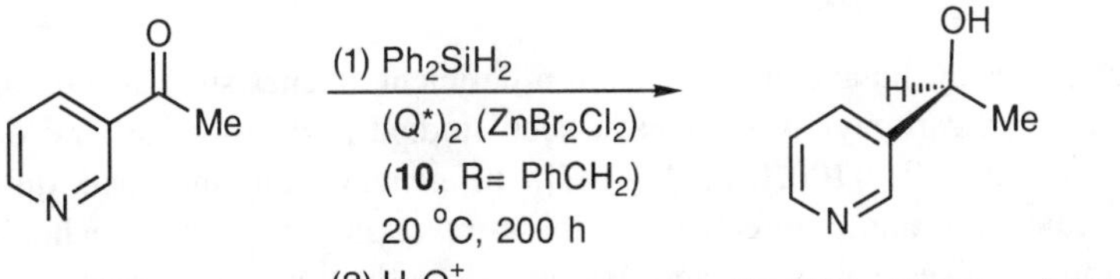

O
‖
(pyridin-3-yl)–C–Me   (1) Ph$_2$SiH$_2$ →   (pyridin-3-yl)–CH(OH)–Me

58

(Q*)$_2$ (ZnBr$_2$Cl$_2$)
(**10**, R= PhCH$_2$)
20 °C, 200 h
(2) H$_3$O$^+$

59 (15%, 53.5% op) [95]

Electroreduction [18] (with chiral quat as the supporting electrolyte) has been compared with chemical reduction (NaBH$_4$) in the presence of chiral quats for ketone (up to 28% op) and imine (up to 22% op, Scheme 10) reductions [98]. Interestingly, for ketone reduction, the major enantiomer formed in the electroreduction was generally of opposite absolute configuration to that formed in the chemical reduction. Additional ketone reductions catalyzed by the *Cinchona* [62,71,106,124,135], *Ephedra* [66,71,106,136–143], crown ether [96,98], and other [62,136,139,140, 144] systems have been reported. Other imine reductions promoted by the *Cinchona* [97], *Ephedra* [97], and other [97] catalyst systems have been studied.

Scheme 10

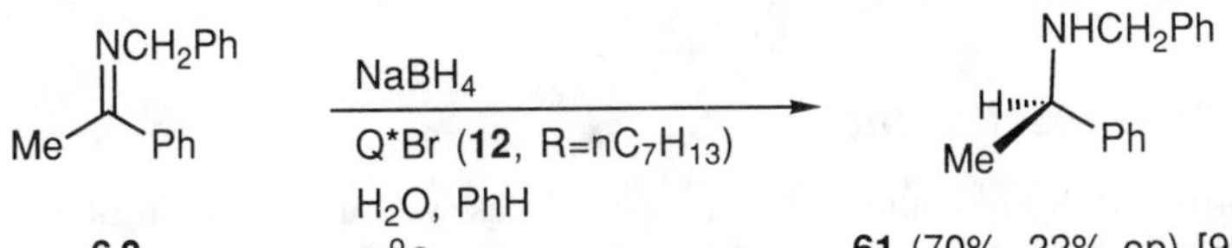

NCH$_2$Ph
‖
Me–C–Ph   NaBH$_4$ →   Me–CH(NHCH$_2$Ph)–Ph

60

Q*Br (**12**, R=nC$_7$H$_{13}$)
H$_2$O, PhH
0 °C

61 (70%, 22% op) [98]

8.3.2.2. Other Reductions

A recent single example of enantioselective hydrodehalogenation of the dichlorobenzazepinone **62** in the presence of a chiral *Cinchona* quat (Scheme 11) is interesting [28]. However, the best asymmetric induction (25% ee) was obtained under optimized conditions with the parent alkaloid, cinchonine, as the catalyst rather than its quaternized derivative. Several reviews concerning enantioselective hydrogenation by PTC have appeared [26–28].

Scheme 11

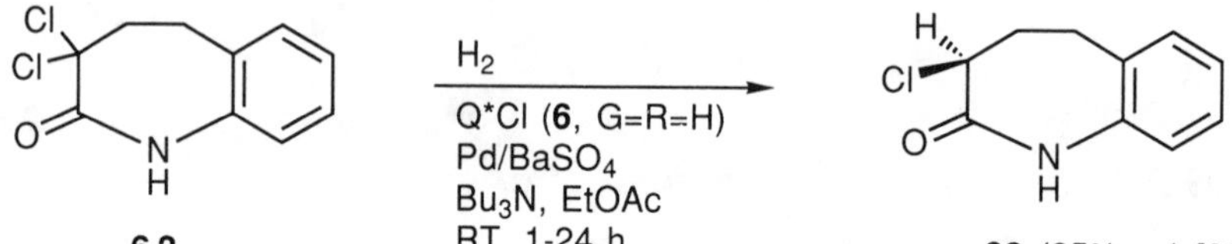

Cl, Cl, O, N, H **6 2**   → H_2 Q*Cl (**6**, G=R=H) Pd/BaSO$_4$ Bu$_3$N, EtOAc RT, 1-24 h   →   H, Cl, O, N, H **63** (25% ee) [99]

8.3.3. Oxidation

8.3.3.1. Epoxidation

Epoxidations of cyclic and acyclic electron-deficient alkenes such as quinones and α,β-unsaturated carbonyl systems have been studied in detail by several different groups (Scheme 12) [100,122,145–157]. In general, quinine- and quinidine-derived quats are more effective in these reactions than the cinchonine- or cinchonidine-derived catalysts, and asymmetric inductions are higher in toluene than those in CH$_2$Cl$_2$ [122]. The epoxidation of cyclohexenone to form **67** uses an interesting bisquaternary catalyst (**16**). A mechanism that rationalizes the origin of the asymmetric induction in these epoxidations has been proposed [19]. Epoxidations other than those involving peroxide-type derivatives have not given promising results [153,158]. Additional epoxidations catalyzed by the *Cinchona* [159], *Ephedra* [63,101], crown ether [102,160], and other [158] systems have been reported.

8.3.3.2. α-Hydroxylation of Ketones

Ketones were converted by chiral PTC to optically active α-hydroxyketones with excellent asymmetric induction using molecular oxygen (Scheme 13) [103]. Of the several different classes of quats examined (*Cinchona*, *Ephedra*, and other) the best catalyst was *p*-trifluoromethylcinchoninium bromide (**3**, G = H, R = CF$_3$).

8.3.3.3. Other Oxidations

Glycol formation by oxidation of styrene [105] as well as oxidation of prochiral phosphines to the optically active phosphine oxides [104] by chiral PTC gave only low asymmetric inductions.

Scheme 12

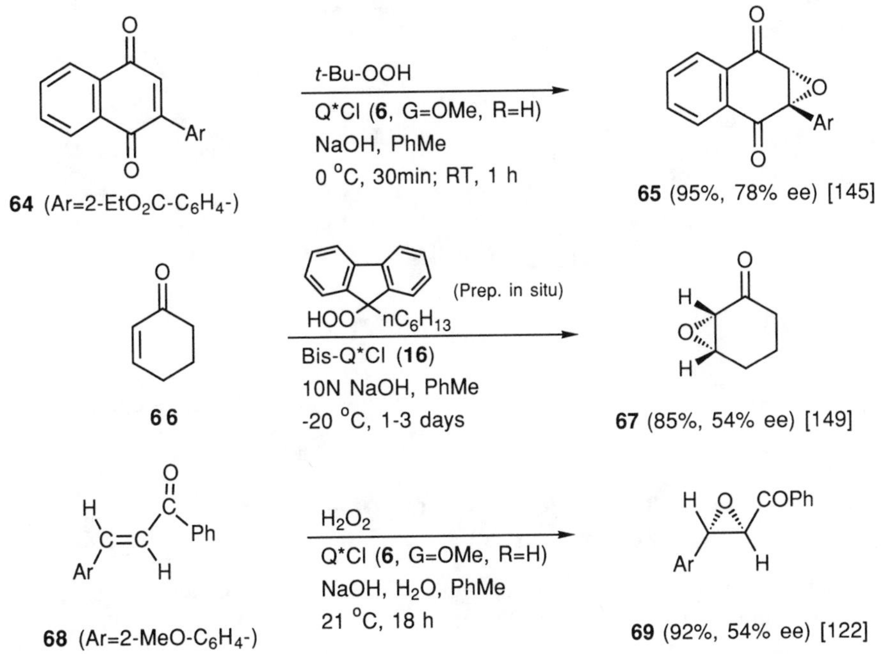

64 (Ar=2-EtO$_2$C-C$_6$H$_4$-)

t-Bu-OOH
Q*Cl (**6**, G=OMe, R=H)
NaOH, PhMe
0 °C, 30min; RT, 1 h

65 (95%, 78% ee) [145]

HOO nC$_6$H$_{13}$ (Prep. in situ)

66

Bis-Q*Cl (**16**)
10N NaOH, PhMe
-20 °C, 1-3 days

67 (85%, 54% ee) [149]

68 (Ar=2-MeO-C$_6$H$_4$-)

H$_2$O$_2$
Q*Cl (**6**, G=OMe, R=H)
NaOH, H$_2$O, PhMe
21 °C, 18 h

69 (92%, 54% ee) [122]

Scheme 13

70

O$_2$
Q*Br (**3**, G=H, R=CF$_3$)
50% NaOH, PhMe
RT, 5 h

71 (95%, 79% ee) [103]

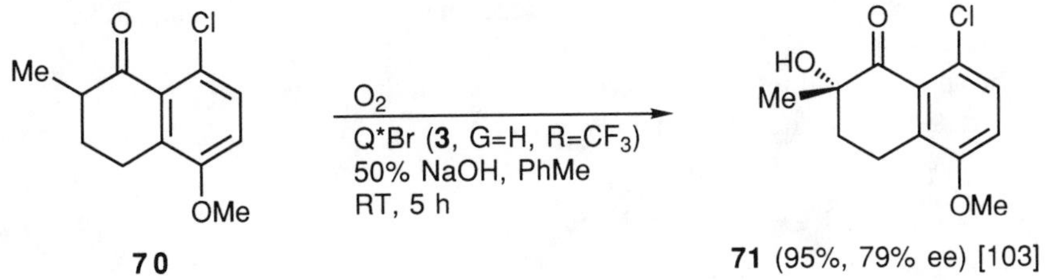

8.3.4. Carbon–Heteroatom Bond Formation by Substitution

In general, transformations by substitution involving a carbon atom and a hetero-atom have given, at best, low asymmetric inductions. Therefore, only a brief discussion and references to specific classes of catalysts used is given in this section.

8.3.4.1. Carbon–Nitrogen Bond Formation

Several groups have studied the synthesis of optically active α-amino acids from the inexpensive and readily available α-halo esters by displacement of the halides with phthalimide in the presence of chiral *Cinchona* catalysts (Scheme 14) [86,106,124,

161,162]. Early studies, using chiral, nonracemic starting material, showed that this reaction occurred with partial inversion of configuration and likely involved a kinetic resolution [161]. Other studies using the *Cinchona* [163] and *Ephedra* [106] catalysts have been reported.

Scheme 14

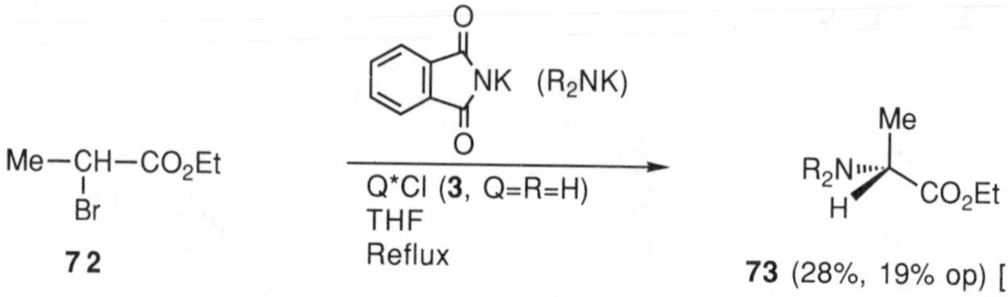

Me—CH—CO₂Et
|
Br

7 2

Q*Cl (3, Q=R=H)
THF
Reflux

73 (28%, 19% op) [161]

8.3.4.2. Carbon–Oxygen and Carbon–Sulfur Bond Formation

O-Alkylation of racemic secondary alcohols (a kinetic resolution) in the presence of a chiral nonracemic unfunctionalized quat, (*S*)-Et₃NCH₂CH(Me)Et Br, with asymmetric inductions up to 48% ee was reported, but these results could not be reproduced [55]. Such catalysts would not be capable of the multipoint interaction between the catalyst and reactants in the transition state that are thought to govern the stereochemistry of these types of reactions [56]. Other *O*-alkylations with *Cinchona* [86] and *Ephedra* [86] catalysts have been studied.

The addition of thiophenol to cyclohexenone in the presence of a *Cinchona* quat gave the Michael adduct with 36% op in 85% yield [107]. Other C—S bond formations with *Cinchona* [86,124,164,165] and *Ephedra* [86,107,127] catalysts have been reported.

8.3.4.3. Carbon–Halogen Bond Formation

Early chlorination of various alkenes gave only low optical rotations in the isolated products using either *Ephedra* or *Cinchona* quats [108].

8.4. Conclusion and Future Prospects

Catalytic, enantioselective synthesis by phase transfer catalysis is a field of great potential which is still in its infancy. As many of the examples cited above demonstrate, continued advancement will depend, in large part, on the careful and systematic study of these reactions in order to understand the details of the asymmetric induction step as well as the other processes that occur during the reaction. With such an in-depth appreciation, it will be possible to design substrates, catalysts, and reaction systems that will lead to both high chemical yield and high levels of asymmetric induction.

Acknowledgements

The author expresses his appreciation to all past and current members of his research group. The special contribution of Shengde Wu, who is responsible for most of the asymmetric PTC chemistry in the group to date, is gratefully acknowledged. He also thanks the National Institutes of Health (GM 28193), the North Atlantic Treaty Organization, and Eli Lilly and Company for financial support.

References

1. Brändström, A. *Preparative Ion Pair Extraction;* Apotakarsocieten/Hässle: Läkemedel, Sweden, 1974.

2. Weber, W. P.; Gokel, G. W. *Phase Transfer Catalysis in Organic Synthesis;* Springer-Verlag: Berlin, 1977.

3. Starks, C. M.; Liotta, C. *Phase Transfer Catalysis: Principles and Techniques;* Academic Press: Orlando, FL, 1978.

4. Caubère, P. *Le transfert de phase et son utilisation en chimie organique;* Masson: Paris, 1982.

5. Dehmlow, E. V.; Dehmlow, S. S. *Phase Transfer Catalysis,* 2nd ed.; Verlag Chemie: Weinheim, 1983.

6. (a) Keller, W. E. *Phase-Transfer Reactions. Fluka Compendium;* Thieme: Stuttgart, 1986; Vol. 1.
   (b) Keller, W. E. *Phase-Transfer Reactions. Fluka Compendium;* Thieme: Stuttgart, 1987; Vol. 2.
   (c) Keller, W. E. *Phase-Transfer Reactions. Fluka Compendium;* Thieme: Stuttgart, 1990; Vol. 3.

7. *Phase Transfer Catalysis (ACS Symposium Series: 326);* Starks, C. M. (Ed.); American Chemical Society: Washington, DC, 1987.

8. Goldberg, Y. *Phase Transfer Catalysis: Selected Problems and Applications*; Gordon and Breach Science Publishers, Philadelphia, PA, 1992.

9. Brändström, A. *Adv. P. Org. Chem.* **1977,** *15,* 267–330.

10. Sjöberg, K. *Aldrichimica Acta,* **1980,** *13,* 55–58.

11. Montanari, F.; Landini, D.; Rolla, F. *Top. Curr. Chem.* **1982,** *101,* 147–200.

12. Starks, C. M. *Is. J. Chem.* **1985,** *26,* 211–215.

13. Freedman, H. H. *Pure Appl. Chem.* **1986,** *58,* 857–868.

14. Rabinovitz, M.; Cohen, Y.; Halpern, M. *Angew. Chem., Int. Ed. Engl.* **1986,** *25,* 960–970.

15. Makosza, M.; Fedorynski, M. *Adv. Catal.* **1987,** *35,* 375–422.

16. Bram, G.; Galons, H.; Labidalle, S.; Loupy, A.; Miocque, M.; Petit, A.; Pigeon, P.; Sansoulet, J. *Bull. Soc. Chim. Fr.* **1989,** 247–251.

17. Kong, F. *Huaxue Tongbao,* **1985,** 35–42; *Chem. Abstr.* **1986,** *105,* 41955v.

18. Tallec, A. *Bull. Soc. Chim. Fr.* **1985,** 743–761.

19. Wynberg, H. In *Topics in Stereochemistry;* Eliel, E. L.; Wilen, S.; Allinger, N. L. (Eds.); Wiley: New York, 1986; Vol. 16, pp. 87–129.

20. Dolling, U.-H.; Hughes, D. L.; Bhattacharya, A.; Ryan, K. M.; Karady, S.; Weinstock, L. M.; Grabowski, E. J. J. In *Phase Transfer Catalysis (ACS Symposium Series: 326);* Starks, C. M. (Ed.); American Chemical Society: Washington, DC, 1987; pp. 67–81.

21. Zhang, J.; Wu, Y. *Huaxue Tongbao,* **1987,** *27,* 18–23; *Chem. Abstr.* **1988,** *108,* 203978w.

22. Dolling, U.-H.; Hughes, D. L.; Bhattacharya, A.; Ryan, K. M.; Karady, S.; Weinstock, L. M.; Grenda, V. J.; Grabowski, E. J. J. In *Catalysis of Organic Reactions (Chem. Ind.: 33);* Rylander, P. N.; Greenfield, H.; Augustine, R. L. (Eds.); Dekker: New York, 1988; pp. 65–86.

23. Oare, D. A.; Heathcock, C. H. In *Topics in Stereochemistry;* Eliel, E. L.; Wilen, S. H. (Eds.); Wiley: New York, 1989; Vol. 19, pp. 227–403.

24. Chen, Z.; Zeng, Z. *Huaxue Shiji,* **1989,** *11,* 243–247, 228; *Chem. Abstr.* **1989,** *111,* 231422r.

25. Baba, N. *Okayama Daigaku Nogakubu Gakujutsu Hokoku,* **1990,** *75,* 31–45; *Chem. Abstr.* **1990,** *113,* 210919j.

26. Blaser, H.-U.; Müller, M. *Stud. Surf. Sci. Catal.* **1991,** *59,* 73–92.

27. Blaser, H.-U.; Jalett, H. P.; Monti, D. M.; Baiker, A.; Wehrli, J. T. In *Structure-Activity and Selectivity Relationships in Heterogeneous Catalysis;* Grasselli, R. K.; Sleight, A. W. (Eds.); Elsevier: Amsterdam, 1991; Vol. 67, pp. 147–155.

28. Blaser, H.-U. *Tetrahedron: Asymmetry,* **1991,** *2,* 843–866.

29. Stoddart, J. F. In *Topics in Stereochemistry;* Eliel, E. L.; Wilen, S. H. (Eds.); Wiley: New York, 1987; Vol. 17, pp. 207–288.

30. Lehn, J.-M. *Angew. Chem., Int. Ed. Engl.* **1988,** *27,* 90–112.

31. Cram, D. J. *Angew. Chem., Int. Ed. Engl.* **1988,** *27,* 1009–1020.

32. Pedersen, C. J. *Angew. Chem., Int. Ed. Engl.* **1988,** *27,* 1021–1027.

33. *Cation Binding by Macrocycles: Complexation of Cationic Species by Crown Ethers;* Inoue, Y.; Gokel, G. W. (Eds.); Dekker: New York, 1990.

34. Dijkstra, G. D. H.; Kellogg, R. M.; Wynberg, H.; Svendsen, J. S.; Marko, I.; Sharpless, K. B. *J. Am. Chem. Soc.* **1989,** *111,* 8069–8076.

35. Dijkstra, G. D. H.; Kellogg, R. M.; Wynberg, H. *J. Org. Chem.* **1990,** *55,* 6121–6131.

36. Aglietto, M.; Chiellini, E.; D'Antone, S.; Ruggeri, G.; Solaro, R. *Pure Appl. Chem.* **1988,** *60,* 415–430.

37. Cram, D. J.; Sogah, D. Y. *J. Am. Chem. Soc.* **1985,** *107,* 8301–8302.

38. Mijs, W. J.; Addink, R. *Recl. Trav. Chim. Pays-Bas,* **1991,** *110,* 526–542.

39. Rosini, C.; Salvadori, P. *Chim. Ind. (Milan)* **1990,** *72,* 195–208.

40. Jacques, J.; Collet, A.; Wilen, S. H. *Enantiomers, Racemates, and Resolutions;* Wiley-Interscience: New York, 1981.

41. de Min, M.; Levy, G.; Micheau, J. C. *J. Chim. Phys.* **1988,** *85,* 603–619.

42. Leusen, F. J. J.; Bruins Slot, H. J.; Noordik, J. H.; van der Haest, A. D.; Wynberg, H.; Bruggink, A. *Recl. Trav. Chim. Pays-Bas,* **1991,** *110,* 13–18.

43. Kellogg, R. M. *Angew. Chem., Int. Ed. Engl.* **1984,** *23,* 782–794.

44. Fendler, J. H. *Membrane Mimetic Chemistry: Characterizations and Applications of Micelles, Microemulsions, Monolayers, Bilayers, Vesicles, Host-Guest Systems, and Polyions;* Wiley-Interscience: New York, 1982.

45. Konno, K.; Kitahara, A.; El Seoud, O. A. *Surf. Sci. Ser.* **1987,** *23,* 185–231.

46. O'Donnell, M. J.; Bennett, W. D.; Wu, S. *J. Am. Chem. Soc.* **1989,** *111,* 2353–2355.

47. Prices listed in Charts 1 and 2 are for 0.1 mol of parent alkaloids from Aldrich Chemical Company, 1992.

48. The pseudoenantiomeric catalysts derived from the alkaloid pairs cinchonine/cinchonidine and quinidine/quinine (**1/4** or **2/5**, respectively) are related as diastereomers but are enantiomeric with respect to the front "working face" defined by carbons 8 and 9.

49. McIntosh, J. M.; Acquaah, S. O. *Can. J. Chem.* **1988**, *66*, 1752–1756.

50. McIntosh, J. M. *Tetrahedron Lett.* **1979**, 403–404.

51. McIntosh, J. M. *Can. J. Chem.* **1980**, *58*, 2604–2609.

52. McIntosh, J. M. *Tetrahedron*, **1982**, *38*, 261–266.

53. McIntosh, J. M.; Khan, M. A.; Delbaere, L. T. J. *Can. J. Chem.* **1982**, *60*, 1073–1077.

54. McIntosh, J. M. *J. Org. Chem.* **1982**, *47*, 3777–3779.

55. Verbicky, J. W., Jr.; O'Neil, E. A. *J. Org. Chem.* **1985**, *50*, 1786–1787. Reference 56 questions this paper.

56. Dehmlow, E. V.; Sleegers, A. *J. Org. Chem.* **1988**, *53*, 3875–3877.

57. Hughes, D. L.; Dolling, U.-H.; Ryan, K. M.; Schoenewaldt, E. F.; Grabowski, E. J. J. *J. Org. Chem.* **1987**, *52*, 4745–4752.

58. Dehmlow, E. V.; Singh, P.; Heider, J. *J. Chem. Res., Synop.* **1981**, 292–293.

59. Mikhlina, E. E.; Yakhontov, L. N. *Khim. Geterotsikl. Soedin.* **1984**, 147–61; *Chem. Abstr.* **1984**, *100*, 209536g.

60. Singh, P.; Arora, G. *Indian J. Chem.* **1986**, *25B*, 1034–1037.

61. Dehmlow, E. V.; Knufinke, V. *J. Chem. Res., Synop.* **1989**, 224–225, 400.

62. Esikova, I. A.; Serebryakov, É. P. *Izv. Akad. Nauk SSSR, Ser. Khim.* **1989**, 1836–1843; *Chem. Abstr.* **1990**, *112*, 54595x.

63. (a) Hiyama, T.; Mishima, T.; Sawada, H.; Nozaki, H. *J. Am. Chem. Soc.* **1975**, *97*, 1626–1627; **1976**, *98*, 641. Reference 69 questions this paper.

64. Hiyama, T.; Sawada, H.; Tsukanaka, M.; Nozaki, H. *Tetrahedron Lett.* **1975**, 3013–3016. Reference 70 questions this paper.

65. Fiaud, J.-C. *Tetrahedron Lett.* **1975**, 3495–3496. Reference 58 questions this paper.

66. Massé, J. P.; Parayre, E. R. *J. Chem. Soc., Chem. Commun.* **1976**, 438–439. References 58 and 71 question this paper.

67. Chen, J.-T.; Chen, Y.-H.; Sheng, H.-Y. *Youji Huaxue*, **1988**, *8*, 164–166; *Chem. Abstr.* **1989**, *110*, 24259g. Reference 72 questions this paper.

68. Zhu, Z.; Yu, L. *Beijing Shifan Daxue Xuebao, Ziran Kexueban*, **1989**, 63–67; *Chem. Abstr.* **1990**, *113*, 24466j. Reference 73 questions this paper.

69. Hernandez, O.; Bhatia, A. V.; Walker, M. *J. Liq. Chromatogr.* **1983**, *6*, 1475–1489.

70. Dehmlow, E. V.; Lissel, M.; Heider, J. *Tetrahedron*, **1977**, *33*, 363–366.

71. Colonna, S.; Fornasier, R. *J. Chem. Soc., Perkin Trans. 1*, **1978**, 371–373.

72. Reference 67 describes the chiral PTC benzylation of the benzaldehyde imine of glycine ethyl ester (**44**, Ar = Ph, R¹ = H, R² = Et) using catalyst Q*Cl (**6**, G = R = H) in an S/L PTC system of KOH–K$_2$CO$_3$ in CH$_2$Cl$_2$ at 30°C to give, following hydrolysis, D-phenylalanine in 61.6% yield and 89.9% optical purity. We have not been able to obtain the reported level of induction upon repeating this reaction in our laboratory.

73. Reference 68 describes the chiral PTC benzylation (with 4-O$_2$N-C$_6$H$_4$CH$_2$Cl) of the benzophenone imine of glycine ethyl ester (**40**, ethyl instead of *t*-butyl ester) using catalyst Q*Cl (**3**, G = R = H)

in an S/L PTC system of KOH–K$_2$CO$_3$ in CH$_2$Cl$_2$ at 20°C for 6 hours to give the benzylated product in 51% yield and 96.6% optical purity. We have not been able to obtain the reported level of induction upon repeating this reaction in our laboratory.

74. Liotta, C. L.; Burgess, E. M.; Ray, C. C.; Black, E. D.; Fair, B. E. In *Phase Transfer Catalysis (ACS Symposium Series: 326);* Starks, C. M. (Ed.); American Chemical Society: Washington, DC, 1987; pp. 15–23.

75. Mason, D.; Magdassi, S.; Sasson, Y. *J. Org. Chem.* **1990,** *55,* 2714–2717.

76. Moberg, R.; Bökman, F.; Bohman, O.; Seigbahn, H. O. G. *J. Am. Chem. Soc.* **1991,** *113,* 3663–3667.

77. Dolling, U.-H.; Davis, P.; Grabowski, E. J. J. *J. Am. Chem. Soc.* **1984,** *106,* 446–447.

78. Lipkowitz, K. B.; Cavanaugh, M. W.; Baker, B.; O'Donnell, M. J. *J. Org. Chem.* **1991,** *56,* 5181–5192.

79. Vicent, C.; Jiménez-Barbero, J.; Martín-Lomas, M.; Penadés, S.; Cano, F. H.; Foces-Foces, C. *J. Chem. Soc., Perkin Trans. 2,* **1991,** 905–912.

80. Vicent, C.; Bosso, C.; Cano, F. H.; de Paz, J. L. G.; Foces-Foces, C.; Jiménez-Barbero, J.; Martín-Lomas, M.; Penadés, S. *J. Org. Chem.* **1991,** *56,* 3614–3618.

81. Struchkova, M. I.; Él'perina, E. A.; Abylgaziev, R. I.; Serebryakov, E. P. *Izv. Akad. Nauk SSSR, Ser. Khim.* **1989,** 2492–2500; *Chem. Abstr.* **1990,** *112,* 217371n.

82. Pochapsky, T. C.; Stone, P. M.; Pochapsky, S. S. *J. Am. Chem. Soc.* **1991,** *113,* 1460–1462.

83. Scott, J. W. In *Topics in Stereochemistry;* Eliel, E. L.; Wilen, S. H. (Eds.); Wiley: New York, 1989; Vol. 19, pp. 209–226.

84. Crosby, J. *Tetrahedron,* **1991,** *47,* 4789–4846.

85. Enantiomeric excess (% ee) is the percent of the major enantiomer minus that of the minor enantiomer, while optical purity (% op) is the ratio, expressed as a percent, of the optical rotation of a mixture of enantiomers to that of the pure enantiomer.

86. Julia, S.; Ginebreda, A.; Guixer, J.; Tomás, A. *Tetrahedron Lett.* **1980,** *21,* 3709–3712.

87. Saigo, K.; Koda, H.; Nohira, H. *Bull. Chem. Soc. Jpn.* **1979,** *52,* 3119–3120.

88. Gasparski, C. M.; Miller, M. J. *Tetrahedron,* **1991,** *47,* 5367–5378.

89. Soai, K.; Watanabe, M. *J. Chem. Soc., Chem. Commun.* **1990,** 43–44.

90. Stoddart, J. F. *Biochem. Soc. Trans.* **1987,** *15,* 1188–1191.

91. Nerinckx, W.; Vandewalle, M. *Tetrahedron: Asymmetry,* **1990,** *1,* 265–276.

92. Loupy, A.; Sansoulet, J.; Zaparucha, A.; Merienne, C. *Tetrahedron Lett.* **1989,** *30,* 333–336.

93. Cram, D. J.; Sogah, G. D. Y. *J. Chem. Soc., Chem. Commun.* **1981,** 625–628.

94. Banfi, S.; Cinquini, M.; Colonna, S. *Bull. Chem. Soc. Jpn.* **1981,** *54,* 1841–1843.

95. Rubina, K. I.; Goldberg, Y. S.; Shymanska, M. V.; Lukevics, E. *Appl. Organomet. Chem.* **1987,** *1,* 435–439; *Chem. Abstr.* **1989,** *110,* 39049n.

96. Shida, Y.; Ando, N.; Yamamoto, Y.; Oda, J.; Inouye, Y. *Agric. Biol. Chem.* **1979,** *43,* 1797–1799.

97. Horner, L.; Skaletz, D. H. *Justus Liebigs Ann. Chem.* **1977,** 1365–1409.

98. Horner, L.; Brich, W. *Justus Liebigs Ann. Chem.* **1978,** 710–716.

99. Blaser, H.-U.; Boyer, S. K.; Pittelkow, U. *Tetrahedron: Asymmetry,* **1991**, *2,* 721–732.

100. Harigaya, Y.; Yamaguchi, H.; Onda, M. *Chem. Pharm. Bull.* **1981**, *29,* 1321–1327.

101. Mazaleyrat, J. P. *Tetrahedron Lett.* **1983**, *24,* 1243–1246.

102. Dehmlow, E. V.; Sauerbier, C. *Justus Liebigs Ann. Chem.* **1989**, 181–185.

103. Masui, M.; Ando, A.; Shioiri, T. *Tetrahedron Lett.* **1988**, *29,* 2835–2838.

104. Bourson, J.; Goguillon, T.; Jugé, S. *Phos. Sulfur,* **1983**, *14,* 347–356.

105. Inoue, K.; Noguchi, H.; Hidai, M.; Uchida, Y. *Yukagaku,* **1980**, *29,* 397–401; *Chem. Abstr.* **1981**, *94,* 102469t.

106. Julia, S.; Ginebreda, A.; Guixer, J.; Masana, J.; Tomás, A.; Colonna, S. *J. Chem. Soc., Perkin Trans. 1,* **1981**, 574–577.

107. Colonna, S.; Re, A.; Wynberg, H. *J. Chem. Soc., Perkin Trans. 1,* **1981**, 547–552.

108. Julia, S.; Ginebreda, A. *Tetrahedron Lett.* **1979**, 2171–2174.

109. Bhattacharya, A.; Dolling, U.-H.; Grabowski, E. J. J.; Karady, S.; Ryan, K. M.; Weinstock, L. M. *Angew. Chem., Int. Ed. Engl.* **1986**, *25,* 476–477.

110. Belokon', Y. N.; Maleev, V. I.; Savel'eva, T. F.; Garbalinskaya, N. S.; Saporovskaya, M. B.; Bakhmutov, V. I.; Belikov, V. M. *Izv. Akad. Nauk SSSR, Ser. Khim.* **1989**, 631–635; *Chem. Abstr.* **1990**, 112:36400a.

111. Belokon', Y. N.; Maleev, V. I.; Videnskaya, S. O.; Saporovskaya, M. B.; Tsyryapkin, V. A.; Belikov, V. M. *Izv. Akad. Nauk SSSR, Ser. Khim.* **1991**, 126–134; *Chem. Abstr.* **1991**, *115,* 50229v.

112. O'Donnell, M. J.; Wu, S. *Tetrahedron: Asymmetry,* **1992**, *3,* 591–594.

113. O'Donnell, M. J.; Wu, S. *Unpublished results.*

114. Lee, T. B. K.; Wong, G. S. K. *J. Org. Chem.* **1991**, *56,* 872–875.

115. Chen, B.-H.; Ji, Q.-E. *Huaxue Xuebao,* **1989**, *47,* 350–354; *Chem. Abstr.* **1989**, *111,* 194508a.

116. An, X.-X.; Ding, M.-X. *Huaxue Xuebao,* **1991**, *49,* 507–512; *Chem. Abstr.* **1991**, *115,* 158668b.

117. Valli, V. L. K.; Sarma, G. V. M.; Choudary, B. M. *Indian J. Chem.* **1990**, *29B,* 481–482.

118. Palladium-catalyzed carbonylation of α-methylbenzyl bromide in the presence of an optically active phosphine under phase transfer conditions (CH_2Cl_2, aqueous NaOH, and hexadecyltrimethylammonium bromide) gave 2-phenylpropionic acid in up to 64% ee with 9% chemical yield (42% ee with 65% chemical yield). The reaction was proposed to involve a kinetic resolution. Arzoumanian, H.; Buono, G.; Choukrad, M.; Petrignani, J.-F. *Organometallics,* **1988**, *7,* 59–62.

119. (a) Julia, S.; Ginebreda, A. *Afinidad,* **1980**, *37,* 194–196. (b) Tong, Y.-J.; Ding, M.-X. *Youji Huaxue,* **1990**, *10,* 464–470; *Chem. Abstr.* **1991**, *114,* 101294b.

120. Delee, E.; Jullien, I.; Le Garrec, L.; Loupy, A.; Sansoulet, J.; Zaparucha, A. *J. Chromatogr.* **1988**, *450,* 183–189.

121. Colonna, S.; Hiemstra, H.; Wynberg, H. *J. Chem. Soc., Chem. Commun.* **1978**, 238–239.

122. Wynberg, H.; Greijdanus, B. *J. Chem. Soc., Chem. Commun.* **1978**, 427–428.

123. Hermann, K.; Wynberg, H. *J. Org. Chem.* **1979**, *44,* 2238–2244.

124. Colonna, S.; Annunziata, R. *Afinidad,* **1981**, *38,* 501–502.

125. Conn, R. S. E.; Lovell, A. V.; Karady, S.; Weinstock, L. M. *J. Org. Chem.* **1986**, *51,* 4710–4711.

126. Cui, J.; Yu, L. *Synth. Commun.* **1990**, *20*, 2895–2900.

127. Annunziata, R.; Cinquini, M.; Colonna, S. *Chem. Ind. (London)* **1980**, 238.

128. Alonso-López, M.; Martín-Lomas, M.; Penadés, S. *Tetrahedron Lett.* **1986**, *27*, 3551–3554.

129. Alonso-López, M.; Jimenez-Barbero, J.; Martín-Lomas, M.; Penadés, S. *Tetrahedron*, **1988**, *44*, 1535–1543.

130. Vicent, C.; Martín-Lomas, M.; Penadés, S. *Tetrahedron*, **1989**, *45*, 3605–3612.

131. Aoki, S.; Sasaki, S.; Koga, K. *Tetrahedron Lett.* **1989**, *30*, 7229–7230.

132. Takasu, M.; Wakabayashi, H.; Furuta, K.; Yamamoto, H. *Tetrahedron Lett.* **1988**, *29*, 6943–6946.

133. Raguse, B.; Ridley, D. D. *Aust. J. Chem.* **1984**, *37*, 2059–2071.

134. Goldberg, Y.; Abele, E.; Bremanis, G.; Trapenciers, P.; Gaukhman, A.; Popelis, J.; Gomtsyan, A.; Kalvins, I.; Shymanska, M.; Lukevics, E. *Tetrahedron*, **1990**, *46*, 1911–1922.

135. Kwee, S.; Lund, H. *Bioelectrochem. Bioenerg.* **1980**, *7*, 693–698; *Chem. Abstr.* **1981**, *94*, 204217b.

136. Colonna, S.; Fornasier, R. *Synthesis*, **1975**, 531–532.

137. Balcells, J.; Colonna, S.; Fornasier, R. *Synthesis*, **1976**, 266–267.

138. Massé, J. P.; Parayre, E. *Bull. Soc. Chim. Fr. II*, **1978**, 395–400.

139. Kinishi, R.; Nakajima, Y.; Oda, J.; Inouye, Y. *Agric. Biol. Chem.* **1978**, *42*, 869–872.

140. Kinishi, R.; Uchida, N.; Yamamoto, Y.; Oda, J.; Inouye, Y. *Agric. Biol. Chem.* **1980**, *44*, 643–648.

141. Li, S.; Meng, S.; Jian, H. *Beijing Gongye Daxue Xuebao*, **1987**, *13*, 117–120; *Chem. Abstr.* **1988**, *109*, 148995t.

142. Passarotti, C.; Bandi, G. L.; Fossati, A.; Valenti, M.; Dal Bo, L. *Boll. Chim. Farm.* **1990**, *129*, 195–198; *Chem. Abstr.* **1991**, *114*, 228584b.

143. Sarkar, A.; Rao, B. R. *Tetrahedron Lett.* **1991**, *32*, 1247–1250.

144. Innis, C.; Lamaty, G. *Nouv. J. Chim.* **1977**, *1*, 503–509.

145. Harigaya, Y.; Yamaguchi, H.; Onda, M. *Heterocycles*, **1981**, *15*, 183–185.

146. Takahashi, H.; Kubota, Y.; Miyazaki, H.; Onda, M. *Chem. Pharm. Bull.* **1984**, *32*, 4852–4857.

147. Onda, M.; Li, S.; Li, X.; Harigaya, Y.; Takahashi, H.; Kawase, H.; Kagawa, H. *J. Nat. Prod.* **1989**, *52*, 1100–1106.

148. Baba, N.; Oda, J.; Kawaguchi, M. *Agric. Biol. Chem.* **1986**, *50*, 3113–3117.

149. Baba, N.; Kawahara, S.; Hamada, M.; Oda, J. *Bull. Inst. Chem. Res., Kyoto Univ.* **1987**, *65*, 144–146; *Chem. Abstr.* **1988**, *109*, 170140g.

150. Baba, N.; Oda, J.; Kawahara, S.; Hamada, M. *Bull. Inst. Chem. Res. Kyoto Univ.* **1989**, *67*, 121–127; *Chem. Abstr.* **1990**, *113*, 23546y.

151. Helder, R.; Hummelen, J. C.; Laane, R. W. P. M.; Wiering, J. S.; Wynberg, H. *Tetrahedron Lett.* **1976**, 1831–1834.

152. Wynberg, H. *Chimia*, **1976**, *30*, 445–451.

153. Hummelen, J. C.; Wynberg, H. *Tetrahedron Lett.* **1978**, 1089–1092.

154. Marsman, B.; Wynberg, H. *J. Org. Chem.* **1979**, *44*, 2312–2314.

155. Wynberg, H.; Marsman, B. *J. Org. Chem.* **1980**, *45*, 158–161.

156. Pluim, H.; Wynberg, H. *J. Org. Chem.* **1980**, *45*, 2498–2502.

157. Snatzke, G.; Wynberg, H.; Feringa, B.; Marsman, B. G.; Greydanus, B.; Pluim, H. *J. Org. Chem.* **1980**, *45*, 4094–4096.

158. Annunziata, R. *Synth. Commun.* **1979**, *9*, 171–178.

159. Domagala, J. M.; Bach, R. D. *J. Org. Chem.* **1979**, *44*, 3168–3174.

160. Banfi, S.; Colonna, S.; Julia, S. *Synth. Commun.* **1983**, *13*, 1049–1052.

161. Julia, S.; Ginebreda, A.; Guixuer, J. *J. Chem. Soc., Chem. Commun.* **1978**, 742–743.

162. Yu, L.; Wang, F. *Gaodeng Xuexiao Huaxue Xuebao*, **1987**, *8*, 336–340; *Chem. Abstr.* **1988**, *108*, 112909h.

163. Wakabayashi, T.; Watanabe, K. *Chem. Lett.* **1978**, 1407–1410.

164. Annunziata, R.; Cinquini, M.; Colonna, S. *J. Chem. Soc., Perkin Trans. 1*, **1980**, 2422–2424.

165. Ahuja, R. R.; Bhole, S. I.; Bhongle, N. N.; Gogte, V. N.; Natu, A. A. *Indian J. Chem.* **1982**, *21B*, 299–303.

Asymmetric Reactions with Chiral Lewis Acid Catalysts

Keiji Maruoka and Hisashi Yamamoto

Department of Applied Chemistry
Nagoya University
Chikusa, Nagoya, Japan

9.1. Introduction

Lewis acids of the BF_3, $AlCl_3$, $SnCl_4$, and $TiCl_4$ types have been used for a variety of carbon–carbon bond formations in organic synthesis. Most often stoichiometric amounts of Lewis acids are employed. Undoubtedly, an efficient way to perform enantioselective carbon–carbon bond formation would be to use catalytic amounts of a chiral Lewis acid in these reactions. In all cases, the complexation of a Lewis acid with a carbonyl functionality would activate the system; therefore, a single chiral Lewis acid has the potential to be a common catalyst for different carbon–carbon bond forming reactions [1].

In 1979 Koga et al. described menthoxyaluminum dichloride as a chiral Lewis acid catalyst in the Diels–Alder reaction, and an enantiomeric excess of 72% was realized [2]. In 1987 Koga and others described the stereochemical relationships between the structure of chiral catalysts and the absolute configuration of the Diels–Alder adducts [3]. The transition state proposed constitutes an anti-*s*-trans configuration of the complex of the aluminum reagent with the carbonyl functionality (eq. 1).

In 1983 Danishefsky described the hetero Diels–Alder reaction of aldehydes and activated dienes (Danishefsky's diene) catalyzed by the chiral shift agent $Eu(hfc)_3$ with moderate asymmetric induction (58% ee) [4] (eq. 2).

Cyclocondensation of an *l*-menthoxydiene afforded the corresponding facial isomer with high enantiomeric purity (eq. 3). Thus, chiral auxiliary–chiral catalyst combinations can be used to synthesize enantiomerically pure pyrones, including some saccharides [5].

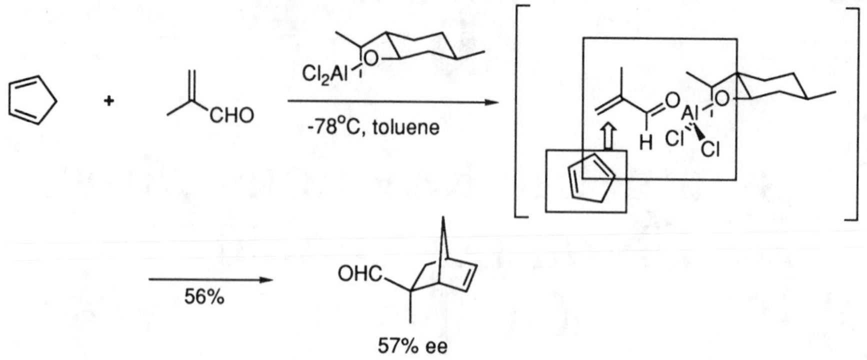

(1)

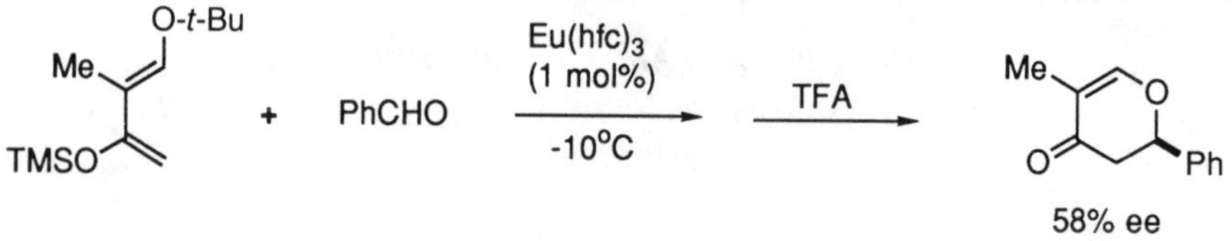

56%

57% ee

(2)

58% ee

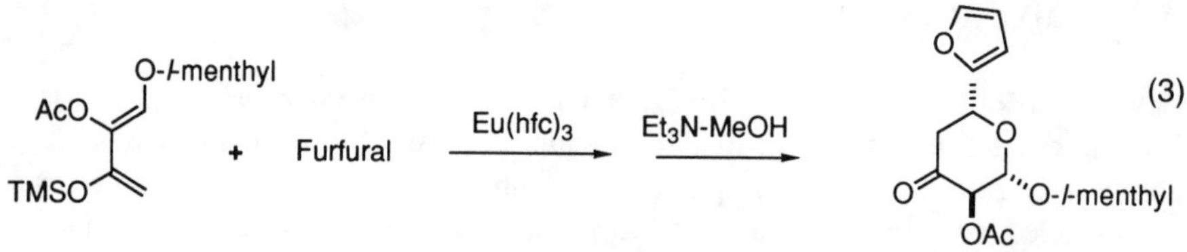

(3)

In 1985 we described the enantioselective cyclization of prochiral unsaturated aldehydes using a chiral zinc reagent derived from dimethylzinc and binaphthol [6] (eq. 4). Although the reagent is rather unstable, the system is flexible and has great potential for further work in catalyst design.

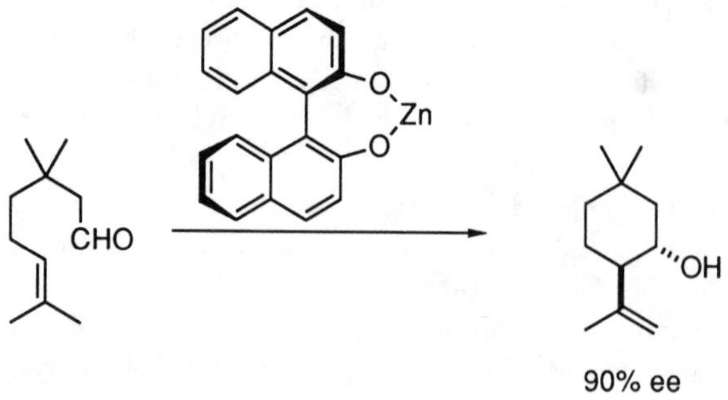

(4)

90% ee

From the aforementioned pioneering work, there is no doubt that a carefully designed chiral Lewis acid has a vast potential for the asymmetric synthesis of carbon skeletons. Choice of the proper metal and design of suitable chiral ligand may be the most important elements for obtaining successful results. What follows is a description of the design of chiral Lewis acids since 1985.

9.2. Chiral Aluminum Reagents

In view of the well-established capacity of aluminum reagents to enhance the reaction rate and selectivity of various organic reactions [7], the utilization of a chiral aluminum catalyst should, in principle, lead to asymmetric induction. However, until recently little work had appeared in the area of asymmetric synthesis with chiral aluminum reagents. In 1979 Koga and coworkers showed that alkoxy-aluminum dichlorides are effective chiral catalysts under certain circumstances [2]. These aluminum catalysts, prepared in situ from optically active terpene alcohols (*l*-menthol, D-neomenthol, and *l*-borneol) and ethylaluminum dichloride, promote the asymmetric Diels–Alder reaction of methacrolein and cyclopentadiene to furnish exo adducts in 25–72% ee. Later, the stereochemical relationship between enantioface selection, the structure of chiral ligands, and the substituents on the aluminum atom was proposed by this group [3].

The chiral mono(alkoxy)aluminum chloride and bis(alkoxy)aluminum chloride reagents, which were generated similarly in situ from 1–2 equiv. of ethylaluminum dichloride, chiral diols **3–7**, and sulfonamide **8**, exhibited high enantioselection in the asymmetric Diels–Alder reaction between crotonoyloxazolidinone **1** and cyclopentadiene [8] (eq. 5). However, synthetic application has been rather limited, since such chiral aluminums must be employed as stoichiometric reagents and only for the specific dienophile **1**. Attempted cycloaddition with acryloyl analog of **1** resulted in a marked loss of enantioselectivity (17–36% ee).

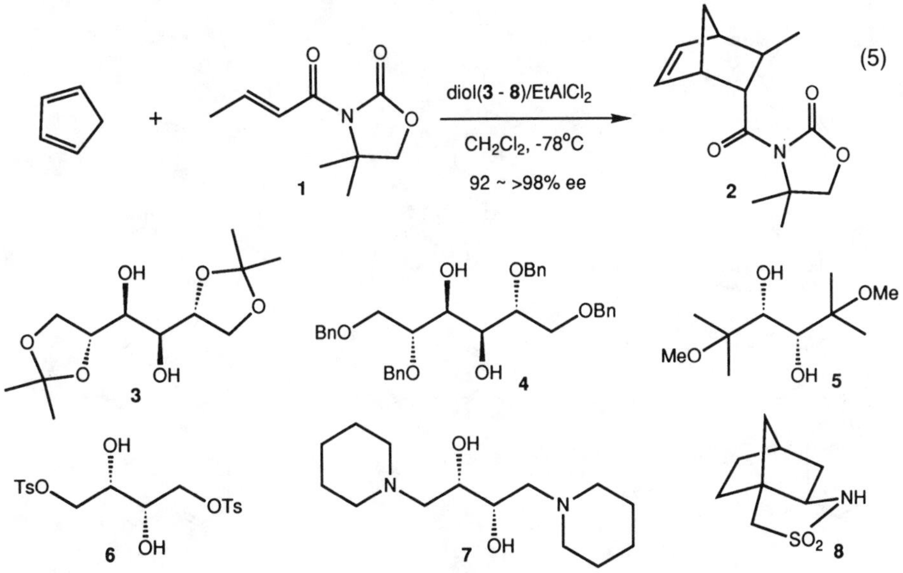

Conformationally rigid cyclic aluminum reagents **9** and **10** can be made from diethylaluminum chloride and the corresponding bidentate ligands, and these were applied to the asymmetric aldol reaction and the formation of cyanohydrin with moderate asymmetric induction [9] (eq. 6).

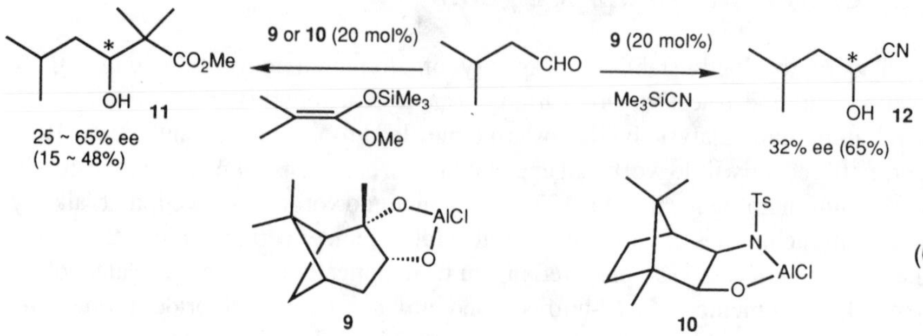

$$(6)$$

The first reliable chiral aluminum reagents of types (R)-**13** and (S)-**13** were devised for enantioselective activation of carbonyl groups using the concept of diastereoselective activation of carbonyl moieties with the exceptionally bulky organoaluminum reagents MAD and MAT [10]. The sterically hindered, enantiomerically pure (R)-(+)-3,3'-bis(triarylsilyl)binaphthol (R)-**14** required for the preparation of (R)-**13** can be synthesized in two steps from (R)-(+)-3,3'-dibromobinaphthol by bistriarylsilylation and subsequent intramolecular 1,3-rearrangement of the triarylsilyl groups [11]. Reaction of (R)-**14** in toluene with trimethylaluminum produced the chiral organoaluminum reagent (R)-**13** quantitatively (Scheme 1). Its molecular weight, found cryoscopically in benzene, corre-

Scheme 1

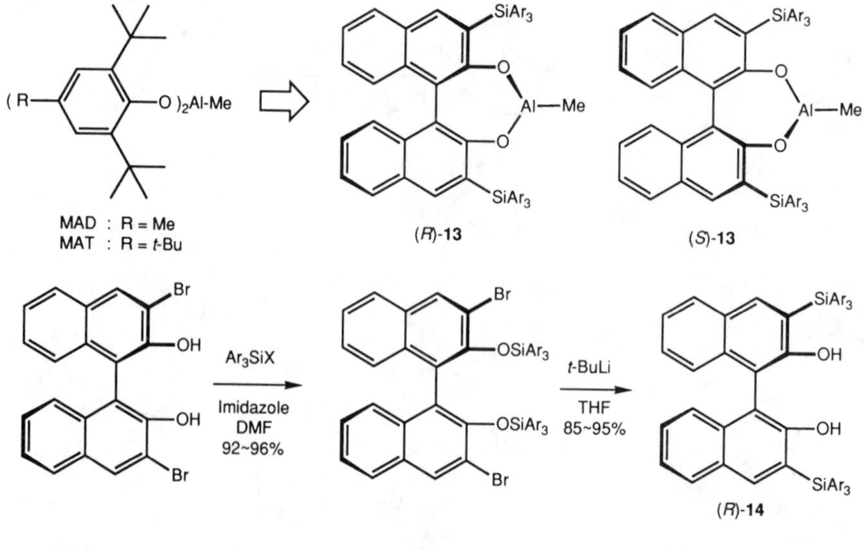

sponds closely to the value calculated for the monomeric species **13**. The chiral organoaluminum reagents (R)-**13** and (S)-**13** were shown to be highly effective as chiral Lewis acid catalysts in the asymmetric hetero Diels–Alder reaction [12]. Reaction of various aldehydes with activated dienes under the influence of a catalytic amount of **13** (5–10 mol %) at −20°C, after exposure of the resulting hetero Diels–Alder adducts to trifluoroacetic acid, gave predominantly cis-dihydropyrone (**16**) in high yield with excellent enantioselectivity (eq. 7). The enantioface differentiation of prochiral aldehydes is controllable by fine-tuning the size of the trialkylsilyl moiety in **13**, thereby allowing the rational design of the catalyst for asymmetric induction. In fact, switching the triarylsilyl substituent (Ar = Ph or 3,5-Xylyl) to the tert-butyldimethylsilyl or trimethylsilyl group led to a substantial loss of enantio as well as cis selectivity in the hetero Diels–Alder reaction of benzaldehyde and activated diene **15**. In marked contrast with this, the chiral organoaluminum reagent derived from trimethylaluminum and (R)-(+)-3,3'-dialkylbinaphthol (alkyl = H, Me, or Ph) was employed only as a stoichiometric reagent and gave less satisfactory results in both the reactivity and enantioselectivity of the hetero Diels–Alder reaction.

$$(R)\text{-}13 : \quad Ar = Me \qquad : 64\% \text{ ee, } 72\% \ (53:47)$$
$$Ar_3 = t\text{-BuMe}_2 \quad : 84\% \text{ ee, } 91\% \ (69:31)$$
$$Ar = Ph \qquad : 95\% \text{ ee, } 87\% \ (92:8)$$
$$Ar = 3,5\text{-Xylyl} : 97\% \text{ ee, } 93\% \ (97:3)$$

(7)

An interesting method for the preparation of chiral aluminum reagents has appeared recently. The chiral organoaluminum reagent (R)-**13** or (S)-**13** can be generated in situ from the corresponding racemate (±)-**13** by diastereoselective complexation with certain chiral ketones [13] (Scheme 2). Among several terpene-derived chiral ketones, 3-bromocamphor was found to be the most satisfactory. The hetero Diels–Alder reaction of benzaldehyde and 2,4-dimethyl-1-methoxy-3-trimethylsiloxy-1,3-butadiene (**15**) with 0.1 equiv. of (±)-**13** (Ar = Ph) and d-bromocamphor at −78°C gave rise to cis adduct **16** as the major product with 82% ee. Although the extent of asymmetric induction is not yet as satisfactory as that (95% ee) with the enantiomerically pure **13** (Ar = Ph), one recrystallization of **16** (82% ee) from hexane gave enantiomerically pure **16**, thereby enhancing the practicality of this method. This study demonstrates the potential for broad application of chiral catalysts generated in situ by way of diastereoselective complexation in asymmetric synthesis.

The enantioselective activation of carbonyl groups with the chiral aluminum (R)-**13** or (S)-**13** also enabled the asymmetric ene reaction of electron-deficient

Scheme 2

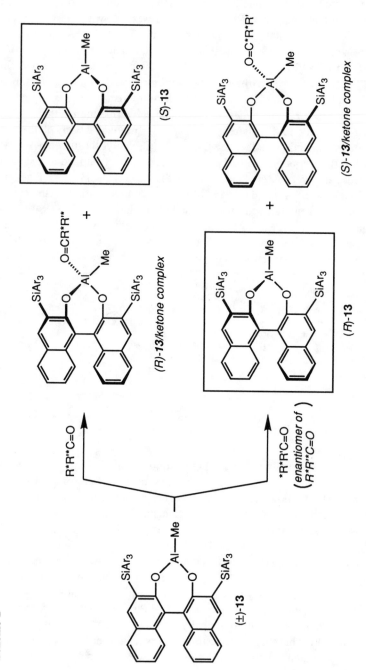

418

aldehydes with various alkenes [14] (eq. 8). In the presence of powdered 4A molecular sieves, the chiral aluminum reagent (R)-**13** or (S)-**13** can be utilized as a catalyst without any loss of enantioselectivity.

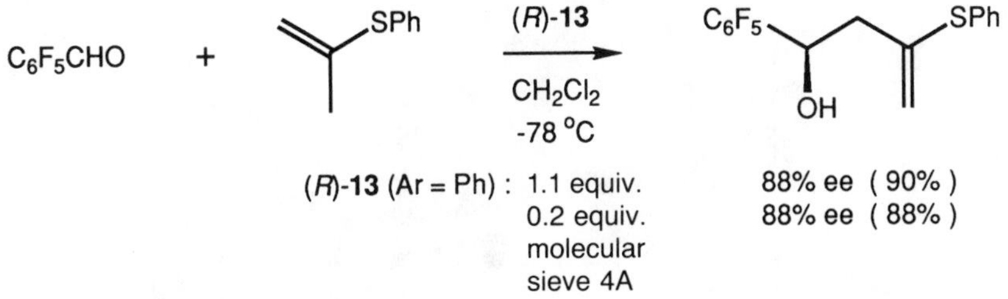

$$C_6F_5CHO \quad + \quad \text{(alkene with SPh)} \xrightarrow[\substack{CH_2Cl_2 \\ -78\,^{\circ}C}]{(R)\text{-}13} \text{(product with } C_6F_5, OH, SPh)$$

(8)

(R)-**13** (Ar = Ph) : 1.1 equiv.   88% ee (90%)
                          0.2 equiv.   88% ee (88%)
                          molecular
                          sieve 4A

The concept of the enantioselective activation of carbonyl groups with the bulky chiral aluminums (R)-**13** and (S)-**13** has been further extended to the enantioselective activation of an ether oxygen, allowing for the first successful example of the asymmetric Claisen rearrangement of allylic vinyl ethers **17**, as illustrated in Scheme 3 and Table 1 [15]. This method provides a facile asymmetric synthesis of various acylsilanes **18** or **19** (X = SiR_3) and acylgermanes **18** and **19** (X = $GeMe_3$) with high enantiomeric purity. Among the various trialkylsilyl substituents of **13**, use of the bulkier t-butyldiphenylsilyl group results in the highest enantioselectivity. Conformational analysis of the two possible chairlike transition state structures of an allyl vinyl ether substrate (**17**) reveals that the chiral organoaluminum reagent **13** can discriminate between these two conformations, **A** and **B**, which differ from each other only in the orientation of α-methylene groups of ethers.

Notably, the asymmetric Claisen rearrangement of cis-allylic α-(trimethylsilyl)vinyl ethers with (R)-**13** produced optically active acylsilanes with the same

Scheme 3

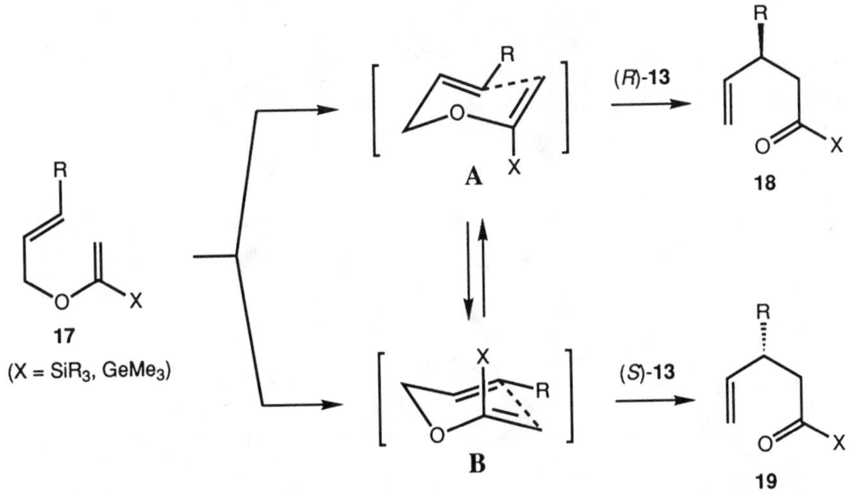

Table 1 Asymmetric Claisen Rearrangement of Allylic Vinyl Ethers (17)

Allylic Vinyl Ether	Catalyst (R)-13	Reaction Conditions (°C, h)	Claisen Product	Yield (%)	% ee[a]
17 (R = Ph, X = SiMe$_3$)	Ar$_3$ = t-BuMe$_2$	−40, 0.1; −20, 24	**18** (R = Ph, X = SiMe$_3$)	22	14
17 (R = Ph, X = SiMe$_3$)	Ar = Ph	−40, 0.1; −20, 8	**18** (R = Ph, X = SiMe$_3$)	86	80
17 (R = Ph, X = SiMe$_3$)	Ar$_3$ = t-BuPh$_2$	−40, 0.1; −20, 3	**18** (R = Ph, X = SiMe$_3$)	99	88
17 (R = Ph, X = SiMe$_3$)	Ar = Ph[b]	−40, 0.1; −20, 5	**18** (R = Ph, X = SiMe$_3$)	85	80
17 (R = Ph, X = SiMe$_2$Ph)	Ar = Ph	−78, 0.1; −40, 16	**18** (R = Ph, X = SiMe$_2$Ph)	65	85
17 (R = Ph, X = SiMe$_2$Ph)	Ar$_3$ = t-BuPh$_2$	−78, 0.1; −40, 8	**18** (R = Ph, X = SiMe$_2$Ph)	76	90
17 (R = cyclohexyl, X = SiMe$_3$)	Ar = Ph	−40, 0.1; −20, 4	**18** (R = cyclohexyl, X = SiMe$_3$)	79	61
17 (R = cyclohexyl, X = SiMe$_3$)	Ar$_3$ = t-BuPh$_2$	−40, 0.1; −20, 4	**18** (R = cyclohexyl, X = SiMe$_3$)	84	71
17 (R = trans-cinnamyl, X = SiMe$_3$)	Ar = Ph	−40, 0.1; −20, 10	**18** (R = trans-cinnamyl, X = SiMe$_3$)	40	60
17 (R = Ph, X = GeMe$_3$)	Ar = Ph	−78, 0.1; −40, 15	**18** (R = Ph, X = GeMe$_3$)	73	91
17 (R = Ph, X = GeMe$_3$)	Ar$_3$ = t-BuPh$_2$	−78, 0.1; −40, 16	**18** (R = Ph, X = GeMe$_3$)	68	93

[a] Determined by capillary gas–liquid chromatographic analysis after conversion of the products to the corresponding acetals of $(2R, 3R)$-butanediol.
[b] (S)-**13** was used as the catalyst.

absolute configuration as those from *trans*-allylic α-(trimethylsilyl)vinyl ethers [16] (eq. 9).

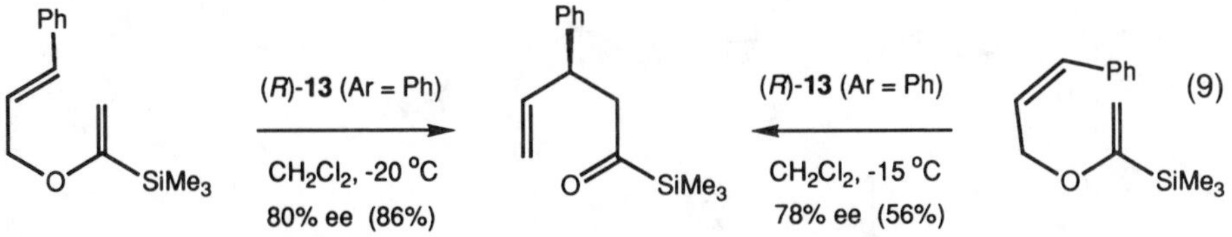

Another reliable chiral aluminum reagent **20** (R = Me or *i*-Bu) was prepared in situ by the reaction of chiral bis(trifluoromethanesulfonamide) with trimethyl aluminum or diisobutylaluminum hydride and used to catalyze asymmetric Diels–Alder reactions [17]. The chiral aluminum **20** (R = Me) catalyzed the cycloaddition of 5-(benzyloxymethyl)-1,3-cyclopentadiene (**21**) with *N*-acryloyl-1,3-oxazolidin-2-one (**22**), yielding the valuable prostaglandin intermediate **23** with 95% ee (eq. 10). Simple recrystallization brings the purity up to 100% ee, showing the outstanding practical utility of this method.

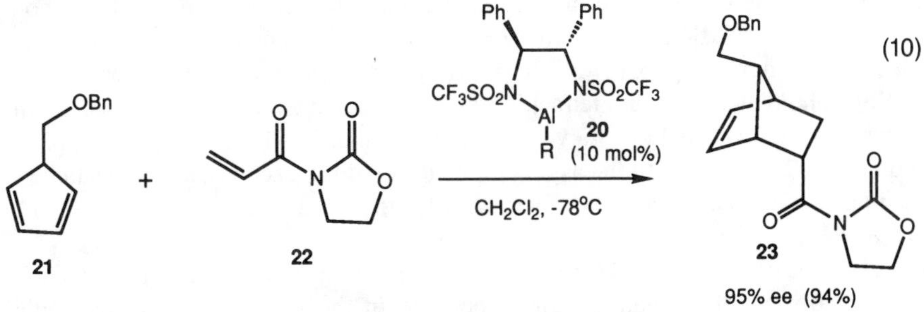

An efficient kinetic resolution of a racemic epoxide has been accomplished with a binaphthol-modified lithium aluminate (**24**), which is generated in situ from (*R*)-binaphthol, dimethylaluminum chloride, and lithium butoxide. Exposure of a racemic epoxide **25** to 0.75 equiv. of **24** induced an intramolecular acetalization, and the remaining epoxide (*R*)-**25** was obtained in 20% yield with better than 98% ee [18] (Scheme 4). The enantiomerically pure **25** was further transformed to C_{16}-juvenile hormone.

The stereoselective cationic polymerization of propylene oxide has been developed with a chiral aluminum alkoxide initiator, prepared by the reaction of (*R*)-(−)-3,3-dimethyl-1,2-butanediol with aluminum hydride [19]. This initiator was found to be highly reactive in combination with equimolar zinc chloride.

9.3. Chiral Titanium Reagents

Several types of chiral titanium reagent have been developed for different asymmetric C—C bond forming reactions, including the Diels–Adler, ene, and alkylation reactions, as well as cyanohydrin formation.

Scheme 4

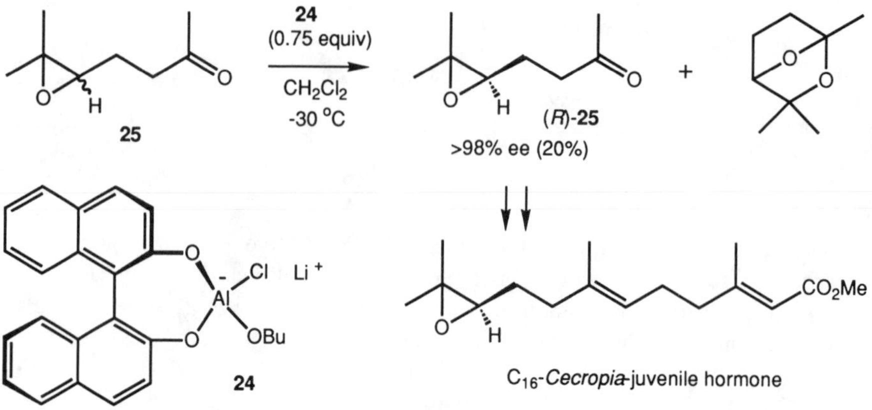

Attempts to prepare a binaphthol-modified dichlorotitanium reagent **26** (X = Cl) often led to the undesired formation of ring-opened oligomeric materials, as confirmed by ^{13}C NMR analysis. Reetz et al. successfully prepared the monomeric dichlorotitanium complex (S)-**26** (X = Cl) by treatment of the dilithio derivative of (S)-binaphthol with titanium tetrachloride. This reagent effects the asymmetric synthesis of cyanohydrin **12** (85% yield) with 82% ee or less [9] (see eq. 6). Application of (S)-**26** (X = Cl) to the asymmetric Diels–Alder reaction of cyclopentadiene and methacrolein at −78°C, however, afforded the exo adduct (see eq. 1) with only 16% ee.

The dihalotitanium (R)-**26** (X = Cl or Br) can also be generated from chiral binaphthol and dihalotitanium diisopropoxide in the presence of 4A molecular sieves. This reagent catalyzed the asymmetric glyoxylate–ene reaction to furnish α-hydroxy esters of biological and synthetic importance [20] (eq. 11). The use of molecular sieves has proven to be essential for the in situ preparation of the chiral catalyst **26** but not for the ene reaction.

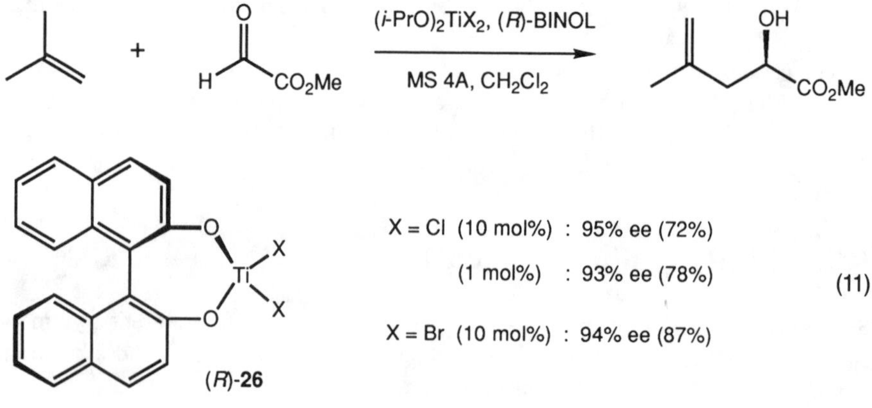

X = Cl (10 mol%) : 95% ee (72%)

(1 mol%) : 93% ee (78%)   (11)

X = Br (10 mol%) : 94% ee (87%)

Another way of preparing dichlorotitanium reagents in situ involves mixing bis-trimethylsilyl ethers of the chiral diol **3** and (*R*)-3,3'-diphenylbinaphthol with titanium tetrachloride. These reagents, employed stoichiometrically in the asymmetric Diels–Alder reaction between crotonoyloxazolidinone **1** and cyclopentadiene, gave endo adduct **2** (86–99% yield) with 96–98% ee or more [8].

The chiral monochlorotitanium reagents **27** and **28**, which are derived in situ from chlorotitanium triisopropoxide and the corresponding chiral diols by azeotropic removal of isopropanol, were found to exhibit a moderate asymmetric induction (42–50% ee) in the asymmetric Diels–Alder reaction of cyclopentadiene and methyl acrylate [21] (eq. 12).

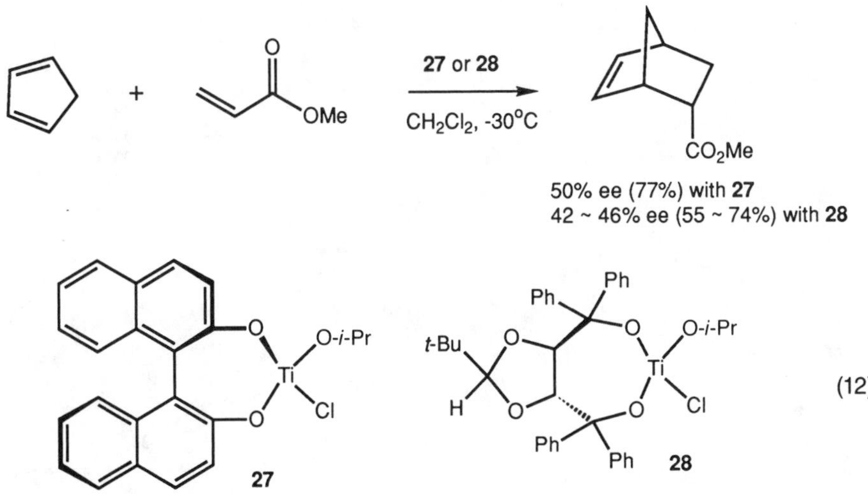

50% ee (77%) with **27**
42 ~ 46% ee (55 ~ 74%) with **28**

(12)

A promising chiral titanium reagent was devised recently by Narasaka et al. [22]. The chiral titanium reagent **29**, which can readily be prepared by mixing the chiral 1,4-diol **30** and dichlorotitanium diisopropoxide at room temperature, can be used as a catalyst for various asymmetric reactions in the presence of 4A molecular sieves. Addition of 4A molecular sieves is indispensable, since they shift the equilibrium of the mixture toward the formation of the chiral titanium reagent **29** (eq. 13). Asymmetric Diels–Alder reaction between 1,3-oxazolidin-2-one of fumaric acid (**31**) and isoprene in toluene–petroleum ether (1:1) proceeds under the influence of a catalytic amount of **29** to furnish cycloadduct **32** with 94% ee (eq. 14). Here, a profound solvent effect was observed, and use of mesitylene, chlorofluorocarbons, or a mixture of toluene and petroleum ether gave higher enantioselectivity than toluene alone.

The titanium reagent **29** has been successfully applied to asymmetric [2 + 2] cycloadditions [23] (eq. 15), ene reactions [24], (eq. 16), and cyanohydrin formation [25] (eq. 17) with high asymmetric induction. The [2 + 2] cycloadduct **33** is convertible, in enantiomerically pure form, to a carbocyclic analog (**34**) of a four-membered nucleoside, oxetanocin A (**35**).

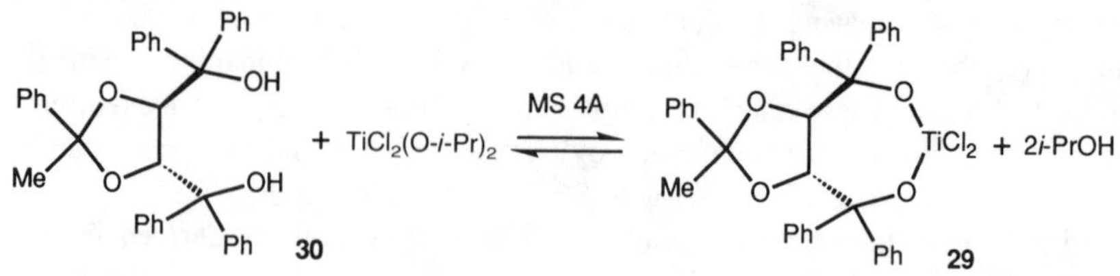

$$\textbf{30} + TiCl_2(O\text{-}i\text{-}Pr)_2 \xrightleftharpoons[MS\ 4A]{} \textbf{29} + 2i\text{-PrOH} \quad (13)$$

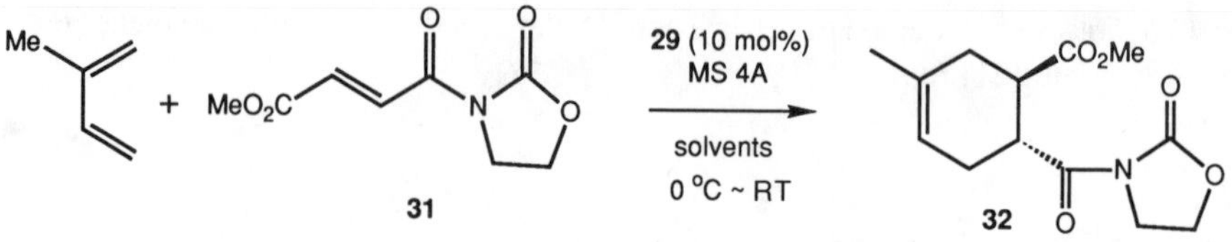

Solvent effect : benzene   41% ee
                  toluene   60% ee
                  mesitylene   93% ee   (14)
                  CFCl$_3$ 92% ee
                  toluene/PE (4:1)   80% ee
                  toluene/PE (1:1)   94% ee

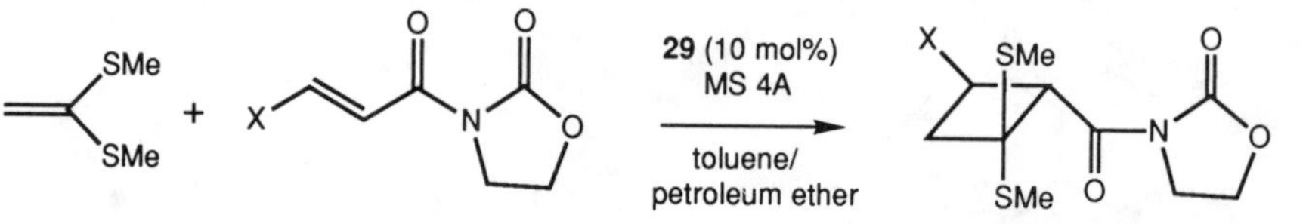

$X = CO_2Me$: 98% ee (96%)
$X = H$: 88% ee (78%)

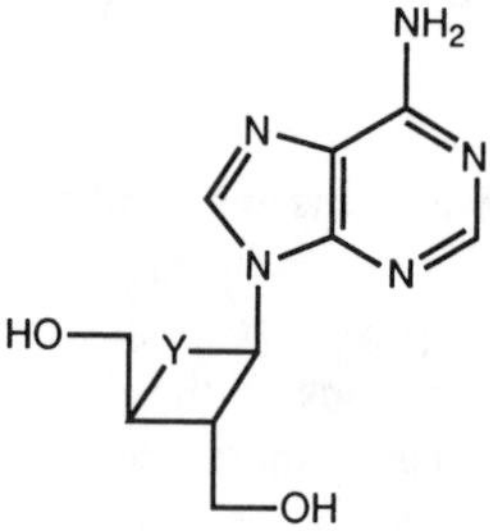

$\textbf{34}$: $Y = CH_2$
$\textbf{35}$: $Y = O$ (Oxetanocin A)

$$(15)$$

$$(16)$$

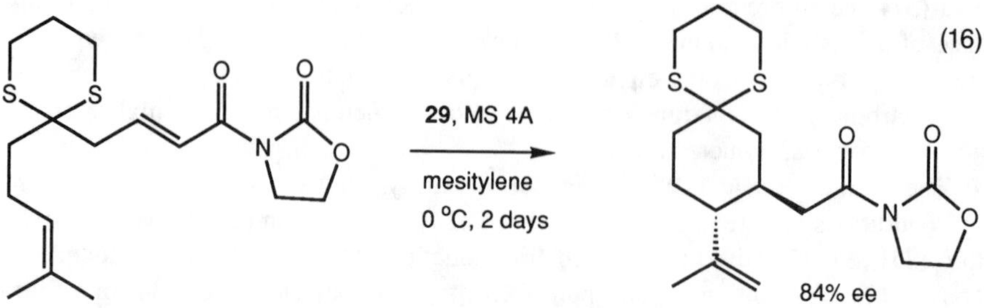

84% ee

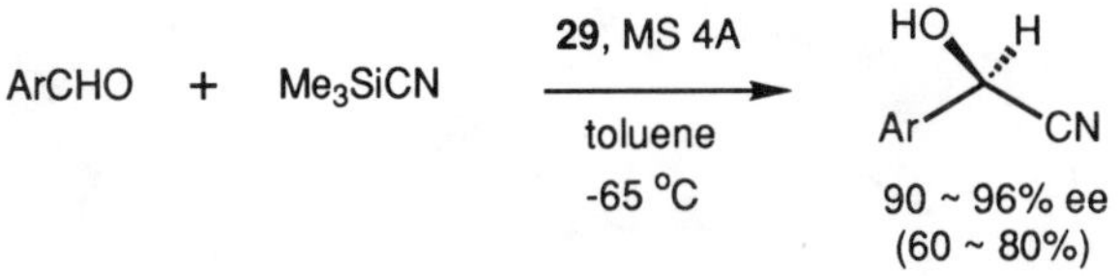

$$\text{ArCHO} + \text{Me}_3\text{SiCN} \xrightarrow[\substack{\text{toluene} \\ -65\,^\circ\text{C}}]{\textbf{29}, \text{MS 4A}} \begin{array}{c} \text{HO} \quad \text{H} \\ \diagdown\!\!\diagup \\ \text{Ar} \quad \text{CN} \\ 90 \sim 96\%\ \text{ee} \\ (60 \sim 80\%) \end{array} \qquad (17)$$

Another characteristic application of the chiral titanium catalyst **29** is the asymmetric solvolysis of racemic S-(2-pyridyl) thioesters of α-arylcarboxylic acid **36** [26] (eq. 18). The large rate difference ($k_R/k_S = 29$–42) between the enantiomers provides a highly efficient method for kinetic resolution of α-arylcarboxylic acid derivatives.

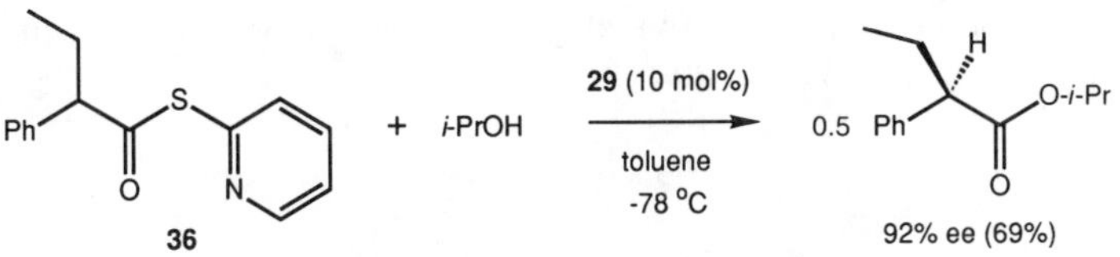

$$(18)$$

92% ee (69%)

A chiral titanium complex derived from a 1:1:1 ratio of the 1,4-diol **30**, titanium tetrachloride, and titanium tetraisopropoxide has proven to be very effective for the asymmetric [2 + 2] cycloaddition of substituted styrenes and 1,4-benzoquinones [27] (eq. 19).

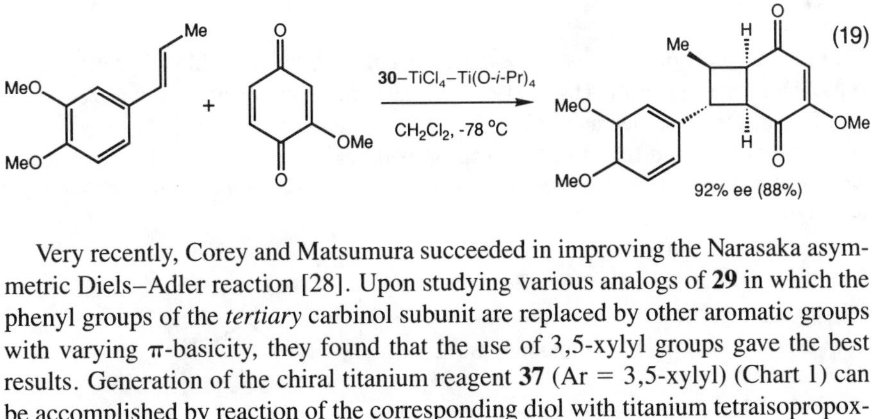

$$(19)$$

92% ee (88%)

Very recently, Corey and Matsumura succeeded in improving the Narasaka asymmetric Diels–Adler reaction [28]. Upon studying various analogs of **29** in which the phenyl groups of the *tertiary* carbinol subunit are replaced by other aromatic groups with varying π-basicity, they found that the use of 3,5-xylyl groups gave the best results. Generation of the chiral titanium reagent **37** (Ar = 3,5-xylyl) (Chart 1) can be accomplished by reaction of the corresponding diol with titanium tetraisopropoxide and subsequent treatment with silicon tetrachloride. The asymmetric cycloaddition of 5-((benzyloxy)methyl)-1,3-cyclopentadiene (**21**) with N-acryloyl-1,3-oxazolidin-2-one (**22**) was carried out in the presence of a catalytic amount of **37** (Ar = 3,5-xylyl) in toluene at -30°C producing the cycloadduct **23** (83%) with 95% ee (see eq. 10).

A chiral titanium catalyst **38**, derived in situ from (R)-bis(trifluoromethanesulfonamide) and titanium tetraisopropoxide, accelerates the asymmetric alkylation of benzaldehyde with diethylzinc, producing (S)-1-phenyl-1-propanol with high enan-

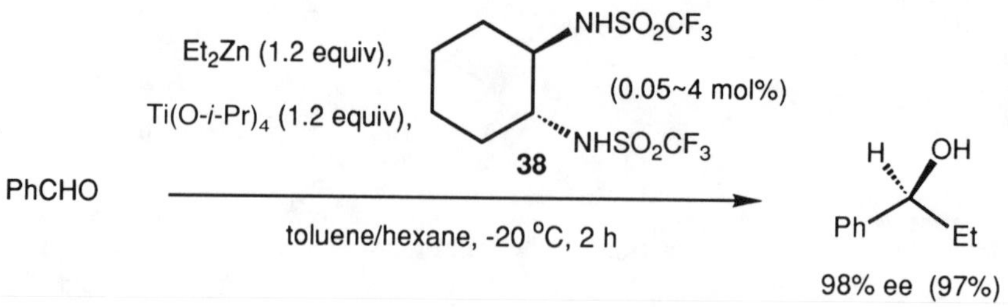
Et2Zn (1.2 equiv),
Ti(O-i-Pr)4 (1.2 equiv),
NHSO2CF3
(0.05~4 mol%)
NHSO2CF3
38
PhCHO
toluene/hexane, -20 °C, 2 h
H OH
Ph Et
98% ee (97%)

(20)

tioselectivity [29] (eq. 20). The introduction of the chiral sulfonamide auxiliary onto titanium increases Lewis acidity and markedly accelerates the alkylation reaction. An NMR study suggests that alkyltitanium is generated in situ from diethylzinc and the titanate **38**. This type of enantioselective alkylation can also be effected with chiral titanium catalysts such as **39** and **40** [30] (Chart 1).

Chart 1

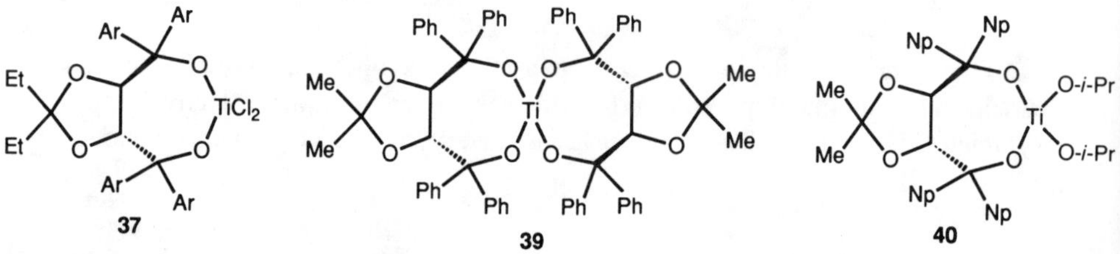
37
39
40

The acyclic dipeptide ester, 2-hydroxy-1-naphthylideneimino-(*S*)-valinyl-(*S*)-tryptophane methyl ester **41**, whose terminal amino group was derivatized as a Schiff base, was designed as a chiral ligand for a titanium catalyst [31]. An equimolar mixture of **41** and titanium tetraisopropoxide catalyzed enantioselective hydrocyanation of benzaldehyde to afford optically active (*R*)-mandelonitrile (88% yield) with 88% ee (eq. 21).

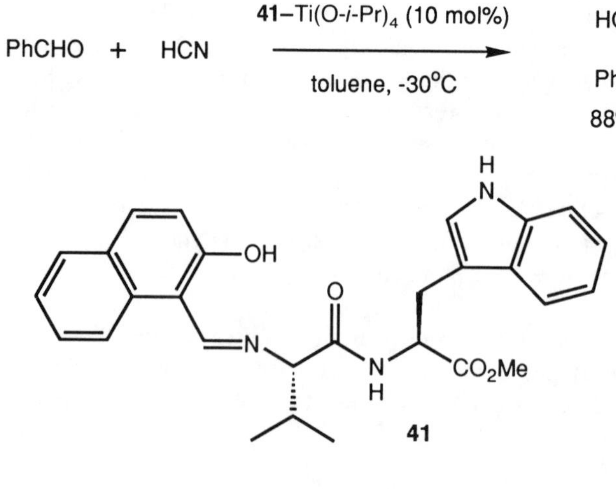
PhCHO + HCN
41–Ti(O-i-Pr)4 (10 mol%)
toluene, -30°C
HO H
Ph CN
88% ee (88%)
OH
O
N
N
H
CO2Me
41

(21)

9.4. Chiral Boron Reagents

9.4.1. Chiral (Acyloxy)borane (CAB) Reagent I

The remarkable reactivity of borane toward carboxylic acids over esters is one of the conspicuous characteristics of this compound, which is rarely seen in other hydride reagents. An acyloxyborane is recognized as an initial intermediate. The carbonyl group in this molecule, which is essentially a mixed anhydride, is activated by the electronegative nature of the trivalent boron atom. This fact led us to examine whether the acyloxyborane could be attacked by a molecule other than a simple hydride. If the acyloxyborane is reactive enough to cause functionalization of the acid moiety, various applications can be anticipated. Thus, acyloxyboranes derived from unsaturated acids were investigated for reactivity in the Diels–Alder reaction [32]. The reaction proceeded smoothly with the following features: (1) progress of the reaction was satisfactory even with a catalytic amount of borane; (2) addition of excess cyclohexanol to the reaction media remarkably reduced the reactivity of acyloxyborane, probably as a result of the irreversible exchange of the acyloxy group with the alcohol in the borane complexes, as well as the reduced reactivity of the resulting alkoxy-substituted acyloxyborane; and (3) monoalkoxysubstituted acyloxyborane is still sufficiently reactive to produce a Diels–Alder product.

With these experimental findings, it became of interest to evaluate the effect of chiral auxiliaries by introducing them into the acyloxyborane intermediate. Reaction of the monoacylated tartaric acid with 1 equiv. of the borane–THF complex in dichloromethane at 0°C should result in the formation of a chiral acyloxyborane intermediate. Acrylic acid (10-fold excess) was then added to this catalyst solution at 0°C, the mixture was cooled to −78°C and treated with cyclopentadiene. After the usual workup and chromatographic separation, a Diels–Alder adduct (78% ee, R) was isolated in 93% yield (a mixture of endo and exo isomers in a ratio of 96:4) (Scheme 5). The extent of asymmetric induction largely depended on the acyl moiety of tartaric acid derivatives [32].

Scheme 5

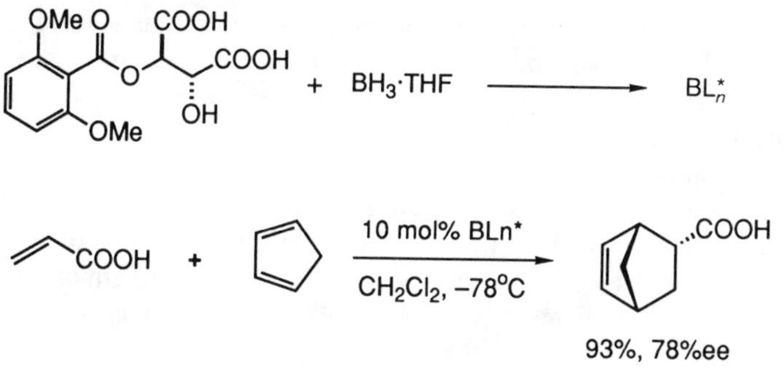

93%, 78%ee

In this process, the acid moiety of the (acyloxy)borane is activated by the electronegative nature of the trivalent boron atom. Conversely, however, the boron atom of the (acyloxy)borane itself should be activated by the electron-withdrawing acyloxy groups. Thus, it is worth considering that (acyloxy)borane derivatives may have enough Lewis acidity to catalyze certain reactions. In fact, the (acyloxy)borane used for the reaction in Scheme 5 was shown to be an excellent catalyst for the Diels–Adler reaction of unsaturated aldehydes (Scheme 6).

Scheme 6

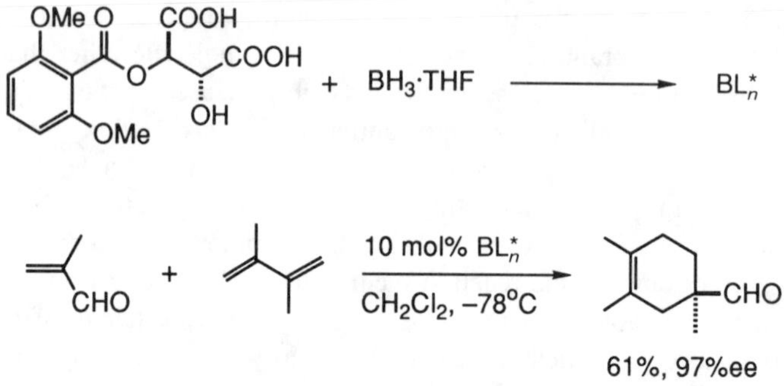

In a typical experiment, to a stirred suspension of monoacylated tartaric acid in dichloromethane was added BH_3–THF at 0°C. Methacrolein and cyclopentadiene were successively introduced to this catalyst solution at −78°C, and the mixture was stirred at the same temperature for 3 hours. The Diels–Adler adduct was isolated after the usual workup in 85% yield (endo/exo = 11:89): The major enantiomer was shown to have an R configuration with 96% ee [33].

This process is quite general and is applicable for the reaction of various dienes and aldehydes with high enantioselectivity. Some examples are listed in Chart 2, which reveals a striking feature of the process: namely, the α-substituent on the dienophile increases enantioselectivity (acrolein vs. methacrolein), while β-substitution dramatically decreases it (crotonaldehyde). In the case of a substrate having substituents at both α- and β-positions, high enantioselectivity was observed; thus the effect of the α-substituent overcomes that of the β-substituent. This trend apparently relates to the structure of the transition state of this reaction.

From a mechanistic standpoint, the actual structures of the chiral boron catalyst and its complex with a dienophile are of considerable interest. In a series of investigations using several kinds of tartaric acid derivative, we found that the boron atom might form a five-membered ring structure with the α-hydroxy acid moiety of tartaric acid, and the remaining carboxyl group might not bond to the boron atom (Fig. 1). Thus, in the reaction of **A** with BH_3–THF, the evolution of only 2.2–2.3 equiv. of hydrogen was observed at 0°C for 15 minutes. Furthermore, the tartaric acid derivative of type **B** revealed a comparable enantioselectivity to the catalyst derived from **A**.

The intramolecular Diels–Adler reaction of 2-methyl-(E,E)-2,7,9-decatrienal catalyzed by the CAB catalyst proceeds with the same high diastereo- and enantioselectivities [34] (eq. 22).

Chart 2 Diels–Alder reaction catalyzed by CAB.

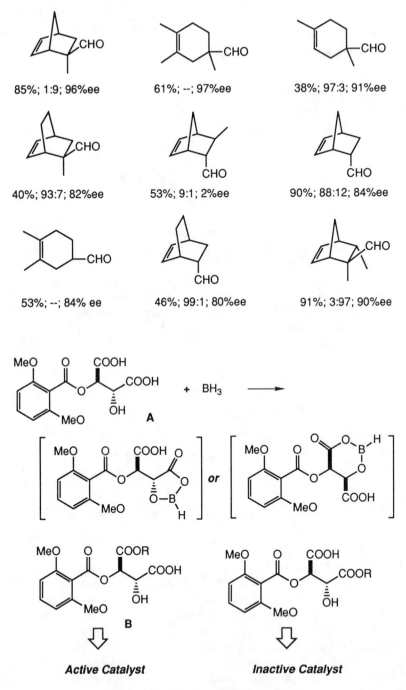

CHO
85%; 1:9; 96%ee
CHO
61%; --; 97%ee
CHO
38%; 97:3; 91%ee
CHO
40%; 93:7; 82%ee
CHO
53%; 9:1; 2%ee
CHO
90%; 88:12; 84%ee
CHO
53%; --; 84% ee
CHO
46%; 99:1; 80%ee
CHO
91%; 3:97; 90%ee

MeO
O
COOH
COOH
O
OH
MeO
A
+ BH3
or
B
Active Catalyst
Inactive Catalyst
COOR

Figure 1 Structure of CAB catalyst.

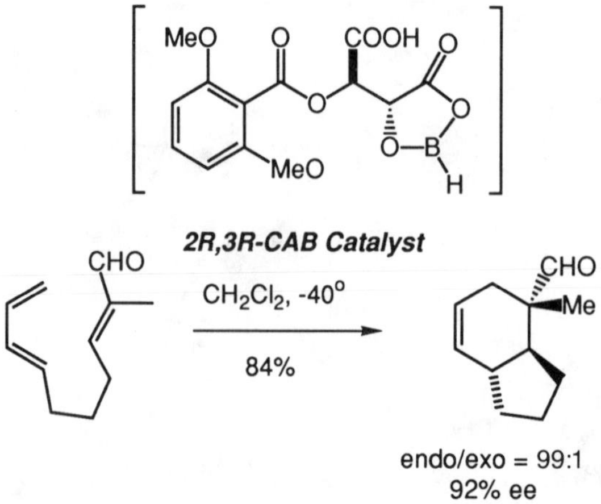

2R,3R-CAB Catalyst

$$CH_2Cl_2, -40°$$

84%

endo/exo = 99:1
92% ee

(22)

A chiral (acyloxy)borane (CAB) complex was shown to be an excellent catalyst for the Mukaiyama condensation of simple chiral enol silyl ethers of ketones with various aldehydes (Scheme 7). This CAB-catalyzed aldol process allows the adducts to be formed in a highly diastereo- and enantioselective manner (up to 96% ee) under mild reaction conditions (Chart 3). The reactions are catalytic: that is, 20 mol % of catalyst is sufficient for efficient conversion, and the chiral source is recoverable and reusable [35].

Scheme 7

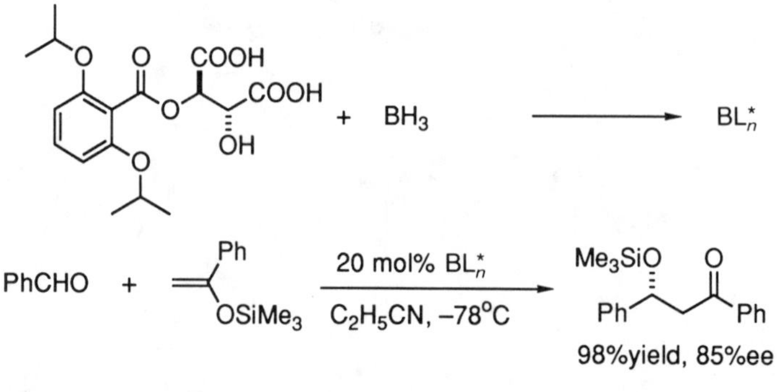

$$PhCHO + \quad \quad \quad \xrightarrow[\text{C}_2\text{H}_5\text{CN}, -78°\text{C}]{20 \text{ mol\% BL}_n^*} $$

98%yield, 85%ee

The relative stereochemistry of the major adducts was assigned as *erythro*, and predominant *re*-face attack of enol ethers at the aldehyde carbonyl carbon was confirmed in cases where a natural tartaric acid derivative was used as the Lewis acid ligand. Use of the unnatural form of tartaric acid as the chiral source afforded the other enantiomer, as expected. Almost perfect asymmetric inductions were achieved in the *erythro* adducts, reaching 96% ee, although a slight reduction in both enantio- and diastereoselectivities was observed in the reactions with saturated aldehydes. It is noteworthy that regardless of the stereochemistry of the starting enolsilyl ethers generated from diethyl ketone, *erythro*-aldols were obtained with high selectivity from these reactions (Scheme 8). These unprecedentedly high

Chart 3 Asymmetric aldol reaction catalyzed by chiral (acyloxy)borane (CAB) complex (e:t = erythro:threo).

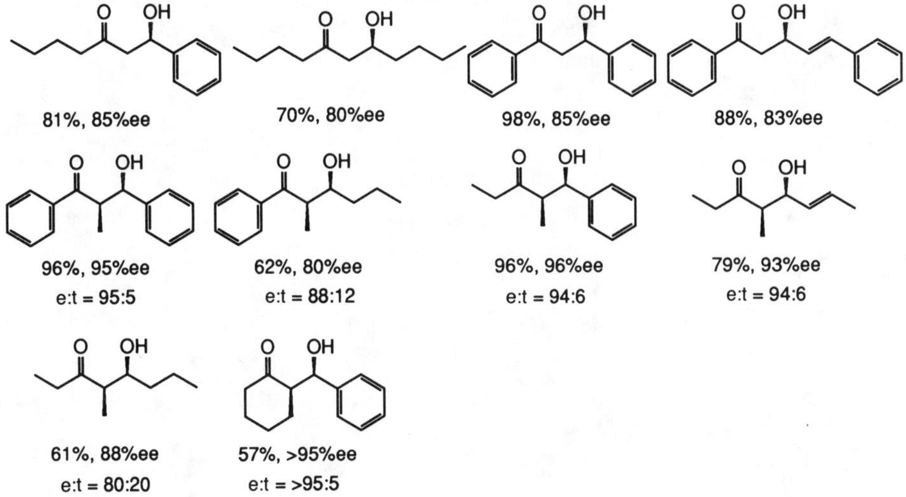

81%, 85%ee

70%, 80%ee

98%, 85%ee

88%, 83%ee

96%, 95%ee
e:t = 95:5

62%, 80%ee
e:t = 88:12

96%, 96%ee
e:t = 94:6

79%, 93%ee
e:t = 94:6

61%, 88%ee
e:t = 80:20

57%, >95%ee
e:t = >95:5

erythro selectivities, together with the lack of dependence on the stereochemistry of the silyl ethers in the CAB-catalyzed reactions, are fully consistent with Noyori's TMSOTf-catalyzed aldol reactions of acetals [36] and, thus, may reflect an acyclic extended transition state mechanism postulated for the latter reactions. Judging from the product configurations, the CAB catalyst (from natural tartaric acid) should effectively cover the *si* face of the carbonyl upon its coordination and thus the selective approach of nucleophiles from the *re face* should result. This mechanism is consistent and in good agreement with the results of previously described CAB-

Scheme 8

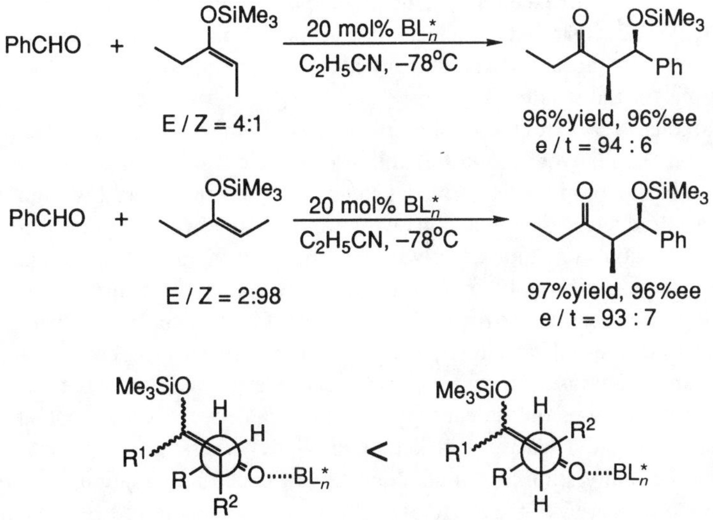

PhCHO +   [OSiMe3, E/Z = 4:1]   20 mol% BL_n^* / C_2H_5CN, −78°C   →   [product]   96%yield, 96%ee   e/t = 94 : 6

PhCHO +   [OSiMe3, E/Z = 2:98]   20 mol% BL_n^* / C_2H_5CN, −78°C   →   [product]   97%yield, 96%ee   e/t = 93 : 7

catalyzed Diels–Alder reactions. It therefore follows that the sense of asymmetric induction of CAB-catalyzed reactions is the same for all the aldehydes examined.

A catalytic asymmetric aldol-type reaction of ketene silyl acetals with achiral aldehydes also proceeded smoothly with the CAB catalyst to furnish *erythro*-β-hydroxy esters with high enantiomeric purities [37] (eq. 23).

$$\text{RCHO} + \underset{R^1}{\overset{OR^2}{\diagup}}\text{OSiMe}_3 \xrightarrow[\text{C}_2\text{H}_5\text{CN}]{20 \text{ mol\% BL}_n^*} R\underset{R^1}{\overset{OH}{\diagup}}\text{COOR}^2 \qquad (23)$$

63%; 84%ee

49%; 76%ee

83%; 79:21; 92%ee

57%;79:21; 88%ee

97%; 96:4; 97%ee

86%; 95:5; 94%ee

The sensitivity of this reaction to the substituents of the starting ketene acetals is remarkable. The reactions of ketene silyl acetals derived from more common ethyl esters were totally stereorandom, giving equivalent amounts of *erythro* and *threo* isomers with moderate enantioselectivities. Benzyl esters also exhibited similar results but in somewhat improved chemical yields. In sharp contrast with this, the use of ketene silyl acetals generated from phenyl esters led to good diastereo- and enantioselectivities in excellent chemical yields. The reason for the observed substituent effect is not clear, but the undesirable secondary interactions between electron-rich ketene silyl acetals derived from alkyl esters and the Lewis acid may be responsible for the results [37].

Analogous to the previous results of enol silyl ethers of ketones, unsubstituted ketene silyl acetals were found to exhibit lower levels of stereoregulation. On the other hand, the propionate derived ketene silyl acetals showed a high level of asymmetric induction. The reactions with aliphatic aldehydes, however, resulted in a slight reduction in enantioselectivity and chemical yields. With the phenyl ester derived ketene silyl acetals *erythro* adducts predominate, but the selectivities are moderate in most cases relative to the reactions of ketone enol silyl ethers. Notable exceptions are the α,β-unsaturated aldehydes, which have revealed excellent diastereo- and enantioselectivities. The observed erythro selectivities and *re*-face attack of nucleophiles on the carbonyl carbon of aldehydes are consistent with the aforementioned aldol reactions of ketone enol silyl ethers.

Asymmetric allylation is a valuable means of constructing chiral functionalized structures, and therefore many chiral allyl–metal reagents have been designed and synthesized [38]. Although some of them have exhibited good to excellent enantio- and diastereoselectivities in reactions with achiral aldehydes, there is no method yet available for a catalytic process.

The CAB catalyst was found to be a powerful activator for the Sakurai–Hosomi allylation of aldehydes to furnish homoallylic alcohols with excellent enantiomeric excess [39] (Scheme 9).

Scheme 9

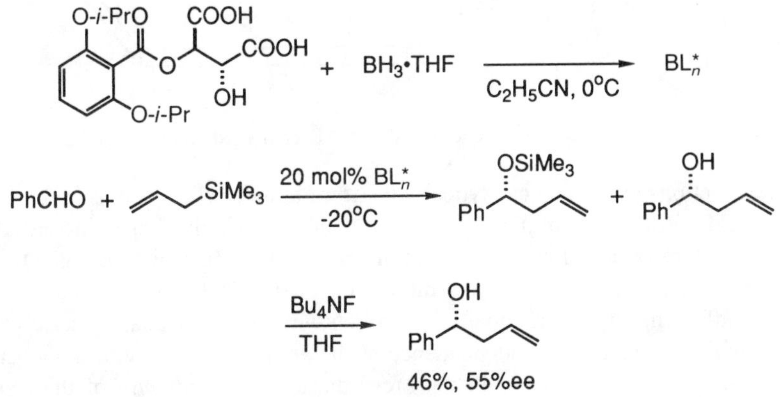
O-i-PrO
COOH
COOH
OH
O-i-Pr
+ BH3•THF
C2H5CN, 0°C
BLn*
PhCHO +
SiMe3
20 mol% BLn*
-20°C
OSiMe3
Ph
+
OH
Ph
Bu4NF
THF
OH
Ph
46%, 55%ee

A catalyst solution was prepared by reacting borane-THF complex with mono(2,6-diisopropoxy)benzoyltartaric acid in dry propionitrile at 0°C. Reaction of the achiral aldehydes with allysilanes was promoted by this catalyst (20 mol %) at −78°C to produce homoallylic alcohols with the enantioselectivities shown in Scheme 10.

Scheme 10

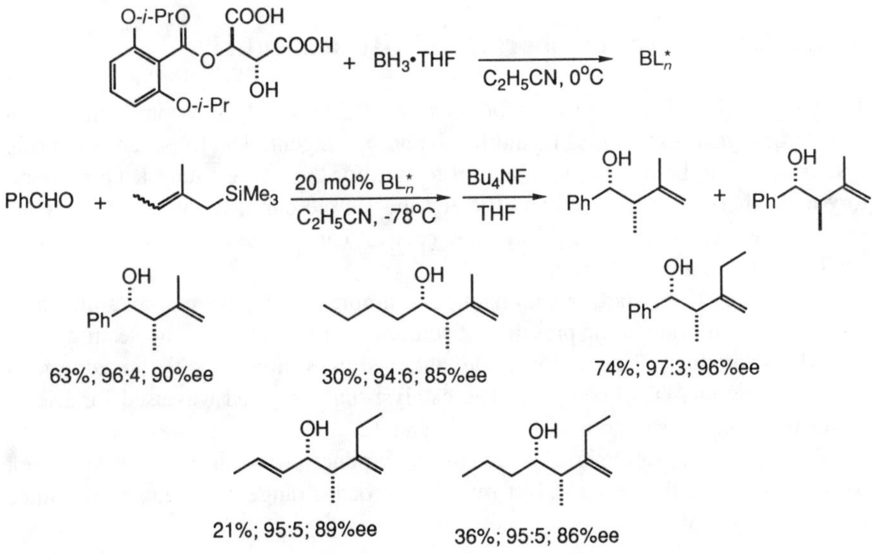
O-i-PrO
COOH
COOH
OH
O-i-Pr
+ BH3•THF
C2H5CN, 0°C
BLn*
PhCHO +
SiMe3
20 mol% BLn*
C2H5CN, -78°C
Bu4NF
THF
OH
Ph
+
OH
Ph
OH
Ph
63%; 96:4; 90%ee
OH
30%; 94:6; 85%ee
OH
Ph
74%; 97:3; 96%ee
OH
21%; 95:5; 89%ee
OH
36%; 95:5; 86%ee

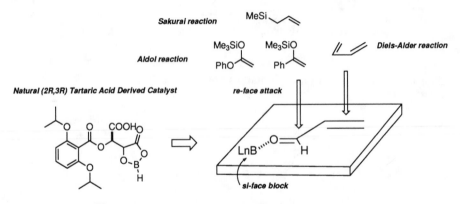

Figure 2 Stereochemistry of CAB-catalyzed reactions.

The reactions proceeded catalytically to afford homoallylic alcohols in modest to good yields. Alkyl substitution at the olefin moiety of allylsilanes increased the reactivity, permitting a lower reaction temperature with improved asymmetric induction. γ-Alkylated allylsilanes exhibited excellent diastereo- and enantioselectivities, affording *erythro*-homoallylic alcohols with a higher enantiomeric purity. Of particular interest is the independence of the *erythro* selectivity of the reactions on allylsilane stereochemistry: that is, regardless of the geometry of the starting allysilanes, the predominant isomer in this reaction has the *erythro* configuration. The observed preference for both relative and absolute configurations for the adduct alcohols from the (2R,3R)-ligand–borane reagent was predicted on the basis of an extended transition state model similar to that for the CAB catalyzed aldol reaction.

Summarizing these three different carbon–carbon bond formation processes, the CAB catalyst derived from natural tartaric acid can effectively cover the *re* face of carbonyl and the reagent attacks from the *si* face [40] (Fig. 2).

9.4.2. Chiral (Acyloxy)borane (CAB) Reagent II

The characteristic feature of the aforementioned CAB catalyst system is the use of an α-hydroxycarboxylic acid ligand for the boron reagent. The five-membered ring system seems to be the major structural feature for the active catalyst. Our laboratory and others have reported an efficient and simple chiral boron catalyst for the Diels–Alder reaction [41,42]. This new catalyst can be prepared from the readily available sulfonamide of amino acids.

Preparation of the starting sulfonamide is simple: the exposure of an amino acid to a sulfonyl chloride in the presence of sodium hydroxide affords the desired white crystalline product. The sulfonamide obtained was treated with an equimolar amount of borane–THF complex. The catalyst thus obtained was used for Diels–Alder reactions as shown in Schemes 11 and 12.

Although the enantiomeric excess of the products from these reactions is not particularly high, the new catalyst reveals a broader range of applicability: Since

Scheme 11

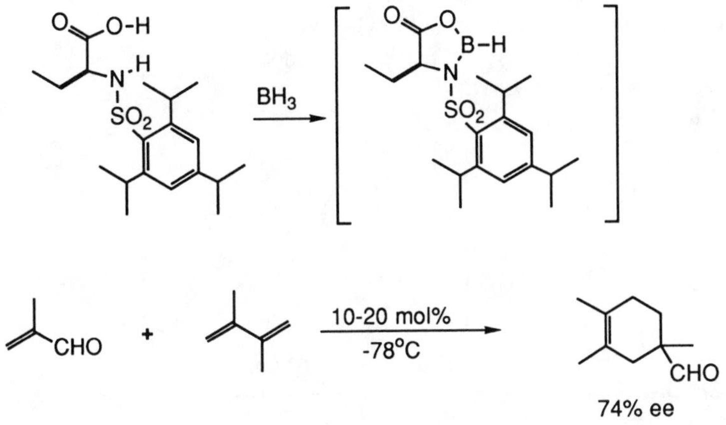

74% ee

Scheme 12

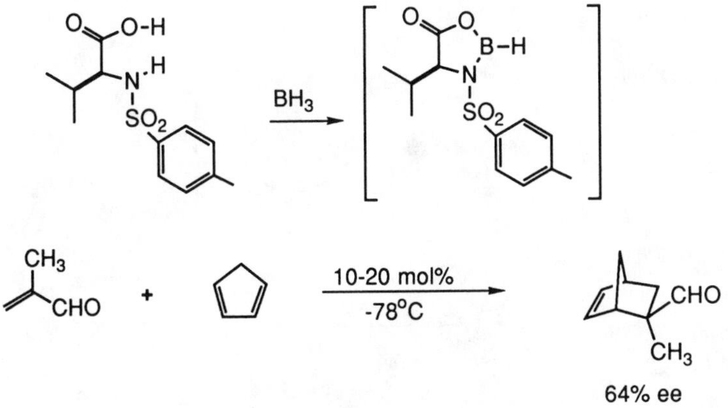

64% ee

(24)

PhCH₂CH₂CHO +

R = Me: 96% ee
R = H: 62% ee

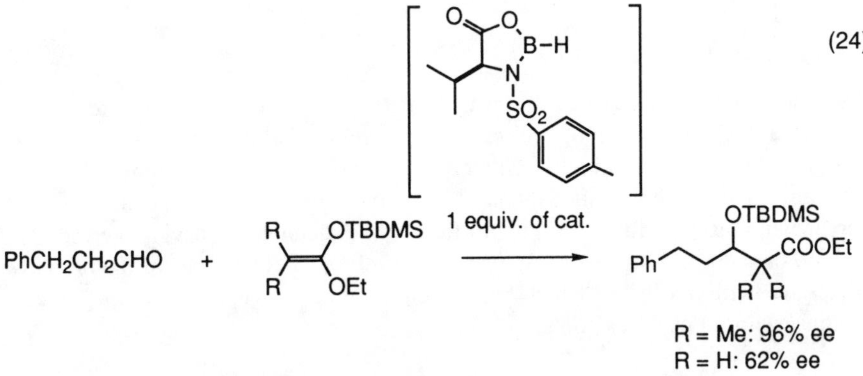

D-amino acids are commercially available, the other enantiomer can also be easily prepared. Furthermore, enantioselectivity of the reaction increases with the increased bulkiness of the benzenesulfonyl group.

Recently, the same catalyst was utilized in aldol [43] (eqs. 24 and 25) and Diels–Adler reactions [44] (eq. 26)

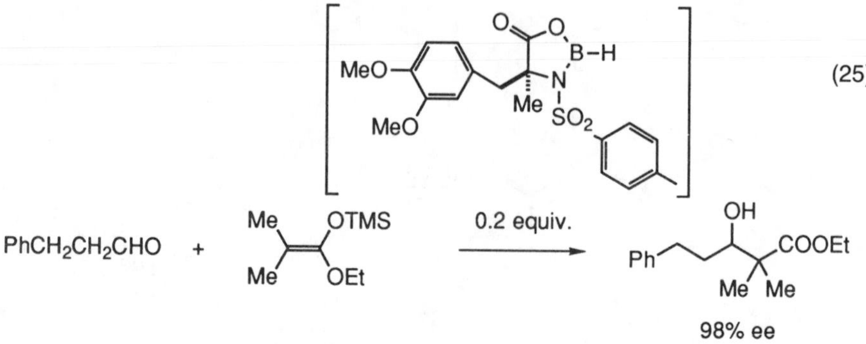

$$\text{(25)}$$

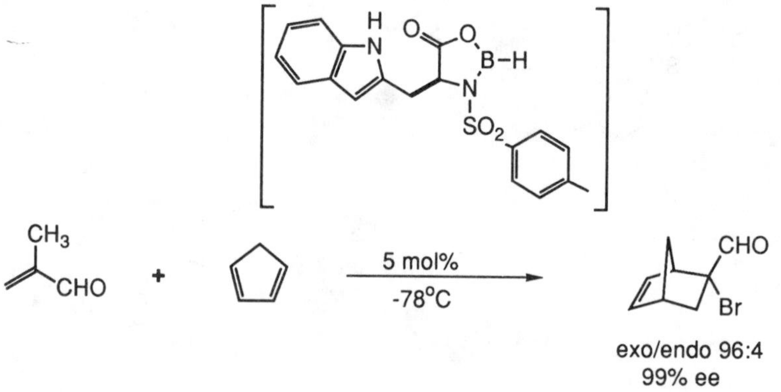

98% ee

$$\text{(26)}$$

exo/endo 96:4
99% ee

9.4.3. Boron Catalysts of Other Types

Alkyldichloroborane (Scheme 13) was prepared by hydroboration and resolution [45]. Diels–Alder reaction of cyclopentadiene and methyl crotonate with this catalyst gave the adduct with high asymmetric induction. The reaction proceeded with the conformation shown in Scheme 13. In this conformation, the edge of the naphthalene moiety blocks the bottom face of the dienophile, leaving the top face open to the approach of dienes. In fact, the absolute configurations of the products consistent with those predicted by the conformational analysis of the methyl crotonate–boron complex (Scheme 13).

Scheme 13

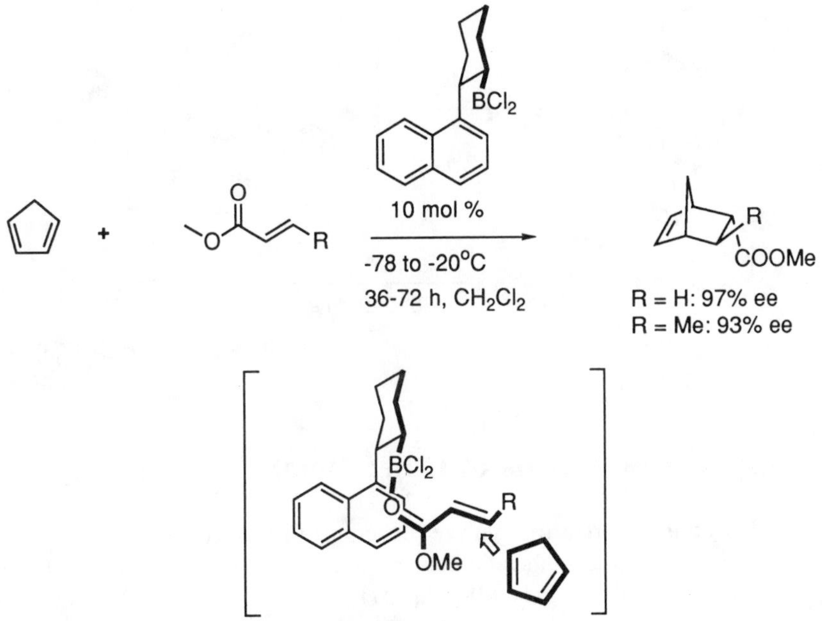

R = H: 97% ee
R = Me: 93% ee

The boron reagent derived from a chiral prolinol derivative gave a Diels–Alder adduct with good to excellent asymmetric induction [46] (Scheme 14).

Scheme 14

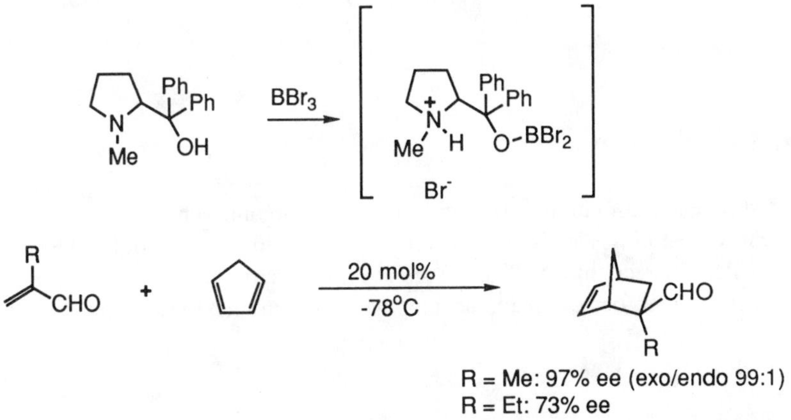

R = Me: 97% ee (exo/endo 99:1)
R = Et: 73% ee

An interesting C_3 symmetric tetradecacyclic diborate was prepared and the structure was determined by X-ray analysis. The catalyst proved to be useful for the Diels–Alder reaction [47] (Scheme 15).

Scheme 15

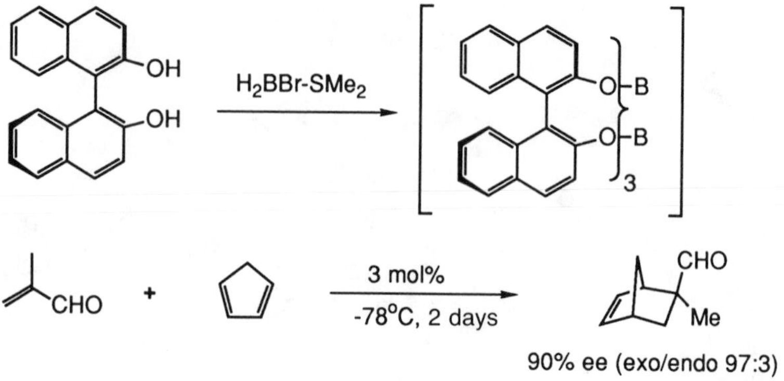

90% ee (exo/endo 97:3)

9.5. Chiral Lewis Acids of Other Metals

Mukaiyama et al. reported tin-catalyzed reactions that proceeded with great efficiency. Their system is a combination of several reagents, the most important being the structure of the chiral ligand [48] (eq. 27).

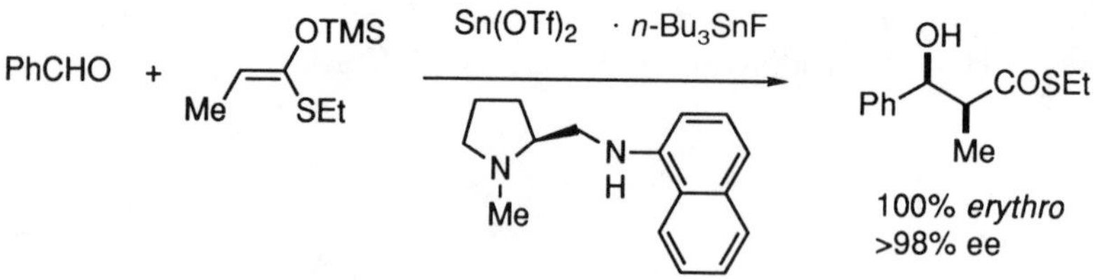

100% *erythro*
>98% ee (27)

9.6. Conclusion

The vast potential for chiral Lewis acid catalysis in organic synthesis is increasingly apparent. Chiral Lewis acids are probably an equally important tool for the chemistry of molecular recognition. Thus, it should not be surprising if research on chiral Lewis acids undergoes extensive development over the next few years.

References

1. Narasaka, K. *Synthesis*, **1991**, *1*, and references cited therein.

2. Hashimoto, S.-I.; Komeshima, N.; Koga, K. *J. Chem. Soc., Chem. Commun.* **1979**, 437.

3. Takemura, H.; Komeshima, N.; Takahashi, I.; Hashimoto, S.-I.; Ikota, N.; Tomioka, K.; Koga, K. *Tetrahedron Lett.* **1987**, *28*, 5687.

4. Bednarski, M.; Maring, C.; Danishefsky, *Tetrahedron Lett.* **1983**, *24*, 3451.

5. Bednarski, M.; Danishefsky, *J. Am. Chem. Soc.* **1983,** *105,* 6968.

6. Sakane, S.; Maruoka, K.; Yamamoto, H. *Tetrahedron Lett.,* **1985,** *26,* 5535.

7. Reviews: (a) Mole, T.; Jeffrey, E. A. *Organoaluminum Compounds;* Elsevier: Amsterdam, 1972. (b) Bruno, G. *The Use of Aluminum Alkyls in Organic Synthesis;* Ethyl Corporation: Baton Rouge, LA, 1970, 1973, and 1980. (c) Maruoka, K.; Yamamoto, H. *Angew. Chem., Int. Ed. Engl.* **1985,** *24,* 668.

8. Chapius, C.; Jurczak, J. *Helv. Chim. Acta,* **1987,** *70,* 436.

9. Reetz, M. T.; Kyung, S.-H.; Bolm, C.; Zierke, T. *Chem. Ind.* **1986,** 824.

10. (a) Maruoka, K.; Itoh, T.; Yamamoto, H. *J. Am. Chem. Soc.* **1985,** *107,* 4573. (b) Maruoka, K.; Itoh, T.; Sakurai, M.; Nonoshita, K.; Yamamoto, H. *J. Am. Chem. Soc.* **1988,** *110,* 3588.

11. Maruoka, K.; Itoh, T.; Araki, Y.; Shirasaka, T.; Yamamoto, H. *Bull. Chem. Soc. Jpn.* **1988,** *61,* 2975.

12. Maruoka, K.; Itoh, T.; Shirasaka, T.; Yamamoto, H. *J. Am. Chem. Soc.* **1988,** *110,* 310.

13. Maruoka, K.; Yamamoto, H. *J. Am. Chem. Soc.* **1989,** *111,* 789.

14. Maruoka, K.; Hoshino, Y.; Shirasaka, T.; Yamamoto, H. *Tetrahedron Lett.* **1988,** *29,* 3967.

15. (a) Maruoka, K.; Banno, H.; Yamamoto, H. *J. Am. Chem. Soc.* **1990,** *112,* 7791. (b) Maruoka, K.; Banno, H.; Yamamoto, H. *Tetrahedron: Asymmetry,* **1991,** *2,* 647.

16. Maruoka, K.; Yamamoto, H. *Synlett,* **1991,** 793.

17. Corey, E. J.; Imwinkelried; Pikul, S.; Xiang, Y. B. *J. Am. Chem. Soc.* **1989,** *111,* 5493.

18. (a) Naruse, Y.; Esaki, T.; Yamamoto, H. *Tetrahedron Lett.* **1988,** *29,* 1417. (b) Naruse, Y.; Esaki, T.; Yamamoto, H. *Tetrahedron,* **1988,** *44,* 4747.

19. (a) Haubenstock, H.; Panchalingam, V.; Odian, G. *Makromol. Chem.* **1987,** *188,* 2789. (b) Kasperczyk, J.; Dworak, A.; Jedlinski, Z. *Makromol. Chem., Rapid Commun.* **1981,** *2,* 663.

20. (a) Mikami, K.; Terada, M.; Nakai, T. *J. Am. Chem. Soc.* **1989,** *111,* 1940. (b) Mikami, K.; Terada, M.; Nakai, T. *J. Am. Chem. Soc.* **1990,** *112,* 4949. (c) Mikami, K.; Shimizu, M.; Nakai, T. *J. Org. Chem.* **1991,** *56,* 2952. (d) Mikami, K.; Terada, M.; Sawa, E.; Nakai, T. *Tetrahedron Lett.* **1991,** *32,* 6571. For asymmetric cycloadditions using this chiral chiral titanium reagent, see (e) Mikami, K.; Terada, M.; Motoyama, Y.; Nakai, T. *Tetrahedron: Asymmetry,* **1991,** *2,* 643. (f) Terada, M.; Mikami, K.; Nakai, T. *Tetrahedron Lett.* **1991,** *32,* 935.

21. Seebach, D.; Beck, A. K.; Imwinkelried, R.; Roggo, S.; Wonnacott, A. *Helv. Chim. Acta,* **1987,** *70,* 954.

22. (a) Narasaka, K.; Inoue, M.; Okada, N. *Chem. Lett.* **1986,** 1109. (b) Narasaka, K.; Inoue, M.; Yamada, T. *Chem. Lett.* **1986,** 1967. (c) Narasaka, K.; Inoue, M.; Yamada, T.; Sugimori, J.; Iwasawa, N. *Chem. Lett.* **1987,** 2409. (d) Iwasawa, N.; Hayashi, Y.; Sakurai, H.; Narasaka, K. *Chem. Lett.* **1989,** 1581. (e) Narasaka, K.; Iwasawa, N.; Inoue, M.; Yamada, T.; Nakashima, M.; Sugumori, J. *J. Am. Chem. Soc.* **1989,** *111,* 5340.

23. Hayashi, Y.; Narasaka, K. *Chem. Lett.* **1989,** 793. (b) Ichikawa, Y.; Narita, A.; Shiozawa, A.; Hayashi, Y.; Narasaka, K. *J. Chem. Soc., Chem. Commun.* **1989,** 1919.

24. Narasaka, K.; Hayashi, Y.; Shimada, S. *Chem. Lett.* **1988,** 1609.

25. (a) Narasaka, K.; Yamada, T.; Minamikawa, H. *Chem. Lett.* **1987,** 2073. (b) Minamikawa, H.; Hayakawa, S.; Yamada, T.; Iwasawa, N.; Narasaka, K. *Bull. Chem. Soc. Jpn.* **1988,** *61,* 4379.

26. Narasaka, K.; Kanai, F.; Okudo, M.; Miyoshi, N. *Chem. Lett.* **1989,** 1187.

27. Engler, T. A.; Letavic, M. A.; Reddy, J. P. *J. Am. Chem. Soc.* **1991,** *113,* 5068.

28. Corey, E. J.; Matsumura, Y. *Tetrahedron Lett.* **1991**, *32*, 6289.

29. (a) Yoshioka, M.; Kawakita, T.; Ohno, M. *Tetrahedron Lett.* **1989**, *30*, 1657. (b) Yoshioka, H.; Kawakita, T.; Yoshioka, M.; Kobayashi, S.; Ohno, M. *Tetrahedron Lett.* **1989**, *30*, 7095.

30. (a) Seebach, D.; Behrendt, L.; Felix, D. *Angew. Chem., Int. Ed. Engl.* **1991**, *30*, 1008. (b) Schmidt, B.; Seebach, D. *Angew. Chem., Int. Ed. Engl.* **1991**, *30*, 1321.

31. Mori, A.; Nitta, H.; Kudo, M.; Inoue, S. *Tetrahedron Lett.* **1991**, *32*, 4333.

32. Furuta, K.; Miwa, Y.; Iwanaga, K.; Yamamoto, H. *J. Am. Chem. Soc.* **1988**, *110*, 6254.

33. Furuta, K.; Shimizu, S.; Miwa, Y., Yamamoto, H. *J. Org. Chem.* **1989**, *54*, 1481.

34. Furuta, K.; Kanematsu, A.; Yamamoto, H. *Tetrahedron Lett.* **1989**, *30*, 7231.

35. Furuta, K.; Maruyama, T., Yamamoto, H. *J. Am. Chem. Soc.* **1991**, *113*, 1041.

36. Noyori, R.; Murata, S.; Suzuki, M. *Tetrahedron*, **1981**, *37*, 3899.

37. Furuta, K.; Maruyama, T.; Yamamoto, H. *Synlett*, **1991**, 439.

38. For example, see: (a) Brown, H. C.; Randad, R. S.; Bhat, K. S.; Zaidlewicz, M. Racherla, U. S. *J. Am. Chem. Soc.* **1990**, *112*, 2389. (b) Roush, W. R.; Hoong, L. K.; Palmer, M. A. J.; Park, J. C. *J. Org. Chem.* **1990**, *55*, 4109, and references cited therein.

39. Furuta, K.; Mouri, M.; Yamamoto, H. *Synlett*, **1991**, 561.

40. The CAB catalyzed Diels–Alder reaction thus should proceed via anti-*s*-trans configuration of Lewis acid–acrolein complex, and this is inconsistent with reports by Houk et al. using ab initio calculations: Loncharich, R. J.; Schwartz, T. R.; Houk, K. N. *J. Am. Chem. Soc.* **1987**, *109*, 14.

41. Takasu, M.; Yamamoto, H. *Synlett*, **1990**, 194.

42. (a) Sartor, D.; Saffrich, J.; Helmchen, G. *Synlett*, **1990**, 197. (b) Sartor, D.; Saffrich, J.; Helmchen, G.; Richards, C. J.; Lambert, H. *Tetrahedron: Asymmetry*, **1991**, *2*, 639.

43. (a) Kiyooka, S.; Kaneko, Y.; Komura, M.; Matsuo, H.; Nakano, M. *J. Org. Chem.* **1991**, *56*, 2276. (b) Parmee, E. R.; Tempkin, O.; Masamune, S. *J. Am. Chem. Soc.* **1991**, *113*, 9365.

44. Corey, E. J.; Loh, T.-P. *J. Am. Chem. Soc.* **1991**, *113*, 8966.

45. Hawkins, J. M.; Loren, S. *J. Am. Chem. Soc.* **1991**, *113*, 7794.

46. Kobayashi, S.; Murakami, M.; Harada, T.; Mukaiyama, T. *Chem. Lett.* **1991**, 1341.

47. Kaufmann, D.; Boese, R. *Angew. Chem., Int. Ed. Engl.* **1990**, *29*, 545.

48. Mukaiyama, T.; Uchiro, H.; Kobayashi, S. *Chem. Lett.* **1989**, 1001.

Epilogue

After reading through all the chapters, readers have surely acquired an updated understanding of *catalytic asymmetric synthesis* achieved with man-made chiral catalysts. It is evident that those man-made chiral catalysts, bearing much simpler and much smaller ligands than the proteins of naturally occurring enzymes, can efficiently create molecules with extremely high enantiomeric purities. The chiral catalysts have a beautiful C_2 symmetry in some cases and in other cases a fascinating dissymmetry. It is breath-taking to realize that such simple and beautiful small molecules can compete, practically and efficiently, with highly sophisticated enzymes that nature has created. This is a great encouragement for synthetic organic chemists to continue and to expand their efforts to design and develop highly efficient chiral catalysts. The current situation and future direction of the "chiral drugs" arena is spawning a new technology called "chirotechnology," and "chirotechnology" industries may well emerge in the near future in a manner similar to the biotechnology industries. It is obvious that *catalytic asymmetric synthesis* will take a central role in these developments.

From a practical point of view, the impacts of Sharpless oxidation, Noyori–Takaya's second-generation asymmetric hydrogenation, Sharpless dihydroxylation, and Jacobsen's epoxidation have been tremendous in recent years. Sharpless dihydroxylation and Jacobsen's epoxidation were licensed to Sepracor, one of the newly spawned "chirotechnology" firms, forming the basis of a supply of new enantiomeric intermediates. I should emphasize that any of the catalytic asymmetric synthesis discussed in this book could become an industrial process not only in "chirotechnology" industries, but also in the pharmaceutical, chemical and agricultural industries when its development reaches a certain efficiency. Asymmetric isomerization was successfully applied to the large-scale commercial synthesis of

441

l-menthol by using the "Takasago process," and other applications of this type of asymmetric catalytic process will emerge in the future. Many promising results have accumulated for asymmetric cyclopropanations, from which many useful applications can be envisioned. This is especially true in the light of the successful commercial asymmetric synthesis of a silastatin component using an Aratani catalyst ("Sumitomo process"). Asymmetric carbon–carbon bond forming reactions have already been widely used in research laboratories, and it is only a matter of time before some of these reactions become important commercial processes, especially in "chirotechnology" industries. Asymmetric syntheses catalyzed by chiral Lewis acids have rapidly gained significant attention in the synthetic community, and these processes have a very high potential for commercial applications. Asymmetric hydrosilylation has a somewhat limited scope in commercial applications, but certain highly enantioselective reactions cannot be achieved by other methods and, more important, this "chirotechnology" may well be applicable for the production of silicon-based chiral "new materials." Asymmetric carbonylation is still a great challenge for synthetic chemists, but there are encouraging results that surely provide hints for the development of practical processes that will have a significant impact on "chirotechnology." Asymmetric phase transfer reactions are in an early stage of development, but these reactions obviously have great potential as commercial processes. I would like to remind readers that most of the inventions described in *Catalytic Asymmetric Synthesis* were made in academic institutions and were immediately secured by industries for development as commercial processes. Accordingly, it is evident that creative basic research in universities is crucial for the birth and growth of innovative "chirotechnology."

It is noteworthy that catalytic asymmetric synthesis has made significant advances in terms of substrate–catalyst interactions and the chiral recognition of substrate structures. Knowles's epoch-making "Monsanto process," established in early 1970s for the asymmetric synthesis of L-DOPA, was based on the interactions between the multifunctionalized substrate *N*-acetyldehydroamino acid and a chiral diphosphine–rhodium catalyst. However, second-generation asymmetric hydrogenation based on chiral diphosphine–ruthenium catalysts can now be applied to simple acrylic acids with exceptionally high enantioselectivity, as demonstrated for the asymmetric synthesis of (*S*)-naproxen, a potent anti-inflammatory drug. Sharpless oxidation of allylic alcohols, extensively developed in the 1980s, required hydroxyl functionality to anchor the substrates to a chiral titanium catalyst. In sharp contrast with this, the recently developed Sharpless dihydroxylation and Jacobsen's epoxidation work extremely well with *unfunctionalized* olefins. Synthetic organic chemists should now have confidence in the rational design of a chiral catalyst that brings about extremely high enantioselectivity without the assistance of a huge protein backbone.

I would like to emphasize the importance of detailed mechanistic studies for accurate understanding of asymmetric induction steps as well as key catalytic species, since these studies have been and will be essential for breakthroughs in *catalytic asymmetric synthesis*. Although this book does not include asymmetric organozinc reactions catalyzed by chiral amino alcohols, I should point out that a very intriguing "asymmetric amplification" phenomenon was observed in this reaction. This finding indicates the extremely interesting possibility of developing catalytic asymmetric processes in which enantiomerically pure catalysts are *not* necessary. For example, even when the enantiomeric purity of an (*S*)-alkylzinc complex is 30% (i.e., *S*/*R* = 65:35), the (*S*,*S*)-dimer (30% concentration) can promote the reaction as if it were the sole active reagent, since the (*R*,*S*)-isomer (70% concentration) is virtually inactive under the reaction conditions. Although this phenomenon does not have generality, it certainly warrants further active investigations. When only low enantioselectivities are observed using man-made chiral catalysts, synthetic organic chemists tend to think that their design is insufficient and the system is too simple. However, detailed mechanistic study has shown that in many cases it is not the efficacy of the designed catalyst system, but other coexisting catalyst species in the system, that are responsible for the poor results. Once the well-designed catalyst species has been generated selectively or coaxed to work selectively, excellent results are obtained. This is a very important message to synthetic organic chemists.

Although I put emphasis on the remarkable efficacy of man-made chiral catalysts, it is not my intention to undermine the use of enzymatic and biological methods to produce enantiomerically pure compounds. Enzymes are surely excellent chiral catalysts for asymmetric organic transformations. The catalytic asymmetric organic reactions complied in this book and those promoted by enzymes or microorganisms will complement each other and both are essential for the development of "chirotechnology."

I sincerely hope that this book attracts the interests of a broad range of synthetic organic and medicinal chemists, especially among the younger generation in both academia and industry, so that many talented young chemists will introduce creative ideas into this fascinating area of research, ones that will promote significant future advances in *catalytic asymmetric synthesis*.

Finally, I thank Professor Scott M. Sieburth, my colleague at Department of Chemistry, State University of New York at Stony Brook, for his valuable advice on editing this book.

Iwao Ojima
March 1993

Appendix

List of Chiral Ligands

The list of chiral ligands is provided in this Appendix. Typical chiral ligands are summarized for each chapter, and references given are taken from the corresponding References section of the chapter.

Chapter 1

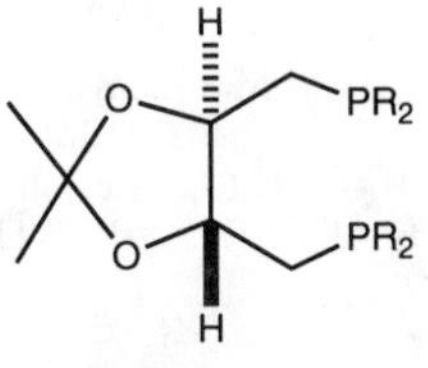

(**R,R**)-**DIOP**: R = Ph [6]
(**R,R**)-**MODDIOP**: R = 2,5-dimethyl-4-methoxyphenyl [86a]

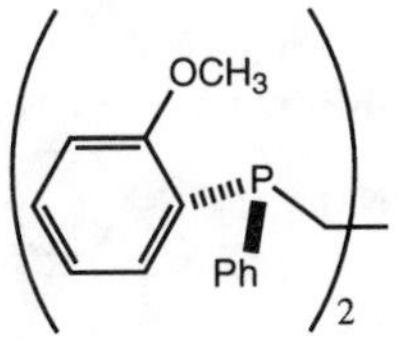

(**R,R**)-**DIPAMP** [7b]

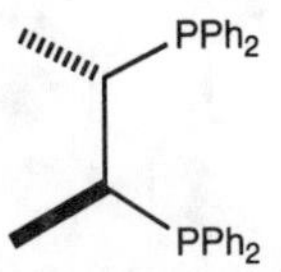

(**S,S**)-**CHIRAPHOS** [10]

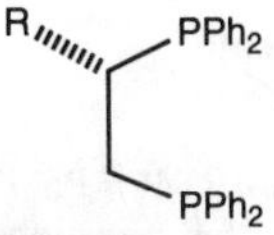

(**S**)-**PROPHOS**: R = Me [11]
(**S**)-**PHENPHOS**: R = Ph [12]
(**S**)-**CYCPHOS**: R = cyclohexyl [13a]

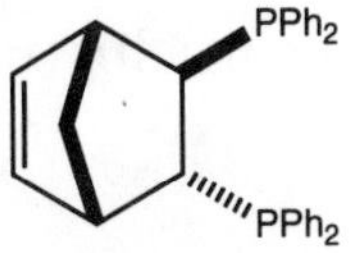

(**R,R**)-**NORPHOS** [14]

(*Continued*)

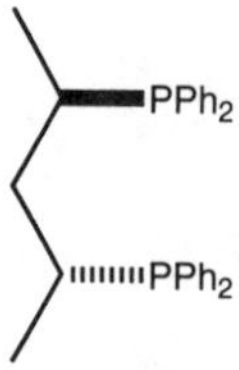

(R,R)-BDPP [15]

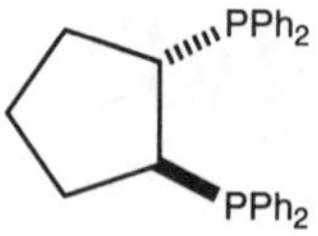

(S,S)-DPCP [16]

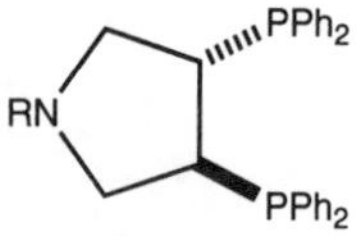

(S,S)-R-PYRPHOS [17]

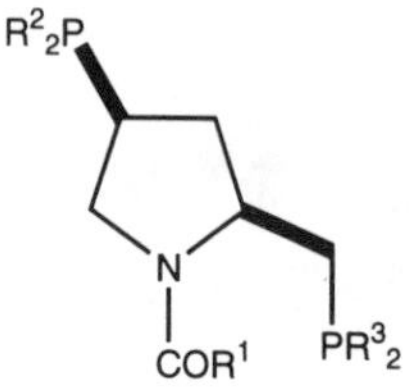

(S,S)-BPPM: R^1 = O-t-Bu, R^2 = R^3 = Ph [18]
(S,S)-Ph-CAPP: R^1 = NHR, R^2 = R^3 = Ph [19]
(S,S)-MCCPM: R^1 = NHMe, R^2 = cyclohexyl, R^3 = Ph [126]
(S,S)-BCPM: R^1 = O-t-Bu, R^2 = cyclohexyl, R^3 = Ph [125b]
(S,S)-PCCPM: R^1 = NHPh, R^2 = cyclohexyl, R^3 = Ph [126]
(S,S)-MCPM: R^1 = OMe, R^2 = cyclohexyl, R^3 = Ph [126]
(S,S)-PCPM: R^1 = OPh, R^2 = cyclohexyl, R^3 = Ph [126]
(S,S)-MCCXM: R^1 = NHMe, R^2 = cyclohexyl, R^3 = $C_6H_3(3,5-Me_2)$ [120]

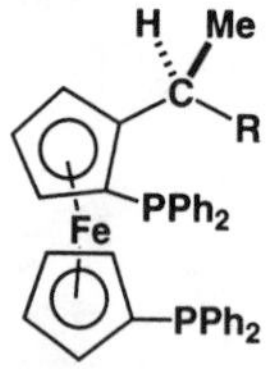

(R,S)-BPPFA: R = NMe_2 [9b]
(R,S)-BPPFOH: R = OH [9b]

R = $NMeCH_2CH_2N$⟨ ⟩ [91a]

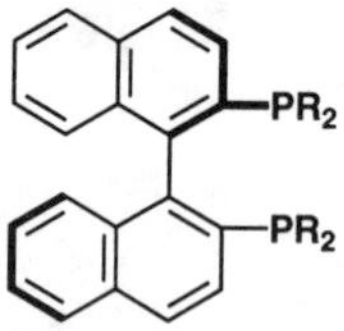

(R)-BINAP: R = Ph [20]
(R)-TolBINAP: R = C_6H_4(4-Me) [20c]
(R)-m-TolBINAP: R = C_6H_4(3-Me) [130]
(R)-m-XylylBINAP: R = $C_6H_3(3,5-Me_2)$ [130]
(R)-3,5-Bu$_2$BINAP: R = $C_6H_3(3,5-(t-Bu)_2)$ [130]

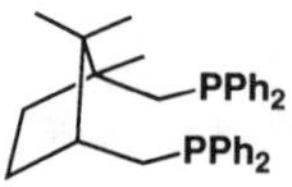

CAMPHOS [21]

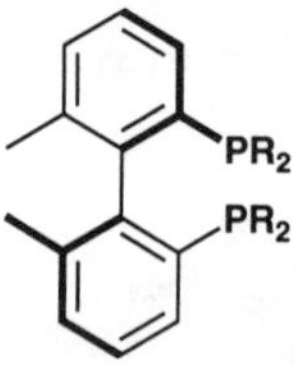

(R)-BIPHEMP: R = Ph [22]
(R)-BICHEP: R = cyclohexyl [23]

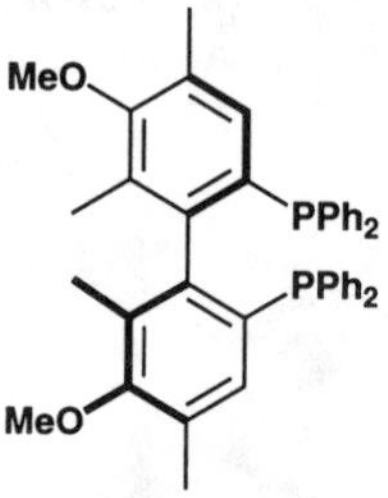

(R)-BIMOP [111]

(*Continued*)

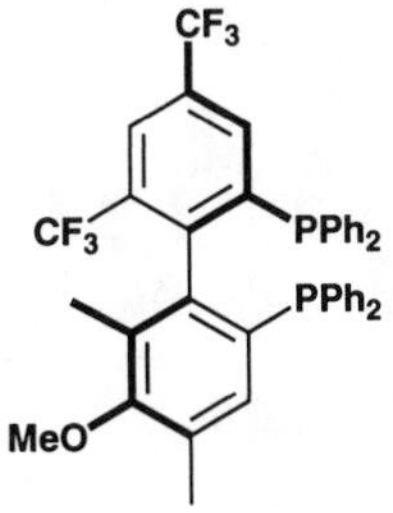

(*R*)-FUPMOP [111]

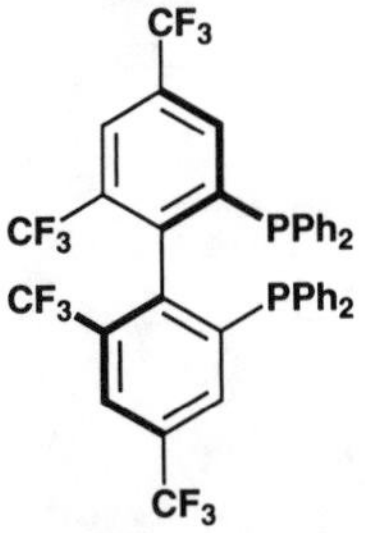

(*R*)-BIFUP [111]

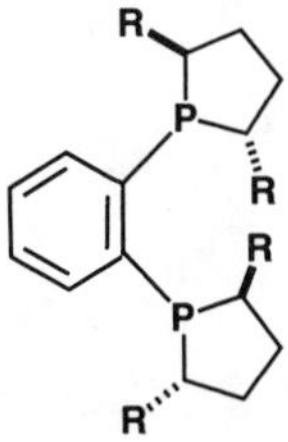

Me-DuPHOS: R = Me [24]
Et-DuPHOS: R = Et [24]

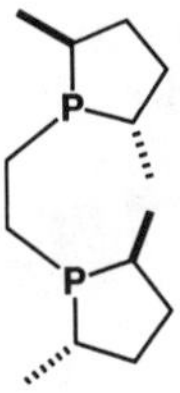

[24]

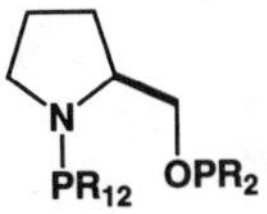

(*S*)-**ProNOP**: $R^1 = R^2 = Ph$ [25]
(*S*)-**Cp,Cy-ProNOP**:
     R^1 = cyclopentyl, R^2 = cyclohexyl [127a]

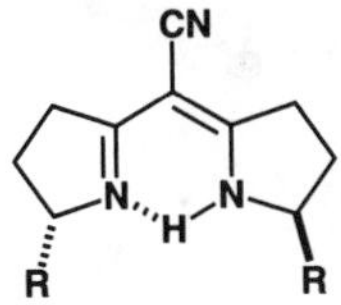

[29b]

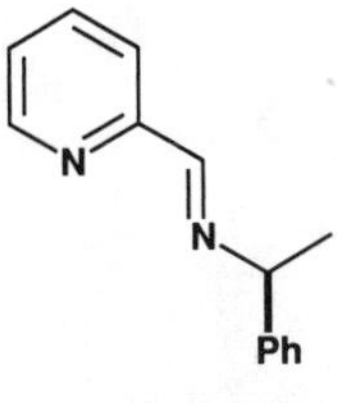

(*S*)-**PPEI** [30]

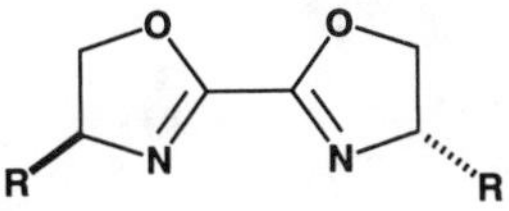

[31]

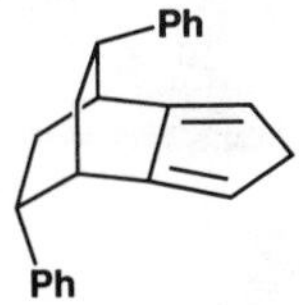

[32]

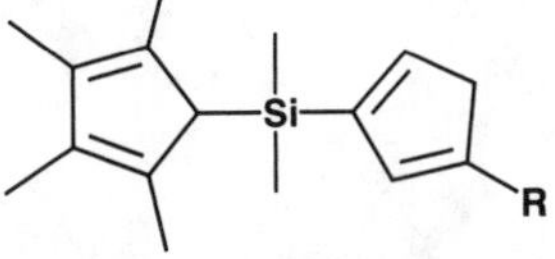

R = neomenthyl [105]
R = menthyl [105]

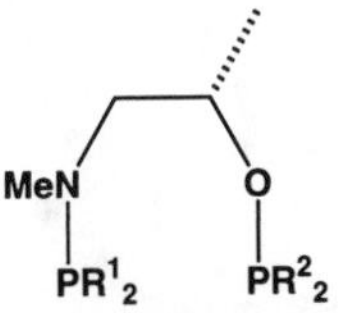

(*S*)-**Cy-isoAlaNOP**: $R^1 = R^2$= cyclohexyl [127a]
(*S*)-**Cp-isoAlaNOP**: $R^1 = R^2$= cyclopentyl [127a]
(*S*)-**Cp,Cy-isoAlaNOP**:
     R^1 = cyclopentyl, R^2 = cyclohexyl [127a]

Chapter 2

For the chiral ligands in this chapter, see the Ligand List for Chapter 1.

Chapter 3

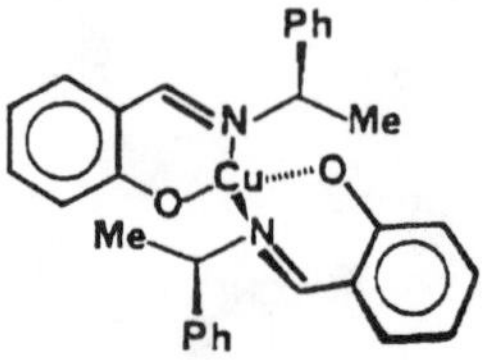

1 [2]

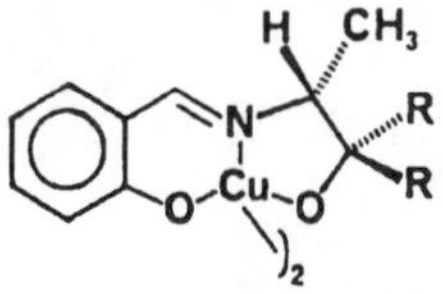

2: R = 5-*t*-Bu-2-*n*-C$_8$H$_{17}$-C$_6$H$_3$- [6]

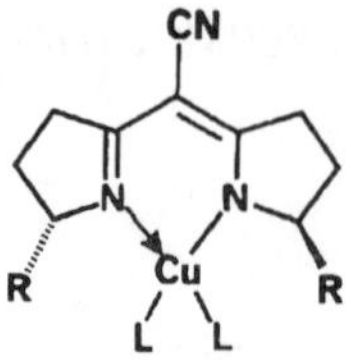

6: R = HO-CMe$_2$- [39]

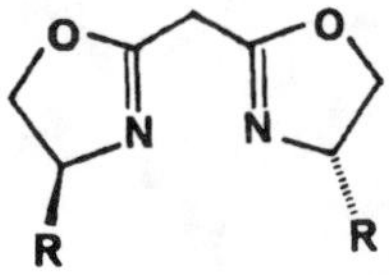

8: R = *t*-Bu [37]

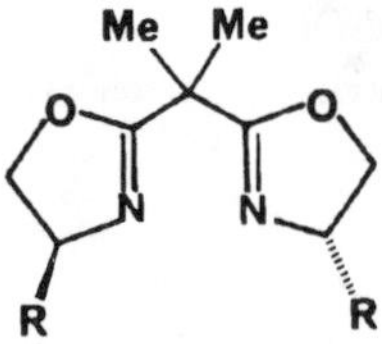

10: R = *t*-Bu [38]

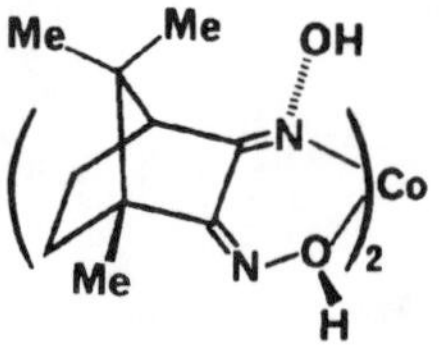

cqdH [48]

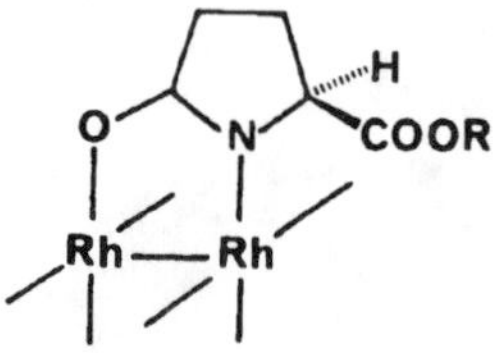

MEPY [9]

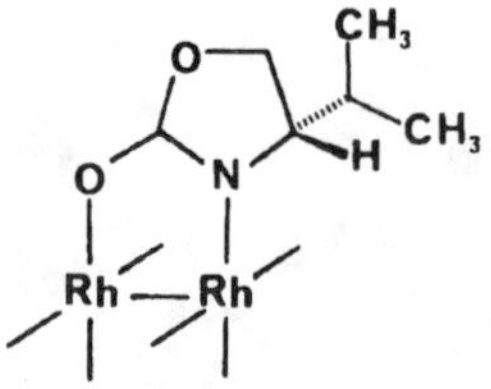

IPOX [9]

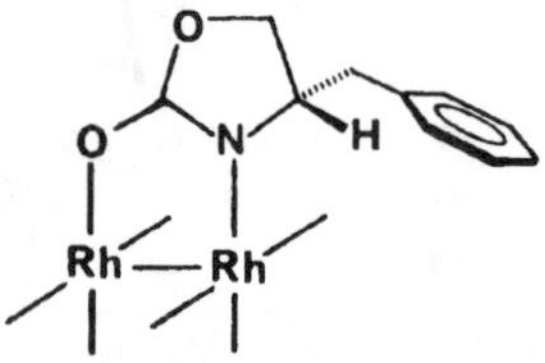

BNOX [9]

Chapter 3 (*Continued*)

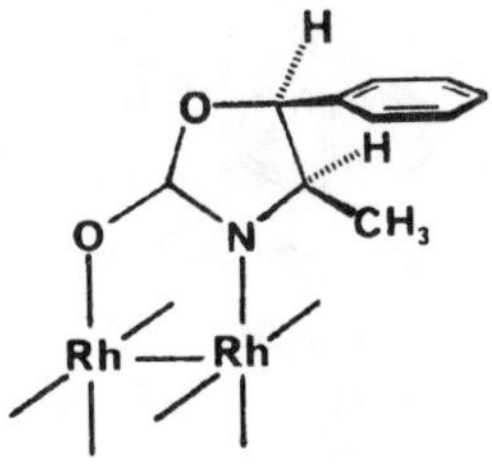

MPOX [9]

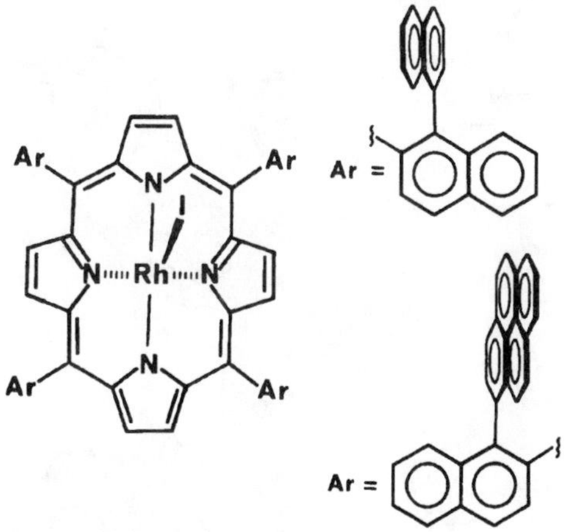

Ar = [58]

[58]

Ar =

Chapter 4.1

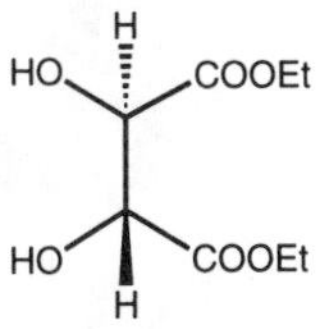

(+)-**DET** 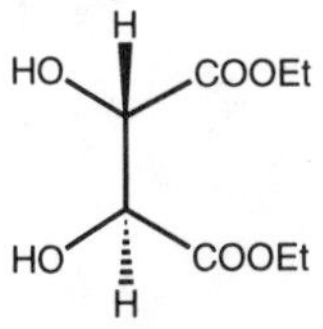   (−)-**DET** 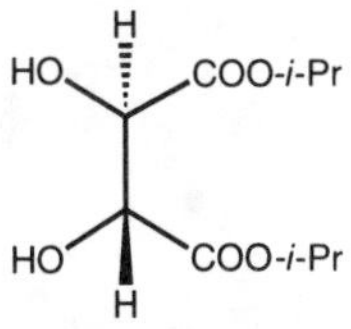   (+)-**DIPT** 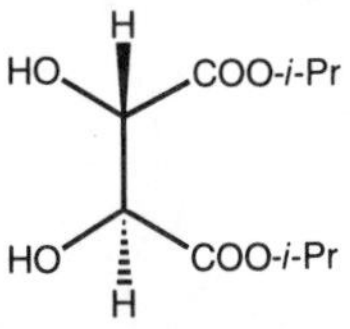   (−)-**DIPT**

These tartrates are all commercially available.

Chapter 4.2

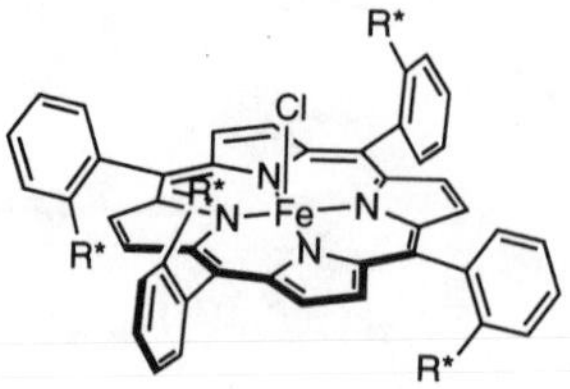

1: R* =

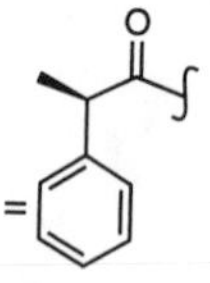

[4]

2: R* =

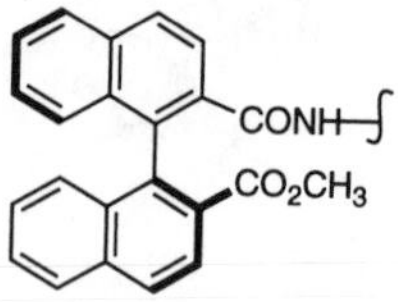

[4]

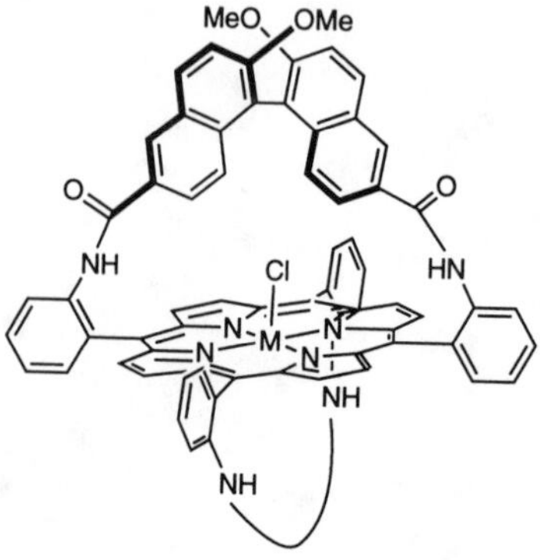

4: M = Fe [7]
5: M = Mn [7]

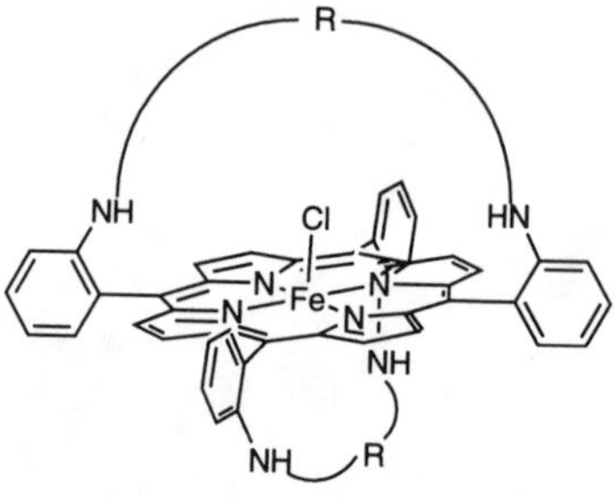

6: R = 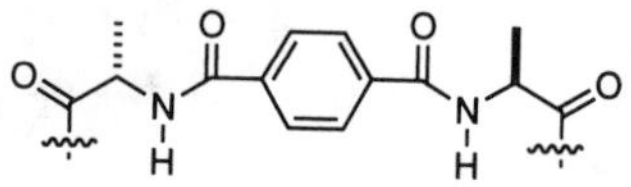   [8]

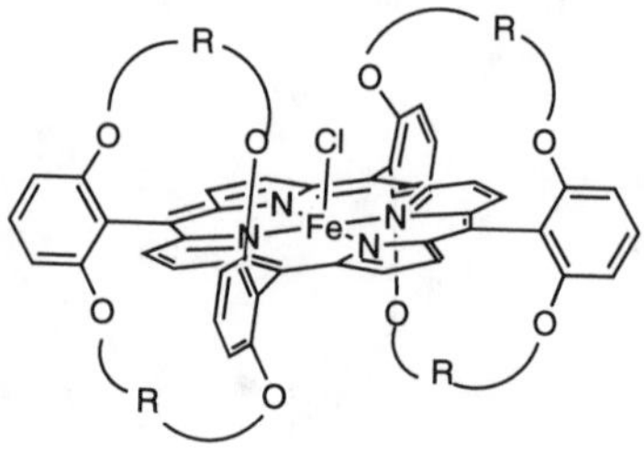

7: R = 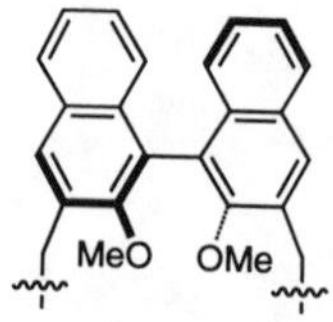   [9]

8: R = 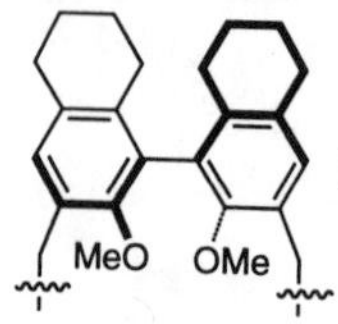   [9]

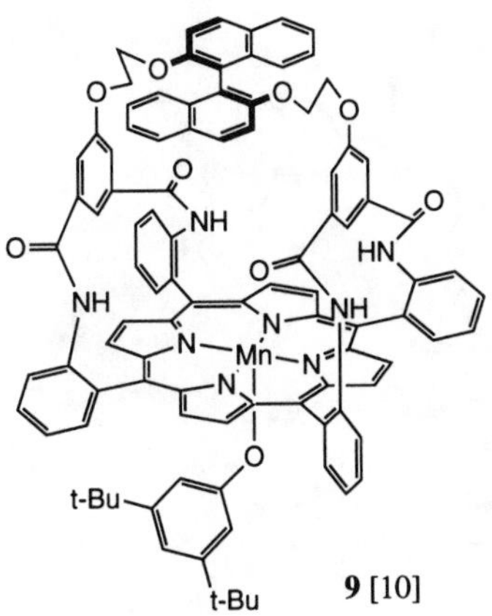

9 [10]

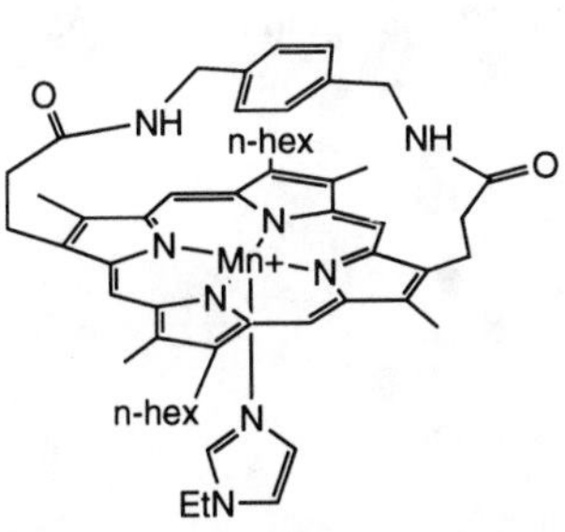

10 [11]

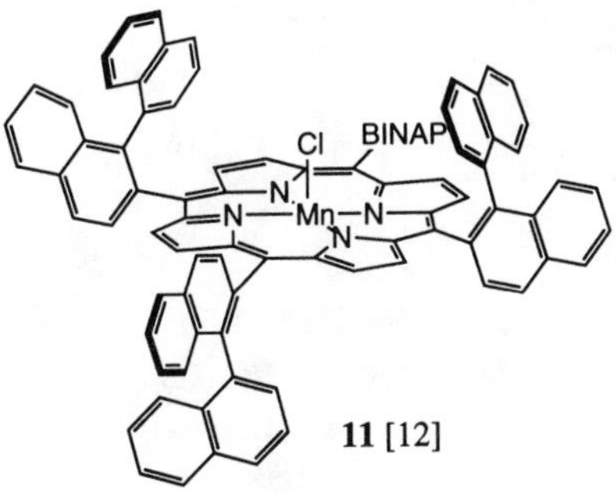

11 [12]

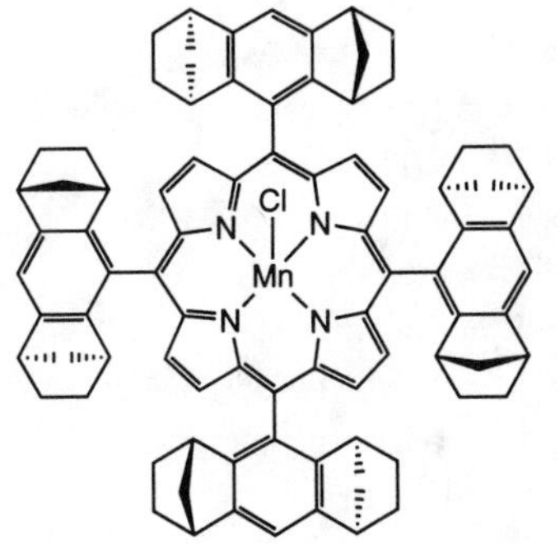

12 14]

Ligands with manganese, **13–65** [18–22]

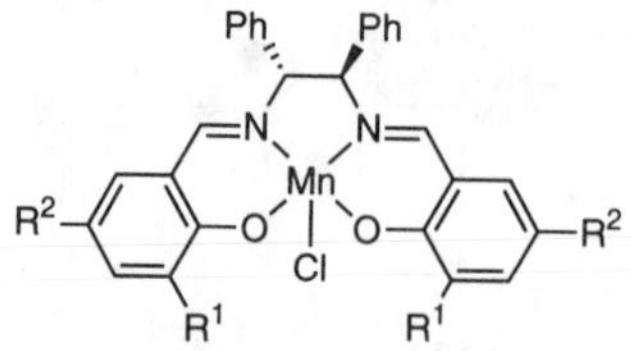

13 $R^1 = H, R^2 = H$
14 $R^1 = t\text{-Bu}, R^2 = H$
15 $R^1 = t\text{-Bu}, R^2 = Cl$
16 $R^1 = t\text{-Bu}, R^2 = OMe$
17 $R^1 = 1\text{-methylcyclohexyl}, R^2 = Me$
18 $R^1 = C(Ph)_2Me, R^2 = H$
19 $R^1 = C(Et)_2Me, R^2 = H$
20 $R^1 = Me_3Si, R^2 = H$
21 $R^1 = t\text{-Bu}, R^2 = Me$
22 $R^1 = 9\text{-methyl-9-fluorenyl}, R^2 = Me$
23 $R^1 = 1\text{-adamantyl}, R^2 = Me$
24 $R^1 = R^2 = Br$
25 $R^1 = t\text{-Bu}, R^2 = NO_2$

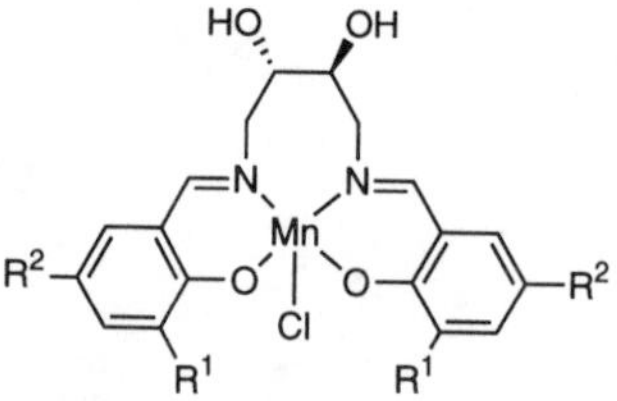

26 $R^1 = R^2 = H$
27 $R^1 = H, R^2 = NO_2$
28 $R^1 = t\text{-Bu}. R^2 = Me$

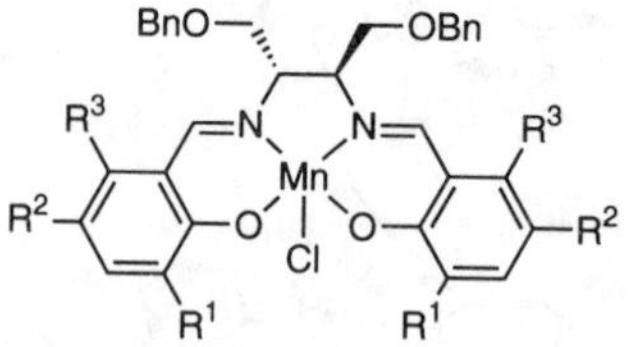

29 $R^1 = R^2 = R^3 = H$
30 $R^1 = R^3 = H, R^2 = t\text{-Bu}$
31 $R^1 = R^3 = H, R^2 = NO_2$
32 $R^1 = R^2 = Br, R^3 = H$
33 $R^1 = H, R^2, R^3 = \text{-CH=CH-CH=CH-}$
34 $R^1 = t\text{-Bu}, R^2 = Me, R^3 = H$

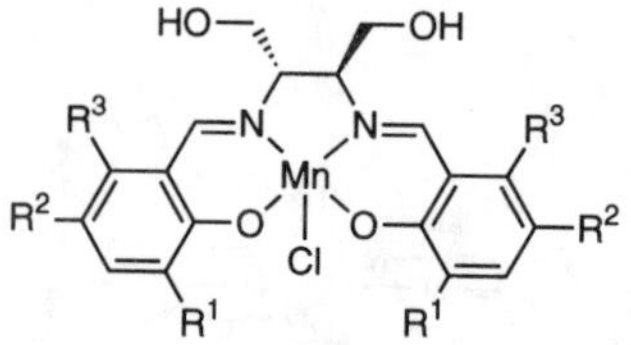

35 $R^1 = R^2 = R^3 = H$
36 $R^1 = R^3 = H, R^2 = t\text{-Bu}$
37 $R^1 = R^3 = H, R^2 = NO_2$
38 $R^1 = R^2 = Br, R^3 = H$
39 $R^1 = H, R^2, R^3 = \text{-CH=CH-CH=CH-}$
40 $R^1 = t\text{-Bu}, R^2 = Me, R^3 = H$

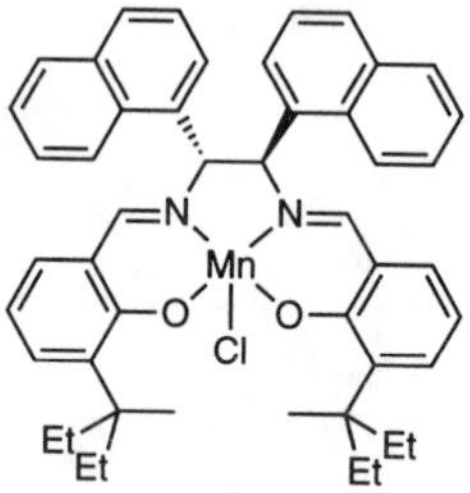

41

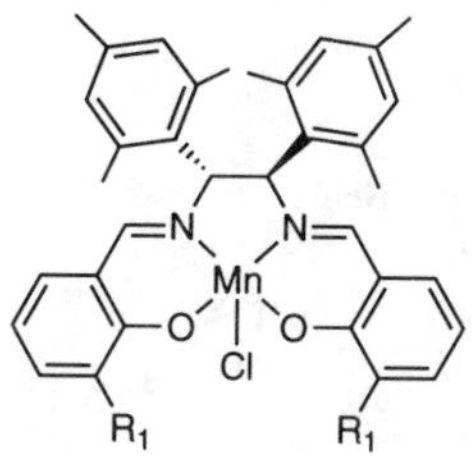

42 $R^1 = t$-Bu
43 $R^1 = C(Ph)_2Me$

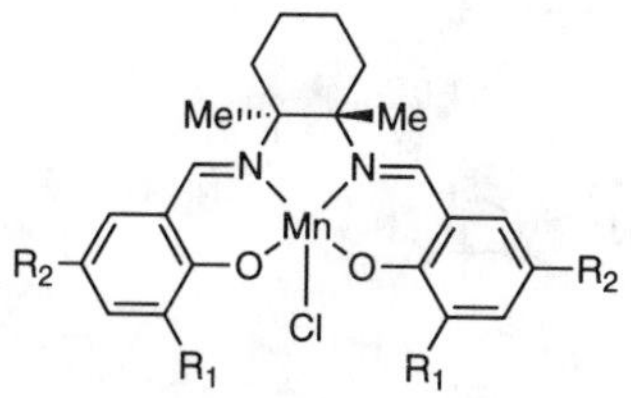

44 R = H
45 R = Ph (unstable)

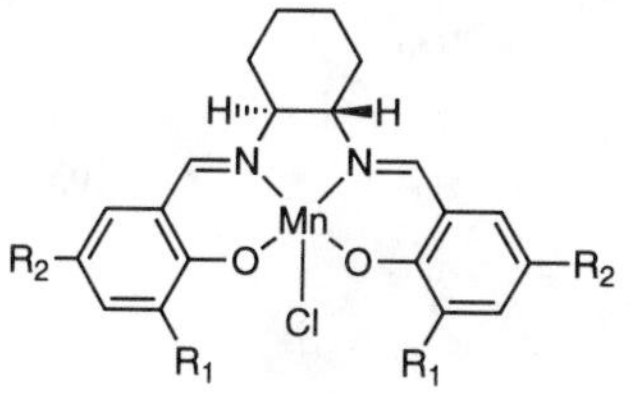

46

47 $R^2 = Me$
48 $R^2 = t$-Bu

49 $R^1 = R^2 = H$
50 $R^1 = C(Ph)_2Me, R^2 = H$
51 $R^1 = R^2 = t$-Bu
52 $R^1 = t$-Bu, $R^2 = Me$
53 $R^1 = $ 1-methylcyclohexyl, $R^2 = Me$

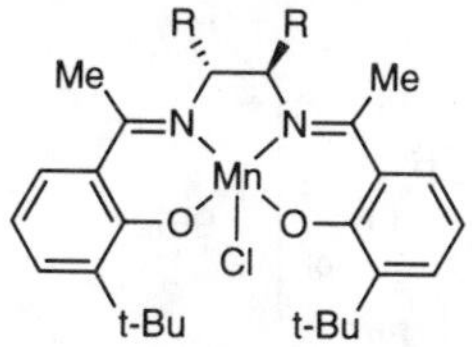

54 $R^1 = H, R^2 = H$
55 $R^1 = t$-Bu, $R^2 = Me$
56 $R^1 = t$-Bu, $R^2 = Cl$
57 $R^1 = t$-Bu, $R^2 = H$
58 $R^1 = t$-Bu, $R^2 = OMe$
59 $R^1 = t$-Bu, $R^2 = NO_2$
60 $R^1 = $ 1-methylcyclohexyl, $R^2 = Me$
61 $R^1 = $ 9-methyl-9-fluorenyl, $R^2 = Me$
62 $R^1 = $ 1-adamantyl, $R^2 = Me$
63 $R^1 = R^2 = t$-Am
64 $R^1 = R^2 = Br$
65 $R^1 = R^2 = t$-Bu

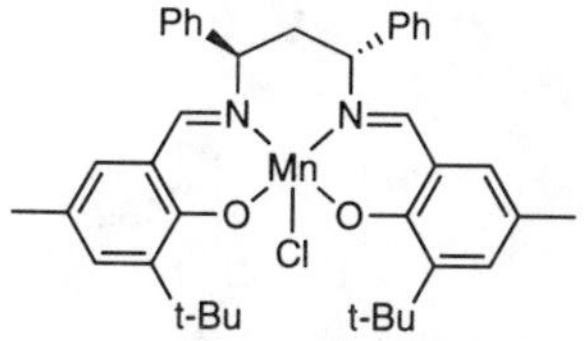

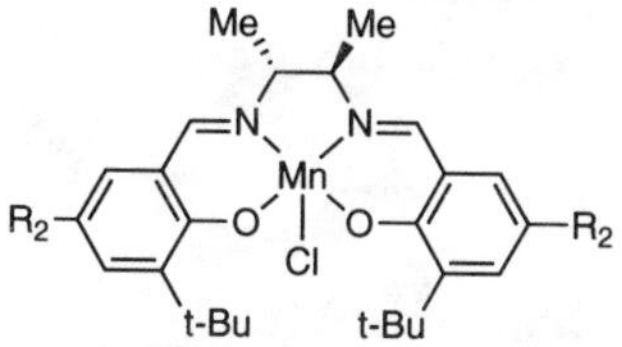

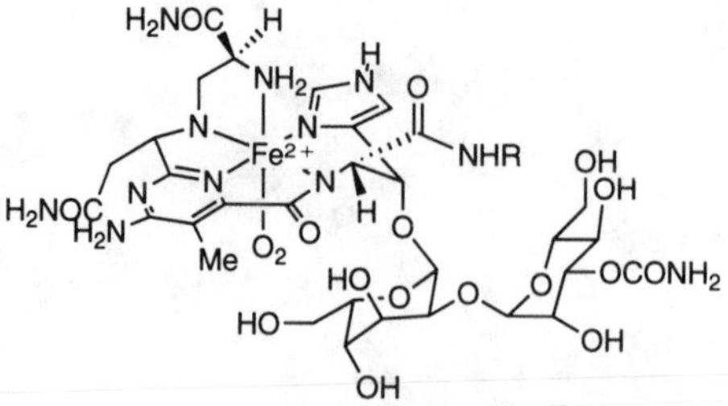

100 [41]

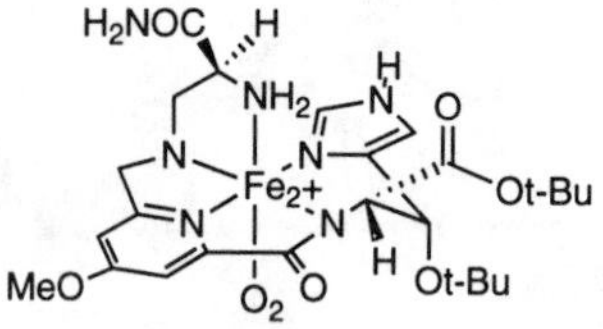

101 [41]

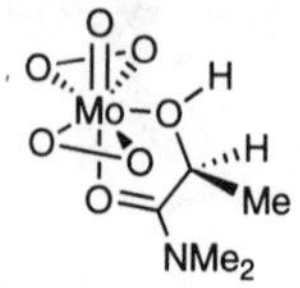

102 [43]

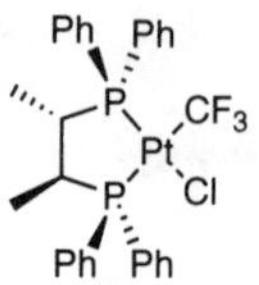

103 [47]

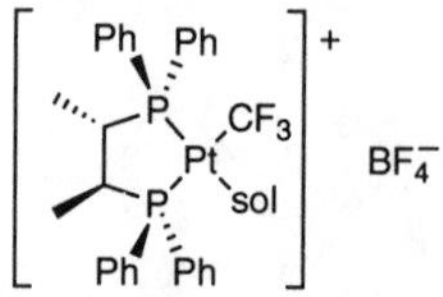

104 [47]

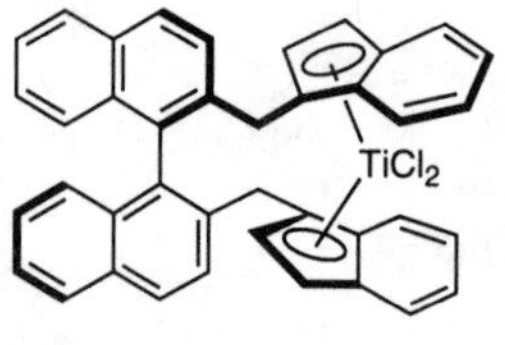

105 [48]

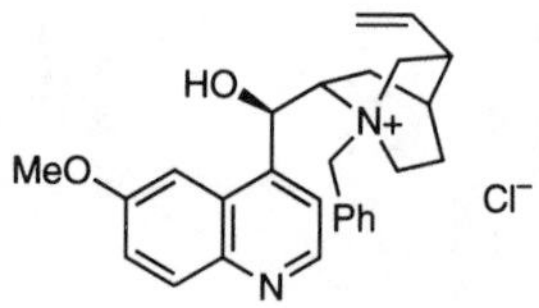

122 [60]

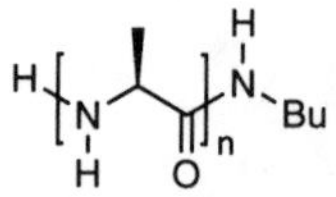

124 [69]

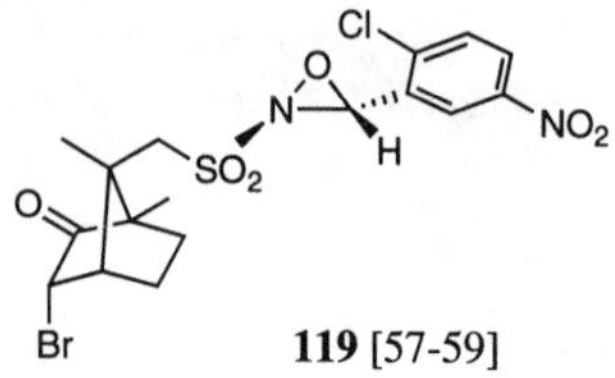

119 [57-59]

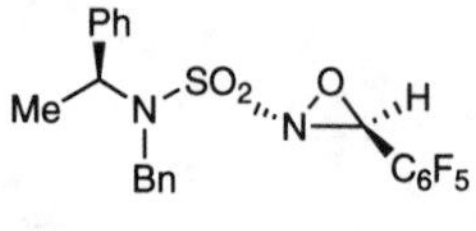

120 [57-59]

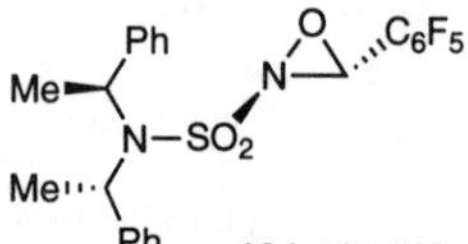

121 [57-59]

Chapter 4.3

For the chiral ligands in this chapter, see the Ligand Lists for Chapters 4.1 and 4.2.

Chapter 4.4

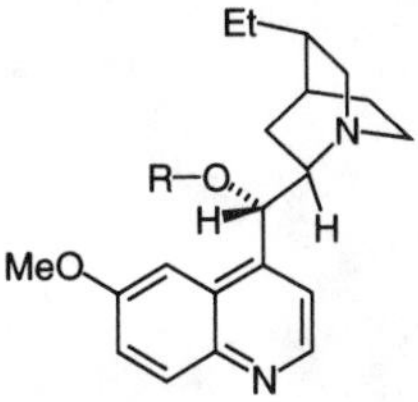

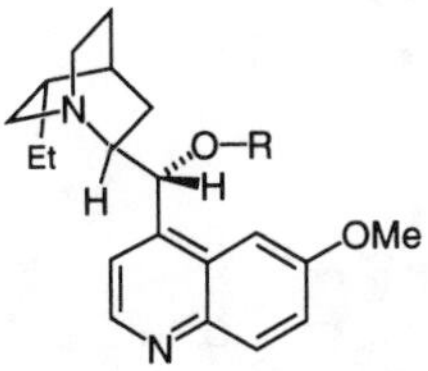

DHQD: R = H [39]

DHQD-OAc: R = —ĈCH₃ [17]

DHQD-CLB: R = —Ĉ—⟨⟩—Cl [17]

(DHQD)₂-PHL: R = —⟨N–N⟩—ODHQD [29]

R = —⟨MeO⟩ [41]

DHQD-PHN: R = —⟨⟩ [41]

DHQD-MEQ: R = —⟨Me⟩ [41]

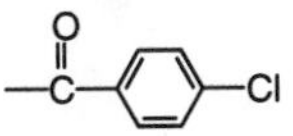

DHQD-IND: R = —Ĉ—N⟨⟩ [57b]

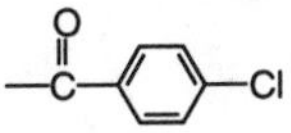

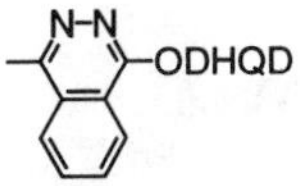

DHQ: R = H [39]

DHQ-OAc: R = —ĈCH₃ [17]

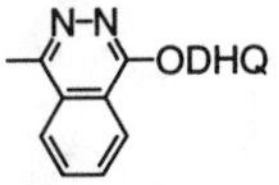

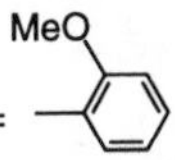

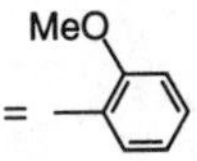

DHQD-CLB: R = —Ĉ—⟨⟩—Cl [17]

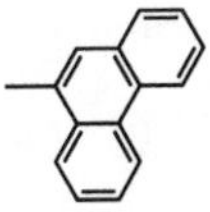

[(DHQ)₂-PHL: R = —⟨N–N⟩—ODHQ [29]

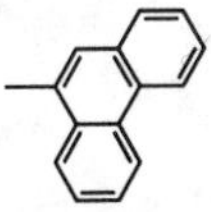

R = —⟨MeO⟩ [41]

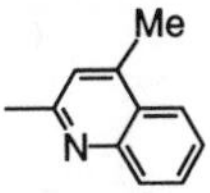

DHQ-PHN: R = —⟨⟩ [41]

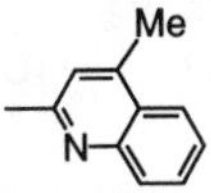

DHQ-MEQ: R = —⟨Me⟩ [41]

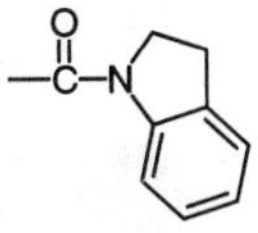

DHQ-IND: R = —Ĉ—N⟨⟩ [57b]

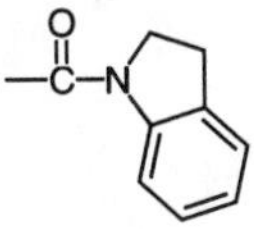

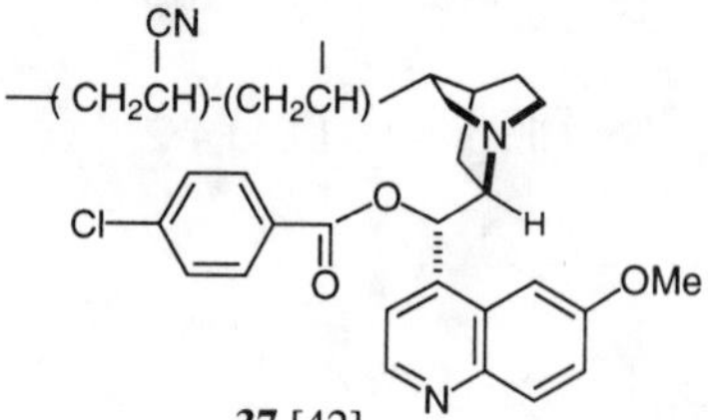

37 [42]

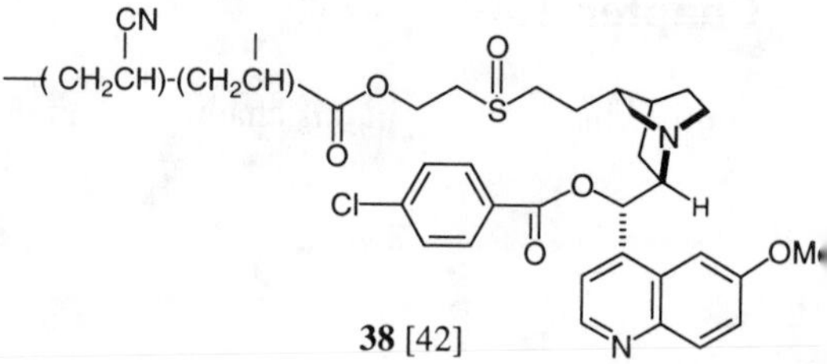

38 [42]

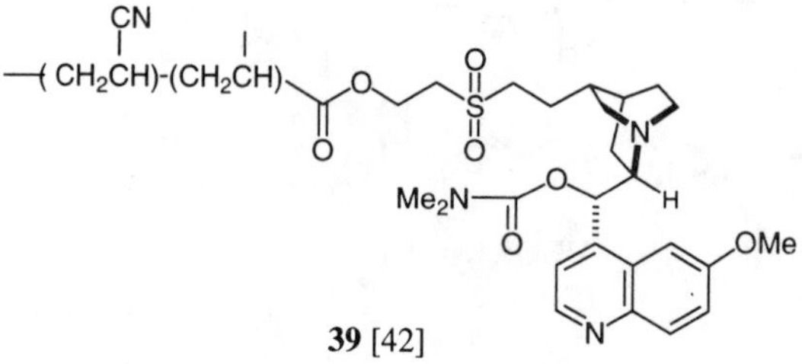

39 [42]

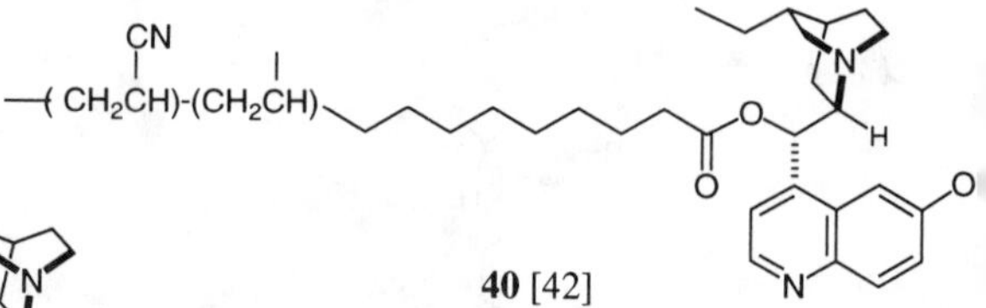

40 [42]

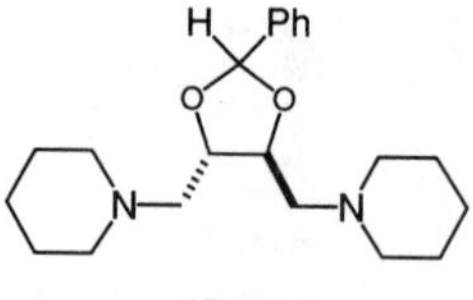

17 [20]

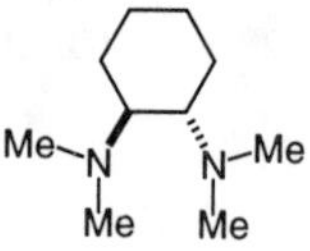

18 [21]

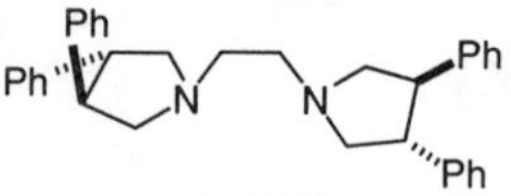

19 [22]

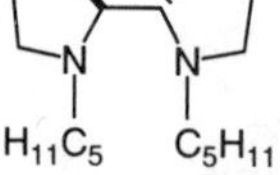

20 [23]

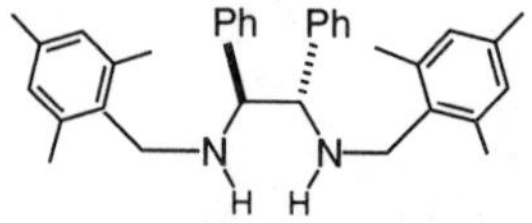

21 [24]

Chapter 5

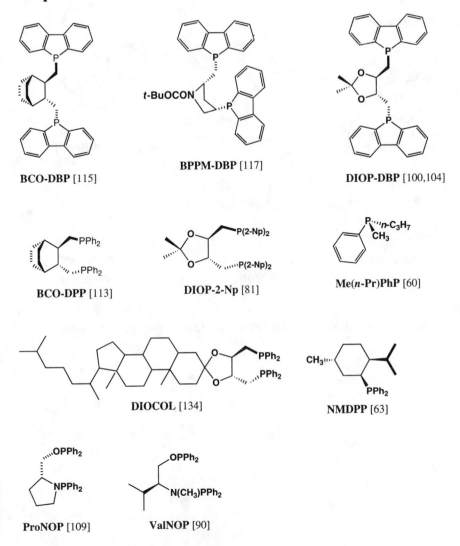

BCO-DBP [115]

BPPM-DBP [117]

DIOP-DBP [100,104]

BCO-DPP [113]

DIOP-2-Np [81]

Me(*n*-Pr)PhP [60]

DIOCOL [134]

NMDPP [63]

ProNOP [109]

ValNOP [90]

For **BDPP, BPPM, BINAP, CHIRAPHOS,** and **DIOP,** see the Ligand List for Chapter 1.

Chapter 6

Chiral Phosphine Ligands

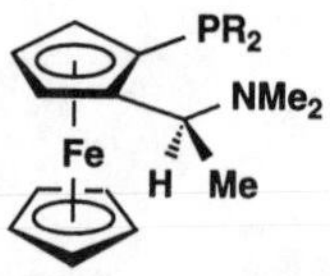

PPFA: R = Ph [44]
MPFA: R = Me [48]

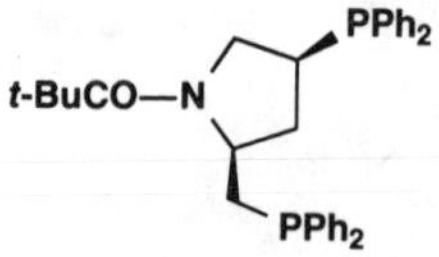

PPPM [51]

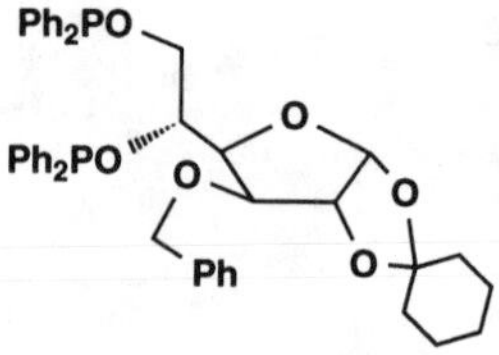

Glucophinite [22]

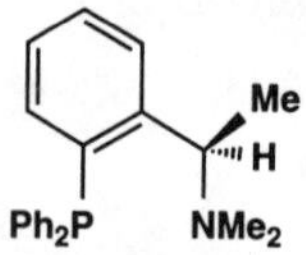

AMPHOS [6]

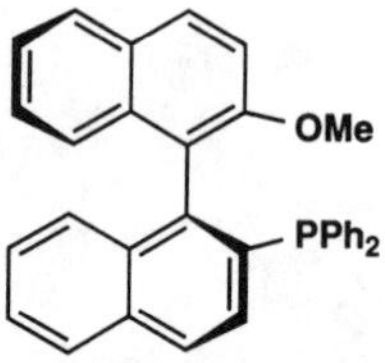

MOP [17]

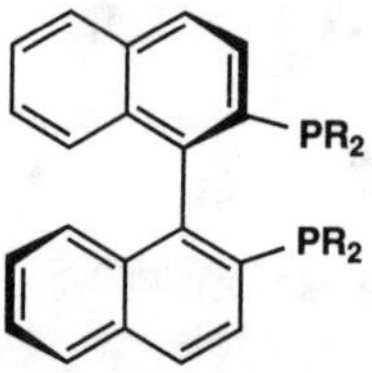

Cy-BINAP [20]
R = cyclohexyl

For **DIOP, NORPHOS,** and **BINAP,** see the Ligand List for Chapter 1.

Chiral Nitrogen Ligands

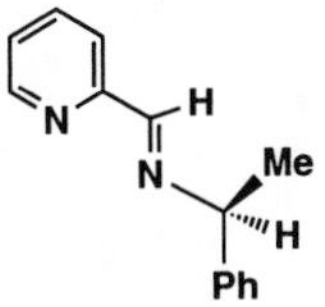

11 [23]

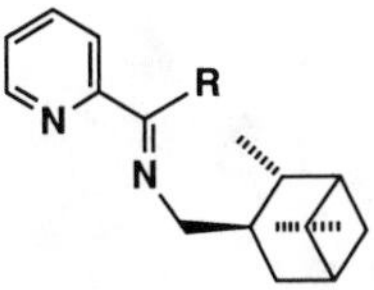

12: R = H [23]
13: R = Me [23]
14: R = Ph [23]

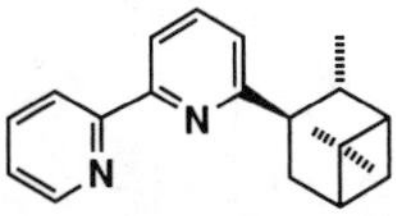

15 [8]

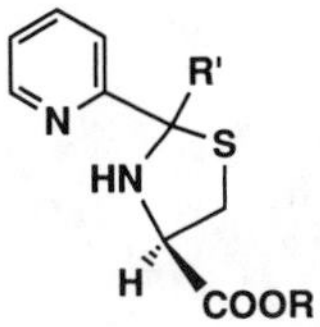

Pythia-(Me,H): R' = Me,H [9]
Pythia-(Me,Me): R,R' = Me,Me [9]
Pythia-(Et,H): R,R' = Et,H [9]

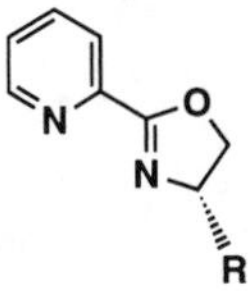

Pymox–*i*-Pr(S) R = *i*-Pr [11,25]
Pymox–*t*-Bu R = *t*-Bu [11,25]

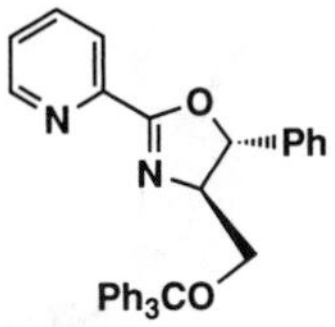

21 [27]

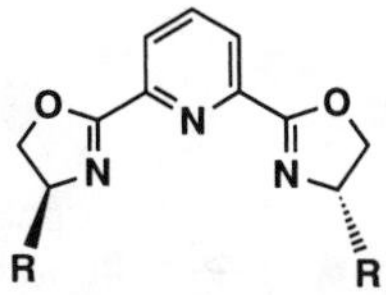

Pybox–*i*-Pr(S,S) R = *i*-Pr [11,12]
Pybox–*t*-Bu(S,S) R = *t*-Bu [11,12]
Pybox–BZ R = CH$_2$Ph [11,12]

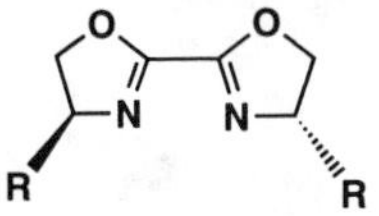

25 R = *i*-Pr [27]
26 R = CH$_2$Ph [27]

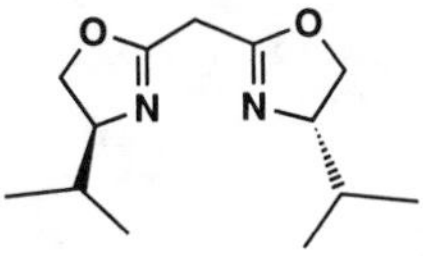

27 [27]

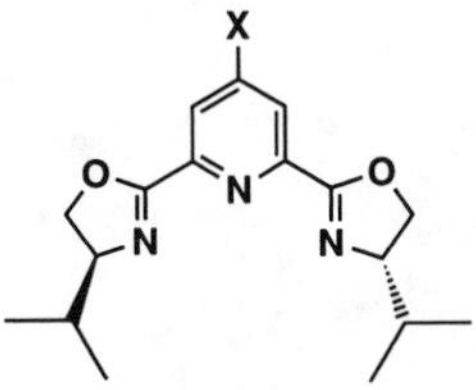

4-Me$_2$N–Pybox–*i*-Pr(S,S) X = Me$_2$N
4-MeO–Pybox–*i*-Pr(S,S) X = MeO
4-Cl–Pybox–*i*-Pr(S,S) X = Cl

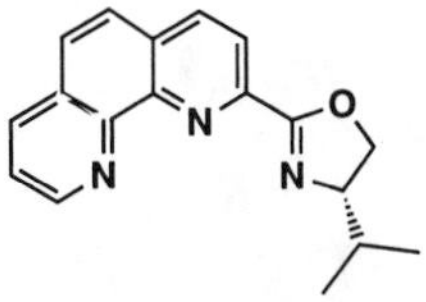

31 [29]

Chapter 7.1

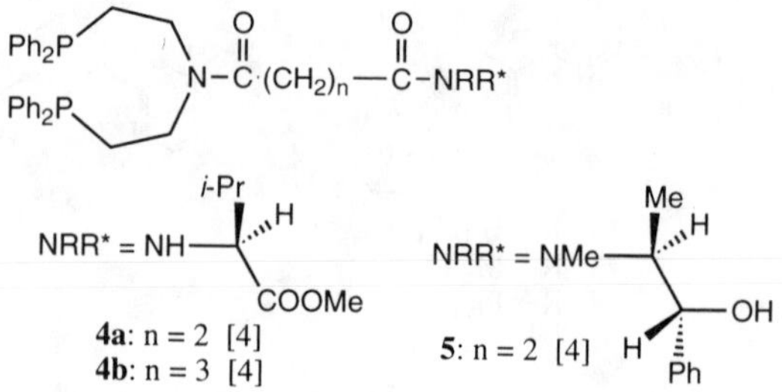

4a: n = 2 [4]
4b: n = 3 [4]

5: n = 2 [4]

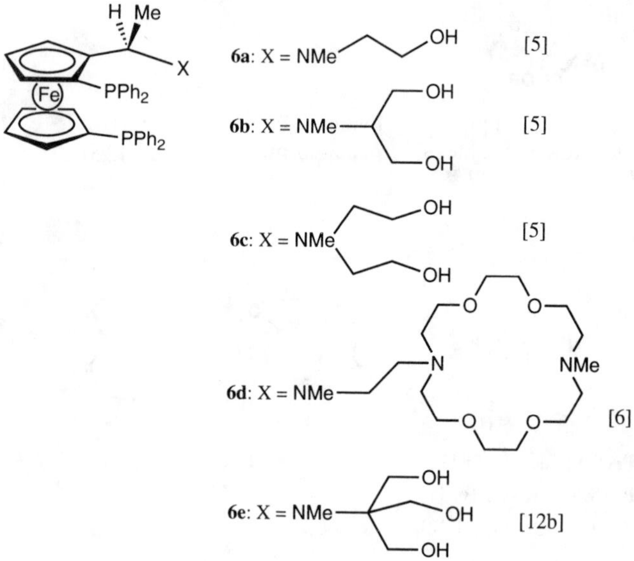

6a: X = NMe ⌒⌒OH   [5]

6b: X = NMe   [5]

6c: X = NMe   [5]

6d: X = NMe   [6]

6e: X = NMe   [12b]

6g: X = Me [12a]

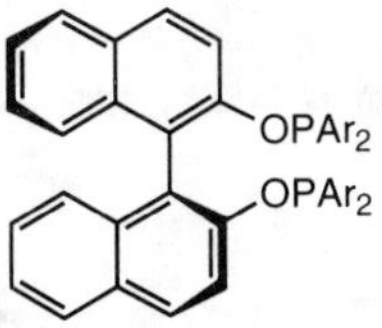

(S)-**BINAPO** (**15a**): Ar = Ph [11]
         15b: Ar = 3,5-(Me$_3$Si)$_2$C$_6$H$_3$ [11]

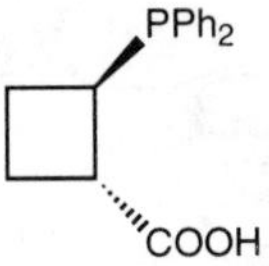

26a [15]

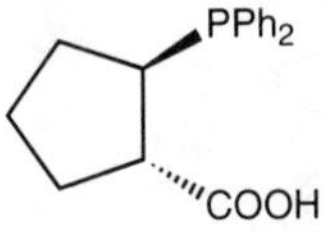

26b [15]

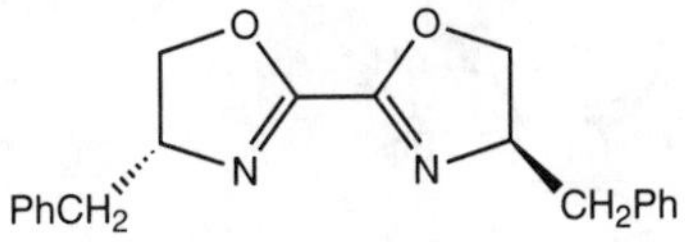

30 [16]

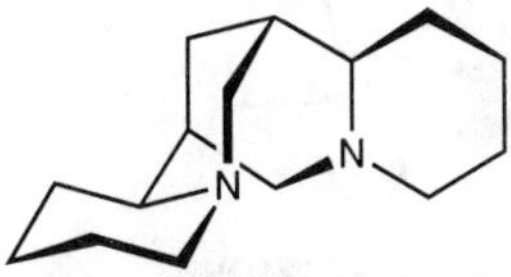

31 [17]

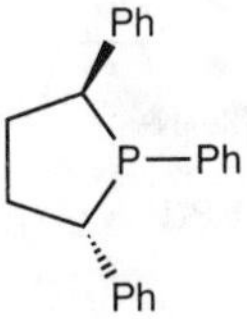

33 [19]

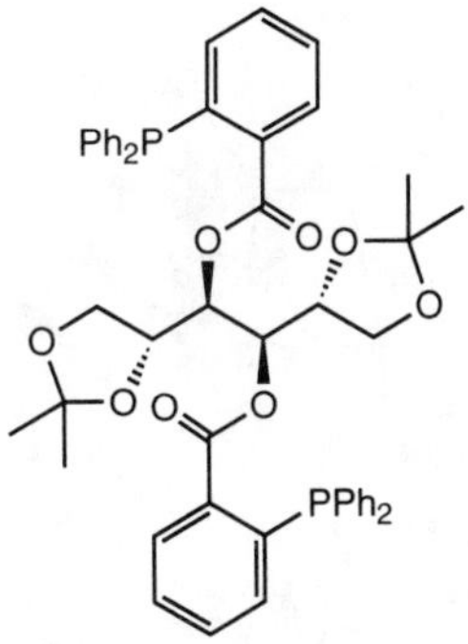

87 [41]

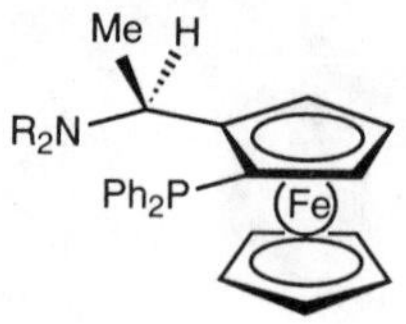

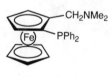

(S)-(R)-**PPFA (99a)**: NR$_2$ = NMe$_2$ [45a,b]

99b: NR$_2$ = N   [45b]

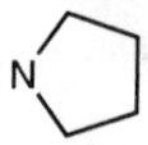

99c: NR$_2$ = N 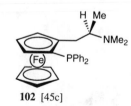 [45b]

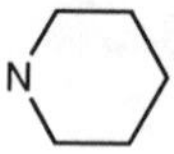

100 [45a,b] 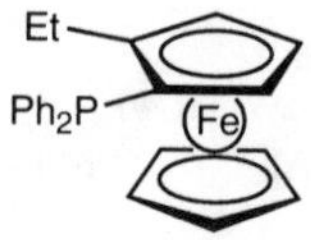   **101** [45a,b]   **102** [45c]

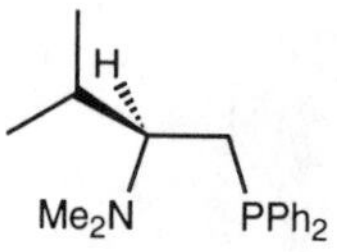

VALPHOS [47] 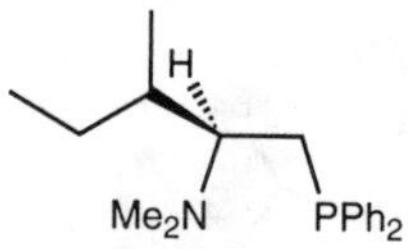   **ILEPHOS** [47] 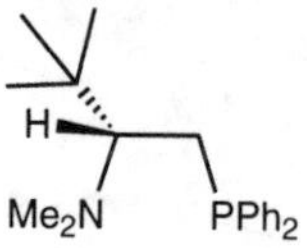 t-**LEUPHOS** [47]

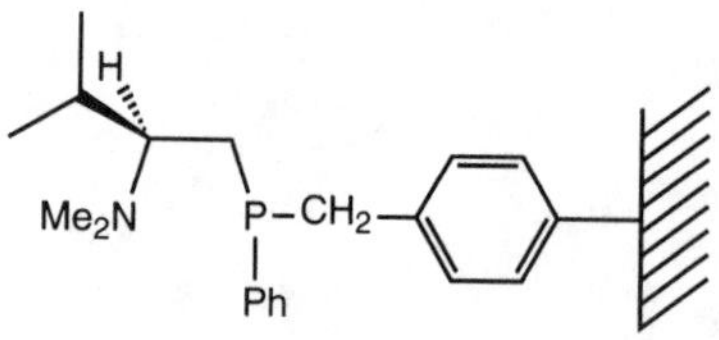

104 [48]

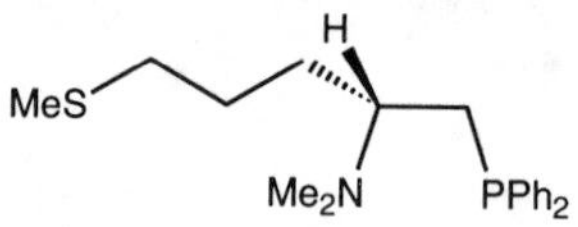

105a [49a]

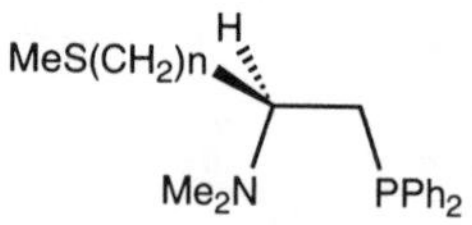

105b: n = 1 [49b]
105c: n = 2 [49b]

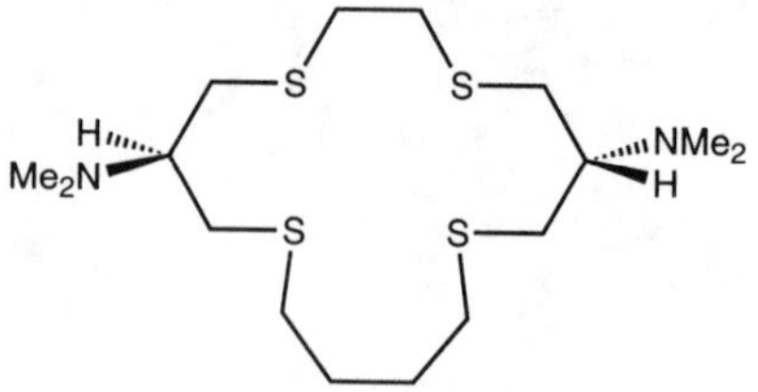

106 [50]

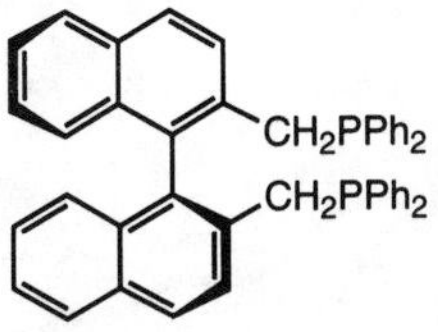

107 [53]

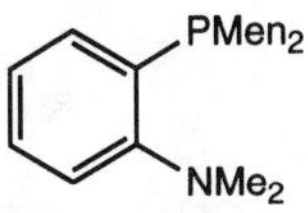

108 [54]

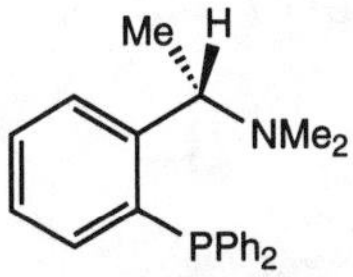

109 [55]

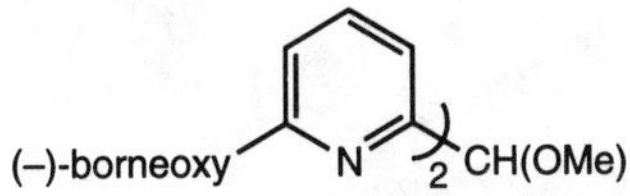

(–)-borneoxy

110 [56]

$Ph_2PCH_2CH_2P$

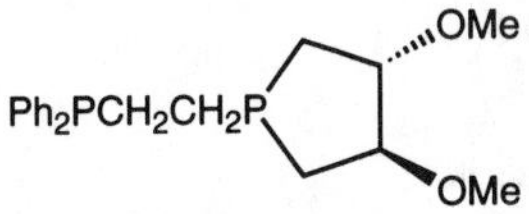

111 [57]

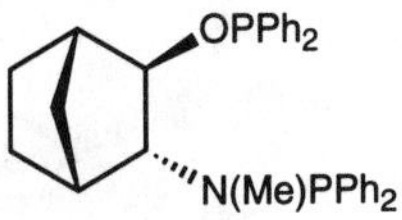

112 [58]

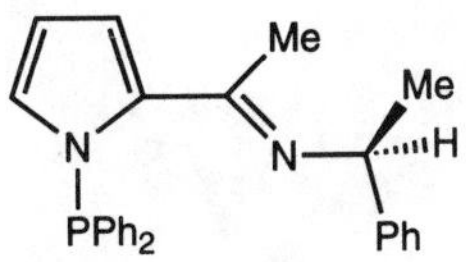

113 [59]

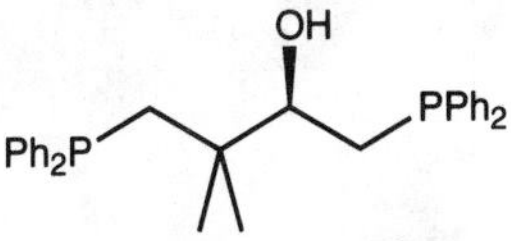

114 [60]

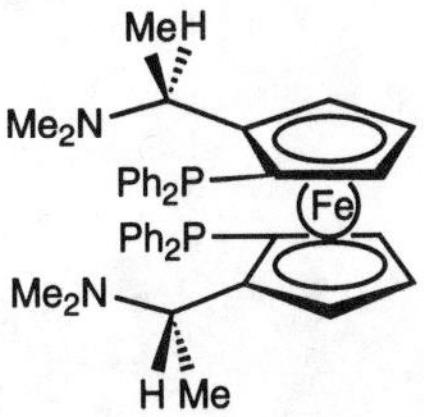

120 [63]

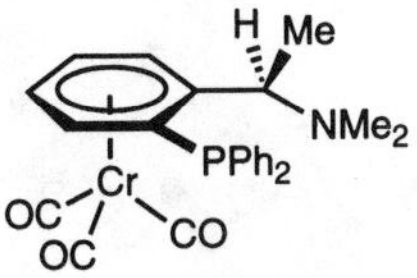

121 [64]

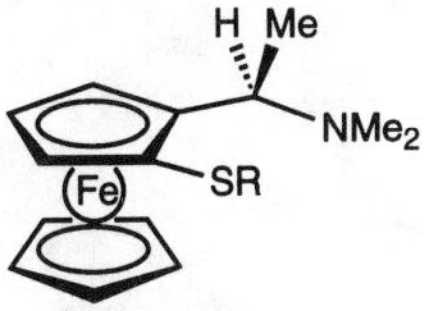

132 [70a]

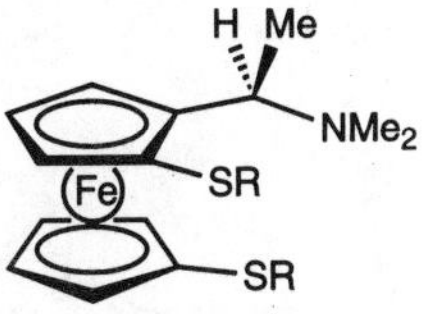

133 [70b]

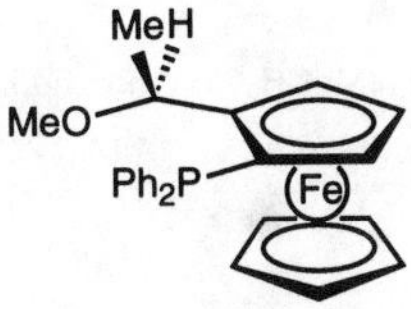

138 [73,74]

For **DIOP (1)**, **BPPFA (6f)**, **BINAP (16)**, **CHIRAPHOS (17)**, **NORPHOS (24)**, **NMDPP (32)**, **ProNOP (35)**, **BIPHEMP (38)**, **DPCP (39)**, and **PROPHOS (60)**, see the Ligand List for chapter 1.

Chapter 7.2

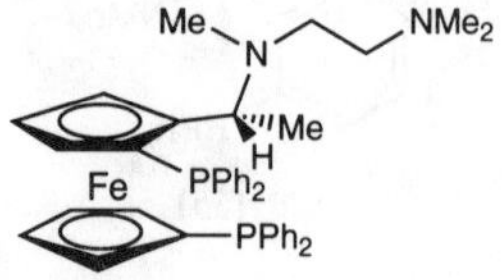

(R)-(S)-**BPPF**NMeCH$_2$CH$_2$NMe$_2$ [4] 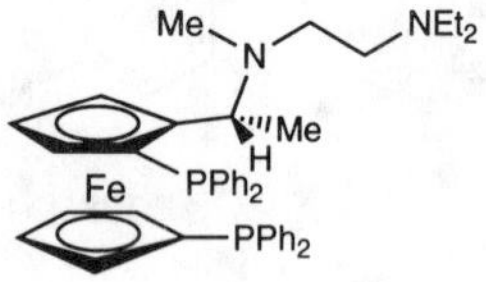   (R)-(S)-**BPPF**NMeCH$_2$CH$_2$NEt$_2$ [4] 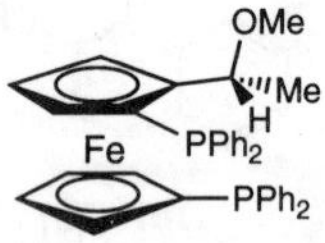   (R)-(S)-**BPPF**OMe [4]

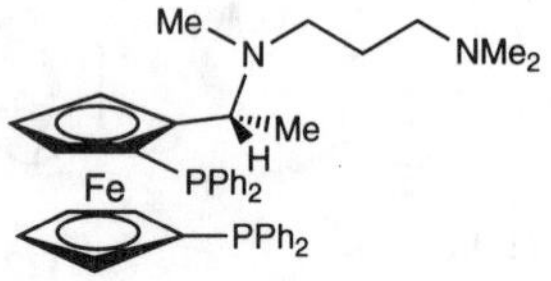

(R)-(S)-**BPPF**NMe(CH$_2$)$_3$NMe$_2$ [4] 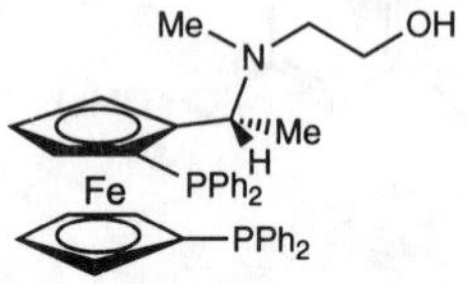   (R)-(S)-**BPPF**NMeCH$_2$CH$_2$OH [4]

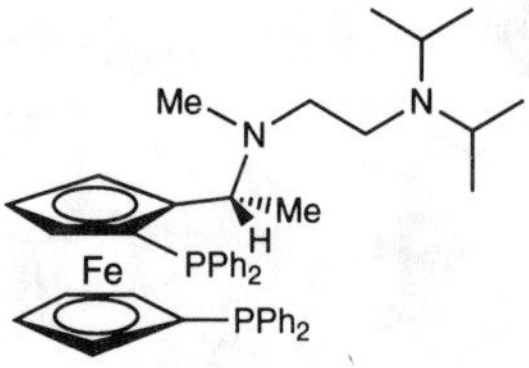

(R)-(S)-**BPPF**NMeCH$_2$CH$_2$N(i-Pr)$_2$ [5] 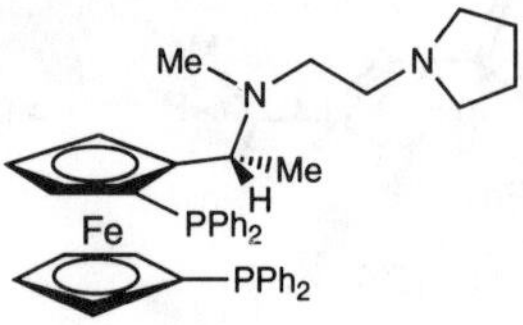   (R)-(S)-**BPPF**NMeCH$_2$CH$_2$(pyrrolidine) [5]

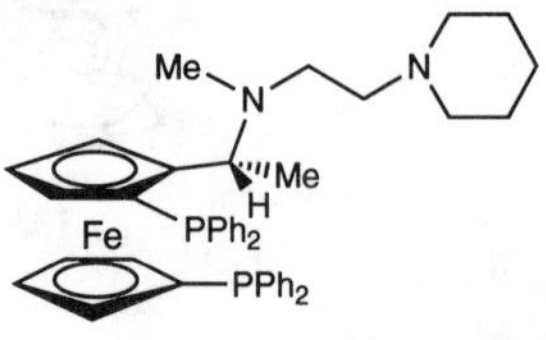

(R)-(S)-**BPPF**NMeCH$_2$CH$_2$(piperidine) [5] 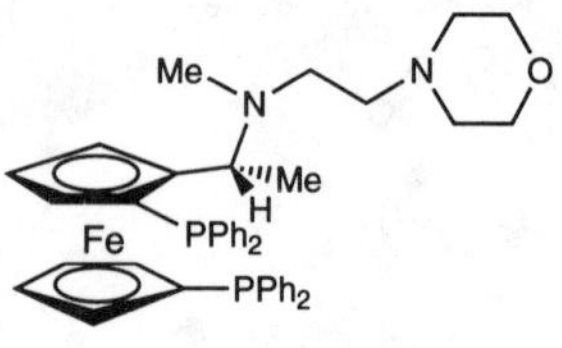   (R)-(S)-**BPPF**NMeCH$_2$CH$_2$(morpholine) [5]

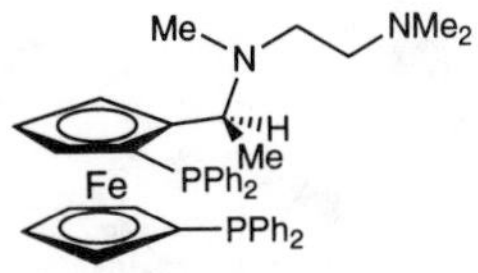

(S)-(S)-**BPPFNMeCH$_2$CH$_2$NMe$_2$** [23]

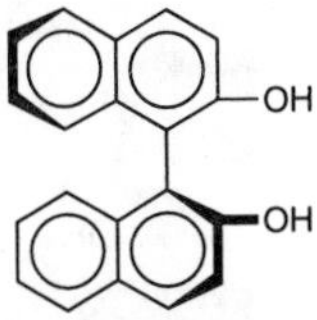

(S)-**Binaphthol**

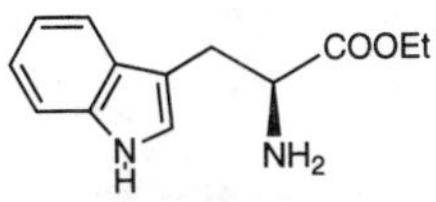

Tryptophan Ethyl Ester

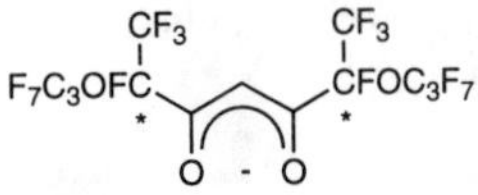

DPPM [30]

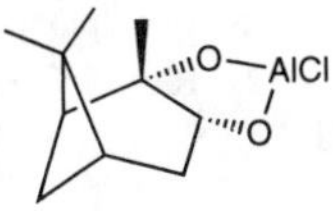

[33]

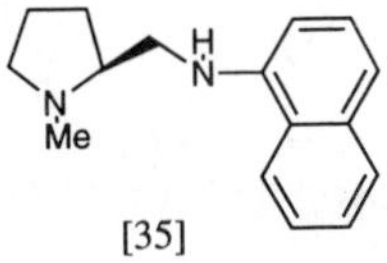

[35]

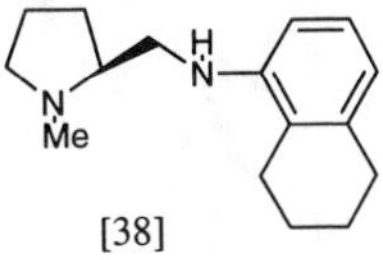

[38]

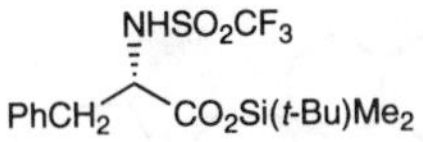

[42]

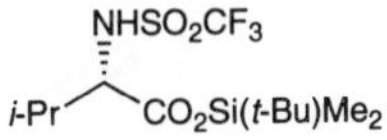

[42]

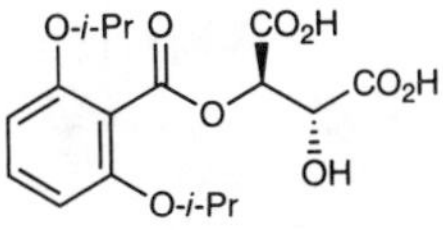

[43]

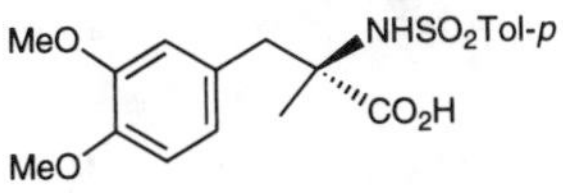

[45]

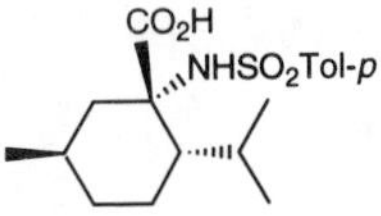

[45]

For **BPPFA**, **CHIRAPHOS**, **DIOP**, **TolBINAP**, and **NORPHOS**, see the Ligand List for Chapter 1.

Chapter 8

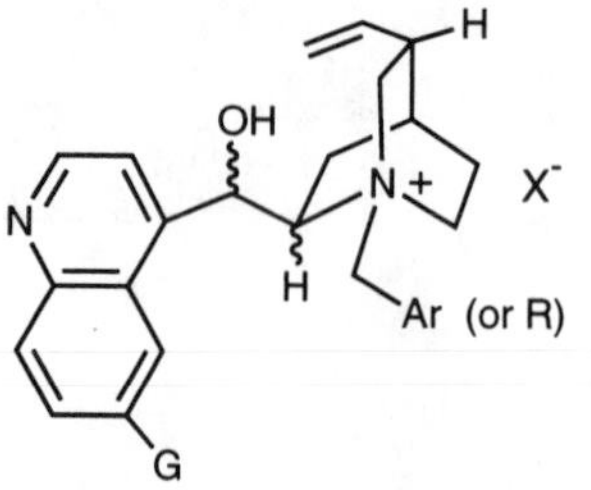

3　Ar = 4-CF$_3$C$_6$H$_4$, G = H [77]
6　Ar = Ph, G = MeO [145]

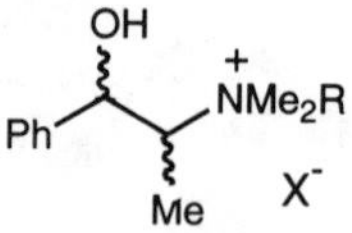

8　R = PhCH$_2$, X = Cl [89]
10　R = PhCH$_2$ X = Cl [92]
12　R = n-C$_7$H$_{15}$, X = Br [98]

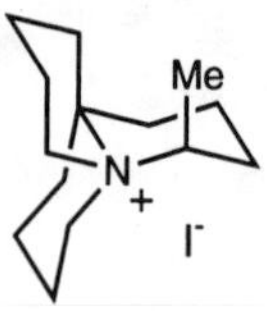

13 [49]

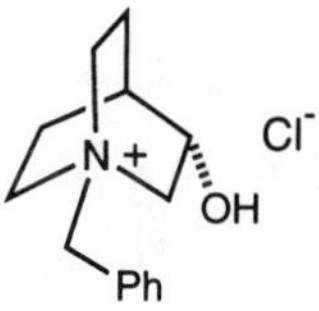

14 [62]

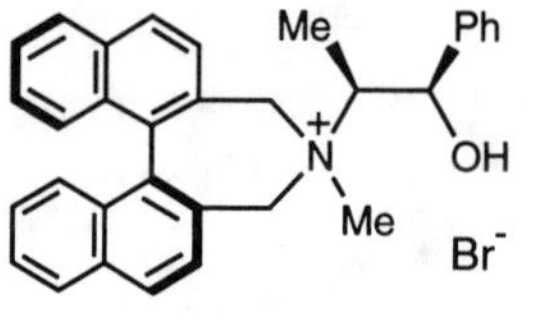

15 [101]

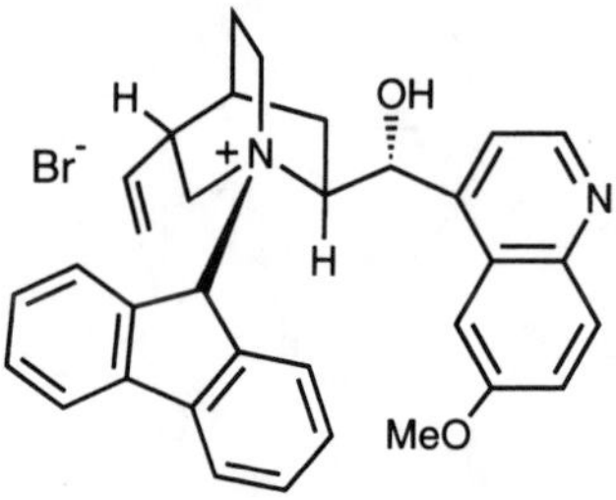

17 [148]

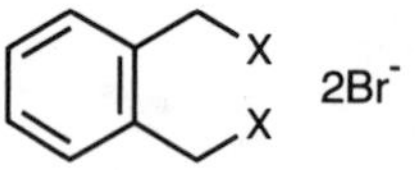

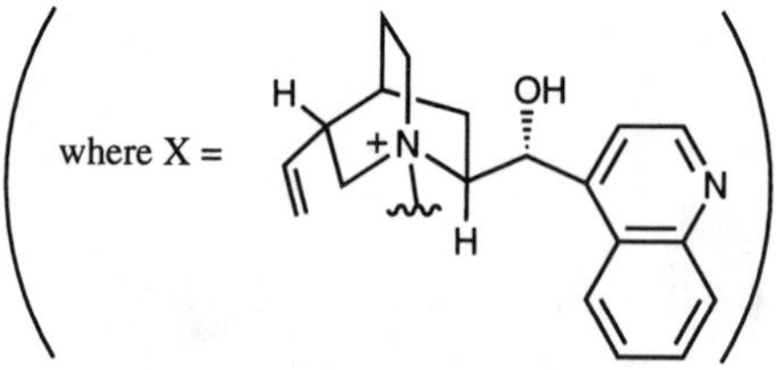

16 [148]　where X =

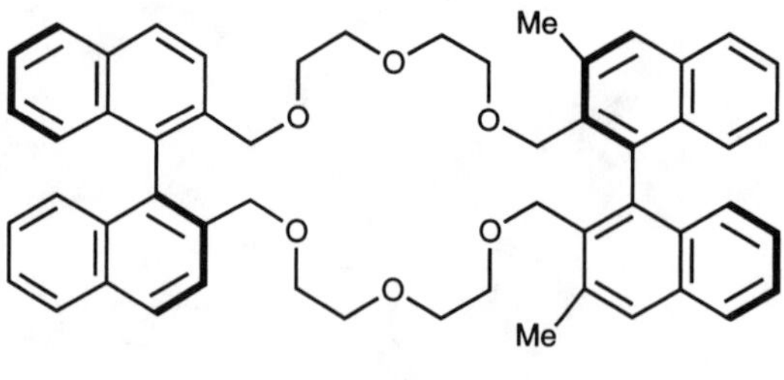

18 [93]

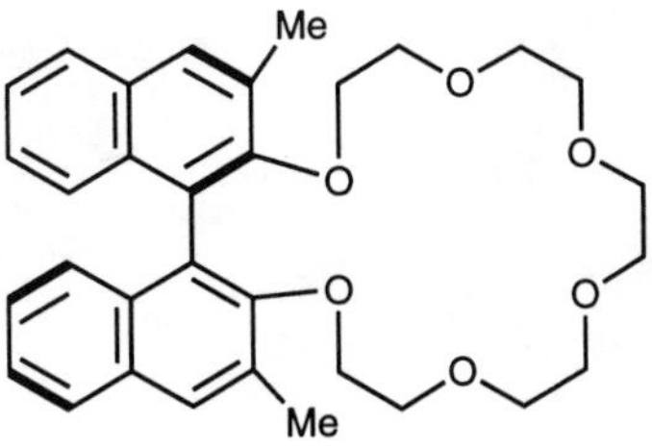

19 [93]

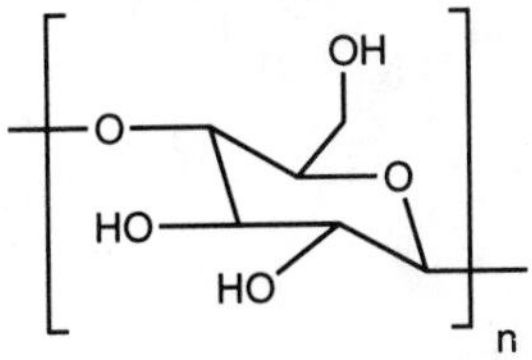

20 [160]

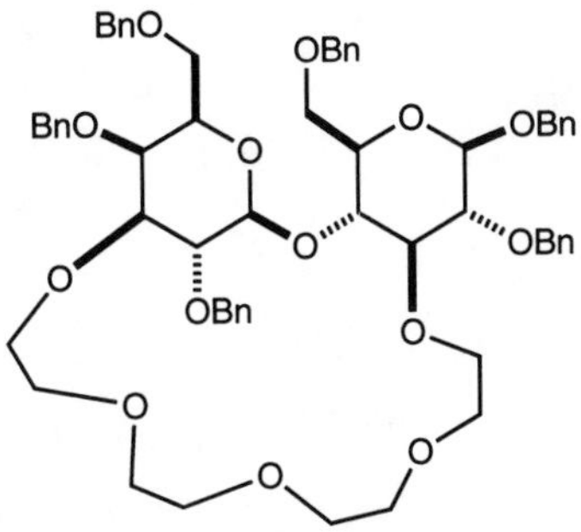

21 [79]

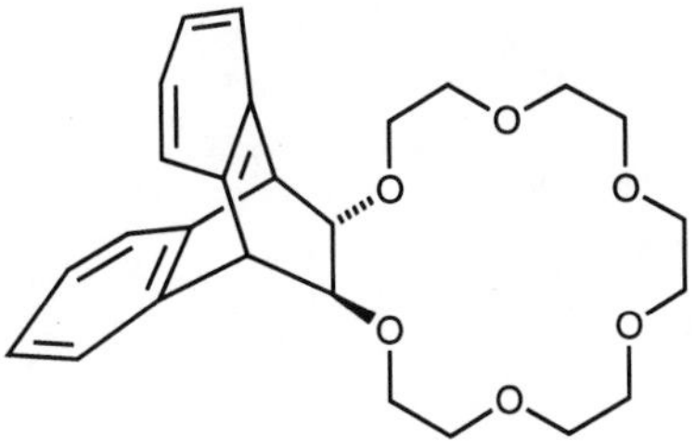

22 [132]

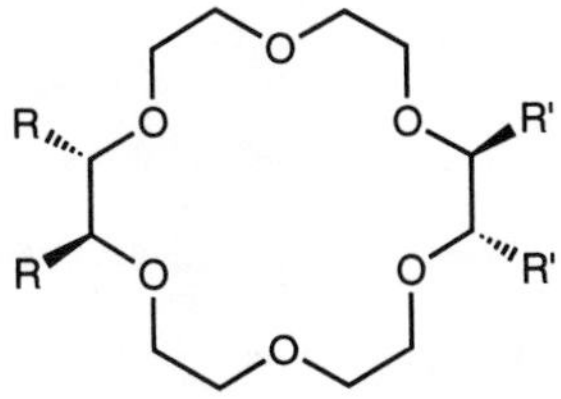

23 R = R' = Ph [90]
24 R = H, R' = Me [131]

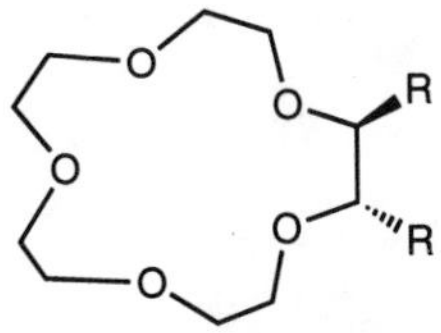

25 R = (2,4-Cl$_2$C$_6$H$_3$)CO$_2$CH$_2$ [102]
26 R = (2,4-Cl$_2$C$_6$H$_3$)CO$_2$CH$_2$ [102]

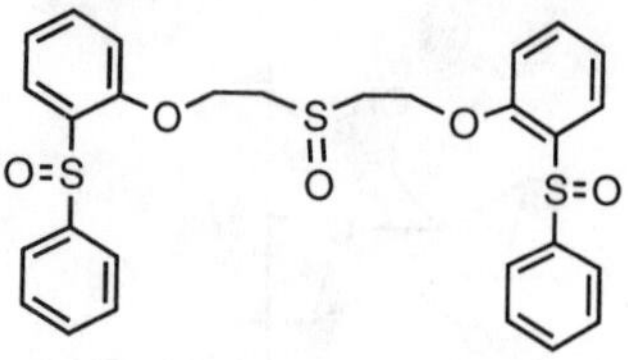

27 [94]

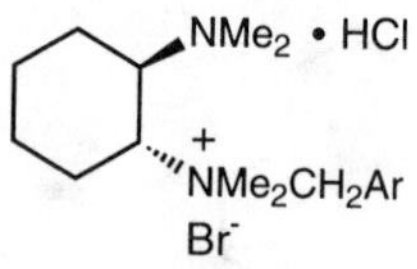

28 [87]

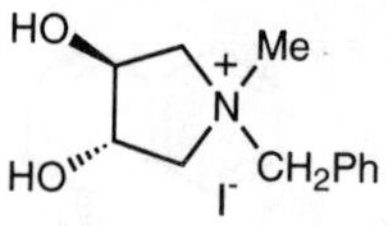

29 [117]

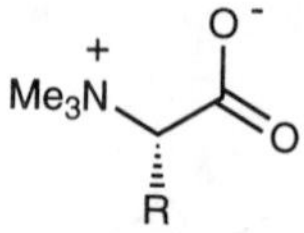

30 [134]

Chapter 9

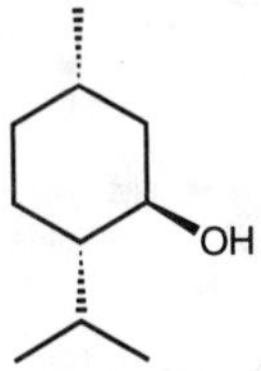

l-Menthol

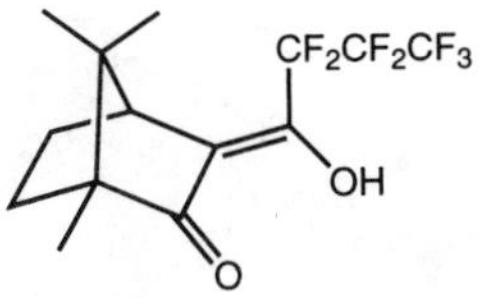

[4]

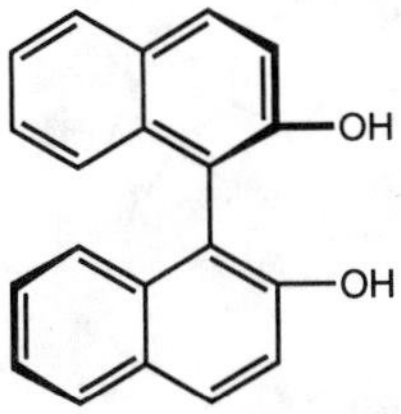

(*R*)-**Binaphthol**

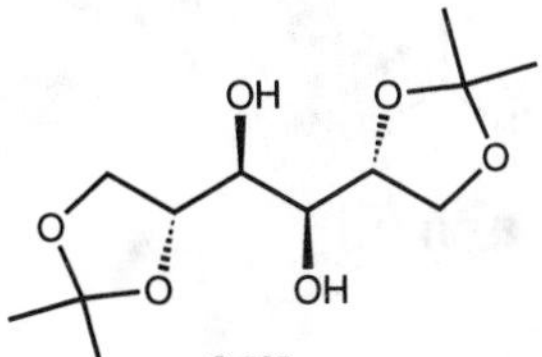

3 [8]

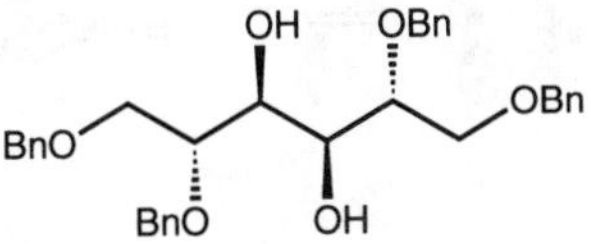

4 [8]

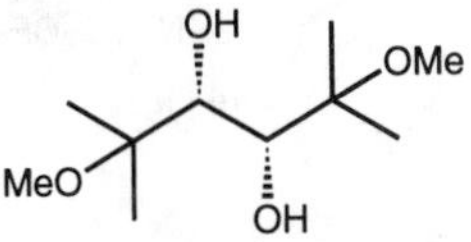

5 [8]

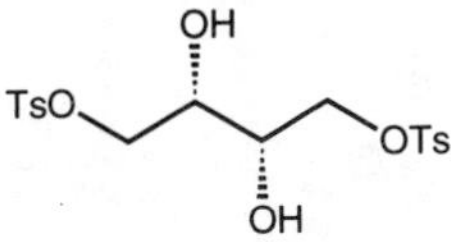

6 [8]

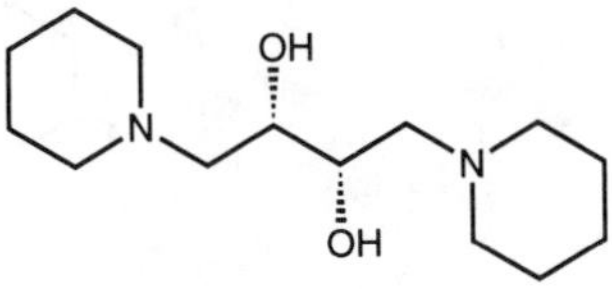

7 [8]

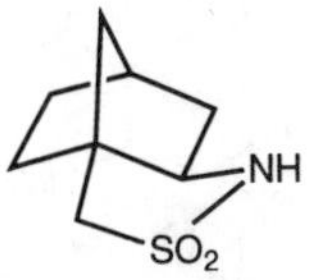

8 [8]

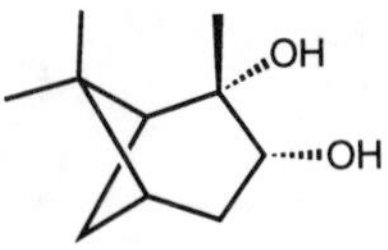

9 [9]

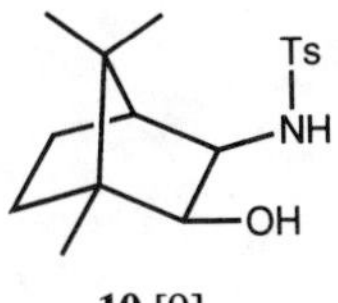

10 [9]

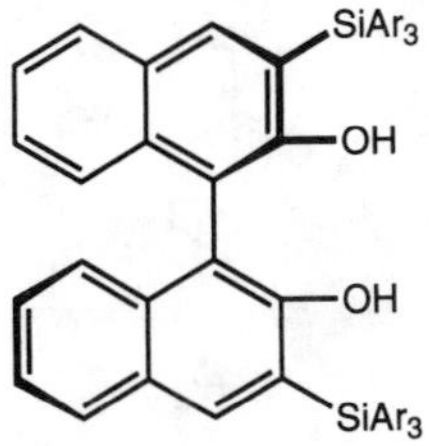

14 [11]

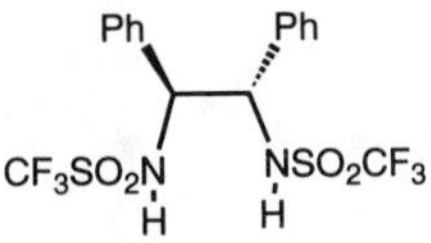

[17]

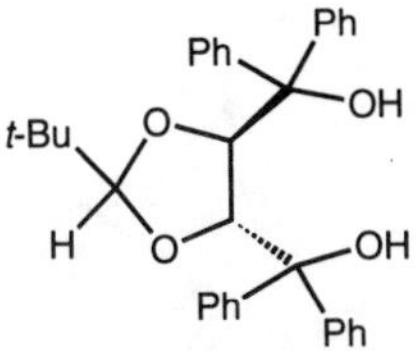

[21]

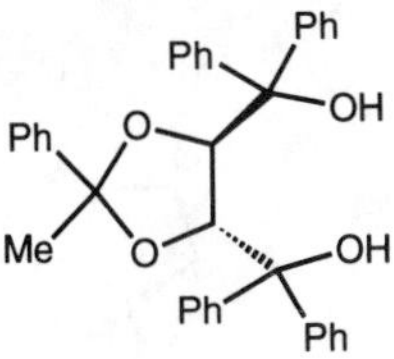

[22]

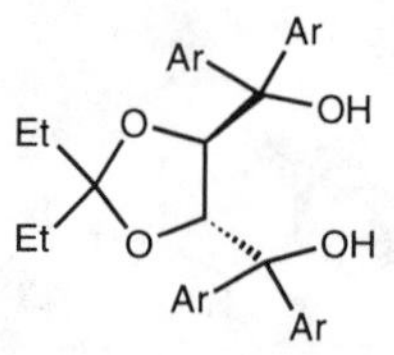

[28]

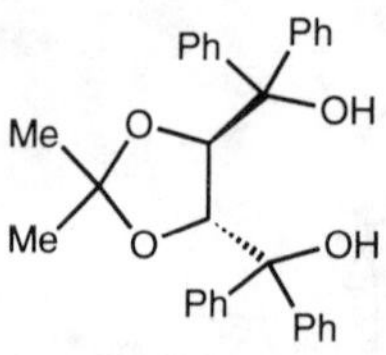

[30]

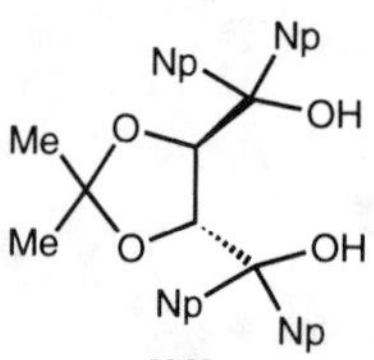

[30]

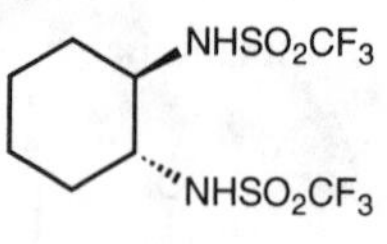

38 [29]

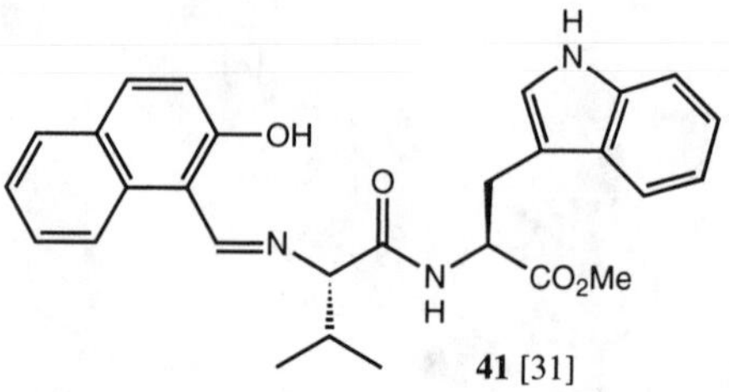

41 [31]

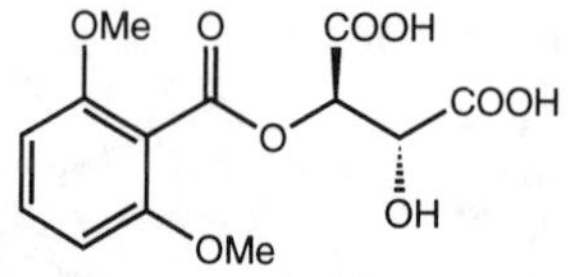

[32]

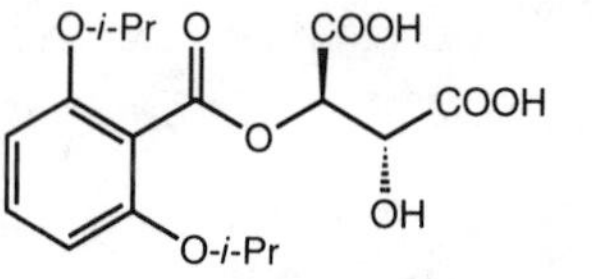

[35]

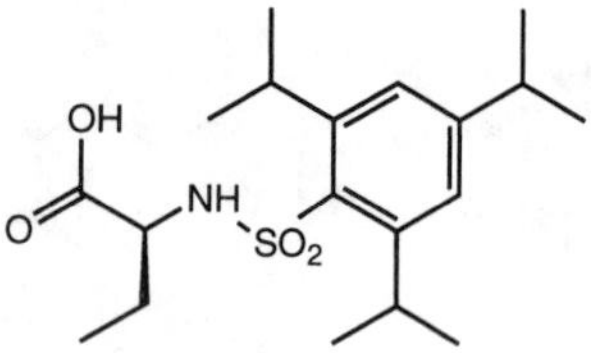

[41]

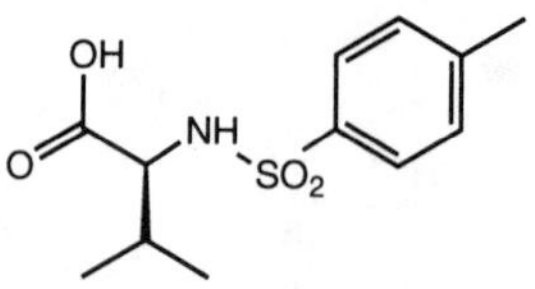

[23]

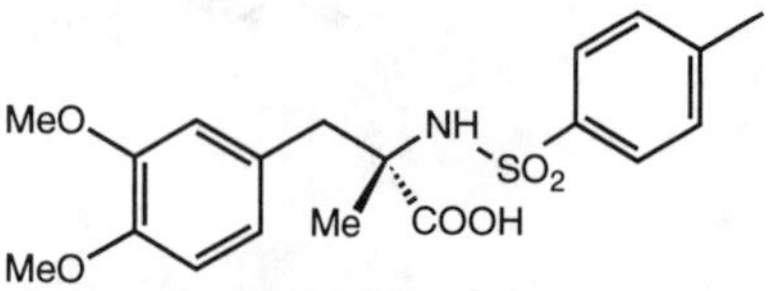

[43b]

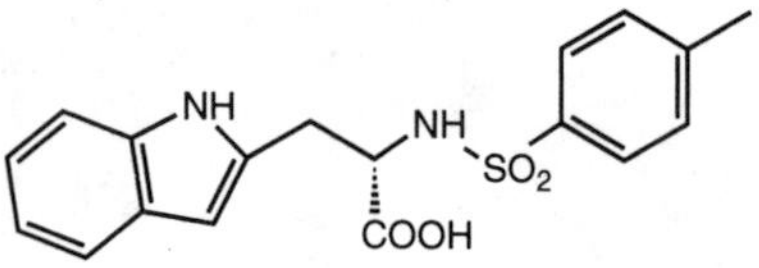

[44]

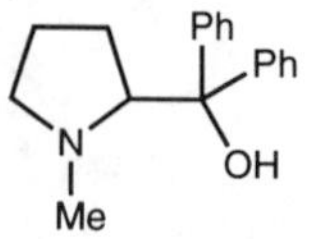

[46]

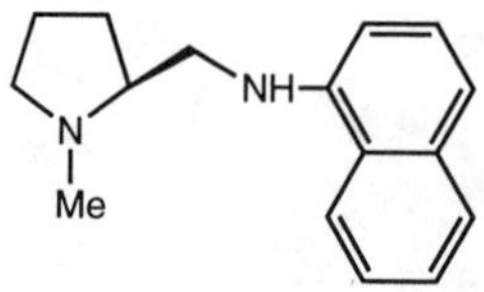

[48] (See also the Ligand List for Chapter 8.)

Index

Acetoxyhalogenation, regioselective, 265
α-tocopherol, 17
Aluminum catalyst, chiral
  for asymmetric aldol reaction, 380, 416
  for asymmetric Claisen rearrangement,
    419–21
  for asymmetric Diels-Alder reaction,
    415, 421
  for asymmetric ene reaction, 419
  for asymmetric hetero Diels-Alder reac-
    tion, 417
Aratani catalyst
  for asymmetric cyclopropanation, 68, 91
2-arylpropionic acids, 14, 289, 340, 356
Asymmetric aldol reactions, 367–88
  mechanism of, 370
  of α-isocyanocarboxylates, 367–77
  of nitromethane, 377–78
  promoted by Lewis acid catalysts, 379–
    86
Asymmetric allylic substitution reactions,
  325–50
  forming a chiral center in allylic sub-
    strates, 331–50
  forming a chiral center in nucleophiles,
    328–31

σ-π-σ mechanism of, 327
  type I and type II reactions of, 326
Asymmetric carbon–carbon bond forming
  reactions, 323–88
Asymmetric carbonylation, 273–302
  asymmetric hydrocarbonylation, 286–89
  asymmetric hydroformylation, 277–86
  enantiomer discriminating reactions of,
    291–94
Asymmetric cyclopropanation, 63–99
  intermolecular reactions, 67–89
  intramolecular reactions, 89–96
  mechanism of, 65, 69, 73, 79, 82
  of alkynes, 94–96
Asymmetric dihydroxylation (AD), 227–71
  by olefin substitution pattern, 245–58
  catalytic cycles of, 235, 238
  comparison with asymmetric epoxida-
    tion, 266
  double diastereoselectivity, 258–61
  general features of, 233–45
  mnemonic scheme of enantioselectivity,
    244
  synthetic applications of, 261–66
Asymmetric epoxidation (AE)
  of allylic alcohols, 103–58

Asymmetric epoxidation (AE) (*cont.*)
  compatibility of functional groups, 107
  enantioselectivity principles of, 105–7
  fundamental elements of, 104–8
  kinetic resolution with, 132–41
  mechanisms of, 144–48
  reaction variables of, 108–12
  of homoallylic alcohols, mono-, bis-, tris-, 142–44
  of unfunctionalized olefins, 159–202
    ligand design for, 172–77
    mechanisms of, 179–81
    synthetic applications of, 183–89
    with transition metal based catalysts, 160–93
    with organic oxidants, 193–99
  of conjugated ketones, 196–99
Asymmetric Grignard cross-coupling reactions, 350–62
  mechanism of, 351
Asymmetric hydrogenation, 1–39
  mechanism of, 8, 15
  of allylic and homoallylic alcohols, 15–18
  of α,β-unsaturated carbonyl compounds, 12–15
  of β-keto esters, 28–30
  of dehydropeptides, 9
  of enamides, 9–10
  of enol derivatives, 11
  of imines, 31–32
  of ketones, 20–31
  of *N*-acylaminoacrylic acids, 6–9
  of olefins, 6–19
  of simple ketones, 30–31
  of simple olefins, 19
  origin of enantioselectivity, 8
Asymmetric hydrosilylation, 303–22
  for the synthesis of chiral silicon compounds, 319–20
  mechanism of, 309–11
  of α,β-unsaturated ketones, 312
  of imines, 314
  of keto esters, 313
  of ketones, 304–12
  of olefins, 315–19
Asymmetric isomerization
  commercial manufacture with, 51–54
  mechanism of, 58–59
  of allylamines, 41–61
  of allylic alcohols, 58
  process development of, 46–51
  scope and limitation of, 54–59
Asymmetric oxidation of sulfides, 203–26
  applications of, 211–12
  mechanism of, 208
  with enzymes, 220–23
  with heterogeneous catalysts, 216–17
Asymmetric phase transfer reactions, 389–411
  alkylation, 395–98
  asymmetric induction step of, 397, 400
  carbon-heteroatom bond formation, 403
  1,2-carbonyl addition, 398–99
  cyclopropanation, 401
  mechanism of, 389–91
  Michael addition, 399–401
  oxidation, 402
  reduction, 401
Asymmetric reactions with chiral Lewis acid catalysts, 413–40
  aldol reaction, 416, 430–34, 435, 436, 438
  alkylation, 425, 426
  Claisen rearrangement, 419
  cyanohydrin formation, 425, 426
  [2+2] cycloaddition, 424, 425
  Diels-Alder reaction, 415, 421, 423, 424, 427–30, 435, 436, 437, 438
  ene reaction, 419, 422, 424
  hetero Diels-Alder reaction, 417
  kinetic resolution, 421–22

"Basket handle" porphyrin, Mansuy's, 163
Benzomorphans, 10
BINAP ligand
  commercial synthesis of, 46–47
Bleomycin (BLM), 189
Borane catalysts, acyloxy- (CAB), chiral
  for asymmetric aldol reaction, 384–86, 430–32, 435–36
  for asymmetric allylation, 433–34
  for asymmetric Diels-Alder reaction, 427–30, 435, 436
  stereochemistry of the reactions with, 434

Borane catalysts, alkyldichloro-, chiral
  for asymmetric Diels-Alder reaction, 437
Borate catalysts, mono-, di-, chiral
  for asymmetric Diels-Alder reaction,
    437–38
Bovin serum albumin (BSA)
  for asymmetric sulfoxidation, 218–20
Brevitoxin, precursors of, 128

Catalyst-substrate adduct
  X-ray crystal structure of, 7–8
Chiral ligands
  for asymmetric aldol reaction, 368, 370,
    375
  for asymmetric allylic substitution reac-
    tion, 328, 329, 330, 332, 333, 337,
    338, 339, 345, 349
  for asymmetric dihydroxylation, 230,
    241
  for asymmetric epoxidation, 169–71
    design of, 172–79
  for asymmetric Grignard cross-coupling
    reaction, 336, 355, 357, 358, 360,
    361
  for asymmetric hydrogenation, 1–3
  for asymmetric hydrosilylation, 305–6,
    309
  list of, 445–70
"Chiral wall" porphyrin, 165
Chrysanthemic acid esters, 64, 68, 73
Cilastatin, 68
Cinchona alkaloids
  as catalysts for asymmetric dihydroxyla-
    tion, 230, 241
  for asymmetric PTC reactions, 391
Cinchona quat catalysts
  for asymmetric PTC reactions, 392
(+)-citronellal, 17, 41, 51, 52
(−)-citronellol, 52
(+)-citronellol, 53
Cobalt catalysts
  for asymmetric cyclopropanation, 78–80
  for asymmetric isomerization, 41, 44
Copolymerization, stereoselective
  of olefins with carbon monxide, 289–91
Copper catalysts
  for asymmetric cyclopropanation, 67–
    77, 89–91
Cromakalim, 183, 187

Crown ethers, chiral
  for asymmetric phase transfer reaction,
    393
Curcumene, 356
Cyclodextrins
  for asymmetric sulfoxidation, 218

11-deoxyanthracyclinones, precursors of,
  128
Dihydrochrysanthemolactone, 89
Dihydroquinidine
  for asymmetric dihydroxylation, 230
Dihydroquinine
  for asymmetric dihydroxylation, 230
2,2-dimethylcyclopropanecarboxylic acid
  ethyl ester of, 64, 68
3,7-dimethyl-1-octanal, 53
Diol activation, 261
Dioxiranes, chiral
  for asymmetric epoxidation, 194
Dioxo(diloato)Os(VI) complex with
  dihydroquinine ester
  X-ray crystal structure of, 231
Dirhodium(II) catalysts, chiral
  for asymmetric cyclopropanation, 80–86
Dolichols, 17
D_4 symmetric metaloporphyrin, 165
Dynamic kinetic resolution, 28–30

Enantioface discriminating carbonylation,
  275
Enantiomer discriminating carbonylation,
  291
Enantiomer discrimination
  of racemic aziridines by carbonylation,
    293
  of racemic halides by carbonylation, 292
  of racemic olefins by carbonylation, 291
Enantioselective hydrocarbalkoxylation,
  287
Enantioselective hydrocarbonylation, 286
Enantioselective hydrocarboxylation, 287
Enantioselective hydroformylation, 277
Enantiotopic group discrimination
  in lactone formation, 294
Ephedra alkaloids
  for asymmetric PTC reactions, 391
Ephedra quat catalysts
  for asymmetric PTC reactions, 392

Epinephrine hydrochloride, 4
3,6-epoxyauraptene, intermediate to, 128
Europium catalyst, Eu(DPPM), chiral
    for asymmetric aldol reaction, 379

Ferricyanide/potassium carbonate
    as cooxidant for dihydroxylation, 234–
    37
FK506, intermediate for, 21
Flavins, chiral
    as catalysts for asymmetric oxidation,
    217
Fluoxetine hydrochloride, 25

Geraniol, 16, 132
Glycidol, 115
Gold catalysts, chiral
    for asymmetric aldol reaction, 367–75

Homogeraniol, 16
Hydroformylation, mechanism of, 277
Hydroperoxyimine, chiral
    for asymmetric epoxidation, 194
(−)-7-hydroxycitronellal, 53
1-hydroxysqualene, 130

Ibuprofen, 289, 356
Iridium catalysts, chiral
    for asymmetric hydrogenation, 30–32
Iron-porphyrin catalysts
    for asymmetric oxidation, 214–16
(−)-isopulegol, 51, 52
Isoquinoline alkaloids, 10

Jacobsen's epoxidation, 159–202
Juvenile hormone, C_{16}-Cecropia-, 422

L-allose, 120, 122
LAT$_4$, intermediate to, 189, 190
Lanthanum alkoxides, chiral
    for asymmetric nitroaldol reaction, 377–
    78
Laudanosine, 10
Levamisole, 25
Lewis acid catalysts, chiral
    for asymmetric aldol reaction, 379–86
"Loaded" titanium-tartrate catalyst
    proposed structure of, 148

Manganese-porphyrin catalysts
    for asymmetric oxidation, 214–16
Marmine, intermediate to, 128
"Matching and mismatching" di-
    astereoselectivity, 258
MeBmt, a component of cyclosporine, 372
(−)-menthol, 17, 51, 52
    industrial production of, 51
Mephenboxalone, 25
7-methoxycitronellal, 53
1β-methylcarbapenems, 18
4-methylene-5α-cholestan-3β-ol, 138
Microbiological oxidation
    of sulfoxides, 221–22
Models for cytochrome P-450, 160
Monsanto process, 442
Montmorillonite
    as catalyst for asymmetric oxidation, 216
Morphinans, 10
Mukaiyama condensation, 430

Naproxen, 14, 289, 340
Narasaka asymmetric Diels-Alder reaction,
    425
N-cyclohexylgeranylamine, 43
Nerol, 16
Nickel catalysts, chiral
    for asymmetric allylic substitution reac-
        tion, 339–40, 343–44
    for asymmetric Grignard cross-coupling
        reaction, 351–58, 360–61
N-methylmorpholine-Noxide (NMO)
    as cooxidant for dihydroxylation, 231,
        238–39
N,N-diethylgeranylamine, 41, 43
N,N-diethylnerylamine, 41, 43
Norreticurine, 10

Osmium tetroxide
    for dihydroxylation of olefins, 228
Oxaziridines, chiral
    for asymmetric epoxidation, 195, 196
Oxetanocin A, 424
Oxo transfer catalysts
    energy diagram for, 182
    for asymmetric epoxidation
        inorganic peroxide-based, 191–93
        porphyrin-based, 160–66
        salen-based, 166–77

Palladium catalysts, chiral
  for asymmetric allylic substitution reaction, 328–39, 340–50
  for asymmetric Grignard cross-coupling reaction, 351–52, 357, 358–59
  for asymmetric hydrocarbonylations, 287–89
  for asymmetric hydrosilylation, 315–19
  for enantiomer discriminating carbonylations, 291, 294
  for stereoselective copolymerization, 289–91
Pantolactone, 26
Pauson-Kand reaction, 295
Permethrinic acid, 64
(+)-peroxycamphoric acid
  for asymmetric epoxidation, 194
Phaltz (semicorrin) catalyst
  for asymmetric cyclopropanation, 73, 91
Phase transfer catalysis (PTC), 389
Phenylephrine hydrochloride, 25
"Picnic basket" porphyrin, Collman's, 163
Platinum catalysts, chiral
  for asymmetric hydroformylation, 279, 282–84
Polymer-linked tartrate esters
  for asymmetric epoxidation, 111
Polymer-supported polyamino acids
  for asymmetric epoxidation, 198
Polymer-supported Pt and Rh catalysts
  for asymmetric hydroformylation, 286
Polyrenols, 17
Poly[(S)-alanine] catalyst
  for asymmetric epoxidation, 197
Prostagrandin, an intermediate to, 421

Quaternary ammonium compounds, "quats," chiral
  for asymmetric phase transfer reaction, 391, 392, 393

Reiterative two-carbon extension cycle, 120
Rhodium catalysts, chiral
  for asymmetric aldol reaction, 380
  for asymmetric hydroformylation, 279–81, 284
  for asymmetric hydrogenation, 3–4

  for asymmetric hydrosilylation, 303–15, 319–20
  for asymmetric isomerization, 43–46, 48–51
  for cyclopropanation, 65–67, 80–87, 90–96
  for enantiomer discriminating carbonylation, 293
Ruthenium catalysts, chiral
  for asymmetric hydrogenation, 4–6

Sakurai-Hosomi allylation, 433
Sakurai reaction, 434
(Salen)Mn(III) complexes, chiral
  for asymmetric epoxidation, 167–72
Salsolidine, 10
Samarium complexes
  for asymmetric hydrogenation, 19
Sharpless dihydroxylation, 227–71
Sharpless oxidation, 103–58
Silver catalysts, chiral
  for asymmetric aldol reaction, 376
Sphingosine, D-*threo*-, D-*erythro*-, 371
(S)-propranolol hydrochloride, 25
Statines, 22
"Strapped" porphyrin, Inoue's, 163
Sulcatol, 312
Sulfonylation, regioselective, 264
Sumitomo process, 68

Takasago process, 42
Tartrate esters
  for asymmetric epoxidation, 110
Taxol side chain, 186, 189
Taxotère sude chain, 188
Tetrahydrofarnesol, 16
Tetrahydropapaveline, 10
$Ti_2(dibenzyltartramide)_2(OR)_4$
  X-ray structure of, 146
Tin catalysts, chiral
  for asymmetric aldol reaction, 381–82, 438
Titanium catalysts, chiral
  for asymmetric alkylation, 426
  for asymmetric cyanohydrin formation, 425, 426
  for asymmetric [2+2] cycloaddition, 424, 425

Titanium catalysts, chiral (*cont.*)
  for asymmetric Diels-Alder reaction,
    423–24
  for asymmetric ene reaction, 422, 424
  for asymmetric epoxidation, 104–57
  for kinetic resolution, 425
Titanium—excess tartrate combination
  for asymmetric oxidation, 210–11
Titanium—Schiff base complexes
  for asymmetric oxidation, 212–14
Titanium tartrate catalysts
  for asymmetric epoxidation, 104–57
Titanium tartrate—water combination
  for asymmetric oxidation, 204–10

Titanocene catalysts, chiral
  for asymmetric hydrogenation, 19
Trethoquinol, 10
"Twin coronet" porphyrin, Naruta's, 163

Vanadium—Schiff base complexes
  for asymmetric oxidation, 214
Verbenol, 138
Virantmycin, intermediate to, 128

Water-soluble catalysts, 7, 31

Zinc catalyst
  for asymmetric aldol reaction, 379, 383